THE INTERNATIONAL SERIES
OF
MONOGRAPHS ON PHYSICS

GENERAL EDITORS

W. MARSHALL   D. H. WILKINSON

H. S. W. MASSEY, E. H. S. BURHOP
AND H. B. GILBODY

ELECTRONIC AND IONIC IMPACT PHENOMENA

SECOND EDITION
IN FOUR VOLUMES

VOLUME I

Collision of Electrons with Atoms

BY H. S. W. MASSEY AND
E. H. S. BURHOP

OXFORD
AT THE CLARENDON PRESS
1969

Oxford University Press, Ely House, London W. 1

GLASGOW NEW YORK TORONTO MELBOURNE WELLINGTON
CAPE TOWN SALISBURY IBADAN NAIROBI LUSAKA ADDIS ABABA
BOMBAY CALCUTTA MADRAS KARACHI LAHORE DACCA
KUALA LUMPUR SINGAPORE HONG KONG TOKYO

© OXFORD UNIVERSITY PRESS 1969

PRINTED IN GREAT BRITAIN
AT THE UNIVERSITY PRESS, OXFORD
BY VIVIAN RIDLER
PRINTER TO THE UNIVERSITY

PREFACE TO THE SECOND EDITION

THE immense growth of the subject since the first edition was produced has raised many problems in connection with a new edition. Apart from the sheer bulk of new material, the interconnections between different parts of the subject have become very complex, while the sophistication of both theoretical and experimental techniques has greatly increased.

It became clear at the outset that it was no longer possible to attempt a nearly comprehensive treatment. As sources of data on cross-sections, reaction rates, etc., for use in various applications are now available and are becoming more comprehensive, it seemed that in the new edition the emphasis should be on describing and discussing experimental and theoretical techniques, and the interpretation of the results obtained by their use, rather than on compilation of data. Even so, a greater selectivity among the wide range of available material has been essential. Within these limitations the level of the treatment has been maintained roughly as in the first edition, although some allowance has been made for the general increase in the level of sophistication.

When all these considerations were taken into account it became clear that the new edition would be between four and five times larger than the first. To make practicable the completion of the task of writing so much against the rate of production of new results, it was decided to omit any discussion of phenomena occurring at surfaces (Chapters V and IX of the first edition) and to present the new edition in four volumes, the correspondence with the first edition being as follows:

Second edition	First edition
Volume I	Chapters I, II, III
Volume II	Chapters IV, VI
Volume III	Chapter VII
Volume IV	Chapters VIII, X

In this way Volumes I and II deal with electron impact phenomena: Volume I with electron–atom collisions, and Volume II with electron–molecule collisions. Volume II also includes a detailed discussion of photo-ionization and photodetachment, which did not appear in the first edition.

Volumes III and IV deal, in general, with collisions involving heavy particles. Thus, Volume III is concerned with thermal collisions involving neutral and ionized atoms and molecules, Volume IV with higher energy collisions of this kind. In addition, recombination is included in this volume as well as a description of collision processes involving slow positrons and muons, which did not appear in the first edition.

Because of the complicated mesh of cross-connections many difficult decisions had to be made as to the place at which a new technique should be described in detail. Usually it was decided to do this in relation to one of the major applications of the technique rather than attempting, in a wholly artificial way, to avoid forward references at all stages.

In covering such a wide field in physics, and indeed also in chemistry, acute difficulties of notation are bound to arise. The symbol k for wave number is now so universally used in collision theory that we have been so impious as to use κ instead of k for Boltzmann's constant. Unfortunately k is also used very widely by physical chemists to denote a rate constant. In some places we have adhered to this but elsewhere, to avoid confusion with wave number, we have substituted a less familiar symbol. Another unfamiliar usage we have employed is that of f for oscillator strength to distinguish it from f for scattered amplitude. Again, we have been unfashionable in using F instead of E for electric field strength because of risk of confusion with E for energy.

No attempt has been made to adopt a set of symbols of universal application throughout the book, although we have stuck grimly to k for wave number and Q for cross-section, as well as e, h, κ, and c.

We have tried not to be too pedantic in choice of units, though admitting to a predilection for eV as against kcal/mol. When dealing with phenomena of strongly chemical interest we have at times used kcal/mol, but always with the value in eV in brackets.

The penetration of the work into chemistry, though perhaps occurring on a wider front, is no deeper than before. The deciding factor has always been the complexity of the molecules involved in the reactions under consideration.

In order to complete a volume it was necessary, at a certain stage, to close the books, as it were, and turn a blind eye to new results coming in after a certain date—unless, of course, they rendered incorrect anything already written. The closing date for Volumes I and II was roughly early 1967, for Volume III mid-1968, and for Volume IV about six months later. Notes on later advances over the whole field will be included at the end of Volume IV.

H. S. W. M.
E. H. S. B.
H. B. G.

London
February 1969

PREFACE TO THE FIRST EDITION

THERE are very many directions in which research in physics and related subjects depends on a knowledge of the rates of collision processes which occur between electrons, ions, and neutral atoms and molecules. This has become increasingly apparent in recent times in connection with developments involving electric discharges in gases, atmospheric physics, and astrophysics. Apart from this the subject is of great intrinsic interest, playing a leading part in the establishment of quantum theory and including many aspects of fundamental importance in the theory of atomic structure. It therefore seems appropriate to describe the present state of knowledge of the subject and this we have attempted to do in the present work.

We have set ourselves the task of describing the experimental techniques employed and the results obtained for the different kinds of collision phenomena which we have considered within the scope of the book. While no attempt has been made to provide at all times the detailed mathematical theory which may be appropriate for the interpretation of the phenomena, wherever possible the observations have been considered against the available theoretical background, results obtained by theory have been included, and a physical account of the different theories has been given. In some cases, not covered in *The theory of atomic collisions*, a more detailed description for a particular theory has been provided. At all times the aim has been to give a balanced view of the subject, from both the theoretical and experimental standpoints, bringing out as clearly as possible the well-established principles which emerge and the obscurities and uncertainties, many as they are, which still remain.

It was inevitable that some rigid principles of exclusion had to be practised in selecting from the great wealth of available material. It was first decided that phenomena involving the collisions of particles with high energies would not be considered, and that other phenomena associated with the properties of atomic nuclei such as the behaviour of slow neutrons would also be excluded. It was also natural to regard work on chemical kinetics as such, although clearly involving atomic collision phenomena, as outside the scope of the book, but certain of the more fundamental aspects are included. Phenomena involving neutral atoms or molecules only have otherwise been included on an equal footing with those involving ions or electrons. A further extensive class of phenomena have been excluded by avoiding any discussion of collision processes occurring within solids or liquids, confining the work to processes occurring in the gas phase or at a gas–solid interface. Among the latter phenomena electron diffraction at a solid surface has been rather arbitrarily excluded as it is a subject already adequately dealt with in other texts. Secondary electron emission and related effects are, however, included.

By limiting the scope of the book in this way it has just been possible to provide a fairly comprehensive account of the subjects involved. It is perhaps too much to hope that even within these limitations nothing of importance has been missed, but it is believed that the account given is fairly complete. Extensive tables of observed and theoretical data have been given throughout for reference purposes and the extent to which the data given are likely to be reliable has been indicated. Every effort has been made to provide a connected and systematic account but it is inevitable that there will be differences of opinion as to the relative weight given to the various parts of the subject and to the different contributions which have been made to it.

We are particularly indebted to Professor D. R. Bates for reading and criticizing much of the manuscript and for many valuable suggestions. Dr. R. A. Buckingham has also assisted us very much in this direction while Dr. Abdelnabi has checked some of the proofs. We also wish to express our appreciation of the remarkable way in which the Oxford University Press maintained the high standard of their work under the present difficult circumstances.

H. S. W. M.
E. H. S. B.

London
August 1951

ACKNOWLEDGEMENTS
VOLUMES I AND II

We must express our indebtedness to a number of people whose help has been invaluable. First among these we would like to thank Mrs. J. Lawson for her assistance, so cheerfully and ably afforded, in preparing the diagrams and tables, obtaining and checking references, and in many other ways. In these tasks her husband, through the use of the computer which he so skilfully constructed, and otherwise, has provided valuable help. The task of transforming indecipherable manuscript to typescript has fallen to Mrs. M. Harding, who has remained unwaveringly in good humour throughout, no matter how bulky the material became. We are grateful for the speed and quality of the work she has done and for her assistance in many other details of checking, etc.

We have had the benefit of many discussions with colleagues, including particularly Professors M. J. Seaton, L. Castillejo, I. C. Percival, and J. B. Hasted, and Drs. A. Burgess, R. F. Stebbings, S. Zienau, D. W. O. Heddle, and R. G. W. Keesing. Dr. G. Peach and Professor L. Castillejo have read the proofs of some of the theoretical chapters and made a number of valuable suggestions. Dr. R. W. Lunt has similarly assisted with some of the chapters of Volume II. Professors J. B. Hasted, I. C. Percival, P. Burke, and Dr. A. Burgess have helped by providing us with advance information on the results of work in which they were engaged. We wish to express our thanks to all of these as well as to Professor L. O. Brockway, Drs. L. S. Bartell, R. P. Madden, and P. Marmet for providing us with photographic illustrations from their original work.

Finally, it is a pleasure to express our appreciation of the work done by the Clarendon Press and by the printers in the speed of publication and the excellence of the format. Their assistance in checking was of great value in handling such a bulk of material.

CONTENTS

1. THE PASSAGE OF ELECTRONS THROUGH GASES: TOTAL COLLISION CROSS-SECTION, ITS DEFINITION AND MEASUREMENT
   1. Classification of collisions ... 1
   2. The concept of collision cross-section ... 2
      2.1. Analysis of the total collision cross-section ... 5
      2.2. Differential cross-sections ... 5
      2.3. Note on nomenclature ... 6
   3. Broad features and fine structure in the variation of cross-sections with electron energy ... 6
   4. Experimental methods for observation of the broad features of the total cross-section ... 7
      4.1. Ramsauer's method ... 7
      4.2. Measurement of collision cross-sections using atomic beams ... 9
         4.2.1. Method depending on observation of scattered electrons ... 10
         4.2.2. Method depending on observation of atomic beam attenuation ... 12
         4.2.3. Method depending on observation of electron beam attenuation ... 17
   5. Experimental methods for observation of the fine structure of total cross-sections ... 19
      5.1. Use of spherical energy analyser ... 20
      5.2. Use of the retarding potential difference method ... 21
   6. Observed total collision cross-sections of atoms ... 24
      6.1. Broad features of observed cross-sections ... 24
      6.2. Observed fine structure in total cross-sections of atoms ... 31
         6.2.1. Introductory—the first observations of fine structure ... 31
         6.2.2. Atomic hydrogen ... 31
         6.2.3. Helium ... 35
         6.2.4. Neon ... 40
         6.2.5. Argon, krypton, and xenon ... 41
         6.2.6. Mercury ... 43
      6.3. Summarizing remarks ... 44

2. SWARM EXPERIMENTS WITH ELECTRONS IN ATOMIC GASES—MOMENTUM-TRANSFER CROSS-SECTION
   1. Diffusion of electrons through gases ... 45
      1.1. Momentum-transfer (diffusion) cross-section ... 47
      1.2. The electron velocity distribution, drift, and random velocity ... 48
         1.2.1. Proof of velocity distribution formula ... 49

1.3. The mean energy and drift velocity	52
1.4. Allowance for energy of motion of the gas atoms	53
2. Measurement of the characteristic energy of a diffusing swarm of electrons	53
3. Measurement of drift velocity	58
3.1. The electrical shutter method	58
3.2. Hornbeck's method	60
3.3. Use of proportional counters for drift velocity measurement	61
3.4. Errors in drift velocity measurements due to diffusion	62
3.5. Measurement of drift velocity in a glow discharge	64
4. Drift in combined electric and magnetic fields	65
5. Microwave afterglow methods for measuring momentum-transfer cross-sections for collisions with thermal electrons	67
5.1. Resonant cavity method	71
6. Cyclotron resonance method for measuring momentum-transfer cross-sections at very low electron energies	74
6.1. Effect of a magnetic field on the resonant frequency of a cavity containing plasma	78
7. Results of measurements of transport properties and their analysis in terms of momentum-transfer cross-sections	79
7.1. Results for helium	81
7.2. Results for neon	84
7.3. Results for argon, krypton, and xenon	87
7.4. Results for caesium	93

3. THE EXPERIMENTAL ANALYSIS OF THE CROSS-SECTIONS FOR IMPACT OF ELECTRONS WITH ATOMS AND IONS—IONIZATION CROSS-SECTIONS

1. Introduction	97
2. The electrical measurement of ionization cross-sections	98
2.1. Measurement of the apparent total ionization cross-section	99
2.1.1. Measurement of the ion current, i_+	100
2.1.2. Measurement of the electron current, i_e	101
2.1.3. Measurement of the electron path-length, l	101
2.1.4. Measurement of the concentration, N, of scatterers	102
2.1.5. Reliability of ionization cross-section measurements	103
2.2. Analysis of positive ion current	104
2.3. The crossed-beam method for the measurement of total apparent ionization cross-section	106
2.4. Ionization cross-section measurements with high energy resolution	112
2.4.1. Use of a high-resolution electron beam	112
2.4.2. The retarding potential difference method	114
2.4.3. Morrison's method of studying fine structure in ionization cross-sections	116

CONTENTS

2.4.4. Problems involved in experimental study of threshold behaviour	118
2.4.5. The use of crossed-beam methods for studying ionization near the threshold	119
2.5. Observed results—measurement of ionization cross-sections	123
2.5.1. The total ionization cross-section—variation with energy	125
2.5.2. The cross-sections for single and multiple ionization	125
2.5.3. Detailed structure in the form of variation of Q_i with electron energy	127
2.5.3.1. The increase of ionization cross-section near the threshold	131
2.5.3.2. Fine structure of ionization cross-section near the threshold	134
2.5.3.3. Multiple ionization through the Auger effect	140
2.5.3.4. Production of metastable ions	141
3. The ionization of positive ions by electron impact	144
3.1. The measurement of the ionization cross-section	145
3.1.1. The crossed-beam method	145
3.1.2. The ion-trap method	149
3.2. Results of measurements of ionization cross-sections of positive ions	151
4. The detachment of electrons from negative ions by electron impact	153
4.1. The experimental method	153
4.2. Results of measurements of the detachment cross-sections of H^- by electrons	158
5. Inner-shell ionization of atoms by electron impact	158
5.1. The measurement of inner shell ionization cross-sections	160
5.2. Results of the measurements	165
5.3. Double inner-shell ionization	167

4. THE EXPERIMENTAL ANALYSIS OF THE CROSS-SECTIONS FOR IMPACT OF ELECTRONS WITH ATOMS AND IONS—CROSS-SECTIONS FOR PRODUCTION OF EXCITED ATOMS

1. Optical measurement of cross-sections for excitation	169
1.1. Principle of the method	169
1.2. Principles involved in absolute measurement of cross-sections—polarization of emitted radiation	171
1.3. The measurement of optical excitation functions and polarizations	172
1.3.1. Measurement of optical excitation functions	172
1.3.2. The measurement of the polarization of impact radiation	178
1.3.3. Measurements using gas or vapour in bulk	179
1.3.4. Crossed-beam methods	193

1.4. Observed results of measurement of optical excitation functions and polarizations ... 196
    1.4.1. Results of measurements in atomic hydrogen ... 196
    1.4.2. Results of measurements in helium ... 196
        1.4.2.1. Optical excitation functions for electron energies greater than 35 eV ... 197
        1.4.2.2. Optical excitation functions for electron energies less than 35 eV ... 207
        1.4.2.3. Polarization ... 214
    1.4.3. Results of measurements in mercury vapour ... 216
    1.4.4. Results of measurements in other gases and vapours ... 223
        1.4.4.1. Inert gases ... 223
        1.4.4.2. Alkali metals ... 223
        1.4.4.3. Cadmium ... 227
1.5. Simultaneous ionization and optical excitation ... 228

2. Direct measurement of excitation functions for metastable states of atoms ... 231
  2.1. The experimental methods ... 231
    2.1.1. Quenching of metastable states ... 231
    2.1.2. Direct measurement of metastable flux ... 234
        2.1.2.1. Electron ejection from surfaces by metastable atoms ... 234
        2.1.2.2. Detection of metastable atoms by ionizing collisions of the second kind ... 237
        2.1.2.3. Detection of auto-ionizing metastable atoms ... 239
    2.1.3. Methods combining quenching and direct detection of metastable atoms ... 242
    2.1.4. Optical methods for measuring metastable atom production rates ... 245
        2.1.4.1. The optical absorption method ... 246
        2.1.4.2. The anomalous dispersion method ... 248
  2.2. Results of the measurements ... 250
    2.2.1. Excitation of the $2s$ metastable states of atomic hydrogen ... 250
    2.2.2. Excitation of the metastable states of He ... 253
    2.2.3. Excitation of metastable states of other inert gases ... 257
    2.2.4. Excitation of the auto-ionizing $(1s\,2s\,2p)\,^4P$ state of lithium ... 261

3. The measurement of cross-sections for collisions of electrons with excited or ionized atoms. ... 263
  3.1. Collisions with metastable atoms ... 263
    3.1.1. Superelastic collisions ... 263
    3.1.2. Total cross-sections ... 267
  3.2. Measurement of cross-section for excitation of He^+ ions to the $2s$ state by an electron-ion crossed-beam method ... 270

5. THE EXPERIMENTAL ANALYSIS OF THE CROSS-SECTIONS FOR IMPACT OF ELECTRONS WITH ATOMS AND IONS—ANALYSIS OF ENERGY AND ANGULAR DISTRIBUTIONS

1. Introduction	277
2. Determination of excitation cross-sections from electron energy loss studies	277
2.1. Diffusion through a gas of electrons with energy sufficient to produce inelastic collisions	278
2.2. Measurement of excitation cross-sections by the diffusion method	280
2.3. The electron-trap method	284
2.4. Detection of superelastic collisions	288
3. Evidence about inelastic cross-sections derived from swarm experiments	290
3.1. Ionization and excitation coefficient	290
3.1.1. Measurement of the ionization coefficient	291
3.1.2. Study of microwave gas discharge breakdown	293
3.1.3. Measurement of drift velocity at large F/p values	296
3.2. Calculation of the velocity distribution function when inelastic collisions are important	299
3.3. Interpretation of experimental results	301
3.3.1. Application to helium	301
3.3.2. Application to other rare gas atoms	307
3.3.3. Application to caesium	309
4. The energy loss of electrons scattered through a fixed angle	309
4.1. Methods of velocity analysis of the scattered electrons	310
4.2. Results of energy loss measurements	312
4.2.1. Resonances in inelastic scattering	316
5. Measurement of the angular distributions of the scattered electrons	321
5.1. Types of apparatus	322
5.1.1. The electron source	322
5.1.2. The monochromator	324
5.1.3. The analyser and collector	324
5.2. Apparatus for special conditions	328
5.2.1. Scattering of very slow electrons	328
5.2.2. Scattering at 180°	328
5.2.3. Scattering at small angles	329
5.2.4. Scattering by metal vapours	331
5.2.5. Crossed-beam method for studying angular distribution of electrons scattered in atomic hydrogen	332
5.3. Observed angular distributions	332
5.3.1. Elastic scattering	332
5.3.2. Inelastic scattering	337
5.3.3. Excitation of resonant states—differential cross-sections	345

6. Spin polarization of electrons following elastic scattering		349
6.1. Measurement of the spin polarization of the scattered electrons		353
6.2. Results of the spin polarization measurements		355
7. Spin-exchange collisions		357
7.1. Atomic-beam method		358
7.2. Optical-pumping method		363
7.3. Use of spin-exchange collisions to produce a polarized electron beam		369

6. ELECTRON COLLISIONS WITH ATOMS—THEORETICAL DESCRIPTION—GENERAL AND SEMI-EMPIRICAL THEORY OF ELASTIC SCATTERING

1. Subdivision of the theoretical problem	373
2. Elastic scattering—semi-empirical 'optical model' approach	376
2.1. The static field of an atom	376
2.2. The Hartree and Hartree–Fock field of an atom	376
2.3. The statistical model	378
3. Quantum theory of scattering by a centre of force	379
3.1. Total cross-section	379
3.2. Angular distribution of elastically scattered electrons—differential cross-section	381
3.3. Proof of quantum formulae	383
3.4. The variation of the phase shifts with energy and angular momentum	385
3.5. Effective range expansions for $k^{2l+1}\cot\eta_l$	388
3.6. Relation of the phase shifts to number of bound states	390
3.7. 'Classical' approximation for the phase η_l	391
3.8. Variational methods for determining the phase η_l	393
3.9. Scattering by a Coulomb field	396
3.10. Scattering by a modified Coulomb field	399
4. Application to calculation of the broad features of elastic cross-sections of atoms	401
4.1. The Ramsauer–Townsend effect	401
4.2. Similar behaviour of the heavier rare gases	405
4.3. Behaviour of neon and helium	407
4.4. Large cross-sections for alkali metals	407
4.5. Similarity of behaviour of chemically similar atoms	408
4.6. Angular distribution for scattering of low-energy electrons by rare gas atoms	410
5. Pressure shift of the high series terms of the alkali metals—relation to the elastic cross-section of the perturbing atoms for very low-energy electrons	414
6. Relativistic effects including spin polarization	418

7. ELECTRON COLLISIONS WITH ATOMS—THEORETICAL DESCRIPTION—BORN'S APPROXIMATION

1.	Born's approximation and the scattering of electrons by hydrogen atoms	426
2.	Scattering by a helium ion—the Coulomb–Born approximation	428
3.	Generalization to scattering by complex atoms	429
4.	Application to elastic scattering—angular distributions at high energy	429
5.	Inelastic collisions	434
	5.1. The cross-sections for excitation and ionization	434
	5.2. Relation to optical transition probabilities—generalized oscillator strengths	438
	5.2.1. Optical transition probabilities and oscillator strengths and cross-sections	438
	5.2.2. Sum rules for oscillator strengths	441
	5.2.3. Alternative forms for the transition matrix element	443
	5.2.4. Relation to Born's collision theory	444
	5.2.5. Alternative forms for Born matrix elements	445
	5.3. Impact parameter formulation	445
	5.3.1. Excitation by a time-dependent perturbation	445
	5.3.2. Excitation by a moving centre of force	447
	5.3.3. Relation to Born's approximation	448
	5.3.4. Excitation of optically allowed transitions	450
	5.3.5. Application to transitions between closely coupled states	452
	5.4. Electron exchange	453
	5.4.1. The Born–Oppenheimer approximation	454
	5.4.2. Ochkur's approximation	455
	5.5. Theoretical limit to the magnitude of collision cross-sections	457
	5.6. Application of Born's first approximation and its range of validity	458
	5.6.1. Excitation of atomic hydrogen	458
	5.6.2. Ionization of atomic hydrogen	461
	5.6.3. Relative probabilities of different types of collision	470
	5.6.4. Atomic hydrogen—comparison with observation	471
	5.6.5. Excitation and ionization of He^+ and other hydrogen-like ions	472
	5.6.6. Excitation of helium—calculated cross-sections	477
	5.6.7. Excitation of helium—comparison with observation—optically allowed transitions	480
	5.6.8. Ionization of helium	486
	5.6.9. Excitation of optically forbidden transitions in helium	491
	5.6.10. Summarizing remarks—the applicability of Born's approximation to inelastic collisions in helium	494
	5.6.11. Applications to other atoms and ions	495
	5.6.12. Angular distribution of the totality of inelastically scattered electrons	498
	5.6.13. Detachment of electrons from H^- ions by electron impact	499

CONTENTS

8. ELECTRON COLLISIONS WITH ATOMS—THEORETICAL DESCRIPTION—ANALYTICAL THEORY FOR SLOW COLLISIONS

1. Introduction — 501

2. Collisions with hydrogen atoms—elastic scattering at energies below the excitation threshold — 502
   2.1. The generalized variational method — 502
   2.2. The eigenfunction expansion — 503
      2.2.1. Analysis of elastic scattering in terms of phase shifts — 504
      2.2.2. The forward intensity of elastic scattering—dispersion relations — 505
      2.2.3. The truncated eigenfunction expansion approximation — 506
   2.3. Allowance for distortion of the atom by the incident electron—Temkin's analysis of s-scattering. — 509
   2.4. Allowance for distortion of the atom by the incident electron—the polarized-orbital and exchange-adiabatic approximations — 510
   2.5. Use of many-parameter variational trial functions — 514
   2.6. Results obtained by various methods—the zero order phase shifts $\eta_0^\pm$ — 514
   2.7. The first and second order phase shifts $\eta_1^\pm$, $\eta_2^\pm$ — 517
   2.8. Comparison with observation — 518
   2.9. Dispersion relations and the elastic scattering of electrons by atomic hydrogen — 519

3. Spin-exchange collisions — 521

4. Collision with hydrogen atoms—elastic and inelastic collisions with electrons at energies above the excitation threshold — 524
   4.1. The generalized variational method — 524
   4.2. Distorted-wave method — 525
   4.3. The excitation of the $2s$ and $2p$ states of hydrogen — 527
   4.4. The excitation of the $2s$–$2p$ transition — 531
   4.5. The excitation of higher discrete states — 533
   4.6. Ionization — 533

5. Collisions with He$^+$ ions — 534
   5.1. Elastic scattering below the excitation threshold — 534
   5.2. Excitation of $2s$ and $2p$ states of He$^+$ — 536

6. Collisions with helium atoms — 538
   6.1. Elastic scattering — 539
   6.2. Excitation of the 2-quantum states—elastic collisions with metastable atoms — 549

7. Collisions with Ne, Ar, Kr, Xe atoms — 553
   7.1. Elastic scattering — 553
      7.1.1. Effective range analysis — 553
      7.1.2. Application of the exchange-adiabatic approximation — 555

7.2. Angular distribution of inelastically scattered electrons—diffraction effects	558
8. Collisions with alkali metal atoms	560
8.1. Application of the s–p close-coupling approximation to elastic scattering	562
8.2. Spin-exchange cross-sections	568
8.3. Excitation of the ns–np (resonance) transitions	568
9. Collisions with atoms and ions with incomplete outer p shells	570
9.1. Introductory	570
9.2. Collisions with neutral oxygen, carbon, and nitrogen atoms	574
9.3. Collisions with ionized atoms	577
9.4. Transitions between fine structure levels	580
10. The theory of the polarization of impact radiation	581
10.1. Inclusion of electron spin	583
10.2. Effects of nuclear spin	586
11. Classical theory and electron–atom collisions	587

9. RESONANCE PHENOMENA—THRESHOLD BEHAVIOUR   594

1. Resonance phenomena and auto-ionization	595
1.1. Perturbation theory involving the interaction of a discrete state and a continuum	597
1.1.1. Cross-sections for excitation of unbound states from the ground state	602
1.2. Interaction between one discrete state and two or more continua	604
1.2.1. Cross-sections for excitation of continuum states, of energy E, from the ground state	607
1.3. Extension to scattering of electrons with non-zero angular momentum	610
1.4. Variation of resonance parameters along a Rydberg series	611
2. Summary of resonance effects and qualitative comparison with experimental results in electron scattering by atoms	612
3. Accurate calculation of resonance parameters—relation to coupled equations	615
3.1. Application to the calculation of resonance energies and level widths associated with the doubly-excited states of helium based on the $2s$ and $2p$ states of He^+.	619
3.2. Comparison with experiment	627
4. Resonance energies and level widths of doubly-excited states of H^-	629
5. Resonance energies and level widths of doubly and triply excited states of He^-	638
6. Doubly-excited states of Ne^-	640
7. Doubly-excited states of Ar^-, Kr^-, and Xe^-	641
8. Doubly-excited states of Hg^-	641
9. One-body or shape resonances	642

10. The S, T, R, and M matrices—behaviour of cross-sections at thresholds ... 645

    10.1. The two-channel S, T, R, and M matrices for states of zero angular momentum ... 645

    10.2. Generalization to coupling between states of different angular momentum ... 653

    10.3. Generalization to any number of channels ... 654

    10.4. Modifications introduced by the presence of a Coulomb field ... 655

    10.5. Generalization of the quantum-defect method to coupled channels ... 657

    10.6. Threshold effects of orbital degeneracy ... 660

    10.7. Quantum defect extrapolation for collisions with He^+ ... 663

    10.8. Threshold law for ionization ... 663

AUTHOR INDEX ... 1

SUBJECT INDEX ... 8

1

THE PASSAGE OF ELECTRONS THROUGH GASES: TOTAL COLLISION CROSS-SECTION, ITS DEFINITION AND MEASUREMENT

1. Classification of collisions

A NUMBER of different effects may result from the encounter of an electron with a gas atom or molecule. These may first be distinguished as elastic, inelastic, superelastic, or radiative.

In an elastic collision no energy exchange takes place between the internal motion of the atom and the electron. The electron loses some energy in such encounters, but only because the finite ratio of the mass m of the electron to the mass M of the atom results in the transfer of some velocity to the centre of mass of the atom or molecule as a whole. As a result, a fraction of order m/M of the initial kinetic energy of the electron is lost in an elastic collision (cf. Chap. 2, § 1.1). This is always less than 10^{-3}, so may be neglected for many purposes. Except when otherwise stated, we shall assume henceforward that an electron loses no energy in an elastic collision with an atom.

In an inelastic collision some kinetic energy is lost by the electron in exciting internal motion in the atom. Further classification of such cases may be carried out by distinguishing the state of internal motion excited. First, the broad distinction may be made between ionizing and non-ionizing impacts, depending on whether or not sufficient energy is transferred to lead to ejection of one or more electrons from the atom. Non-ionizing inelastic collisions will involve excitation of distinct atomic states, so that specification of the state excited provides still further classification. Ionizing impacts may be analysed further in terms of the number and energy of the electrons ejected from the atom and in terms of excitation of the ions.

Superelastic collisions can occur only between an electron and an excited atom and are such that the electron gains energy from the internal motion of the atom. This is clearly possible only if the energy of the internal motion of the atom before the impact is not already a minimum, i.e. the atom is excited. Important instances of collisions of this type occur between electrons and metastable atoms.

Collisions may also occur in which electromagnetic radiation is emitted. They are essentially inelastic as far as the electron is concerned, but differ in that the whole or part of the additional energy is lost as radiation. If the loss is great enough, the electron may be captured to form a negative ion.

Provided the electron has sufficient energy to produce an inelastic collision it is roughly true that the proportion of impacts is evenly divided between elastic and inelastic types (cf. Chap. 7, § 5.6.3). Collisions involving radiation are rare compared with other types of encounter, but are of importance in certain phenomena (see Chaps. 14 and 15). We must now discuss how to represent quantitatively the chance that an electron should make an encounter of any specified type.

2. The concept of collision cross-section

Consider a parallel beam of electrons of homogeneous velocity passing through a hypothetical gas consisting of solid spherical atoms of cross-sectional area Q cm². If there are N such atoms per cm³ the chance that an electron will make a collision in moving a small distance δx cm through the gas will be $NQ\delta x$. Regarding any such impact as removing an electron from the beam, the amount of the beam current strength lost in traversing a distance δx from a point P will be given by

$$\delta I = NQI_\mathrm{P}\,\delta x,$$

where I_P is the current strength at P. Hence, on integration, we have, if P is at a distance x from O,

$$I_\mathrm{P} = I_0\,\mathrm{e}^{-NQx}.$$

If I_P/I_0 is measured as a function of x and N, Q may thus be found.

Now we can imagine an experiment carried out in an actual gas in which similar measurements are made. A beam of electrons of homogeneous velocity is fired into the gas. In traversing a certain distance x in the gas a fraction of the electrons, proportional to x, will be deviated from their original paths and/or lose energy by collisions with the gas atoms. All electrons so affected are considered to be lost from the beam. If measurements are made of the rate at which the current remaining in the beam varies with x, it will again be found that

$$I = I_0\,\mathrm{e}^{-\alpha x},$$

where α can be regarded as an absorption coefficient of the gas for the electron beam. If p is the gas pressure in torr, then

$$N = 2{\cdot}7\times 10^{19} p/760$$

and we may derive from α an effective collision cross-section Q, where

$$Q = \alpha/N = 2\cdot 81 \times 10^{-17} \alpha/p \,\text{cm}^2. \tag{1}$$

This quantity is called the total collision cross-section of the gas atoms for electrons of the beam velocity.

As far as rate of collision is concerned the actual gas behaves towards electrons of the particular velocity just as a hypothetical gas of rigid spheres of cross-section Q would do. We must be careful, however, of carrying this analogy too far. The actual gas atoms are not rigid spheres with defined boundaries—the force between an electron and an atom will fall off continuously with distance and not drop suddenly to zero at some definite separation. This raises at once an important point concerning the independence of Q, defined as above, of the nature, and particularly the angular resolving power, of the measurements involved. We have defined loss from the beam as occurring whenever an electron is deviated from its path or loses energy or both. But, on the classical theory, as long as some field exists between an electron and an atom some deviation will occur. In this case the true effective cross-section of an atom for an electron would be infinite and the observed value would depend strongly on the ability of the apparatus used to distinguish very small deviations of electrons from their original paths. In these circumstances our definition of total absorption cross-section would be theoretically meaningless. However, when allowance is made for quantum uncertainty effects, it turns out that a finite value of the total effective cross-section is to be expected, provided the force between an atom and an electron falls off at large separations r faster than r^{-3}. Provided a certain minimum resolving power is achieved, this finite value can be determined in principle by experiments of the type discussed. The reason for this can be seen from the following argument due to Mott.†

To relate the classical and quantal descriptions of the scattering of an electron through a certain angle we must regard the electron as represented by a wave-packet that spreads as it moves. Two conditions must be satisfied before the two descriptions of the scattering give essentially the same result. Not only must the wavelength of the electron be small compared with the closest distance of approach of the electron to the scattering atom, but the spread of the electron wave-packet must not be so large as to mask a deflexion through the angle concerned.

Consider an electron wave-packet travelling with velocity v_x in the x-direction so that its centre would, in the absence of deviation, pass at

† MOTT, N. F., *Proc. R. Soc.* **A127** (1930) 658.

a distance y from the scattering centre. The classical orbit will only be closely followed if the breadth Δy of the packet is very small compared with y. According to the uncertainty principle, this breadth Δy will be associated with an uncertainty Δv_y in the transverse velocity by the relation
$$\Delta y\, \Delta v_y \simeq h/m,$$
where h is Planck's constant and m the electron mass. We must therefore have
$$y \gg h/m\, \Delta v_y. \qquad (2)$$
The existence of the uncertainty Δv_y means that the wave-packet has a spreading angle $\Delta v_y/v_x$. It is a further necessary condition for the classical description of the particular collision to be valid that this spreading should be small compared with the deviation α which the electron would suffer according to classical mechanics. Thus
$$\alpha \gg \Delta v_y/v_x. \qquad (3)$$
Combining (2) and (3) we must have
$$\alpha y \gg h/mv_x. \qquad (4)$$
Now, except for scattering fields with potential falling off more slowly than the inverse square of the distance at large distances, αy tends to zero with α. This means that, for any wavelength h/mv of the electrons, the classical description fails at sufficiently small angles. Deviations less than this are not observable and, as it is the contribution from very small deviations that makes the classical value of Q infinite, the quantal value remains finite. In other words, increase of resolving power in an experimental apparatus does not lead to indefinite increase of the observed Q, for resolution is limited in any case by intrinsic uncertainties.

Experimental evidence in support of the quantum viewpoint can be derived, either from a study of the variation of the observed Q with experimental resolving power or of the distribution in angle of electrons scattered from atoms. Agreement of measurements of Q by observers using a wide variety of apparatus also provides strong support for the correctness of the quantum viewpoint.

Although the quantity Q we have defined has a definite meaning provided the resolving power of the apparatus is sufficiently high, it is very important in making measurements to ensure that this condition is satisfied. The higher the electron velocity the higher the required resolving power, and in the study of the passage of positive ions through gases (see Vol. IV) the requirement is very much more severe.†

† As an example of the type of resolution required of an apparatus for the study of electron collisions, in order to obtain an accuracy of 1 per cent in the measured total

There is one other important consequence of the gradual decrease of the scattering field with distance as contrasted with the rigid sphere case—the effective cross-section must be expected to vary with electron velocity.

2.1. *Analysis of the total collision cross-section*

So far we have defined the total effective cross-section for all types of collision presented by a gas atom towards electrons of a given velocity v. No attention has been paid to specifying quantities that determine separately the rate of elastic and the various kinds of inelastic, superelastic, and radiative collisions. To do this we may assign probabilities $P_0(v)$, $P_n(v)$, etc., which represent the chance that a collision of an electron of velocity v with a gas atom should be elastic or inelastic, involving excitation of the nth state of the atom, etc., respectively. These quantities can be expected to vary with the electron velocity. The cross-section $P_0(v)Q$ is then defined as the effective cross-section for elastic collisions of electrons of velocity v with the gas atoms concerned, $P_n(v)Q$, that for inelastic collisions involving excitation of the nth state, and so on. We then have

$$Q = P_0(v)Q(v) + \sum P_n(v)Q(v) = Q_0 + \sum Q_n,$$

where we have written Q_0, Q_n, etc., respectively for the individual cross-sections. Experiments designed to measure Q_0, Q_n, etc., respectively can be conceived in principle and have to some extent been carried out in practice. They will be discussed in some detail in Chapters 2–5, 10–13, 14, and 15.

2.2. *Differential cross-sections*

So far we have paid no attention to the specification of the distribution in angle of electrons undergoing a particular type of collision. Consider for the moment elastic collisions. The elastic cross-section Q_0 may be further broken down as follows. Let $p(\theta)\sin\theta\,d\theta d\phi$ be the probability that, in an elastic collision, the electron is scattered into the solid angle $d\Omega\,(=\sin\theta\,d\theta d\phi)$. Then $p(\theta)Q_0\sin\theta\,d\theta d\phi$ is called the differential cross-section for elastic scattering into the solid angle $d\Omega$. It is usually written

$$I_0(\theta)\sin\theta\,d\theta d\phi$$

collision cross-section, the maximum angle of scattering that an electron may undergo without being lost to the beam must not exceed 11° for 1-V electrons, 6·5° for 10-V electrons, 2·3° for 100-V electrons, 0·85° for 1000-V electrons, and 0·2° for 10 000-V electrons. These figures are practically independent of the nature of the scattering material.

so we have
$$p(\theta) = I_0(\theta)/Q_0 \tag{5}$$

and
$$Q_0 = \int_0^\pi \int_0^{2\pi} I_0(\theta)\sin\theta \, d\theta d\phi. \tag{6}$$

In the same way we may define differential cross-sections for inelastic collisions. Thus
$$I_n(\theta)\sin\theta \, d\theta d\phi$$
is the differential cross-section for scattering into the solid angle $d\Omega$ in an inelastic collision involving excitation of the nth state of the atom.

In terms of the differential cross-section the condition for the finiteness of Q_0 requires the existence of the integral in (6). This will be convergent if $I(\theta)$ does not increase as rapidly as $1/\theta^2$ for small θ. The relation of this to the law of force producing the scattering will be discussed in more detail in Chapter 16, § 5.4.2.

2.3. *Note on nomenclature*

In order to avoid confusion we shall throughout this book employ the term 'total collision cross-section' to mean the sum of the cross-sections for every type of collision. The term 'total elastic (inelastic) cross-section' will be used for the cross-sections Q_0 (Q_n) in contradistinction to the corresponding differential cross-sections. Under certain conditions, as, for example, when the electron energy is insufficient to excite the atoms, the total collision cross-section will be practically equal to the total elastic cross-section, but this will not be true in general.

3. Broad features and fine structure in the variations of cross-sections with electron energy

The total collision cross-sections of atoms towards electron impact vary with electron energy in complicated ways. It is convenient to distinguish between the comparatively gradual variations in which a change of electron energy of the order of a few eV or so is required to produce much change in the cross-section and the much more rapid variations occurring within an electron energy range of less than 0·1 eV that arise from many-body resonance effects.

This distinction is useful both from the point of view of theoretical interpretation and the design of experiments. To observe the gradual variations it is not necessary to employ specially refined methods for producing electron beams with a high degree of energy homogeneity. Equally, it is not necessary to resort to such a refined theory to interpret these gradual variations because they are much less dependent on the many-body character of the phenomena. Most earlier work was really concerned with the study of the gradual variations or, more correctly perhaps, with the variation with mean energy of the cross-section averaged over a finite energy spread that largely obscures the fine detail of the energy variation. It is only recently that techniques have developed so far as to make detailed study of the fine structure possible.

Accordingly we shall first describe experiments that yield information on the broad features of the magnitude and energy variation of the cross-section and then discuss the techniques used to study the fine structure. A similar distinction will be made throughout, both in dealing with experimental and with theoretical aspects of the subject. It is of interest to note that the distinction has been well made in nuclear physics for many years—the gradual variations manifest in the form of broad peaks in the variation of mean cross-sections with collision energy or with mass number of the target nucleus are referred to as 'shape' resonances to distinguish them from the narrow resonances that are a prominent feature of nuclear collision phenomena.

4. Experimental methods for observation of the broad features of the total cross-section

4.1. *Ramsauer's method*

The first experiments on the absorption of electrons in gases were made as long ago as 1903 by Lenard,† and in 1916 Akesson‡ obtained evidence that, in certain gases, slower electrons were more penetrating than the faster. Quantitative absolute measurements of total collision cross-sections as a function of electron velocity really date, however, from the introduction of Ramsauer's§ method. It is a very direct method, the principle of which is illustrated in Fig. 1.1.

Electrons from a source F were accelerated to the desired velocity before passing through the slit S_1. In Ramsauer's original apparatus F was a zinc plate and the electrons were ejected photo-electrically. By means of a magnetic field perpendicular to the plane of the paper, those electrons that suffered no collision described a circular path through the slits S_1–S_7 and entered a collector C. Electrons that were elastically scattered from the beam failed to pass through the slits, while those that suffered inelastic collisions, even without deflexion, moved in a new

---

† LENARD, P., *Annln Phys.* **12** (1903) 714.
‡ AKESSON, N., *Årsskr. Univ. lund.* N.F. **12** (1916) 29.
§ RAMSAUER, C., *Annln Phys.* **64** (1921) 513; **66** (1921) 546.

circular path of smaller radius in the magnetic field and so failed to pass through the succeeding slits. In Ramsauer's apparatus the beam defined by the slit system had a mean diameter of 20 mm and was 1 mm wide and 8 mm high.

The procedure adopted for this measurement was as follows. With a pressure p_1 torr in the apparatus the currents i_1 to C alone and j_1 to B and C together were measured by an electrometer. If x is the length of

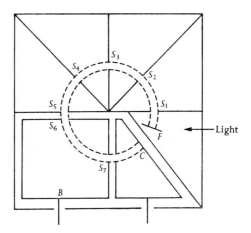

Fig. 1.1. Ramsauer's apparatus for measurement of collision cross-sections.

the path between slits S_6 and S_7, $i_1 = j_1 \mathrm{e}^{-\alpha p_1 x}$, where α is the absorption coefficient at a pressure of 1 torr.

If i_2, j_2 are similar currents when the pressure is p_2 torr,

$$i_2 = j_2 \mathrm{e}^{-\alpha p_2 x},$$

so that $\qquad (p_1 - p_2)\alpha x = \ln(j_1 i_2 / i_1 j_2).$

α, and hence Q, may thus be determined from (1).

Provided the slits are narrow enough this method should give the true total collision cross-section, as no type of encounter fails to be detected, except for the small number of elastic collisions through very small angles.

Ramsauer's apparatus has not only been used for measuring the total absorption cross-section for electrons in gases and vapours but has also been adapted by Brode† for measurements in metal vapours. For this purpose the whole of the collision chamber was designed so that it could be maintained at such temperatures that the vapour pressure of the

† Brode, R. B., Phys. Rev. 34 (1929) 673.

metal was of the order of 10^{-5} torr, convenient for the experiment. This was made possible by constructing the metal parts of tantalum and using a hot-filament source for the electrons. In Brode's apparatus the beam had a mean diameter of 30 mm and was 1 mm wide and 10 mm high. A single collecting chamber was used, the electron emission from the source being assumed to remain constant at the different pressures of vapour in the apparatus.

There is no difficulty in principle in employing Ramsauer's method with an electron source that provides a beam of much greater homogeneity in energy so as to investigate the fine structure in the cross-section-energy variation. An apparatus of this type is described in § 5.

4.2. *Measurement of collision cross-sections using atomic beams*

A method such as Ramsauer's is suitable for observing the total cross-section of a particular atom only if, at ordinary temperatures, the gaseous phase composed of such atoms is monatomic. However, a knowledge of the total cross-section of atomic hydrogen as a function of electron energy is of great interest, since this is the simplest system from the point of view of theoretical interpretation. Another example is atomic oxygen, the collision properties of which are important in the interpretation of upper atmospheric phenomena. In recent years methods have been developed in which the scattering atoms form an atomic beam that crosses the electron beam. The products of the collision are then detected and analysed. In these experiments, which are of importance in the measurement of many types of collision cross-section (see Chap. 3, §§ 2.3, 2.4.5, 3, and 4, and Chap. 4, § 3.2), the number density of particles in the beam is frequently considerably less than the molecules of residual or recombined gas. To pick out the wanted signal due to collisions with atoms in the beam, either the atomic or electron beam is chopped mechanically at an audio-frequency. The wanted signal is amplified by a narrow-band amplifier tuned to the modulation frequency of the atomic beam and measured by a phase-sensitive detector.

Although the particular aim is the measurement of cross-sections for atoms that do not normally exist uncombined in the gas phase, the methods are nevertheless applicable to the study of collisions in monatomic gases and vapours and offer certain advantages in dealing with experimentally unpleasant substances such as the alkali metal vapours.

Three different techniques have been developed. One, which is of wider application to the study of the cross-sections for different types of collision, depends on the measurement of the current of scattered electrons. The other two, which are concerned only with the measurement of total cross-sections, rely on measurement of the attenuation either of the atomic beam or of the electron beam due to collisions with the other crossed beam.

4.2.1. *Method depending on observation of scattered electrons.* Technical details of the first method are more fully discussed in § 2.3 of Chapter 3

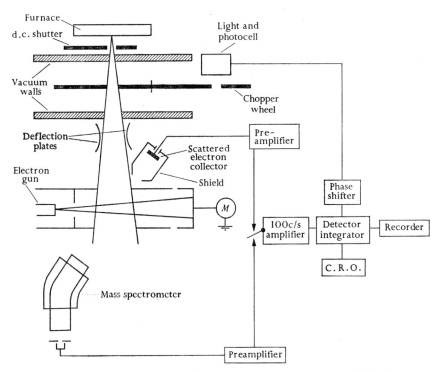

FIG. 1.2. Schematic diagram of the arrangement of apparatus used by Fite, Brackmann, and Neynaber for observation of elastic scattering of electrons by atomic hydrogen.

in relation to the measurement of ionization cross-sections that provides the basic information on which other applications depend. In this section we describe briefly its specific application by Brackmann, Fite, and Neynaber[†] to the determination of the total collision cross-section of atomic hydrogen. Fig. 1.2 is a schematic diagram of the apparatus used. Hydrogen gas was introduced into a furnace consisting of a tungsten

[†] BRACKMANN, R. T., FITE, W. L., and NEYNABER, R. H., *Phys. Rev.* **112** (1958) 1157. An earlier experiment in which the atomic-beam method was applied to measure the total cross-section of atomic hydrogen was conducted by BEDERSON, B., MALAMUD, H., and HAMMER, J., *Bull. Am. Phys. Soc.* Ser. II, **2** (1957) 172.

tube operated at a temperature hot enough to produce atomic hydrogen by thermal dissociation. The atomic hydrogen beam that emerged from an orifice in the side of the tungsten tube was chopped by the chopper wheel at a frequency of 100 c/s. An oxide-coated cathode was used in an electron gun to produce a d.c. electron beam that crossed the modulated atomic beam at right angles and was collected by a cylindrical bucket electrode. The detector for the scattered electrons consisted of a disc placed so as to collect electrons scattered into a cone of semi-apex angle of 45° and with its axis perpendicular to the direction of the electron beam. The disc was connected to a one-valve preamplifier and both were placed inside a closed metal shield containing a circular aperture through which the scattered electrons entered. The electronic circuit arrangements for the amplification of the scattered signal were similar to those described in Chapter 3, § 2.3.

Measurements were made of the current scattered into the collector for electrons in the energy range 1–10 eV. The use of atomic hydrogen gave rise to an apparent increase of the cathode work-function as determined from retarding potential measurements, amounting to 0.4 eV over a time of about 1 h. This caused a drift in the electron energy, resulting in some uncertainty in the cross-section measurements below 3 eV.

The hydrogen furnace was first operated at a temperature T_r (°K) so low that the dissociation of the hydrogen was negligible, and under these conditions the scattered electron signal was S_r. At the working temperature T where the dissociation fraction is D the total signal

$$S = S_1 + S_2,$$

where S_1 and S_2 are the contributions to the signal due respectively to the atoms and molecules of the beam. Then

$$S_2 = (1-D)S_0,$$

$$S_1 = \sqrt{(2)} D (Q_A/Q_M) S_0,$$

where $S_0 = S_r(T_r/T)^{\frac{1}{2}}$ is the signal that would have been observed at this temperature if dissociation had not occurred and Q_A, Q_M are the scattering cross-sections of the hydrogen atom and molecule respectively, so that

$$\frac{Q_A}{Q_M} = \frac{1}{\sqrt{(2)}D}\left\{\frac{S}{S_r}\left(\frac{T}{T_r}\right)^{\frac{1}{2}} + D - 1\right\}. \tag{7}$$

The mass spectrometer enabled the relative numbers of atomic and molecular hydrogen ions produced in the beam to be determined, and from the known ratio of ionization cross-sections of atomic and molecular hydrogen (see Chap. 3, § 2.3) the dissociation fraction, D, could be estimated.

Since in the experiment of Brackmann et al.† scattered electrons were collected over an angular range around 90°, the estimate of the total cross-section required a knowledge of the angular distribution of the scattered electrons. In the energy range used in the experiment the scattering is predominantly isotropic, arising from electrons with zero angular momentum relative to the nucleus of the target atom. The small contribution due to other electrons (10 per cent) was estimated from the calculations of Bransden, Dalgarno, John, and Seaton.‡

4.2.2. *Methods depending on observation of atomic beam attenuation.* Measurements of total collision cross-sections in lithium, sodium, and potassium in the energy range up to a few eV using the second atomic-beam method have been made by Perel, Englander, and Bederson.§ In their measurements the number of collisions was determined by measuring the reduction in intensity of the atomic beam as a result of recoil following electron collisions. The intensity of the atomic beam was estimated from the surface ionization current when it impinged on a hot wire made of platinum–tungsten alloy (92 per cent platinum, 8 per cent tungsten). In these experiments the atomic beam was not modulated but measurements were made using an electron beam modulated at a frequency of 30 c/s. The tuned detector technique was used to study the component of this frequency in the atomic beam current. By using this recoil method it was possible to obtain the total cross-section directly without making assumptions about the angular distribution of the scattered electrons as is necessary in the method of Brackmann et al.,† although some care had to be taken to define the minimum electron scattering angle that could be detected (about 10° in these experiments). In these measurements the apparatus used contained two ovens, one for producing an atomic beam of the alkali metal being studied and one for producing an atomic beam of potassium. The cross-sections at each energy were compared with those for potassium. The absolute cross-section for potassium was obtained by normalizing to the absolute value obtained by Brode‖ at 2 eV.

† loc. cit.
‡ BRANSDEN, B. H., DALGARNO, A., JOHN, T. L., and SEATON, M. J., *Proc. phys. Soc.* **71** (1958) 877.
§ PEREL, J., ENGLANDER, P., and BEDERSON, B., *Phys. Rev.* **128** (1962) 1148.
‖ loc. cit.

Sunshine, Aubrey, and Bederson† have applied essentially the same method to measure the total cross-section for atomic oxygen. Fig. 1.3 shows the general arrangement of their experiment. Oxygen gas enters a source chamber from which it issues as a molecular beam. By applying an r.f. discharge to the chamber the oxygen is partly dissociated and issues as a mixed atomic and molecular beam. This beam is crossed by an electron beam which is chopped at a frequency of 50 c/s. Arrangements are also made to chop the main beam at the same frequency so that comparable measurements may be made by selective amplification and phase-sensitive detection both for the main beam and for the residual beam after scattering. Detection is carried out by ionizing the beam as it passes through an exit slit and then mass-analysing the ions produced.

Omitting for the moment the modifications due to the phase-sensitive detection, the observations and analysis required to obtain the total cross-section of atomic oxygen are as follows.

Let I_0 be the incoming flux of oxygen molecules to the source chamber. Then, if the source temperature is T, the number of such molecules with velocity between v and $v+dv$ is given by

$$I_m(v)\, dv = 2I_0\, v^3 \alpha_m^{-4}\, e^{-v^2/\alpha_m^2}, \tag{8}$$

where $\alpha_m = (2\kappa T/M_m)^{\frac{1}{2}}$, M_m being the mass of an O_2 molecule. The chance of ionization of a molecule entering the detector leading to production of O_2^+ and O^+ ions respectively will be B_i^m/v, fB_m^i/v respectively, where B_i^m and f are constants. Hence the currents J_0^m, J_0^a of O_2^+ and O^+ ions observed with the full beam when the discharge is off are given by

$$J_0^m = 2I_0\, \alpha_m^{-4} B_i^m i_m(2), \tag{9}$$

$$J_0^a = 2I_0\, \alpha_m^{-4} f B_i^m i_m(2), \tag{10}$$

where
$$i_m(2) = \int v^2 e^{-v^2/\alpha_m^2}\, dv. \tag{11}$$

When the discharge is on, the beam will be partially dissociated. If D is the fractional dissociation the currents J_d^m, J_d^a of O_2^+ and O^+ ions observed with the full beam are given by

$$J_d^m = 2I_0\, \alpha_a^{-4}(1-D) B_i^m i_m(2), \tag{12}$$

$$J_d^a = 2I_0\{2D\alpha_a^{-4} B_i^a i_a(2) + (1-D)\alpha_m^{-4} f B_i^m i_m(2)\}. \tag{13}$$

From (9) and (10),
$$J_0^a/J_0^m = f. \tag{14}$$

and from (12) and (13),

$$J_d^a - f J_d^m = 4I_0\, D\alpha_a^{-4} B_i^a i_a(2). \tag{15}$$

† SUNSHINE G., AUBREY, B. B., and BEDERSON, B., *Phys. Rev.* **154** (1967) 1.

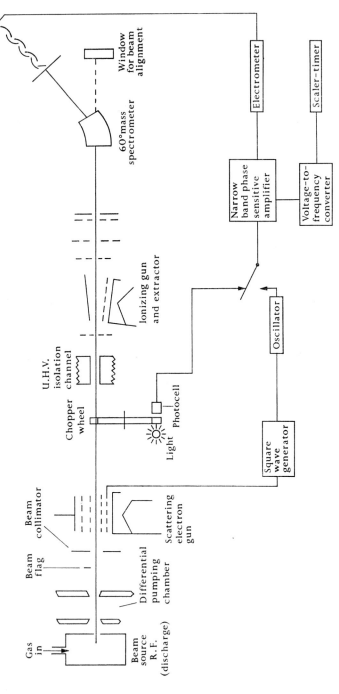

FIG. 1.3. Schematic diagram of the arrangement of apparatus used by Sunshine, Aubrey, and Bederson for observation of the total cross-section of atomic oxygen for electrons.

A similar analysis may be carried out for the flux scattered from the beam. The chance of a molecule of velocity v being scattered can be written B_s^m/v where B_s^m is a constant, the chance being proportional to the time the molecule spends in passing across the electron beam. Similarly for an atom of the same velocity the chance is B_s^a/v. Denoting the scattered fluxes corresponding to J_0^m, J_0^a, J_d^m, J_d^a respectively by S_0^m, S_0^a, S_d^m, S_d^a we have

$$S_0^m = 2I_0 \alpha_m^{-4} B_i^m B_s^m i_m(1), \tag{16}$$

etc., where
$$i_m(1) = \int v e^{-v^2/\alpha_m^2} \, dv. \tag{17}$$

Just as for (15) we have

$$S_d^a - f S_d^m = 4I_0 D \alpha_a^{-4} B_i^m B_s^a i_a(1), \tag{18}$$

so that
$$B_s^a = \frac{S_d^a - f S_d^m}{J_d^a - f J_d^m} \frac{i_a(1)}{i_a(2)} = \frac{S_d^a - f S_d^m}{J_d^a - f J_d^m} \alpha_a. \tag{19}$$

In terms of the total collision cross-section Q_a for atoms

$$B_s^a = Q_a i/eh, \tag{20}$$

where i/e is the number of electrons/s in the electron beam and h is the path length followed by an atom in traversing this beam.

It follows that, from measurements of S_d^a, S_d^m, J_d^a, J_d^m, J_0^a, and J_0^m, Q_a may be obtained. In the course of these measurements the molecular cross-section Q_m may also be obtained from

$$B_s^m = S_0^m \alpha_m / J_0^m = Q_m i/eh. \tag{21}$$

When phase-sensitive detection is used the detected signal is proportional to the input signal multiplied by the cosine of the phase angle between it and an arbitrary phase reference signal. Thus (9) is replaced by

$$J_0^m = 2I_0 \alpha_m^{-4} B_i^m \int_0^\infty v^2 \cos\{\phi_1 + 2\pi\nu l_1/v\} e^{-v^2/\alpha_m^2} \, dv, \tag{22}$$

and (16) by

$$S_0^m = 2I_0 \alpha_m^{-4} B_i^m B_s^m \int_0^\infty v \cos\{\phi_2 + 2\pi\nu l_2/v\} e^{-v^2/\alpha_m^2} \, dv, \tag{23}$$

where l_1 is the distance from the beam chopper to the detector and l_2 from the electron-scattering gun to the detector, and ν is the modulation frequency. ϕ_1 and ϕ_2 are the arbitrary phase differences that are set to maximize the signal. Carrying through the appropriate analysis, evaluating numerically the integrals that occur, it is found that, with the particular apparatus used

$$Q_\mathrm{a} = 1\cdot026 \frac{he\alpha_\mathrm{a}}{i}\left\{\frac{S_\mathrm{d}^\mathrm{a}-fS_\mathrm{d}^\mathrm{m}}{J_\mathrm{d}^\mathrm{a}-fJ_\mathrm{d}^\mathrm{m}}\right\}, \qquad (24)$$

$$Q_\mathrm{m} = 1\cdot064 \frac{he\alpha_\mathrm{m}}{i} \frac{S_0^\mathrm{m}}{J_0^\mathrm{m}}. \qquad (25)$$

The ratio
$$\frac{Q_\mathrm{a}}{Q_\mathrm{m}} = 0\cdot964 \frac{S_\mathrm{d}^\mathrm{a}-fS_\mathrm{d}^\mathrm{m}}{J_\mathrm{d}^\mathrm{a}-fJ_\mathrm{d}^\mathrm{m}} \frac{J_0^\mathrm{m}}{S_0^\mathrm{m}} \frac{\alpha_\mathrm{a}}{\alpha_\mathrm{m}} \qquad (26)$$

is independent of the geometry of the beam systems and is likely to be more accurately determined.

The apparatus was arranged in four separately pumped vacuum chambers—the source chamber operating at 6×10^{-5} torr, the scattering chamber at $1\cdot5\times10^{-7}$ torr, the differential pumping chamber between them at 4×10^{-6} torr with 3×10^{-9} torr in the detector chamber.

Oxygen dissociation was produced in a quartz discharge tube operated at about 30 Mc/s. To assist dissociation 35 per cent of hydrogen was admixed with the oxygen. Usually the fractional dissociation was about 25 per cent. The source slit was 0·125 in high and 0·007 in wide. It is important that no appreciable fraction of excited atoms or molecules should reach the interaction region. Some reassurance is provided by the fact that the measured molecular cross-sections were the same whether or not the discharge was operating (see, however, Chap. 13, § 4.4.2 and Chap. 14, § 7.5.2). Furthermore, no evidence for production of O ^1S or ^1D metastable atoms has been found by magnetic resonance experiments[†] in a beam produced by a similar r.f. discharge.

The electron gun produced an electron beam of $0\cdot125\times1\cdot5$-in cross-section. Collimation was ensured by applying a uniform magnetic field of about 1000 gauss along the path of the beam. Retarding potential analysis showed that the energy spread of the beam between half maxima was about 0·7 eV.

The detection system was of the type described by Aberth.[‡] To isolate the ultra-high vacuum of the detector chamber from the rest of the vacuum system the entrance channel to the chamber was of cross-sectional area $0\cdot040\times0\cdot4$ in.

The angular resolving power, as far as the atomic beam is concerned, defined as that angle for which 50 per cent of all scattering events are observable, was about 10^{-3} rad. In terms of electron scattering angle

[†] BRINK, G., private communication.
[‡] ABERTH, W., *Rev. scient. Instrum.* **34** (1963) 928.

this amounts to an effective resolution, averaged over the Maxwellian distribution of velocities, ranging from 16° at 1 eV to 10° at 12 eV for atomic oxygen. The respective values for O_2 were 18·5° and 11°.

Discussion of the results obtained is deferred to § 6.1.

4.2.3. *Method depending on observation of electron beam attenuation.* A further variant of the crossed-beam technique applied to atomic oxygen is that used by Neynaber, Marino, Rothe, and Trujillo.† They measured the total cross-section from the attenuation of the electron,

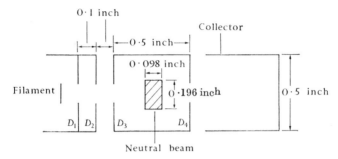

FIG. 1.4. Illustrating the geometry of the electron gun used by Neynaber, Marino, Rothe, and Trujillo in their experiments on the total cross-section of atomic oxygen for electrons. The diameters of the electron beam holes D_1, D_2, D_3, and D_4 were 0·08, 0·08, 0·100, and 0·120 in respectively.

instead of the atom, beam. In other respects the arrangement was very similar to that described above. An r.f. discharge produced partial dissociation of the oxygen, and the mixed molecular and atomic beam was detected by ionization and a mass analyser in the form of a 60° sector mass spectrograph. The neutral beam was chopped mechanically at 101·8 c/s so that selective a.c. detection could be used.

Fig. 1.4 is a diagram of the geometry of the electron gun, all components of which, with the exception of the filament, were kept grounded. The attenuation of the electron beam was measured by the ratio of the current modulated at the chopping frequency to the direct current. With the geometry employed, the angular resolution, measured in terms of the angle of scattering for which the detection efficiency is 50 per cent, was 25°.

If S' and S are the scattered a.c. currents per unit direct electron current with the discharge respectively on and off, and D is the fraction of molecules dissociated, the ratio of the total cross-sections for atomic

† NEYNABER, R. H., MARINO, L. L., ROTHE, E. W., and TRUJILLO, S. M., *Phys. Rev.* **123** (1961) 148.

and molecular oxygen is given by

$$\frac{Q_a}{Q_m} = \frac{(S'/S)-1+D}{D\sqrt{2}}. \qquad (27)$$

D is measured by the ratio of the current of O_2^+ ions observed in the detector with the discharge on to that when it is off. Equation (27) assumes effusive flow conditions. However, to produce sufficient attenuation of the electron beam, the intensity of the neutral beam had to be so high that the pressure in the source exceeded that required for effusive flow. The importance of errors introduced in this way was investigated by carrying out a subsidiary experiment to determine the ratio of the total ionization cross-sections for O and O_2 by the method described in Chapter 3, § 2.3. In one set of experiments the ratio was measured for electrons of 200-eV energy, using the source at high pressure; in another set the source was operated under effusive flow conditions. Ionization cross-section ratios differing by 20 per cent were found and this was used to provide a correction factor for the attenuation measurement of the total cross-section ratio. This is reasonable because (see Chap. 3, § 2.3) the formula for the ionization cross-section ratio is similar to (27) with S'/S replaced by a ratio of total ionization currents produced by electron impact from the neutral beam.

Similar equipment was used to measure the total cross-section of atomic hydrogen,† the only difference being the source of the atomized hydrogen. The discharge was operated at about 0·8 torr and produced about 30 per cent dissociation. In contrast to the oxygen experiments the source temperature tended to rise by about 50° C during the experiment so that a cooling jacket was installed. As for oxygen, correction was made because the flow from the source was not effusive.

A third experiment‡ was also carried out with atomic nitrogen. Opportunity was taken to improve the electron gun so that, at equivalent energies, the electron beam current was six to seven times larger. This made it possible to work with lower gas densities in the atomized beam so that effusive flow conditions prevailed. Experiments with an undissociated nitrogen beam gave a cross-section energy variation consistent with the observations of Normand,§ and the latter were used in normalizing the atomic nitrogen data. The results for hydrogen, oxygen, and nitrogen are described in § 6.

† NEYNABER, R. H., MARINO, L. L., ROTHE, E. W., and TRUJILLO, S. M., *Phys. Rev.* **124** (1961) 135.
‡ Ibid. **129** (1963) 2069.
§ NORMAND, C. E., ibid. **35** (1930) 1217.

5. Experimental methods for observations of the fine structure of total cross-sections

Investigation of fine structure in the energy variation of the total cross-section requires the development of electron beams with a smaller energy spread than those of the experiments described above. Since electrons emitted from a hot filament will have an energy spread of several tenths of an electronvolt due to the temperature and the effect of the drop of potential along the filament, it is essential to use an energy monochromator to select electrons of a narrow band of energies before

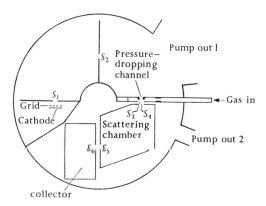

Fig. 1.5. Ramsauer type apparatus used by Golden and Bandel.

they enter the collision region. Golden and Bandel† have improved the resolution in a Ramsauer experiment so that the electron energy spread, between half maxima, is about 0·1 eV at 20 eV. Fig. 1.5 illustrates the experimental arrangement. The chamber was of all-metal construction and was part of a bakeable system within which the background pressure could be reduced to 10^{-9} torr. Contact potential differences were reduced by spraying all metal parts exposed to the electron beam with colloidal graphite. The cathode and energy-selection regions were pumped separately from the rest of the chamber so that background scattering in the selection region was reduced as was also gas cooling of the cathode surface.

The mean radius of the beam trajectory was 2·5 cm and the respective slit dimensions were, referring to Fig. 1.5, $S_1 = S_2 = 0·091 \times 0·500$ cm, $S_3 = 0·076 \times 0·500$ cm, $S_4 = 0·159 \times 0·599$ cm,

$$S_5 = S_6 = 0·401 \times 1·198 \text{ cm}.$$

† GOLDEN, D. E., and BANDEL, H. W., ibid. **138** (1965) A14.

5.1. *Use of spherical energy analyser*

Several investigators have used cylindrical electrostatic monochromators to obtain beams monochromatic to 0·06 to 0·10 eV. These are described in greater detail in Chapter 3, § 2.4 since they have found

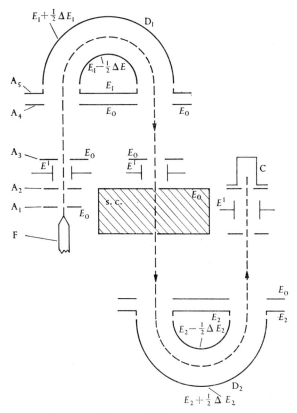

Fig. 1.6. Schematic diagram of the arrangement of apparatus developed by Simpson for observing fine structure in the variation with energy of the transmission of electrons through gases.

most application in the study of ionization cross-sections. Simpson[†] has developed an apparatus for studying total collision cross-sections of electrons in gases in which both monochromator and analyser are identical and consist of concentric spherical electrostatic deflectors as described by Purcell.[‡] These produce a point focus at 180° deflexion. Fig. 1.6 shows a schematic drawing of Simpson's apparatus. Electrons

[†] SIMPSON, J. A., Atomic collision processes, *Proc. 3rd Int. Conf. Phys. electron. atom. Collisions*, p. 128. (North Holland, Amsterdam, 1963).
[‡] PURCELL, E. M., *Phys. Rev.* **54** (1938) 818.

from the hot filament F are accelerated toward the anode A_1, maintained at a potential E_0 ($\simeq 20$ V) relative to F. The electrodes A_3, A_4, A_5 constitute a lens that focused the slit in A_1 on to the entrance of the hemispherical monochromator D_1. At the same time the beam is decelerated by maintaining A_5 at potential E_1 ($\simeq 1.35$ V) and focused after deflexion through 180° in the monochromator by means of the deflexion voltage, ΔE_1. It is then re-accelerated and passed through the scattering chamber, maintained at potential E_0, decelerated again into the second deflector D_2 maintained at potential E_2, and again accelerated before being collected by C. Effects of slit scattering and secondary electron emission are reduced by dispensing with physical slits in the focal planes of the deflectors. At each change of energy a virtual slit is imaged on a physical slit, or vice versa. The inter-electrode gap in the deflectors D_1, D_2 is comparatively large (0·63 cm) compared to their mean radius (2·54 cm). The electrodes E^1, shown in Fig. 1.6 enable the beam to be centred electrostatically.

The scattering chamber s.c. consists of a stainless steel cylinder, 2 cm long, closed at the ends by molybdenum discs containing apertures 0·5 mm in diameter. The gas enters through a 5-mm diameter tube midway between the apertures. The scattering assembly is mounted inside a stainless-steel vacuum envelope and, after baking, pressures as low as 10^{-9} torr can be obtained using a mercury diffusion pump with liquid air trap. Under these high-vacuum conditions the width of the beam at half the maximum energy is $\simeq 0.038$ eV and the collected current 3×10^{-10} A. Great care is taken to shield the whole apparatus magnetically and to bypass all leads into the chamber in order to remove r.f. pick-up. With this apparatus, electrons that have been scattered through an angle greater than 0·02 rad, or lost an energy greater than 0·05 eV, are not collected by C. Accurate measurements of the pressure in the scattering chambers are not possible. The pressure in the vacuum envelope is used to monitor the pressure in the scattering chamber. Thus the apparatus is used to detect fine structure in the variation of total cross-section with electron energy rather than for the accurate absolute measurement of cross-sections.

5.2. *Use of the retarding potential difference method*

Schulz[†] has also carried out transmission experiments with high-energy resolution, as well as making observations of the energy variation of the intensity of electrons scattered through a fixed angle (72°). Fig. 1.7

[†] SCHULZ, G. J., ibid. **136** (1964) A650.

illustrates the arrangement of the apparatus in his transmission experiments. To improve the energy resolution the retarding potential difference (r.p.d.) method (see also Chap. 3, § 2.4.2) could be used.

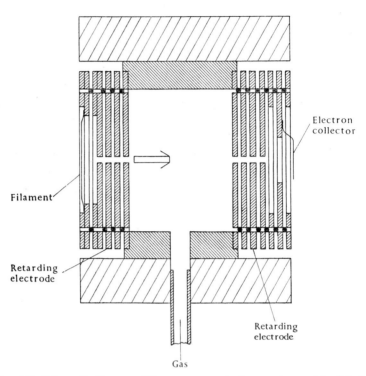

Fig. 1.7. Schematic diagram of the arrangement of apparatus used by Schulz for observing fine structure in the variation with energy of the transmission of electrons through gases.

The principle of this method is as follows. Referring to Fig. 1.8, consider the three-electrode system consisting of a cathode C, an electrode E from which the beam issues, and an intermediate electrode D. E is grounded and C maintained at a potential V_1 below ground, so the energy of the electrons passing through E will be V_1 eV. If D is maintained at a potential $V_2 (< V_1)$ below ground only electrons emitted from C with energy $> e(V_1 - V_2)$ can pass through the slit S in D to issue from E with energy eV_1. Suppose now that V_1 is kept fixed and measurements of the current issuing from E are made for values

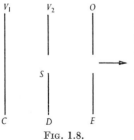

Fig. 1.8.

V_2, $V_2-\Delta V$ of the potential of D. The difference of these two currents gives the contribution from electrons emitted from the filament with energies between $e(V_1-V_2)$ and $e(V_1-V_2+\Delta V)$, i.e. electrons with an energy spread $e\Delta V$. The energy of the issuing beam may be varied while maintaining the constant energy spread if V_1 is varied keeping V_1-V_2 and ΔV constant. In practice this may be applied with ΔV close to 0·1 V.

Electrons from a thoria-coated iridium filament were aligned by a magnetic field and traversed the space between the first three plates that could be used as electrodes for application of the r.p.d. method. They then entered the collision chamber 1·5 cm long containing the scattering gas at about 0·3 torr pressure. Those leaving the chamber were decelerated to nearly zero energy and collected. This deceleration discriminates strongly against the collection of scattered electrons. The electrodes were gold-plated and the tube could be baked at 400° C.

To facilitate the search for fine structure the tube was operated with a.c. A square voltage wave of 0·1 V was applied to the first retarding electrode of the r.p.d. system at a frequency of about 17 c/s and the current measured by a preamplifier. The a.c. signal from the modulated difference current was then amplified and synchronously detected.

Schulz[†] applied this technique to observe fine-structure effects in the total cross-section of atomic hydrogen. The gas containing hydrogen circulated through the collision chamber after passage through a Pyrex bulb in which it was partially dissociated. To reduce loss of atomic hydrogen by recombination on metal surfaces the collision chamber was formed by 5 wires each 0·1 mm in diameter.

It was found convenient to mix water vapour with the incoming gas, partly to enhance dissociation and partly to balance out the decrease with energy of the total cross-section of atomic and molecular hydrogen with that of water vapour which, in the energy range of interest, increases with energy (see Chap. 10, Fig. 10.22 (d)). Fine structure shows up prominently against a nearly uniform background.

No attempt was made to obtain the absolute value of the atomic hydrogen cross-section. It was verified that the fine structure observed was due to H atoms by showing that it remained unaltered when the ionizing hydrogen was mixed either with H_2O, Ne, or both H_2O and Ne or He, and was certainly not due to H_2.

In all these experiments it was important to determine accurately the energies at which rapid variations of the scattering cross-sections

[†] SCHULZ, G. J., *Phys. Rev. Lett.* **13** (1964) 583.

occurred. This was done by comparing these energies with other well-known atomic excitation energies. For example, Schulz calibrated his energy measurements relative to the energy of excitation to the 2^3S state of He at 19·8 eV. Simpson introduced an admixture of argon into the collision chamber and related his measured energies to the energy of ionization to the $^2P_{\frac{1}{2}}$ state of A+ at 15·9 eV.

6. Observed total collision cross-sections of atoms[†]

It has been customary in describing experimental results to give the observed value of α at 1 torr pressure, usually called the 'probability of collision'.

To obtain Q in units of πa_0^2 (where $a_0 = 0.53 \times 10^{-8}$ cm is the radius of the first Bohr orbit of the hydrogen atom, usually taken as the atomic unit of length) we have Q (in units of πa_0^2) $= 0.318\alpha$.

6.1. Broad features of observed cross-sections

We discuss first the results obtained without employing special techniques for producing electron beams with very small energy spread.

In Fig. 1.9 is illustrated the variation of Q, in units of πa_0^2, with electron velocity expressed in $\sqrt{}$(volts), as observed for the three rare gases argon, krypton, and xenon. It is immediately obvious that the observed variation is not what would be expected on classical ideas. The slower the electrons the more effectively should they be scattered by the atomic field. On the contrary, a pronounced maximum is observed for each of the three gases for electrons with energy in the neighbourhood 8–15 eV. The gases are practically transparent to electrons of about 1 eV energy. For even slower electrons the cross-section rises again, the increase being more pronounced the heavier the rarer gas atom.

The remarkable transparency of the heavier rare gas atoms towards electrons of energy 1 eV or so was discovered independently by Ramsauer[‡] and by Townsend and Bailey[§] using a more indirect method (see Chap. 2, § 7.3), so will be referred to as the Ramsauer–Townsend effect. The measurements illustrated in Fig. 1.9 represent the average of those obtained by several observers.[||] Over most of the electron energy range the agreement between the different observers is within

[†] See the summary by KOLLATH, R., *Phys. Z.* **31** (1930) 985.
[‡] RAMSAUER, C., *Annln Phys.* **64** (1921) 513; **66** (1921) 546.
[§] TOWNSEND, J. S. and BAILEY, V. A., *Phil. Mag.* **43** (1922) 593.
[||] RAMSAUER, C., *Annln Phys.* **72** (1923) 345; BRODE, R. B., *Phys. Rev.* **25** (1925) 636; RUSCH, M., *Annln Phys.* **80** (1926) 707; BRÜCHE, E., ibid. **84** (1927) 279; RAMSAUER, C., and KOLLATH, R., ibid. **3** (1929) 536; NORMAND, C. E., *Phys. Rev.* **35** (1930) 1217.

10 per cent, indicating that the true total collision cross-section has been measured.

The total cross-section–velocity curves illustrated in Fig. 1.9 are by no means typical of other atoms. Thus in Fig. 1.10 the observed curves obtained for helium† and for neon‡ are illustrated. It will be noticed

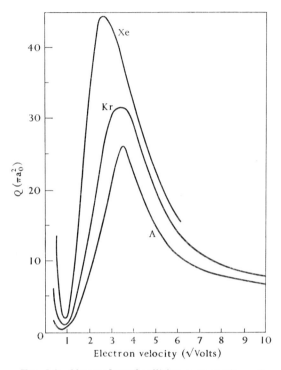

Fig. 1.9. Observed total collision cross-sections of A, Kr, and Xe.

that the cross-sections are very much smaller on the whole than for the other rare gases and do not exhibit such marked variations with electron velocity.

Fig. 1.11 illustrates observed results for atomic hydrogen. The full line curves give the results of Neynaber, Marino, Rothe, and Trujillo,§

† RAMSAUER, C., *Annln Phys.* **66** (1921) 546; BRÜCHE, E., ibid. **84** (1927) 279; RAMSAUER, C. and KOLLATH, R., ibid. **3** (1929) 536; NORMAND, C. E., *Phys. Rev.* **35** (1930) 1217.

‡ RAMSAUER, C., *Annln Phys.* **66** (1921) 546; RUSCH, M., *Phys. Z.* **26** (1925) 748; BRÜCHE, E., *Phys.* **84** (1927) 278; RAMSAUER, C. and KOLLATH, R., loc. cit.; NORMAND, C. E., loc. cit.

§ NEYNABER, R. H., MARINO, L. L., ROTHE, E. W., and TRUJILLO, S. M., *Phys. Rev.* **124** (1961) 135.

using the method of § 4.2.3 that determines directly at each electron energy the ratio of the total cross-sections of atomic and molecular hydrogen. Comparison with the results obtained by Brackmann et al.†

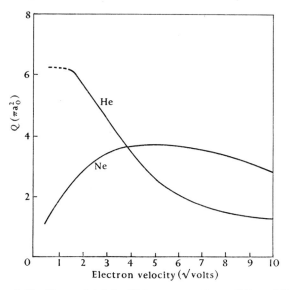

FIG. 1.10. Observed total collision cross-sections of He and Ne.

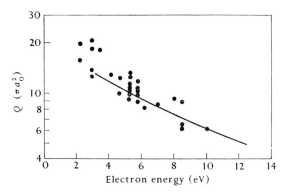

FIG. 1.11. Total collision cross-sections of atomic hydrogen.
● observed by Brackmann, Fite, and Neynaber; ⎯⎯ observed by Neynaber, Marino, Rothe, and Trujillo.

requires some knowledge of the angular distribution of the scattered electrons. Assuming this to be given by the theory as discussed in Chapter 8, § 2 the data of Brackmann et al. give the points shown in Fig. 1.11 which agree fairly well with the results of Neynaber et al. It

† BRACKMANN, R. T., FITE, W. L., and NEYNABER, R. H., *Phys. Rev.* **112** (1958) 1157.

is to be noted that there is no sign of a Ramsauer effect for atomic hydrogen.

Although the accuracy yet attained is not high, the general features of the total cross-section for atomic oxygen are known. Fig. 1.12 compares the results obtained by Sunshine et al.† (see p. 13) with those of Neynaber et al. (see p. 17)‡ for the ratio of the atomic to the molecular cross-section. There is agreement within the uncertainty of the observa-

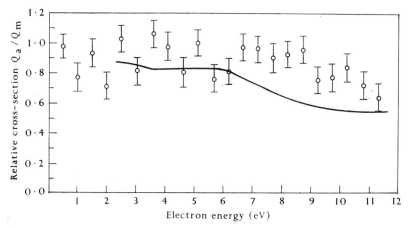

FIG. 1.12. Observations of the ratio of the total collision cross-sections of atomic and molecular oxygen Q_a, Q_m respectively. ⌀ observed by Sunshine, Aubrey, and Bederson; —— observed by Neynaber, Marino, Rothe, and Trujillo (smooth curve fitting the observed data).

tions for energies below 6 eV but at higher energies the results of Sunshine et al. lie definitely higher. Additional uncertainty is introduced in obtaining the absolute atomic cross-section because that for the molecule is not known very accurately. Marino et al. obtained values for O_2 that agree well with those of Brüche§ (see Chap. 10, Fig. 10.21), whereas Sunshine et al. find cross-sections which, while exhibiting much the same energy variation, are all about 25 per cent larger. On the other hand, Sunshine et al. find good agreement with measurements made by Ramsauer and Kollath‖ at energies below 1·5 eV (see Chap. 10, Fig. 10.21).

Fig. 1.13 shows values for atomic oxygen obtained directly by Sunshine et al. by the method described in § 4.2.2. It seems likely that the true cross-section varies with energy as indicated by the line in

† SUNSHINE, G., AUBREY, B. B., and BEDERSON, B., Phys. Rev. **154** (1967) 1.
‡ NEYNABER, R. H., MARINO, L. L., ROTHE, E. W., and TRUJILLO, S. M., ibid. **123** (1961) 148.
§ BRÜCHE, E., Annln Phys. **13** (1927) 1065.
‖ RAMSAUER, C. and KOLLATH, R., ibid. **4** (1930) 91.

Fig. 1.13, but the absolute magnitude is uncertain to at least 30 per cent. Comparison of the result with theory is discussed in Chapter 8, § 9.2.

Fig. 1.14 illustrates the observations of the cross-section for atomic

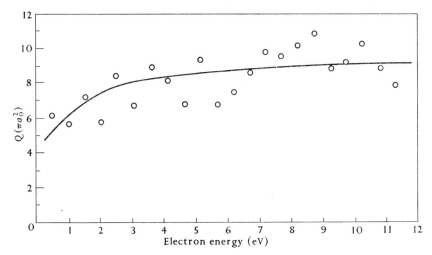

FIG. 1.13. Observed total collision cross-sections of atomic oxygen. ○ observed by Sunshine, Aubrey, and Bederson; —— probable best fit.

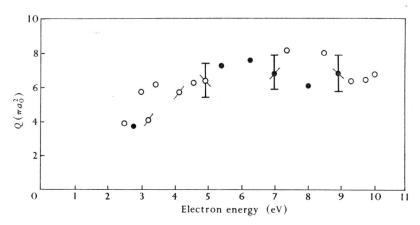

FIG. 1.14. Observed total collision cross-sections of atomic nitrogen. ○ ● ◊ ⌀ ↯ ↟ denote experimental points of increasing reliability.

nitrogen made by Neynaber, Marino, Rothe, and Trujillo† (see § 4.2.3). The experimental uncertainty is rather large, particularly at electron energies below 5 eV. However, the investigators consider that $4 \cdot 5\pi a_0^2$ is

† NEYNABER, R. H., MARINO, L. L., ROTHE, E. W., and TRUJILLO, S. M., *Phys. Rev.* **129** (1963) 2069.

an upper limit to the total cross-section between 1·6 and 2·2 eV. Comparison of these results with theory is discussed in Chapter 8, § 9.2.

Fig. 1.15 illustrates Brode's[†] measurements for zinc, cadmium, and mercury. Apart from some differences in detail, the behaviour of the cross-section velocity curves is generally similar for all three of these chemically similar atoms. It will be noticed that the cross-section increases rapidly at low electron velocities to values considerably in

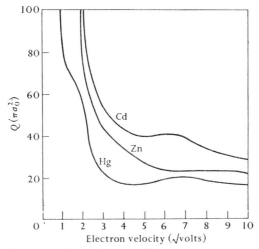

FIG. 1.15. Observed total collision cross-sections of Cd, Zn, and Hg.

excess of those observed for the rare gases. The greatest cross-sections observed, however, are those for the alkali metals. The results of Brode's[‡] measurements for sodium, potassium, and caesium and of the measurements of Perel et al.[§] for lithium are shown in Fig. 1.16. Again the behaviour of the chemically similar set of atoms is much the same, both as regards variation with velocity and the huge magnitude which is attained at certain electron energies. It is noteworthy also that, for each atom, the cross-section attains a sharp maximum for electrons with energy of order 2–3 eV, falls sharply for a small range of lower energy, and then appears, from the observations in sodium and caesium, to begin rising again. Once more we have here a complicated velocity variation. The only other monatomic substance that has been investigated is thallium, and Brode's measurements[||] for this atom are illustrated in Fig. 1.17.

[†] BRODE, R. B., *Proc. R. Soc.* A**125** (1929) 134; *Phys. Rev.* **35** (1930) 504.
[‡] Ibid. **34** (1929) 673. [§] loc. cit., p. 12.
[||] BRODE, R. B., *Phys. Rev.* **37** (1931) 570.

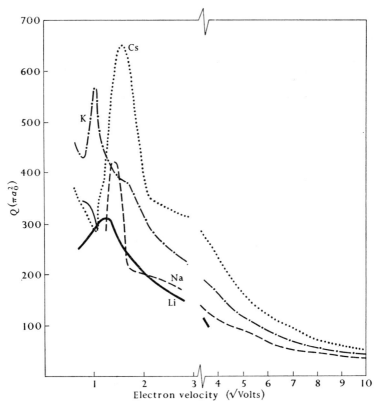

Fig. 1.16. Observed total collision cross-sections of Li, Na, K, and Cs.

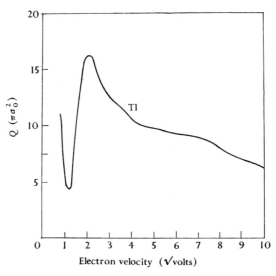

Fig. 1.17. Observed total collision cross-section of Tl.

6.2. *Observed fine structure in total cross-sections of atoms*

6.2.1. *Introductory—the first observations of fine structure.* The results so far discussed all refer to experiments using electron beams with rather a large energy spread (up to 0·5 eV). When nearly monochromatic electron beams with energy spread $\simeq 0\cdot05$ eV are used a characteristic fine structure in the variation of cross-section with energy has been found in some cases, even when the energy of the incident electrons is below the threshold for excitation or ionization of the scattering atom concerned. The first observation of such a fine structure in the elastic scattering cross-section was made by Schulz,† who found a sharp reduction in the elastic scattering of electrons at a fixed angle of 72° in helium at an energy of $19\cdot3\pm0\cdot1$ eV, i.e. 0·5 eV below the threshold of the first excited (2^3S) state. His results are shown in Fig. 1.18 (a).

Shortly afterwards Simpson‡ observed a similar 'resonant' behaviour in the form of a sharp peak in the total cross-section for electron scattering in helium at the same energy and his measurements of the transmitted electron intensity at different energies in helium are shown in Fig. 1.18 (b).

Fleming and Higginson§ also independently observed the same effects using an apparatus primarily designed for the study of inelastic collisions of electrons in helium with energy close to the 2^3S excitation threshold. This method, essentially due to Maier–Leibnitz, is described in Chapter 5, § 2.2. It is of interest to note that, as shown in Fig. 1.18 (c), Golden and Bandel‖ (see Fig. 1.5) were able to observe the resonance as a sharp reduction in the total cross-section.

Fine structure was observed at about the same time for neon, and since then a great amount of data has become available for all the rare gases, for atomic hydrogen, and for mercury. We shall describe these data in some detail.

6.2.2. *Atomic hydrogen.* The first observations of fine structure in the total cross-section of atomic hydrogen were made by Schulz,†† using the equipment described in § 5.2. Fig. 1.19 (a) reproduces his results, which show a marked increase in transmission of electrons with energy centred round $9\cdot7\pm0\cdot15$ eV. As the energy resolution was not better than 0·3 eV much detail is obscured.

Confirmation of the existence of structure at about the same energy

---

† Schulz, G. J., *Phys. Rev. Lett.* **10** (1963) 104.
‡ Simpson, J. A. and Fano, U., ibid. **11** (1964) 158.
§ Fleming, R. J. and Higginson, G. S., *Proc. phys. Soc.* **81** (1963) 974.
‖ Golden, D. E. and Bandel, H. W., *Phys. Rev.* **138** (1965) A14.
†† Schulz, G. J., *Phys. Rev. Lett.* **13** (1964) 583.

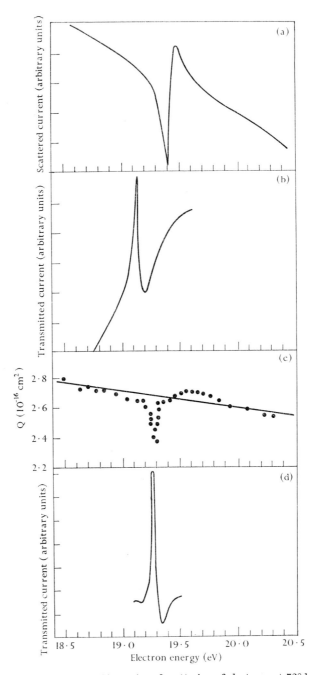

Fig. 1.18. (a) Variation of intensity of scattering of electrons at 72° by helium atoms with electron energy, as observed by Schulz. (b) Variation with electron energy of the transmission of electrons through helium as observed by Simpson. (c) Variation with electron energy of the total cross-section for elastic scattering of electrons by helium atoms, as observed by Golden and Bandel. ●, experimental points. (d) Variation with electron energy of the transmission of electrons through helium as observed by Golden and Nakano.

was provided by the observations of Kleinpoppen and Raible† who used a modulated cross-beam apparatus essentially similar to that described in § 4.2.1. A hydrogen beam, 85 per cent dissociated by passage through a tungsten oven at 2600 °K, was chopped and then crossed by an electron

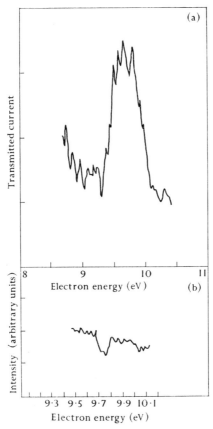

FIG. 1.19. (a) Variation with electron energy of the transmission of electrons through dissociated hydrogen, as observed by Schulz. (b) Variation of intensity of scattering of electrons at 94° by an 85 per cent dissociated hydrogen beam, as observed by Kleinpoppen and Raible.

beam rendered homogeneous in energy to about 0·1 eV by passage through a 127° electrostatic analyser. The intensity of electron scattering through 94° was monitored, as a function of electron energy, by an electron multiplier using phase-sensitive detection. Fig. 1.19 (b) illustrates the observed variation, showing a reduction of intensity over approximately the same energy range as that which Schulz observed in

† KLEINPOPPEN, H. and RAIBLE, V., *Phys. Lett.* **18** (1965) 24.

transmission. Although the width of the resonance is about four times smaller than in Schulz's experiments the resolution was still not high enough to reveal further structure.

Much greater detail was revealed in the experiments of McGowan, Clarke, and Curley.† They studied the variation with energy of the intensity of electrons scattered at 90° to the plane of intersection of a predominantly atomic hydrogen beam and an electron beam. The

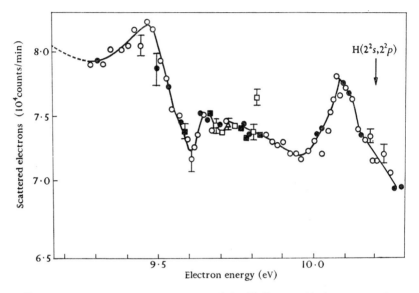

FIG. 1.20. Resonance structure observed by McGowan, Clarke, and Curley in the intensity of scattering of electrons through 90° in atomic hydrogen.
——— mean curve through observed points.

electron beam was rendered homogeneous in energy by means of an electrostatic analyser of the Clarke–Marmet–Kerwin type (Chap. 3, § 2.4). It was estimated from observations of the half-depth of the helium resonance at 19·30 eV that the half-width of the energy distribution was between 0·08 and 0·09 eV. The electron energy scale was calibrated by the position of the minimum in the main helium resonance and also by observing the onset potential for production of H⁺ ions (13·60 eV).

Fig. 1.20 shows the resonance structure observed, exhibiting much more detail than in the earlier experiments and indicating the presence of more than one resonance. One of these, with a maximum at 9·45 and minimum at 9·6 eV, is very clearly defined and there is evidence

† McGowan, J. W., Clarke, E. M., and Curley, E. K., *Phys. Rev. Lett.* **15** (1965) 917.

of a second and much weaker resonance with a minimum near 9·7 eV Comparison with the earlier experiments (see Fig. 1.19) shows that the centre of the distribution of the fine structure effects occurs quite close to that, 9·7±0·5 eV, observed by Schulz. The agreement with the experiments of Kleinpoppen and Raible is not so close but it must be remembered that we can expect the relative importance of different resonances to vary with angle of scattering, and McGowan has shown that when this is taken into account the disagreement largely disappears. This is discussed in Chapter 9, § 4 in connection with a detailed theoretical analysis of the atomic hydrogen data.

6.2.3. *Helium.* A detailed study of the fine structure of the total cross-section for helium has been carried out by Kuyatt, Simpson, and Mielczarek,† using the apparatus described in § 5.1. This was concerned primarily with the location of sharp resonances and not with the absolute magnitude of the cross-section. The method of operation was to apply a slowly varying voltage to change the electron energy in the scattering chamber. This voltage was simultaneously applied to the X-axis of an XY recorder, the Y-axis of which was applied to output of the vibrating reed electrometer that measured the transmitted current. With no scattering gas present, the plots of transmitted current versus electron energy showed a broad maximum due to an electron-optical focusing effect‡ associated with changing the energy. The location of this maximum could be changed by adjusting the various focusing voltages so that fine structure under study could be displayed to advantage.

Fig. 1.21 illustrates two plots of this kind. In (a) the principal resonance near 19·3 eV shows up very prominently as a sharp peak followed by a shallow depression. The small maxima at A and B are considered to be due to excitation of the 2^3S and 2^1S states at 19·818 and 20·614 eV respectively. If this is accepted the peak transmission with the principal resonance occurs at 19·31±0·03 eV, which agrees well with 19·3±0·1 eV found by Schulz for the minimum scattering at 72°.

In plot (b) further, less marked, fine structure is visible. Sometimes this takes the form of a peak followed by a depression, as for the principal resonances, and sometimes the reverse. The location of the resonances

† KUYATT, C. E., SIMPSON, J. A., and MIELCZAREK, S. R., *Phys. Rev.* **138** (1965) A385.
‡ This effect arises as follows. The beam issuing from the monochromator is diverging and the electrodes in the accelerating region form an electron lens which focuses it. Before the accelerating potential is scanned the various voltages in this lens are adjusted to give maximum detector current. However, they are not adjusted so as to maintain these conditions during scanning. Scansion of the accelerating potential in either direction therefore produces a reduction in current through the lens which depends on the electron energy.

with respect to the 2, 3, 4, 5, and 6 quantum singly excited states of helium is indicated on the diagram.

Table 1.1 summarizes the location and character of the fine structure effects observed.

Fig. 1.22 shows the presence of two resonances at considerably higher energies $57 \cdot 1 \pm 0 \cdot 1$ eV and $58 \cdot 2 \pm 0 \cdot 1$ eV.

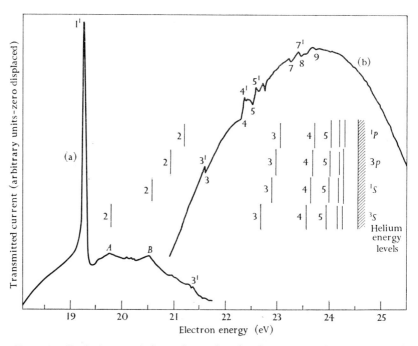

FIG. 1.21. Typical transmission–voltage plots for electrons in helium as observed by Kuyatt, Simpson, and Mielczarek.

The behaviour of the total cross-section for helium at electron energies below 3 eV has recently been the subject of intensive experimental study. This is because of the evidence from earlier investigations of fine structure in this energy range. Thus reference to Fig. 1.23 shows in an expanded energy scale, the observations of Normand† and of Ramsauer and Kollath.‡ Both exhibit rather similar structure though a little displaced in energy. Golden and Bandel§ using the Ramsauer-type apparatus described in § 5 paid special attention to examining the reality of this structure. At first they found some evidence of structure that was reproducible, but when care was taken to avoid following the same

† loc. cit., p. 25.   ‡ loc. cit., p. 25.   § loc. cit., p. 19.

TABLE 1.1

Structure observed in transmission of electrons through helium

Level	Type	Energy (eV)
1′	Peak	19·31±0·01
1	Depression	19·37±0·01
2′	Peak	19·43±0·01
2	Depression	19·47±0·01
A	Break	19·818
B	Break	20·59±0·01
3′	Peak	21·50±0·1
3	Depression	21·55±0·1
4	Depression	22·34±0·02
4′	Peak	22·39±0·02
5	Depression	22·54±0·02
5′	Peak	22·60±0·02
6′	Peak	22·81±0·02
6	Depression	22·85±0·02
7	Depression	22·30±0·02
7′	Peak	23·44±0·02
8	Depression	23·49±0·02
9	Depression	23·75±0·05
9′	Peak	23·82±0·05

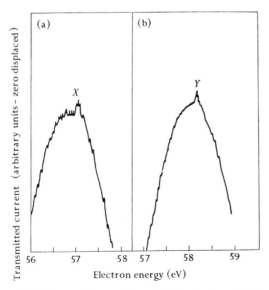

FIG. 1.22. Transmission–voltage plots for electrons in helium showing resonance effects at energies close to 60 eV.

routine in carrying out the observations this degenerated to merely a small scatter in the results. However, they were hampered by the fact that the variation of the total cross-section with electron energy could not be continuously recorded at energies below 1·5 eV. Golden and Nakano† therefore constructed new equipment which, while possessing good energy resolution and sensitivity, permitted continuous recording down to very low electron energies.

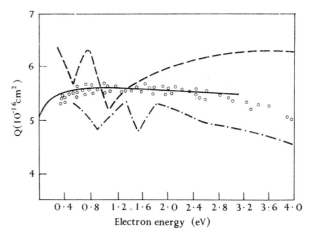

FIG. 1.23. Total collision cross-sections for low-energy electrons in helium observed by Ramsauer and Kollath, Normand, and Golden and Bandel. – – – – Ramsauer and Kollath; –·–·–· Normand; o Golden and Bandel.

Fig. 1.24 illustrates the experimental arrangement. The chamber was of an all-metal construction. For energy resolution a 127° electrostatic analyser of the type designed by Hughes and McMillen‡ was used. By application of suitable voltages to the electrodes of lens 1 a beam of electrons from the gun was focused on the entrance slit of the analyser, which was operated with positive and negative voltages V_1, V_2 (with respect to ground) on the inner and outer electrodes respectively. In a similar way, using lens 2, the emergent beam was focused through the scattering chamber into the collector. Dimensions of the apparatus are as follows:

Grid slit, and slits in lenses 1 and 2	0·039 × 0·394 in
Distance of grid from first slit of lens 1	0·125 in
Lenses 1 and 2	0·3 × 1·25 in

† GOLDEN, D. E. and NAKANO, H., *Phys. Rev.* **144** (1966) 71.
‡ HUGHES, A. L. and MCMILLEN, J. H., ibid. **39** (1932) 585.

Distance of lens from respective entrance and exit
of analyser   0·25 in
Radii of analyser plates   1·969, 2·362 in
Differential pumping slit   0·050 × 0·500 in
Scattering chamber and electron collector slits   0·060 × 0·500 in
Scattering chamber   1·00 in long
Distance of scattering chamber from electron
collector slits   0·150 in

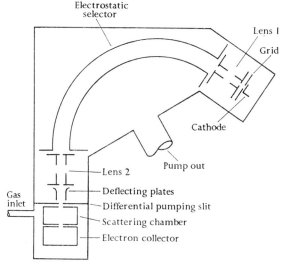

FIG. 1.24. Apparatus used by Golden and Nakano to search for fine structure in the total cross-section of helium for impact with low-energy electrons.

An ultra-high vacuum system was used capable of an ultimate pressure less than 10^{-9} torr.

The energy of the electron beam was determined by making retarding potential measurements on the collector and the half-width was found to be about 0·05 eV. To vary the electron energy, the voltage applied to the scattering chamber was varied.

Fig. 1.18 (d) shows a recorder tracing of the transmitted current in passing through the main helium resonance, and it gives a peak at 19·27±0·08 eV with half-width 0·05±0·005 eV. To test the ability of the apparatus to observe resonances at low electron energies the elastic resonances in nitrogen (Chap. 10, § 3.5.2) that occur over an energy range from 1·4 to 3·8 eV were looked for and clearly observed. No evidence of any fine structure was found in helium between about 0·1 and 19·3 eV, despite the fact that the sensitivity for detection of cross-section changes was such that a proportional change higher than 2×10^{-4} could have been detected.

O'Malley† has suggested that the fine structure found in the earlier experiments may have been due to the presence of N_2 and O_2 as impurities. In any case it now seems very unlikely that there is any fine structure in the helium total cross-section in this low-energy region.

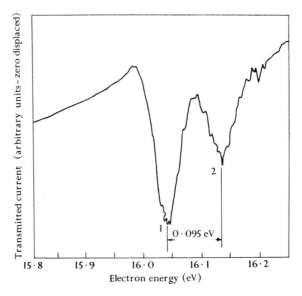

Fig. 1.25. Transmission–voltage plots for electrons in neon.

6.2.4. *Neon.* The first resonance was observed in neon near 16·0 eV by Schulz‡ and this was shown by Simpson§ to be double. Fig. 1.25 reproduces a transmission–voltage plot obtained by Kuyatt *et al.*‖ that shows this very clearly. In contrast to the principal resonance in helium, those in neon appear predominantly as deep depressions. The separation of the two is 0.095 ± 0.002 eV. Table 1.2 lists the structural features observed and their location in energy. The voltage scale was calibrated by first comparing with the position of the helium resonance and then with the inelastic thresholds associated with the $^3P_2$ and $^3P_0$ states at 16·619 and 16·715 eV respectively. These were consistent within 0·05 eV.

† O'Malley, T. F., *Phys. Rev.* **130** (1963) 1020.
‡ loc. cit., p. 31.
§ loc. cit., p. 31.   ‖ loc. cit., p. 35.

TABLE 1.2

Fine structure observed in transmission of electrons through neon

Feature	Energy (eV)
1	16·04±0·02
2	16·135±0·02
A	16·62±0·03
B	16·715 (calibrating point)
3	18·18±0·05
4	18·29±0·05
5	18·46±0·03
6	18·56±0·03

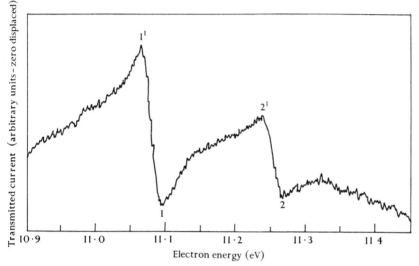

FIG. 1.26. Transmission–voltage plot for electrons in argon.

TABLE 1.3

Fine structure observed in transmission of electrons through argon

Feature		Energy (eV)
1'	Peak	11·064±0·002
1	Depression	11·094
2'	Peak	11·235±0·002
2	Depression	11·267±0·003

6.2.5. *Argon, krypton, and xenon.* Fig. 1.26 illustrates a typical plot† showing the presence of two resonances about 0·5 eV below the first excited state of argon. These are located as given in Table 1.3.

† KUYATT, C. E., SIMPSON, J. A. and MIELCZAREK, S. R., loc. cit., p. 35.

Typical results for krypton[†] are shown in Fig. 1.27. In this case the only clear feature is a peak at 9·45 eV followed by a depression at 9·48±0·01 eV.

For xenon more detail was found as given in Table 1.4.

TABLE 1.4

Fine structure observed in transmission of electrons through xenon

Feature		Energy (eV)
1'	Peak	7·74
1	Depression	7·77±0·01
2	Depression	9·02±0·04
3'	Peak	9·33±0·02
3	Depression	9·40±0·02
A		9·45±0·02—threshold for excitation of $5p^5(^2P^0_{\frac{1}{2}})6s$
4'	Peak	10·71±0·02
4	Depression	10·76±0·02
5'	Peak	10·81±0·02
5	Depression	10·86±0·02

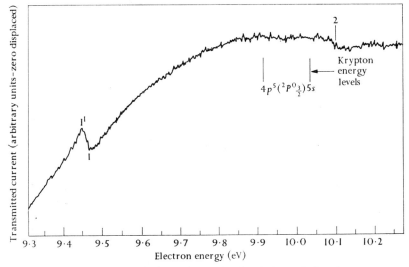

FIG. 1.27. Transmission–voltage plot for electrons in krypton.

In his transmission measurements (§ 5.2) Schulz,[‡] while observing fine structure in krypton and xenon close to the principal resonance features referred to above, did not resolve the peaks and depressions at all clearly, presumably because of poorer electron-energy resolution.

[†] KUYATT, C. E., SIMPSON, J. A. and MIELCZAREK, S. R., loc. cit., p. 35.
[‡] loc. cit., p. 31.

Table 1.5

Fine structure observed in transmission of electrons through mercury

Feature	Energy (eV)	Feature	Energy (eV)
1	4.07 ± 0.01	7	8.22 ± 0.05
2	4.30	8	8.83 ± 0.05
A	4.68 ± 0.02	9	8.99 ± 0.05
3	4.89 ± 0.01	10	9.75 ± 0.05
4	7.81 ± 0.05	11	10.29 ± 0.05
5	7.94 ± 0.05	12	10.58 ± 0.05
6	8.14 ± 0.05	13	10.88 ± 0.08

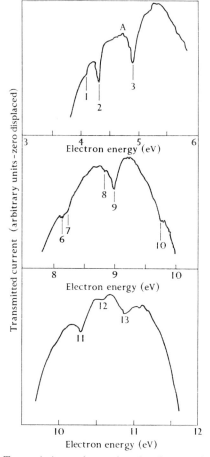

FIG. 1.28. Transmission–voltage plots for electrons in mercury.

6.2.6. Mercury. The features observed in transmission of electrons through measuring vapour are listed in Table 1.5. Typical plots showing these features are shown in Fig. 1.28.

6.3. *Summarizing remarks*

We have now described both the broad features and the fine structure in the energy variation of total cross-sections of atoms for electron impact. There is clearly a great deal to be understood theoretically even as far as the broad features are concerned. We have only to refer to the Ramsauer–Townsend effect for the heavier rare gases and the large peaked cross-sections for the alkali metal atoms. It will appear from the theoretical discussion given in Chapter 6 that these features can be interpreted in terms of a model in which the scattering atom is a source of a central potential acting on the electron.

The fine structure cannot be interpreted in this way. It must be supposed to arise through resonant capture of electrons to form unstable negative ions from which they are subsequently ejected. The energies of these negative ion states will lie closely below those of excited states of the atoms concerned, and this explains why the fine structure described in § 6.2 usually occurs at an energy not far below an excitation threshold. The resonance effects appear in a variety of other ways and further experimental results concerning them are discussed in Chapter 3, § 2.5.3, Chapter 4, § 1.4.2.2, and Chapter 5, § 5.3.3. A detailed theoretical interpretation is given in Chapter 9. Resonance phenomena associated with collisions of electrons with molecules are discussed in Chapters 10–13.

2

SWARM EXPERIMENTS WITH ELECTRONS IN ATOMIC GASES—MOMENTUM-TRANSFER CROSS-SECTION

1. Diffusion of electrons through gases

WE now consider the diffusion of a swarm of electrons through a gas at pressure p under the influence of a constant electric field of strength F. This is of interest, not only from its application to electric discharge and other phenomena, but also because it can provide valuable information about the collision cross-sections of atoms and molecules towards slow electrons. It is particularly useful in determining the mean energy loss suffered by a slow electron on collision with a gas molecule. The electrons concerned may have too small an energy to produce any electronic excitation within the molecule but may excite vibration or rotation. In an atomic gas the only loss will occur in so-called elastic collisions and may be calculated from a knowledge of the atomic weight of the gas. Nevertheless, it is of importance to study diffusion in such cases to show that the method is on a sure foundation. We therefore discuss first the experimental and theoretical researches that have been carried out concerning the diffusion of slow electrons through the rare gases. These studies are of historical interest also inasmuch as Townsend and Bailey[†] observed in this way the anomalous behaviour of slow electrons in argon as long ago as 1921, independently of Ramsauer's investigations.[†] It will be shown that the two sets of results are entirely compatible. In Chapter 11 the application of the diffusion method to the study of energy losses due to excitation of molecular vibration and rotation, and in Chapter 12 to negative ion formation, will be described.

The principle of the method that may be used to determine the mean free path and mean energy loss for collision is as follows. As the electrons diffuse they gain energy from the field but, in the steady state, this is balanced by the energy lost in collision with the gas atoms. In this steady state the swarm will possess a definite mean drift velocity u in the direction of the field and also a definite mean kinetic energy ϵ. A very simple approximate argument shows how these quantities are

† loc. cit., p. 24.

related to the collision properties of the gas atoms towards the electrons.

We suppose that the mean free path of the electrons in the gas has a constant value l and that the fraction of its energy that an electron loses in a collision with a gas atom is also constant and equal to a small quantity λ. Then, if c is the mean velocity of the random motion of the electron, the number of collisions made in traversing a distance x in the direction of the field will be cx/ul since the actual length of path will be xc/u. The energy lost in these collisions will be $\lambda \epsilon cx/ul$. Hence, for a steady state,
$$\lambda \epsilon cx/ul = Fex, \qquad (1\,\text{a})$$
the energy gained from the field. As $\epsilon = \tfrac{1}{2}mc^2$ approximately, where m is the mass of an electron, we have
$$\lambda c^3/u = 2Fel/m. \qquad (1\,\text{b})$$

A second relation follows by considering the mean distance traversed in the direction of the field in the interval δt between collisions. Assuming all directions of motion after the collision to be equally probable, then this distance will be $u\,\delta t$, so that
$$\tfrac{1}{2}Fe(\delta t)^2/m = u\,\delta t.$$
Apart from a numerical factor of order unity, δt may be taken as l/c, giving
$$2uc = Fel/m. \qquad (2)$$

Since l is inversely proportional to the gas pressure p at a fixed temperature T, both c and u are functions of F/p. Also, if c and u are measured for a fixed value of F/p, then l and λ can be obtained. Early observations were carried out at room temperature so F/p referred unambiguously to this temperature. It is now often more convenient to consider transport coefficients such as c and u as functions of F/N, where N is the number of atoms per cm³. We then have F/p in V cm⁻¹ torr⁻¹ at 300° K $= 3\cdot 22 \times 10^{16} F/N$ V cm². In most cases we shall show quantities as functions both of F/p at 300° K and of F/N.

In practice l will not be a constant and the measurements of c and u will give certain mean values of l and λ over the velocity distribution of the electrons. We cannot therefore hope to derive, from experiments on electron swarms, information as definite about collision cross-sections as may be obtained from experiments of the Ramsauer type. It is possible, however, to work with electrons of lower energy than in the more direct experiments and to obtain information about the energy losses of electrons on collision with molecules which cannot be got in any other way.

To analyse the diffusion phenomena in more detail in relation to collision cross-sections, we shall consider a swarm of such small mean energy that inelastic collisions with the gas atoms do not occur. The only loss of energy in elastic collisions therefore arises from the finite value of the ratio of the mass m of an electron to that M of a gas atom. On the other hand, we shall suppose that the mean energy of the atoms is very much smaller than that of the electrons. We also suppose that the electron concentration in the stream is so small that interaction between the electrons can be ignored. Under these conditions it becomes possible, given the results of Ramsauer's experiments and also of measurements on the angular distribution of elastically scattered electrons, to compute both the drift velocity and the mean kinetic energy as functions of F/p. These may then be compared with values observed by methods described in Chapter 1, § 4 and good agreement will be found. This enables one to apply the technique with confidence to molecular gases in which the mean energy loss λ is not known beforehand and in which attachment may occur (see Chaps. 11 and 13).

1.1. *Momentum-transfer (diffusion) cross-section*

With the assumptions made, the fraction of its energy lost per impact by an electron that is scattered through an angle θ is given by

$$2(1-\cos\theta)m/M \tag{3}$$

if $(m/M)^2$ is ignored.

If $p(\theta)\sin\theta\, d\theta d\phi$ is the probability that, on collision, the electron be scattered into the solid angle $d\Omega$ about θ, the mean fractional energy loss per collision will be

$$2(m/M)\int_0^\pi\int_0^{2\pi}(1-\cos\theta)p(\theta)\sin\theta\, d\theta d\phi. \tag{4}$$

Referring to Chapter 1, §§ 2.1 and 2.2 we have, in terms of the total and differential cross-sections Q_0 and $I_0\, d\Omega$ for elastic scattering,

$$p(\theta) = I_0(\theta)/Q_0,$$

so the mean fractional energy loss per collision becomes

$$2(m/M)Q_d/Q_0,$$

where
$$Q_d = \int_0^\pi\int_0^{2\pi} I_0(\theta)(1-\cos\theta)\sin\theta\, d\theta d\phi, \tag{5}$$

and is usually referred to as the momentum-transfer cross-section.†

† Q_d is sometimes referred to as the diffusion cross-section on account of its importance in the discussion of diffusion phenomena (see Chap. 16, § 3.1).

Thus the fractional amount of energy lost by an electron in traversing a distance x in a gas containing N atoms/cm^3 is $2(m/M)NQ_\mathrm{d}x$, just as if the total elastic cross-section Q_0 were replaced by Q_d and the fractional energy lost per collision by $2m/M$. Thus, for purely elastic collisions, we should take $l = 1/NQ_\mathrm{d}$ and $\lambda = 2m/M$ in (1) and (2). It must be remembered, however, that in general Q_d will be a function of electron energy.

Q_d differs appreciably from Q_0 only when there is a pronounced concentration of scattering in either the backward or forward directions.

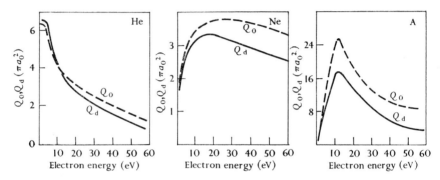

Fig. 2.1. Comparison of momentum transfer (diffusion) cross-sections with total collision cross-section for He, Ne, and A. ---- observed total cross-section; —— derived momentum transfer cross-section.

If $I_0(\theta)$ is a constant, independent of θ, then Q_d and Q_0 are equal. It will be shown in Chapter 6, § 3.2 that, for sufficiently slow electrons, the scattering is independent of angle so, in such cases, Q_d need not be distinguished from Q_0. At higher energies there is an appreciable difference as may be seen by reference to Fig. 2.1, in which Q_d and Q_0 are compared for argon, neon, and helium. In obtaining Q_d the observed angular distributions, described in Chapter 5, § 5.3, have been used.

If the mean energy ϵ_g of the gas molecules cannot be ignored the expression (3) for the mean fractional energy loss per collision must be modified by multiplication by the factor $1-\epsilon_\mathrm{g}/\epsilon$. If the electrons themselves have a Maxwellian distribution of mean energy $\bar{\epsilon}$ the corresponding factor becomes modified to $\tfrac{4}{3}(1-\epsilon_\mathrm{g}/\bar{\epsilon})$.

1.2. *The electron velocity distribution, drift, and random velocity*

Consider the steady state of a swarm of electrons diffusing through a gas containing N atoms/cm^3, of mass M, with momentum transfer cross-section Q_d, under the action of a uniform electric field F in the direction of the x-axis. Let $f(\xi, \eta, \zeta)\, \mathrm{d}\xi \mathrm{d}\eta \mathrm{d}\zeta$ be the fraction of electrons with velocities lying between $\xi, \xi+\mathrm{d}\xi;\ \eta, \eta+\mathrm{d}\eta;\ \zeta, \zeta+\mathrm{d}\zeta$. Then it may be shown that,† ignoring $(m/M)^2$,

where
$$f(\xi, \eta, \zeta) = f_0(v) + (\xi/v)f_1(v),$$
$$v^2 = \xi^2 + \eta^2 + \zeta^2 = 2\epsilon/m,$$

$$\left.\begin{aligned} f_0(v) &= A \exp\!\left\{-(6m/M)\int_0^\epsilon \epsilon\, \mathrm{d}\epsilon/\epsilon_l^2\right\} \\ f_1(v) &= -(\epsilon_l/mv)\, \mathrm{d}f_0/\mathrm{d}v \end{aligned}\right\}, \tag{6}$$

and A is such that
$$4\pi \int_0^\infty f_0(v) v^2\, \mathrm{d}v = 1.$$

ϵ_l is given by $Fe/NQ_\mathrm{d} = Fel$, the energy gained in falling through a free path in the direction of the field, and is in general a function of ϵ. If, however, Q_d is effectively constant for all electron energies of importance, then
$$\left.\begin{aligned} \pi\Gamma(\tfrac{3}{4})f_0(v) &= (3m/M)^{\frac{3}{4}}(m/2\epsilon_0)^{\frac{3}{2}}\exp\{-(3m/M)(\epsilon/\epsilon_0)^2\} \\ f_1(v) &= 6(m\epsilon/M\epsilon_0)f_0(v) \end{aligned}\right\}, \tag{7}$$

where ϵ_0 is written for the now constant value of Fel.

The effect of the variability of Q_d on the distribution function is illustrated in Fig. 2.2. Here four distributions are given, all with the same mean energy. One is simply the form (7), while the other three are those for helium, neon, and argon, using the general result (6) and the cross-sections Q_d illustrated in Fig. 2.1. It will be seen that the Ramsauer–Townsend effect in argon leads to quite a pronounced deficiency of the higher energy electrons.

1.2.1. *Proof of velocity distribution formula*

We follow the method of Morse, Allis, and Lamar.‡ Consider an element $\mathrm{d}\gamma = \mathrm{d}\xi \mathrm{d}\eta \mathrm{d}\zeta$ of the velocity space. The representative points of the electrons in this space are given by their velocity components. In the steady state the number of representative points that leave this element per second due to the applied field must be equal to the net number entering it due to collisions.

The number $c\, \mathrm{d}\gamma$ leaving per second due to the applied field will be given by

$$\frac{\mathrm{d}\xi}{\mathrm{d}t}\frac{\partial f}{\partial \xi}\,\mathrm{d}\gamma = \frac{eF}{m}\frac{\partial f}{\partial \xi}\,\mathrm{d}\gamma. \tag{8}$$

The number $a\, \mathrm{d}\gamma$ leaving the element per second due to collisions will be given simply by
$$a\, \mathrm{d}\gamma = NQ_0\, vf\, \mathrm{d}\gamma,$$

† DRUYVESTEYN, M. J., *Physica* **10** (1930) 61; **1** (1934) 1003; DAVYDOV, V. B., *Phys. Z. Sowjun.* **8** (1935) 59; MORSE, P. M., ALLIS, W. P., and LAMAR, E. S., *Phys. Rev.* **48**; (1935) 412. ‡ loc. cit.

where N is the number of gas molecules/cm³ and Q_0 is the total elastic cross-section. This must be subtracted from the number b entering the element due to collisions to give
$$c = b-a.$$

The calculation of b involves the momentum transfer cross-section. Since in a collision electrons lose a fraction $2(m/M)(1-\cos\theta)$ of their initial velocity,

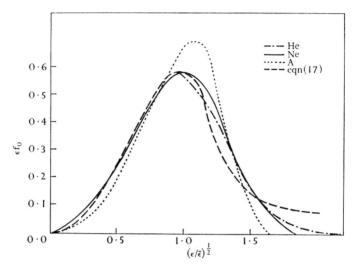

FIG. 2.2. Electron velocity distribution in a diffusing electron swarm, inelastic collisions being ignored. The function ϵf_0, where f_0 is defined in (6) and ϵ is the electron energy, is plotted as a function of $(\epsilon/\bar{\epsilon})^{\frac{1}{2}}$, where $\bar{\epsilon}$ is the mean energy. The curves for He, Ne, and A have been obtained using the diffusion cross-sections illustrated in Fig. 2.1. The curve (----) has been obtained from (7) using a constant diffusion cross-section.

those which, after collision, find their representative points in the element $d\gamma$ must, before collision, have had a velocity
$$v' = v\{1+(1-\cos\theta)(m/M)\},$$
$(m/M)^2$ being ignored. These representative points must have come from an element
$$d\gamma' = d\gamma(v'/v)^3.$$
Hence
$$b\,d\gamma = Nv' \int_0^{2\pi}\int_0^{\pi} I_0(v',\theta)f(v',\xi')\sin\theta\,d\theta d\phi d\gamma'$$
$$= N(v'^4/v^3) \int_0^{2\pi}\int_0^{\pi} I_0(v',\theta)f(v',\xi')\sin\theta\,d\theta d\phi d\gamma.$$

We may, therefore, write
$$b-a = (N/v^3)\int_0^{2\pi}\int_0^{\pi} \{v'^4 f(v',\xi')I_0(v',\theta) - v^4 f(v,\xi)I_0(v,\theta)\}\sin\theta\,d\theta d\phi.$$

2.1 MOMENTUM-TRANSFER CROSS-SECTION

To simplify this, we expand in powers of m/M, retaining only the first two terms, to give

$$b-a = (N/v^3) \int_0^{2\pi}\int_0^\pi \left[v^4\{f(v,\xi')-f(v,\xi)\}I_0(v,\theta) + \right.$$
$$\left. + (m/M)(1-\cos\theta)v\frac{\partial}{\partial v}\{v^4 f(v,\xi')I_0(v,\theta)\}\right]\sin\theta\, d\theta d\phi. \quad (9)$$

At this stage we must make some assumption as to the form of $f(v,\xi)$. We expand it as
$$f(v,\xi) = f_0(v) + (\xi/v)f_1(v), \quad (10)$$

ignoring higher order harmonics in ξ/v. In doing this, it is supposed that the effect of the electric field on the distribution function is small, a result that will be justified at the end.

We have then

$$\int_0^{2\pi}\int_0^\pi [\{f(v,\xi')-f(v,\xi)\}I_0(v,\theta)]\sin\theta\, d\theta d\phi$$
$$= f_1(v)\int_0^{2\pi}\int_0^\pi (\cos\omega'-\cos\omega)I_0(v,\theta)\sin\theta\, d\theta d\phi, \quad (11)$$

where $\cos\omega = \xi/v$, $\cos\omega' = \xi'/v$. From the geometry of the collision
$$\cos\omega' = \cos\omega\cos\theta + \sin\omega\sin\theta\cos(\phi-\chi),$$

so that the integral (11) becomes
$$-f_1(v)Q_d(v)\cos\omega.$$

Again, treating $f_1(v)$ as small,
$$\left(\frac{m}{M}\right)\int_0^{2\pi}\int_0^\pi (1-\cos\theta)\frac{\partial}{\partial v}\{v^4 f(v,\xi')I_0(v,\theta)\}\sin\theta\, d\theta d\phi$$
becomes
$$\left(\frac{m}{M}\right)\frac{d}{dv}\{v^4 f_0(v)Q_d\}.$$

Thus we have
$$b-a = -NQ_d\, vf_1\cos\omega + (mN/Mv^2)\frac{d}{dv}(v^4 Q_d f_0). \quad (12)$$

Returning now to c we have, using the form (10) for f,
$$c = \left(\frac{eF}{m}\right)\left\{\frac{\partial f_0}{\partial \xi} + \frac{f_1}{v} + \xi\frac{\partial}{\partial \xi}\left(\frac{f_1}{v}\right)\right\}.$$

Now
$$\frac{\partial f}{\partial \xi} = \frac{\xi}{v}\frac{df}{dv},$$

so
$$c = \left(\frac{eF}{m}\right)\left\{\frac{\xi}{v}\frac{df_0}{dv} + \frac{f_1}{v} + \frac{\xi^2}{v}\frac{d}{dv}\left(\frac{f_1}{v}\right)\right\}.$$

Since we are ignoring the effect of all harmonics beyond the first we replace ξ^2 in this expression by its spherical average $\frac{1}{3}v^2$ and obtain

$$c = \left(\frac{eF}{m}\right)\left\{\cos\omega\frac{df_0}{dv} + \frac{1}{3v^2}\frac{d(v^2 f_1)}{dv}\right\}. \quad (13)$$

Equating separately the terms in c and $b-a$ with and without $\cos\omega$ gives now

$$\left(\frac{eF}{m}\right)\frac{df_0}{dv} = -NQ_\mathrm{d} vf_1, \tag{14}$$

$$\left(\frac{eF}{3m}\right)\frac{d(v^2 f_1)}{dv} = (mN/M)\frac{d}{dv}(v^4 Q_\mathrm{d} f_0). \tag{15}$$

Integrating (15) gives

$$\tfrac{1}{3}eFf_1 = 2(m/M)N\epsilon Q_\mathrm{d} f_0 + \mathrm{const}/v^2, \tag{16}$$

where $\epsilon = \tfrac{1}{2}mv^2$. This equation represents the energy balance, the term on the left-hand side being proportional to the energy gained from the field, the term involving Q_d to that lost by collision. In a steady state the constant should be taken equal to zero.

Writing now
$$\epsilon_l = eF/NQ_\mathrm{d},$$
we have
$$f_1 = 6(m\epsilon/M\epsilon_l)f_0.$$

Substitution in (14) gives

$$\frac{df_0}{dv} = -3(m^3 v^3/M\epsilon_l^2)f_0. \tag{17}$$

This can be integrated to give

$$f_0 = A\,\exp\!\left\{-(6m/M)\int_0^\epsilon \epsilon\,d\epsilon/\epsilon_l^2\right\}, \tag{18}$$

A being a normalizing constant which is given by

$$\int f_0\,d\gamma = 1.$$

1.3. *The mean energy and drift velocity*

We have for the mean energy

$$\bar{\epsilon} = \tfrac{1}{2}m\iiint (\xi^2+\eta^2+\zeta^2)f(\xi,\eta,\zeta)\,d\xi d\eta d\zeta$$

$$= 2\pi m\int_0^\infty v^4 f_0(v)\,dv \tag{19a}$$

$$= 0{\cdot}427(M/m)^{\frac{1}{2}}\epsilon_0, \quad \text{when } Q_\mathrm{d} \text{ is constant.} \tag{19b}$$

The root mean square velocity c is therefore given by

$$c = 0{\cdot}924(M\epsilon_0^2/m^3)^{\frac{1}{4}}. \tag{20}$$

Similarly we have, for the drift velocity u,

$$u = \iiint \xi f(\xi,\eta,\zeta)\,d\xi d\eta d\zeta$$

$$= \tfrac{4}{3}\pi\int_0^\infty v^3 f_1(v)\,dv \tag{21}$$

$$= 0{\cdot}634_5(m/M)^{\frac{1}{4}}(2\epsilon_0/m)^{\frac{1}{2}}, \quad \text{when } Q_\mathrm{d} \text{ is constant.} \tag{22}$$

It is of interest now to compare the formulae for c^3/u and cu with those we obtained by very crude arguments. From (20) and (22) we have, treating Q_d as constant,

$$c^3/u = 0.880 M\epsilon_0/m^2, \tag{23}$$

$$2cu = 1.658\epsilon_0/m. \tag{24}$$

As λ is to be taken as $2m/M$ and $\epsilon_0 = Fel$, we see that the more refined treatment introduces factors 0.880 and 1.658 on the right-hand sides of the equations.

After describing the methods used to measure c and u we shall discuss the comparison of the observed values for the rare gases with the predictions of the above formulae, using the cross-sections Q_d of Fig. 2.1.

1.4. *Allowance for energy of motion of the gas atoms*

For small values of F/p the mean energy of the diffusing electrons will not be much greater than that of the gas atoms. Allowance must therefore be made for the effect of the velocity of the gas atom on the probability of a particular energy loss being suffered by an electron in a collision. It may be shown that, if T_g is the temperature of the gas, the required correction can be made by replacing f_0 on the right-hand side of (17) by

$$f_0 + \frac{\kappa T_g}{mv}\frac{df_0}{dv}. \tag{25}$$

If f_0 is Maxwellian about the temperature T_g, (25) vanishes as it must do because under such circumstances collisions can have no effect on the distribution.

Following through the same analysis as above, but with this modification, gives, in place of (18),

$$f_0(v) = A\exp\left\{-(6m/M)\int_0^\epsilon \epsilon\, d\epsilon\Big/\left(\epsilon_l^2 + \frac{6m}{M}\kappa T_g\epsilon\right)\right\}, \tag{26}$$

so that, if ϵ_l is constant and equal to ϵ_0,

$$f_0(v) = A\left(1 + \frac{6m}{M}\frac{\kappa T_g\epsilon}{\epsilon_0^2}\right)^{(M\epsilon_0^2/6m\kappa^2 T_g^2)}\exp(-\epsilon/\kappa T_g). \tag{27}$$

Also

$$f_1(v) = \left(\frac{6m}{M}\frac{\epsilon}{\epsilon_0}\right)\left(1 + \frac{6m}{M}\frac{\kappa T_g\epsilon}{\epsilon_0^2}\right)^{-1} f_0. \tag{28}$$

2. Measurement of the characteristic energy of a diffusing swarm of electrons

In 1925 Townsend[†] observed the lateral spread of a swarm of electrons as it diffused under the action of a uniform field. From such observations the ratio D/u of the diffusion coefficient to the drift velocity may be obtained. It is shown below that, if the collision frequency is independent of electron velocity,

$$D/u = \tfrac{2}{3}\bar{\epsilon}/eF, \tag{29}$$

[†] TOWNSEND, J. S., *Motion of electrons in gases* (Clarendon Press, Oxford, 1925); *J. Franklin Inst.* **200** (1925) 563.

where $\bar{\epsilon}$ is the mean energy. Although this relation does not hold exactly in any other case it is often quite a good approximation (see, however, Fig. 2.24) so that for many purposes Townsend's method gives the mean energy. We shall refer to eFD/u as the characteristic energy ϵ_k.

The principle of Townsend's method is illustrated in Fig. 2.3. An electron stream which has already acquired a steady drift motion through a gas at a pressure p in a uniform electric field of strength F, enters the diffusion chamber through an orifice A in the upper electrode 1. It continues to diffuse through this chamber under the same uniform field and is collected on a receiving electrode 5. This electrode consists of an inner portion B and outer portions C_1, C_2 so that the currents reaching B and C_1, C_2 may be derived as follows.

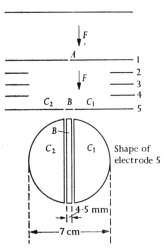

FIG. 2.3. Illustrating the principle of Townsend's method for measuring D/u.

The diffusion coefficient D for the swarm is defined such that, in the absence of any electric field, the local velocity of flow at any point within the swarm is given by
$$n\mathbf{v} = -D\,\mathrm{grad}\,n, \tag{30}$$
where n is the number of electrons/cm³ at that point. In doing this we assume that the electron energy distribution is not appreciably affected by the sideways diffusion. The effect of the electric field will now be to superpose the drift velocity $\mathbf{u}$ so that
$$n\mathbf{v} = -D\,\mathrm{grad}\,n + n\mathbf{u}. \tag{31}$$
The motion must also satisfy the equation of continuity
$$\mathrm{div}\,n\mathbf{v} = 0,$$
so that
$$\mathrm{div}(-D\,\mathrm{grad}\,n + n\mathbf{u}) = 0,$$
or
$$\nabla^2 n = (1/D)\mathbf{u}.\mathrm{grad}\,n. \tag{32}$$
In our case the field is taken along the x-direction so
$$\nabla^2 n = \frac{u}{D}\frac{\partial n}{\partial x}. \tag{33}$$

By solving this equation with the appropriate boundary conditions the value of n at each point of the receiving electrode may be determined. The fraction R of the current received on the inner portion B, as a function of D/u and the geometry of the apparatus, may now be determined by integration.

The diffusion coefficient, D, is given by

$$D = \frac{4\pi}{3n} \int_0^\infty (v^3 f_0/Q_\mathrm{d})\, dv \qquad (34)$$

as in the kinetic theory of gases (Chap. 16, § 3.1), while

$$u = \frac{4\pi}{3} \int_0^\infty v^3 f_1\, dv = -\frac{4\pi}{3} \frac{eF}{nm} \int_0^\infty \frac{v^2}{Q_\mathrm{d}} \frac{df_0}{dv}\, dv, \quad \text{using (14).} \qquad (35)$$

If the collision frequency is constant, Q_d is inversely proportional to v and

$$\frac{D}{u} = -(m/eF) \int_0^\infty v^4 f_0\, dv \bigg/ \int_0^\infty v^3 \frac{df_0}{dv}\, dv$$

$$= (\tfrac{2}{3}/eF) \int_0^\infty \epsilon v^2 f_0\, dv \bigg/ \int_0^\infty v^2 f_0\, dv$$

$$= \tfrac{2}{3}\bar{\epsilon}/eF. \qquad (36)$$

In the experiments of Townsend† the distance between the electrodes 1 and 5 was 4 cm. Guard rings such as 2, 3, 4, in Fig. 2.3 were included to ensure uniformity of the applied electric field. The orifice A consisted of a rectangular slot 1·5 cm long and 2 mm wide. The shape and dimensions of the electrodes B, C_1, and C_2 are illustrated in Fig. 2.3. Very small currents of the order 10^{-12} A were received on electrode 5 and a special induction balance was devised to measure them with accuracy. No precautions were taken to ensure good vacuum conditions, the apparatus not being constructed of materials that would permit of outgassing by heating. Nevertheless no important errors seem to have been introduced in this way except in some cases at the smaller values of F/p (see Fig. 2.21).

The first experiments in which modern vacuum techniques were applied were those of Huxley and Zaazou,‡ and since then most of the earlier measurements of Townsend and his collaborators have been repeated, employing the best available techniques. Most of this work has been directed towards extension of the measurements to such low values of F/p that the mean electron energy is not appreciably in excess of thermal. For this purpose it is necessary to eliminate stray electric fields as well as to ensure a high degree of uniformity of the field in the

† loc. cit., p. 53.
‡ HUXLEY, L. G. H. and ZAAZOU, A. A., Proc. R. Soc. A196 (1949) 402.

drift space. In general the principle adopted has still been that of Townsend.

As an example of precision technique we may take the work of Crompton, Elford, and Gascoigne† and of Warren and Parker.‡ Both employed a circular geometry for the collector system as indicated in Fig. 2.4, which

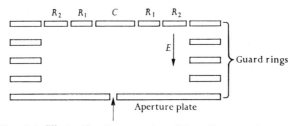

Fig. 2.4. Illustrating the geometry of the collector system used by Warren and Parker in their measurement of D/u.

TABLE 2.1

Electrode dimensions in apparatus used to measure mean electron energy

Dimension	Crompton, Elford, and Gascoigne (cm)	Warren and Parker (cm)
Drift distance	10	8·9
Central collector disc C, radius	0·5	0·615
First annular ring R_1, outer radius	3	0·21
Second annular ring R_2, outer radius	..	1·79
Aperture radius	0·1	0·13
Guard ring inner radius	4·5	3·81
Guard ring spacing	0·5	1·2

shows the apparatus of Warren and Parker.‡ The latter authors introduced two concentric annular electrodes, R_1 and R_2, as well as the central disc C, and measured the currents received at all three. The dimensions of the electrode systems used are given in Table 2.1. The choice of tube diameter was dictated by the desire to carry out observations over a wide range of gas temperature requiring immersion of the whole apparatus in a cold or hot bath. The next important requirement is that of field uniformity in the drift space, which dictates the dimensions of the guard ring system. If the geometry is accurately preserved Crompton *et al.* calculate that in their equipment the field within an axial cylinder of 6 cm diameter would remain uniform to within 0·05 per cent.

† Crompton, R. W., Elford, M. T., and Gascoigne, J., *Aust. J. Phys.* **18** (1965) 409.
‡ Warren, R. W. and Parker, J. H., *Phys. Rev.* **128** (1962) 2661.

To maintain the geometry, their guard rings, which were made of copper, were separated by glass spacers 0·05 cm thick with a tolerance of ±0·003 cm. All metal surfaces exposed to the drift space were gold-coated to reduce the effect of contact potential differences. The gap between the central collector disc and the annulus was kept as low as 0·003 cm so that penetration of stray fields into the drift space should be small. Warren and Parker took similar precautions using a spacing of 0·25 cm but with overlapping electrodes.

In both experiments the materials used were limited to glass, copper, and stainless steel. Warren and Parker worked under ultra-high vacuum conditions with a pressure of order 10^{-9} torr, but Crompton *et al.* avoided baking-out their apparatus because of the risk of changing electrode spacings, etc. However, they used a 5-litre/s getter-ion pump so as to avoid contamination by mercury vapour and minimize that due to hydrocarbons. Their working pressure was about 10^{-7} torr.

Crompton *et al.* used a hot filament source for their electrons and were able to check the operation of the equipment by using a Kunsman source for K$^+$ ions and verifying that, for the smallest values of F/p, the ions were in thermal equilibrium with the molecules in hydrogen gas. They found that the effect of small potential differences, less than 0·2 V between the disc and annular collectors, can lead to serious errors and it was necessary to carry out the current measurements so that no such differences arose.

Warren and Parker found evidence from their observations of the presence of field inhomogeneities in the drift space due to departures from the ideal geometry. They allowed for this by introducing an empirical relation between the current ratio and the parameter $\beta = 3cF^2d/4 = uFd/2D$, where d is the drift distance. This relation differed from the theoretical one only slightly and gave results which, while still preserving the essential requirement that D/u should be a function of F/p alone and tend to $\frac{3}{2}\kappa T$ as $F/p \to 0$, were much more consistent.

A further correction had to be introduced in their measurements owing to the use of a photoelectric source for the electrons. Some electrons were produced within the drift region from ultra-violet light scattered through the aperture. To correct for this the current to the collector was measured with the light source on but with the cathode held so positive with respect to the aperture that no electrons from the cathode could reach it. This current was subtracted from readings made in the course of normal operation.

Warren and Parker measured D/u in various gases down to values of F/p as low as $1{\cdot}5\times 10^{-4}$ V cm^{-1} torr^{-1}. The lower limit of F/p in the observations of Crompton *et al.* was somewhat higher (6×10^{-3} V cm^{-1} torr^{-1}), though about fifty times smaller than the lowest values used by Townsend.

Results of measurements of D/u in the rare gases are discussed in § 7 in connection with their analysis in terms of momentum-loss cross-sections. Observations in molecular gases are discussed in Chapter 11, § 7 and Chapter 13, §§ 1.7, 3.8, 4.4.5, 5.3, and 8.

3. Measurement of drift velocity

3.1. *The electrical shutter method*

This operates on the shutter principle, as follows. Let S_1, S_2 be two shutters placed at different levels in the diffusing electron swarm. At regular intervals these shutters open for a short interval of time so that electrons may pass through. They are synchronized in phase so that both are open at the same time. Electrons will succeed in passing through both shutters only if they traverse the distance between them in an integral number of cycles. By observing the variation of the current passing through S_2 as a function of the frequency of shutter operation, the drift velocity may be determined.

In the adaptation of this method by Bradbury and Nielsen† to the measurement of electron drift velocities, the shutters consisted of an electron filter of the type introduced by Loeb and Cravath‡ (see Chap. 19, § 2 for an alternative shutter used for measuring positive ion mobilities). This is a grid of copper between alternate wires of which a high-frequency potential difference may be applied to sweep electrons, passing through the grid, to one or other set of wires. If the magnitude of the alternating potential difference is reduced all the electrons arriving at the grid are not collected by it—those that reach the grid at a time when the alternating field is nearly zero will pass through it. Two such devices with synchronized alternating field functioned as the shutters S_1, S_2. Electrons that travelled between the filters in a whole number of half-cycles were able to pass through the second filter. In practice the frequency was fixed and F/p varied to give the value appropriate to this drift velocity, this being repeated for various frequencies.

The grids used by Bradbury and Nielsen were made of copper wires of 0·08 mm diameter mounted 1·0 mm apart on mica frames and were

† BRADBURY, N. E. and NIELSEN, R. A., *Phys. Rev.* **49** (1936) 388; **50** (1936) 950; **51** (1937) 69.
‡ LOEB, L. B. and CRAVATH, A. M., ibid. **33** (1929) 605; **48** (1935) 684.

maintained at a distance 5·93 cm apart. The alternating potential applied to the grids by a radio-frequency oscillator could be varied between 0 and 200 V and the frequency between 10^4 and 10^7 c/s. Six guard rings maintained a uniform field in the space through which the electrons diffused. The whole apparatus was sealed in a Pyrex glass tube with a graded quartz seal. This was to admit ultra-violet light for the production of electrons by photo-emission. The apparatus could be baked out and pumped down to a very low pressure (10^{-7} torr) before admitting the gas under investigation. There can be no doubt that effects due to impurities were negligible in this work.

More recent measurements made with this technique have introduced features specially designed to extend the observations down to values of F/p small enough for the mean energy of the electrons to be effectively equal to that of the gas atoms. Under these conditions it is important to eliminate stray field effects as completely as possible. As a step towards this Phelps, Pack, and Frost† replaced the alternating potential applied to the grids by narrow rectangular pulses but they found, in the course of a study of drift in helium that, for low F/p, the observed drift time between the grids was no longer proportional to the drift distance. Correction was made for these end effects by measuring the drift times t_1 and t_2 for two different drift distances d_1 and d_2 respectively and taking the drift velocity as given by $(d_1-d_2)/(t_1-t_2)$.

In a second set of experiments two further changes were made.‡ The photoelectric electron source was pulsed so that the times t_1 and t_2 now referred to drift times from the cathode to the two grids, and there was no need to vary the grid separation to eliminate the end effects. To reduce field penetration effects still further the grid transmission was varied in a different way. Whereas in the earlier methods a d.c. bias was applied between alternate sets of grid wires so as to reduce transmission to 5 per cent and transmission was then restored by application of rectangular pulses, in the new method the d.c. bias was removed and a pulsed voltage applied to reduce transmission by collecting electrons arriving at the grid. Fig. 2.5 illustrates the variation of the current received in the two modes of operation by a collecting electrode, after passage through both grids, as a function of the time delay between the pulse applied to either grid and the light pulse. In the unbiased mode the time of travel is determined by the appearance of a minimum in the collector current instead of a maximum as in the earlier modes.

† PHELPS, A. V., PACK, J. L., and FROST, L. S., *Phys. Rev.* **117** (1960) 470.
‡ PACK, J. L. and PHELPS, A. V., ibid. **121** (1961) 798.

The advantage of the unbiased mode is that no voltage is applied between grid wires during the period between pulses. When used no end effects were found, the drift times being strictly proportional to the drift distance.

In these experiments the grids were of 3-mil gold-plated molybdenum wires spaced 140 mil apart. The drift tube was constructed of metal

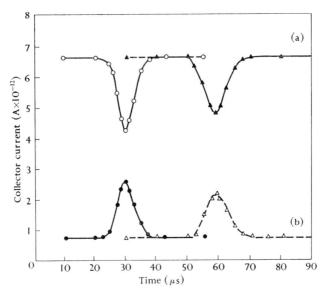

FIG. 2.5. Form of the variation with time of the current received at a collecting electrode in the drift velocity experiments of Pack and Phelps: (a) with d.c. bias; (b) without d.c. bias.

glass and fired lava with gold-wire gasket vacuum seals. All the electrodes, apart from the grids, were gold-plated and mounted on fired lava supports shielded with gold-plated Advance metal to minimize effects due to charging up. The drift tube was operated under ultra-high vacuum conditions at an ultimate background pressure, after baking, of 10^{-9} torr.

Measurements were made, not only at room temperature $300\pm3°$ K, but also at $77°$, $195°$, and $373°$ K by immersion in a suitable bath. They covered a range of F/p from 10^{-4} to 1 V cm^{-1} torr^{-1}.

3.2. Hornbeck's method

The principle of this method, which was first introduced by Hornbeck† and which has been applied particularly to measure the drift velocities of

---

† HORNBECK, J. A., *Phys. Rev.* **83** (1951) 374.

positive ions (see Chap. 19, § 2), is very simple. A group of photo electrons is released by a flash of light from a photocathode and is then allowed to drift in a uniform electric field until collection at an anode. The drifting electrons produce a current in an external resistor and the time variation of this current is recorded on an oscilloscope. From this the transit time of the electron between cathode and anode may be obtained and hence the drift velocity.

As used by Bowe[†] for measuring the drift velocities of electrons in various gases the electrodes were parallel plates $7\frac{1}{2}$ in in diameter separated accurately by 1 in. The photocathode was a disc of flat Pyrex glass coated with tin. Photoelectrons were produced by a light pulse of $\sim 0.4\,\mu$s duration, from a flash tube, that passed through a quartz window and perforated anode to reach the cathode. Special precautions were taken to avoid the use of materials likely to evolve gases. After obtaining a vacuum of 2×10^{-6} torr the chamber was baked at 80° C for a few days while outgassing the calcium turnings in the purifier at 500° C. After admission of the gas it was purified by forced circulation over the calcium.

Fig. 2.6 illustrates typical shapes of current pulses. The time of transit was taken as the time between the midpoint of the leading and trailing edge. The minimum error in measurement of transit time was estimated at 2·5 per cent and the over-all accuracy at 5 per cent. As we shall see in § 7 the results obtained by Bowe[†] for the rare gases differ from those of other observers, using different methods, by considerably more than this.

3.3. *Use of proportional counters for drift velocity measurement*

Bortner, Hurst, and Stone[‡] have developed a pulse method, originally suggested by Stevenson,[§] for measuring drift velocities of electrons, which is particularly suitable for studying gases commonly used in ionization chambers and counters. A plane geometry is used, the drift velocity being measured from the time taken for electrons to drift between two parallel plates P_1, P_2 a distance d apart. A uniform electric field is maintained between the plates and also between P_1 and a parallel plate P_3 on the opposite side to P_2. Electrons, produced close to P_3 by a collimated source of alpha particles, are drawn towards P_1, which includes two slots S_1 and S_2. Electrons that reach S_2 produce a pulse in

[†] BOWE, J. C., ibid. **117** (1960) 1411.
[‡] BORTNER, T. E., HURST, G. S., and STONE, W. G., *Rev. scient. Instrum.* **28** (1957) 103.
[§] STEVENSON, A., ibid. **23** (1952) 93.

a proportional counter C_1, while some of those that pass through S_1 drift along the field to reach a slot S_3, aligned so that $S_1 S_2$ is perpendicular to the plates P_1, P_2. Those electrons that pass through S_3 activate a second proportional counter C_2. The time between the pulses recorded on C_1 and C_2 is the time taken for electrons to drift the distance d between P_1 and P_2.

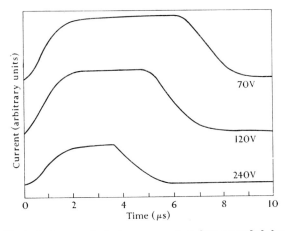

Fig. 2.6. Typical shapes of electron current pulses recorded by Bowe in experiments by Hornbeck's method for measuring electron drift velocities in helium. The helium pressure was 112 torr and the chamber voltages that determined the applied electric field are indicated in each case.

3.4. *Errors in drift velocity measurements due to diffusion*

Lowke† has made a detailed analysis of the principal errors involved in the measurement of drift velocity by the techniques described in §§ 3.1 and 3.2.

In the method of Nielsen and Bradbury it is assumed that the time required for the centre of an undisturbed pulse to travel the distance between the shutters is $1/f$, where f is the lowest frequency of the applied attenuating field for which the transmission through the second shutter S_2 is a maximum. There are a number of reasons why this will not be exactly correct. The principal ones are as follows.

(*a*) Back-diffusion of electrons will produce absorption of some electrons from each pulse by S_1 after it is shut, thus altering a little the position of maximum electron density in the pulse.

(*b*) Absorption of electrons by the second shutter before it opens also disturbs the pulse.

† Lowke, J. J., *Aust. J. Phys.* **15** (1962) 39.

(c) The presence of a diffusion current means that the frequency corresponding to maximum electron current is not equal to that corresponding to maximum electron density at S_2.

(d) Variation of the number of electrons in each pulse with frequency.

(e) The decay of the maximum electron density of a pulse while it is passing through S_2 causes the current/frequency curves to be unsymmetrical about each current maximum.

(f) If the mean velocity of agitation of the electrons varies along the length of the pulse errors will be introduced because the transmission of the shutters increases with electron velocity.

(g) The geometry of the electron source.

Lowke[†] analysed the contributions from all of these with the exception of (f), assuming that the shutters are open for a very short time interval. He found that the relative errors introduced are of order D/ud where D is the diffusion coefficient and u the drift velocity. This means that the absolute error is of order D/d and is minimized by use of a long path-length and high pressure so that D is small. The effect of using a point source with small collecting electrode is to increase the relative error as compared with a plane source by $2D/ud$. No estimate was made of the likely error arising from (f).

Experiments were carried out to verify these conclusions. The shutters, which were separated by a distance of 6 cm, were constructed by threading Nichrome wire of diameter 0·003 in through holes pierced in mica so that the distance between adjacent grid wires was 0·5 mm. Guard rings were placed every 0·5 cm to ensure a uniform electric field between the shutters.

Fig. 2.7 illustrates the observed variation of drift velocity in helium with gas pressure for $F/p = 0.4$ V cm^{-1} torr^{-1} at 20° C. The results follow quite closely the calculated variation for a plane source, so the expected effects are clearly shown. More detailed observations in nitrogen showed the same effect.

Similar errors due to diffusion arise in the modified procedure introduced by Phelps, Pack, and Frost.[‡] However, in their experiments and those of Pack and Phelps,[§] the difference in times to drift over two distances was observed. Since the absolute error is independent of distance this means that it is largely cancelled by this procedure.

The errors to be expected from diffusion in Hornbeck's method are also of the same order.

† loc. cit., p. 62.   ‡ loc. cit., p. 59.   § loc. cit., p. 59.

Measurements of drift velocity for the rare gases using the diffusion methods are described and discussed in relation to momentum-transfer cross-sections in § 7. Observations for molecular gases are discussed in Chapter 11, § 7 and Chapter 13, §§ 1.7, 3.8, 4.4.5, 5.3, and 8.

3.5. *Measurement of drift velocity in a glow discharge*

Phelps, Pack, and Frost† have sought to extend drift velocity measurements to higher values of F/p (in the range 2–40 V cm^{-1} torr^{-1}) by measuring the current, voltage gradient, and electron concentration in the positive column of a glow discharge in helium maintained in a Vycor

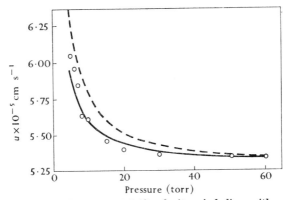

Fig. 2.7. Variation of the apparent drift velocity u in helium with gas pressure. ○ measured by Lowke using the Bradbury–Nielsen method; ——— calculated for a plane electron source; – – – – a point electron source.

tube 30 cm long, 13 mm diameter, in the pressure range 0·25–3 torr. Currents of up to 10 mA were used in the discharge. The field was estimated by measuring the floating potentials of two 3-mil tungsten wire probes placed 15 cm apart. The electron concentration n_e was estimated by measuring the frequency shift of a rectangular resonant cavity surrounding a section of the positive column. Measurements were confined to electron densities in the range 2×10^9 to 4×10^{10} per cm^3—sufficiently small to justify the application of the theory of § 1.

In measurements of this kind great care is needed to ensure the absence of plasma oscillations, and in the measurements of Phelps, Pack, and Frost the discharge current was found to have less than a 5 per cent ripple when observed on an oscilloscope with an amplifier and cable which passed frequencies from 10 to 10^4 c/s. It is difficult to be sure, however, that oscillations of higher frequency may not have been present, thus modifying the velocity distribution.

† loc. cit., p. 59.

4. Drift in combined electric and magnetic fields

If a uniform magnetic field H is applied in a direction perpendicular to the electric field F the stream is deviated in the steady state through an angle $\theta = \arctan(\beta Hu/F)$, where β is a quantity close to unity which depends on the variation of the momentum-transfer cross-section with velocity. If this cross-section is independent of velocity, $\beta = 1\cdot 06$. At first the measurement of θ was regarded as determining the drift velocity u, the uncertainty in β being regarded as small, but now it is regarded as affording, through the determination of β, a further means of obtaining data about momentum-transfer cross-sections.

The full theory has been given by Allis and Allen.† We outline their method for the case in which the mean energy of the gas atoms may be neglected.

Their derivation follows on very similar lines to that involved in the determination of the energy distribution function. Using the same notation as in § 1.2.1 and taking the magnetic field along the z-axis we have

$$c = \frac{d\xi}{dt}\frac{\partial f}{\partial \xi} + \frac{d\eta}{dt}\frac{\partial f}{\partial \eta}, \tag{37}$$

with

$$\frac{d\xi}{dt} = \frac{eF}{m} + \frac{eH}{m}\eta,$$

$$\frac{d\eta}{dt} = -\frac{eH}{m}\xi.$$

Following (10) we now write

$$f(v,\xi,\eta) = f_0(v) + (\xi/v)f_1(v) + (\eta/v)f_2(v). \tag{38}$$

Carrying out the same procedure as before gives

$$c = \frac{eF}{m}\left\{\frac{\xi}{v}\frac{df_0}{dv} + \frac{1}{3v^2}\frac{d(v^2 f_1)}{dv}\right\} + \frac{eH}{m}\left\{\frac{\eta}{v}f_1 - \frac{\xi}{v}f_2\right\}, \tag{39}$$

$$b - a = -NQ_d(f_1\xi + f_2\eta) + (mN/Mv^2)\frac{d}{dv}(v^4 Q_d f_0). \tag{40}$$

Equating separately to zero those terms in $c - b + a$ which are independent of, and those which are proportional to, ξ, η respectively, gives

$$(eF/mv^2)\frac{d}{dv}(v^2 f_1) = (3m/Mv^2)\frac{d}{dv}(v^4 f_0/l), \tag{41}$$

$$(eF/mv)\frac{df_0}{dv} - (eH/mv)f_2 = -f_1/l, \tag{42}$$

$$(eH/mv)f_1 = -f_2/l, \tag{43}$$

where l is $1/NQ_d$ as before.

† Allis, W. P. and Allen, H. W., *Phys. Rev.* **52** (1937) 703.

These equations may be easily integrated to give

$$\ln f_0 = -(3m/M)\left\{\int \left(\frac{2\epsilon}{\epsilon_l^2} + \frac{H^2}{mF^2}\right) d\epsilon\right\} + \ln A, \qquad (44)$$

$$f_1 = (6m\epsilon/M\epsilon_l)f_0, \qquad (45)$$

$$f_2 = -(3mvH/MF)f_0. \qquad (46)$$

As before, the constant A is to be chosen so that $\iiint f_0\, d\xi d\eta d\zeta = 1$.

We note that the equation for f_0 may be written in the form

$$\ln f_0 = -(3m^2/Me^2F^2)\int (\nu^2 + \omega_H^2)\, d\epsilon + \ln A, \qquad (47)$$

where ν is the collision frequency $NQ_d v$ for electrons of energy ϵ and $\omega_H = eH/m$ is the Larmor frequency for electrons in the field H.

The components u_x, u_y of the drift velocity are now given by

$$u_x = \tfrac{4}{3}\pi \int_0^\infty v^3 f_1(v)\, dv, \qquad u_y = \tfrac{4}{3}\pi \int_0^\infty v^3 f_2(v)\, dv. \qquad (48)$$

When Q_d and hence l is a constant these integrals may be expanded in a power series in $(2elH^2/F)(3/Mm)^{\frac{1}{2}}$ to give

$$u_y/u_x = 3Hu/2^{\frac{3}{2}}F = 1\cdot 06 Hu/F, \qquad (49)$$

u being the drift velocity in the absence of the magnetic field and terms of higher order in H being neglected.

This relation was used extensively by Townsend and Bailey[†] and their collaborators to measure 'drift' velocities in a number of gases, assuming that the ratio $\beta = u_y F/u_x Hu$ is effectively unity. However, the departure of β from unity is now significant, so attention is concentrated on the measurement of β as a further transport coefficient known as the *magnetic deflexion coefficient*. Fu_y/Hu_x is sometimes known as the *magnetic drift velocity*. The apparatus of Townsend and Bailey was the same as that used for the measurement of mean energies. In addition to the uniform electric field, a uniform magnetic field of strength H was applied in a direction perpendicular to the electric field and parallel to the straight sides of the electrode B (Fig. 2.3). The magnetic field was adjusted so that the current arriving on the electrode C_1 was equal to that arriving on B and C_2. In this case the centre of the stream fell on the narrow slot between C_1 and B. If d is the distance between the plane of the slit A and the plane of B and $2b$ is the width of B, we then have

$$u_y/u_x = b/d = \beta uH/F. \qquad (50)$$

Huxley and Zaazou[‡] employed a modification to this procedure. They used a circular geometry as in Fig. 2.4. Their receiving electrode was split into two semicircular portions that were disconnected. With a given magnetic field H applied parallel to the common diameter the

---

[†] See § 7.3 of this chapter and §§ 8.1, 8.2, and 8.3 of Chapter 11.
[‡] HUXLEY, L. G. H. and ZAAZOU, A. A., *Proc. R. Soc.* A**196** (1949) 402.

currents received on the two halves are unequal. The ratio R_1 is a function of $\tan\theta$ and $Fd/(D/u)$, which may be calculated. Having measured D/u by the method described in § 2, it follows that $\tan\theta$ may be obtained from R_1.

This method has been developed particularly by Jory,[†] who has used very much the same apparatus as that of Crompton and Elford[‡] in the measurement of D/u, except that the current can be separately measured to each half of the receiver electrode.

5. Microwave afterglow methods for measuring momentum-transfer cross-sections for collisions with thermal electrons

Referring to § 1.3 we see that the drift velocity of a swarm of electrons diffusing under the action of a uniform steady electric field F is given by

$$u = \frac{4\pi}{3} \int_0^\infty v^3 f_1(v)\, dv, \tag{51}$$

or, using (14), by

$$u = -\tfrac{4}{3}\pi \frac{eF}{m} \int_0^\infty \frac{v^3}{\nu} \frac{df_0}{dv}\, dv, \tag{52}$$

where ν is the collision frequency $NQ_d v$ and f_0 is the spherically symmetrical term in the expansion of the electron velocity distribution in momentum space. The specific electrical conductivity σ of the gas in which the swarm is diffusing will then be given by nue/F, where n is the number of electrons per cm³. Hence

$$\sigma = -\tfrac{4}{3}\pi n \frac{e^2}{m} \int_0^\infty \frac{v^3}{\nu} \frac{df_0}{dv}\, dv. \tag{53}$$

This analysis has been extended by Margenau[§] and Phelps, Fundingsland, and Brown[∥] to calculate the complex conductivity of a plasma to which an alternating field of angular frequency ω is applied. They find that

$$\sigma = \sigma_\mathrm{r} + i\sigma_\mathrm{i}$$
$$= -\frac{4\pi}{3} \frac{ne^2}{m} \int_0^\infty \frac{\nu - i\omega}{\nu^2 + \omega^2} v^3 \frac{df_0}{dv}\, dv, \tag{54}$$

so that

$$\sigma_\mathrm{r}/\sigma_\mathrm{i} = -\frac{1}{\omega} \int_0^\infty \frac{\nu}{\nu^2+\omega^2} v^3 \frac{df_0}{dv}\, dv \bigg/ \int_0^\infty \frac{1}{\nu^2+\omega^2} v^3 \frac{df_0}{dv}\, dv. \tag{55}$$

[†] Jory, R. L., *Aust. J. Phys.* **18** (1965) 237.   [‡] loc. cit., p. 56.
[§] Margenau, H., *Phys. Rev.* **69** (1946) 508.
[∥] Phelps, A. V., Fundingsland, O. T., and Brown, S. C., ibid. **84** (1951) 559.

If $\omega \gg \nu$ this simplifies still further to give

$$\sigma_r/\sigma_i = -\frac{1}{\omega} \int_0^\infty \nu v^3 \frac{df_0}{dv} dv \bigg/ \int_0^\infty v^3 \frac{df_0}{dv} dv. \qquad (56)$$

These relations may be used to determine the collision frequency for electrons of nearly thermal energies, using applied fields of microwave frequencies that certainly satisfy the condition $\omega \gg \nu$.

Margenau has shown further that in a gas at temperature T

$$\ln f_0 = -\int_0^v \left\{ \frac{\kappa T_g}{m} + \frac{Me^2 F_m^2}{6m^3} \frac{1}{\nu^2 + \omega^2} \right\}^{-1} v \, dv, \qquad (57)$$

where F_m is the amplitude of the alternating electric field. This reduces to (26) when $\omega \to 0$ and $F_m^2 = 2F^2$. On the other hand, if $\omega \gg \nu$ the distribution is Maxwellian with a temperature, T_e, given by

$$T_e = T_g + Me^2 F_m^2 / 6\kappa m^2 \omega^2. \qquad (58)$$

The proof of these results may be outlined as follows. Using the notation of § 1.2.1, since the field $F = F_m \cos \omega t$ is an r.f. field of angular frequency ω, f will now depend on t explicitly so that

$$c = b - a - \partial f/\partial t.$$

We have then, in place of (14) and (15),

$$(eF_m \cos \omega t/m) \frac{\partial f_0}{\partial v} = -\nu f_1 - \frac{\partial f_1}{\partial t}, \qquad (59)$$

$$(eF_m \cos \omega t/3m) \frac{\partial}{\partial v}(v^2 f_1) = (m/M) \frac{\partial}{\partial v}\left\{ v^3 \nu \left(f_0 + \frac{\kappa T_g}{m\nu} \frac{\partial f_0}{\partial v} \right) \right\} - \frac{\partial f_0}{\partial t}, \qquad (60)$$

where $\nu \, (= NQ_d v)$ is the collision frequency.

To proceed further f_0 and f_1 may be expanded in a Fourier series in $\genfrac{}{}{0pt}{}{\cos}{\sin} \omega t$. The leading term in f_0 is independent of t while that in f_1 is

$$f_1 = f_1^c \cos \omega t + f_1^s \sin \omega t. \qquad (61)$$

Ignoring other terms we have now from (59)

$$\frac{eF_m}{m} \cos \omega t \frac{\partial f_0}{\partial v} = -\nu(f_1^c \cos \omega t + f_1^s \sin \omega t) + \omega(f_1^c \sin \omega t - f_1^s \cos \omega t),$$

so that $f_1^s = (\omega/\nu) f_1^c$, and

$$f_1^c = -\frac{eF_m}{m} \frac{\nu}{\nu^2 + \omega^2} \frac{\partial f_0}{\partial v}.$$

It follows from (61) that

$$f_1 = \frac{eF_m}{m} \left[\mathrm{Re}\left\{ -\left(\frac{\nu - i\omega}{\nu^2 + \omega^2} \right) \right\} e^{i\omega t} \right] \frac{\partial f_0}{\partial v}. \qquad (62)$$

Putting f_1 in (51), the phase difference between the current in the drifting swarm of electrons and the applied electric field leads in the usual way to a complex conductivity of the plasma given by (54) above.

To obtain the leading term of f_0 we substitute (62) for f_1 in (60) to give

$$\frac{\partial}{\partial v}\left[\left\{ \frac{e^2 F_m^2}{3m^2} v^2 \cos \omega t \, \mathrm{Re}\left(\frac{-\nu + i\omega}{\nu^2 + \omega^2} e^{i\omega t} \right) - \frac{\kappa T_g \nu v^2}{M} \right\} \frac{\partial f_0}{\partial v} - \frac{m}{M} v^3 \nu f_0 \right] = 0. \qquad (63)$$

2.5 MOMENTUM-TRANSFER CROSS-SECTION

Averaging over a period $2\pi/\omega$ yields now

$$\frac{df_0}{dv} + \alpha(v) f_0 = 0,$$

where
$$\alpha(v) = mv \bigg/ \left(\kappa T_\mathrm{g} + \frac{e^2 F_m^2}{6m^2} \frac{M}{v^2 + \omega^2} \right). \tag{64}$$

The formula (57) then follows immediately.

If the momentum-transfer cross-section Q_d is independent of v as is nearly the case for helium (see Fig. 2.16) and $\omega \gg \nu$ then (56) gives

$$\sigma_\mathrm{r}/\sigma_\mathrm{i} = -\tfrac{8}{3}(NQ_\mathrm{d}/\omega)(2\kappa T_\mathrm{e}/m\pi)^{\frac{1}{2}}. \tag{65}$$

It follows that, for this case, there is no difficulty in deriving Q_d from the measured values of $\sigma_\mathrm{r}/\sigma_\mathrm{i}$. Moreover, by carrying out measurements at different gas temperatures and such fixed values of F_m that $T_\mathrm{e} \simeq T_\mathrm{g}$ it is possible to verify the assumption of constant Q_d. It is also possible, in principle, to apply a further check by varying F_m to change T_e but this may introduce considerable complications in practice.

The analysis of observed data is more complicated and less definite when Q_d varies appreciably with electron velocity. Two procedures have been applied. If observations can be carried out over a sufficient range of gas temperatures with $T_\mathrm{e} \simeq T_\mathrm{g}$ the results may be represented in terms of a power series of the form

$$\sigma_\mathrm{r}/\sigma_\mathrm{i} = -\sum_{j=-3}^{l} a_j \bar{v}^j, \tag{66}$$

where $\bar{v}$ is the root mean square velocity of the electrons.

Using (56), we then have

$$Q_\mathrm{d}(v) = \frac{\omega \Gamma(\tfrac{5}{2})}{N} \sum_{j=-3}^{l} \frac{a_j v^{j-1}}{\Gamma(\tfrac{5}{2} + \tfrac{1}{2}j)}. \tag{67}$$

This may be applied at gas pressures for which $\omega \gg \nu$.

The alternative procedure is to work in the pressure range in which ω and ν are comparable so that (55) must be used. $\sigma_\mathrm{r}/\sigma_\mathrm{i}$ is then measured over a considerable pressure range with T_g and ω fixed. The collision frequency is assumed to have the form $av^h p$, where p is the gas pressure, so that (55) becomes

$$\sigma_\mathrm{r}/\sigma_\mathrm{i} = -\gamma \int_0^\infty \frac{y^{\frac{1}{2}h+\frac{3}{2}}}{y^h + \gamma^{-2}} e^{-y} \, dy \bigg/ \int_0^\infty \frac{y^{\frac{3}{2}} e^{-y}}{y^h + \gamma^{-2}} \, dy, \tag{68}$$

with
$$\gamma = (2\kappa T_\mathrm{g}/m)^{\frac{1}{2}h} ap/\omega. \tag{69}$$

γ/p is chosen to give the best fit between the observed values of $(\sigma_\mathrm{r}/\sigma_\mathrm{i})p$ as a function of $\sigma_\mathrm{r}/\sigma_\mathrm{i}$ for $\sigma_\mathrm{r} \simeq \sigma_\mathrm{i}$ for some values of h.

Even if neither of these procedures may be adopted a mean value of Q_d may be obtained by applying the formula (65), but this gives a complicated average value. If the collision frequency does not vary rapidly with velocity, however, the mean will not be very different from that for electrons with a Maxwellian velocity distribution and in that case may be used for comparison with mean cross-sections derived from experiments at higher velocities.[†]

A further complication that must be allowed for if the electron concentration is not very small is the contribution to the collision frequency arising from electron collisions with positive ions so that the observed value of ν is given by
$$\nu = \nu_{ea} + \nu_{ei},$$
where ν_{ea} is the desired frequency of collisions with atoms, ν_{ei} that with positive ions. It may be shown that
$$\nu_{ei} = C_1 n_e T_e^{-\frac{3}{2}} \ln(C_2 T_e^{\frac{3}{2}} n_e^{-\frac{1}{2}}),$$
where C_1 and C_2 are constants while n_e, T_e are the electron concentration and temperature. For a limited range of variation of n_e
$$\nu_{ei} = n_e C(T_e),$$
where $C(T_e)$ is practically independent of n_e. If p_0 is the gas pressure (referred to 0° C)
$$\nu/p_0 = (\nu_{ea}/p_0) + \alpha C(T_e) n_e(0)/p_0,$$
where $n_e(0)$ is the electron concentration at the axis of the cavity and α is a constant that can be calculated from the known distribution of electrons in the tube when ambipolar diffusion prevails. For a cylindrical tube $\alpha = 0.675$.

Since ν_{ea}/p_0 should be a constant for given T_e a plot of ν/p_0 against $n_e(0)/p_0$ should be linear with slope $\alpha C(T_e)$. The intercept on the ν/p_0 axis is then equal to ν_{ea}/p_0.

Fig. 2.8 illustrates such plots for neon, observed by Chen,[‡] the conditions being such that $\omega \gg \nu$.

The first application of this method was made by Margenau and Adler,[§] who used microwave conductivity measurements to determine the mean free path of electrons in the positive column of a mercury discharge. Phelps, Fundingsland, and Brown[||] were the first to carry

---

[†] A discussion of the effect of the velocity of the collision frequency on the average obtained using different experimental methods is given by FEHSENFELD, F. C., *J. chem. Phys.* **39** (1963) 1653.
[‡] CHEN, C. L., *Phys. Rev.* **135** (1964) A627.
[§] MARGENAU, H. and ADLER, F. P., *Phys. Rev.* **79** (1950) 970.
[||] loc. cit. p. 67.

out observations for thermal electrons using plasma contained in a microwave cavity. At about the same time Varnerin† applied the method by measuring the propagation constant for microwaves in a rectangular wave guide containing the plasma.

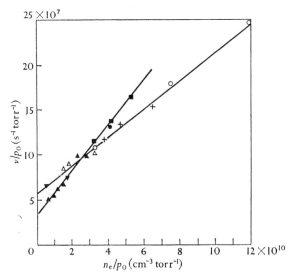

FIG. 2.8. Variation of ν/p_0 with n_e/p_0 in afterglow experiments in neon at two temperatures, 305° and 480° K. The observations were made at the following values of pressure p in torr: ■ 3·66, ● 3·78, ▼ 6·52, ▲ 9·60, ○ 0·28, + 4·49, △ 8·15.

5.1. *Resonant cavity method*

The resonant cavity method will be discussed in some detail as other measurements of considerable importance for the study of reactions between electrons, ions, and neutral atoms and molecules may be made using this method (see Chap. 12, § 7.5.5, Chap. 13, § 4.4.3 and 11.2, Chap. 18, § 5, Chap. 19, §§ 2, 6, Chap. 20, § 3). The change, $\Delta\omega_0$ in resonant frequency of a microwave cavity produced by the presence of a discharge is given by $\Delta\omega_0 = An_e/\omega_0$, where ω_0 is the resonant frequency of the empty cavity and n_e the electron concentration in the discharge. A is a constant that depends on the distribution of electron density and electric field in the cavity.‡ The change in the quality factor Q of the cavity depends upon the ratio of the real and imaginary parts of the electron conductivity of the plasma.

The most direct method of measuring the change in resonant frequency and Q of the cavity and thence the ratio σ_r/σ_i involves measuring the

† VARNERIN, L. J., ibid. **84** (1951) 563.
‡ See, for example, ROSE, D. J. and BROWN, S. C., *J. appl. Phys.* **23** (1952) 1028.

resonance curve of the cavity with and without the presence of the plasma. This method is most sensitive, however, when the cavity is tightly coupled to the measuring signal so that there is a danger of the discharge being disturbed by it.

To avoid this a method was developed by Gould and Brown[†] that has the advantage of being most sensitive when the coupling to the cavity is very weak. They introduced a microwave signal of power P_i into the cavity and measured the ratio of the power P_t in the transmitted signal to P_i.

Gould and Brown then show that

$$\eta P_i/P_t = 1+x^2/\alpha^2 \equiv \gamma, \qquad (70)$$

where
$$x = 2(\omega-\omega_0'),$$

ω being the circular frequency of the signal and ω_0' the resonant frequency of the cavity in the presence of the discharge. η and α depend on the admittance of the cavity. Samples of the incident and output signal were introduced to the push-pull input of an oscilloscope. The frequency ω was varied round the resonant frequency ω_0' at which P_t was a maximum and the signal levels to the push-pull input adjusted for a very small or zero deflexion. Under these conditions

$$x = 0, \qquad \eta = 1, \qquad \gamma = 1.$$

The attenuation in the receiver for the transmitted power was then decreased by an accurately measured amount and the frequency adjusted away from resonance to restore the signal to zero. Two values ω_1, ω_1' of ω on either side of resonance give a null signal. Then

$$x = 2|\omega_0'-\omega_1| = 2|\omega_1'-\omega_0'|,$$

while the value of γ corresponding to this value of x was measured by the ratio of the attenuations for the resonance and off-resonance conditions. A plot of γ against x^2 gave a straight line as expected, the slope of the line providing a measure of $1/\alpha^2$.

The difference $\Delta\alpha$ in the value of α, with and without the presence of the discharge, is then given by

$$\Delta\alpha = 2(\sigma_r/\sigma_i)\Delta\omega_0. \qquad (71)$$

In their measurements of collision frequency of electrons in helium using this method Gould and Brown[‡] used a rectangular cavity with

---

[†] GOULD, L. and BROWN, S. C., *J. appl. Phys.* **24** (1953) 1053.
[‡] GOULD, L. and BROWN, S. C., *Phys. Rev.* **95** (1954) 897.

dimensions 7·16, 7·88, and 6·48 cm designed to resonate in its three fundamental modes at wavelengths of 9·5, 10·0, and 10·5 cm. Breakdown was produced by a 9·5-cm pulse obtained from a magnetron of 100-W peak power. The pulse duration was varied between 0·1 and 5 ms and the repetition rate from 20 to 120 c/s. The 10·5-cm mode was used for measurement. Power from a c.w. tunable magnetron operating in this wavelength band was fed into the cavity and the proportion transmitted measured in the manner described above. In order not to disturb the discharge, the electric field introduced into the cavity by the measuring signal never exceeded 0·1 V cm^{-1} throughout the cavity. The transmitted power ratio was measured by a transient receiver operative for a time from 20–100 μs. This could be delayed in time relative to the breakdown pulse so that the conductivity of the afterglow could be measured at different times after the breakdown pulse.

Measurements of the conductivity ratio σ_r/σ_i were made at temperatures of 77°, 195°, 300°, and 400° K. Gould and Brown were unable to heat the cavity to higher temperatures because this released impurities to a serious extent as judged by the inconsistency of the experimental results. However, they carried out a further remarkable experiment in which they used a c.w. magnetron operating in the 10-cm mode to produce an electric field within the cavity great enough to increase the mean electron energy according to (58) by an order of magnitude, although complications were introduced because the energy input was no longer uniform over the plasma. Gould and Brown were able to allow for this assuming Q_d to be nearly constant. In this way it was possible to extend the measurements up to electron temperatures of 25 000° K.

Fig. 2.9 shows the observed variation of the conductivity ratio σ_r/σ_i with electron temperature. The solid line shows the expected variation of σ_r/σ_i if the momentum-transfer cross-section has a constant value of $5·1 \times 10^{-16}$ cm^2, in good agreement with the value of $5·3 \times 10^{-16}$ cm^2 derived in the earlier measurements of Phelps et al.[†] by the same technique.

Since these early experiments a number of other investigations have been carried out in helium and other gases. The results obtained are discussed, in connection with other methods of determining momentum-transfer cross-sections at low electron energies, in § 7 for the rare gases and caesium and in Chapter 11, § 7 and Chapter 13, §§ 1.7, 3.8, 4.4, 5, 6, and 8 for molecular gases.

† PHELPS, A. V., FUNDINGSLAND, O. T., and BROWN, S. C., Phys. Rev. 84 (1951) 559.

6. Cyclotron resonance method for measuring momentum-transfer cross-sections at very low electron energies

Consider a plasma containing free electrons in a uniform steady magnetic field H. If a microwave electric field $E\cos\omega t$ is applied at right angles to the magnetic field, power will be absorbed by the electrons. The variation of the absorbed power will be of resonance form about the

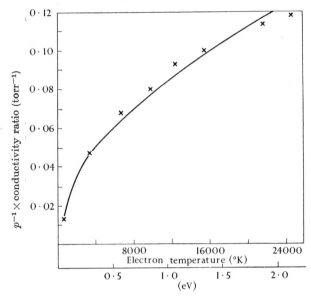

FIG. 2.9. Observed variation with electron temperature of $\sigma_r/p\sigma_i$ where σ_r/σ_i is the ratio of the real to the imaginary electrical conductivity of a helium afterglow and p is the helium pressure, as observed by Gould and Brown.

cyclotron angular frequency $\omega_H = eH/m$ of the electrons and the width of the resonance curve will depend on the collision frequency of the electrons in the plasma. This provides a further way[†] of studying the momentum-transfer cross-sections of atoms and molecules for electrons of nearly thermal energy, which possesses certain advantages.

Because it is a resonance method it is of high sensitivity so that only quite small electron concentrations are required: for some atoms such as caesium thermal ionization is an adequate source, and this has the advantage of ensuring that the electron velocity distribution is Maxwellian. Also, as the shape of the absorption line depends on the way in which the collision frequency varies with electron velocity it is possible in principle to obtain information about the form of this variation from measurements at a single electron temperature.

† KELLY, D. C., MARGENAU, H., and BROWN, S. C., *Phys. Rev.* **108** (1957) 1367.

The formula for the power absorption as a function of ω may be obtained by an extension† of the method used by Margenau to derive the velocity distribution function for electrons in an oscillating electric field $F_m \cos \omega t$ (see § 5). In the presence of the magnetic field, which we take to be in the z-direction, we choose for the harmonic expansion of the distribution function the same form as in § 4:

$$f(v,t) = f_0(v,t) + (\xi/v)f_1(v,t) + (\eta/v)g_1(v,t). \tag{72}$$

It is then found that

$$\ln f_0(v,t) = -\int_0^v \frac{mv \, dv}{\kappa T + MB/6} + \text{const}, \tag{73}$$

where

$$B = \frac{e^2 F_m^2}{2m^2}\left\{\frac{1}{\nu^2+(\omega-\omega_H)^2} + \frac{1}{\nu^2+(\omega+\omega_H)^2}\right\}, \tag{74}$$

ν being the collision frequency for electrons of velocity v. When $\omega_H = 0$ this reduces to (57) and when $\omega = 0$ and $F_m^2 = 2F^2$ to (47).

The current density is given by

$$\mathbf{J} = -\frac{4\pi e n}{3}\int_0^\infty \mathbf{G}\frac{\partial f_0}{\partial v}v^3 \, dv, \tag{75}$$

where n is the electron density and

$$\mathbf{G} = \frac{e}{m}\text{Re}\left\{\frac{\mathbf{F}_m(\nu+i\omega)+\boldsymbol{\omega}_H \times \mathbf{F}_m}{\omega_H^2+(\nu+i\omega)^2}e^{i\omega t}\right\}. \tag{76}$$

$\boldsymbol{\omega}_H$ is a vector in the direction of $\mathbf{H}$ and of magnitude equal to the cyclotron frequency eH/m. $-G(\partial f_0/\partial v)$ reduces to (62) when $\omega_H \to 0$.

The average power absorbed is

$$\bar{P} = \langle \mathbf{J} \cdot \mathbf{F}_m \cos \omega t \rangle \tag{77}$$

where $\langle \rangle$ denotes a time average over a cycle $2\pi/\omega$. Using (73)–(77) we have

$$\bar{P} = \frac{2\pi m n}{3}\int_0^\infty \nu B v^3 \frac{\partial f_0}{\partial v} \, dv \tag{78}$$

where B is given by (74).

In the special case in which the collision frequency and hence B is independent of v we have

$$\bar{P} = 2\pi m n \nu B \int_0^\infty v^2 f_0 \, dv = \frac{e^2 F_m^2 n \nu}{4m}\left\{\frac{1}{\nu^2+(\omega-\omega_H)^2} + \frac{1}{\nu^2+(\omega+\omega_H)^2}\right\}. \tag{79}$$

The conditions may be readily chosen so that $\omega_H^2 \gg \nu^2$ and we have

$$\bar{P} = \frac{e^2 F_m^2 n \nu}{4m}\{\nu^2+(\omega-\omega_H)^2\}^{-1}. \tag{80}$$

The line half-width is then clearly equal to the constant value of ν.

If ν varies with v the shape of the resonance curve may be distorted considerably from the Lorentz form (80). For example, Fig. 2.10 shows the cyclotron resonance absorption line shapes for the case of

$$\nu = Av^\lambda \quad (\lambda = -2, -1, 0, 1, 2).$$

† ibid.

Assuming a variation of the form

$$\nu = \nu_0 + av^\lambda,$$

the constants ν_0, a, λ may be determined to give the best fit to the observed shape.

To obtain accurate results by this method it is necessary to work under conditions such that other sources of line-broadening are negligible. These include Doppler, Stark, and plasma effects and magnetic field

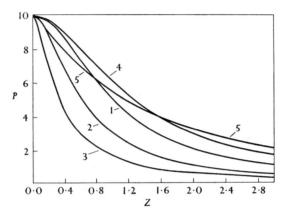

Fig. 2.10. Illustrating the dependence of the shape of a cyclotron resonance absorption line on the form of the variation of the electron collision frequency ν with electron velocity v. P is the power absorption normalized to unit maximum. Z is the ratio $(\omega - \omega_H)/\nu$ where ω_H is the angular frequency at resonance. Curves 1, 2, 3, 4, and 5 refer to collision frequencies which vary as v^λ with $\lambda = -2, -1, 0, 1,$ and 2 respectively.

inhomogeneities. However, electron concentrations much less than 10^9 per cm^3 may be used because of the high sensitivity of the method, and, at such low concentrations, Stark and plasma effects are unimportant. At low electron energies the Doppler effect is also negligible. Special care must be taken, however, to reduce magnetic field inhomogeneities to less than 0·1 per cent of the main field over the measuring volume.

Fig. 2.11 shows the apparatus used by Fehsenfeld[†] to study collision frequencies in helium, argon, oxygen, and CO_2 by the cyclotron resonance method. The gas, which was ionized by application of r.f. power in the discharge cavity, flowed through a Vycor tube into the detection cavity which was located between the pole-faces of an electromagnet. The detection cavity formed part of a microwave spectrometer in which microwave power from a klystron was fed into the cavity and the strength of the signal reflected from the cavity measured. The power

† FEHSENFELD, F. C., J. chem. Phys. **39** (1963) 1653.

in the reflected signal, passing through a minimum at the cyclotron frequency, provided a measure of the shape of the cyclotron-resonance absorption of the probing microwave signal. The frequency of the klystron was locked to the resonant frequency of the microwave cavity by the automatic frequency control of the spectrometer, thus ensuring that only the real part of the r.f. conductivity was measured.

The length of the tube connecting the discharge and detection cavities could be varied between 3 and 35 cm, corresponding to a flow time between the cavities of from 0·1 to several milliseconds, and allowing

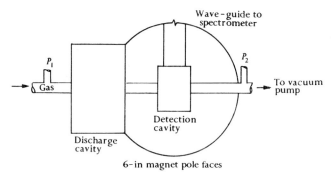

Fig. 2.11. Arrangement of apparatus used by Fehsenfeld for cyclotron resonance experiments.

sufficient time for plasma thermalization and electron density decay in the afterglow.

Pressure measurements were made by the gauges P_1 and P_2 and the pressure in the detection cavity (of the order of 5 torr) calculated from the dimensions of the flow tube, assuming laminar viscous flow. Both rectangular and cylindrical detection cavities were used with identical results, provided the microwave power was kept low enough and sufficient afterglow time allowed for thermalization and electron density decay.

To check that the equipment was operating successfully it was necessary to show that the line shape was unaffected by the power level of the microwave probe, thus indicating that no appreciable heating of the electrons in the cavity was taking place. It was also shown that the line-width of the cyclotron resonance was proportional to the gas pressure in the cavity. The cyclotron resonance method has also been used to measure electron collision frequencies in caesium vapour, ionized by thermal ionization.†

† MEYERAND, R. G. and FLAVIN, R. K., private communication.

The results obtained for helium and argon are discussed in § 7 and compared with those obtained by other methods.

6.1. *Effect of a magnetic field on the resonant frequency of a cavity containing plasma*

The change $\Delta\omega_0$ in the resonant frequency of a microwave cavity due to the presence of a plasma is proportional to the imaginary part of the complex conductivity, i.e. to

$$\frac{4\pi}{3}\frac{ne^2}{m}\int_0^\infty \frac{\omega}{\nu^2+\omega^2} v^3 \frac{\partial f_0}{\partial v}\, dv \qquad (81)$$

in the absence of a magnetic field (see (54)).

In the presence of a magnetic field the interaction between a microwave field and a plasma has been formulated by Allis† in terms of a high-frequency conductivity tensor, whose component transverse to the field can be written

$$\sigma_t = \frac{4\pi}{3}\frac{ne^2}{m}\int_0^\infty \frac{\nu+i\omega}{\omega_H^2+(\nu+i\omega)^2}\frac{\partial f_0}{\partial v} v^3\, dv \quad \text{(compare eqns (75) and (76))}. \qquad (82)$$

The imaginary part of σ_t is

$$-\frac{4\pi}{3}\frac{ne^2}{m}\int_0^\infty \frac{\omega(\nu^2+\omega^2-\omega_H^2)}{\{\nu^2+(\omega+\omega_H)^2\}\{\nu^2+(\omega-\omega_H)^2\}}\frac{\partial f_0}{\partial v} v^3\, dv. \qquad (83)$$

For a cylindrical microwave resonator operating in the TE mode the change in resonant frequency is proportional to (83) instead of (81). If ν is independent of v (83) vanishes for

$$\omega_H^2 = \nu^2+\omega^2. \qquad (84)$$

Thus by studying the effect of a magnetic field on the resonant frequency of a cavity containing a discharge plasma and measuring the magnitude of the magnetic field for which the resonant frequency ω is equal to that of the cavity in the absence of plasma, the collision frequency can be derived from (84).

In general where ν depends on v there is still some value of ω_H for which (83) vanishes and a value of ν averaged over a complicated velocity distribution can be obtained. This method has been applied by Hirshfield and Brown‡ to determine the momentum-transfer cross-section in helium.

---

† ALLIS, W. P., *Handb. Phys.* (ed. Flügge) vol. xxi, p. 383 (Springer-Verlag, Berlin, 1956).
‡ HIRSHFIELD, J. L. and BROWN, S. C., *J. appl. Phys.* **29** (1958) 1749.

The gas was contained in a quartz bottle connected to the vacuum system and placed axially within the cavity. A discharge was produced by using a pulsed magnetron to excite the TE_{111} mode of the cavity at at 3250 mc. Observations were made in the afterglow after sufficient time had elapsed to enable the electrons to reach thermal equilibrium. Transmission curves for the TE_{011} mode of the cavity round 4200 mc were measured for different magnetic fields and different times in the afterglow.

Fig. 2.12 shows oscilloscope traces of the amplitude of the microwave signal transmitted through the cavity for different times in the afterglow (and thence for different electron densities): (a) for $H = 0$, (b) for $H = H'$, the value for zero cavity frequency shift, and (c) for $H > H'$.

If $\nu/\omega_0 \ll 1$ and $\Delta\omega = \omega_0 - \omega_H \ll \nu$, where ω_0 is the resonant frequency of the cavity without plasma, the condition for the vanishing of (83) can be written

$$\int_0^\infty (1 - 2\omega_0 \Delta\omega/\nu^2) v^3 \frac{\partial f_0}{\partial v} \, dv = 0. \tag{85}$$

Writing $\nu = Ap$, where p is the gas pressure in the cavity and $\Delta H = H' - H_0 = m\Delta\omega/e$ it is seen that for different gas pressures (85) is satisfied for values H' of the magnetic field such that ΔH is proportional to p^2. Fig. 2.13 shows the measured ΔH plotted against p^2 for helium. The straight lines represent the curves expected for the different values of P_c, the probability of collision in cm^2/cm^3 torr. The results of the experiment are consistent with $P_e = 20\pm1$ cm^2/cm^3 torr, corresponding to a constant cross-section $Q_d = 5\cdot6\pm0\cdot3\times10^{-16}$ cm^2.

7. Results of measurements of transport properties and their analysis in terms of momentum-transfer cross-sections

No perfectly straightforward procedure may be outlined for determining the momentum-transfer cross-section and its variation with electron energy from observations of drift velocity, characteristic energy, and other transport quantities as functions of F/p, the ratio of electric field strength to gas pressure. A trial-and-error procedure must be adopted in which a first approximation is arrived at by examination of the different data, including the beam results at the lowest energies. This is then substituted in the appropriate formulae to obtain a first approximation to the velocity distribution function and thence to the transport quantities. From comparison with the observed data the assumed momentum-transfer cross-section is modified and the procedure repeated until a satisfactory fit is obtained with all the data. If this is not possible then it is likely that some of the data are unreliable. Sometimes it may be clear which data are suspect and consistency obtained when these are ignored.

For a given gas the velocity distribution function $f_0(\epsilon)$ is a function only of the gas temperature T and of $F/NQ_\mathrm{d}(\epsilon_0)$, where $Q_\mathrm{d}(\epsilon_0)$ is the momentum-transfer cross-section at some reference energy ϵ_0 and N is the concentration of gas atoms. It follows that the drift velocity u and

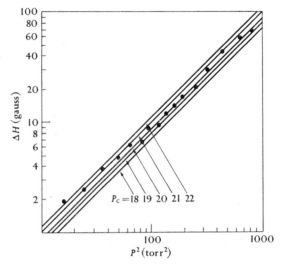

FIG. 2.13. Variation of ΔH with p^2 for helium as observed by Hirshfield and Brown. Straight lines represent results expected for different values of P_c.

characteristic energy ϵ_k are also functions only of T and $F/NQ_\mathrm{d}(\epsilon_0)$. Hence if measurement of u and ϵ_k for fixed T show a variation as $(F/N)^s$ then $Q_\mathrm{d}(\epsilon_0)$ is proportional to $\epsilon_k^{1/s}$ and $u^{1/s}$ and so is determined most accurately when s is large.

As for the relative effectiveness of data on u and on ϵ_k we note that, if the collision frequency varies as $\epsilon^{s+\frac{1}{2}}$ then, provided $\epsilon_k \gg \kappa T$,

$$u \propto \{F/NQ_\mathrm{d}(\epsilon_0)\}^{1/s+1}, \tag{86 a}$$

$$\bar{\epsilon}_k \propto \{F/NQ_\mathrm{d}(\epsilon_0)\}^{2/s+1}. \tag{86 b}$$

Under these circumstances Q_d can be determined more precisely from data on ϵ_k than on u, assuming these data to be equally reliable. However, there seems to be some belief that measurements of ϵ_k are more liable to systematic error, so analysis is usually based in the first instance on u.

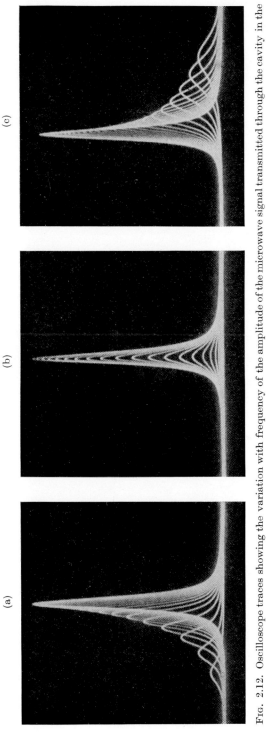

FIG. 2.12. Oscilloscope traces showing the variation with frequency of the amplitude of the microwave signal transmitted through the cavity in the experiments of Hirshfield and Brown. In each photograph the various traces are for different times in the afterglow. (a) $H = 0$; (b) $H = H'$, the value for zero shift of cavity frequency; (c) $H > H'$.

7.1. *Results for helium*

Figs. 2.14 and 2.15 show observations of drift velocities and of characteristic energies of electrons in helium at two gas temperatures, 77° and 293–300° K as functions of F/p_{300}.

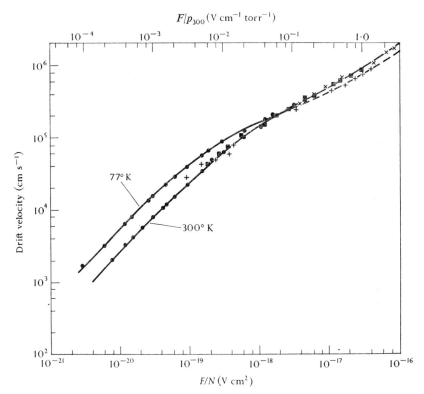

FIG. 2.14. Drift velocities for electrons in helium. Observed: ● Pack and Phelps; ■ Crompton, Elford, and Jory; + Bowe; × Nielsen; —— calculated from momentum-transfer cross-sections given by curves A and B (Fig. 2.16); –––– calculated from momentum-transfer cross-sections given by curve C (Fig. 2.16).

It will be seen that, while there is good agreement between the observations made by Pack and Phelps[†] using the method described in § 3.1 and by Crompton, Elford, and Jory[‡] using an improved version of the method of Bradbury and Nielsen, there is a considerable disagreement with those of Bowe[§] made by Hornbeck's method. Much earlier results obtained by Nielsen,[||] which are not shown in Fig. 2.14, also agree well with those of Pack and Phelps.

[†] PACK, J. L. and PHELPS, A. V., *Phys. Rev.* **121** (1961) 798.
[‡] CROMPTON, R. W., ELFORD, M. T., and JORY, R. L., *Aust. J. Phys.* **20** (1967) 369.
[§] BOWE, J. C., *Phys. Rev.* **117** (1960) 1411. [||] NIELSEN, R. A., ibid. **50** (1936) 950.

The modern data shown on the characteristic energy ϵ_k, which are due to Warren and Parker† and to Crompton, Elford, and Jory, agree well with each other and also with the much earlier measurements of Townsend and Bailey.‡

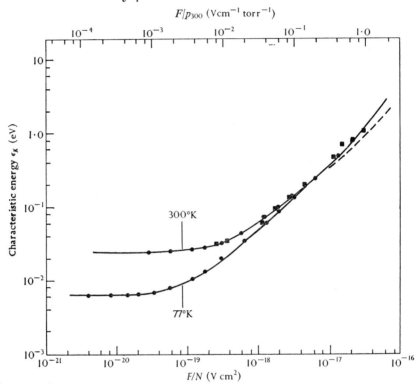

FIG. 2.15. Characteristic energies ϵ_k for electrons in helium. Observed: ● Warren and Parker; ■ Crompton, Elford, and Jory; —— calculated from momentum-transfer cross-sections given by curves A and B (Fig. 2.16); ---- calculated from momentum-transfer cross-sections given by curve C (Fig. 2.16).

Fig. 2.16 shows the momentum-transfer cross-sections Q_d derived from the drift velocity data. Curve A was derived by Crompton, Elford, and Jory from their own data, and curve B by Frost and Phelps§ from the data of Pack and Phelps. Curve C was derived by Frost and Phelps from the data of Bowe, excluding observations for $F/p < 0.03$ V cm^{-1} torr^{-1}, as to fit these would have required much too rapid a variation of Q_d with electron energy.

† WARREN, R. W. and PARKER, J. H., ibid. **128** (1962) 2661.
‡ TOWNSEND, J. S. and BAILEY, V. A., *Phil. Mag.* **46** (1923) 657 and **44** (1922) 1033.
§ FROST, L. S. and PHELPS, A. V., *Phys. Rev.* **136** (1964) A1538.

To discriminate between these results for Q_d they may be used to calculate the characteristic energy. When this is done it is found (see Fig. 2.15) that both the cross-sections of curves A and B in Fig. 2.16 give agreement within experimental error with the observations. On the other hand, the cross-sections of curve B give values for ϵ_k that are definitely below the observed values. The accuracy of the observations is not sufficient to discriminate between A and B.

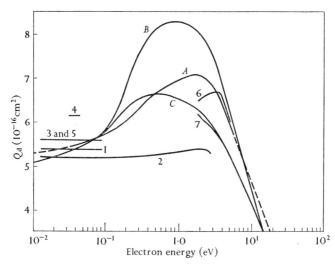

Fig. 2.16. Momentum-transfer cross-section for collisions of electrons with helium atoms.

A, B, C forms derived from analysis of drift velocity data. (1)–(7) derived from the respective experiments of Phelps, Fundingsland, and Brown; Gould and Brown; Chen, Leiby, and Goldstein; Fehsenfeld; Hirshfield and Brown; Ramsauer and Kollath; and Golden and Bandel.

Crompton, Elford, and Jory have also calculated the so-called 'magnetic drift velocity', $(F/H)u_y/u_x$, using their derived momentum-transfer cross-section (curve B), and compared it with their observations made with the equipment described in § 4 and with those of Townsend and Bailey made much earlier. Although the agreement is not unsatisfactory it is not so close as for ϵ_k, possibly due to less precision in the observations.

Further evidence about the cross-sections is available from microwave observations. In Fig. 2.16 the results obtained by Phelps et al.,[†] for thermal electrons, using the cavity-resonance method, are shown as well as those of Gould and Brown[‡] who extended the observations to

[†] PHELPS, A. V., FUNDINGSLAND, O. T., and BROWN, S. C., *Phys. Rev.* **84** (1951) 559.
[‡] GOULD, L. and BROWN, S. C., ibid. **95** (1954) 897.

higher mean electron energies by using microwave heating. Observations obtained by other methods that are also shown are those of Chen, Leiby, and Goldstein,[†] who used the microwave interaction technique of Anderson and Goldstein,[‡] of Hirshfield and Brown[§] from the effect of a magnetic field on the resonance frequency of a cavity containing helium plasma, and of Fehsenfeld[||] from the cyclotron-resonance method. Finally, the momentum-transfer cross-sections derived from the total cross-section measurements of Ramsauer and Kollath[††] and of Golden and Bandel[‡‡] are shown.

Quite good agreement exists between most of these data and the cross-sections (curves A and B) derived from the drift velocity and characteristic energy data. It must be remembered that the analysis of the microwave observations of Gould and Brown in the superthermal region is difficult and not too much weight should be attached to the fact that in this region they appear to give rather low cross-sections.

A further source of information on the cross-section at very low electron energies is the pressure shift produced in the high series terms of the alkali metals. The principle of this method is discussed and the results obtained described in Chapter 6, § 5.

Finally, comparison of these results with those of theoretical calculations is discussed in Chapter 8, § 6.1 and will be seen to reveal a not unsatisfactory position.

7.2. Results for neon

The analysis of results for neon is less satisfactory than for helium, partly because the cross-section at the low-energy limit is very small.

Fig. 2.17 illustrates the drift velocity measurements at 77° and 300° K. There is quite good agreement between the observations of Pack and Phelps[§§] at 300° K and the much earlier measurements of Nielsen[||||] at 293° K. As in helium, Bowe's results[†††] fall appreciably lower for F/p greater than 0·2 V cm^{-1} torr^{-1}. Frost and Phelps[‡‡‡] found that it was not possible to fit these data both at 77° and 300° K with the same cross-sections to within 30 per cent, so more reliable information comes from the microwave and pressure shift (Chap. 6, § 5) methods.

[†] CHEN, C. L., LEIBY, C. C., and GOLDSTEIN, L., ibid. **121** (1961) 1391.
[‡] ANDERSON, J. M. and GOLDSTEIN, L., ibid. **102** (1956) 933.
[§] HIRSHFIELD, J. L. and BROWN, S. C., ibid. **122** (1961) 719.
[||] FEHSENFELD, F. C., *J. chem. Phys.* **39** (1963) 1653.
[††] RAMSAUER, C. and KOLLATH, R., *Ann. Phys.* **12** (1932) 529.
[‡‡] GOLDEN, D. E. and BANDEL, H. W., *Phys. Rev.* **138** (1965) A14.
[§§] loc. cit., p. 81.   [||||] NIELSEN, R. A., ibid. **50** (1936) 950.
[†††] loc. cit., p. 81.   [‡‡‡] loc. cit., p. 82.

Chen† measured the microwave conductivity of a decaying neon plasma at temperatures ranging from 200° to 600° K. After removing the contribution to the collision frequency from collisions with ions, as described in § 5, p. 70, he obtained the mean collision frequency per electron-neon collision as a function of temperature shown in Fig. 2.18.

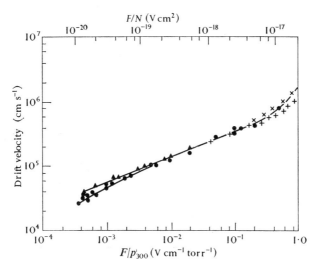

FIG. 2.17. Observed drift velocities for electrons in neon.
● Pack and Phelps (300° K); ▲ Pack and Phelps (77° K);
+ Bowe (300° K); × Nielsen (293° K).

Close agreement is found with the earlier thermal observations of Phelps et al.,‡ using the same technique. Over this temperature range it was found that a good fit could be obtained if the momentum-transfer cross-section is given by either

$$Q_d = 2\cdot54 \times 10^{-16} V^{\frac{1}{2}} \text{ cm}^2 \tag{87}$$

or
$$= 1\cdot07 \times 10^{-17} + 2\cdot17 \times 10^{-16} V^{\frac{1}{2}} \text{ cm}^2, \tag{88}$$

where V is the electron energy in eV. These give the curves shown in Fig. 2.18.

Gilardini and Brown§ extended their observations to higher energies by using microwave heating. Analysis of their results gives the cross-section–energy curve shown in Fig. 2.19.

† CHEN, C. L., *Phys. Rev.* **135** (1964) A627.
‡ PHELPS, A. V., FUNDINGSLAND, O. T., and BROWN, S. C., ibid. **84** (1951) 559.
§ GILARDINI, A. L. and BROWN, S. C., ibid. **105** (1957) 31.

Some observations at 77° K have been carried out by Dougal and Goldstein† using the microwave technique and by Marshall, Karrigan, and Goldstein‡ by the cyclotron-resonance method. These results

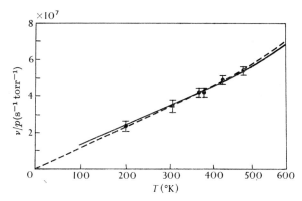

Fig. 2.18. Variation with gas temperature of the electron collision frequency ν in a neon afterglow. ● observed by Chen; ▲ observed by Phelps, Fundingsland, and Brown; —— calculated assuming Q_d to be given by (88); – – – – calculated assuming Q_d to be given by (87);

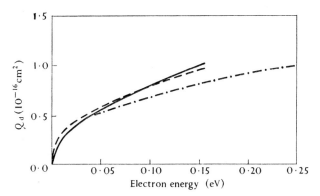

Fig. 2.19. Momentum-transfer cross-section for electrons in neon. – · – · – · from observations of Gilardini and Brown; —— from (87); – – – – from (88).

suggest somewhat larger cross-sections at these very low energies than would be expected from extrapolation of Chen's results.

As we shall see from the discussion of Chapter 8, § 7.1.1 empirical theory suggests that the cross-sections shown in Fig. 2.19 are not far

---

† DOUGAL, A. A. and GOLDSTEIN, L., ibid. **109** (1958) 615.
‡ Private communication to C. L. Chen.

from correct although there is considerable uncertainty about the low-energy limits.

7.3. Results for argon, krypton, and xenon

Special interest attaches to the early drift velocity measurements in argon. They were made by Townsend and Bailey† by the method described on p. 66 on the assumption that the factor β in (50) is unity.

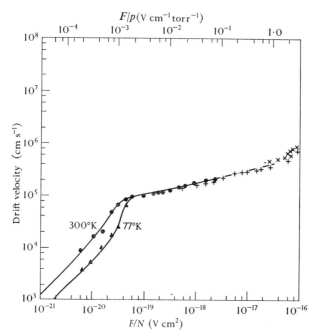

FIG. 2.20. Drift velocities for electrons in argon at 77° and 300° K. Observed: ● Pack and Phelps (300° K); ▲ Pack and Phelps (77° K); + Bowe (300° K); × Nielsen (293° K; —— calculated by Frost and Phelps.

Using the formulae (1) and (2) with $\lambda = 2m/M$ the free path l was derived and found to have a maximum at a mean energy, measured by the Townsend technique described in § 2, near 0·4 eV. This discovery was made at much the same time as Ramsauer's observations by the beam method described in Chapter 1, § 4.1 and so is referred to as the Ramsauer–Townsend effect.

Observations of the drift velocity and characteristic energy for electrons in argon at 77° and 300° K are illustrated in Figs. 2.20 and 2.21 respectively. The agreement between the earlier observations for

† TOWNSEND, J. S. and BAILEY, V. A., *Phil. Mag.* **43** (1922) 593; **44** (1922) 1033.

both u (Nielsen†) and ϵ_k (Townsend and Bailey‡) with the much later results of Pack and Phelps§ for u and Warren and Parker‖ for ϵ_k is quite good except possibly for u at the larger F/p. Bowe's†† results for u deviate somewhat more.

Frost and Phelps‡‡ have analysed these results. Using the momentum-transfer cross-section as a function of electron energy shown in Fig. 2.22

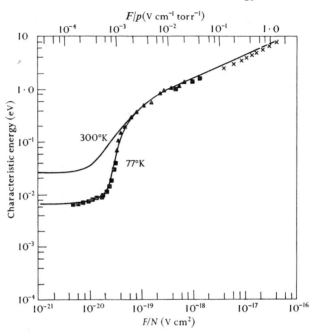

FIG. 2.21. Characteristic energies for electrons in argon at 77° and 300° K. Observed: ■ Warren and Parker (77° K); ▲ Warren and Parker (84° K); × Townsend and Bailey (288° K); —— calculated by Frost and Phelps.

they calculated u and ϵ_k, which are compared with the observations in Figs. 2.20 and 2.21. In this case, because of the rapid decrease in cross-section as the Ramsauer–Townsend minimum is approached from the low-energy side, ϵ_k is more sensitive than u to the assumed cross-section. Both ϵ_k and u are very insensitive for $\epsilon_k > 0.7$ eV. For this region further sensitivity was obtained by fitting the measurements of drift velocity in a mixture of 10 per cent H_2 with 90 per cent A (see Chap. 13,

† NIELSEN, R. A., *Phys. Rev.* **50** (1936) 950.
‡ TOWNSEND, J. S. and BAILEY, V. A., loc. cit.
§ PACK, J. L. and PHELPS, A. V., *Phys. Rev.* **121** (1961) 798.
‖ WARREN, R. W. and PARKER, J. H., loc. cit., p. 56.
†† BOWE, J. C., loc. cit., p. 81.
‡‡ FROST, L. S. and PHELPS, A., loc. cit., p. 82.

§ 1.7.) The cross-sections derived by Frost and Phelps are between 30 and 40 per cent of those derived by Bowe from his data.

The microwave data are available for comparison. The early measurement of Phelps *et al.*,† gave a thermal cross-section below 10^{-16} cm², which is clearly inconsistent with the other data and suggests that in these experiments the electrons were not in thermal equilibrium with the gas atoms. Fehsenfeld‡ found a more consistent value, indicated in Fig. 2.22, by the cyclotron-resonance method.

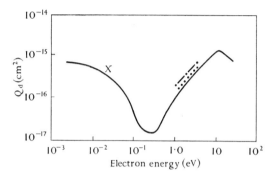

FIG. 2.22. Momentum-transfer cross-sections for electrons in argon. ——— derived by Frost and Phelps; .... from total cross-sections observed by Ramsauer and Ramsauer and Kollath; –·–·–· derived by Bowe; × observed by Fehsenfeld.

The momentum-transfer cross-sections, derived from the total cross-section observations of Ramsauer§ and of Ramsauer and Kollath‖ by using observed angular distributions of scattering, are also shown in Fig. 2.22. There is a reasonable degree of consistency and, when account is taken of the data from the pressure shift in alkali-metal spectra and of semi-empirical theory (see Chap. 8, § 7.1.1), the position is not unsatisfactory.

The existence of the Ramsauer–Townsend effect causes the energy distribution of drifting low-energy electrons to depart very markedly from the Maxwellian and Druyvesteyn forms. Fig. 2.23 shows three distributions calculated by Frost and Phelps with the momentum-transfer cross-sections shown in Fig. 2.22. For low values of F/p ($< 2 \times 10^{-3}$ V cm⁻¹ torr⁻¹) the most probable electron energies are near to thermal and to that of the Ramsauer–Townsend minimum. As F/p increases

† PHELPS, A., FUNDINGSLAND, O. T., and BROWN, S. C., loc. cit., p. 85.
‡ FEHSENFELD, F. C., *J. chem. Phys.* **39** (1963) 1653.
§ RAMSAUER, C., loc. cit., p. 25.
‖ RAMSAUER, C. and KOLLATH, R., loc. cit., p. 25.

the rapid rise of the cross-section from the minimum as the electron energy increases causes the distribution to fall very rapidly at the high-energy end. Because these distributions are so narrow, the characteristic energy ϵ_k is much larger than the mean electron energy $\bar{\epsilon}$ as indicated on the distributions shown in Fig. 2.23. This is to be contrasted with $\epsilon_k = \frac{2}{3}\bar{\epsilon}$ for a Maxwellian distribution. One result is that the coefficient β, relating the drift velocity to the angle of deflexion of an electron beam (see p. 66) drifting in combined perpendicular electric and

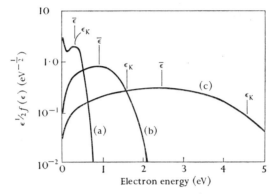

FIG. 2.23. Energy distribution functions $\epsilon^{\frac{1}{2}}f(\epsilon)$ for electrons in argon at 300° K for the following values of F/N: (a) 5×10^{-20} V cm^{-2}; (b) $1 \cdot 04 \times 10^{-18}$ V cm^{-2}; (c) $1 \cdot 04 \times 10^{-17}$ V cm^{-2}. Value of the mean and characteristic energies $\bar{\epsilon}$ and ϵ_k are indicated on each distribution.

magnetic fields F and H respectively, departs very markedly from unity. This is shown in Fig. 2.24, which gives $\beta = F\tan\theta/uH$ as a function of F/p, calculated using velocity distributions obtained from (44), (45), and (46) with the momentum-transfer cross-section derived by Frost and Phelps. It will be seen that β reaches a maximum as high as 3·7 for $F/p \simeq 3$ V cm^{-1} torr^{-1}. The early observations of Townsend and Bailey shown in Fig. 2.24 reveal this effect quite clearly although the maximum is not as large as calculated.

Similar analysis applies to krypton and xenon. Fig. 2.25 shows drift velocities for these two gases observed by Pack, Voshall, and Phelps[†] and by Bowe[‡] compared with those calculated by Frost and Phelps[§] from the momentum-transfer cross-sections shown in Figs 2.26 (a) and (b).

Comparison may be made with the observations of Chen[||] who measured the microwave conductivity of decaying plasma established

[†] PACK, J. L., VOSHALL, R. E., and PHELPS, A. V., *Phys. Rev.* **127** (1962) 2084.
[‡] loc. cit., p. 81. [§] loc. cit., p. 82.
[||] CHEN, C. L., *Phys. Rev.* **131** (1963) 2550.

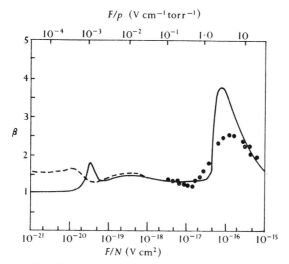

FIG. 2.24. The magnetic deflection factor $\beta = F \tan\theta/uH$ for argon. ---- calculated (300° K); —— calculated (77° K); ● observed (Townsend and Bailey) (293° K).

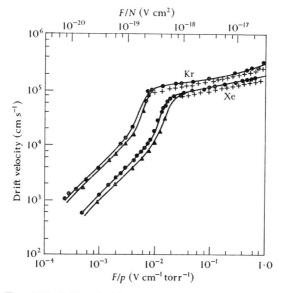

FIG. 2.25. Drift velocities for electrons in krypton and xenon. Observed: ● 300 °K (Pack, Voshall, and Phelps); ▲ 195° K (Pack, Voshall, and Phelps); + 293° K (Bowe); —— calculated (Frost and Phelps).

in mixtures of helium with the respective heavy rare gases. This ensured that the electrons reached thermal equilibrium before observations were made, as was checked by comparison of the microwave noise emitted by the plasma with that of a standard noise source. It was found that the observations, which covered a gas temperature range from 200°

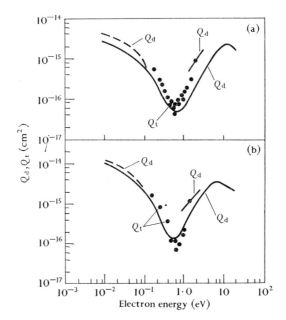

FIG. 2.26. Momentum-transfer cross-sections for electrons in krypton and xenon: (a) krypton, (b) xenon; —— derived by Frost and Phelps; – – – – observed by Chen; —●— derived by Bowe; ●●● total cross-sections observed by Ramsauer.

to 600° K, could be well represented by taking the momentum-transfer cross-section to be of the form

$$Q_d \times 10^{15} \text{ cm}^2 = a - bV^{\frac{1}{2}} + cV,$$

where V is the electron energy in eV, and a, b, c have the values

$a = 6\cdot 56, \quad b = 27\cdot 9, \quad c = 31\cdot 4 \quad$ for krypton,

$a = 19\cdot 1, \quad b = 83, \quad c = 94 \quad$ for xenon.

These cross-sections are shown in Fig. 2.26. Below 0·1 eV they are, for krypton, about 40 per cent above those derived by Frost and Phelps from the drift velocity data. The latter differ by a fraction of 2 to 4 from those found by Bowe from his analysis of his own drift velocity data, which differ by 20–30 per cent from those of Pack et al. For xenon similar discrepancies in the same sense are also found.

Momentum-transfer cross-sections obtained from the observations of Ramsauer and of Ramsauer and Kollath are also shown in Fig. 2.26 and are no less compatible with the cross-section derived by Frost and Phelps and by Chen than they are with each other.

These results will be further discussed in Chapter 8, § 7.1.1 in conjunction with semi-empirical theory and the data obtained from observations of the pressure shift in alkali metal spectra.

7.4. *Results for caesium*

A great amount of attention has been paid to the determination of Q_d for caesium at thermal and near-thermal energies. This is because of the importance of caesium seeding as a means of producing gas of sufficient electrical conductivity for use in the proposed technique for direct conversion of heat to electricity. Despite the effort put in, it cannot be said that at the time of writing (January 1967) we have any valuable information about Q_d. This is perhaps not surprising because the experiments are difficult.

As long ago as 1933 Boeckner and Mohler† measured the drift velocity of electrons in the positive column of a caesium discharge. This they did by combining measurements of the electric field across the column with determinations of electron concentration and temperature obtained from Langmuir probe characteristics. Results were extrapolated to zero electron concentration to remove effects due to collisions with ions. Their data have been re-analysed by Nolan and Phelps‡ who used more recent vapour pressure temperature data to determine the caesium pressure and allowed for errors in the Langmuir probe data that were corrected in a later paper by Mohler.§ The net result is a collision frequency over the electron energy range from 0·22 to 0·40 eV, of $1·1 \times 10^{-6}$ cm^3 s^{-1}, giving a momentum-transfer cross-section as shown in Fig. 2.27.

Chen and Raether‖ determined the cross-section by microwave conductivity in afterglows in pure caesium and in mixtures with helium. In the latter case the contribution to the collision frequency from collisions with helium atoms was subtracted, it being assumed that Q_d for such collisions has the constant value $5·3 \times 10^{-16}$ cm^2 for

† BOECKNER, C. and MOHLER, F. L., *Bur. Stand. J. Res.* **10** (1933) 357.
‡ NOLAN, J. F. and PHELPS, A. V., *Phys. Rev.* **140** (1965) A792.
§ MOHLER, F., *Bur. Stand. J. Res.* **17** (1936) 849.
‖ CHEN, C. L. and RAETHER, M., *Phys. Rev.* **128** (1962) 2679.

electron energies between 0·04 and 2 eV. Results obtained for the mean momentum-transfer cross-section are shown in Fig. 2.27, and it will be seen that quite good agreement is obtained between the results from the pure vapour and the mixture. It seems likely that the rapid relaxation in the pure caesium was due to diffusion cooling, a phenomenon discussed in Chapter 19, § 2.3.

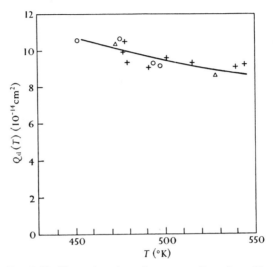

FIG. 2.27. Momentum-transfer cross-sections for collisions of electrons with caesium atoms observed by Chen and Raether. + from pure caesium; ○, △ deduced from observations on mixtures with helium partial pressures of 1·82 and 1·35 torr (at 0° C) respectively; —— calculated from the empirical formula (89).

Chen and Raether analysed their data assuming that Q_d has the form
$$10^{-14} Q_d = aV^{-1} + bV^{-\frac{1}{2}} + c, \tag{89}$$
where V is the electron energy in eV. They found from a least-squares fit that
$$a = 2·805, \quad b = -13·55, \quad c = 23·0,$$
giving the cross-section shown in Fig. 2.28.

Evidence of a strong energy-dependence of the cross-section as in (89) was provided by carrying out a microwave cross-modulation experiment (Chap. 11, § 6) in a helium–caesium mixture. According to (89) the collision frequency in the mixture will have a minimum value for a particular electron temperature T_e, so that at this temperature the attenuation of a microwave passing through the gas should also pass through a minimum. The electron temperature was raised by introducing a rectangular pulse of 20 μs duration, 8650 Mc/s frequency and 120 mW power, about 15 μs after the sensing pulse of 60 μs duration, 9410 Mc/s frequency and 10 μW power. The sensing pulse was picked up at the far end of the discharge tube through a ferrite isolator and two microwave filters, which rejected and absorbed the disturbing wave. In this way the occurrence of a bump in the transmission record was observed, showing the existence of a minimum collision frequency.

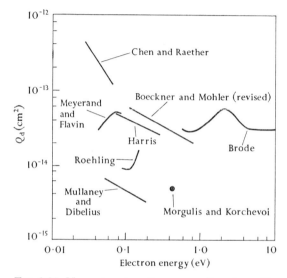

FIG. 2.28. Momentum-transfer cross-sections for collisions of electrons with caesium atoms, derived from different experiments. The total cross-section as measured by Brode is also included.

Meyerand and Flavin[†] applied the cyclotron-resonance method over a temperature range from 300° to 650° C and found that the mean cross-section passed through a minimum about midway in this range. Assuming that the collision frequency varies as $a+bV^\lambda$, where V is the energy of the electrons in eV, they determined a, b, and λ to provide the best fit to their data. In this way they derived the cross-sections shown in Fig. 2.28.

Apart from the isolated observation by Morgulis and Korchevoi,[‡] which was made by the same method as that of Boeckner and Mohler,

[†] MEYERAND, R. G. and FLAVIN, R. K., *Atomic collision processes*, ed. McDOWELL, M. R. C., p. 59 (North Holland, Amsterdam, 1964) (Proceedings of the Third International Conference on the Physics of Electronic and Atomic Collisions, London, 1963).

[‡] MORGULIS, N. D. and KORCHEVOI, Y. P., *Zh. tekh. Fiz.* 32 (1962) 900; *Soviet Phys. tech. Phys.* 7 (1962) 655.

the other data shown in Fig. 2.28 have all been obtained by measurement of electrical conductivity as a function of temperature, either in caesium-seeded inert gases (Harris†) or in pure caesium (Roehling‡). In these cases no attempt has been made to analyse the data in detail to obtain Q_d. It has been assumed in plotting Harris's results that the collision frequency is constant over the experimental energy range, while it has been assumed that Roehling's results for the variation of an average Q_d with temperature reproduces that of Q_d with energy taken as the mean energy. The total cross-section measured by Brode,§ using beam technique for energies greater than 0·8 eV, is also shown in Fig. 2.28. The only thing that is clear is that below 0·8 eV we have no reliable information about the momentum-transfer cross-section. More experiments are clearly required, particularly as no help can be expected from theory (see Chap. 8, § 8).

† HARRIS, L. P., *J. appl. Phys.* **34** (1963) 2958.
‡ ROEHLING, D., *Adv. Energy Conversion* **3** (1963) 69.
§ BRODE, R. B., *Phys. Rev.* **34** (1929) 673.

3

THE EXPERIMENTAL ANALYSIS OF THE CROSS-SECTIONS FOR IMPACT OF ELECTRONS WITH ATOMS AND IONS—IONIZATION CROSS-SECTIONS

1. Introduction

IN Chapter 1 we have described the methods that have been used for measuring the total collision cross-sections of atoms towards electrons. For electron energies less than E_1-E_0—the difference in energy between the lowest excited state and the ground state of the atom—the total cross-section is equal to the elastic cross-section, contributions from radiative encounters being negligible (see Chap. 15). At higher energies, however, excitation and ionization of the atoms can occur and the total collision cross-section includes the sum of the cross-sections for all these processes in addition to the elastic cross-section, i.e.

$$Q = Q_0 + \sum_s Q_s + \int_0^{\epsilon_{\max}} Q_\epsilon \, d\epsilon. \tag{1}$$

The cross-section Q_s is that for excitation of the sth state of the atom and $Q_\epsilon \, d\epsilon$ that for an ionizing collision in which the energy of the ejected electron lies between ϵ and $\epsilon + d\epsilon$.

We are concerned now with the analysis, by suitable experiments, of the total cross-section into individual cross-sections. Furthermore, an individual cross-section such as Q_s may be analysed further in the form

$$Q_s = 2\pi \int_0^\pi I_s(\theta) \sin\theta \, d\theta, \tag{2}$$

where $I_s(\theta) \, d\omega$, the differential cross-section (see Chap. 1, § 2.2), defines the angular distribution of the electrons that are scattered in producing collision excitation of an atom to the sth state.

For various reasons that will be given later, it is not possible to make a really complete analysis that would involve knowledge, for a given atom, of $I_s(\theta)$ for all s, θ, and electron energies. Cross-sections for inelastic collisions of electrons with atoms may be measured by studying the ions or excited atoms produced, or by studying the energy losses of the scattered electrons. In the present chapter measurements by the former method of ionization cross-sections, including cross-sections for multiple ionization, for the further ionization of positive ions and the detachment of electrons from negative ions are described. Chapter 4 describes measurements of a similar kind of cross-sections for atomic excitation. Measurements of ionization and excitation cross-sections obtained from studies of the inelastically scattered electrons are described in Chapter 5, which also discusses the analysis of energy and angular distributions for electrons scattered after elastic and inelastic impacts with atoms.

The theoretical interpretation of the data obtained is discussed in Chapters 7, 8, and 9.

2. The electrical measurement of ionization cross-sections

The chance that an electron of energy E, in passing a small distance δl through a gas, will undergo an ionizing collision with a gas atom is $NQ_i \, \delta l$, where N is the number of gas atoms/cm³ and Q_i is the total cross-section for ionization

$$Q_i = \int_0^{\epsilon_{\max}} Q_\epsilon \, d\epsilon. \tag{3}$$

These ionizing collisions give rise to a positive ion current i_+. The ratio of the strength of this current to that of the incident beam i_e will be given by
$$i_+/i_e = NQ_i \, \delta l, \tag{4}$$
provided that only one atomic electron is ejected in each collision. As this ratio may be measured directly, it follows that Q_i may readily be determined. It is, of course, important to remember that the ratio of collected positive ion current to incident electron current will only be equal to $NQ_i \, \delta l$ when the ratio is small.

If an electron is sufficiently energetic there will be a finite chance that it will knock off more than one electron from an atom in an ionizing collision. Thus if $Q_i^{(1)}$, $Q_i^{(2)}$,... are the cross-sections for single, double,... ionization, the observed positive ion current will be given by

$$i_+ = N \, \delta l (Q_i^{(1)} + 2Q_i^{(2)} + ...) i_e \tag{5}$$

as doubly-charged ions give a double contribution to the collected current. To determine the separate cross-sections it is now necessary to measure further the relative values $Q_i^{(2)}/Q_i^{(1)}$, $Q_i^{(3)}/Q_i^{(1)}$,..., etc. This may be done by a charge-to-mass-ratio analysis of the positive ion current arriving at the collector. We shall call the sum $Q_i^{(1)} + 2Q_i^{(2)} + ...$

the apparent cross-section Q_i^a for ionization. It will equal the true cross-section only at electron energies below the threshold for double ionization (usually several times the energy for single ionization), but as single ionization is usually much more probable than multiple ionization, the difference between the apparent and true cross-sections is rarely very great.

Many writers express their results in terms of quantities other than the ionization cross-section. Sometimes the 'probability of ionization', P_i is used, and this is defined as the fraction of the total number of electron collisions with the gas molecules being studied that give rise to ionization, i.e. $P_i = Q_i/Q$. Since the calculation of P_i requires a knowledge of the total collision cross-section, or alternatively of the mean free path of the electrons in the gas being studied, it is a quantity less directly associated with the actual experimental observations than Q_i.

Another quantity that has been used by many authors to express their results is the number, α_i, of positive ions formed by each electron in traversing 1 cm of path at a pressure of 1 torr. This is equal to nQ_i^a, where n is the number of gas molecules per cm³ at a pressure of 1 torr. Just as in Chapter 1, § 4, we have

$$Q_i^a = 2 \cdot 81 \times 10^{-17} \alpha_i \text{ cm}^2.$$

2.1. *Measurement of the apparent total ionization cross-section*

One of the first quantitative measurements of the apparent ionization cross-section of electrons in gases, Q_i^a, was made by Compton and van Voorhis† using He, Ne, A, Hg, H_2, N_2, and HCl. In their work, however, difficulty was experienced in preventing electrons scattered from the main beam reaching the ion-collecting electrode. In order to prevent this a large negative potential had to be applied to the collector. The resulting electric field in the region of the electron beam in the collision chamber gave rise to a 40 per cent spread in the energy of the electrons producing ionizing collisions.

To overcome this difficulty Tate and co-workers‡ used a longitudinal magnetic field of strength a few hundred gauss to confine the scattered electrons to a region close to the beam so that they could not reach the ion collector.§ A typical arrangement due to Smith is shown in Fig. 3.1.

† COMPTON, K. T. and VAN VOORHIS, C. C., *Phys. Rev.* **26** (1925) 436.
‡ JONES T., ibid. **29** (1927) 822; SMITH, P. T., ibid. **36** (1930) 1293; **37** (1931) 808; TATE, J. T. and SMITH, P. T., ibid. **39** (1932) 270; BLEAKNEY, W., ibid. **34** (1929) 157; **35** (1930) 1180; **36** (1930) 1303; BLEAKNEY, W. and SMITH, P. T., ibid. **49** (1936) 402.
§ Compton and van Voorhis had experimented with such a field but abandoned it because their collected positive ion current was irregular and was sensitive to the

Electrons from the filament K, after acceleration through the collimating slits S_1 and S_2, which were maintained at fixed potentials, entered the collision chamber. The energy of the electrons as they passed through the chamber between the plates P_2 and P_3 was determined by the potential of S_2. Positive ions produced along the path of the electrons were attracted to P_2, a rectangular plate 6×2 cm by a small potential difference between P_3 and P_2. The collector plate P_2, whose long dimension was parallel to the electron beam, was surrounded by a guard ring device G.

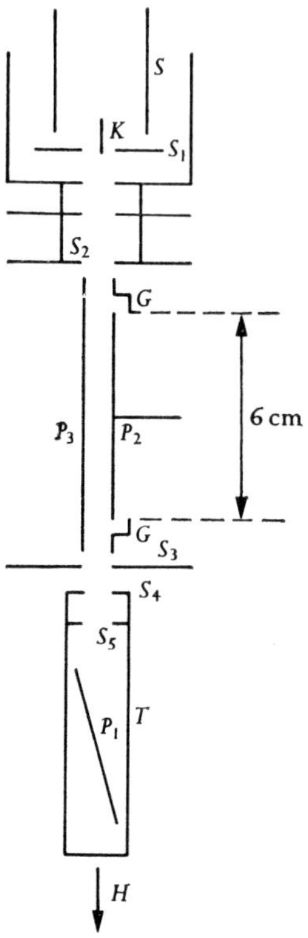

The electron beam was collected by the electrode P_1, which could be held at a positive potential of 400 V relative to the cage T.

Evaluation of the ionization cross-section Q_i^a from (5) requires accurate measurements of i_+, i_e, N and δl. We discuss now the factors influencing the accurate measurement of each of these quantities in an apparatus of the Smith type.

2.1.1. *Measurement of the ion current* i_+. It is necessary to ensure collection on P_2 of all the positive ions produced between plates P_2 and P_3. In order to verify that this condition is satisfied the current to the ion collector is measured as a function of the collector potential. In a monatomic gas the longitudinal magnetic field enables saturation ion currents to P_2 to be obtained with a field of the order of 5 V cm^{-1} between P_2 and P_3. Ions produced by molecular dissociation may be formed with appreciable kinetic energy however (see Chap. 12, § 7), so that much larger fields (30 V cm^{-1}) may be needed to ensure complete ion collection in studying the ionization of molecules (see Chap. 12, § 7.2).

FIG. 3.1. Smith's apparatus for measurement of Q_i^a.

The presence of the longitudinal magnetic field, by preventing the escape of secondary electrons from the ion collector, helps to ensure that the measured current to P_2 gives the true ion current.

The guard-ring device G ensures that the ions collected by P_2 are produced from a carefully defined path-length of the beam very nearly equal to the long dimension of P_2. By maintaining a large difference of

magnetic field. Jones showed that these effects disappeared when steps were taken to eliminate secondary electron emission.

potential between the electron collector P_1 and its surrounding cage T any ions produced in T are swept up and do not enter the ionization chamber. This is an important consideration when studying the ionization of molecular gases. In the same way ions produced in the collimator region can be prevented from entering the ionization chamber if the potential of the last slit S_2 is above that of the preceding slits. This may not always be possible, however, when studying ionization by low-energy electron beams.

Spurious ionization, due to secondary electrons arising from slit bombardment by the electron beam or from gas ionization, has to be reduced as far as possible. The former can be reduced by a suitable arrangement of electrode potentials, with the ionization chamber having the lowest potential of the whole system. The effect of the latter can be reduced by operating at as low a gas pressure as feasible.

2.1.2. *Measurement of the electron current, i_e.* The longitudinal magnetic field prevents spread of the electron beam so that the total electron current producing the ionization is collected by P_1, apart from electrons scattered out of the beam. These can be reduced by making the product of gas pressure and path length as small as convenient. The collection of secondary electrons ejected in the ionizing collisions or through slit bombardment has to be made small. Low-pressure operation practically eliminates the former. Detailed studies of currents collected by the various electrodes under different operating conditions may be required in order to verify the absence of the latter.

Above all, however, P_1 and T must be operated at a sufficiently high potential (several hundred volts) relative to the ionization chamber to ensure the absence of secondary electron emission or of electron reflection from the collector.

2.1.3. *Measurement of electron path length l.* The exact value of l is also subject to uncertainty owing to the helical path of the electrons in the collimating magnetic field, H. The maximum possible path corresponding to motion on a helix of diameter d mm equal to that of the defining hole is given by

$$l_{\max} = l(1 + 1 \cdot 10 \times 10^{-4}\, d^2 H^2/V), \tag{6}$$

where V is the energy of the electrons in eV and H the magnetic field strength in gauss. Asundi† pointed out, however, that electrons would need to receive transverse velocity components that are unrealistically large in order for them to move in a helix of diameter a few mm, the

† ASUNDI, R. K., *Proc. phys. Soc.* **82** (1963) 372.

diameter of the defining hole of a typical apparatus. In general, the correction will be determined by the energy of transverse motion (V_T) of the electron, and is then given by

$$l_{max} = l(1+0.92\, V_T/V), \qquad (7)$$

independent of the magnetic field.

Transverse velocity components may be imparted to the electrons owing to the electrostatic lens effect of the accelerating apertures, but these are negligible in collimating magnetic fields of the strength normally employed in the Tate and Smith type of apparatus. Using a different type of electron gun, however, Boksenberg† showed that variations of electron path-length of up to 10 or 15 per cent could be expected for electrons in the energy range from 20 to 300 eV. Scattering of the electrons by the gas in the ionization chamber can produce appreciable transverse velocity components. The estimate of this effect is very sensitive to the angular distribution of the scattered electrons, since even a very small fraction scattered through 90° can have a significant effect on the average path-length of the electrons in the gas. Systematic differences near the threshold between measurements of the ionization cross-sections of the rare gases made by Smith‡ and by Rapp and Englander-Golden§ have been attributed by Kieffer and Dunn‖ to the fact that the former measurements, which were carried out at higher pressure, were subject to error owing to neglect of a correction to l due to this effect.

2.1.4. *Measurement of the concentration, N, of scatterers.* Uncertainty also arises in the measurement of the atom or molecule concentration, N. A McLeod gauge has usually been employed for absolute measurement of the gas pressure. If p_M, T_M are respectively the pressure in torr and temperature in °K of the gas in the gauge and p_I, T_I the corresponding quantities in the ionization chamber, N has been obtained from the relations

$$p_I/p_M = (T_I/T_M)^{\frac{1}{2}}, \qquad (8\,\text{a})$$

$$N = 3.535 \times 10^{16} p_I (273/T_I). \qquad (8\,\text{b})$$

It is necessary to introduce a cold trap between the McLeod gauge and the apparatus and this leads to unreliable readings of the McLeod gauge, the mercury stream between the gauge and the trap acting like a diffusion pump.††

† BOKSENBERG, A., Thesis, University of London (1961). ‡ loc. cit., p. 99.
§ RAPP, D. and ENGLANDER-GOLDEN, P., *J. chem Phys.* **43** (1965) 1464.
‖ KIEFFER, L. J., and DUNN, C. H., *Rev. mod. Phys.* **38** (1966) 1.
†† The possibility of this effect was pointed out by GAEDE, W., *Annln Phys.* **46** (1915) 354. The unreliable behaviour of the gauge was found by ISHII, H. and NAKAYAMA, K.,

Schram et al.,† following a suggestion of Ishii and Nakayama,‡ cooled the walls of the McLeod gauge just above the reservoir. With this arrangement they found increased pressure readings ranging from 1 per cent for H_2 to 14 per cent for Xe. It is not certain, however, that the entire uncertainty was eliminated by this means. Rapp and Englander-Golden§ used a McLeod gauge for pressure measurements only in the case of H_2, where the error is expected to be small. In each case the gas was introduced into the ionization chamber through a leak from a reservoir. Under conditions of effusive flow they showed that the quasi-steady state pressure in the ionization chamber should be proportional to the reservoir pressure, independent of the gas. Knowing therefore the chamber pressure for H_2 it could be estimated for any other gas from the reservoir pressure. The possibility of other systematic errors in this method cannot be excluded however.

Absolute measurements of ionization cross-sections of metal vapours have been made using an apparatus of the Tate and Smith type with the vapour density in the ionization chamber obtained from vapour pressure–temperature tables.‖ Some of the early work is undoubtedly in error owing to uncertainty in the relation between vapour pressure and temperature. To obtain accurate results it is necessary to know the temperature of the ionization chamber very accurately. For example, for Zn a temperature uncertainty of 1 °K causes an error in vapour-pressure estimate of 8 per cent. On the other hand, Heil and Scott†† have measured the ionization cross-section of Cs using an apparatus of this type, but determining the vapour density by means of a surface ionization detector. Uncertainty in the efficiency of the detector may then affect the accuracy of the results obtained.

2.1.5. *Reliability of ionization cross-section measurements.* It has been seen above that while the true ionization current should be proportional to the gas or vapour pressure, some of the disturbing effects are proportional to higher powers of the pressure. A measurement of i_+ as a function of pressure is necessary to detect effects due to gas scattering of the electron beam, ionization by secondary electrons

---

Proc. 8th natn. Vac. Symp., 1961, p. 519, and has been confirmed by MEINKE, C. and REICH, G., *Vacuum* **13** (1963) 579, and by ROTHE, E. W., *J. Vac. Sci. Technol.* **1** (1964) 66.

† SCHRAM, B. L., DE HEER, F. S., VAN DER WIEL, M. J., and KISTEMAKER, J., *Physica* **31** (1965) 94.   ‡ loc. cit., p. 102.   § loc. cit., p. 102.

‖ Hg: BLEAKNEY, W., *Phys. Rev.* **35** (1930) 139; SMITH, P. T., ibid. **37** (1931) 808; HARRISON, H., Thesis (Catholic University of America Press Inc., Washington, D.C., 1956). Zn, Cd: POTTIE, R. F., *J. chem. Phys.* **44** (1966) 916.

†† HEIL, H. and SCOTT, B., *Phys. Rev.* **45** (1966) 2791.

ejected in a primary gas ionization process, and other multiple collision effects. i_+ should also be measured as a function of i_e to verify the absence of space charge or other non-linear effects.

A systematic study of spurious instrumental effects that make accurate measurements of ionization cross-sections difficult has been made by Marmet and Morrison.† They observed time-dependent effects, as a result of which the detailed form of the energy dependence of the ionization cross-section depends on the time that has elapsed after admission of the sample gas to the apparatus. They concluded that

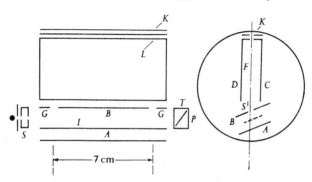

FIG. 3.2. Bleakney's apparatus for analysis of the ion current.

these and other anomalies may be related to ion formation at the walls of the ionization chamber. The effects were overcome by the use of higher ion extraction fields and by appropriate design of the ion chamber to avoid extraction of ions formed on the wall.‡

2.2. Analysis of positive ion current

To complete the determination of the true cross-section for single, double,... ionization it is necessary to perform a charge-to-mass-ratio analysis of the positive ions produced. The first systematic study of the relative probabilities of multiple ionization processes were made by Bleakney,§ using the apparatus illustrated schematically in Fig. 3.2.

Just as in Smith's apparatus, electrons were collimated by a magnetic field and the ions produced in the chamber I were attracted to the plate B by means of a small potential difference between A and B. In Bleakney's work the first slit S had dimensions 1×4 mm and a tungsten filament was stretched lengthwise behind it. This ensured that the

† MARMET, P. and MORRISON, J. D., *J. chem. Phys.* **36** (1962), 1238.
‡ MORRISON, J. D., ibid. **40** (1964), 2488.
§ BLEAKNEY, W., *Phys. Rev.* **34** (1929) 137; **35** (1930) 1180; **36** (1930) 1303; BLEAKNEY, W. and SMITH, P. T., ibid. **49** (1936) 402.

collimated electron beam was in the form of a ribbon, having dimensions approximately defined by the first slit.

A slit S' having dimensions $0 \cdot 25 \times 60$ mm was cut in the plate B, parallel to the direction of the electron beam. A sample of ions attracted to B passed through this slit into the analysing chamber F. The charge to mass ratio of the ions emerging from the slit S' was measured by passing them between the condenser plates C and D, maintained at different potentials. Under the action of crossed electric and magnetic fields ions describe trochoids. In Fig. 3.2 let y refer to a direction parallel to the condenser plates and x to a direction perpendicular to them. If a particle of charge q and mass M passes through S' moving in the y-direction with velocity v_0 at time $t = 0$, the equation of its path in the condenser will be

$$x = (M/qH)(v_0 - F/H)\{1 - \cos(qHt/M)\},$$
$$y = (Ft/H) + (M/qH)(v_0 - F/H)\sin(qHt/M), \qquad (9)$$

where F and H are the electric and magnetic field strengths respectively.

For a particular ion to be collected by the electrode K, x must vanish identically, the condition for which can be written $q/M = F^2/2VH^2$, where V is the potential through which the ion has fallen before entering the analyser.

Thus by measuring the current to K as the ratio F/H was changed, the relative numbers of singly, doubly, ... charged ions produced in I could be determined.

The plates A and B in Bleakney's apparatus were set at a slight angle to the x-direction to allow for the curvature of the ion path before reaching the slit S'.

The Bleakney trochoidal type of analyser has the advantage that it can be operated under conditions of high collection efficiency. Boksenberg[†] used an arrangement corresponding effectively to a slitless spectrometer in which he concluded that ions formed without initial kinetic energy were being collected with 100 per cent efficiency. More conventional forms of mass spectrometer commonly employed to analyse the charge carried by the positive ions suffer from the disadvantage of a small collection efficiency, which may depend on q/M and even on the energy of the electron beam. The former dependence may arise if the trajectories of ions in the extraction and focusing electric and magnetic fields depend on q/M. The latter dependence may arise from changes

† loc. cit., p. 102.

with energy of the electron beam configuration. In addition care must be taken to allow for any variation with q/M of the efficiency of the ion detector employed. Early experiments with an electron multiplier detector gave incorrect results on multiple ionization owing to the failure to allow for the variation of detection efficiency with the q/M value of the ions. Further, since multiple processes may be of importance in producing multiple ionization, it is necessary to verify that the observed relative cross-sections are independent of the pressure in the ionization chamber and of the beam current.

The ion currents corresponding to highly ionized atoms may be very feeble and if the M/q value corresponds to that of residual gas ions the background correction may be large. Schram[†] used ^{86}Kr instead of the more abundant ^{84}Kr in studying the ionization of Kr. In this way he was able to avoid confusion with residual gas ion peaks at M/q values of 28, 14, and 12. Similarly, Fox's[‡] measurements on helium ionization were carried out in ^{3}He to avoid confusion with H_2^+ and D^+.

If i^{n+} is the positive ion current carried by ions with n-fold charge and $Q_i^{(n)}$ the cross-section for their production then

$$Q_i^{(n)} = \frac{i^{n+}}{\sum_{n=1}^{n_{\max}} i^{n+}} \frac{Q_i^a}{n}, \qquad (10)$$

where Q_i^a is the known apparent total ionization cross-section under the conditions of the experiment, provided it can be assumed that the collection efficiency is the same for all the ions.

Mass spectrometer methods are of particular importance in studying the products of molecular ionization. We return to the discussion of their use in Chapter 12, § 7.3, where apparatus primarily designed for the study of molecular ionization but also useful for the investigation of the multiple ionization of atoms is described.

2.3. *The crossed-beam method for the measurement of total apparent ionization cross-section*

The method of Tate and Smith is not applicable to determine the ionization cross-section of a free atom that is chemically unstable. In such cases an atomic beam technique is used in which a beam of electrons is fired across a beam containing atoms whose ionization cross-section is to be measured. If i_e is the electron current that crosses the beam, i_+ the total ionization current when all the ions produced in the beam are

[†] SCHRAM, B. L., *Physica* **32** (1966) 197.
[‡] Fox, R. E., *Advances in mass spectrometry*, p. 397 (Pergamon Press, London, 1959).

collected, l the length of the electron path across the beam, and N the number of atoms per cm³ in the portion of the beam traversed by the electrons, then the apparent ionization cross-section is again given by (5).

The crossed-beam technique was in fact first developed for the study of Q_i^a of atoms such as the alkali metals, where it was convenient to use an oven to produce a beam of the vapour atoms or molecules. Quite apart from the need to obtain a sufficient concentration of the atoms being studied, the crossed-beam method possesses some particular advantages such as the elimination of wall effects and of multiple collision processes. The early experiments in which measurements of Q_i^a were attempted using this method† ran into difficulties owing to the very large correction required for ions formed in the residual gas of the apparatus, some of which could enter the ion collector. Many useful investigations have been carried out, however, using this method in conjunction with a mass spectrometer to study relative cross-sections for the formation of a particular ion with electrons of different energies.‡

In recent experiments,§ and following the work of Branscomb and Fite‖ on the photo-detachment of H^- (see Chap. 15, § 2), the difficulty of the background correction has been overcome by modulating the atomic beam by means of a mechanical 'beam chopper' in the form of a rotating toothed wheel as described already in relation to the measurement of elastic scattering cross-sections (Chap. 1, § 4.2). The beam ionization is then produced at the beam modulation frequency so that, by using a narrow band amplifier tuned to the modulation frequency to detect the ionization current, the signal-to-noise ratio can be greatly enhanced and reliable measurements of Q_i^a obtained. However, the passage of the modulated atomic beam through the collision chamber produces a rise in background gas pressure and thence a spurious signal of the modulation frequency. This signal reaches its maximum at the end of the modulated atomic beam pulse and is out of phase with the desired signal. It can be discriminated against by use of a phase-sensitive

† SMYTH, H. D., *Proc. R. Soc.* A**102** (1922) 293; VON HIPPEL, A., *Annln Phys.* **87** (1928) 1035; DITCHBURN, R. W. and ARNOT, F. L., *Proc. R. Soc.* A**123** (1929) 516; FUNK, H., *Annln Phys.* **4** (1930) 149.

‡ DIBELER, V. H. and REESE, R. M., *J. chem. Phys.* **31** (1959) 282; BRINK, G. O., *Phys. Rev.* **127** (1962), 1204; KANEKO, Y., *J. phys. Soc. Japan* **16** (1961) 2288; **18** (1963) 1822; FIQUET-FAYARD, F. and LAKMANI, M., *J. Chim. phys.* (1962) 1050; ZIESEL, J. P., ibid. **62** (1965) 328.

§ See, for example, BOYD, R. L. F. and GREEN, G. W., *Proc. phys. Soc.* **71** (1958) 351; FITE, W. L. and BRACKMANN, R. T., *Phys. Rev.* **112** (1958) 1141; ibid. **113** (1959) 815; ROTHE, E. W., MARINO, L. L., NEYNABER, R. H., and TRUJILLO, S. M., ibid. **127** (1962) 582; SMITH, A. C. H., CAPLINGER, E., NEYNABER, R. H., ROTHE, E. M., and TRUJILLO, S. M., ibid. **127** (1962) 1647.

‖ BRANSCOMB, L. M. and FITE, W. L., ibid. **93** (1954) 651.

detector with an arrangement for varying the phase of the detector to maximize the wanted signal as described in Chapter 1, § 4.2.2, p. 15.

Fig. 3.3 shows a schematic diagram of the apparatus used by Fite and his associates for studying ionization cross-sections of hydrogen and oxygen atoms. For atomic hydrogen the source was a tube 3 in long and $\frac{3}{16}$ in diameter, made by rolling tungsten foil of thickness 0·001 in

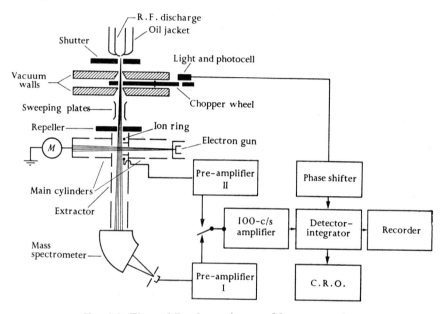

Fig. 3.3. Fite and Brackmann's crossed-beam apparatus.

into a tube with walls composed of six layers of the foil and closed at either end with plugs of tungsten wadding. Thermal dissociation occurred when the furnace was heated to a temperature of about 2000 °C by passing a heavy current through the foil.

For atomic oxygen, Fite and Brackmann[†] used an r.f. discharge tube source. It was constructed of $\frac{3}{8}$-in glass tubing surrounded by a cooling jacket through which oil circulated. The length of the discharge region was about 6 in. The r.f. discharge (frequency 13 Mc/s) was provided by an oscillator coupled capacitatively by means of electrodes on the outside of the cooling jacket. Under optimum conditions of pressure (a few hundred torr), temperature (near room temperature), and discharge power (about 10 W), the neutral beam emerging through a hole of diameter 1 mm at the end of the discharge tube was about 30 per cent

† FITE, W. L. and BRACKMANN, R. T., ibid. **113** (1959) 815.

dissociated, contained only small traces of water vapour and nitrogen, and had at least 97 per cent of the beam particles in their ground state.†

After leaving the source the beam was collimated by means of slits through which it entered and left the second chamber containing the rotating toothed wheel used to modulate the beam at the required frequency.‡

The atomic beam then passed through a pair of sweeping plates across which a difference of potential was maintained so that any charged particles were swept out of the beam before it entered the ionization chamber where it was crossed by a homogeneous electron beam from the electron gun. The atomic beam was 4 cm long and in the collision region had a cross-section of about 4×6 mm. The electron beam was focused by electrostatic lenses in the gun and collimated by means of apertures to ensure that all electrons were passing through the neutral beam and all ions formed came from electrons whose current was being measured by the electron collector. The total ionization produced in the beam was measured by collecting it on an ion ring maintained at a potential carefully chosen to collect all the ions, including those formed by dissociative ionization. The relative ionization carried by atomic and molecular ions was measured by extracting them through a cylinder, maintained at a suitable potential. Ions of a given mass to charge ratio were focused by the mass spectrometer on to the detector.

The apparatus was divided into three vacuum chambers containing respectively the source, chopper wheel, and the main apparatus, each with separate differential pumping. A shutter between the source and the chopper wheel enabled the direct beam to be blocked out without seriously changing the conditions of gas flow between the various chambers.

The signal to the ion ring or the detector was passed through a plate follower preamplifier and then successively into a tuned amplifier and a phase-sensitive detector. The latter was locked to a signal taken directly from the chopper wheel by a light and photocell.

If $Q_{i(1)}$ is the cross-section for ionization of atomic hydrogen, i.e. $e+H \to e+e+H^+$, and $Q_{i(2)}$ that for molecular ionization without

† In their measurements of the ionization cross-sections Rothe *et al.* (loc. cit., p. 107) and Boksenberg (loc. cit., p. 102) used an r.f. discharge source for both H and O.

‡ A modulation frequency of 100 c/s was used in the experiments of Fite and co-workers. Other investigators have used frequencies of 10 c/s, or even 2 c/s. The frequency should be sufficiently high so that the Fourier components of the frequency spectrum of the background pressure fluctuations at the modulation frequency are negligible.

dissociation, i.e. $e+H_2 \rightarrow e+e+H_2^+$, the ratio $Q_{i(1)}/Q_{i(2)}$ can be determined readily from the crossed-beam measurements.

Let S_r be the measured H_2^+ current in the mass spectrometer when the furnace is operated at a temperature T_r, so low that the dissociation can be neglected and S_0 the H_2^+ current that would have been observed at the working temperature T if no dissociation had occurred. Owing to the shorter time spent by the molecules in traversing the beam at the higher temperature,
$$S_0 = S_r(T_r/T)^{\frac{1}{2}}. \tag{11 a}$$

Let α be the ratio of the currents in the mass spectrometer peaks corresponding to H^+ and H_2^+ respectively with the source at temperature T_r. Even under these conditions a peak due to H^+ will be observed as a result of the dissociative ionization of H_2 (see Chap. 13, § 1.5).

Let S_1 and S_2 be the actual mass spectrometer currents corresponding to H^+ and H_2^+ with the source at the working temperature T and let D be the fractional dissociation of the beam at this temperature. Then
$$S_2 = (1-D)S_0, \tag{11 b}$$
$$S_1 = \alpha S_2 + 2^{\frac{1}{2}} D S_0 (Q_{i(1)}/Q_{i(2)}), \tag{11 c}$$
where the second term in (11 c) arises from direct atomic ionization. Using (11 a, b, c)
$$Q_{i(1)}/Q_{i(2)} = \frac{S_1 - \alpha S_2}{2^{\frac{1}{2}}\{S_r(T_r/T)^{\frac{1}{2}} - S_2\}}. \tag{12}$$

For H_2 the absolute cross-section $Q_{i(2)}$ has been determined independently (see Chap. 13, § 1.4.1), so that $Q_{i(1)}$ can be derived from the crossed-beam measurements.

For measurements using the r.f. discharge source a similar analysis gives
$$Q_{i(1)}/Q_{i(2)} = \frac{1}{2^{\frac{1}{2}}}\left(\frac{S_1 - S_1^0}{S_2 - S_2^0} + \frac{S_1^0}{S_2^0}\right), \tag{13}$$
where S_1, S_1^0 are the currents in the mass spectrometer peak corresponding to atomic ions with the discharge source on and off respectively and S_2, S_2^0 the similar currents corresponding to the molecular peak.†

Smith et al.‡ used a crossed-beam method to measure the ionization cross-section of atomic nitrogen. The source used was the afterglow of a pulsed d.c. discharge since there is evidence§ that the only active species present in appreciable concentration in the low-pressure after-

---

† This expression neglects a small correction due to the heating of the gas by the discharge amounting to about 3 °K per 10 W of discharge power in Fite's experiments with O_2. ‡ loc. cit., p. 107.
§ BERKOWITZ, L., CHUPKA, W. A., and KISTIAKOWSKY, B. B., J. chem. Phys. 25 (1956) 457; TANAKA, Y., JURSA, A. S., LEBLANC, F. J., and INN, E. C. Y., Planet, Space Sci. 1 (1958) 7.

glow of nitrogen is the ground state nitrogen atom. The method used by Fite and Brackmann, in which the ionization cross-sections for the atomic and molecular gases are compared, assumes the collection efficiency for both to be the same. This implies that the momentum transfer from electron to atom or molecule in the ionizing collision does not affect the collection efficiency. This is by no means obvious since the momentum transfer may be appreciable in comparison with the beam momentum in the case of beams of thermal energy, although Fite and Brackmann gave indirect evidence that the assumption was justified.

The modulated crossed-beam method has been applied to the measurement of ionization cross-sections of the alkali and alkaline earth metals.† An oven source was used and the absolute cross-sections derived directly from (5) instead of from comparison with the known ionization cross-sections of a stable molecule. To determine the beam density N of the atomic beam, the flux R of the particles in the beam was measured, using surface ionization on an oxidized tungsten wire of 0·002 in diameter.

If P_s is the surface ionization probability of beam atoms on the hot wire, v the mean velocity of the particles in the beam, and i_{sd} the ion current produced by the surface detector,

$$Q_i = C_1 \frac{v P_s}{i_e} \frac{i_+}{i_{sd}}, \qquad (14)$$

where C_1 is a geometric factor determined by the oven slit area, the defining slit width, the size of the surface detector wire, and the distances from oven to the electron beam and the surface detector wire.

The same surface ionization measurement enabled v to be estimated from the law for effusive flow through the oven slit and the constants in the vapour pressure–temperature relation for the material being studied but without requiring knowledge of the source temperature. A subsidiary experiment gave P_s a value close to 1 under the conditions of operation of the detector.

Peterson‡ has carried out crossed-beam measurements of ionization cross-sections using a neutral beam formed by charge exchange from a charged beam of positive ions of energy in the range 1–5 MeV. The method was applied to argon, neon, and nitrogen (see also Chap. 12, § 7.2). The current of neutral atoms was estimated by measuring the

† BRINK, G. O., *Phys. Rev.* **127** (1962) 1204; **134** (1964) A345; MCFARLAND, R. H. and KINNEY, J. D., ibid. **137** (1965) A1058.

‡ *Atomic collision processes*, ed. MCDOWELL, M. R. C., p. 465 (North-Holland, Amsterdam, 1964) (Proceedings of the Third International Conference on the Physics of Electronic and Atomic Collisions, London, 1963).

energy given up when they were brought to rest in a calorimeter. The observed absolute apparent ionization cross-sections for argon and neon were, however, substantially greater than those obtained for the same gases using a static method, suggesting that some of the atoms in the neutral beam may be formed in excited states in these cases (see § 2.5.3.4).

2.4. *Ionization cross-section measurements with high-energy resolution*

The study of the variation of the ionization cross-section in the neighbourhood of the threshold is important for the accurate determination of ionization potentials. A fine structure of the energy variation of the ionization has also been observed, especially in the region just above the ionization threshold. It is related to indirect modes of ionization of the atoms and to the formation of ions in excited states. For the study of phenomena such as these, special precautions are needed to ensure that the cross-section measurements relate to ionization by electrons with a very small spread in energy.

2.4.1. *Use of a high-resolution electron beam.* The most direct method of studying the fine details of the variation of ionization cross-section with electron energy involves the use of energy analysis of the incident electron beam.

The first experiments of this kind were those of Lawrence,[†] who used a magnetic field to provide an energy analysis of the incident electrons. They were extended by Nottingham,[‡] using an apparatus that is a development of that used earlier by Lawrence with the whole apparatus in the magnetic field. This makes it difficult to define the exact electron path in order to obtain absolute measurements.

Several more recent investigations have employed electrostatic analysis to produce very homogeneous electron beams for studies of this kind.[§] Fig. 3.4 shows the arrangement used by Clarke. Electrons from the filament A were accelerated through a potential difference of 3 V and entered the electrostatic selector through a slit in plate S. The selector plates C and B were maintained at potentials of 1·5 and 4·5 V respectively relative to A, and focused the electrons on to the exit plate of the selector after deflexion through 127°. A voltage applied to the ionization chamber D determined the energy with which they bombarded gas molecules entering the ionization chamber from E. The ions produced were analysed by a mass spectrometer whose accelerating

[†] LAWRENCE, E. O., *Phys. Rev.* **28** (1926) 947.
[‡] NOTTINGHAM, W. B., ibid. **55** (1939) 203.
[§] CLARKE, E. M., *Can. J. Phys.* **32** (1954) 764; FONER, S. M. and NALL, R. H., *Phys. Rev.* **122** (1961) 512.

slit system is shown at F, H, and G. With this arrangement the spread of the electron energies at half intensity was 0·3 eV for a beam current of 10^{-8} A in the energy range of 7–60 eV.

Marmet and Kerwin† sought to reduce the energy spread further without reducing the beam current. They showed that the principal obstacle in the way of improving the performance of the electrostatic analyser arose from the space charge due to electrons reflected from the surfaces and from scattering in the gas atmosphere of the mass

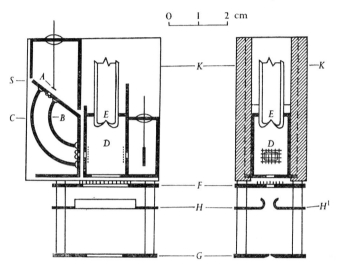

Fig. 3.4. Clarke's apparatus for studying ionization cross-sections under high resolution.

spectrometer ion source. Accumulation of electrons that built up a space charge in the region of the entrance slit of the selector was avoided by protecting the deflector plates with transparent tungsten mesh grids of 90 per cent transparency. A different solution had to be found to the problem of avoiding the build-up of space charge behind the entrance slit to the ion chamber. Marmet and Kerwin showed that the reflection coefficient of slow electrons depended strongly on the degree of polish of a metal surface. They finally developed a complicated procedure for making a durable surface of low reflection coefficient (20 per cent) in contrast to the usual values of 60–70 per cent for normal metal surfaces. The inside walls of the ionization chamber were made of this surface, which the authors refer to as 'electron velvet'. Fig. 3.5 shows the appearance of these inside surfaces. The procedure used for preparing electron

† MARMET, P. and KERWIN, L., *Can. J. Phys.* **38** (1960) 787.

velvet is described by the authors as follows: 'Fine aluminium wire (0·5 mm) is cleaned and plated with 0·01 mm copper. Thousands of 3- or 6-in lengths of plated wire are inserted in an aluminium cylinder of the same length and about 4 cm in diameter. The whole is compressed to approximately an oval section with a pressure of about 600 kg/cm, squeezing the wires tightly together. The end of the cylinder is then worked flat and polished, when the fine network of copper plating may be seen on the aluminium surface formed by the ends of the squeezed wires. The surface is placed for a few seconds in sodium hydroxide, which dissolves a little aluminium and leaves the copper plating (tubes) in slight relief. The surface is then heavily plated with copper (0·5 mm). This end of the cylinder is then sliced off to form a plate about 2 mm thick, which is ground and polished on both sides to a thickness of about 1 mm. It is then immersed in sodium hydroxide until all of the aluminium has been dissolved, leaving the fine copper tubes, about 1 mm long and 0·5 mm in diameter, anchored to their plated copper base. The sheet is then plated with gold.'

With this technique it was possible to reduce the energy spread of the electrons at half intensity to 0·04 eV for a current of about 10^{-9} A emerging from the analyser, and to determine the position of characteristic structure of the ionization curve for argon to an accuracy of 0·01 eV on the energy scale.

2.4.2. *The retarding potential difference method.*[†] Fox and his co-workers at the Westinghouse Research Laboratories have developed the retarding potential difference (r.p.d.) method for obtaining ionization cross-sections relevant to a very small spread of electron energy.[‡] The principle of their apparatus is shown in Fig. 3.6. Electrons emitted from filament 1 are subjected to a retarding potential difference V_R between 1 and the slit 2 so that only electrons emitted with energy greater than eV_R get accelerated through the slit 3 (at potential V_a relative to 2) into the ionization chamber, and there is a sharp cut-off on the low-energy side of the spectrum of electrons entering the chamber. The retarding potential is then changed by the small amount δV_R without changing the potentials of any of the other electrodes, so that those electrons possessing initial energies greater than eV_R will have their kinetic energy in the ionization chamber unaffected by the potential of slit 2. The change in current collected by the electron collector 5 then gives the number of

[†] See also Chapter 1, § 5.2.
[‡] Fox, R. E., Hickam, W. M., Grove, D. J., and Kjeldaas, T., *Rev. scient. Instrum.* **26** (1955) 1101. The same method was also used by Frost, D. C. and McDowell, G. A., *Proc. R. Soc.* A232 (1955) 227.

Fig. 3.5. Electron 'velvet' ionization chamber of Marmet and Kerwin.

electrons with energy between eV_R and $e(V_R+\delta V_R)$ and the change in ionization measures the number of ions produced by electrons in this range. The slits 2 and 3 are gold-plated to reduce contact potentials. An electrode 4, charged to a positive potential V_P relative to the rest of the ionization chamber, pushes the ions through a slit in the side of the ionization chamber, from where they are accelerated into a mass spectrometer. This ion push-out field and the electron beam are pulsed out-of-phase so that there is no field in the ion chamber when the electron beam

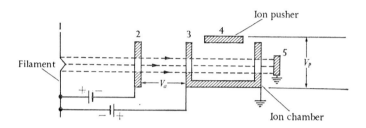

FIG. 3.6. Retarding potential difference apparatus of Fox and collaborators.

is passed through. The pulsing of the electron beam is achieved by means of an electrode between 1 and 2. It is estimated that an effective electron energy spread of 0·06 eV at half maximum can be obtained using the r.p.d. method. The r.p.d. method has received wide application. Burns† has modified it so that the small difference potential, δV_R, is applied to the retarder plate as a 13-cycle square wave that produces a 13-cycle modulation of the ion beam. After amplification and rectification this gives a signal that is proportional to the number of ions formed by a beam of electrons that should be uniform in energy to within $e(\delta V_R)$ eV.

Difficulties have been experienced in the application of the r.p.d. method however. Under certain conditions, near the ionization threshold, the difference ionization current signals may actually take negative values. Simpson‡ has pointed out that anomalous effects are to be expected when magnetic fields are used in electron energy analysis, owing to periodic focusing effects. Anderson, Eggleton, and Keesing,§ in a detailed calculation for a particular r.p.d. geometry in a magnetic field, have been able to reproduce conditions in which electric field penetration through the aperture in the retarding electrode leads to

† BURNS, J. F., *Atomic collision processes*, ed. McDOWELL, M. R. C., p. 451 (North Holland, Amsterdam, 1964).
‡ SIMPSON, J. A., *Rev. scient. Instrum.* **32** (1961) 1283.
§ ANDERSON, N., EGGLETON, P. P., and KEESING, R. G. W., ibid. **38** (1967) 924.

negative values of the ionization difference current. It arises from a dependence on the applied retarding potential of the shape of the high-energy side of the electron energy distribution. The build-up of electron space charge in front of the retarding electrode aperture might also be expected to produce anomalous effects.

Burns† has found that these difficulties arise with the r.p.d. method under rather abnormal operating conditions. Undoubtedly it is desirable that studies of the fine structure of the dependence of Q_i^a on electron energy, particularly near threshold, should be repeated using homogeneous electron beams obtained by electrostatic or magnetic analysis. Most detailed work of this kind has so far employed the r.p.d. method, however, and in the few cases where comparison can be made with measurements using magnetic or electrostatic analysis the agreement is satisfactory. The r.p.d. results are also consistent with known spectroscopic data for the cases studied. In these circumstances, therefore, it seems justified to accept a great many of the results using the r.p.d. method, especially at energies well above the ionization threshold, E_i. In any case, many anomalies persist in measurements of ion currents in mass spectrometers that have nothing to do with the r.p.d. method.

2.4.3. *Morrison's method of studying fine structure in ionization cross-sections.* Morrison and his collaborators‡ have developed methods of studying fine structure in the energy dependence of ionization cross-sections in the presence of an energy spread of the electron beam.

This fine structure is related to the onset of new ionization processes. The accuracy with which the threshold potentials concerned can be determined depends on the way the ionization cross-section for the process increases above the threshold. This variation is discussed in § 2.5.3.1 and in Chapter 9, § 10.8. For n-fold ionization Q_i is expected to be proportional to $(E-E_i)^n$, very close to the threshold.

The onset of single ionization processes should be characterized by a sharp change of slope in the apparent total ionization cross-section variation with energy, and these should be easily distinguished. The threshold for higher-order ionization is more difficult to determine. For n-fold ionization, however, according to the theory referred to above, the $(n-1)$th differential of the ionization cross-section curve should exhibit a sharp change of slope at threshold. Dorman and Morrison§ analysed their observed form of variation of the ionization current, $i(V)$, for

† BURNS, J. F., loc. cit., p. 115.
‡ MORRISON, J. D., *J. chem. Phys.* **21** (1953) 1767; DORMAN, F. H. and MORRISON, J. D., ibid. **34** (1961) 578.
§ loc. cit.

electrons of nominal energy eV, assuming the validity of the threshold ionization theory and taking account of the electron energy spread.

If $w(U)\,dU$ is the number of electrons with energy between $Ve+U$ and $Ve+U+dU$, they write

$$i(V) = \int_{E_i}^{\infty} |M(E)|^2 P(E) w(E-Ve)\,dE, \tag{15}$$

where $E = Ve+U$. Neglecting the variation with energy of $M(E)$ over a small range of energy near the threshold, differentiation of (15) with respect to V gives

$$\frac{di(V)}{dV} = \int_{E_i}^{\infty} \frac{\partial w(E-Ve)}{\partial V} P(E)|M(E)|^2\,dE$$

$$= -e \int_{E_i}^{\infty} \frac{\partial w(E-Ve)}{\partial E} P(E)|M(E)|^2\,dE. \tag{16}$$

Integrating by parts,

$$\frac{di(V)}{dV} = [-ew(E-Ve)P(E)|M(E)|^2]_{E_i}^{\infty} +$$

$$+ \int_{E_i}^{\infty} ew(E-Ve) \frac{dP(E)}{dE} |M(E)|^2\,dE. \tag{17}$$

If $P(E_i) = 0$, the first term on the right-hand side vanishes since clearly $w(E-V_e) \to 0$ as $E \to \infty$.

Provided $P(E_i)$ and all its derivatives up to and including order $n-1$ vanish at $E = E_i$, as will be the case for n-fold ionization if

$$P(E) \propto (E-E_i)^n,$$

the process can be repeated and leads to the expression

$$\frac{d^n i(V)}{dV^n} = e^n \int_{E_i}^{\infty} w(E-Ve) \frac{d^n P(E)}{dE^n} |M(E)|^2\,dE, \tag{18}$$

which is constant if $P(E) \propto (E-E_i)^n$. For the $(n+1)$th derivative, however,

$$\frac{d^{n+1} i(V)}{dV^{n+1}} = w(E_i - Ve) \left(\frac{d^n P(E)}{dE^n}\right)_{E=E_i} |M(E_i)|^2 e^{n+1} \tag{19}$$

since
$$\frac{d^{n+1} P(E)}{dE^{n+1}} = 0.$$

This shows that the $(n+1)$th derivative of the ionization cross-section with respect to nominal electron energy is closely related to the electron energy distribution. It is in fact proportional to a curve having the shape of this distribution, reversed on the energy scale and with a sharp cut-off at $V = V_i = E_i/e$. Thus if Fig. 3.7 (a) is the shape of the electron energy distribution, $w(U)$, Fig. 3.7 (b) shows the shape that would be expected for the $(n+1)$th differential curve of $i(V)$.

Morrison and Nicholson† have analysed double ionization cross-sections along these lines. They recorded directly the first derivative

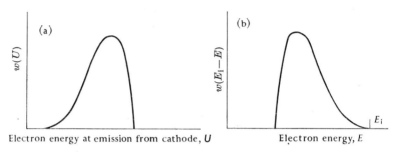

Fig. 3.7. (a) Energy spread of electrons from cathode. (b) Shape of $(n+1)$th differential curve of $i(V)$.

$\Delta I/\Delta V$ of the ionization current where $\Delta V = 0.1$ eV. Dorman and Morrison‡ tried to distinguish auto-ionization processes from the onset of new direct ionization processes in the analysis of the fine structure of ionization curves (see § 2.5.3.2). The onset of an auto-ionization process produces a step in the ionization curve itself, the onset of a single direct ionization process a step in the first differential ionization curve, a direct double ionization process a step in the second differential ionization curve, and so on. This method has, however, achieved only limited success.

2.4.4. *Problems involved in experimental study of threshold behaviour.* The accurate measurement of appearance potentials and of structure in ionization cross-section curves near the threshold is fraught with considerable difficulties of an instrumental nature. Spurious effects may give rise to erroneous estimates of appearance potentials, or distort the shape of the ionization cross-section curve.

Some of these difficulties have been summarized by Fox, Hickman, Grove, and Kjeldaas.§

The electron collector should be kept at the same potential as the

† Morrison, J. D. and Nicholson, A. J. C., *J. chem Phys.* **31** (1959) 1320.
‡ loc. cit., p. 116.
§ Fox, R. E., Hickman, W. M., Grove, D. J., and Kjeldaas, T., *Rev. scient. Instrum.* **26** (1955) 1101.

ionization chamber. If it is made positive relative to the chamber, ions formed between the end of the chamber and the collector may be driven back into the collection region, producing a 'tailing' of the apparent ionization curve near the threshold, and leading to estimates of the appearance potentials that are too low.

A further difficulty is caused by the formation, at the point where the electron beam strikes the collector, of an insulating spot, attributed to the breakdown of surface contaminants, such as oil from the vacuum pumps. Local surface potentials may be built up at such a spot to such an extent that the electrons are reflected back into the collection region and may produce further ionization. This effect is largest for low-energy electrons and can lead to distortion of the shape of the curve near the threshold. If it sets in under particular bombardment conditions it could cause the appearance of spurious breaks. Fox *et al.* observed spurious effects of this kind commonly to set in after some time (about 50 h) of operation of a new source. The effect is made more serious by the collimating magnetic field of the ion source that ensures that the reflected electrons return to the collection region.

The collimating magnetic field may also introduce tailing in the ionization curves near the threshold by introducing a spread of effective path-length of electrons through the ionization chamber, depending on their component of motion transverse to the magnetic field. Instability of the beam in traversing the slit system arising from a complex interaction of the electrons with surface charges in the presence of the magnetic field may introduce marked distortion of the curves. It is very sensitive to the detailed alignment of the magnetic field in the region of the slits.

Change of contact potential due to surface contamination may have the result that electrodes no longer define planes of constant potential. The building-up of space charge effects either from the beam itself or as a result of secondary electron emission at various parts of the apparatus may introduce further electric field irregularities, giving rise to distortion of the shape of the ionization curve and so 'tailing' near the threshold. Such effects are particularly serious when the r.p.d. method is employed, since they may have the result that the cut-off by the retarding electrode is not sharp and the energy difference band may not be entirely determined by the change, ΔE_{R}, in the retarding electrode potential.

2.4.5. *The use of crossed-beam methods for studying ionization near the threshold.* McGowan and his associates[†] have developed the use of

[†] See, for example, MCGOWAN, J. W. and CLARKE, E. M., *Phys. Rev.* **167** (1968) 43.

crossed-beam techniques for investigating the behaviour of the ionization cross-section near the threshold with high electron energy resolution. In their apparatus not only could the ions be collected and analysed, but also the scattered electrons could be analysed in energy and angle.

Fig. 3.8 illustrates the general arrangement of their experiment. An atomic beam is crossed by an electron beam of very closely prescribed

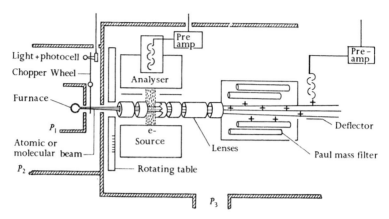

FIG. 3.8. General arrangement of the apparatus used by McGowan and his associates for studying ionization near the threshold.

energy. The ions produced in the interaction region travel on with the atomic beam and are then accelerated into a Paul quadrupole mass filter and analysed.

The general arrangement of the electron spectrometer is shown in Fig. 3.9. The electron beam is energy-selected by a 127° electrostatic analyser of Marmet–Clarke–Kerwin type (§ 2.4.1) in which the analysing field is applied through a potential difference of around 1·05 V between the inside and outside grids G_0 and G_1 while the plates P_0 and P_1 are biased (at potentials between 0 and 40 V) so as to collect electrons which have penetrated the grids. Energy selection is carried out at an energy near 0·5 eV, and the issuing electrons are then accelerated to the desired energy before entering the interaction region that is defined by P_C and a number of shields such as S_A and S_C. Energy analysis of electrons scattered through any angle between −95° and +35° can be carried out by a second similar electrostatic analyser that can be rotated through this angular range.

To minimize the reflection of low-energy electrons gold black is deposited on all surfaces that see electrons. This proved to be easy to apply in practice.

The atomic beam is defined in cross-sectional area by a rectangular slit 5 × 3 mm placed a short distance in front of the interaction region. It can be chopped mechanically so as to make phase-sensitive detection and hence increased signal-to-noise ratio possible. The electron beam is defined by a slit 8 mm long and 1 mm wide, parallel to the axis of the

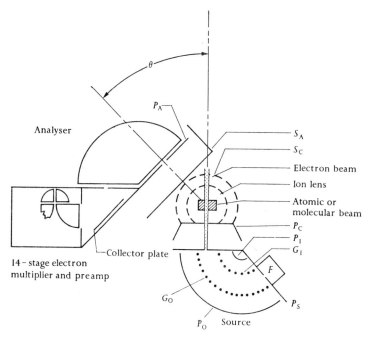

Fig. 3.9. General arrangement of the electron spectrometer in the experiments of McGowan and his associates (see Fig. 3.8).

atomic beam. Measurement of the actual breadth of the beam in the interaction region was carried out with a sliding probe; it was found to be close to 2 mm and remained approximately the same at the entrance slit of the analyser.

It is of the greatest importance in precise studies of threshold behaviour to be sure of the shape and width of the electron energy distribution. Assuming the shape to be gaussian the measured energy spread at half-peak height should be given by

$$\Delta = (\Delta_s^2 + \Delta_a^2)^{\frac{1}{2}}, \qquad (20)$$

where Δ_s, Δ_a are the full widths at half-intensity characteristic of the selector and analyser electron monochromators. To verify this and to determine Δ_s and Δ_a, Δ_s was first fixed by applying a definite field in the selector. The issuing electron beam was then passed directly into the analyser and the shape of the collected current distribution determined with a number of different fields applied between the analyser grids. It was verified that the shapes were gaussian and the width Δ was plotted as a function of the applied field in the analyser. This was carried out for potentials between the grids ranging from 0·5 to 8 V. Non-linear extrapolation to zero potential difference then gives Δ_s. This was found to be 0·05 eV, which is consistent with the fact that in all experiments Δ was kept below 0·07 eV, the value which could be obtained if both Δ_s and Δ_a were each 0·05 eV. No high-energy tail was observed in the distribution. This is very important, as a spurious extension of the ionization function below the threshold could arise if such a high-energy tail were present.

Further checks were made of the electron energy distribution by measuring the breadths of the elastic scattering resonances in He (see Chap. 1, § 6.2.3) and H (Chap. 1, § 6.2.2). The results obtained suggest that the width at half maximum was as low as 0·04 eV.

In collecting the ions, allowance must be made for the lateral momentum communicated to the atom in the electron impact that varies with the electron energy. Because of this it is necessary to collect at least a constant fraction of the ions with an efficiency independent of the electron energy. The potential applied to the first lens in the ion extraction system must be sufficient to ensure uniform collection but not so strong as to penetrate into the interaction region. It is also important to choose the potential so that the collection efficiency is not increased at the cost of too great an increase in signal-to-noise ratio through collection of ions formed in background gas. For this reason it was found best to work at 95 per cent collection as judged by the ratio to the total current produced at saturation.

In carrying out experiments extreme stability of all components is an important requirement. Thus, during a single run to measure an ionization efficiency curve, lasting for 2 h or more, the electron current must be constant to better than 4 per cent. This is because current fluctuations are associated with energy fluctuations that are far more serious in confusing the interpretation of data.

This technique has been applied to investigate ionization threshold behaviour in the molecular gases H_2, N_2, and O_2 and also to atomic hydrogen. We discuss the results of the latter experiment in § 2.5.3.1. The observations in the molecular gases are described and discussed in Chapter 13, § 1.4 (H_2), § 3.4.2 (N_2), and § 4.3.2 (O_2).

2.5. Observed results—measurement of ionization cross-sections

The discussion of the earlier sections makes clear the great care that has to be taken in order to obtain reliable results for the absolute measurement of ionization cross-sections. Even for relative cross-section measurements at different electron energies or for different degrees of ionization of the same atom at the same energy, spurious effects may introduce systematic errors.

A searching and critical analysis of all ionization cross-section measurements so far reported has been made by Kieffer and Dunn.† On the basis of this analysis they compiled a series of figures showing the best available data on ionization cross-sections. The investigations used for the compilation of this data are listed in Table 3.1, which also includes some investigations published subsequently. In this chapter we can present only a selection of these data, and for further information the reader is referred to the article of Kieffer and Dunn. The figures reproduced here show that, despite the care taken in the selection of data, systematic errors remain, so that few ionization cross-sections can be considered as determined to better than 20 per cent and in some cases the uncertainties are much larger.

TABLE 3.1

Measurements of ionization cross-sections of neutral atoms that appear in the Kieffer–Dunn compilation

Author (year)	Species	Type of apparatus	Absolute (A) or Relative (R) measurement
Jones (1927)[1]	Hg	Parallel plate total ion collection	A
Bleakney (1930)[2]	Hg	Parallel plate total ion collection with mass spectrometer	A
Smith (1930)[3]	He, Ne, A	Parallel plate total ion collection	A
Smith (1931)[4]	Hg	Parallel plate total ion collection	A
Liska (1934)[5]	He, Hg	Parallel plate geometry, small solid angle for collection of ions	R
Harrison (1956)[6]	He, Hg	Parallel plate total ion collection, r.f. mass spectrometer	A
Fite and Brackmann (1958)[7]	H	Crossed beam with mass spectrometer	R

† KIEFFER, L. J. and DUNN, C. H., *Rev. mod. Phys.* **38** (1966) 1.

TABLE 3.1 (cont.)

Author (year)	Species	Type of apparatus	Absolute (A) or Relative (R) measurement
Fite and Brackmann (1959)[8]	O	Crossed beam with mass spectrometer and total ion collector	R
Tozer and Craggs (1960)[9]	A, Kr, Xe	Lozier tube with cylindrical draw-out tube (see Chap. 12, § 7.4)	R
Boksenberg (1961)[10]	H, O	Mass spectrometer (H) with total ion collector (O)	R
Rothe et al. (1962)[11]	H, O	Crossed beam with total ion collection	R
Asundi and Kurepa (1963)[12]	He, Ne, A, Kr, Xe	Parallel plate, total ion collection	A
Peterson (1963)[13]	N	Fast crossed beam with mass spectrometer, total ion collection	A
Rapp and Englander-Golden (1965)[14]	He, Ne, A, Kr, Xe	Parallel plate, total ion collection	A
McFarland and Kinney (1965)[15]	Li, Na, K, Rb, Cs	Crossed beam, total ion collection	A
Schram et al. (1965)[16] Schram† et al. (1966)[17]	He, Ne, A, Kr, Xe	Parallel plate total ion collection	A
Heil and Scott† (1966)[18]	Cs	Parallel plate, total ion collection	A

† This paper appeared after the Kieffer–Dunn compilation was published.

(1) JONES, T. J., *Phys. Rev.* **29** (1927) 822.
(2) BLEAKNEY, W., ibid. **35** (1930) 139.
(3) SMITH, P. T., ibid. **36** (1930) 1293.
(4) SMITH, P. T. ibid. **37** (1931) 808.
(5) LISKA, J. W., ibid. **46** (1934) 169.
(6) HARRISON, H., Thesis, Catholic University of America Press Inc., Washington D.C. (1956).
(7) FITE, W. L. and BRACKMANN, R. T., *Phys. Rev.* **112** (1958) 1141.
(8) FITE, W. L. and BRACKMANN, R. T., ibid. **113** (1959) 815.
(9) TOZER, B. A. and CRAGGS, J. D., *J. Electron. Control* **8** (1960) 103.
(10) BOKSENBERG, A., Thesis, University of London (1961).
(11) ROTHE, E. W., MARINO, L. L., NEYNABER, R. H., and TRUJILLO, S. M., *Phys. Rev.* **125** (1962) 582.
(12) ASUNDI, R. K. and KUREPA, M. V., *J. Electron. Control* **15** (1963) 41.
(13) PETERSON, J. R., *Atomic collision processes*, ed. MCDOWELL, M. R. C., p. 465 (North Holland, Amsterdam, 1964) (Proceedings of the Third International Conference on the Physics of Electronic and Atomic Collisions, London, 1963).
(14) RAPP, D. and ENGLANDER-GOLDEN, P., *J. chem. Phys.* **43** (1965) 1464.
(15) MCFARLAND, R. H. and KINNEY, J. D., *Phys. Rev.* **137** (1965) A1058.
(16) SCHRAM, B. L., DE HEER, F. J., VAN DER WIEL, M. J., and KISTEMAKER, J., *Physica*, **31** (1965) 94.
(17) SCHRAM, B. L., MOUSTAFA, H. R., SCHUTTEN, J., and DE HEER, F. J., ibid. **32** (1966) 734.
(18) HEIL, H. and SCOTT, B., *Phys. Rev.* **145** (1966) 279.

Some of the data that do not appear in the compilation of Kieffer and Dunn illustrate particular features about the structure of the ionization cross-section variation near the threshold or about indirect processes of ionization. These will be discussed in later sections.

2.5.1. *The total ionization cross-section—variation with energy.* The variation of the total ionization cross-section with energy has a characteristic form, rising from the ionization potential V_i to a maximum at an electron energy of from three to seven times V_i and then decreasing. Fig. 3.10 shows the dependence of Q_i on electron energy for H, He, Na, A, Xe, and Hg in the electron energy range up to 1000 eV. For He, A, Xe, and Hg these measurements have been extended up to 20 keV. The experimental points obtained by different observers are given. These indicate that although there is general agreement about the shape of the curves the absolute values of Q_i obtained in the various experiments differ by amounts up to 50 per cent, indicating the existence of systematic errors far in excess of those estimated by the individual authors.

There are significant differences between the variation of Q_i for different materials. For example, the potential $V_{\max}$, at which Q_i passes through a maximum varies from $3V_i$ for Na to $9V_i$ for Xe. Q_i decreases to half its maximum value at an electron accelerating potential of $13V_i$ for Na and $50V_i$ for Hg. The curves for the alkali metals rise to their maximum value considerably more rapidly then those for the rare gases. The curves are smooth for H, He, Na, and Hg, but those for A and Xe show an inflection before reaching the maximum. The curves presented here are a characteristic sample. For a full summary of all the available data the reader is referred to the review article of Kieffer and Dunn.†

2.5.2. *The cross-sections for single and multiple ionization.* Analysis of the q/M values of the ions enables cross-sections for the production of ions with a specified charge to be derived. Fig. 3.11 shows cross-sections for the production of different multiply charged argon ions obtained by Fox.‡ The general shape of the variation of $Q_i^{(n)}$ with electron energy is similar to that for total ionization but there is much more detailed structure. Unfortunately this structure is not entirely independent of instrumental conditions (see § 2.2). Fox observed variations of up to 15 per cent in the ratio of the cross-sections for single ionization of argon at 50 and 200 eV, as the voltage used to extract the A$^+$ ions increased from 50 to 200 V, the ion source conditions being otherwise kept constant.

† loc. cit., p. 123.
‡ Fox, R. E., *J. chem. Phys.* **33** (1960) 200.

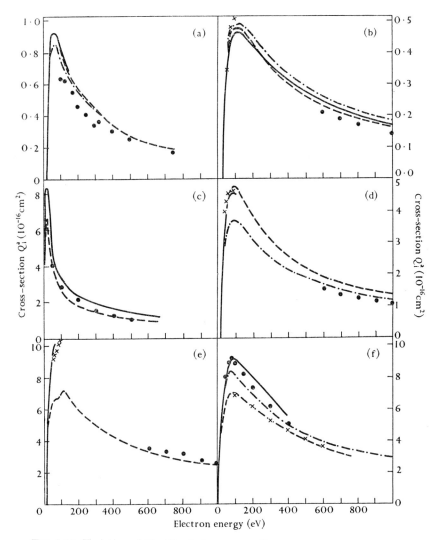

FIG. 3.10. Variation of Q_i^a with electron energy for
(a) H, ——— Fite and Brackman (1)†, - - - - - Boksenberg (2)†, ●●●● Rothe et al. (3)†.
(b) He, ××× Asundi and Kurepa‡, —·—·— Rapp and Englander–Golden (4)†, ——— Smith (5)†, - - - - Harrison (6)† ●●●● Schram et al. (7)†.
(c) Na, ——— Tate and Smith (18)†, - - - - Brink (12)†, ●●●● McFarland and Kenney (13)†.
(d) A, - - - - Smith (5)†, —·—·— Rapp and Englander–Golden (4)†, ●●●● Schram (7)†, ××× Tozer and Craggs§, ——— Asundi and Kurepa‡.
(e) Xe, - - - - Rapp and Englander–Golden (4)†, ●●●● Schram (7)†, ××× Tozer and Craggs§, ——— Asundi and Kurepa‡.
(f) Hg, ——— Jones (24)†, - - - - Smith (22)† ●●●● Bleakney (23)†, —·—·— Harrison (6)†, ××× Liska (8)†.

† The numbers refer to the references in Table 3.2, pp. 129–30.
‡ ASUNDI, R. K. and KUREPA, M. V., J. Electron. Control 15 (1963) 41.
§ TOZER, B. A. and CRAGGS, J. D., ibid. 8 (1960) 103.

The main features of the structure are independent of instrumental conditions, however, and can be related to an indirect process of ionization, such as auto-ionization, discussed in § 2.5.3.

Table 3.2 summarizes some features of the results of measurements of absolute ionization cross-sections of singly and multiply charged atomic

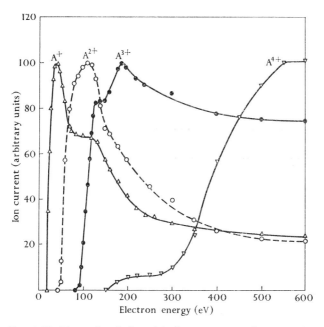

FIG. 3.11. Form of variation with electron energy of cross-section for multiple ionization of A. The curves are arbitrarly normalized to 100 at the maximum cross-section.

ions. In this table V_i is the ionization potential, V_{max} the energy at which Q_i has its maximum value, $Q_{i(max)}$. Further data, for ionization of molecules, are discussed in Chapter 13.

2.5.3. *Detailed structure in the form of variation of Q_i with electron energy.* The general shape of the variation of Q_i has been discussed in § 2.5.2. Superimposed on this general shape, however, is a detailed finer structure characteristic of each ion. The behaviour of the ionization curve near the threshold is of interest and appears to depend on the charge of the ion. The threshold region is especially rich in fine structure that can be interpreted as due to the production of the ions in excited states or to ionization occurring via excited auto-ionizing levels of the neutral atom. Complicated structure is also observed, however, at

TABLE 3.2

Ion	V_i (eV)	V_{max} (eV)	$Q_i(V_{max})$ $\times 10^{-16}$ cm²	$Q_i(500\text{ eV})$ $\times 10^{-16}$ cm²	$Q_i(1000\text{ eV})$ $\times 10^{-16}$ cm²	$Q_i(10\,000\text{ eV})$ $\times 10^{-16}$ cm²	Reference
H⁺	13·54	56	0·70	0·22			(1)
		63	0·66				(2)
				0·21			(3)
He⁺	24·46	126	0·37	0·22	0·14		(4)
		110	0·35	0·20	0·13		(5)
		100	0·35	0·19	0·125		(6)
				0·18	0·11	0·17	(7)
						0·17	(8)
He²⁺	78	500	0·0012	0·0012			(9)
		560	0·0012	0·0012	0·0007		(6)
		500	0·0019	0·0019	0·0010		(10)
				8×10^{-4}	$5\cdot 2\times 10^{-4}$	$4\cdot 7\times 10^{-5}$	(11)
Li⁺							(12)
		14	5·2	0·9			(13)
N⁺	14·4	77	2·95	1·65			(14)
		100	1·40	0·65			(15)
O⁺	13	89	1·52	0·87			(16)
				0·81			(3)
		100	2·36				(2)
Ne⁺	21·5	150	0·78	0·58	0·36		(5)
		187	0·84	0·62	0·40		(4)
		160	0·64	0·44	0·29	0·05	(7)
Ne²⁺	62·6	250	0·045	0·035			(17)
				0·019	0·0105	0·0125	(11)
Ne³⁺	126·6			0·0031			(17)
				0·001	0·0006	0·00008	(11)
Ne⁴⁺	223·6			$1\cdot 9\times 10^{-5}$	$1\cdot 5\times 10^{-5}$	$6\cdot 5\times 10^{-6}$	(11)
Ne⁵⁺	349·9				$1\cdot 7\times 10^{-6}$	$3\cdot 5\times 10^{-6}$	(11)
Na⁺	5·12	16	8·4	1·1			(12)(13)
		16	6·9	1·1			(18)
Na²⁺	52	500	0·18	0·18			(18)
A⁺	15·76	90	3·21	1·60	0·90		(17)
		92	2·72	1·31	0·83		(4)
				1·24	0·76	0·14	(7)
A²⁺	43·4	120	0·31	0·12			(17)
				0·071	0·040	0·0077	(19)
A³⁺	84·3	190	$1\cdot 1\times 10^{-2}$	$1\cdot 0\times 10^{-2}$			(17)
				$1\cdot 71\times 10^{-2}$	$0\cdot 66\times 10^{-2}$	$0\cdot 20\times 10^{-2}$	(19)
A⁴⁺	144			$1\cdot 03\times 10^{-3}$	$1\cdot 3\times 10^{-3}$	$0\cdot 36\times 10^{-3}$	(19)
A⁵⁺	219			$0\cdot 80\times 10^{-4}$	$1\cdot 51\times 10^{-4}$	$0\cdot 44\times 10^{-4}$	(19)
A⁶⁺	310				$1\cdot 22\times 10^{-5}$	$0\cdot 82\times 10^{-5}$	(19)
A⁷⁺					$0\cdot 58\times 10^{-6}$	$0\cdot 75\times 10^{-6}$	(19)
K⁺		10	8·4	2·1			(12)(13)
K²⁺	30	130	0·65	0·32			(18)
Kr⁺	14·00	80	5·0				(20)
		80	3·8	1·9	1·2		(4)
				1·75	1·15	0·18	(7)

TABLE 3.2 (cont.)

Ion	V_i (eV)	V_{max} (eV)	$Q_i(V_{max})$ $\times 10^{-16}$ cm²	$Q_i(500$ eV$)$ $\times 10^{-16}$ cm²	$Q_i(1000$ eV$)$ $\times 10^{-16}$ cm²	$Q_i(10\,000$ eV$)$ $\times 10^{-16}$ cm²	Reference
Kr^{2+}	38·6	100	0·5				(18)
				0·150	0·077	0·016	(19)
Kr^{3+}	75·5	250	0·042				(18)
				0·055	0·035	0·0096	(19)
Kr^{4+}				0·012	0·0083	0·0026	(19)
Kr^{5+}				$2·7 \times 10^{-3}$	$2·8 \times 10^{-3}$	$0·92 \times 10^{-3}$	(19)
Kr^{6+}					$6·0 \times 10^{-4}$	$2·5 \times 10^{-4}$	(19)
Kr^{7+}					$6·0 \times 10^{-5}$	$10·13 \times 10^{-5}$	(19)
Kr^{8+}					$0·55 \times 10^{-6}$	$2·8 \times 10^{-5}$	(19)
Kr^{9+}						$0·4 \times 10^{-5}$	(19)
Rb^+		19	8·9	1·6			(19)
Rb^{2+}		90	2·2	0·70			(15)(19)
Rb^{3+}				0·27			(15)(19)
Xe^+	12·13	115	4·8	2·65	1·68		(4)
				2·85	1·8	0·25	(7)
Xe^{2+}	33·3	100	0·75	0·62			(18)
				0·295	0·217	0·043	(19)
Xe^{3+}	65·5	140	0·24	0·265			(18)
				0·116	0·097	0·017	(19)
Xe^{4+}	111			0·104			(18)
				0·034	0·031	0·0056	(19)
Xe^{5+}	171			0·042			(18)
				0·0095	0·0083	0·0019	(19)
Xe^{6+}				$1·12 \times 10^{-3}$	$2·18 \times 10^{-3}$	$0·90 \times 10^{-3}$	(19)
Xe^{7+}					$4·2 \times 10^{-4}$	$4·7 \times 10^{-4}$	(19)
Xe^{8+}					$0·77 \times 10^{-4}$	$2·55 \times 10^{-4}$	(19)
Xe^{9+}						$0·60 \times 10^{-4}$	(19)
Xe^{10+}						$1·55 \times 10^{-5}$	(19)
Xe^{11+}						$5·0 \times 10^{-6}$	(19)
Xe^{12+}						$1·45 \times 10^{-6}$	(19)
Cs^+	3·83	33	9·4	2·9			(18)
			8·0	2·9			(13)
		25	5·2				(21)
Cs^{2+}	22·7	100	1·1	0·25			(18)
Cs^{3+}	73·5	125	0·16	0·126			(18)
Cs^{4+}	117	300	5·1	0·046			(18)
Cs^{5+}	200	500	0·04	0·04			(18)
Cs^{6+}	270	600	0·015				(18)
Hg^+	10·4	75	5·0	2·2			(22)
			5·8	2·3	1·4	0·27	(6)
			6·4				(23)(24)
				2·2	1·3	0·18	(8)
Hg^{2+}	29·2	115	0·90	0·26			(23)
			0·73	0·36	0·22	0·043	(6)
Hg^{3+}	63·4	210	0·20	0·044			(23)
			0·24	0·013	0·0063	0·0012	(6)
Hg^{4+}	122		0·048	0·024			(23)
				0·026	0·014	0·0027	(6)
Hg^{5+}				0·0081	0·0063	0·0021	(6)

(1) FITE, W. L. and BRACKMANN, R. T., *Phys. Rev.* **112** (1958) 1141.
(2) BOKSENBERG, A., Thesis, University of London (1961).
(3) ROTHE, E. W., MARINO, L. L., NEYNABER, R. H., and TRUJILLO, S. M., *Phys. Rev.* **125** (1962) 582.
(4) RAPP, D. and ENGLANDER-GOLDEN, P., *J. chem. Phys.* **43** (1965) 1464.
(5) SMITH, P. T., *Phys. Rev.* **36** (1930) 1293.
(6) HARRISON, H., Thesis, Catholic University of America Press Inc., Washington, D.C. (1956).

(7) SCHRAM, B. L., DE HEER, F. J., VAN DER WIEL, M. J., and KISTEMAKER, J., *Physica* **31** (1964) 94.
(8) LISKA, J. W., *Phys. Rev.* **46** (1934) 169.
(9) BLEAKNEY, W. and SMITH, L. G., ibid. **49** (1936) 402.
(10) STANTON, H. E. and MONAHAN, J. E., ibid. **119** (1960) 711.
(11) SCHRAM, B. L., BOERBOOM, A. J. H., and KISTEMAKER, J., *Physica* **32** (1966) 185.
(12) BRINK, G. O., *Phys. Rev.* **127** (1962) 1204.
(13) MCFARLAND, R. H. and KINNEY, J. D., ibid. **137** (1965) A1058.
(14) PETERSON, J. R., in *Atomic collision processes*, ed. MCDOWELL M. R. C., p. 465 (North Holland, Amsterdam, 1964).
(15) SMITH, A. C. H., CAPLINGER, E., NEYNABER, R. H., ROTHE, E. W., and TRUJILLO, S. M., *Phys. Rev.* **127** (1962) 1647.
(16) FITE, W. L. and BRACKMANN, R. T., ibid. **113** (1959) 815.
(17) BLEAKNEY, W., ibid. **36** (1930) 1303.
(18) TATE, J. T. and SMITH, P. T., ibid. **46** (1934) 773.
(19) SCHRAM, B. L., *Physica* **32** (1966) 197.
(20) ASUNDI, R. K. and KUREPA, M. V., *J. Electron. Control* **15** (1963) 41.
(21) HEIL, H. and SCOTT, B., *Phys. Rev.* **145** (1966) 279.
(22) SMITH, P. T., ibid. **37** (1931) 808.
(23) BLEAKNEY, W., ibid. **35** (1930) 139.
(24) JONES, T. J., ibid. **29** (1927) 822.

In addition the following references give relative ionization cross-sections and illustrate the detailed fine structure of ionization cross-section curves.

FOX, R. E., HICKAM, W. M., and KJELDAAS, T., *Phys. Rev.* **89** (1953) 555 (Kr^+, Xe^+).
CLARKE, E. M., *Can. J. Phys.* **32** (1954) 764 (Xe^{2+}).
HICKAM, W. M., *Phys. Rev.* **95** (1954) 703 (Zn^+, Cd^+, Hg^+).
FROST, D. C. and MCDOWELL, C. A., *Proc. R. Soc.* **A232** (1955) 227 (Kr^+).
CLOUTIER, G. G. and SCHIFF, H. I., *J. chem. Phys.* **31** (1959) 793 (A^+).
DIBELER, V. H. and REESE, R. M., ibid. 282 (Na^+, Na^{2+}, Na^{3+}).
KERWIN, L., MARMET, P., and CLARKE, E. M., *Advances in mass spectrometry*, p. 522 (Pergamon Press, London 1959).
FOX, R. E., ibid., p. 397 (He^+, He^{2+}, Xe^+, Xe^{2+}, Xe^{3+}, Xe^{4+}, Xe^{5+}, Xe^{6+}, Xe^{7+}).
BLAIS, N. C. and MANN, J. B., *J. chem. Phys.* **33** (1960) 100 (Cu^+, Ag^+, Au^+, Au^{2+}).
FOX, R. E., ibid. 200 (A^+, A^{2+}, A^{3+}, A^{4+}, Kr^+, Kr^{2+}, Kr^{3+}, Kr^{4+}).
MARMET, P. and KERWIN, L., *Can. J. Phys.* **38** (1960) 787 (A^+).
BURNS, J. F., *Nature, Lond.* **192** (1961) 651 (Kr^+).
DORMAN, F. H. and MORRISON, J. D., *J. chem. Phys.* **35** (1961) 575 (Xe^+, Kr^+); ibid. **34** (1961) 578 (Xe^{2+}, Kr^{2+}).
FONER, S. N. and NALL, B. H., *Phys. Rev.* **122** (1961) 512 (A^+, Kr^+, Xe^+).
FOX, R. E., *J. chem. Phys.* **35** (1961) 1379 (He^+, Ne^+, A^+, Hg^+).
KANEKO, Y., *J. phys. Soc. Japan* **16** (1961) 1587 (A^+, Kr^+); ibid. **16** (1961) 2288 (Na^+, Na^{2+}, K^+, K^{2+}, Mg^+, Mg^{2+}).
BRINK, G. O., *Phys. Rev.* **127** (1962) 1204 (Li^+, Na^+, Na^{2+}, K^+, K^{2+}, Rb^+, Rb^{2+}).
FIQUET-FAYARD, F. and LAKMANI, M., *J. Chim. phys.* **59** (1962) 1052 (K^+, K^{2+}, K^{3+}, K^{4+}, K^{5+}, Ca^+, Ca^{2+}, Ca^{3+}, Ca^{4+}, A^+, A^{2+}, A^{3+}, A^{4+}, A^{5+}).
KANEKO, Y. and KANOMATA, I., *J. phys. Soc. Japan* **18** (1963) 1822 (Ca^+, Ca^{2+}).
MCGOWAN, J. W. and KERWIN, L., *Can. J. Phys.* **41** (1963) 1535 (A^{2+}).
BURNS, J. F., *Atomic collision processes*, ed. MCDOWELL, M. R. C., p. 451 (Xe^+, Kr^+) (North Holland, Amsterdam 1964).
MORRISON, J. D., *J. chem. Phys.* **40** (1964) 2488.
STUBER, F. A., ibid. **42** (1965) 2639 (Na^+, Ne^{2+}, Ne^{3+}, Ne^{4+}, Ne^{5+}, A^+, A^{2+}, A^{3+}, A^{4+}, A^{5+}, A^{6+}, Kr^+, Kr^{2+}, Kr^{3+}, Kr^{4+}, Kr^{5+}, Kr^{6+}, Kr^{7+}, Kr^{8+}, Xe^+, Xe^{2+}, Xe^{3+}, Xe^{4+}, Xe^{5+}, Xe^{6+}, Xe^{7+}, Xe^{8+}, Xe^{9+}).
ZIESEL, J. P., *J. Chim. phys.* **62** (1965) 328 (Ne^+, Ne^{2+}, Ne^{3+}, Ne^{4+}, Na^+, Na^{2+}, Na^{3+}, Na^{4+}, Mg^+, Mg^{2+}, Mg^{3+}, Mg^{4+}).
HEIL, H. and SCOTT, B., *Phys. Rev.* **145** (1966) 279 (Cs^+).

electron energies far above the threshold in the case of multiply charged ions, and it appears possible to interpret this in terms of Auger processes following primary inner-shell ionization.

2.5.3.1. *The increase of ionization cross-section near the threshold.* For electrons of energy E, Wannier† and Geltman‡ have written for the ionization cross-section $Q_i(E) = |M(E)|^2 P(E)$, where $M(E)$, the matrix element for the process concerned, varies very slowly with energy while $P(E)$, a factor proportional to the volume of phase space available for the electron in the final state, is responsible for almost the whole variation of the cross-section with energy. This implies that the interaction time is sufficiently long, so that a statistical equilibrium between the different degrees of freedom is reached in the final states. This is justified very close to the threshold where the emergent electrons are moving very slowly, but it is difficult to estimate the extent of the range of energy over which it should be valid.

The factor $P(E)$ will depend on the number of electrons among which the available energy in the final state is divided. For single ionization the most elaborate and detailed quantum mechanical analysis (see Chap. 9, § 10.8) gives $P(E)$ proportional to $E-E_i$ where E_i is the threshold energy. An earlier classical calculation by Wannier gave $(E-E_i)^{1.127}$, and a good deal of effort has been directed towards discrimination between these two threshold laws.

There are certain basic difficulties in carrying out decisive experiments to this end. Thus, as pointed out above, there is no knowledge of the range from the threshold throughout which the threshold law is applicable. Once the energy resolution becomes comparable with 0·03 eV, the mean energy of thermal motion of gas molecules at room temperature and also of thermal photons, further difficulties arise. Excitation of upper states lying around 0·03 eV below the true ionization threshold could lead to ion production through thermal ionization of these highly excited atoms. It is necessary to discriminate against this possibility by carrying out experiments in a low temperature bath. Such experiments have not yet been done.

Fig. 3.12 shows results obtained for the threshold behaviour of the ionization cross-section for atomic hydrogen by McGowan and his associates§ using the technique described in § 2.4.5. Above a certain energy, about 0·1 eV from the apparent threshold, the cross-section

---

† WANNIER, G. H., *Phys. Rev.* **100** (1955) 1180.
‡ GELTMAN, S., ibid. **102** (1956) 171.
§ McGOWAN, J. W. and CLARKE, E. M., loc. cit., p. 119.

increases linearly. If this linear portion is extrapolated towards the threshold, allowing for an energy spread of 0·05 eV in the electron beam, a low-energy tail is obtained that is appreciably less prominent than that actually observed. Special precautions, described in § 2.4.5, were taken to minimize spurious contributions to energy spread but it does not follow that a real departure from linear behaviour as indicated means that the linear threshold law for direct single ionization

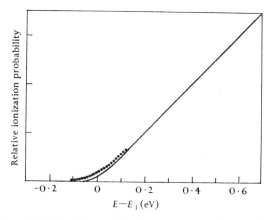

FIG. 3.12. Variation with electron energy close to the threshold of the cross-section for ionization of atomic hydrogen by electron impact. •••• observed by McGowan et al.; —— derived assuming a linear variation with slope tangential to the observed results at the higher energies and folding in the energy distribution in the electron beam used in the experiments.

is not valid. Thus the assumption of its validity over a range up to 0·6 eV or more above the threshold may be incorrect, in which case the electron energy distribution should be folded into a line of different slope extending only a short distance above the threshold. Again some of the observed tail may have arisen through multiple processes involving initial excitation of bound states lying only just below the ionization limit.

For n-fold ionization ($n > 1$) the threshold behaviour is expected to be as $(E-E_i)^n$. However, the presence of fine structure near the threshold makes it difficult to test; for example, Fig. 3.13 (a) shows the data obtained by Fox† for Kr^{3+}, plotted in different ways. The quantity $\{Q_i(Kr^{3+})\}^{1/n}$ is plotted against incident electron energy for $n = 1, 2$, and 3. It is clear that the data very close to the threshold can be made consistent with any of these values of n, provided it is postulated that breaks

† Fox, R. E., J. chem. Phys. **33** (1960) 200.

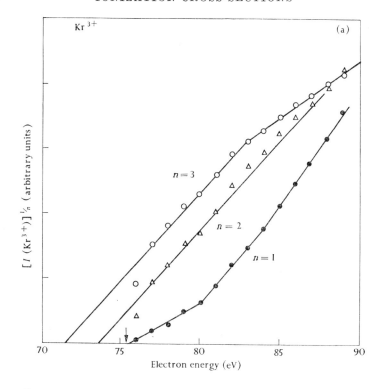

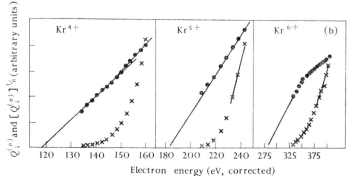

FIG. 3.13. (a) $(Q_i^{(3)})^{1/n}$ for Kr plotted against E for $n = 1, 2, 3$ (Fox).
(b) ●—$(Q_i^{(n)})^{1/n}$ for Kr plotted against E for $n = 4, 5, 6$. × $Q_i^{(n)}$ for Kr plotted against E for $n = 4, 5, 6$ (Dorman and Morrison).

may occur in the cross-section at energies a little above the threshold owing to the onset of new processes. On the other hand, the data of Dorman and Morrison† on Kr^{4+}, Kr^{5+}, Kr^{6+}, shown in Fig. 3.13 (b) indicate linearity for the electron energy dependence of $\{Q_i(Kr^{4+})\}^{\frac{1}{4}}$,

† DORMAN, F. H. and MORRISON, J. D., *J. chem. Phys.* **36** (1961) 1407.

$\{Q_i(\mathrm{Kr}^{5+})\}^{\frac{1}{5}}$, $\{Q_i(\mathrm{Kr}^{6+})\}^{\frac{1}{6}}$, while the plots of the cross-sections themselves against electron energy are far from linear. The ionization energy obtained by extrapolation of Q_i itself usually appears to be closer to that determined spectroscopically, however, than is the case for the $\{Q_i\}^{1/n}$ plot. This is seen for example from Fox's data in Fig. 3.13 (a); Table 3.3 shows values of the ionization potential for a number of cases evaluated by direct extrapolation or by the $\{Q_i\}^{1/n}$ plot, compared

TABLE 3.3

Ionization potential for multiple ionization obtained by different methods

Ion	Spectroscopic	Ionization energy (eV)		
		Direct extrapolation		nth root extrapolation
		Fox	Dorman and Morrison	Dorman and Morrison
A^{2+}	43·38	43·4		43·3
A^{3+}	84·28	84·8		84·0
A^{4+}	144·07	150·0		..
Kr^{2+}	38·56	38·45	38·5	38·1
Kr^{3+}	75·46	75·6	77	73·3
Xe^{2+}	33·33	33·3	33·3	33·5
Xe^{3+}	65·5	66·2	65·4	64·8
Xe^{4+}	111	110	111	107
Xe^{5+}	187	170	171	160
Hg^{2+}	29·18	..	29·8	29·0
Hg^{3+}	63·38	..	68·5	63·5

with spectroscopic values. The spectroscopic data for higher ionized atoms are less well known than for the neutral or singly ionized cases. The variation of the ionization cross-section near threshold like $(E-E_i)^n$ for n-fold ionization does not appear yet to have been fully established.

2.5.3.2. *Fine structure of ionization cross-section near the threshold.* The presence of structure in the ionization curve for Hg close to the threshold was first observed by Lawrence† and confirmed by Nottingham.‡ Q_i was found to rise to a distinct maximum at 10·8 eV, 0·4 eV above the threshold, and definite structural details were observed up to 16 eV.

Fig. 3.14 (a) shows the ionization cross-section curve obtained by Nottingham near the threshold using a magnetic analyser to produce a homogeneous electron beam. Fig. 3.14 (b) shows the curve obtained

† LAWRENCE, E. O., *Phys. Rev.* **28** (1926) 947.
‡ NOTTINGHAM, W. B., ibid. **55** (1939) 203.

by Fox,† normalized to the absolute cross-section measured by Nottingham at 12 eV. The agreement is satisfactory, demonstrating that it is possible to achieve conditions that give reliable results using the r.p.d. method.

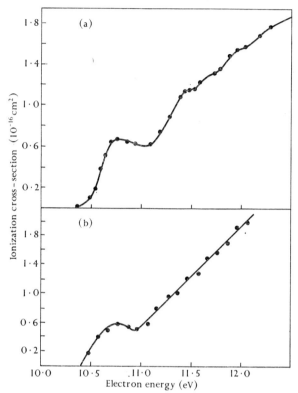

FIG. 3.14. Variation of Q_i with E near threshold for Hg.
(a) Nottingham, (b) Fox.

Fine structure near the threshold is a common feature of ionization cross-section curves. Fig. 3.15 shows the variation of Q_i near threshold for Xe, observed by Burns‡ using the r.p.d. method. The two traces represent independent sets of measurements (displaced on the ion current scale) and give an idea of the reproducibility of the measurements. The threshold energies for the production of Xe⁺ in the $^2P_{\frac{3}{2}}$ and $^2P_{\frac{1}{2}}$ states are indicated. It is clear that the marked change of slope at electron energy 13·43 eV (1·3 eV above threshold) is to be

† Fox, R. E., *J. chem. Phys.* **35** (1961) 1379.
‡ BURNS, J. F., *Atomic collision processes*, ed. McDOWELL, M. R. C., p. 451 (North Holland, Amsterdam, 1964).

attributed to the production of Xe⁺ ions in the $^2P_{\frac{1}{2}}$ metastable state. Further structure is also evident, however, giving changes of slope at 12·5 and 12·9 eV. Burns interpreted these as due to the excitation of

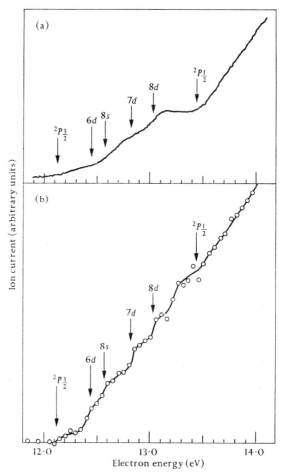

Fig. 3.15. Variation of Q_i with E near threshold for Xe and relation of structure with energies of auto-ionizing Beutler levels. (a) Burns, (b) Morrison.

auto-ionizing states of Xe. Such states were identified by Beutler† in a study of the absorption spectra of A, K, and Xe (see Chap. 14, § 7.3).

He observed a number of strong lines in the spectra, which he resolved into two series corresponding to transitions between the configurations

$$p^6(^1S_0) - p^5(^2P_{\frac{1}{2}})md(S_1) \quad \text{and} \quad p^6(^1S_0) - p^5(^2P_{\frac{1}{2}})ms(S_2),$$

† BEUTLER, H., *Z. Phys.* **93** (1935) 177.

the final states having an energy above the $^2P_{\frac{3}{2}}$ state of the ion. The energies of the Beutler levels corresponding to different values of m are shown in Fig. 3.15 and there is seen to be some correlation.

Fig. 3.15 shows for comparison the results obtained for Xe by Morrison,† using an electrostatic analyser to obtain electrons homo-

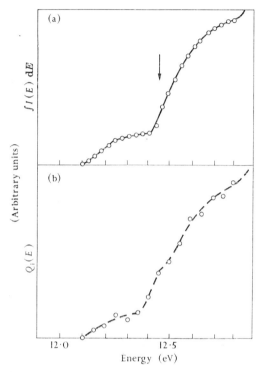

FIG. 3.16. Ionization cross-section $Q_i(E)$ for Xe near threshold (b) (Nicholson) compared with integrated photo-ionization efficiency curve for same material (a).

geneous in energy. The main details observed by Burns are visible in Morrison's results but the latter observed additional structure, well correlated with the known positions of the Beutler levels. Morrison has stressed the close correlation between the structure of the ionization cross-section curves near threshold and the measurements of photo-ionization efficiency, $I(E)$ (see also Chap. 13, §§ 1.4.2 and 4.3.2).

Fig. 3.16 (b) shows the results obtained by Nicholson‡ for Xe. Ionization directly to the ground state of the ion and indirectly through the two lowest auto-ionizing states is clearly distinguishable. In Fig. 3.16 (a)

† MORRISON, J. D., *J. chem. Phys.* **40** (1964) 2488.
‡ NICHOLSON, A. J. C., ibid. **39** (1963) 954.

the integrated photo-ionization efficiency (see Chap. 14, § 8.1)

$$\int_{E_i}^{E} I(E)\,dE$$

is plotted against E and compared with the form of $Q_i(E)$ for electron ionization obtained by Morrison, revealing a definite correlation.

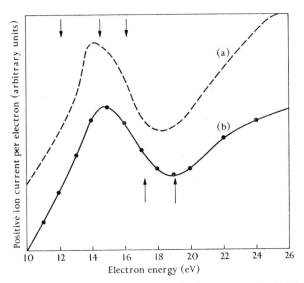

Fig. 3.17. Ionization cross-section (Q_i) for Cs near threshold. (a) Heil and Scott, (b) Tate and Smith.

The importance of auto-ionization processes for the interpretation of the variation of Q_i near the threshold appears to have been stressed first by Fox, Hickam, and Kjeldaas.† Foner and Nall,‡ using a homogeneous electron beam from an electrostatic analyser, observed structure at 1·33 eV above the ionization threshold, corresponding to $Xe^+(^2P_{\frac{3}{2}})$ production, and at 0·70 eV above the threshold, corresponding to the excitation of the strongest auto-ionizing state observed by Burns. They also observed structure at 2·05 eV above the threshold outside the range of observation in Burns' experiment.

Similar structure has been observed by Heil and Scott§ in the region just above the threshold for Cs ionization, using a Tate and Smith type of apparatus. Fig. 3.17 shows their measured absolute cross-section for different electron energies. The arrows above the curve

---

† Fox, R. E., Hickam, W. M., and Kjeldaas, T., *Phys. Rev.* **89** (1953) 555.
‡ Foner, S. N. and Nall, B. H., ibid. **122** (1961) 512.
§ Heil, H. and Scott, B., ibid. **145** (1966) 279.

show the positions of the three strongest auto-ionizing levels obtained from measurements of the Cs absorption spectrum by Beutler and Guggenheimer.† The two arrows below the curve indicate the thresholds for ionization to the following pairs of excited states of the ion: $^3P_2$ and $^3P_1$ at 17·3 eV and $^3P_0$ and $^1P_1$ at 19·1 eV. Tate and Smith's‡ relative measurements of Cs ionization, given on the same figure, show an almost identical structure.

In Fig. 3.17 the Tate–Smith curve is normalized to the absolute measurements of McFarland and Kinney,§ using the crossed-beam

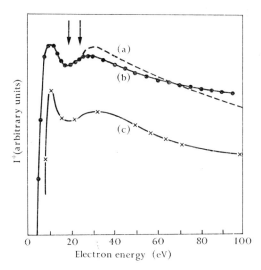

Fig. 3.18. Structure of the ionization curve for K.
(a) Tate and Smith, (b) Kaneko, (c) Brink.

technique. The absolute values obtained by Heil and Scott are only about half as large as those of McFarland and Kinney. The discrepancy is attributable no doubt to difficulties in the estimation of the Cs vapour density.

Evidence of the importance of auto-ionization processes in influencing the structure of the ionization curves is provided by the measurements for K. Fig. 3.18 shows the results obtained by Tate and Smith,∥ Kaneko†† (r.p.d. method), and Brink‡‡ (crossed-beam method). All three measurements reproduce the characteristic double maximum in the region below

---

† BEUTLER, H. and GUGGENHEIMER, K., *Z. Phys.* **88** (1934) 25.
‡ TATE, J. T. and SMITH, P. T., *Phys. Rev.* **46** (1934) 773.
§ McFARLAND, R. H. and KINNEY, J. D., ibid. **137** (1965) A1058.
∥ loc. cit., p. 130.   †† KANEKO, Y., *J. phys. Soc. Japan* **16** (1961) 2288.
‡‡ BRINK, G. O., *Phys. Rev.* **127** (1962) 1204.

30 eV. The arrows show respectively the positions of the auto-ionizing level $3p^5 4s^2$ of K at an excitation potential of 18·8 eV and the position of the first excited state ($^3P_2$) of K$^+$ at 24·5 eV (20·15 eV above the ground ($^1S_0$) state). As pointed out by Brink, the onset of the process responsible for the second maximum seems to be associated with excitation of the auto-ionizing state.

The ionization curve for Rb$^+$ shows a similar double maximum† that can be interpreted as due to the excitation of the corresponding auto-ionizing level of Rb at 15·5 eV.

2.5.3.3. *Multiple ionization through the Auger effect.* Structure has been observed in the dependence on the electron energy of the cross-section for multiple ionization of a number of atoms. Fig. 3.19 shows ionization cross-section curves for A^{3+}, A^{4+}, K^{3+}, K^{4+}, Ca^{3+} and Ca^{4+}.‡ The structure in these cases is well above the threshold. There appear in fact to be two thresholds, one corresponding to the energy just necessary to eject the electrons from the atom and the other at a considerably higher energy, evidently corresponding to the onset of another process. The effect was first pointed out by Fox§ for Cs^{2+}, Xe^{2+}, and A^{4+}, and he suggested that the enhanced ionization at higher energies could be attributed to inner shell ionization followed by multiple ionization as a result of Auger type reorganization processes. The second thresholds in the curves of Fig. 3.19 are seen to be close to the threshold for ionization in one of the L shells. For argon the second process could be thought of as taking place through the following cascade,

$$A^+_{L_1} \to A^{2+}_{L_{2,3}, M_1} \to A^{3+}_{M_1 M_1 M_{2,3}}. \tag{21}$$

The usual two-electron Auger process $M_1 - M_{2,3} M_{2,3}$ is probably endothermic but the three-electron process $M_1 M_1 - M_{2,3} M_{2,3} M_{2,3}$ is energetically possible. Similar considerations apply to the other cases illustrated in Fig. 3.19. Ziesel‖ has observed similar structure for Na^{5+}, Ne^{4+} in the neighbourhood of the K-shell ionization potential for these elements. It is difficult to account for the high degree of ionization in terms only of a cascade of exothermic two-electron Auger processes. It appears likely, however, that additional electrons may be lost in 'shake-off' processes, either in the primary ionization process or in the first stage of the Auger cascade.

† BRINK, G. O., loc. cit., p. 139.
‡ FIQUET-FAYARD, F. and LAHMANI, M., *J. Chim. phys.* **59** (1962) 1050.
§ Fox, R. E., *Advances in mass spectrometry*, p. 397 (Pergamon Press, London, 1959); *J. chem. Phys.* **33** (1960) 200.
‖ ZIESEL, J. R., *J. Chim. phys.* **62** (1965) 328.

2.5.3.4. *Production of metastable ions.* Some of the structure of ionization cross-section curves just above threshold has been interpreted (§ 2.5.3.2) in terms of the formation of ions in excited states. Direct evidence for the production of ions in long-lived metastable states has been obtained by Hagstrum,† who studied the ejection of secondary

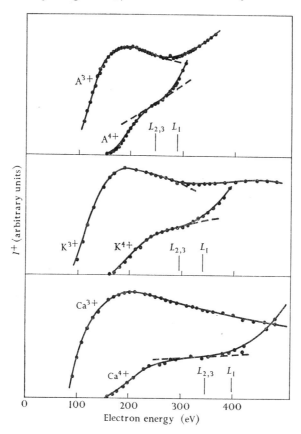

FIG. 3.19. Structure of $Q_i^{(n)}$ curves for A^{3+}, A^{4+}, K^{3+}, K^{4+}, Ca^{3+} and Ca^{4+} (Fiquet-Fayard and Lahmani).

electrons by the impact on a clean molybdenum target of beams of A^+, Kr^+, and Xe^+ of known energy as a function of the energy of the electron beam used to produce the ions in a Nier-type source. Fig. 3.20 shows the number of secondary electrons observed as a function of the electron beam energy. The ionization potentials of the rare gas atoms are indicated in the figure and it is seen that a marked increase of secondary electron emission occurs between V_1 and V_2. The energies of the four

† HAGSTRUM, H. D., *Phys. Rev.* **104** (1956) 309.

known metastable states ($^4D_{\frac{7}{2}}$, $^4F_{\frac{9}{2}}$, $^4F_{\frac{7}{2}}$, $^2F_{\frac{7}{2}}$) of A$^+$, Kr$^+$, and Xe$^+$, also shown on the figure, are seen to be close to the electron beam energy at which enhancement of the secondary emission occurs. Hagstrum therefore attributes the enhancement to the onset of Auger-type ejection of secondary electrons from molybdenum by the metastable atoms.†

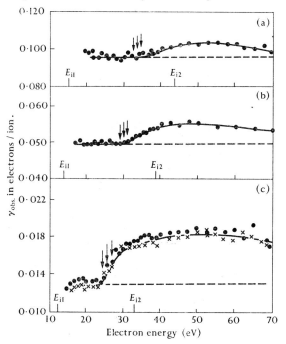

FIG. 3.20. Efficiency of ejection of secondary electrons from molybdenum surface by (a) A$^+$, (b) Kr$^+$, and (c) Xe$^+$ beams of given energy as a function of energy of electrons used to produce ions, illustrating effect of metastable ion formation (Hagstrum). The arrows correspond to the energies required to form the $^4D_{\frac{7}{2}}$, $^4F_{\frac{9}{2},\frac{7}{2}}$ and $^2F_{\frac{7}{2}}$ states.

No similar enhancement was observed for Ne$^+$ or He$^+$. In Ne$^+$ there are no similar metastable states but for He$^+$ the metastable $2\,^2S_{\frac{1}{2}}$ state exists. Hagstrum assumed, therefore, that the proportion of metastable ion formation in helium is too small to detect. From his measurements Hagstrum estimated that the maximum proportion of ions formed in the metastable state in argon, krypton, and xenon was 2 per cent. The longest transit time of the ions through the apparatus was 30 μs. The expected lifetime of these metastable ionized states is greater than 1 s unless two-quantum decay mechanisms are important.‡

† loc. cit., p. 141.
‡ See, for example, BREIT, G. and TELLER, E., *Astrophys. J.* **91** (1940) 215.

The formation of metastable A⁺ ions has been observed also by McGowan and Kerwin† in a mass spectrometer. The energy of these metastable ions was just below the ionization threshold of A^{2+} and they were detected through their conversion to A^{2+} in the drift region of the mass spectrometer through the pressure dependent process:

$$A^{+m}+A \to A^{2+}+A+e. \tag{22}$$

Direct evidence for the formation of singly and doubly ionized argon atoms in auto-ionizing metastable states has been obtained by Daly,‡ using a rather similar apparatus to that of Hagstrum but with the addition of rough electrostatic analysis of the beam incident on the target. This was achieved by deflecting the beam on to the target by maintaining the latter at a negative potential that could be varied up to 40 kV.

Daly used a novel form of detector for the ion beams in this experiment.§ Secondary electrons, ejected from the target by impact of the positive ions, were accelerated through a potential of some tens of kV on to a scintillator and the resultant pulses detected by a photomultiplier.

He observed two types of ion beam deflected through the magnetic spectrometer at the A^{4+} position. One of the beams hit the target detector when the potential applied to the latter was -35 kV and the variation of current in the beam with the energy of electrons in the source is given by curve (1) of Fig. 3.21. It had a threshold of 145 V and the normal shape of an ionization curve.

As the potential applied to the target fell the currents in the photomultiplier fell too, but rose to a maximum again when the target potential was -17 kV, corresponding to the potential expected for a beam of A^{2+} entering the detector. The ionization current for this beam is given by curve (2) of Fig. 3.21 and the threshold energy after excitation of the beam in the detector was 44·5 eV, about 1 eV above the threshold for direct excitation of A^{2+} (curve (3) of Fig. 3.21). The maximum cross-section for this beam, however, occurred at an electron energy of 60 eV compared with 150 eV for A^{2+}. The phenomenon was independent of pressure in the apparatus and could not be interpreted in terms of gas collisions in the magnetic analysis system.

Daly interpreted the anomalous A^{4+} beam as due to the formation in the source of metastable singly-ionized A^{+m}. After acceleration it

† McGowan, J. W. and Kerwin, L., *Can. J. Phys.* **41** (1963) 1535.
‡ Daly, N. R., *Proc. phys. Soc.* **85** (1965) 897.
§ Daly, N. R., *Rev. scient. Instrum.* **31** (1960) 264.

entered the chamber C with momentum half that of A^{4+}. While passing through C it underwent auto-ionization through the process

$$A^{+m} \rightarrow A^{2+} + e,$$

and passed through the magnet with half the momentum, half the charge, and thence the same trajectory, as a beam of A^{4+} ions. The maximum cross-section for production of A^{+m} was estimated to be $1 \cdot 3 \times 10^{-20}$ cm² and its lifetime, obtained by studying its intensity as a

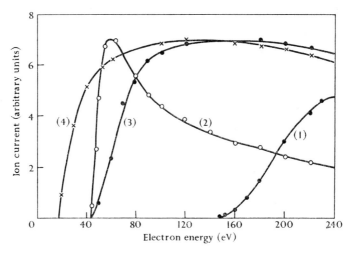

FIG. 3.21. Evidence for production of metastable A^+ ions in experiments of Daly. (1) i^+ v. E_e for A^{4+} beam. (2) i^+ v. E_e for production of metastable A^+ ions. (3) i^+ v. E_e for A^{2+} beam. (4) i^+ v. E_e for A^+ beam.

function of the repeller voltage used to extract it from the Nier source, as 14 ns. These metastable ions had energy above the ionization energy of A^{2+} and were thus different from those observed by McGowan and Kerwin and by Hagstrum.

Similar anomalous behaviour corresponding to the passage of the A^{3+} beam into the detector was attributed to the formation of a metastable state of A^{2+}. Metastable states were also observed for Xe^+ and Xe^{2+}.

3. The ionization of positive ions by electron impact

The production of multiply ionized atoms in a series of single ionization processes is of importance in astrophysics and in dense plasmas, and we describe now the measurements that have been performed to measure ionization cross-sections of ions.

3.1. The measurement of the ionization cross-section

3.1.1. The crossed-beam method. The first direct measurements of the ionization cross-sections of ions were made in He+ by Dolder, Harrison, and Thonemann,† using a crossed-beam method. Crossed-beam measurements of ionization cross-sections of positive ions have also been carried out by Latypov et al.‡ for rare gas ions and mercury, and by Lineberger et al.§ for alkali metal ions. Fig. 3.22 shows a simplified plan of the apparatus used by Dolder et al. Ions of energy 5 keV from the source S were deflected by the magnet M_1 through an aperture in plate P placed at a position to accept a beam of the ions. This target

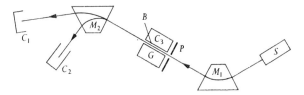

Fig. 3.22. Simplified plan of apparatus of Dolder, Harrison, and Thonemann.

beam was crossed in the interaction space B by an electron beam from the gun G to collector C_3 and some ions were further ionized to produce doubly charged ions. The momentum transfer in the ionization process is small compared with the primary ion beam momentum, so that the doubly charged ions continued along with the beam and entered the second magnet M_2 where the two types of ion were deflected into the collectors C_1 and C_2 respectively.

With a charged target beam it is much easier to obtain absolute cross-section measurements using the crossed-beam method than in the case of the ionization of neutral atoms. The distributions of current in the electron and ion beams were monitored, using a shutter constructed in the form of a right-angled bracket so that both beams could be intercepted simultaneously. The simultaneous measurement of the beam distributions is important because the crossing electron beam, by partially neutralizing the space charge of the positive ion beam, modifies its spatial distribution.

† DOLDER, K. T., HARRISON, M. F. A., and THONEMANN, P. C., Proc. R. Soc. **A264** (1961) 367.
‡ LATYPOV, Z. Z., KUPRIYANOV, S. E., and TUNITSKII, N. N., Zh. éksp. teor. Fiz. **46** (1964) 833; Soviet Phys. JETP **19** (1964) 570.
§ LINEBERGER, W. C., HOOPER, J. W., and McDANIEL, E. W., Phys. Rev. **141** (1966) 151; HOOPER, J. W., LINEBERGER, W. C., and BACON, F. M., ibid. **141** (1966) 165.

If i_+, i_e are the currents of ions and electrons crossing the interaction space, i_{2+} the current of doubly charged ions produced by electron impact, h the height of the ion beam, and V, v the ion and electron velocities respectively, the cross-section Q_i^+ for ionization of the ions is given by

$$Q_i^+ = \frac{i_{2+}}{i_+ i_e} \frac{hevV}{2(v^2+V^2)^{\frac{1}{2}}} F, \qquad (23)$$

where the factor
$$F = \frac{\int_0^h i_+(z)\,dz \int_0^h i_e(z)\,dz}{h \int_0^h i_+(z)i_e(z)\,dz}, \qquad (24)$$

which measures the degree of overlap of the beams, was obtained from the measurements with the bracket slit.

Background effects necessitate a number of corrections in order to obtain the true current i_{2+} from the measured current, $i_{2+}(\text{meas})$. A small current $i_{2+}(0)$ to the collector C_2 may be present due to leakage current driven by thermal and contact potentials even when neither beam is present. The correction for this current was appreciable in the experiments of Lineberger et al. using alkali metal ions.

Production of the doubly charged ion can also occur by collision with molecules of the background gas in a reaction of the type:

$$\text{He}^+ + \text{X} \rightarrow \text{He}^{2+} + \text{X}' + \text{e}, \qquad (25)$$

where X represents a molecule of the background gas. In addition, in the case of helium an additional background was due to the production of metastable He atoms when the He$^+$ beam hit metal surfaces inside the apparatus. These metastable atoms, drifting into the collector C_2, gave rise to a small negative current. The current $i_{2+}(I)$ has therefore to be measured with the ion beam passing through the apparatus but without the crossing electron beam. Owing to space charge effects this background current changes in the presence of the crossing electron beam. The electron beam current had therefore to be kept small to reduce the error due to this effect.

In Lineberger's experiments a current $i_{2+}(e)$ was observed with no ion beam but with the electron beam on. This current, attributed to stray electrons entering C_2, could be reduced by means of external magnets.

The true current i_{2+} to be used in (23) is then given by

$$i_{2+} = \{i_{2+}(\text{meas}) - i_{2+}(I)\} - \{i_{2+}(\text{e}) - i_{2+}(0)\}. \qquad (26)$$

Dolder et al. allowed for these background effects by an ingenious arrangement of pulsed beams.

Voltage pulses applied to a deflector plate deflected the He⁺ beam out of the interaction space 5000 times per second to produce a square-topped wave form for the ion current with duty cycle $\frac{1}{2}$. Similarly, pulses applied to the control grid of the electron gun gave square-topped pulses of electron current of the same repetition rate but with a duty cycle $\frac{1}{4}$. Two modes of pulsing were used. In the coincidence mode (Fig. 3.23) the pulses were so phased that ions and electrons crossed

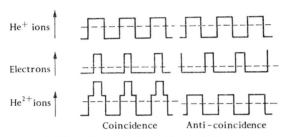

FIG. 3.23. Wave form of pulses of ion and electron currents in coincidence and anticoincidence modes.

the interaction space simultaneously. The time-averaged current $i_{2+}(c)$ collected by a vibrating reed electrometer is given by

$$i_{2+}(c) = \tfrac{1}{4}i_{2+} + i_{2+}(0) + \tfrac{1}{4}i_{2+}(e) + \tfrac{1}{2}i_{2+}(I). \tag{27a}$$

In the anticoincidence mode (Fig. 3.23) the pulses were so phased that ions and electrons crossed out of phase and the collector current $i_{2+}(a)$ is given by

$$i_{2+}(a) = i_{2+}(0) + \tfrac{1}{4}i_{2+}(e) + \tfrac{1}{2}i_{2+}(I) \tag{27b}$$

and
$$i_{2+}(c) - i_{2+}(a) = \tfrac{1}{4}i_{2+}, \tag{28}$$

i_{2+} being the steady current of doubly charged ions that would have been observed in steady operation. Substitution in (23) then gives Q_{i+} provided i_+, i_e refer to the amplitude of the ion and electron current pulses. Dolder et al. corrected (28) for loss of ions in travelling from the interaction space to the collector.

The modulated crossed-beam method in these measurements serves a different purpose from that served in the case of measurements of the ionization cross-section of neutral atoms. In that case the main background effects were not modulated, so that by using a detector tuned to pick up signals of the modulation frequency this type of background could be eliminated. For ionization of positive ions, however, the main background effects are modulated with the same frequency as the beam modulation frequency—direct double ionization of the background gas is not such an important factor. Nevertheless the modulation technique was used in the experiments of Dolder et al. because the outgassing effect of the electron beam increases the background contribution from process (25). The background current $i_{2+}(I)$, measured under steady conditions, is less than the actual background during the experiment when the electron beam is flowing. By modulating at a sufficiently large frequency the background gas pressure assumes a steady value that is the same whether or not the electron current is flowing, so that a realistic background correction can be made.

In their experiments on the alkali metal ions, obtained from hot coated filament sources, Lineberger et al. relied on operation under very high vacuum conditions, with small electron beam currents to reduce the background corrections. In their apparatus the pressure with both beams operating was 10^{-8} torr. Even so, the background corrections were large, as is shown by the following figures, given as typical of a run with lithium.

$$i_+ = 2 \cdot 00 \times 10^{-7} \text{ A}, \qquad i_e = 2 \cdot 00 \times 10^{-3} \text{ A},$$
$$i_{2+}(\text{meas}) = 2 \cdot 22 \times 10^{-15} \text{ A}, \qquad i_{2+}(0) = -0 \cdot 65 \times 10^{-15} \text{ A},$$
$$i_{2+}(I) = 1 \cdot 44 \times 10^{-15} \text{ A}, \qquad i_{2+}(e) = -2 \cdot 14 \times 10^{-15} \text{ A}.$$

In Na and K, where it was possible to operate at lower electron currents, the background correction was smaller.

Similarly, in the experiments of Dolder et al. on Ne+ with background gas pressures of 10^{-7} torr the operating conditions were such that

$$i_{2+}(c) = 3 \cdot 01 \times 10^{-14} \text{ A}, \qquad i_{2+}(a) = 2 \cdot 34 \times 10^{-14} \text{ A}.$$

Latypov et al. used continuous beams but their vacuum conditions (background pressure 10^{-6}–10^{-7} torr) were not so good as in the experiments of Lineberger et al.

In all experiments in which the target atoms are ionized it is difficult to ensure that all the ions are initially in their ground state. In the crossed-beam experiments discussed here the distance between the ion source and the intersection region ensures that the only excited states to survive in the beam are metastable states with lifetime greater than 10^{-6} s. Latypov, Kupriyanov, and Tunitskii[†] found that the threshold for ionization of Kr+ and Xe+ depended on the energy of the electrons used to produce the Kr+ and Xe+ beams. For example, when Xe+ ions formed by electrons of energy 30 eV were ionized, the threshold

† loc. cit., p. 145.

energy for the production of Xe^{2+} was 20 eV, in agreement with the known single ionization potential of the Xe^+ ion (21·2 eV). On the other hand, when the Xe^+ ions were formed by electrons of energy 60 eV, the threshold energy for Xe^{2+} production was approximately 10 eV. The shape of the ionization cross-section curve was markedly different in the two cases. This strongly suggests the presence of metastable Xe^+ ions in the beam in the second case.

The ground state of Ne^+ and He^+ consists of two levels, $^2P_{\frac{3}{2}}$ and $^2P_{\frac{1}{2}}$, separated by less than 0·1 eV, however, and the measured cross-sections refer to ions in a mixture of these two states in an unknown proportion, even when no higher energy metastable states are excited.

3.1.2. *The ion-trap method.* Space charge effects give rise to a negative potential inside an electron beam. Ions formed by collision of the electrons within the beam with gas atoms are then trapped if their kinetic energy is less than the depth of this potential well. A Nier type mass spectrometer source was used by Plumlee[†] to trap Hg ions within the electron beam. He observed relative proportions of Hg ions with one to five charges to be 100: 62: 42: 22: 1. The corresponding proportions observed for Hg-ion production in single collisions at the same electron energy (80 eV) is 100: 14: 0·45: 0: 0. Evidently a considerable proportion of the ionization in Plumlee's source occurred in successive step processes of the type $Hg^{n+}+e \rightarrow Hg^{(n+1)+}+2e$.

Baker and Hasted[‡] have used a Nier source operated as an ion trap to study ionization potentials for step ionization processes of positive ions and the behaviour of the cross-sections for such processes near threshold. For an electron beam of energy $V_e e$ and current of strength i_e, uniform over a radius a, the well depth in volts seen by a positive ion at radius r can be written, if the quantities are expressed in mks units:

$$V = -\frac{1}{4\pi\epsilon_0}\left(\frac{m}{2V_e e}\right)^{\frac{1}{2}}\frac{i_e r^2}{a^2}. \qquad (29)$$

For $i_e = 50\ \mu A$, $V_e = 30$ V, $a = 0·5$ mm, the coefficient of r^2 in (29) is 55·2 $V^{-1}\ cm^{-2}$, so that cylindrical potential traps of depth $> 0·1$ V are readily obtainable. The longitudinal trapping potential is of the order of 1 V. The kinetic energy of ions at formation ($\frac{3}{2}\kappa T = 0·04$ eV) is small enough to ensure that many ions are trapped inside the electron beam. The ion concentration within the beam builds up until equilibrium conditions are reached when the rate of extraction of ions by the extractor field is equal to their rate of production.

[†] PLUMLEE, R. H., *Rev. scient. Instrum.* **28** (1957) 830.
[‡] BAKER, F. A. and HASTED, J. B., *Phil. Trans. R. Soc.* **261**A (1966) 33.

Baker and Hasted operated at gas pressures of the order of 10^{-6} torr and relied on the longitudinal magnetic field of the Nier source to confine their electron beam. They obtained trapping times up to 10^{-4} s and electron current densities of about 10^{-4} A/cm². Redhead† operated at lower pressures ($\sim 5 \times 10^{-8}$ torr) and added potential barriers of 25 V at both ends of the ionization region to improve the axial trapping of the ions.

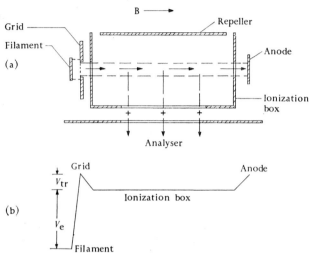

FIG. 3.24. (a) Electrode arrangement for ion-trap method. (b) Axial distribution of potential (Redhead).

Fig. 3.24 (a) shows the electrode arrangement used by Redhead.† The grid and anode were connected together and maintained at a positive potential (V_{tr}) relative to that of the ion chamber (V_{e}). The field for the extraction of the ions was provided by the repeller plate at the back of the chamber. The axial distribution of potential is shown in Fig. 3.24 (b). Redhead was able to build up the trapped ion concentration to obtain space-charge neutralized electron current densities of 5×10^{-2} A/cm² with a trapping time up to 0·5 s.

If the source is operated with values of V_{e} below the threshold for further ionization of the doubly charged ions, or for direct double ionization of the neutral gas atoms, the ratio i_{2+}/i_{+} of the current of doubly to singly charged ions is shown to be

$$i_{2+}/i_{+} = C_0 Q_{2i} E, \tag{30}$$

where C_0 is a constant and E the 'trap efficiency factor' is relatively

† REDHEAD, P. A., Can. J. Phys. **45** (1967) 1791.

insensitive to V_e. The ion trap is useful mainly in measurements of thresholds for ionization processes and for the study of structure in ionization cross-section curves. More recent developments[†] of the trap use separate electron beams for the production of ions and for their further ionization, and under these circumstances E (eqn (30)) may be independent of electron energy so that the form of the ionization cross-section dependence on energy can be determined. It appears difficult to adapt the method for absolute measurements however.

Ionization may occur singly or through multiple processes, but the two types of processes can be separated by studying the form of variation of i_{2+} with the electron current i_e.

3.2. Results of measurements of ionization cross-sections of positive ions

Fig. 3.25 shows the results of measurements of ionization cross-sections of He^+, Ne^+, N^+, Li^+, K^+, Na^+, using the crossed-beam method. The general shape of the variation of Q_i^+ with V_e is similar to that for the ionization of neutral atoms.

Results have also been given by Latypov et al.[‡] for the ionization cross-sections of Ne^+, A^+, A^{2+}, Kr^+, Kr^{2+}, Xe^+, Xe^{2+}, Hg^+, Hg^{2+}. In general it appears that the maximum cross-sections for ionization of singly and doubly charged ions are of the same order of magnitude as those for ionization of the neutral atom.

Fig. 3.26 (a) shows the results obtained by Baker and Hasted for the ionization of Ne^+ and A^+, using the ion-trap method. These curves suggest an approximately linear rise of cross-section from threshold. (The curvature at threshold is evidently due to the electron energy spread.)

The curves obtained by Redhead for the production of multiply charged rare gas ions in the ion trap showed considerable structure. For A^{2+}, Kr^{2+}, and Xe^{2+} the appearance potential fell below the spectroscopic value for ionization of the singly charged ion. Fig. 3.26 (b) shows the normalized ion current for these ions plotted against $V_e - V_B$ where V_B is the spectroscopic ionization threshold. The ionization peak below the threshold is attributed to ionization in two stages through the excitation of an intermediate metastable state, $A^{+(m)}$, etc. in the processes

$$\left.\begin{array}{l} e + A^+ \rightarrow e + A^{+(m)} \\ e + A^{+(m)} \rightarrow 2e + A^{2+} \end{array}\right\}. \tag{31}$$

[†] HASTED, J. B., private communication.
[‡] LATYPOV, Z. Z., KUPRIYANOV, S. E., and TUNITSKII, N. N., loc. cit., p. 145.

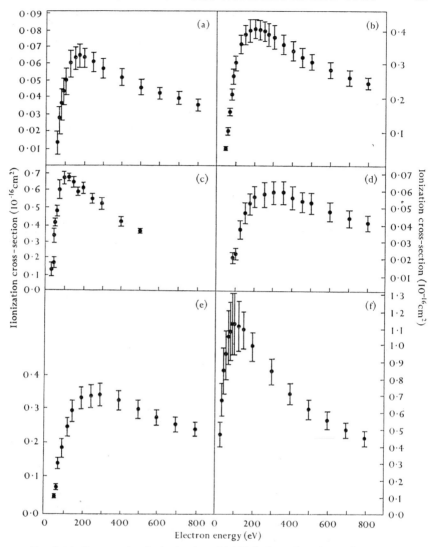

Fig. 3.25. Cross-section for ionization of (a) He$^+$, (b) Ne$^+$, and (c) N$^+$ (Harrison et al.), and (d) Li$^+$, (e) Na$^+$, (f) K$^+$ (Lineberger et al.).

The threshold for this process should then be the excitation energy of the metastable ion. The metastable levels observed by Redhead for A$^+$, Kr$^+$, and Xe$^+$ were, within experimental error, in agreement with the ^4D or ^4F levels determined spectroscopically and also with the values obtained by Hagstrum† from measurements of Auger electron emission from metals caused by the impact of metastable ions.

† See § 2.5.3.4.

4. The detachment of electrons from negative ions by electron impact

4.1. *The experimental method*

The detachment of electrons from negative ions by electron impact is a similar process to neutral atom ionization. The cross-section for the process

$$e + H^- \rightarrow H + 2e \tag{32}$$

has been measured using modulated crossed electron and H^- beams by

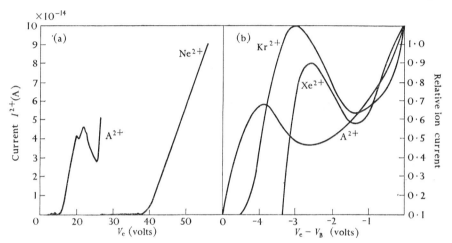

Fig. 3.26. Cross-section for ionization of (a) Ne^+, A^+ (Baker and Hasted), (b) A^+, Kr^+, Xe^+ (Redhead). V_B is the threshold potential for direct ionization of the ion.

Tisone and Branscomb† and by Dance, Harrison, and Rundel.‡ Fig. 3.27 shows the H^- source used by Tisone and Branscomb. Electrons emitted from the tungsten filament were accelerated towards the anode and collimated by an axial magnetic field of about 500 gauss produced by external magnets. The anode plate and cathode holder were of soft iron so that the field was uniform near the centre of the source. Ammonia was introduced into the source and H^- ions produced in the discharge. The positive potential attracted these ions through the anode slit. After magnetic analysis the H^- beam passed through the interaction space where it was intersected by an electron beam of variable energy.

Fig. 3.28 (a) shows a schematic drawing of the arrangement used by Dance *et al.* and Fig. 3.28 (b) shows a view of a section of their apparatus

† TISONE, G. and BRANSCOMB, L. M., *Phys. Rev. Lett.* **17** (1966) 236.
‡ DANCE, D. F., HARRISON, M. F. A., and RUNDEL, R. D., *Proc. R. Soc.* A**299** (1967) 525.

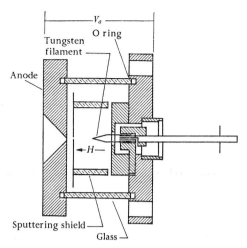

Fig. 3.27. H⁻ source used by Tisone and Branscomb.

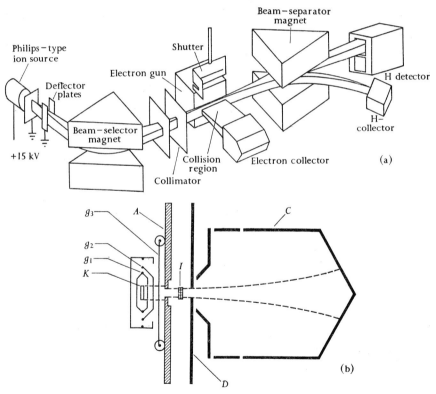

Fig. 3.28. (a) Schematic drawing of, and (b) section through apparatus of Dance *et al.* for studying detachment cross-section of H⁻ by electron impact.

through the electron beam. The electrons were accelerated toward the anode A through the grids g_1, g_2, g_3. g_2 and g_3 were biased positive with respect to A so that secondary electrons formed at g_2, g_3 and A could not enter the collision region. After intersecting the ion beam I they passed through the defining plate D into the collector C. After passing the interaction space the H⁻ beam (of energy several keV) was deflected magnetically into a collector. The neutral H atoms formed by detachment were incident on a copper-beryllium plate from which they ejected secondary electrons. In the experiment of Dance et al., this plate formed the first dynode of a 17-stage Venetian-blind type particle multiplier, operated as a single-particle counting device. In the experiment of Tisone and Branscomb a Daly-type detector was used.† The secondary electrons were accelerated through a potential difference of 10–15 kV on to a scintillator, and the photons produced transferred to the photomultiplier tube by a light pipe. The amplified time-integrated output of the photomultiplier constituted the signal.

With the electron beam on, but without the ion beam, photons produced by electron collisions with residual gas atoms or with electrodes of the gun may give a background secondary electron current in the detector. With the ion beam on, but without the electron beam, neutrals could be produced by stripping of the H⁻ beam on defining slits or residual gas atoms.

Both these sources of background may be compensated by suitable modulation of the two beams. A more troublesome source of background, however, arises from space-charge neutralization effects due to the interaction of the two beams. This may alter the trajectories of the ions after passing through the interaction region, changing the accompanying flux of neutrals produced by stripping of H⁻ on defining slits. This background cannot be distinguished from true signal by modulation techniques. The size of the effect must depend on the velocity of the ions. Conditions were chosen so that the measured detachment cross-section was independent of ion-beam energy in the experiment of Dance et al. Tisone and Branscomb measured the detachment cross-sections of H⁻ and D⁻ beams of the same energy and chose operating conditions in which the two cross-sections were in agreement.

Tisone and Branscomb used different electron modulation frequency (ω_e) and ion modulation frequency (ω_i) with $\omega_e/2\pi = 20$ kc/s and $\omega_i/2\pi = 50$ c/s, so that $\omega_e \gg \omega_i$. This condition enabled the elimination of pressure modulation effects from electron bombardment. The signal

† p. 143.

from the photomultiplier of the neutral detector was passed through two lock-in phase-sensitive amplifiers used in series as shown diagrammatically in Fig. 3.29. The electron modulating frequency was synchronized to the second amplifier.

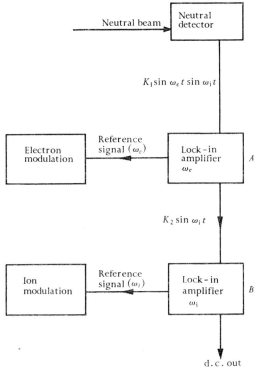

FIG. 3.29. Block diagram of phase-sensitive amplifiers used in experiment of Tisone and Branscomb.

Just as in equation (23), the signal current is proportional to the product $i_I i_e$ so that, expressing i_I, i_e as Fourier sums, the leading terms present in the signal from the detector can be written:

$$S_i = \left(\frac{4}{\pi}\right)^2 \frac{Q^d(v^2+V^2)^{\frac{1}{2}}}{e^2 v V} i_e i_I F G_d \sin \omega_e t \sin \omega_i t +$$
$$+ G_d(A_e \sin \omega_e t + A_I \sin \omega_i t). \quad (33)$$

v, V, F, i_e are as in equation (23), i_I is the current in the negative ion beam, Q^d the cross-section for detachment, G_d the gain of the neutral detector, and A_e, A_I the total amplitudes of the background signals at frequencies ω_e, ω_i.

The component of frequency ω_i is blocked by the first amplifier and that of frequency ω_e by the second amplifier so that only the first term contributes to the signal through both amplifiers. The output signal is given by

$$S_0 = \left(\frac{4}{\pi}\right)^2 G_1 G_2 G_d F i_e i_I \frac{Q^d(v^2+V^2)^{\frac{1}{2}}}{e^2 vV},\qquad(34)$$

where G_1, G_2 are the gains of the first and second amplifiers. The detection system was calibrated, using negative ions incident on the copper beryllium plate. A subsidiary experiment, using a negative ion beam containing a known concentration of neutral H atoms produced by photodetachment, suggested that the secondary electron emission

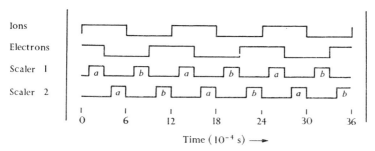

FIG. 3.30. Wave forms of ion and electron beams and scaler gates in experiment of Dance et al.

coefficient for H atoms and H⁻ ions of the same energy were approximately equal.

Dance et al. used their detector to count individual H atoms. Both electron and ion beams were pulsed at a frequency of 1·67 kc/s, 90 degrees out of phase. The output pulses from the detector were fed in parallel to two scalers, gated so that counting takes place over two portions, a, b, of each pulse. The wave forms of the ion and electron beams and scaler gates are shown in Fig. 3.30. If R is the true signal and B_I, B_e, and B_d the backgrounds associated with the ion beam, the electron beam, and the electronic noise, scaler 1 measures $(R+B_I+B_e+B_d)$ in gate a and B_d in gate b: scaler 2 measures (B_I+B_d) in gate a and (B_e+B_d) in gate b. The true signal R is then given by

$$\{\text{scaler 1}-\text{scaler 2}\}_{\text{gate }a}-\{\text{scaler 2}-\text{scaler 1}\}_{\text{gate }b}.$$

They measured the efficiency of detection of H⁺ and H⁻ ions of the same energy with their detector and found them to be equal within 3 per cent and both always greater than 90 per cent. They took then for the efficiency of detection of H atoms of the same energy the mean of the H⁻ and H⁺ detection efficiencies.

4.2. Results of measurements of the detachment cross-sections of H^- by electrons

Fig. 3.31 shows the results obtained by Dance *et al.* and by Tisone and Branscomb. The errors shown on the figure refer to random errors only. In both cases careful estimates have been made of the maximum systematic errors that could be present. These were estimated as approximately 15 per cent by Dance *et al.* and as 45 per cent by Tisone

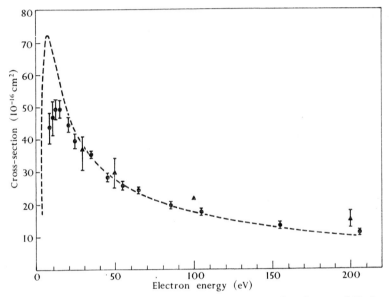

Fig. 3.31. Results of measurement of cross-sections for detachment of H^- by electron impact. (●) Dance, Harrison, and Rundel. (▲) Tisone and Branscomb. The dotted line is a theoretical curve (see Chap. 7, § 5.6.13)

and Branscomb, so that the results of the two experiments are in agreement. The general shape of the variation of Q^d with electron energy is similar to that for ionization. The electron affinity of H^- is 0·75 eV so that the maximum cross-section appears to be reached at an energy of approximately twenty times the threshold energy, relatively much further above threshold than for ionization of neutral atoms.

There appears to be a hint of structure in the cross-section near 30 eV but further study would be needed to establish its reality.

5. Inner-shell ionization of atoms by electron impact

In the ionizing collisions discussed so far we have considered ionization by ejection of one or more of the outer electrons from an ion or atom. A considerable body of data is available, however, on the ionization of an atom in an inner shell, from a study of the efficiency of excitation of characteristic X-radiation by electron impact.

The vacant state in the inner shell can be filled either by a radiative transition from a higher shell or by a radiationless reorganization (Auger effect). The ratios of the intensities of the X-radiation arising from transitions between different outer levels and a given inner level have been thoroughly explored for many elements.[†] The probability that the vacant state will be filled by a radiative transition, the fluorescence yield, has also been measured for the innermost shells of many elements.[‡] In principle, therefore, to determine the cross-section for a given inner-shell ionization it is only necessary to measure the intensity of one of the X-ray lines that results from the filling of the vacancy. For example, the cross-section $Q_i^K(E)$ for K-shell ionization of an atom by electrons of energy E may be derived from measurements of the intensity of the $K\alpha$-radiation emitted as a result of the impact of electrons of this energy on the appropriate target. The number, $N_{K\alpha}$ of quanta of this radiation emitted from a target of thickness t and containing N atoms per cm^3 and on which N_e electrons of energy E are incident is given by

$$N_{K\alpha} = N_e\, Nt Q_i^K(E) \varpi_K\, p_{K\alpha}\, \epsilon, \qquad (35)$$

where ϖ_K is the K-shell fluorescence yield, $p_{K\alpha}$ the proportion of radiative transitions to the K-shell that give rise to the emission of $K\alpha$-radiation, and the correction factor ϵ allows for the target efficiency discussed below. The practical application of these principles, however, encounters many difficulties. As might be expected, the measurement of the absolute value of the cross-section is even more difficult than that of its relative values at different electron energies.

Owing to the complicated nature of the processes occurring at the target of an ordinary X-ray tube, most of the X-ray data available using ordinary thick targets are too complex to allow even of an estimate of the way the cross-section for inner-shell ionization of the atoms concerned varies with electron energy (the K-ionization function). It is necessary to use thin targets, a few hundred Å thick, when studying inner-shell ionization by electrons of energy a few tens of keV. For the excitation of shorter wavelength characteristic radiation, involving the use of more energetic electrons, thicker targets may be employed. With such effectively thin targets the corrections to be made for secondary

[†] See, for instance, COMPTON, A. H. and ALLISON, S. K., *X-rays in theory and experiment*, pp. 640–6 (van Nostrand, New York 1935).

[‡] BURHOP, E. H. S., *The Auger effect and other radiationless transitions* (Cambridge University Press, 1953).

modes of inner-shell ionization are not large and can be estimated with sufficient accuracy. Careful measurements of relative inner-shell ionization cross-sections at different energies have been made by Webster et al.† and by McCue.‡ Absolute values of the K-shell ionization cross-section have been measured for silver by Clark§ and for nickel by Smick and Kirpatrick,|| so that by combining the two types of measurements absolute K-shell ionization cross-sections for incident electrons of energy in the range up to 175 MeV for silver and nickel have been obtained. More recently, absolute measurements of inner-shell ionization cross-sections for the K-shells of copper, silver, and nickel and three L-shells of gold have been made by Green†† for electrons in the energy range up to 50 keV, by Motz and Placious‡‡ for K-shell-ionization of tin and gold by electrons of energy up to 600 keV, and by Hansen, Weigmann, and Flammersfeld§§ for K-shell ionization of zirconium, tin, tungsten, and lead for electrons of energy up to 1440 keV.

5.1. *The measurement of inner-shell ionization cross-sections*

The thin targets used in these experiments were either in the form of thin foils or of very thin layers deposited by vacuum evaporation on a beryllium base. An element of low atomic number such as beryllium was chosen as the underlying metal, in order to reduce the contribution to the X-radiation from the thin target due to fluorescent excitation by primary X-radiation produced in the backing material. Great care was taken to obtain uniform films and it was found that uniformity could be achieved by carrying out the evaporation sufficiently rapidly.

It was necessary to go to some trouble to avoid effects that are usual with hot cathode X-ray tubes, viz. the deposition of layers of tungsten and carbon on the target. Such deposition was avoided by using a carefully designed vacuum line and a set of cold traps. In experiments such as those of Webster and his colleagues in which an X-ray spectrometer was used to isolate the radiation being measured, it was necessary to design the geometry of the slit system so as to ensure that radiation from the whole of the focal spot entered the spectrometer system, since

---

† WEBSTER, D. L., HANSEN, W. W., and DUVENECK, F. B., *Phys. Rev.* **43** (1933) 839; POCKMAN, L. T., WEBSTER, D. L., KIRKPATRICK, P., and HARWORTH, K., ibid. **71** (1947) 330. ‡ MCCUE, J. J. G., ibid. **65** (1944) 168.
§ CLARK, J. C., ibid. **48** (1935) 30.
|| SMICK, A. E. and KIRKPATRICK, P., ibid. **67** (1945) 153.
†† GREEN, G. W., *Proc. 3rd Int. Symp. X-ray Microsc.*, Stanford, 1962 (Academic Press, New York); Thesis, Cambridge, 1962.
‡‡ MOTZ, J. W. and PLACIOUS, R. C., *Phys. Rev.* **136** (1964) A662.
§§ HANSEN, H., WEIGMANN, H., and FLAMMERSFELD, H., *Nucl. Phys.* **58** (1964) 241.

the physical dimensions of the focal spot in an X-ray tube vary with tube potential.

The X-ray relative intensities in the experiments of Webster and his colleagues were measured by means of an ionization chamber. This chamber was made with two compartments connected with opposing potentials, so that the currents due to natural ionization cancelled out. Only one compartment was exposed to the X-rays.

In the absolute intensity measurements of Clark† and of Smick and Kirkpatrick‡ the method of balanced filters developed by Ross§ was used. This has the advantage of separating out monochromatic X-radiation of much greater intensity than is possible by means of the X-ray crystal spectrometer. In this method two filters, composed of materials which are adjacent elements in the periodic system, are used in turn. The thickness of these two filters may be adjusted so that the spectrum of an incident continuous beam of X-radiation after transmission through either filter coincides at all wavelengths, except in the region between the respective K-limits. The difference in ionization produced by the X-radiation passing through these filters must therefore arise from radiation of wavelengths between the two K-limits.

In the experiments of Smick and Kirkpatrick,∥ balanced filters of iron and cobalt were used to isolate the nickel K-radiation. As finally adjusted, the cobalt filter passed 63·9 per cent of the Ni $K\alpha$-radiation, the iron filter passed 6·8 per cent. The transmission of the two filters was appreciably the same, except for a region 0·13 Å wide around the wavelength of the Ni $K\alpha$-radiation. The absolute ionization measurements required a knowledge of the mean X-ray energy expended per ion pair produced in the gas of the ionization chamber (air), and a correction was needed for the absorption of the X-radiation in the window of the ionization chamber.

Green†† used an argon-CO_2 flow proportional counter to detect the X-radiation. The size of the pulse defined the quantum energy of the radiation and the counting rate for pulses of the appropriate size defined its intensity. The counter was 5 cm in diameter and was fitted with a 5-μm Mylar window. The detection efficiency of the counter for radiation of a given quantum energy was found by combining the measured transmission of the window with the calculated absorption of the filling gas.

It was difficult to obtain accurate measurements of the absolute characteristic radiation intensity, using the thin film and the propor-

† loc. cit., p. 160. ‡ loc. cit., p. 160.
§ Ross, P. A., *Phys. Rev.* **28** (1926) 425. ∥ loc. cit., p. 160. †† loc. cit., p. 160.

tional counter, since the intensity was low and scattered electrons striking the inside of the target chamber could produce a significant background radiation from the walls. Absolute measurements of the intensity were therefore made for a solid target of the same material as the film, using the proportional counter. The relative intensity of the characteristic radiation from the film and from the solid target was then measured using a crystal spectrometer. Owing to its high dispersion the crystal spectrometer made it easy to distinguish the background radiation from

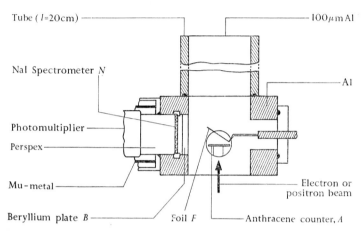

Fig. 3.32. Apparatus of Hansen, Weigmann, and Flammersfeld for studying inner-shell ionization.

the characteristic radiation from the thin target. The absolute intensity of the radiation from the thin target could then be calculated.

Motz and Placious† used a sodium iodide scintillator spectrometer to detect the K-series radiations of tin and gold. They estimated that the efficiency of their detector, including corrections for the photon absorption in the windows of the target chamber and the scintillator and for the escape of the iodine K X-rays from the scintillator, was 0·89 and 0·98 for the Au K- and Sn K-radiations respectively.

In the experiments of Hansen, Weigmann, and Flammersfeld characteristic X-rays were produced by the impact of beta electrons of known energy on a foil, and they employed a sodium iodide scintillator spectrometer to detect them. They compared also the cross-sections for inner-shell ionization by electrons and positrons. Their apparatus is shown in Fig. 3.32. Electrons from a ^{90}Sr source or positrons from a ^{56}Co source

† loc. cit., p. 160.

were analysed by means of a beta-ray spectrometer, which produced beams homogeneous in energy to within 5 per cent. The electron (or positron) beam was then incident on the foil F after passing through a thin anthracene scintillation counter, A. X-rays emitted from the foil nearly perpendicular to the incident-particle beam direction were detected by the sodium iodide scintillation spectrometer N, operated in coincidence with A. Scattered electrons were prevented from reaching N by the 5-mm thick beryllium plate B. Fig. 3.33 shows the peak due to Sn K-radiation (quantum energy 25·2 keV) produced by electrons of energy 530 keV, on a tin foil of surface density 23·6 mg cm^{-2}. The background observed in the absence of the foil and the method of subtracting the background correction to obtain the peak attributable to Sn K-series radiation are also shown in the figure. The ratio of the number of coincidences in the peak due to Sn K-radiation to the number of single counts in A due to particles passing through the foil F in the same time then enables an estimate of the K-ionization cross-section to be made. This ratio was determined by replacing the foil F by a scintillation counter S of the same surface dimensions, and measuring the ratio of the coincidence rate between A and S to the single counting-rate of A.

In deriving the inner-shell ionization cross-section from measurements of this kind careful attention has to be given to the correction for the target efficiency, ϵ, of equation (35). This depends on the following factors:

1. The diffusion of the electron stream in the thin target as a result of which the true mean length of electron path in the target is greater than the target thickness.

2. The back-diffusion of the electrons through the thin target due to the scattering of some electrons through angles greater than 180°. When the thin targets used are mounted on a thick beryllium base most of the back-scattered electrons will re-enter the target from the beryllium. For low-energy electrons, however, back scattering in the thin target itself may be important. The phenomena of diffusion and back-diffusion have been investigated experimentally by Bothe[†] and recently by Thomas.[‡]

3. The energy loss of the electrons in passing through the thin target as a result of which the measured X-ray line intensity is not produced by electrons homogeneous in energy. The correction for this effect,

[†] Bothe, W., *Handb. Phys.* (1st edn) **24** (1927) 18.
[‡] Thomas, M, N., University of Cambridge, Thesis. 1961.

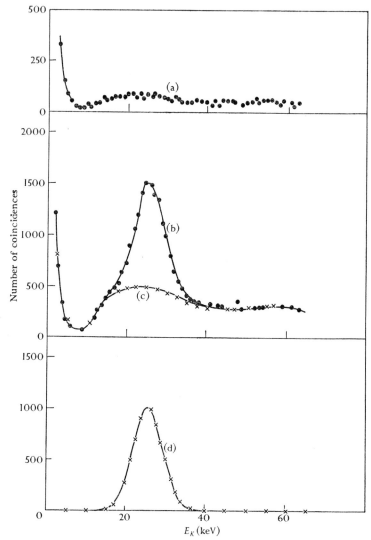

Fig. 3.33. Spectrum of radiation emitted by impact of 530 keV electrons on tin foil, measured by a sodium iodide scintillation spectrometer. (a) Background in absence of foil. (b) Observed spectrum with foil in position. (c) Background correction estimated from (a). (d) SnK peak after subtraction of background.

which can be estimated from the known energy-loss relationship,† is very small except near the threshold.

4. Fluorescent radiation produced in the target by continuous radiation from the beryllium or other backing material. The size of this

† See, for example, WILLIAMS, E. J., Proc. R. Soc. A**130** (1930) 310.

correction can be estimated from a knowledge of the frequency distribution of the radiation emitted from a thick target.

5. Absorption of the characteristic radiation in the thin target itself.

By using targets a few hundred Å thick these corrections can each be reduced to a few per cent with the exception of that due to fluorescent excitation (4), which becomes appreciable for high-energy electrons (above 100 keV). In this case, however, the other corrections become very small so that it is possible to use thicker targets ($\sim$ 1000 Å thick) that have sufficient mechanical strength to permit their use in the form of free foils. Correction (4) does not then arise. A correction is required, however, for fluorescent excitation by bremsstrahlung produced in the foil itself.

5.2. *Results of the measurements*

Fig. 3.34 shows the results of absolute measurements of the K-ionization cross-section Q_i^K in Ni and Cu (Fig. 3.34 (a)), in Ag and Sn (Fig. 3.34 (b)), and in Au (Fig. 3.34 (c)). Also shown in these figures are the results of relativistic and non-relativistic calculation for Ni, Ag, and Hg. In all the figures the cross-section is plotted against E/E_K, E_K being the K-ionization energy. Non-relativistic Born approximation calculations suggest that the curves plotted in this way should all have the same form with the cross-section for a given value of E/E_K proportional to Z^4 (see Chap. 7, § 5.6.5). Relativistic effects are found to be important for nickel and change completely the form of the variation of Q_i^K for Au. The errors quoted for the experimental measurements are approximately 6 per cent for the results of Webster, Kirkpatrick, and their collaborators for Ni and Ag and about 15 per cent for the results of Green for Ni, Cu, and Ag. The errors in the measurements of Motz and Placious for Sn and Au (3–18 per cent) are shown in the figure. The measurements of Webster, Kirkpatrick and collaborators, Green, and Motz and Placious agree reasonably well within their quoted estimated errors. The cross-sections measured by Hansen *et al.*, however, are larger than those of other authors, in cases where comparison is possible, by amounts much larger than their estimated errors of 10–15 per cent.

Some general features emerge from the curves. As the atomic number increases above that of nickel ($Z = 28$) the fall of cross-section past the maximum gets smaller and smaller and for tin ($Z = 50$) appears to have almost disappeared altogether. The K-ionization cross-section is then almost constant over a large range of electron energies. For tungsten, gold, and lead the cross-section is still increasing for $E/E_K = 10$ but

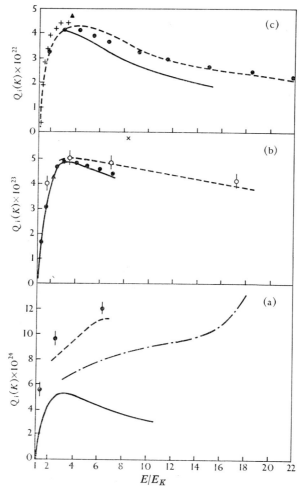

FIG. 3.34. Variation of K-ionization cross-section with E/E_K for Ni, Ag, Sn, W, Au, Pb. (a) ⦁ Au (Motz and Placious). (b) ● Ag (Webster et al., Clark); ▲ Ag (Green); ⦶ Sn (Motz and Placious); × Sn (Hansen et al.). (c) ● Ni (Pockman et al., Smick and Kirkpatrick); ▲ Ni (Green); + Cu (Green). The lines show theoretical cross-sections ——— Burhop† (non-relativistic), - - - - Arthurs and Moiseiwitsch‡ (relativistic), —·—·—·— Perlman§ (relativistic).

according to the measurements of Hansen et al. it falls off for E/E_K between 12 and 20.

Absolute inner-shell ionization cross-sections for the three L-shells of gold have been obtained by Green.‖ These measurements depend

† BURHOP, E. H. S., *Proc. Camb. phil. Soc. math. phys. Sci.* **36** (1940) 43.
‡ ARTHURS, A. M. and MOISEIWITSCH, B. L., *Proc. R. Soc.* **A247** (1958) 550.
§ PERLMAN, H. S. *Proc. phys. Soc.* **76** (1960) 623.   ‖ loc. cit., p. 160.

markedly on a knowledge of the fluorescence yields ϖ_{L_I}, ϖ_{L_II}, ϖ_{L_III} of the three L-shells, which are still not known very well.

Measurements of the shape of the L-shell ionization functions have been made for gold L_II and L_III shells,† tungsten L_I, L_II, and L_III‡ and the silver L_III shell.§ Fig. 3.35 compares the values of these cross-sections measured by Green for Au with the non-relativistic calculations‖ for the neighbouring element, Hg. The experimental curves have been calculated, assuming the following values for the fluorescence yields, $\varpi_{L_\mathrm{I}} = 0{\cdot}13$, $\varpi_{L_\mathrm{II}} = 0{\cdot}51$, $\varpi_{L_\mathrm{III}} = 0{\cdot}55$ and agreement is then obtained

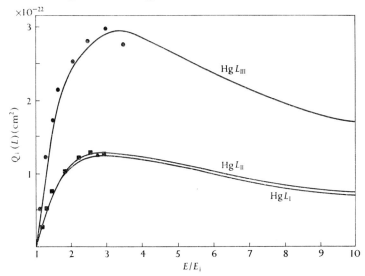

FIG. 3.35. Variation with electron energy of the cross-section for ionization of L-shells of Au. ▲ L_I; ■ L_II; ● L_III (compared with curves calculated for Hg with $\varpi_{L_\mathrm{I}} = 0{\cdot}13$, $\varpi_{L_\mathrm{II}} = 0{\cdot}51$, $\varpi_{L_\mathrm{III}} = 0{\cdot}55$).

with the theoretically calculated cross-sections. The most reliable measurements†† of the fluorescent yields give 0·09 for ϖ_{L_I}, 0·34–0·50 for ϖ_{L_II}, and 0·25–0·3 for ϖ_{L_III}, so that a considerable discrepancy is apparent for L_III ionization.

5.3. *Double inner-shell ionization*

The cross-section for double inner-shell ionization of an atom can be studied and compared with that for a single ionization. Consider a

† WEBSTER, D. L., POCKMAN, L. T., and KIRKPATRICK, P., *Phys. Rev.* **44** (1933) 130.
‡ HUIZINGA, W. J., *Physica* **4** (1937) 317.
§ MCCUE, J. J. G., *Phys. Rev.* **65** (1944) 168.
‖ BURHOP, E. H. S., loc. cit. p. 166.
†† FINK, R. W., JOPSON, R. C., MARK, H., and SWIFT, C. D., *Rev. mod. phys.* **34** (1966) 513.

doubly ionized atom that has lost an electron from both the K- and the L-shell. If such an atom undergoes a transition LL–KL, the X-radiation emitted will be a satellite line of the K_α-radiation. Parratt† and Shaw and Parratt‡ have measured the relative intensities of K_α-satellites as compared with those of the K_α-lines for elements of atomic number 11 to 46. These experiments were not carried out with truly thin targets and so they need correction for the effect of fluorescent excitation by the general radiation produced in the target. Since one might expect

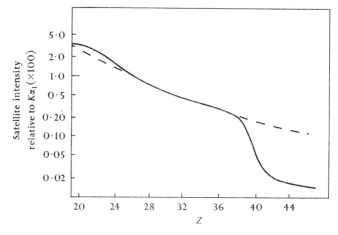

Fig. 3.36. Intensity of K_α satellites relative to K_α intensity, as a function of Z. ——— observed, – – – – calculated.

the double inner-shell photo-ionization of an atom to be very improbable, the effect of the fluorescent excitation would be expected to give too small a value for the intensity of the satellites relative to the parent line.

Fig. 3.36 shows how the observed relative intensity decreases with increasing atomic number. The decrease appears fairly regular up to an atomic number about 38, but for higher atomic numbers there is a sharp decrease in relative intensity. In these experiments the potential differences applied to the X-ray tube were in each case roughly twice the K-ionization potential or a little greater. The curves show, therefore, the form of variation with atomic number of the relative probability of a double KL-ionization to that of a single K-ionization.

† PARRATT, L. G., *Phys. Rev.* **50** (1936) 1.
‡ SHAW, C. H. and PARRATT, L. G., ibid. 1006.

4

THE EXPERIMENTAL ANALYSIS OF THE CROSS-SECTIONS FOR IMPACT OF ELECTRONS WITH ATOMS AND IONS — CROSS-SECTIONS FOR PRODUCTION OF EXCITED ATOMS

WE turn now to the discussion of the measurement of the cross-section of excitation of atoms and ions by electron impact, both to states that lead to the emission of radiation and to metastable states. In the present chapter are described the determination of the cross-section by studying the intensity and polarization of the radiation emitted from the excited atoms or by the direct measurement of the number of atoms excited to metastable states. In Chapter 5 we describe methods of measuring cross-sections for atomic excitation by studying the energy and angular distribution of the scattered electrons.

1. Optical measurement of cross-sections for excitation
1.1. *Principle of the method*

The passage of an electron beam of sufficient energy through a gas produces excited atoms. These atoms dispose of their surplus energy by radiation, and it is possible, in principle, to determine the cross-sections for excitation of the different levels by measurement of the absolute intensities of the emitted spectral lines for a given electron beam current.

Let n_j be the number of excited atoms produced per cm^3 by the beam. The current strength in the beam is supposed so small that collisions between excited atoms and a beam electron may be neglected. In the same way the gas pressure is supposed so low that collisions between excited and normal atoms, which may lead to transfer of excitation, as well as effects due to imprisonment of resonance radiation, may also be ignored. We then have

$$\frac{dn_j}{dt} = -A_j n_j + NQ_j v n_e + \sum A_{ij} n_i. \tag{1}$$

In this equation A_{ij} is the probability/s of the radiative transition from the ith to the jth level (see Chap. 7, § 5.2.1), $A_j = \sum A_{jk}$, the suffix k

referring to levels of lower energy than that specified by j. Q_j is the cross-section for excitation of the jth level, v is the electron velocity, and n_e, N are the respective numbers of electrons and gas atoms/cm³. In equilibrium we have

$$n_j = (NQ_j v n_e + \sum A_{ij} n_i)/A_j. \tag{2}$$

The total number of quanta of the radiation emitted, per unit time, per unit length of the electron beam, with frequency corresponding to the transition $j \to k$ is

$$J_{jk} = A_{jk} n_j S, \tag{3}$$

where S is the cross-sectional area of the beam. From (2) we have finally

$$J_{jk} = \frac{A_{jk}}{A_j}\left(NQ_j \frac{i}{e} + \sum_i J_{ij}\right), \tag{4}$$

where

$$A_{jk}/A_j = J_{jk}\Big/\sum_k J_{jk} \tag{5}$$

and

$$i = vSen_e \tag{6}$$

is the current strength in the beam. It follows that, if measurements are made of the absolute intensities of the radiation emitted in transitions both to and from the jth state of the atom, then the cross-section Q_j may be derived. Similarly, if relative intensity measurements are made for these lines at different electron energies, the form of the variation of Q_j with energy may be derived. Relative magnitudes of the Q_j for different j may also be obtained from the relative intensities of all the lines involved.

What is usually measured in optical experiments of this kind is the relative variation with energy of the intensity of a number of chosen spectral lines. This is called the *optical excitation function* of the particular line concerned. It gives the relative variation with energy of J_{jk} which is not the same as that of Q_j unless the contribution due to transitions from upper levels can be ignored. If sufficient lines have been measured these cascade contributions can be allowed for as described above, but in most experiments this has not been so and resort has had to be made to theoretical estimates of the A_{jk}. Similarly, absolute measurements are usually given in terms of an effective cross-section Q_{jk} which is defined by

$$J_{jk} = NQ_{jk} i/e, \tag{7}$$

so that, in terms of (4),

$$Q_{jk}\left(1 - \frac{\sum J_{ij}}{\sum J_{jk}}\right) = \frac{A_{jk}}{A_j} Q_j. \tag{8}$$

Again, Q_j can only be deduced if the ratios of the quantities J_{ij}, J_{jk}

are obtainable from the observed data either directly or with assistance from theory.

The method is incapable of giving information about the cross-section for excitation of metastable levels. Such cross-sections have been studied by detecting the metastable atoms through ionizing collisions of the second kind with other gas atoms† or through the ejection of Auger electrons when they impinge on metal surfaces‡ (see § 2.1.2).

1.2. *Principles involved in absolute measurement of cross-sections—polarization of emitted radiation*

In practice observations are made of the intensity of radiation emitted per unit solid angle in a particular direction, making an angle ϑ, usually 90°, with that of the beam. By comparison with a standard tungsten lamp, absolute values of this differential cross-section, which we shall call $q_{jk}(\vartheta)$, may be obtained. It is necessary to consider the variation of q_{jk} as a function of ϑ to obtain Q_{jk}.

As the radiation is of dipole type the angular distribution will be of the form
$$q_{jk}(\vartheta) = a_{jk}(1-p\cos^2\vartheta), \tag{9}$$
where p and a_{jk} are independent of ϑ but are functions of electron energy. It follows that
$$Q_{jk} = \int q_{jk}\,d\omega = \frac{4\pi}{3}(3-p)a_{jk},$$
so, for observations at 90°,
$$Q_{jk} = 4\pi(1-\tfrac{1}{3}p)q_{jk}(\tfrac{1}{2}\pi). \tag{10}$$

The quantity p is a measure of the degree of polarization of the beam. Thus if $I_\|$, $I_\perp$ are the intensities of the radiation emitted at 90° with electric vectors respectively parallel and perpendicular to the direction of the electron beam,
$$p = \frac{I_\| - I_\perp}{I_\| + I_\perp}. \tag{11}$$

The percentage polarization of the radiation emitted at 90° is therefore given by $P = 100p$. There is no difficulty in also deriving the corresponding percentage polarization for radiation emitted at an angle ϑ. It is given by
$$P(\vartheta) = \frac{P\sin^2\vartheta}{1-P\cos^2\vartheta/100}. \tag{12}$$

It follows from (10) that the absolute cross-section Q_{jk} can only be derived if the polarization and the intensity per unit solid angle of the

† ČERMAK, V., *J. chem. Phys.* **44** (1966) 3774.
‡ OLMSTED, J., NEWTON, A. S., and STREET, K., *J. chem. Phys.* **42** (1965) 2321.

radiation emitted normal to the beam are both measured. In practice the correction due to the polarization is usually not much greater than 10 per cent and often considerably smaller. As the accuracy of absolute measurement of q_{jk} is usually poorer than this, the measurement of the polarization is not especially important from this point of view. On the other hand, the knowledge of the polarization is useful as it provides further information about the excitation process, and for this reason particularly there has been a considerable revival of interest in its precise measurement.

1.3. *The measurement of optical excitation functions and polarizations*

In order that true optical excitation functions and polarizations be observed it is necessary that attention be paid to a number of experimental requirements.

1.3.1. *Measurement of optical excitation functions.* The gas pressure and electron beam intensity must be low enough to eliminate secondary sources of population or depopulation of different excited states. Thus resonance radiation will be trapped in the gas unless the gas pressure is very low. If the jth state combines optically with the ground state then on the right-hand side of (1) an additional term $f_j A_{j0} n_j$ should be added where f_j is the fraction of quanta radiated from $j \to 0$ transitions that are absorbed in the gas. Also (3) becomes

$$J_{jk} = A_{jk}(1 - \delta_{k0} f_j) n_j S, \tag{13}$$

so that in place of (4)

$$J_{jk} = \frac{A_{jk}(1 - \delta_{k0} f_j)}{A_j - A_{j0} f_j} \left(N Q_j \frac{i}{e} + \sum_i J_{ij} \right). \tag{14}$$

The net result of trapping is that A_{j0} is replaced by $(1-f_j)A_{j0}$, which depends on the pressure.

In principle both J_{jk} and J_{ij} may be measured as functions of pressure and Q_j can be determined from (14) in the limit of zero pressure in which $f_j \to 0$. At high pressures f_j may be obtained in the following way.

We define
$$Q'_j = (e/Ni)(A_j - A_{j0})/A_{jk}, \tag{15}$$

and
$$Q^c_j = (e/Ni) \sum_i J_{ij}. \tag{16}$$

Q'_j represents an apparent cross-section for electron excitation of state j. It would be equal to Q_j if no cascade or transfer of excitation in collisions took place and if, for ^1P states that can emit resonance radiation, the pressure were high enough for imprisonment to be complete (i.e. $f_j = 1$). Q_j^c is the contribution to the apparent cross-section arising from cascade processes. Then equation (14) can be rewritten

$$Q_j = Q_j'\left\{1 + \frac{A_{j0}(1-f_j)}{A_j - A_{j0}}\right\} - Q_j^c. \tag{17}$$

J_{jk} and Q_j^c can be measured at different pressures p and the transition probabilities A_{jk} can be calculated so that Q_j' can be obtained from (15), and f_j estimated.

As an example we consider the imprisonment of the resonance radiation emitted by the 3^1P level of helium. Gabriel and Heddle† measured J_{jk} for the 3^1P–2^1S (λ 5016) transition of helium and obtained Q_j' from the quantities A_{jk} tabulated in Table 4.2 (p. 203). Q_j, obtained under low-pressure conditions, is 455×10^{-20} cm². In comparison Q_j^c ($= 22 \times 10^{-20}$ cm² at a pressure of 5×10^{-3} torr) is small and its variation with pressure can be ignored. Equation (17) becomes in this case

$$42\{1 - f(3^1P)\} = \{477/Q'(3^1P)\} - 1. \tag{18}$$

Fig. 4.1 shows $Q'(3^1P)$ as a function of gas pressure p obtained by Gabriel and Heddle.† The variation of $1-f(3^1P)$ with pressure obtained from Fig. 4.1 using equation (18) may be checked by reference to Holstein's theory of the imprisonment of radiation. According to this theory f_j is a function, illustrated in Fig. 4.2, of $k_j\rho$ where ρ is an effective tube radius and

$$k_j = \frac{\lambda_j^3 N}{8\pi^{\frac{3}{2}}} \frac{g_j}{g_0} A_{j0}(M/2RT)^{\frac{1}{2}}, \tag{19}$$

where λ_j is the wavelength of the radiation concerned, g_j and g_0 are the respective statistical weights of the jth and the ground states, M is the molecular weight of the gas, R the gas constant, and T the absolute temperature. The points on Fig. 4.2 are those obtained using Fig. 4.1 and equation (18) and an effective tube radius of 0·56 cm. In Gabriel and Heddle's apparatus the collimating hole defining the electron beam had a diameter of 0·25 cm. The beam passed along the axis of the collision chamber of diameter 2·5 cm.

The application of Holstein's theory to the analysis of measurements of the excitation cross-section for the resonance (n^1P) states of He has been discussed by Phelps.‡ Although it is possible to avoid the need to allow for imprisonment of radiation by working at sufficiently low gas pressures this may lead to emission of radiation at such low intensity as to make measurement difficult. The introduction of photomultipliers

† GABRIEL, A. H. and HEDDLE, D. W. O., *Proc. R. Soc.* A258 (1960) 124.
‡ PHELPS, A. V., *Phys. Rev.* 110 (1958) 1362.

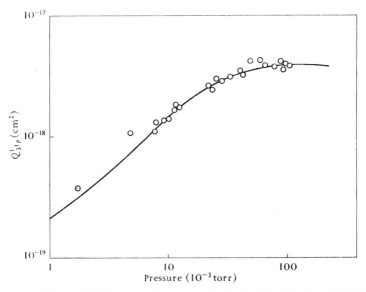

Fig. 4.1. Values of $Q'(3^1P)$ (defined by eqn (15)) obtained by Gabriel and Heddle plotted against helium gas pressure p. At the higher pressures, when imprisonment of resonance radiation is complete $Q'(3^1P)$ tends to the total cross-section for population of the He(3^1P) level directly and by cascade. The correction for collisional excitation transfer (see p. 176) is about 15 per cent at the highest pressure (10^{-1} torr) and is proportional to the pressure.

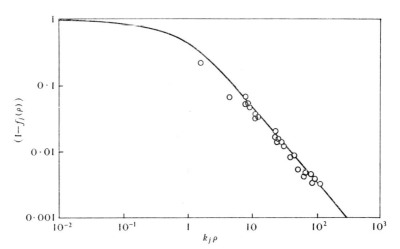

Fig. 4.2. Plot of $1-f_j$ against the parameter k_j according to Holstein's theory of imprisonment. The points are those derived from the measurements of $Q'(3^1P)$ (Fig. 4.1) for He(3^1P) using equation (18) and taking the effective collision tube radius $\rho = 0.56$ cm.

for intensity measurement has made this less likely, but there may still be circumstances in which it is convenient to work under conditions in which f_j is appreciable and to apply the method outlined above to correct for it.

Zapesochnyi and Shimon† have sought to eliminate the effect of self absorption in studying the excitation cross-sections for the resonance

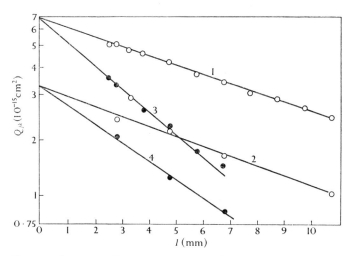

FIG. 4.3. Apparent effective cross-sections for excitation of caesium resonance doublets (Zapesochnyi and Shimon) as a function of distance l between electron beam and exit window. Curves 1 and 3 refer to λ 8521 ($6^2P_{\frac{3}{2}}-6^2S_{\frac{1}{2}}$) at pressures of $8\cdot5\times10^{-5}$ and 3×10^{-4} torr respectively. Curves 2 and 4 refer to λ 8944 ($6^2P_{\frac{1}{2}}-6^2S_{\frac{1}{2}}$) at the same pressures.

doublets of caesium and rubidium by attaching the exit window to a sylphon bellows so that the distance between the electron beam and the exit window could be varied. If l is this distance and Q_{eff} the effective cross-section for photon absorption and if I_0 and I_l are respectively the true intensity of the resonance radiation produced in the collision volume and the intensity after passing a distance l,

$$\ln I_l = \ln I_0 - Q_{\text{eff}} l. \tag{20}$$

The absorption of the collision volume was made negligible by using a flat thin electron beam, 0·4 mm thick. The plot of $\ln I_l$ against l should be linear and extrapolate back to $\ln I_0$ when $l \to 0$.

Fig. 4.3 shows the extrapolation for the two components (λ 8521, λ 8944) of the resonance doublet of caesium for two different pressures,

† ZAPESOCHNYI, I. P. and SHIMON, L. L., *Dokl. Akad. Nauk SSSR* **166** (1966) 320.

showing that the measurements extrapolate to the same value of $\ln I_0$ and thence to the same effective cross-section irrespective of pressure.

Another important pressure-dependent effect arises from transitions between neighbouring excited states occurring on collision between excited and normal atoms. In many cases, involving highly excited states and very small energy differences, the cross-sections may be very much larger than gas-kinetic. The estimation of these excitation transfer cross-sections from observations of the variation of apparent effective excitation cross-sections Q_{jk} with pressure is discussed in Chapter 18. As correction for these effects is much less definite than for imprisonment of radiation, it is desirable for the measurement of true excitation cross-sections to be able to work at sufficiently low pressures for the redistribution of excited states by collision to be unimportant.

Heddle and Lucas[†] have carried out a systematic study of the effect of pressure on the values obtained for the apparent effective excitation cross-section for a number of transitions in helium. These are shown in Fig. 4.4. The maximum pressure that can be used if reliable results are to be obtained depends on the particular transition. For the transition 3^1P–2^1S (Fig. 4.4 (a)) a working pressure of less than 10^{-4} torr is needed, while for the transition 3^3P–2^3S (Fig. 4.4 (d)) reliable results would be obtained with a working pressure as high as 5×10^{-3} torr.

Considerable care must be devoted to the design of the electron source. Thus it is important that the electron beam be nearly homogeneous in energy, particularly at energies not far above threshold in which range the excitation function may vary quite rapidly with energy.

It is equally important that secondary electrons should not be introduced into the collision region in the course of beam measurement or as a result of unsuitable choice of other aspects of the electrode systems.

Furthermore, to avoid errors due to the variation of the electron distribution across a cross-section of the beam it is important to observe the radiation emitted from the whole cross-section.

Determination of absolute cross-sections presents even more difficult problems, as reliable standard monochromatic sources covering the wavelength range concerned must be available.

The earliest measurements of optical excitation functions yielded only qualitative results because of inadequate appreciation of these requirements, but in 1930 Hanle and Schaffernicht[‡] made a detailed study of the excitation of mercury by electrons with energies up to 60 eV

[†] HEDDLE, D. W. O. and LUCAS, C. B., *Proc. R. Soc.* **A271** (1963) 129.
[‡] HANLE, W. and SCHAFFERNICHT, W., *Annln Phys.* **6** (1930) 905.

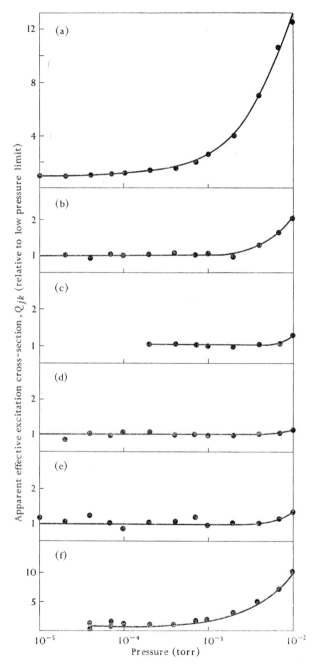

Fig. 4.4. Apparent effective cross-sections for excitation of various helium lines (Heddle and Lucas) as a function of pressure, p. The transitions and electron energies are: (a) 3^1P–2^1S (84 eV); (b) 4^1S–2^1P (87 eV); (c) 4^3S–2^3P (86 eV); (d) 3^3P–2^3S (85 eV); (e) 4^1D–2^1P (35 eV); (f) 4^3D–2^3P (88 eV). Transitions (b) and (f) are enhanced by collisional transfer of excitation.

using a considerably improved technique. In 1932 Thieme† employed the same technique to measure optical excitation functions for helium up to energies as high as 400 eV and his results are still regarded as basic in the analysis of inelastic cross-sections for helium. One of the important features of Thieme's work was the low pressure used (0·005 torr), which gave it an advantage over useful work carried out a year earlier by Lees.‡

Further developments in the experimental study of optical excitation did not begin until over twenty years later with the application of new techniques to improve the homogeneity of electron energy and the light-collecting methods, the replacement of photographic detection by the use of photomultipliers, and the employment of selective amplification in conjunction with imposed modulation to reduce background noise. This work has established beyond doubt that in helium the excitation cross-sections at energies less than 20 eV above the threshold are by no means smooth functions of energy.

A further major development has been the use of the modulated atomic-beam technique (see Chap. 1, § 4.2 and Chap. 3, § 2.3) to determine the excitation functions for the Ly α- and Hα-lines of atomic hydrogen.

1.3.2. *The measurement of the polarization of impact radiation.* The history of the study of the polarization of impact radiation is even more remarkable. In 1927 Skinner and Appleyard§ carried out a systematic study of the polarization of a number of impact-excited lines of mercury. Some similar observations were carried out at about the same time for helium, but at too high a pressure for the results to refer to single collisions. No further work was carried out until in 1957 Lamb and Maiman,∥ in the course of measurements of the fine structure separation of the 3^3P states, measured the excitation cross-section and polarization of the 3^3P–2^3S (λ 3889) transition at a sufficiently low pressure (3×10^{-3} torr) for the results to be valid. The result obtained for the polarization near the threshold, although finite, was considerably smaller than the theoretical expectation. Subsequent investigations†† gave results indicating practically zero polarization near the threshold for most transitions, although non-zero values were obtained for the 3^3P–2^3S (λ 3889)

† THIEME, O., *Z. Phys.* **78** (1932) 412.
‡ LEES, J. H., *Proc. R. Soc.* **A137** (1932) 173.
§ SKINNER, H. W. B. and APPLEYARD, E. T. S., *Proc. R. Soc.* **A117** (1927) 224.
∥ LAMB, W. E. and MAIMAN, T. H., *Phys. Rev.* **105** (1957) 573.
†† HEDDLE, D. W. O. and LUCAS, C. B., *Proc. R. Soc.* **A271** (1963) 129; HUGHES, R. H., KAY, R. B., and WEAVER, L. D., *Phys. Rev.* **129** (1963) 1630; MCFARLAND, R. H. and SOLTYSIK, E. A., *ibid.* **127** (1962) 2090; **128** (1962) 1758, 2222; **129** (1963) 2581.

transition† and for the (4^1D-2^1P) ($\lambda\,4922$) and (5^1D-2^1P) ($\lambda\,4388$) transitions.‡ Considerable effort has been devoted to resolving this difficulty, since the theoretical polarization at the threshold can be derived on very general grounds. Heddle and Keesing§ have shown that the anomaly does not arise if electrons of very homogeneous energy are used to excite the radiation. It is then found that the polarization approaches the theoretical value at threshold but falls sharply at electron energies only about 0·5 eV above the threshold.

Fig. 4.5 shows the effect of the working pressure on the measured values of the polarization.∥ The polarization of the radiation decreases markedly if the working pressure is so high that secondary processes in the formation of the excited state become important. Naturally the maximum permissible working pressure is approximately the same for both excitation cross-section and polarization measurements.

1.3.3. *Measurements using gas or vapour in bulk.* We shall now describe some selected experiments to illustrate the considerations involved, beginning with that of Hanle and Schaffernicht.†† The arrangement of their experiment is illustrated in Fig. 4.6.

The apparatus used is simple in principle. Electrons from an oxide-coated cathode K, after acceleration through 80 V, passed through the slit A_1 and were then retarded to the required energy by a suitable potential on A_2. After suffering collisions in the collision chamber C, they were collected by the Faraday cylinder A_3 at the same potential as A_2, and about 1 cm from it. Atoms within C were excited and emitted radiation in their subsequent reorganization.

The electrode A_2 was electrically connected to the cylinder A_4, which surrounded the whole of the collision space C, thus ensuring that it was free of electrical fields. The inside of the Faraday cylinder was screened by a fine wire mesh to prevent secondary or reflected electrons from entering the collision space. Radiation from this space was passed through the glass envelope of the tube and its intensity measured.

Hanle and Schaffernicht carried out an absolute intensity measurement. After leaving the tube the light was incident on a lens, which brought it to a focus on a photo-cell after passage through a suitable monochromatic filter. In this way a measure could be obtained of the intensity of a single line excited in the collision space C. A mercury

---

† HUGHES, R. H., KAY, R. B., and WEAVER, L. D., loc. cit., p. 178.
‡ MCFARLAND, R. H., *Phys. Rev.* **136** (1964) A1240.
§ HEDDLE, D. W. O. and KEESING, R. G., *Proc. R. Soc.* A**299** (1967) 212.
∥ HEDDLE, D. W. O. and LUCAS, C. B., loc. cit., p. 178.
†† HANLE, W. and SCHAFFERNICHT, W., *Annln Phys.* **6** (1930) 905.

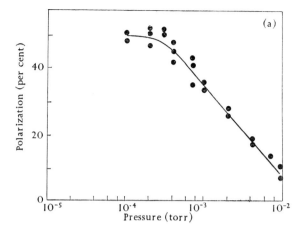
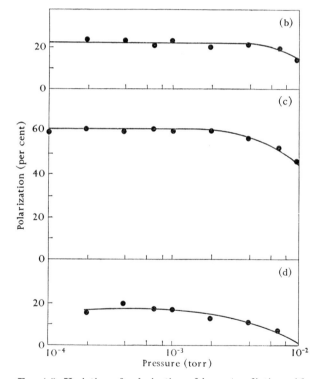

Fig. 4.5. Variation of polarization of impact radiation with pressure in helium (Heddle and Lucas). The transitions and electron energies are: (a) 3^1P-2^1S (34 eV); (b) 3^3P-2^3S (60 eV); (c) 4^1D-2^1P (40 eV); (d) 4^3D-2^3P (33 eV). Collisional transfer causes depolarization in transitions (b) to (d).

lamp whose intensity could be compared with that of a standard Hefner lamp by means of a thermo-element was used to make the measurement absolute. Light from the lamp was focused on the photocell after passing through a neutral filter, the known transmission of which was of suitable magnitude to make the intensity from the mercury lamp comparable with that of the monochromatic radiation reaching the photocell from

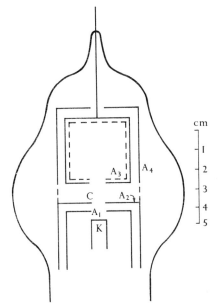

FIG. 4.6. Hanle and Schaffernicht's apparatus for the study of optical excitation cross-sections.

the collision space C. Care had, of course, to be taken to exclude reflected radiation from entering the photocell.

Hanle and Schaffernicht's measurements were made on a number of mercury lines in the visible region. For such lines no errors were encountered due to the resonance absorption of the radiation before leaving the tube. Absolute intensity measurements had previously been carried out by Bricout† for the mercury line λ 2537 (6^1S_0–6^3P_1) in the ultra-violet, but since this line arises from a transition to the ground state of the mercury atom, the possibility of resonance absorption of the radiation before leaving the tube is high and Hanle and Schaffernicht considered that Bricout's measurements were invalidated by this effect.

† BRICOUT, P., *J. Phys. Radium, Paris* **9** (1928) 88.

A general arrangement similar to that described above has been employed by many investigators, except that a spectrograph has been used to isolate the various spectral lines emitted.

An independent measurement of the absolute intensity of the mercury line λ 4358 (6^3P_1–7^3S_1) was made by Fischer.† He used an electron tube of similar design to that of Hanle and Schaffernicht, but isolated the mercury line by means of a spectrograph and compared its intensity photometrically with that in a narrow band of the continuous spectrum around this wavelength emitted from a tungsten incandescent lamp at a temperature of 1700° C. From Wien's law he calculated the energy in the given wavelength band incident on the spectrograph slit from the tungsten lamp and, knowing the dispersive power of the spectrograph, was able to calculate from the photometric data the absolute intensity of the λ 4358 line emitted from the electron tube. For electrons of energy 60 eV Fischer obtained a cross-section for emission of this line of $8 \cdot 04 \times 10^{-18}$ cm², in very good agreement with the independent value of $8 \cdot 25 \times 10^{-18}$ cm² obtained earlier by Hanle and Schaffernicht.

A considerable amount of data on absolute excitation cross-sections has been obtained by Hanle and his collaborators‡ using the absolute intensity of this line of mercury as a standard. The gases studied were present, mixed with mercury vapour, in an apparatus of the type described above (Fig. 4.6). Knowing the partial pressure of the mercury vapour, an intensity measurement of the lines being investigated relative to the λ 4358 line of mercury sufficed to determine their absolute cross-section for excitation.

The arrangement used by Skinner and Appleyard in 1928§ to measure the polarization of a number of lines excited by electron impact in mercury vapour is illustrated in Fig. 4.7.

The radiation emitted from a tube E by the passage of electrons of a definite energy through mercury vapour was focused on the slit S by means of the quartz lens L. It then entered the quartz prism, where it was split into two components by double refraction at the oblique face, the angular separation between the two components being about $\frac{1}{3}°$. The light then entered the quartz prism R of the spectrograph and two images of the slit corresponding to the two components were produced on the photographic plate P. From the relative intensity of these two images the polarization of the radiation could be calculated.

† FISCHER, O., Z. Phys. 86 (1933) 646.
‡ THIEME, O., ibid. 78 (1932) 412; HAFT, G., ibid. 82 (1933) 73.
§ loc. cit., p. 178.

Care had to be taken to prevent spurious effects arising from differential reflection of the two components of the radiation at the prism R that would cause an error up to 25 per cent in the polarization. To avoid this the quartz prism D was cut so that its optic axis was parallel to the direction AB. After the beams had been separated by double reflection the light then emerged from D along the optic axis. Large rotation effects were produced that varied from different parts of the beam and

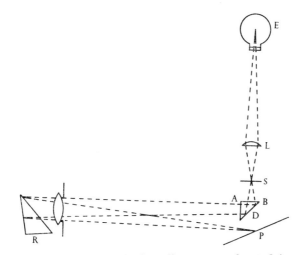

Fig. 4.7. Skinner and Appleyard's apparatus for studying the polarization of impact radiation.

thus the beam was effectively depolarized. Experiments with an unpolarized source at E showed that the error due to apparent polarization produced by the apparatus was always less than 5 per cent.

The most recent investigations of the excitation of mercury have been carried out by Frisch and by Zapesochnyi† and by Jongerius,‡ but in neither case was the polarization investigated. The latter studied not only the excitation of lines in the visible but also in the ultra-violet, and even attempted some observations in the vacuum ultra-violet on the excitation of the $6^1P \to 6^1S$ resonance line at 1849 Å.

The electrode system used by Jongerius is illustrated in Fig. 4.8. The electron gun consisted of a source in the form of an indirectly heated oxide-coated cathode K operated at a temperature of 700° K, an electrode A_1 at potential V_1 relative to the cathode, and a further electrode

† Frisch, S. and Zapesochnyi, I. P., *Dokl. Akad. Nauk SSSR* **95** (1954) 971; *Izv. Akad. Nauk SSSR* **19–11** (1955) 5.
‡ Jongerius, H. M., Thesis, Utrecht (1961).

A_2 at potential V_2. These potentials were adjusted to give a nearly parallel electron beam that was accelerated into the measuring chamber A_3 at potential V_3. The beam passed out of this chamber into the collecting chamber A_4 at potential V_4 to impinge at a 45° angle of incidence on the collecting plate A_5 at potential V_5. By trial it was found that the most suitable choice of potentials (in volts) at the working pressure range

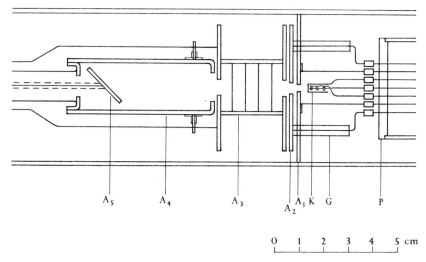

Fig. 4.8. Electrode system used by Jongerius for studying the excitation of mercury impact radiation.

(2×10^{-4} to $1 \cdot 25 \times 10^{-3}$ torr) were $V_1 = 15, V_2 = V_4 = 23, V_5 = 29$. Dimensions of the system are indicated in Fig. 4.8. The beam current was taken as the sum of the currents i_3, i_4, i_5 to A_3, A_4, and A_5 respectively. With the chosen potentials, i_3 ranged from about 0·4 to 0·2 of the total current. Retarding potential measurements showed that the energy width of the beam at half intensity was about 0·4 eV over a range 10–60 μA of beam current and pressures from 9×10^{-4} to $1 \cdot 4 \times 10^{-3}$ torr.

The measurements in the visible region were made as usual by observing at 90° to the beam, using a glass spectrograph as monochromator and measuring the intensity with a photomultiplier. Because of the low beam current and vapour pressures employed the intensity of the light was very low, so that to eliminate the dark current in the multiplier the light beam was mechanically interrupted fifty times a second in order that the wanted signal could be selected by a suitable amplifier (cf. Chap. 3, § 2.3 where a similar method is used in atomic beam collision experiments).

For the ultra-violet measurements the glass spectrograph was replaced by a grating monochromator and recorder, the general arrangement being as shown in Fig. 4.9. The grating could be rotated from outside by means of a drum calibrated in wavelengths so that it was possible to record the spectrum for a fixed electron energy. Alternatively,

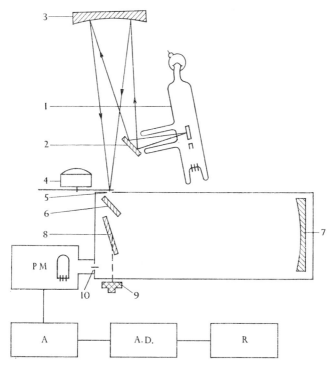

Fig. 4.9. Arrangement of Jongerius for analysing intensity of ultra-violet radiation. 1, excitation tube; 2, 6, flat mirrors; 3, 7, concave mirrors; 4, synchronous motor with light chopper; 5, 10, slits; 8, grating; 9, wavelength drum; PM, photomultiplier and pre-amplifier; A, amplifier; $A.D.$, amplifier-detector; R, recorder.

by means of a circular potentiometer driven by a small synchronous motor, the accelerating voltage could be raised linearly at about 1 V per minute so that, by fixing the wavelength appropriately, the optical excitation functions could be recorded. Correction to such a record had to be made to allow for the variation of beam current with energy.

An interesting effect was observed and corrected for in connection with the dependence of excitation function on beam current and vapour pressure. A typical excitation function exhibits maxima and minima (see Fig. 4.10). It was found that, whereas the intensity at a particular maximum or minimum varies linearly with beam current and vapour pressure over the experimental range (see Fig. 4.11 (a), (b)), the excitation function as a whole is displaced along the energy scale by an amount depending on both these variables (see Fig. 4.10). This was ascribed by Jongerius to space charge effects in the beam, which lead to modification

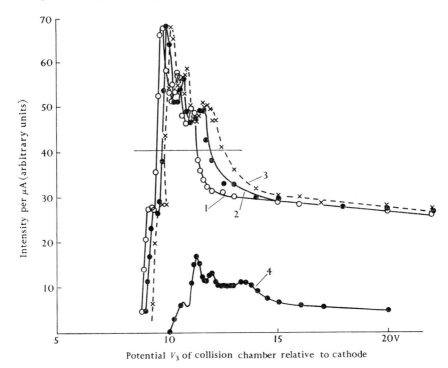

Fig. 4.10. Excitation curves obtained by Jongerius for excitation of the mercury green line (λ 5461) under different operating conditions: (1) beam current $i = 55\ \mu\text{A}$, pressure $p = 1\cdot 25 \times 10^{-3}$ torr. (2) $i = 35\ \mu\text{A}$, $p = 1\cdot 25 \times 10^{-3}$ torr. (3) $i = 17\ \mu\text{A}$, $p = 1\cdot 25 \times 10^{-3}$ torr. (4) $i = 17\ \mu\text{A}$, $p = 2 \times 10^{-4}$ torr.

of the potential in the measuring chamber. To correct for this, results were extrapolated to zero beam current. After this had been done a further constant correction of 2·25 eV to the energy scale was found necessary due to contact potentials.

High resolution measurements of excitation functions of mercury, using a very homogeneous electron beam having an over-all energy spread of the order of 0·5 eV, have also been carried out by Zapesochnyi and Shpenik.†

† ZAPESOCHNYI, I. P. and SHPENIK, O. B., *Dokl. Akad. Nauk SSSR* **160** (1965), 1053; *Soviet Phys. Dokl.* **10** (1965) 140.

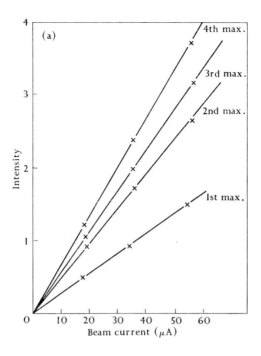
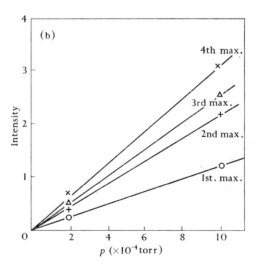

FIG. 4.11. Intensities of first four maxima of the excitation curve of Hg (λ 5461) measured by Jongerius; (a) as function of beam current for a fixed vapour pressure of $1\cdot 25 \times 10^{-3}$ torr; (b) as function of pressure for a fixed beam current.

Smit, Heideman, and Smit† employed a very similar apparatus to that of Jongerius for the measurement of optical excitation functions in helium. The electrode system was essentially the same but the voltages were chosen somewhat differently so as best to fit the conditions. They took V_1 between 15 and 40 V, $V_2 = 6$ V, $V_4 = (V_3-4)$ V $V_5 = (V_3+4)$ V. Very good homogeneity in energy of the electron beam was obtained, the half-width being less than 0·4 V. To eliminate the dark signal background from the photomultiplier they applied a voltage to the measuring cage that varied as shown in Fig. 4.12. The voltage

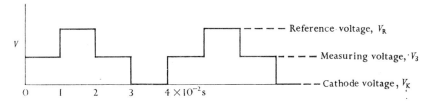

FIG. 4.12. Wave form of voltage applied to measuring cage in experiments of Smit, Heideman, and Smit.

V_R was a standard reference voltage, V_3 the measuring voltage, and V_K the cathode voltage and the duration of each square-topped pulse was 10^{-2} s. The pulses received by the anode of the multiplier when each of the voltages V_R, V_3, V_K were applied could be separately recorded so that, by taking differences, corrections could be made for the dark current of the multiplier and for any changes of sensitivity with time.

At the working pressure ($5 \cdot 3 \times 10^{-3}$ torr) and beam current (15–16 μA) space charge effects were very small in these experiments.

Heddle and Lucas‡ have observed both optical excitation functions and polarizations in helium, using an electrode system similar to that used by Gabriel and Heddle in an earlier investigation. The general arrangement of their experiment is illustrated in Fig. 4.13. In the electron gun the voltages were chosen so as to provide a nearly parallel electron beam through the measuring chamber.

Square wave modulation of the electron beam at frequency $f = 400/3$ c/s was used so that selective amplification could be used to eliminate the dark background signal from the photomultiplier.

The light passed out of the vacuum chamber through a fused silica plate and was collimated by a silica-lithium fluoride achromat. After

† SMIT, C., HEIDEMAN, H. G. M., and SMIT, J. A., *Physica* **29** (1963) 245.
‡ HEDDLE, D. W. O. and LUCAS, C. B., *Proc. R. Soc.* **A271** (1963) 129.

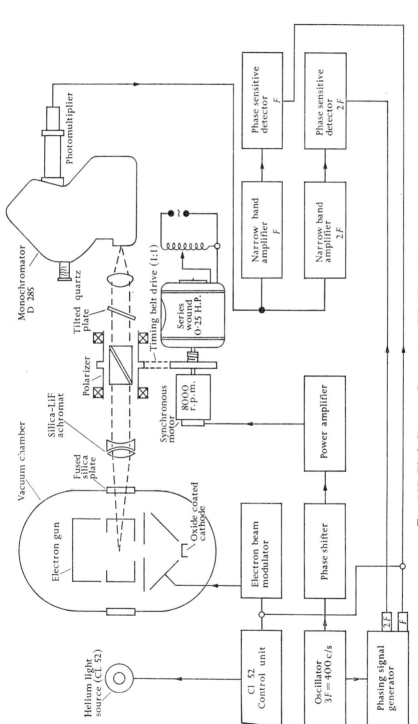

FIG. 4.13. Block diagram of apparatus of Heddle and Lucas.

passage through the polarizer the light was focused on the entrance slit of the monochromator by a silica lens.

An ingenious method was used to determine the polarization. The polarizer was rotated in the light path at a frequency $2f$ so that a signal of this frequency and amplitude, proportional to the difference $I_\parallel - I_\perp$ of the intensities of the polarized components, was superposed on the total light signal modulated at frequency f. It can be shown that the polarization is given by

$$P = \frac{12a_2}{3\pi a_1 + 4a_2} \times 100, \qquad (21)$$

where a_1, a_2 are the amplitudes of the sinusoidal components of frequencies f, $2f$ respectively. These amplitudes were measured by passing the photomultiplier signal through two narrow band amplifiers tuned to these respective frequencies.

For the visible region HN32 polaroid was used as polarizer but for the study of the 3^3P–2^3S transition at 3889 Å it was replaced by a 1-in Glan prism. Correction was made for polarization produced by reflection in the monochromator.

The background pressure, before admission of helium, was about 5×10^{-7} torr and the apparatus was continuously pumped while helium circulated through at the working pressures.

Comprehensive series of measurements of absolute electron excitation cross-sections of helium have been carried out by St. John and his collaborators[†] and by Zapesochnyi.[‡] To eliminate any important contribution from excitation transfer either directly, or indirectly through its effect on cascading, measurements were made at pressures of 10^{-3} torr or less and in the case of ^1P states, whose populations are especially affected by imprisonment of resonance radiation, the working pressures were kept down to 10^{-4} torr by St. John et al. and to 6×10^{-5} torr by Zapesochnyi.

Although the study of the excitation of the 2^1P and 2^3P levels is of special interest no measurements were carried out until very recently, because the radiation that is emitted in transitions to the respective 2^1S and 2^3S levels is in the infra-red, the wavelengths being 20 582 and 10 832 Å respectively. It is true that transitions may take place from 2^1P to the ground 1^1S level, but this is far in the ultra-violet and its

[†] St. John, R. M., Lin, C. C., Stanton, R. L., West, H. D., Sweeney, J. P., and Rinehart, E. A., Rev. scient. Instrum. **33** (1962) 1089; St. John, R. M., Bronco, C. J., and Fowler, R. G., J. opt. Soc. Am. **50** (1960) 28; St. John, R. M., Miller, F. L., and Lin, C. C., Phys. Rev. **134** (1964) A888.

[‡] Zapesochnyi, I. P., Astr. Zh. **43** (1966) 954 (Soviet Astr. **10** (1967) 766).

measurement is again faced with considerable experimental difficulties. However, Jobe and St. John† have been able to measure the intensity of emission of the infra-red radiation, using lead sulphide detectors, even though the most sensitive of these is about 10^{-4} of that obtained in the visible with photomultipliers. The electron beam they used, produced from a pentode-type electron gun, was cylindrical in form, about 1 cm in diameter, and carried a current of from 2 to 10 mA. As in the optical region, absolute calibration was made against a tungsten ribbon standard lamp, the emissivity of tungsten in the infra-red being obtained from the work of de Vos. The light from the lamp was chopped mechanically in phase with the electronically chopped electron beam, so that comparison with the standard could be made either simultaneously with, or immediately following, an observation. Wavelength selection was carried out either by the use of interference filters or grating monochromators.

A further set of experiments designed especially to measure the polarization of impact radiation near the threshold (see § 1.3.2) but which also give optical excitation functions are those of MacFarland and his collaborators.‡ Fig. 4.14 illustrates the electrode system employed in these investigations. The first accelerating electrode, 1, is at $+5$ V relative to the cathode, the second, 2, at $+15$, and the third, 3, and fifth, 5, at the measuring voltage. The fourth, 4, is at the same potential as the first, apart from a square-topped modulation that blocks the beam 100 times a second, so that the usual selective-amplifier technique may be employed in discriminating against background effects in the photomultiplier.

A cylindrical gold grid was placed coaxially within the measuring chamber and insulated from it so that the current to the grid could be measured. In this way an estimate could be made of the intensity of electrons scattered out of the beam.

An axial magnetic field of 10–15 gauss was used to confine the beam. At a pressure of 10^{-7} torr and an accelerating voltage of 25 V this reduced the current collected by the gold grid to 0·5 per cent of the beam current. At a pressure of $2 \cdot 5 \times 10^{-3}$ torr of helium this ratio increased to 6·6 per cent. Magnetic fields for collimation in experiments designed for polarization measurements must be kept small in order to prevent a significant transverse component of the electron velocity. The maximum pitch angle of the helical electron path must be kept small ($\lesssim 10°$).

† JOBE, J. D. and ST. JOHN, R. M., *Phys. Rev.* **164** (1967) 117.
‡ MCFARLAND, R. H. and SOLTYSIK, E. A., *Phys. Rev.* **127** (1962) 2090; **128** (1962) 1758, 2222; **129** (1963) 2581; MCFARLAND, R. H., ibid. **133** (1964) A986; **136** (1964) A1240.

Although all the surfaces seen by the beam were gold-plated there was evidence of surface charging, an effect that increased during the experiment, with the Vacsorb pump in operation. Replacement of the triple carbonate cathode first used, by caesium-impregnated tungsten, reduced the effect, presumably by providing a continuous supply of metal.

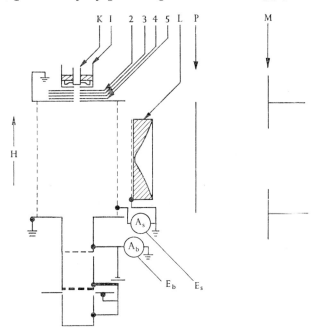

FIG. 4.14. Schematic diagram of apparatus of McFarland and collaborators. K, cathode; 1–5, accelerating electrodes; L, cylindrical lenses; P, polaroid analyser; M, monochromator; E_b, E_s electrometer for beam current and scattered current respectively. A square-wave voltage applied to 4 interrupted the electron beam periodically.

An important new feature of the optical design was the use of a fused quartz cylindrical lens to focus the light on to the slit of the monochromator. This increased the intensity at the photomultiplier by a factor of about 10 so that it was convenient to work under conditions of very low gas pressure and beam current.

A KN-36 polaroid analyser was used to measure the polarization. Correction was made for any polarization by reflection in the monochromator. After baking and before filling with gas a pressure as low as 10^{-9} torr was obtained. During an experiment the three-filter Vacsorb pump continued to operate so as to remove impurities generated during the run.

Judged by the energy resolution of the excitation functions and the sharpness of the threshold behaviour the homogeneity of the electron beam in the measuring chamber was remarkably good, the effective spread being not more than 0·2 eV.

Errors in the form of variation of the excitation cross-section and polarization with electron energy may arise due to variations in the experimental conditions, such as, for example, slow drifts in the amplifier gain, gas pressure, and beam current, and changes in contact potentials during the course of the experiment. Heddle and Keesing,[†] in order to overcome such difficulties, developed an automatic data-logging system that spanned the whole range of electron energies many times in rapid succession during the course of a run, so that any long-term variation in conditions was averaged out.

The apparatus they used was similar to that of Heddle and Lucas.[‡] The collision-chamber potential was obtained from a 10-turn helical potentiometer driven by a reversible stopping motor and was varied in steps of 0·05 V from about 0·5 V below threshold to 2 V above threshold and then in steps of 0·1 V to 4·7 V above threshold. The potential was then reduced by the same intervals and the process repeated some 100 times. Pieces of polaroid sheet oriented in the two principal directions were placed alternately in the light beam. The 'parallel' component was measured in the ascending voltage half-cycle and the 'perpendicular' component in the descending half-cycle. The system was controlled by a mains-driven digital clock. At each voltage the signal was recorded in a scaler for 1·6 s. The count registered was punched on to a paper tape during the next 0·4 s, the scaler cleared, and the collision chamber potential changed. Each complete cycle occupied $4\frac{1}{2}$ min. The data for all counts at a given collision-chamber potential and given plane of polarization were then summed by means of a computer.

The background was determined from the signal corresponding to collision-chamber potential below threshold. This was subtracted from the signal at the various data points to obtain the true signal.

1.3.4. *Crossed-beam methods.* Fite and his associates§ measured the excitation function for the Lyman α-line (1216 Å) of atomic hydrogen by the modulated crossed-beam technique. The apparatus was essentially the same as for the measurement of elastic scattering by atomic hydrogen (Chap. 1, § 4.2) but with a photon detector replacing the

---

[†] loc. cit., p. 179.   [‡] loc. cit., p. 176.
§ FITE, W. L. and BRACKMANN, R. T., *Phys. Rev.* **112** (1958) 1151; FITE, W. L., STEBBINGS, R. F., and BRACKMANN, R. T., ibid. **116** (1959) 356.

electron collector. This detector was a Geiger–Müller counter filled with iodine vapour.

A gas filter was placed in front of the counter. This consisted of a cell with lithium fluoride windows through which a stream of oxygen gas passed. The filter absorbed all radiation in the wavelength range detectable by the counters except for seven 'windows', one of which occurs at about 1216 Å. Ultra-violet radiation excited by electron collisions with the residual gas in the vacuum chamber was strongly attenuated but the cell was transparent to Lyman α-radiation.

Even so the ultra-violet radiation from residual gases at a pressure of 10^{-6} torr was large. The same type of modulation techniques was therefore employed as in other crossed-beam experiments, the atomic hydrogen beam being chopped at a frequency of 100 c/s. The output of the G.–M. counter was treated as an a.c. signal. This produced very large shot-effect noise because one was dealing with random pulses from the counter containing 10^{10} electrons instead of single electrons. By integrating the signal over a time of 40 s, however, a signal to noise ratio of about 15 : 1 could be obtained.

The density of atoms in the atomic beam was monitored by measuring the current of H^+, using the mass spectrometer as in the elastic scattering measurement (Chap. 1, § 4.2). Measurement of the H_2^+ current enabled the H_2 contamination of the beam (usually 3–4 per cent) to be estimated.

The total cross-section Q_1 for excitation of the Lyman α-line is given in terms of the flux $q(\frac{1}{2}\pi)$ of quanta observed per unit solid angle in a direction normal to the electron beam by (see eqn (10))

$$Q_1 = 4\pi(1-\tfrac{1}{3}p)q_1(\tfrac{1}{2}\pi).$$

To measure p it is necessary to make observations of the flux at two angles. For this purpose the photon counter was mounted so that its axis made an angle of 45° with that of the neutral beam, while the electron gun was mounted on a table that permitted rotation about the axis of the beam so that θ could be varied from 45° to 135°. In practice p was determined by observations at 45° and 90° respectively. As explained in § 1.2, $100p$ is the percentage polarization of the radiation emitted at 90°, so that the observations yielded not only the relative optical excitation function but also the polarization, though the accuracy of the determination of the latter was poor (see Fig. 4.15).

An alternative method of determining the relative values of Q_1 at different energies, used as a check, made use of the fact that at the angle ϑ for which $\cos^2\vartheta = \frac{1}{3}$, $\vartheta = 54\cdot5°$ the signal $S(\vartheta)$ received is proportional to Q_1 (compare eqns (9), (10)).

No attempt was made to determine the absolute values of the excitation cross-section but measurements were carried out up to electron energies of 500 eV, high enough for theoretical values given by Born's first approximation to be valid, and this was used for normalization (see Chap. 7, § 5.6.1).

Chamberlain, Smith, and Heddle[†] have used a modulated crossed-beam method to study the variation of the excitation cross-section for Lyman α-radiation near the threshold, using an ionization chamber filled with nitric oxide and having a lithium fluoride window of diameter 16 mm. The excitation energy is 11·6 eV while the threshold for excitation of H_2 to states that produce radiation detectable in the chamber is 13·0 eV, so that in the range of 1·4 eV above threshold background effects were greatly reduced. An estimate was made of the production of Lyman α-radiation due to quenching of the 2s state by static electric or magnetic fields in the interaction volume and found to be negligible.

In 1930 Ornstein and Lindeman[‡] measured excitation functions for the Hα-, Hβ-, and Hγ-lines of atomic hydrogen. They produced a high degree of dissociation of hydrogen in a Wood's tube and allowed the atomized gas to diffuse through a short tube into a region across which an electron beam was passed. The intensity of excitation of the appropriate atomic lines was then observed as a function of electron energy.

Kleinpoppen, Kruger, and Ulmer[§] have applied crossed-beam methods to study the excitation of the Hα-line. This technique was very similar to that of Fite *et al.* but, as they were dealing with visible radiation, they observed the light emission normal to the electron beam isolating the desired wavelength by an interference filter and passing it through a polarizer so that the polarization could also be measured. From these observations the relative cross-sections could be derived by using (10). The measurements have been extended to measure the polarization of the resonance lines of ^{6}Li, ^{7}Li, and ^{23}Na.[||]

A description of the methods used to determine the cross-section for excitation of the 2s metastable states of atomic hydrogen is given in § 2.

[†] Chamberlain, G. E., Smith, S. J., and Heddle, D. W. O., *Phys. Rev. Lett.* **12** (1964) 647.
[‡] Ornstein, L. S. and Lindeman, H., *Z. Phys.* **80** (1933) 525.
[§] Kleinpoppen, H., Kruger, H., and Ulmer, R., *Phys. Lett.* **2** (1962) 78.
[||] Hafner, H., Kleinpoppen, H., and Kruger, H., *Proc. 4th Int. Conf. Phys. electron. atom. Collisions*, 1965, p. 386 (Science Bookcrafters, New York).

1.4. *Observed results of measurement of optical excitation functions and polarizations*

Measurements of the cross-section for excitation of spectral lines by electron impact have been carried out for H, He, Li, Ne, A, Kr, Xe, Na, Cs, Rb, Zn, Cd, Hg, Tl, Ag, and Pb.

1.4.1. *Results of measurements in atomic hydrogen.* The excitation function for the Ly α-line observed by Fite, Stebbings, and Brackmann† is illustrated in Fig. 4.15 (a). Estimates that have been made, using theoretical cross-sections for excitation of higher states, of the contribution to the population of the $2p$ states by optical transitions from such states suggests that this is negligible, so that the observed curve should give correctly the variation with energy of the cross-section for direct excitation of the $2p$ state.

Fig. 4.15 (b) illustrates the rather inaccurate values of the polarization of the emitted radiation obtained in the course of these experiments. The measured values of the cross-section obtained by Chamberlain *et al.*‡ for the excitation of the Ly α-line near the threshold are given in Fig. 4.16. The cross-section rises very steeply to a maximum, decreases to a minimum in about 0·3 eV and then rises again. In this experiment the electron energy distribution was measured by a modulated retarding potential technique and showed a width of 0·35 eV at half-maximum intensities. The dotted curve is the shape expected for the true excitation function obtained from the solid curve and allowing for this electron energy distribution. The solid curve can be expressed by the function

$$Q_1(1s-2p) = 9 \cdot 50(E-11 \cdot 62)^2 - 6 \cdot 00(E-11 \cdot 62) + 7 \cdot 531, \qquad (22)$$

where E is in eV.

Fig. 4.17 shows the observed optical excitation functions for the Hα-, Hβ-, and Hγ-lines. For Hα comparison is made between the earlier result of Ornstein and Lindeman and the later work of Kleinpoppen, Kruger, and Ulmer using the crossed-beam method. The scales are adjusted so that the maxima are of equal magnitude. It will be seen that the agreement is poor. The polarization of the Hα-radiation observed by Kleinpoppen *et al.* is shown in Fig. 4.18.

Theoretical discussion of these results is given in Chapter 7, § 5.6.1 and Chapter 8, § 4.3.

1.4.2. *Results of measurements in helium.* It is convenient to divide consideration of the observed results into two electron energy ranges, above and below about 35 eV respectively. In the higher-energy range

† loc. cit., p. 193.   ‡ loc. cit., p. 195.

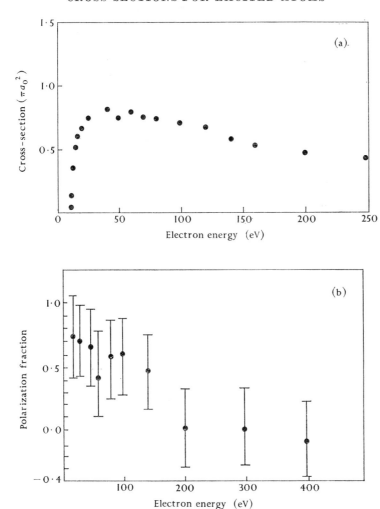

Fig. 4.15. Excitation of Ly α-radiation as measured by Fite, Stebbings, and Brackmann. (a) Excitation cross-section. (b) Polarization fraction, p.

the optical excitation functions and polarization vary smoothly with energy, at least as far as present observations show, and comparison may be made between the results of different observers without paying attention to the energy spread of the electron beam in the measuring chamber. This spread is very important in the lower range in which the excitation functions show structure.

1.4.2.1. *Optical excitation functions for electron energies greater than* 35 eV. The striking feature of the observations is the similarity of

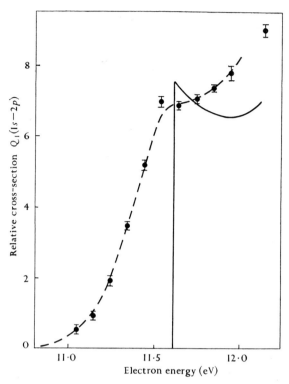

FIG. 4.16. Measurement of Chamberlain, Smith, and Heddle of the relative Ly α-excitation cross-section near the threshold. The broken curve that is fitted to the experimental points is obtained by folding in the electron energy distribution to the excitation function (eqn (22)), shown as a solid line in the figure. The energy scale relates to applied cathode voltage, uncorrected for contact potentials.

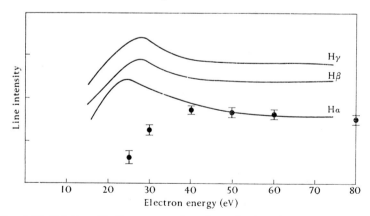

FIG. 4.17. Relative effective cross-sections for excitation of Balmer α-, β-, and γ-radiation (Ornstein and Lindeman). ● Kleinpoppen, Kruger, and Ulmer (Hα only).

the form of the optical excitation functions for lines whose upper states have the same multiplicity and total orbital angular momentum. In fact the results may be represented quite well by writing for the apparent excitation cross-sections for lines of the same series

$$Q_j(E) = M_j q(E), \qquad (23)$$

where j distinguishes a particular line. $q(E)$ is the characteristic excitation function for the series normalized to a maximum value of unity.

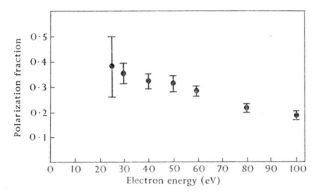

FIG. 4.18. Polarization fraction for Balmer α-radiation (Kleinpoppen, Kruger, and Ulmer).

M_j is a quantity independent of electron energy, E, which determines the scale for the particular line concerned.

Fig. 4.19 shows the forms of $q(E)$ for the ^1S, ^3S, ^1P, ^3P, ^1D, and ^3D series as derived from the observations of St. John, Miller, and Lin.† The main features that are shown here are certainly correct. Thus the cross-sections for excitation of optically allowed transitions (cf. the ^1P series) rise to a maximum at a higher energy (3–4 times the threshold value) than for any other transitions. For transitions associated with a quadrupole moment (the ^1D series) the maximum occurs at about 2–2·5 times the threshold. The excitation function for triplet states (intercombination transitions) are distinguished, not only by possessing maxima very close to threshold, but also by the rapid decrease of the cross-section at high energies. Except for the observations of Jobe and St. John for the excitation of 2^3P, the observed rate of decrease, though rapid, is much slower than expected theoretically (see Chap. 7, Table 7.4). In all the singlet states the rate of decrease at these energies is much slower.

† loc. cit., p. 190.

When we turn to detailed comparison of the excitation functions obtained by different observers we see that considerable discrepancies still exist. This is illustrated in Fig. 4.20, which gives a comparison of results obtained for the excitation of the 3^1P–2^1S, 4^1S–2^1P, and 4^1D–2^1P

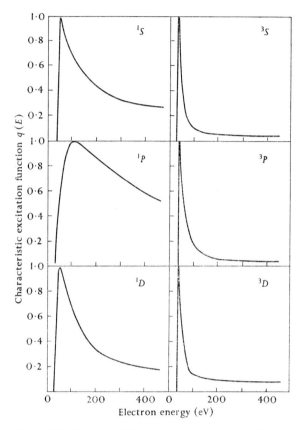

Fig. 4.19. Characteristic excitation functions ($q(E)$ of eqn (23)) for excitation of different types of state in helium.

lines, and also for the corresponding triplet lines. In all cases the magnitudes have been adjusted so that, for the singlet terms, the maximum values agree. It will be seen that in most cases there is a considerable spread of results. For the triplet terms this would be expected, at least at the higher energies, because it is very likely that under these conditions most of the excitation arises from secondary processes as discussed further in Chapter 18. This does not apply to the singlet terms and it

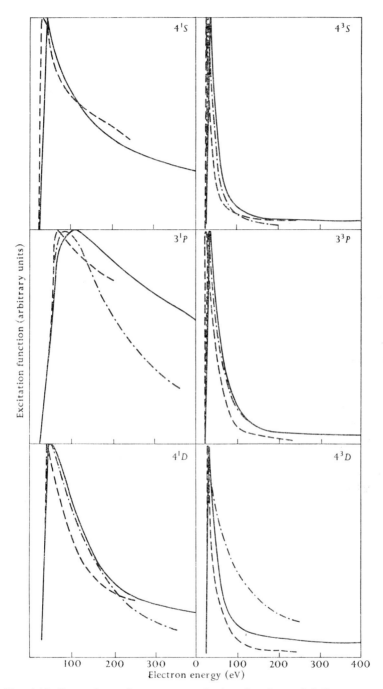

Fig. 4.20. Comparison of apparent excitation functions of helium states observed by different experimenters. ——— St. John, Miller, and Lin; — - — - — Heddle and Lucas; —··—··— McFarland and Soltysik.

may be anticipated that further work using a low working pressure will remove the discrepancies.

Turning now to absolute values of apparent excitation cross-sections, Table 4.1, based on one given by St. John, Miller, and Lin,† gives values obtained for the 4^1S, 3^1P, 4^1D, 4^3S, 3^3P, and 4^3D levels by various authors. These values were obtained from measured cross-sections for the lines (4^1S–2^1P), (3^1P–2^1S), (4^1D–2^1P), (4^3S–2^3P), (3^3P–2^3S), and

TABLE 4.1

Level	4^1S	3^1P	4^1D	4^3S	3^3P	4^3D
Maximum cross-section (10^{-20} cm²)						
Zapesochnyi[1]	24·5	530	28	40	105	14·4
St. John, Miller, and Lin[2]	24	350	17·6	35	97	12·0
Yakhontova[3]	20		17·8	25	83	12·4
Stewart and Gabathuler[4]	27·5	4130	24	37	105	18
Lees[5]		4360	15·1	36	80	15·2
Thieme[6]	28	3660	32	64	1890	23
Cross-section at 108 eV (10^{-20} cm²)						
Gabriel and Heddle[7]	16·5	457	12	4·4	11	4·6
St. John, Miller, and Lin[2]	15	350	12	3·4	15·3	1·64
Zapesochnyi[1]	16·3	530	19	5·0	31	5

(1) ZAPESOCHNYI, I. P., *Astr. Zh.* **43** (1966) 954 (*Soviet Astr.* **10** (1967) 766).
(2) ST. JOHN, R. M., MILLER, F. L., and LIN, C. C., *Phys. Rev.* **134** (1964) A888.
(3) YAKHONTOVA, V. E., *Vest. leningr. gos. Univ.* Ser. Fiz. i Khim. **14** (1959) 27 (Translation 951 by AERE Harwell).
(4) STEWART, D. T. and GABATHULER, E., *Proc. phys. Soc.* **74** (1959) 473.
(5) LEES, J. H., *Proc. R. Soc.* **A137** (1932) 173.
(6) THIEME, O., *Z. Phys.* **78** (1932) 412.
(7) GABRIEL, A. H. and HEDDLE, D. W. O., *Proc. R. Soc.* **A258** (1960) 124.

(4^3D–2^3P), using the theoretical transition probabilities given by Gabriel and Heddle.‡ Values of these probabilities, which are given in Table 4.2, were calculated by Dalgarno and Stewart § for the resonance transitions, while for the other transitions the Coulomb approximation of Bates and Damgaard‖ was used. Even if the absolute values of Table 4.2 are not very reliable it is probable that the relative values are much more so and this is important for the present purpose.

Table 4.1 compares the results obtained by different observers for the peak cross-section and also cross-sections at an accelerating energy of 108 eV for three different groups. All the observers measured light

† loc. cit., p. 190.   ‡ loc. cit., p. 173.
§ DALGARNO, A. and STEWART, A. L., *Proc. phys. Soc.* **76** (1960) 49.
‖ BATES, D. R. and DAMGAARD, A., *Phil. Trans. R. Soc.* **A242** (1949–50) 101.

emitted normally to the electron beam. On the whole the agreement between the results of different observers is reasonably satisfactory. An exception is the case of 3^1P. The large values of Stewart and Gabathuler, of Thieme, and of Lees can be attributed to the fact that they did not work at pressures low enough to eliminate imprisonment of resonance

TABLE 4.2

Radiative transition probabilities in helium

(Units of 10^6 s^{-1})

	2^1P	3^1P	4^1P	5^1P	6^1P	7^1P	8^1P				
1^1S	1780	571	246	127	74·0	46·6	31·0				
2^1S	1·97	13·4	6·81	3·85	2·56	1·60	1·07				
3^1S	18·8	0·25	1·47	0·94	0·57	0·43	0·26				
4^1S	6·60	4·54	0·06	0·30	0·25	0·19	0·14				
5^1S	3·12	2·01	1·49	0·02	0·08	0·07	0·07				
6^1S	1·76	1·07	0·72	0·61	0·01	0·03	0·05				
7^1S	1·21	0·62	0·42	0·30	0·26	0	0·04				
8^1S	0·74	0·41	0·27	0·21	0·15	0·14	0				
								4^1F	5^1F	6^1F	7^1F
3^1D	65·1	0	0·29	0·13	0·09	0·08	0·07	13·8	4·45	2·10	1·22
4^1D	19·3	7·14	0	0·16	0·08	0·05	0·03	0	2·56	1·29	1·04
5^1D	8·89	3·28	1·52	0	0·08	0·04	0·03	0·07	0	0·73	0·43
6^1D	4·94	1·80	0·84	0·47	0	0·04	0·02	0·03	0·05	0	0·26
7^1D	2·63	1·17	0·52	0·29	0·18	0	0·02			0·03	0
8^1D	1·82	0·66	0·38	0·22	0·15	0·14	0				
K_0/P	7·25	1·79	0·70	0·35	0·20	0·12	0·08				
	2^3P	3^3P	4^3P	5^3P	6^3P	7^3P	8^3P				
2^3S	10·2	9·28	5·67	3·08	1·87	1·15	0·79				
3^3S	27·5	1·07	0·71	0·60	0·41	0·29	0·21				
4^3S	9·26	6·42	0·22	0·12	0·14	0·17	0·18				
5^3S	4·33	2·68	2·05	0·06	0·03	0·05	0·07				
6^3S	2·40	1·40	0·90	0·76	0·03	0·01	0·02				
7^3S	1·75	0·86	0·51	0·37	0·36	0·01	0·01				
8^3S	1·18	0·60	0·34	0·25	0·17	0·15	0·01				
								4^3F	5^3F	6^3F	7^3F
3^3D	71·7	0	0·65	0·27	0·06	0·01	0	13·8	4·45	2·10	1·22
4^3D	24·4	6·65	0	0·32	0·16	0·03	0·01	0	2·56	1·29	1·04
5^3D	11·9	3·38	1·27	0	0·16	0·09	0·02	0·07	0	0·73	0·43
6^3D	5·87	1·99	1·05	0·35	0	0·09	0·05	0·03	0·05	0	0·26
7^3D	4·53	1·33	0·60	0·26	0·13	0	0·03			0·03	0
8^3D	3·03	0·98	0·43	0·22	0·10	0·06	0				

radiation, so that their measured cross-sections are abnormally high. The pressure used by Lees was so high (40 μm) that transfer effects were probably also important in his case. The large value found by Thieme for excitation of the 3^3P level is difficult to understand other than as due to calibration errors.

To obtain true cross-sections, corrections have to be made for the effects of polarization and cascading. These corrections vary with electron energy. The polarization correction (eqn (10)) reduces the apparent cross-section, and values based on the measurements of McFarland and Soltysik† (see Fig. 4.29) have been used by St. John *et al.*‡ to correct their data. The correction is zero for S states and in the energy range up to 500 eV varies from 4 to 7 per cent for ^1P levels, 13 to 17 per cent for ^1D levels, 5 per cent for ^3P levels, and 4 to 5 per cent for ^3D levels. The correction was largest for the lowest levels.

The cascade correction reduces the true excitation cross-section still further. It is smaller and more dependent on electron energy for singlet than for triplet levels. St. John *et al.*‡ estimated the following cascade corrections:

3^1S—maximum of 15 per cent at 300 eV.
3^1P—14 per cent at 35 eV dropping to 4 per cent at 100 eV.
^1D levels—maximum in range 5 to 8 per cent at 450 eV.

The cascade corrections for the triplet levels were nearly independent of electron energy. Gabriel and Heddle§ estimated the following values for these at 108 eV:

4^3S, 23 per cent; 5^3S, 7 per cent; 3^3P, 34 per cent; 3^3D, 4 per cent; 4^3D, 17 per cent; 5^3D, 17 per cent.

Fig. 4.21 compares, for the results of St. John *et al.*, the apparent excitation cross-section curves and the true cross-sections after correction for polarization and cascading for excitation of the 3^1S, 3^1P, 3^1D, 3^3S, 3^3P, and 3^3D states of helium.

We consider next the results obtained by Jobe and St. John for excitation of the 2^1P and 2^3P levels using infra-red detectors as described on p. 191. A cascade correction to the apparent excitation functions and absolute apparent excitation cross-sections involved twenty-two states from which cascade population of 2^1P and 2^3P could occur. Because of the relatively low sensitivity of infra-red detectors it was necessary to work at rather higher pressures than in the optical excitation experiments of St. John *et al.*, so that correction to the observed excitation functions for 2^1P had to be made for imprisonment of radiation. This was done by the same method as that used by Gabriel and Heddle.

The remarkable feature of this work is that, for the first time, a triplet excitation function was obtained that fell off as rapidly with electron energy as expected by theory. Referring forward to Fig. 7.29 we see a comparison between the final derived absolute 2^3P excitation function

† McFarland, R. H. and Soltysik, E. A., *Phys. Rev.* **127** (1962) 2090; **128** (1962) 1758.   ‡ loc. cit., p. 190.   § loc. cit., p. 173.

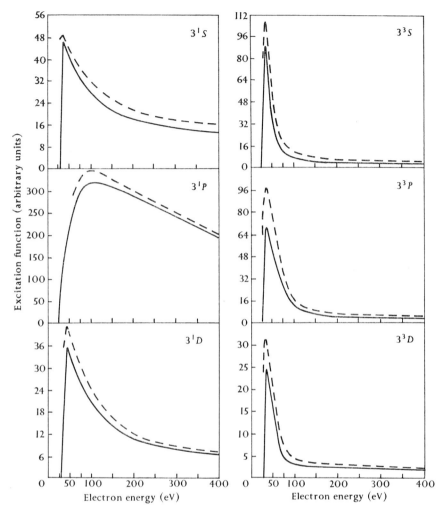

Fig. 4.21. Excitation functions for singlet and triplet states in helium, observed by St. John, Miller, and Lin, showing size of correction for polarization and cascade. —— corrected; — — — uncorrected.

obtained by Jobe and St. John† and the theory that is expected to be valid at electron energies of 200 eV or higher. The agreement at these energies is very good. This is in sharp contrast with the relatively large excitation cross-sections derived for higher triplet states which, at energies of a few hundred eV, are orders of magnitude larger than the theoretical. It seems that this must be due to inadequate allowance for cascade and other indirect means of populating the triplet states.

† Jobe, J. D. and St. John, R. M., loc. cit., p. 191.

The 2^1P excitation cross-section derived by Jobe and St. John is illustrated in Fig. 7.17, again in comparison with theory. The agreement is quite good when it is expected and the disagreement at lower electron energies is equally expected so that one has confidence in the observed results.

TABLE 4.3

Effective principal quantum number, n^*, for terms of helium atoms n

n	^1S	^1P	^1D	^3S	^3P	^3D
1	0·745					
2	1·85	2·01		1·69	1·935	
3	2·855	3·01	2·99	2·695	2·935	2·99
4	3·86	4·01	3·995	3·70	3·92	3·995
5	4·86	5·01	4·995	4·70	4·925	4·995
6	5·865	6·01	5·995	5·70	5·935	5·995
7	6·865	7·01	7·00	6·70	6·925	7·00
8	7·865	8·01	8·00	7·70	7·925	8·00
9	8·865	9·01	9·00	8·71	8·925	9·00
10	9·865	10·01	10·00	9·71	9·925	10·00

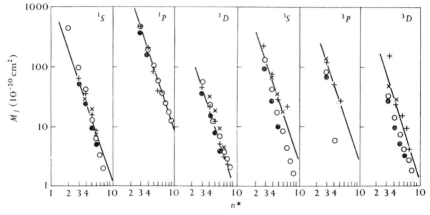

FIG. 4.22. Variation of M_j (eqn (23)) with n^* for excitation of helium states. + Gabriel and Heddle (108 eV); × Thieme (> 30 eV); ● St. John, Miller, and Lin (threshold to 400 eV); ○ Zapesochnyi (maximum values). Straight lines correspond to $M_j \propto n^{*-3}$.

Gabriel and Heddle† found that, for a given type of term the cross-section was proportional to n^{*-3}, where n^* is the effective principal quantum number defined by $n^* = \epsilon^{-\frac{1}{2}}$ where ϵ is the term value in rydbergs. Values of n^* derived from spectroscopic data are given in Table 4.3. This suggests that, if the cross-section can be written in the form (23), then M_j should vary as n^{*-3}. Fig. 4.22 shows that this is

† GABRIEL, A. H. and HEDDLE, D. W. O., *Proc. R. Soc.* A258 (1960) 124.

closely true, both from the results of Gabriel and Heddle and of Thieme. It must be remembered, of course, that a logarithmic scale has been employed and the spread of the data is not so small as it seems.

Using this empirical relation we would have in units 10^{-18} cm^2 $n^{*+3}M_j = 18\cdot9$, 121, 11·6, 31·6, 31·6, 16·3, for ^1S, ^1P, ^1D, ^3S, ^3P, ^3D respectively, the values of n^* being as given in Table 4.3.

A somewhat similar analysis was carried out earlier by Frost and Phelps,† using Thieme's data only. They used the true principal quantum number n in place of n^* and found a less simple relationship of the form $M_j \propto n^{-\gamma}$, where γ varied from 3 to 3·5 for different term systems.

The accuracy of these representations of the helium excitation cross-sections should not be overestimated. While it is likely that the use of the empirical relationship will tend to eliminate inaccuracies in relative values for terms of a given system, the absolute values will not be more accurate than the original observations. Also, while the procedure seems to be effective, for the triplet terms it can only be relied upon to give semi-quantitative results at electron energies very close to the threshold, for reasons already mentioned above. Theoretical discussion of the measurements in helium at the higher energies is given in Chapter 7, §§ 5.6.6, 5.6.7.

1.4.2.2. *Optical excitation functions for electron energies less than 35 eV.* The great improvement in the homogeneity of the electron beams in optical excitation function experiments, particularly in the work of Smit and of McFarland and their collaborators, has yielded convincing information about the detailed structure of the optical excitation functions in the low-energy region.

McFarland‡ used a gun producing an electron beam of energy width 0·2 eV while the corresponding width in the experiments of Smit *et al.*§ was 0·4 eV.

Fig. 4.23 shows the excitation functions obtained for lines with upper ^1S terms. For the line 5048 Å (4^1S–2^1P) (Fig. 4.23 (a)) both experiments revealed a sharp peak very close to the excitation threshold (compare with the behaviour of the cross-section for excitation of the 2^1S metastable state).

The peak is less clear in the measurements of Yakhontova‖ who used a less homogeneous electron beam. Fig. 4.23 (b) and (c) show similar

† FROST, L. S. and PHELPS, A. V., *Westinghouse Research Report* 6-94439-6-R3 (1957).
‡ MCFARLAND, R. H., *Phys. Rev.* **136** (1964) A1240.
§ SMIT, C., HEIDEMAN, H. G. M., and SMIT, J. A., *Physica* **29** (1963) 245.
‖ YAKHONTOVA, V. E., *Vest. leningr. gos. Univ. Ser. Fiz. i Khim.* **14** (1959) 27.

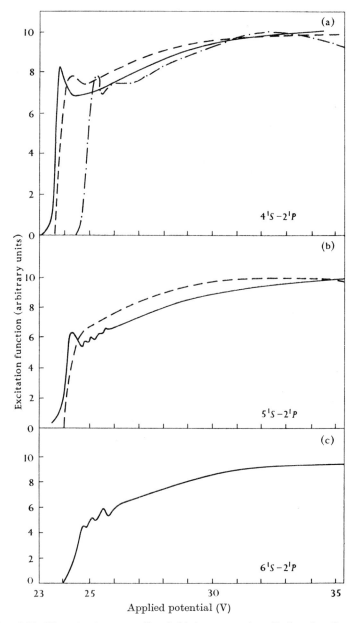

Fig. 4.23. Fine structure near threshold, in apparent excitation functions of helium lines with ^1S upper states, (a) 4^1S–2^1P (λ 5048); (b) 5^1S–2^1P (λ 4438); (c) 6^1S–2^1P (λ 4169). —— Smit; ---- Yakhontova; —·—·— McFarland. Each curve is normalized to unity at the maximum.

structure for the lines 4438 Å (5^1S–2^1P) and 4169 Å (6^1S–2^1P) near threshold. For the latter line the peak near threshold has largely disappeared but quite a complicated structure is still present.

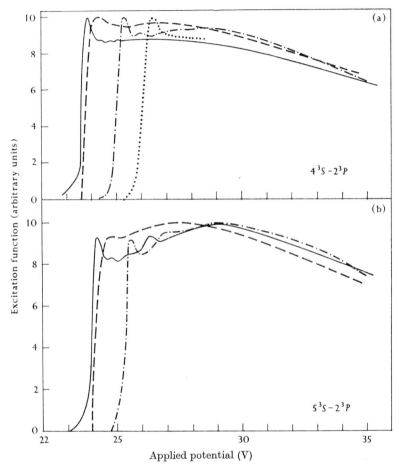

Fig. 4.24. Fine structure near threshold, in apparent excitation functions of helium lines with ^3S upper states. (a) 4^3S–2^3P (λ 4713); (b) 5^3S–2^3P (λ 4121). —— Smit, Heideman, and Smit; – – – – Yakhontova; —·—·—· McFarland; . . . . Heddle and Keesing.

Fig. 4.24 illustrates observed excitation functions for lines with ^3S upper levels. Once again a sharp peak appears very close to the threshold, which is more prominent than the corresponding peaks for the ^1S cases (cf. results for the 2^3S states, Figs. 4.58 and 4.59).

A complex structure is also observed for the excitation of lines with ^1D and ^3D upper states but, as for ^1S, the fluctuations seem to smooth

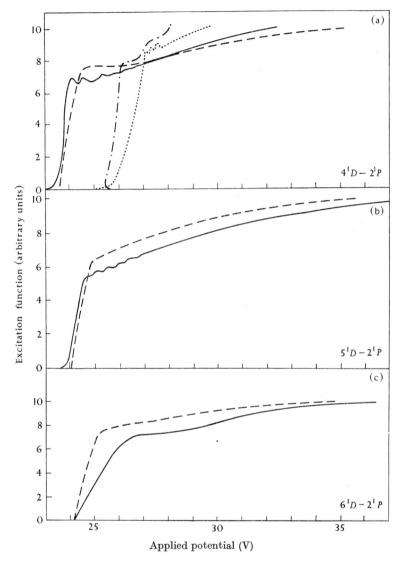

Fig. 4.25. Fine structure near threshold in apparent excitation functions of helium lines with ^1D upper states. (a) 4^1D–2^1P (λ 4922); (b) 5^1D–2^1P (λ 4388); (c) 6^1D–2^1P (λ 4143). —— Smit, Heideman, and Smit; – – – – Yakhontova; —·—·—· McFarland; . . . . Heddle and Keesing.

out as the principal quantum number increases. Fig. 4.25 shows results for lines arising from the respective transitions from 4^1D, 5^1D, and 6^1D–2^1P. For 4^1D–2^1P at 4922 Å a comparison can be made between the results of Smit *et al.*, McFarland, and Heddle and Keesing. The agreement in detail is remarkable and provides very strong confirmation of the reality of the observed structure.

For lines originating on ^3D terms the smoothing that occurs as the principal quantum number increases is most marked. Referring to Fig. 4.26, for the 3^3D–2^3P line at 5876 Å a near-threshold peak and further structure is clearly seen, but this has almost disappeared in the results for the 4^3D–2^2P line at 4471 Å.

No peaks close to the threshold appear for lines originating on ^3P upper states, although McFarland observed structure in the curves. For the ^1P states there is no true peak but a discontinuity in slope just above the threshold.†‡ However, reference to Fig. 4.27 shows that there is reasonably close agreement between the results of different observers, though McFarland finds more detailed structure than do Smit et al. or Heddle and Keesing.†

Finally, in Fig. 4.28 the different results for the 3^3P–2^3S line at 3889 Å are compared. As far as can be judged there does seem to be some agreement in detail between the results of Smit et al. and of McFarland. Yakhontova finds a kink at less than 1 eV above threshold that agrees quite closely with the location of a corresponding fairly pronounced kink in their results. However, the fluctuations are not so marked as in the other cases and it is not yet certain how definite they are. The results of Heddle and Keesing† do not exhibit this kink.

It is of considerable interest to determine whether the structural details found can be accounted for through population from higher states. For the ^1S, ^3S, ^1D, and ^3D states it is very unlikely but for the ^1P and ^3P states some of the structure may arise in this way. In any case the sharp peaks near threshold that appear in so many cases cannot be due to population from above. Thus the peaks arise in lines with S or D upper states. These could be populated only from excited P states but for such states the cross-section rises far more gradually from threshold than for any other states.

The behaviour of the excitation cross-sections for the 4^1D–2^1P (Fig. 4.25) and 4^3D–2^3P (Fig. 4.26) lines have been investigated at energies within a few tenths of an eV of the threshold. These results will be discussed in relation to the theory in Chapter 9, § 5. Other theoretical

---

† HEDDLE, D. W. O. and KEESING, R. G., Atomic collision processes, ed. McDOWELL, M. R. C., p. 179 (North Holland, Amsterdam, 1964) (Proceedings of the Third International Conference on the Physics of Electronic and Atomic collisions).

‡ ZAPESOCHNYI, I. P., Astr. Zh. **43** (1966) 954 (Soviet Astr. **10** (1967) 766).

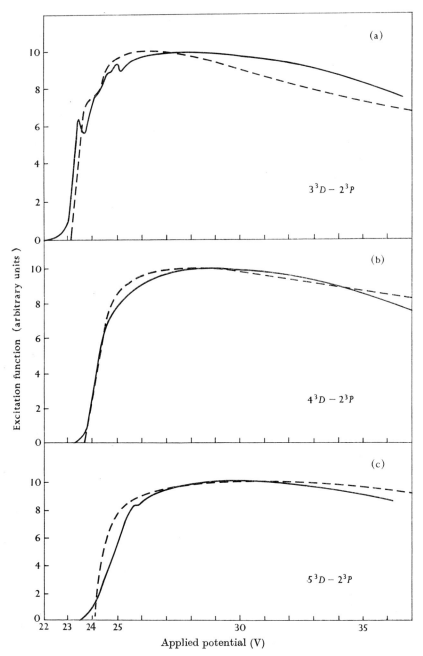

Fig. 4.26. Fine structure near threshold in apparent excitation functions of helium lines with ^3D upper states. (a) 3^3D–2^3P (λ 5876); (b) 4^3D–2^3P (λ 4471); (c) 5^3D–2^3P (λ 4026). —— Smit, Heideman, and Smit; — — — Yakhontova.

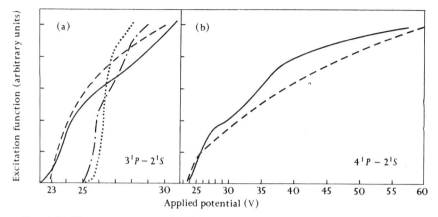

FIG. 4.27. Fine structure near threshold in apparent excitation functions of helium lines with ^1P upper states. (a) 3^1P–2^1S (λ 5016); (b) 4^1P–2^1S (λ 3964). ——— Smit, Heideman, and Smit; – – – – Yakhontova; –·–·–· McFarland; .... Heddle and Keesing.

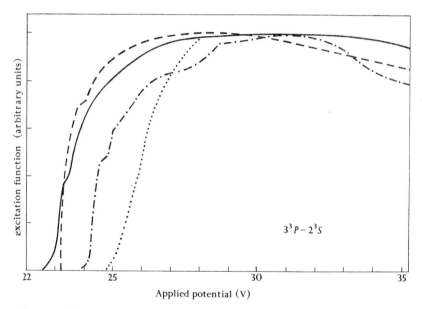

FIG. 4.28. Fine structure near threshold in apparent excitation function of helium line 3^3P–2^3S (λ 3889). ——— Smit, Heideman, and Smit; – – – – Yakhontova; –·–·–· McFarland; .... Heddle and Keesing. (The differences in threshold are probably due to different contact potentials.)

discussion of low energy excitation functions in helium is given in Chapter 8, § 6.2.

1.4.2.3. *Polarization.* Fig. 4.29 compares the results of polarization measurements made by different experimenters[†] on the lines 3^1P–2^1S (λ 5016), 3^3P–2^3S (λ 3889), 4^1D–2^1P (λ 4922), 4^3D–2^3P (λ 4471) in the electron energy range up to 300 eV, using working pressures sufficiently low to produce meaningful results (see Fig. 4.5). There is general agreement between the different authors with respect to the form of variation of the polarization with electron energy although the absolute values obtained vary considerably from author to author.

Of greatest interest are the measurements of the polarization near the threshold. The early observations of Skinner and Appleyard gave a polarization tending to zero at the threshold, a result in clear contradiction with quite general considerations (Chap. 8, § 10). Lamb and Maiman[‡] obtained finite polarization at threshold for the 3^3P–2^3S (λ 3889) line and this was confirmed by Hughes, Kay, and Weaver[§] and McFarland,[||] but in other cases zero threshold polarization continued to be observed. The resolution of this difficulty has presented a severe problem, even employing the most modern techniques to enable work at very low gas pressures and beam currents.

McFarland[††] was the first to obtain results for the 5^1D–2^1P (λ 4388) and 4^1D–2^1P (λ 4922) lines in which the polarization remained high very close to the threshold, in good agreement with theory. His measurements appeared to indicate, however, that the polarization of the 3^3P–2^3S (λ 3889) and 3^1P–2^1S (λ 5016) fell to zero at threshold after passing through maxima of 11 per cent and 18 per cent respectively about 0·3 eV above threshold. McFarland ascribed difficulties encountered in earlier measurements, including those of his own group, as due to elastic scattering of electrons from the beam. This means that part of the light emitted comes from electrons possessing a radial velocity in relation to which the roles of the components $I_\parallel$ and $I_\perp$ are interchanged. Heddle and Keesing,[‡‡] however, were unable to find any significant effect from scattered electrons in their experiments.

[†] HEDDLE, D. W. O. and LUCAS, C. B., *Proc. R. Soc.* A**271** (1963) 129; HUGHES, R. H., KAY, R. B., and WEAVER, L. D., *Phys. Rev.* **129** (1963) 1630; McFARLAND, R. H. and SOLTYSIK, E. H., ibid. **127** (1962) 2090; **128** (1962) 1758, 2222; **129** (1963) 2581.
[‡] LAMB, W. E. and MAIMAN, T. H., *Phys. Rev.* **105** (1957) 573.
[§] HUGHES, R. H., KAY, R. B., and WEAVER, L. D., loc. cit.
[||] McFARLAND, R. H., *Phys. Rev. Lett.* **10** (1963) 397; *Phys. Rev.* **133** (1964) A986.
[††] McFARLAND, R. H., ibid. **136** (1964) A1240.
[‡‡] HEDDLE, D. W. O. and KEESING, R. G. *Atomic collision processes*, ed. McDowell, M. R. C., p. 179 (North Holland, Amsterdam, 1964) (Proceedings of the Third International Conference on the Physics of Electronic and Atomic Collisions, London, 1963).

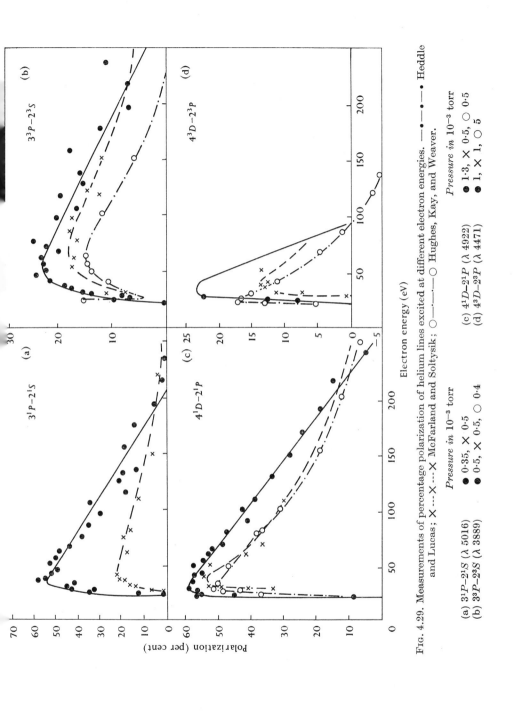

FIG. 4.29. Measurements of percentage polarization of helium lines excited at different electron energies. —·—·— Heddle and Lucas; ×---×--- McFarland and Soltysik; ○—·—·—○ Hughes, Kay, and Weaver.

(a) 3^1P–2^1S (λ 5016)
(b) 3^3P–2^3S (λ 3889)
(c) 4^1D–2^1P (λ 4922)
(d) 4^3D–2^3P (λ 4471)

Pressure in 10^{-3} torr

● 0.35, × 0.5
● 0.5, × 0.5, ○ 0.4

Pressure in 10^{-3} torr

● 1.3, × 0.5, ○ 0.5
● 1, × 1, ○ 5

The polarization measurements of Heddle and Keesing[†] near excitation threshold, using the data-logging method described above to reduce drifts, are shown in Fig. 4.30 for the 4^1D-2^1P (λ 4922), 3^1P-2^1S (λ 5016), 3^3P-2^3S (λ 3889), and 4^3S-2^3P (λ 4713 Å). Also shown are the theoretically expected values at threshold. The agreement is good except for the line λ 5016, the upper level for which (3^1P) can be excited by absorption of resonance radiation. In this case, although a finite value is found at threshold it is considerably less than the theoretical value of 100 per cent. The effective cross-section for the excitation of this line is small and this limits the precision of the polarization measurement. As expected for excitation of S states, the polarization of the 4^3S-2^3P (λ 4713) transition is effectively zero. For comparison, the measurements of McFarland[‡] are shown on the same figure and in addition his measurements of the line 5^1D-2^1P (λ 4388) are also shown in Fig. 4.30 (e).

1.4.3. *Results of measurements in mercury vapour.* The most extensive observations are those of Jongerius,[§] and the transitions that he studied are indicated in the energy-level diagram of the mercury atom shown in Fig. 4.31.

As compared with most earlier work the main additional feature is the detail found in the optical excitation functions because of the improved homogeneity of the electron beam. This is illustrated in Fig. 4.32, in which a number of observed excitation functions for the $7^3S_1-6^3P_2$ line at 5461 Å are compared. Strong support for the validity of the detailed structure is provided from the quite close agreement in this respect between the results of Jongerius,[§] Frisch,[||] and Zapesochnyi and Shpenik.[††] Furthermore, the same details are observed to appear in optical excitation functions for other lines arising from the 7^3S_1 level as shown in Fig. 4.32.[‡‡] As long ago as 1931 Siebertz[§§] observed the excitation function for the $7^3S-6^3P_1$ line that is shown in Fig. 4.32 and it will be seen that it shows most of the detail found nearly twenty years later by Jongerius.

Jongerius made an absolute measurement of the differential cross-section for emission per steradian at 90° of the 4047 Å $7^3S_1-6^3P_0$ line by

[†] HEDDLE, D. W. O. and KEESING, R. G., *Atomic collision processes*, ed. McDowell, M. R. C., p. 179 (North Holland, Amsterdam, 1964) (Proceedings of the Third International Conference on the Physics of Electronic and Atomic Collisions, London, 1963).
[‡] MCFARLAND, R. H., *Phys. Rev.* **136** (1964) A1240.
[§] JONGERIUS, H. M., Thesis, Utrecht (1961).
[||] FRISCH, S., *Proc. VIth Conf. Spectr. Amsterdam*, 1956, p. 350.
[††] ZAPESOCHNYI, I. P. and SHPENIK, O. B., *Dokl. Akad. Nauk SSSR* **160** (1965) 1053; *Soviet Phys. Dokl.* **10** (1965) 140.
[‡‡] SMIT, J. A. and JONGERIUS, H. A., *Appl. scient. Res.* **B5** (1955) 59–62.
[§§] SIEBERTZ, K., *Z. Phys.* **68** (1931) 505.

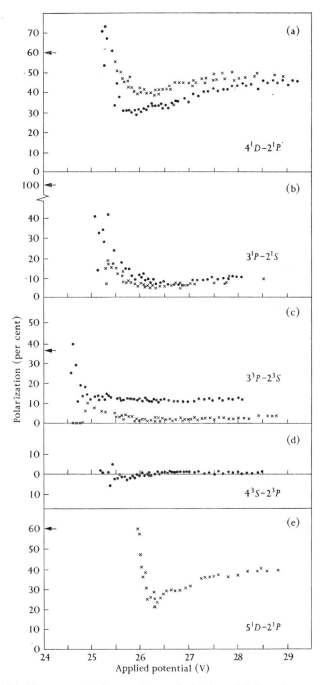

Fig. 4.30. Measurement of percentage polarization of helium lines near the threshold. ● Heddle and Keesing; × McFarland. (a) 4^1D-2^1P (λ 4922); (b) 3^1P-2^1S (λ 5016); (c) 3^3P-2^3S (λ 3889); (d) 4^3S-2^3P (λ 4713); (e) 5^1D-2^1P (λ 4388). In each case the theoretical polarization at threshold is marked by an arrow on the ordinate scale.

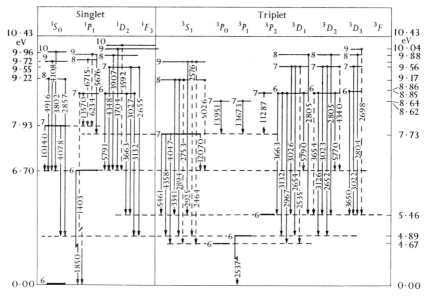

Fig. 4.31. Energy levels of the mercury atom. Energies of levels are given in eV with respect to the ground state. Optical transitions are indicated by arrows.

TABLE 4.4

Differential cross-sections at 90° for excitation of mercury lines by electrons of energy 15 eV

Wavelength (Å)	Diffnl. cross-section at 90° (10^{-19} cm²/sr)	Wavelength (Å)	Diffnl. cross-section at 90° (10^{-19} cm²/sr)	Wavelength (Å)	Diffnl. cross-section at 90° (10^{-19} cm²/sr)
2485	0·4±0·25	3126	2·3±0·3	4047	1·00(±10%)
2537	21±3	3132	3·85±0·45	4078	0·76±0·08
2655–2652	1·5±0·3	3341	0·31±0·03	4108	0·16±0·02
2700	0·05±0·025	3524	0·02±0·01	4348	0·94±0·13
2753–2760	0·04±0·02	3592	0·08±0·01	4358	2·9±0·3
2805	0·26±0·03	3650	6·2±0·07	4916	0·67±0·07
2857	0·04±0·02	3663	0·55±0·1	5461	3·6±0·4
2894	0·26±0·03	3704	0·17±0·02	5676	0·24±0·04
2967	0·51±0·06	3801	0·05±0·01	5770	1·9±0·25
3027–3022	0·93±0·11	3907	0·41±0·04	5791–5790	2·9±0·35

impact of electrons with 15 eV energy, using a tungsten ribbon lamp as comparison standard. A value of $(1\cdot0\pm0\cdot1)\times10^{-19}$ cm²/sr⁻¹ was obtained. The corresponding cross-sections for other visible and near ultra-violet lines observed were determined by intensity measurements relative to that of the 4047 Å line. Her results are tabulated in Table 4.4. Comparison with the results of earlier investigations is difficult

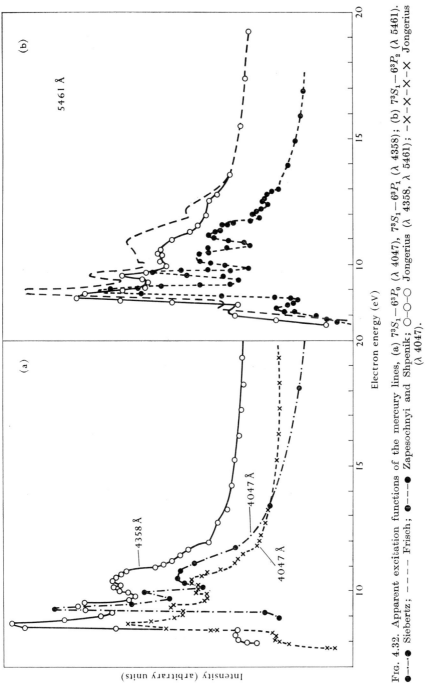

FIG. 4.32. Apparent excitation functions of the mercury lines, (a) $7^3S_1-6^3P_0$ (λ 4047), $7^3S_1-6^3P_1$ (λ 4358); (b) $7^3S_1-6^3P_2$ (λ 5461). •----• Siebertz; ---- Frisch; •---• Zapesochnyi and Shpenik; ○—○—○ Jongerius (λ 4358, λ 5461); -×-×-×-× Jongerius (λ 4047).

because they used 60 eV as the standard electron energy and the complex shape of the excitation functions makes it difficult to transfer to 15 eV or vice versa. Further complication arises from the fact that the earlier observations did not record the detail in the excitation functions. Nevertheless, it does appear that the intensities obtained by Hanle and Schaffernicht† are three or so times larger than those of Jongerius.

It is important to determine the extent to which cascade effects influence the shape of the excitation functions. In many respects a favourable case to study is that of the 2537 Å (6^3P_1–6^1S_0) resonance line. Already in 1933, Ornstein, Lindeman, and Oldeman‡ analysed the different contributions for this line, using data available on the excitation functions of the lines λλ 3131, 3125, 4078, and 4358 arising from transitions to the upper 6^3P_1 state of the resonance line from the 6^3D_1, 6^3D_2, 7^1S_0, and 7^3S_1 states respectively.

This analysis was repeated by Jongerius,§ allowing in addition for population from the 8^3S_1 and $7^3D_{1,2}$ levels by including the relative excitation functions for the respective lines λλ 2894 and 2654. As no allowance was made for resonance trapping of the λ 2537 line only the form of its excitation function could be used. However, the sum of the excitation functions for the cascade transitions exhibits maxima at exactly the same energies as the last three maxima in the observed function for λ 2537. The scale of the latter function could then be chosen so that, when the cascade contribution was subtracted, the function, apart from a weak maximum at 10·5 eV electron energy, decreased monotonically at energies beyond the threshold for cascade excitation. This is shown in Fig. 4.33.

On the assumption that this procedure is correct the maximum value of the differential cross-section for emission at 90° of the λ 2537 line through direct excitation comes out to be $21\pm3\times10^{-19}$ cm²/sr.

Although it seems clear from this analysis that many of the observed maxima in the excitation functions arise from cascade effects, some are certainly characteristics of the cross-sections for direct excitation of the upper level concerned. Thus, referring to Fig. 4.33, we see that a subsidiary maximum occurs in the excitation function for the λ 2537 line at an electron energy below that for the principal maximum. This could not arise from cascade effects.

Analysis of the excitation functions that show maxima cannot be

† HANLE, W. and SCHAFFERNICHT, W., *Annln Phys.* **6** (1930) 905.
‡ ORNSTEIN, L. S., LINDEMAN, H., and OLDEMAN, J., *Z. Phys.* **83** (1933) 171.
§ loc. cit., p. 216.

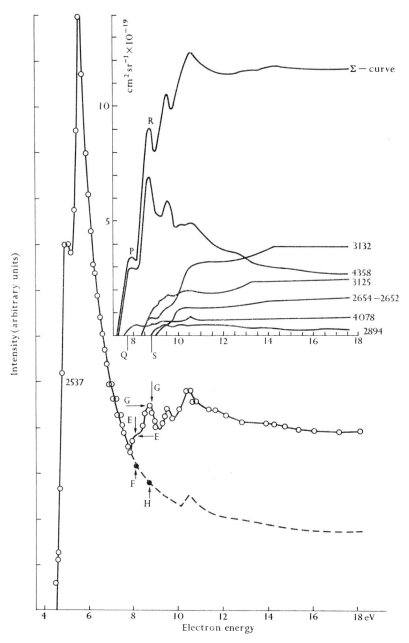

FIG. 4.33. Apparent excitation curve of the Hg line λ 2537 and the sum of the excitation curves of the lines λλ 4358, 4078, 3132, 3125, 2894, and 2654 that populate it by cascade transitions, showing the method of separating the direct and cascade contributions to the λ 2537 excitation curve. The relation $PQ:RS = EF:GH$ was used to extrapolate the 'direct' excitation curve above 8 eV where cascade effects become important.

carried out because of lack of observed data on excitation functions of lines that populate the upper state concerned. Because of the complicated nature of the mercury atom, theoretical results cannot be used as in helium.

A further feature that is a consequence of the high atomic number of mercury is that the LS coupling is no longer a good representation of many spectral terms. The λ 2537 line arises from a level that is of different multiplicity from that of the ground state, so we should at first sight expect the excitation function for the 2^3P_1 level to resemble that for helium ^3P (Fig. 4.19). For electron energies close to the threshold this is the case, but at higher energies it assumes the form characteristic of the excitation of an optically allowed level (^1P of Fig. 4.19). The explanation of this anomaly does not lie in the existence of any secondary mode of excitation (as for the n^3D levels of helium) but in the fact that for a heavy atom, such as mercury, spin-orbit coupling is not negligible and the $^3P_1$ level is not a pure triplet state. This is clear from the fact that the resonance line λ 2537 is one involving an apparent change of multiplicity. The singlet component of the level is responsible for the form of the excitation function at electron energies greater than two or three times the threshold energy.

To derive total cross-sections for excitations of the different lines it is necessary to allow for the polarization (see § 1.2). The very early measurements of Skinner and Appleyard† are shown in Fig. 4.34. These suggest that the correction required for the anisotropy of emission of the different radiations due to polarization, in deriving total cross-sections from the measurements at 90°, is smaller than the remaining inaccuracy of the absolute values.

Fig. 4.35 shows measurements of the polarizations of the Hg lines $\lambda\lambda$ 5791, 5790 (6^1D_2–6^1P_1 and 6^3D_1–6^1P_1) as functions of electron energy obtained by Heideman.‡ On the same figure are shown the apparent excitation curves for $I_\parallel$ and $I_\perp$. The measurements were carried out at a mercury vapour pressure of 10^{-3} torr. The fine structure in the polarization curve, which is correlated with structure in the excitation curves, appeared to be reproducible. Although measurements within 0·3 eV of the threshold were available there were no indications that the polarization tended to zero at threshold. The polarization function of the line λ 5770 (6^3D_2–6^1P_1) was almost the same as for λ 5791 but for the line λ 4358 (7^3S_1–6^3P_1) the polarization was much smaller, as expected.

† loc. cit. p. 178.   ‡ HEIDEMAN, H. G. M., *Phys. Lett.* **13** (1964) 309.

1.4.4. Results of measurements in other gases and vapours

1.4.4.1. Inert gases. Absolute measurements of the excitation cross-sections for the 2p levels of argon, krypton, and xenon have been reported by Zapesochnyi and Feltsan.† The general shape of the excitation curves are similar to those of the ^1P excitation of helium. On passing to heavier atoms, however, the half-width of the maximum of the

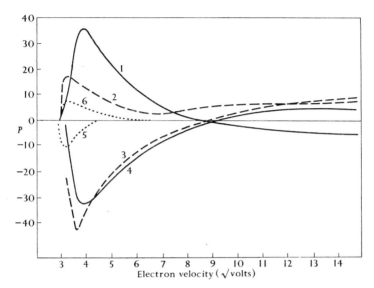

Fig. 4.34. Skinner and Appleyard's measurements of the polarization percentage P, of mercury lines as a function of electron energy. (1) λ 4347 (7^1D_2–6^1P_1); (2) λ 3650 (6^3D_3–6^3P_2); (3) λ 3663 (6^3D_1–6^3P_2); (4) λ 3655 (6^3D_2–6^3P_2); (5) λ 4358 (7^3S_1–6^3P_1); (6) λ 4047 (7^3S_1–6^3P_0).

excitation function decreases and its peak shifts closer to the threshold. Cascade effects (10–50 per cent) are larger than in helium but insufficient data is available to make reliable estimates of these.

1.4.4.2. Alkali metals. In recent years absolute cross-sections have been obtained for the excitation of many lines in the spectra of the alkali metals. Hughes and Hendrickson‡ have measured the excitation function for the λ 6708 ($2p$–$2s$) line of lithium.

Extensive measurements have been made by Volkova for sodium§

† ZAPESOCHNYI, I. P. and FELTSAN, P. V., *Optika Spektrosk.* **20** (1966) 521 (*Soviet Optics and Spectroscopy* **20** (1966) 291).
‡ HUGHES, R. H. and HENDRICKSON, C. G., *J. opt. Soc. Am.* **54** (1964) 1494.
§ VOLKOVA, L. M., *Optika Spektrosk.* **11** (1961) 775 (*Soviet Optics and Spectroscopy* **11** (1961) 420).

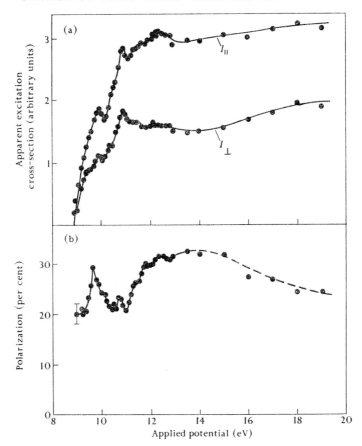

Fig. 4.35. Apparent effective excitation cross-sections and percentage polarization curves for Hg lines λλ 5790/1 (measured by Heideman with $p = 10^{-3}$ torr, $i = 13$ μA). $I_\parallel$ and $I_\perp$ refer to components polarized parallel and perpendicular to electron-beam direction. The vertical line in (b) indicates the statistical error in the polarization measurement near threshold.

and potassium† and by Zapesochnyi and his collaborators for sodium,‡ potassium,§ rubidium,‖ and caesium.††

As in the case of helium, excitation functions for a given series have

---

† VOLKOVA, L. M., ibid. **13** (1962) 849 (*Soviet Optics and Spectroscopy* **13** (1962) 482).

‡ ZAPESOCHNYI, I. P. and SHIMON, L. L., ibid. **19** (1965) 480 (*Soviet Optics and Spectroscopy* **19** (1965) 268).

§ ZAPESOCHNYI I. P., SHIMON, L. L., and SOSNIKOV, A. K., ibid. **19** (1965) 864 (*Soviet Optics and Spectroscopy* **19** (1965) 480).

‖ ZAPESOCHNYI, I. P. and SHIMON, L. L., ibid. **20** (1965) 944 (*Soviet Optics and Spectroscopy* **20** (1965) 525).

†† ZAPESOCHNYI, I. P. and SHIMON, L. L., ibid. **20** (1965) 753 (*Soviet Optics and Spectroscopy* **20** (1965) 421).

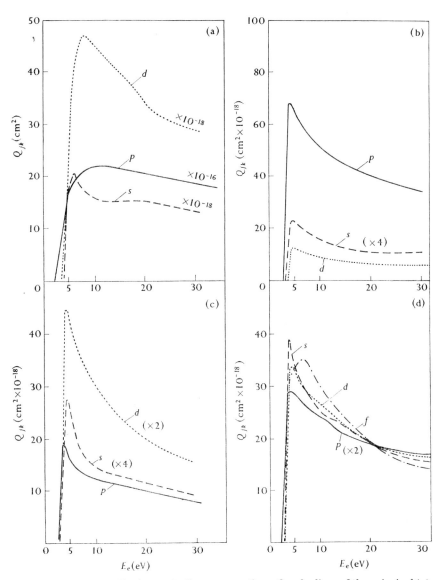

FIG. 4.36. Apparent effective excitation cross-sections, Q_{jk}, for lines of the principal (p), sharp (s), diffuse (d), and fundamental (f) series of the alkali metals measured by Zapesochnyi and co-workers. (a) Sodium: $\lambda\lambda$ 5890–6 (p), 6154–61 (s), 8183–94 (d). (b) Potassium: $\lambda\lambda$ 4044 (p), 6911 (s), 5832 (d). (c) Rubidium: $\lambda\lambda$ 4201 (p), 7408 (s), 7757 (d). (d) Caesium: $\lambda\lambda$ 4555 (p), 7944 (s), 6973 (d), 8079–15 (f).

the same general shape but there are differences between the excitation functions for different series. For example, Fig. 4.36 shows, for Na, K, Rb and Cs, excitation functions for a typical line of the principal (p), sharp (s), and diffuse (d) series corresponding to initial ^2P, ^2S, and ^2D states. For Cs the excitation function for a typical line of the fundamental (f) series corresponding to an initial ^2F state is also shown.

For sodium the excitation curve rises to its maximum less rapidly for the principal series lines (corresponding to an optically allowed $2p$–$2s$ transition) than for the other lines. As the mass of the alkali metal

TABLE 4.5

Element	Excited level	Cross-section (max) ($\times 10^{-18}$ cm^2)	Element	Excited level	Cross-section (max) ($\times 10^{-18}$ cm^2)
Na	$4P$	15	Cs	$8S$	90
	$5S$	35·4		$9S$	40
	$6S$	12·2		$10S$	25
				$7D$	60
Rb	$7S$	34		$8D$	35
	$8S$	10		$9D$	19
	$9S$	3·6		$5F$	35
	$10S$	2·4		$6F$	16
				$7F$	7

atom increases, however, the excitation function, even for this transition, rises more and more sharply from threshold, and the structure in the curve increases.

The contributions from cascade transitions vary considerably between the different alkali metals and the different levels. In those cases where reasonably reliable estimates of the contribution from cascade processes can be made Table 4.5 gives the maximum cross-sections for direct excitation of a number of levels, as derived by Zapesochnyi and his collaborators.

Using the method described in § 1.3 to overcome the effect of resonance absorption, Zapesochnyi and Shimon[†] obtained the absolute excitation curve shown in Fig. 4.37 for the caesium resonance level $6^2P_{\frac{1}{2}}$.

Polarization measurements of radiation from lithium and sodium have been made by Hafner, Kleinpoppen, and Kruger[‡] for the first resonance lines of Li and Na. The polarization expected theoretically

[†] ZAPESOCHNYI, I. P. and SHIMON, L. L., *Dokl. Akad. Nauk SSSR* **166** (1966) 320 (*Soviet Phys. Dokl.* **11** (1966) 44).
[‡] HAFNER, H., KLEINPOPPEN, H., and KRUGER, H., *Phys. Lett.* **18** (1965) 270.

near the threshold depends on the nuclear spin, so that measurements were carried out using the separated ^{6}Li and ^{7}Li isotopes. The results are shown in Fig. 4.38, together with the excitation cross-section curves near the threshold. The polarization tends to a finite value at the threshold in each case and is different for ^{6}Li and ^{7}Li. The observed values are in accord with the theoretical interpretation as discussed in Chapter 8, § 10.2.

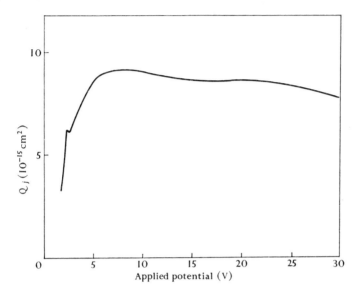

Fig. 4.37. Excitation cross-sections Q_j for the caesium resonance levels, $6^2P_{\frac{1}{2},\frac{3}{2}}$, derived by Zapesochnyi and Shimon from their measurements.

The line λ 4971 (4^2S–2^2P) for natural Li was unpolarized as expected, but the polarization of the Li line λ 6103 (3^2D–2^2P) was found to decrease toward zero near the threshold, contrary to an expected threshold polarization of 47·9 per cent.

1.4.4.3. *Cadmium.* Excitation functions for cadmium lines have been measured by Zapesochnyi and Shevera.† They found excitation functions very similar to those obtained for mercury. Considerable structure was observed. For example, Fig. 4.39 shows the form of excitation function found for the resonance line λ 3261 (5^3P_1–5^3S_0). Some of the maxima are seen to be associated with structure in the excitation curve for the line λ 4800 (6^3S_1–5^3P_1), which populates the 5^3P_1 level

† Zapesochnyi, I. P. and Shevera, V. S., *Dokl. Akad. Nauk SSSR* **141** (1961) 595 (*Soviet Phys. Dokl.* **6** (1962) 1006).

through cascade processes. Several of the maxima are not associated, however, and must represent structure in the direct excitation of the 5^3P_1 level.

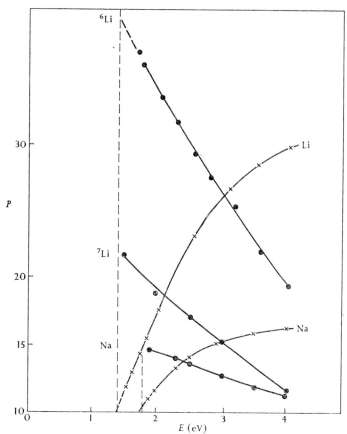

Fig. 4.38. Percentage polarization (P) (●) and apparent excitation cross-section (×) of the component of radiation polarized parallel to the electron-beam direction ($I_\parallel$) for first resonance lines of ^{6}Li, ^{7}Li, and Na. The apparent excitation cross-section is the same for ^{6}Li and ^{7}Li.

1.5. Simultaneous ionization and optical excitation

Some measurements have been made of the excitation curves for lines of the ionized spectrum of the alkali metals and of argon produced in a single collision process between electrons and neutral atoms. These measurements are, of course, closely related to the fine structure observed in the ionization cross-section curves and described in § 2.5.3 of Chapter 3. Since, however, the techniques employed were identical with those for the study of the excitation of the neutral atom, it is more convenient to describe them here. Zapesochnyi and Shimon† have measured absolute excitation cross-sections for 10 lines of NaII, 15 lines of RbII, and 17 lines of CsII, with maximum cross-sections up to 3.5×10^{-18} cm²

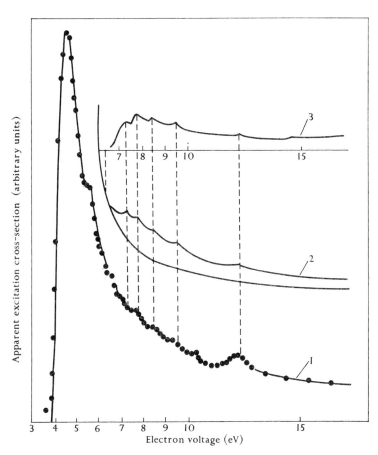

FIG. 4.39. Apparent excitation function for the cadmium resonance line λ 3261 (5^3P_1–5^3S_0) (curve 1) observed by Zapesochnyi and Shevera. The structure is clearly related to that observed for λ 4800 (6^3S_1–5^3P_1) (curve 3). Curve 2 shows the λ 4800 excitation curve superimposed on a curvilinear axis to reproduce the excitation curve observed for λ 3261.

for NaII (λ 3533) at 200 eV, to 7.55×10^{-18} cm² for RbII (λ 4244) at 70 eV, and to 11.9×10^{-18} cm² for CsII (λ 4603) at 30 eV.

Measurements of cross-sections of lines of the AII spectrum by direct excitation from the ground state of the neutral atom have been made by

† ZAPESOCHNYI, I. P. and SHIMON, L. L., *Optika Spektrosk.* **19** (1965) 480 (*Soviet Optics and Spectroscopy*, **19** (1965) 268, 421; **20** (1965) 525, 753, 944).

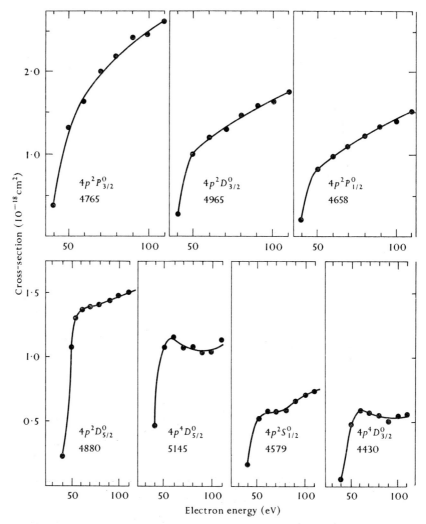

Fig. 4.40. Apparent cross-sections for simultaneous ionization and excitation of argon, measured by Bennett et al. The cross-sections for the excited states of A$^+$ specified were obtained from intensity measurements of the lines of the AII spectrum whose wavelengths (in Å) are given on the curves.

Bennett et al.[†] for lines whose upper levels are of importance in relation to the argon ion laser. Fig. 4.40 shows the apparent excitation cross-sections for seven excited states of A$^+$, derived from measurements of the intensities of the lines specified on the curves.

[†] BENNETT, W. R., MERCER, G. N., KINDLMANN, P. J., WEXLER, B., and HYMAN, H., Phys. Rev. Lett. **17** (1966) 987.

2. Direct measurement of excitation functions for metastable states of atoms

The methods outlined in § 1 are not available for the measurement of excitation functions of metastable states, since radiative transitions from such states do not in general occur. Other methods are available, however, for determining the density of atoms excited to metastable states and these make possible a direct determination of the corresponding excitation functions. We discuss these methods and results obtained using them in this section. In Chapter 5 these results, as well

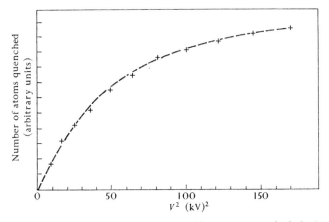

Fig. 4.41. Number of He(2^1S) metastable atoms quenched during transit between two condenser plates as a function of the square of the potential difference V between the plates (Holt and Krotkov).

as those of § 1 on optical excitation, will be compared with the results of indirect methods of measuring these cross-sections by studying the discrete energy losses of the electrons producing the excitation.

2.1. *The experimental methods*

2.1.1. *Quenching of metastable states.* If H atoms in the $2^2s_{\frac{1}{2}}$ metastable state are passed through a region containing an electric field, mixing of the $2^2s_{\frac{1}{2}}$ and $2^2p_{\frac{1}{2}}$ levels takes place. The leakage to the $2^2p_{\frac{1}{2}}$ level leads to the emission of Ly α-radiation until all the metastable atoms are de-excited. Similarly, the metastable 2^1S_0 states of He can be quenched to the 2^1P_0 state in an electric field. Fig. 4.41, given by Holt and Krotkov,† shows how the degree of quenching depends on the square of the electric field in this latter case. In their experiment the quenching electric field was applied across two parallel stainless steel

† HOLT, H. K. and KROTKOV, R., *Phys. Rev.* **144** (1966) 82.

plates 9·84 cm long and 0·053 cm apart and to secure saturation they worked with a potential difference of 12 kV, corresponding to a field of 226 kV/cm.

Experiments have been carried out to measure the cross-section for excitation of the 2s metastable states of atomic H by this method by Stebbings et al.,† using a modulated atomic beam crossed by a steady

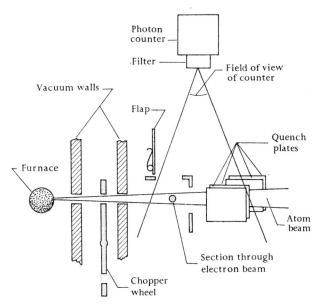

FIG. 4.42. Schematic diagram of arrangement used by Stebbings et al. to study excitation of 2s metastable states of hydrogen.

electron beam, and by Hils et al.‡ using steady crossed atomic and electron beams. In addition some experiments (described in § 2.1.3) have used a combination of quenching and direct metastable detection to study the excitation cross-sections of metastable states of H and He.

Stebbings et al. adapted the method employed by them for the measurement of the cross-section for excitation of the 2p state of atomic hydrogen (see § 1.3.4). Fig. 4.42 illustrates schematically the principle of operation. Essentially it involved a comparison of the cross-sections for production of 2p and 2s excited atoms in terms of the ratio of the intensity of the Ly α-radiation emitted normally from the bombardment region to that of the radiation emitted from the region between plates across which

† STEBBINGS, R. F., FITE, W. L., HUMMER, D. G., and BRACKMAN, R. T., ibid. **119** (1960) 1939.
‡ HILS, D., KLEINPOPPEN, H., and KOSCHMIEDER, H., *Proc. phys. Soc.* **89** (1966) 35.

a quenching electrostatic field was applied. These plates were placed parallel to and on either side of the atom beam that passed between them after traversing the bombardment region.

To make the necessary measurements the G.–M. photon counter with a lithium fluoride window and filled with iodine vapour was mounted on a trolley, so that it could be placed in position to view either the collision or the quenching region. The first measurements were made with the counter directly above the collision region with the flap, shown in Fig. 4.42, raised. This recorded a signal proportional to the cross-section for direct production of the $2p$ state. The counter was then moved so as to view the quenching region directly and the flap was lowered to block out any Ly α-radiation from the collision region. Measurements were made of the Ly α-flux received with the quenching field sufficiently large to quench all the $2s$ atoms passing between the plates, and with zero field. The difference between these two measurements was then taken as the flux arising from total quenching of the metastable atoms. In all cases discrimination against background was achieved by the usual a.c. methods.

To minimize any accidental quenching by stray electric fields in the bombardment region, all the elements of the electron gun were gold-plated and heated to avoid the formation of insulating layers. A uniform magnetic field of 50 gauss was applied in the direction of the electron beam to confine the beam. It also served to prevent scattered electrons from reaching the electrodes in the quenching region and thus giving rise to electrical noise through soft X-ray generation. Care was taken to bias the electron beam collector so that secondary electrons were unable to escape from it, otherwise, being trapped by the magnetic fields, they could travel back to the bombardment region and give rise to spurious signals.

A grounded box enclosed the electron beam. Slots were cut in the box to allow the passage of the atom beam. The size and position of the exit slot was determined from a preliminary experiment in which the angular distribution of the recoiling $2s$ atoms was measured. A detector consisting of quenching plates plus a photon counter, with the usual a.c. discrimination, was used. This detector could be rotated in the plane containing the electron and atom beams.

The electrode system used for quenching consisted of two plates biased symmetrically above and below ground in a position that intercepted a negligible fraction of the emerging Ly α-radiation. To reduce the penetration of this field into the bombardment region a pair of guard plates was added. With this arrangement the field penetrating to this region was less than 3 per cent of the main quenching field.

It was verified that the metastable atom signal was proportional to the strength of the bombarding electron beam, showing that no appreciable quenching by space-charge effects occurred. By using the fact that the Ly α-radiation produced by quenching is isotropic, and knowing the relation of the total cross-section for production of 2p atoms to the differential cross-section at 90°, it is possible to derive the ratio of the total production cross-section for 2s and 2p states from the observations.

The experiment of Hils et al.† used a steady H atomic beam but a similar photon counter detector observing, however, only Ly α-radiation emitted from the quenching region. They observed quenching of the metastable $2s_{\frac{1}{2}}$ atoms due to space charge effects with electron beam currents above 200 μA, but the counting rate was linear with electron current for lower electron currents, indicating that space charge effects were not important at the lower currents at which measurements were carried out. The excitation function was normalized to the values calculated, using Born's approximation in the energy region 200–500 eV in the same manner as used by Lichten and Schultz‡ and described in § 2.1.3 below.

2.1.2. Direct measurement of metastable flux

2.1.2.1. *Electron ejection from surfaces by metastable atoms.* Several experimenters have studied excitation cross-sections of metastable atomic states by using electron emission when metastable atoms strike a metal surface to measure the metastable production rate. Electron emission can occur if the excitation energy exceeds the work function of the surface. The first experiments of this kind were carried out by Dorrestein,§ using the apparatus shown in Fig. 4.43. Electrons emitted from the indirectly heated cathode K were accelerated through the slit system D_1, D_2 along the axis of the copper cylinder C. They were collimated by means of an axial magnetic field produced by the coils M_1, M_2, M_3. The cylinder C contained the gas being studied (He or Ne), at a pressure of about 10^{-3} torr, and metastable atoms were formed along the path of the electron beam. Some of them left C through the side hole H, which was covered with a wire mesh, and were incident on a platinum collecting plate P which could be thoroughly outgassed by means of the heater S. Secondary electrons were ejected from P by the

† loc. cit., p. 232. ‡ LICHTEN, W. and SCHULTZ, S., *Phys. Rev.* **116** (1959) 1132.
§ DORRESTEIN, R., *Physica* **9** (1942) 433, 447.

incident metastable atoms and were collected by the grid G. The cylindrical electrode W surrounding P was positively charged and prevented positive ions reaching the collector.

If i_0 is the electron current passing along C (about 10^{-5} mA in Dorrestein's experiments), n the number of atoms per cm^3, $Q(E)$ the cross-section for the excitation of the metastable state being investigated, the

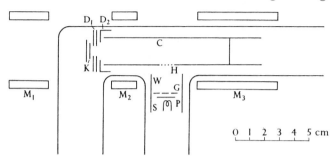

FIG. 4.43. Schematic diagram of Dorrestein's apparatus for measuring excitation cross-sections of metastable atoms by electron impact.

current i_m due to electrons ejected from P by the metastable atoms is given by
$$i_m = i_0 n Q(E) \zeta \rho,$$
where ρ is a factor giving the fraction of the metastable atoms produced in unit path which reach P and ζ is the number of electrons ejected from P per incident metastable atom. ρ can be estimated from geometrical considerations but ζ is not known. If $Q(E)$ can be obtained independently this method may be used to estimate ζ.

Quanta reaching P from C will also eject electrons and may mask the effect of the metastable atoms. To separate the photo-electrons from electrons ejected by metastable atoms, Dorrestein pulsed the current through C by applying a.c. of frequency from 900 to 2×10^5 cycles to the slit D_2. The time taken for a metastable atom of velocity v to reach P is a/v, where a is the distance from H to P. If the frequency is high enough the distribution in energy of the metastable atoms is such that the number incident on P per unit time is practically uniform. The difference between the 'in-phase' and 'out-of-phase' electron current ejected from P therefore measures the effect of radiation.

An elaboration of Dorrestein's method has been used by Schulz and Fox[†] to study excitation of metastable states of He, and by Dowell[‡] for metastable states of He, Ne, A, Kr, and Xe. These experiments used

[†] SCHULZ, G. J. and FOX, R. E., *Phys. Rev.* **106** (1957) 1179.
[‡] DOWELL, J. T., *UCRL Report* 14450 (1965).

the r.p.d. method (see Chap. 3, § 2.4.2) to obtain the excitation produced by an electron beam having a narrow energy spread, and collected metastable atoms on a cylindrical detector surrounding the whole collision region. Fig. 4.44 shows the apparatus used by Dowell. An indirectly heated barium-impregnated tungsten cathode K was used and, after acceleration in the electron gun (electrodes 1 to 6), the electrons passed through the collision chamber, CC, and were collected by the electron collector, EC, which was coated with platinum black to reduce electron reflection. They were collimated by an axial magnetic field of strength 120 gauss. Two cylindrical grids constructed of electroformed gold mesh, forty lines per in, surrounded the electron beam in the collision chamber. The inner of these was connected to the entrance and exit slits of the collision chamber CC. The outer, G, was biased -2 to $+5$ V relative to CC. Metastable atoms diffusing through the grid to the plate M, maintained 10–20 V negative relative to CC and constructed of a non-magnetic Cu–Ni alloy (Advance metal), ejected electrons that were collected by G, which was connected to an electrometer. The background pressure before introduction of the gases being studied was 10^{-8} to 10^{-9} torr.

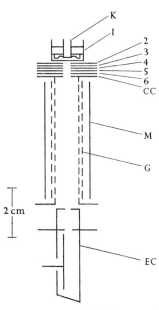

Fig. 4.44. Schematic diagram of Dowell's apparatus for measuring excitation cross-sections of metastable atoms by electron impact.

The total effective energy width of the electron beam was 0·25 eV, and both positive ion onset potentials and electron retardation studies were used for the calibration of energy scales. The operating pressure of the sample gas was in the range 5×10^{-5} to 2×10^{-3} torr and electron beam difference currents between 10^{-9} and 10^{-8} A, linearity of signal with current and pressure being verified in the operating region. The difference current signal due to metastable atoms lay between 5×10^{-13} to 5×10^{-16} A, with typical peak-to-peak noise levels of 5 to 10×10^{-16} A.

In order to obtain absolute excitation cross-sections the electron ejection efficiency ζ must be known independently. For a tungsten detector this has been measured to be approximately 0·3 electrons per metastable atom,† for helium. Usually, however, the method has been employed to study the structure in the metastable excitation function near the threshold, without explicitly separating effects produced by different metastable states. The method also suffers from the disadvantage that photons created by transitions from radiating states can liberate photoelectrons from the metal, although with considerably lower efficiency than for metastable atoms.

Measurements of excitation functions of metastable states of inert gas atoms have also been made by Kuprianov.‡ Crossed-beam methods for measuring metastable excitation cross-sections are generally to be preferred, because the detection region can be separated from the excitation region so that the background due to excitation of radiating states can be greatly reduced. A modulated crossed-beam method has been applied by Olmsted, Newton, and Street§ to measure excitation cross-sections of Ne, A, and Kr, as well as the molecular gases H_2, N_2, and CO (see Chap. 13, § 3.2). The method is similar to that of Stebbings et al., with the exception that metastable detection was accomplished by observing electron ejection from the surface of a silver–magnesium alloy with low work function (4·5 eV).

2.1.2.2. *Detection of metastable atoms by ionizing collisions of the second kind.* A crossed-beam experiment has also been carried out by Čermak,‖ but using a different detection technique, for metastable He. In this experiment the metastable atoms were detected by passing the atomic beam after excitation through a chamber containing argon gas. Ionization of the argon occurs in collisions of the second kind between the metastable helium and the argon in the process (see Chap. 18, § 5)

$$\mathrm{He}(2^3S, 2^1S) + A \rightarrow A^+ + \mathrm{He}(1^1S) + e, \tag{24}$$

the electron energies produced being 4·06 eV for the 2^3S and 4·86 eV for the 2^1S metastable atoms. Retarding potential analysis of the electrons ejected enables the excitation of the 2^3S and 2^1S states to be studied separately.

Fig. 4.45 shows a schematic diagram of Čermak's apparatus. Helium entering the source was formed into an atomic beam by means of a multi-channel tube T_1. It was crossed in the excitation region A by an electron current of 100 μA, collimated by a magnetic field of intensity

---

† HOLT, H. K. and KROTKOV, R., loc. cit. p, 231.
‡ KUPRIANOV, S. E., *Optika Spektrosk.* **20** (1966) 163; *Optics Spectrosc., N.Y.* **20** (1966) 85.
§ OLMSTED, J., NEWTON, A. S., and STREET, K., *J. chem. Phys.* **42** (1965) 2321.
‖ ČERMAK, V., ibid. **44** (1966) 3774.

about 100 gauss. Charged particles were removed from the beam by the electric field between the plates P_1 and P_2. The grid G_s prevented electrons from entering the collision region. The reaction (24) with

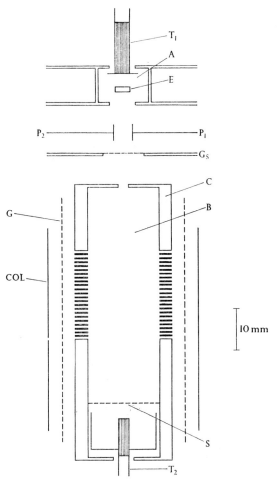

Fig. 4.45. Schematic diagram of Čermak's apparatus for measuring excitation cross-sections of metastable atoms by electron impact.

argon atoms took place in the collision region B, into which argon atoms were introduced through the multi-channel tube T_2. The pressure in B was about 5×10^{-4} torr.

Electrons emitted perpendicular to the long axis of the collision chamber could escape from the field-free region B through the channels in the cylinder C. These channels were coated with colloidal graphite

4.2   CROSS-SECTIONS FOR EXCITED ATOMS   239

to reduce electron reflection. A retarding potential analysis between the grid G and C enabled the energy distribution of the electrons reaching the collector C to be determined. Fig 4.46 (curve 1) shows the current of electrons reaching the collector as a function of the stopping potential between G and C. The differential curve (2) shows clearly the separation between electrons ejected in the de-excitation of 2^3S

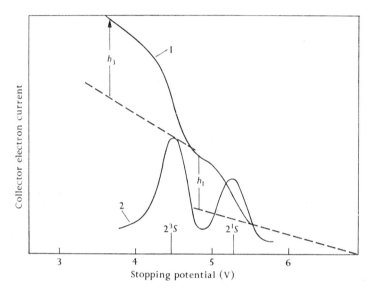

FIG. 4.46. Collected electron current (curve 1) produced by collisions of second kind between A and metastable He atoms in Čermak's experiment. Curve 2 is the differential electron current curve showing the 2^3S and 2^1S signals separated. The dotted lines show the extrapolation of the electron background so that h_1, h_3 are respectively signals due to 2^1S and 2^3S metastable atoms.

and 2^1S metastable atoms. Difficulty was experienced in allowing for the background of electrons that could not be unambiguously assigned to either metastable state. Allowance for this was made by extrapolating the total current curve in the manner shown in Fig. 4.46 so that h_3 and h_1 were taken as the currents due to the 2^3S and 2^1S metastable atoms respectively.

2.1.2.3. *Detection of auto-ionizing metastable atoms.* In § 2.5.3 of Chapter 3 some of the fine structure in ionization cross-section curves near the threshold was attributed to the formation of metastable states of the neutral atom of an energy in excess of the ionization energy, followed by auto-ionization. If such metastable states are formed in an atomic beam they may have a sufficiently long lifetime for them to decay, giving rise to the production of ions sufficiently far from the excitation region to enable the process to be distinguished from direct ionization.

The excitation curve for the formation of the auto-ionizing $(1s\ 2s\ 2p)^4P_{\frac{5}{2}}$ metastable state of lithium has been studied by Feldman and Novick[†] by observing the ions formed after decay. For auto-ionization to occur, the initial discrete and final continuum level must have the same parity and total angular momentum quantum number J. Furthermore, if Russell–Saunders coupling is valid, the separate total orbital and spin quantum numbers must also be conserved. When these conditions are satisfied the lifetime towards auto-ionization is of the order of the time of revolution in a Bohr orbit, 10^{-14} to 10^{-15} s.

We may apply these rules to the consideration of the low-lying auto-ionization states in lithium. These will arise from the $1s\ 2s^2$ and $1s\ 2s\ 2p$ configurations. The former can give rise only to a $^2S_{\frac{1}{2}}$ term from which auto-ionization occurs to a continuum level in which the outgoing electron is in an $s_{\frac{1}{2}}$ state. More interesting possibilities arise from the $1s\ 2s\ 2p$ configurations. The ^2P terms are of no special interest as they may auto-ionize via ejection of a p electron, but the situation is different for the ^4P terms. Of these the $^4P_{\frac{3}{2}}$ and $^4P_{\frac{1}{2}}$ states may auto-ionize through departure from Russell–Saunders coupling, which links them with the $^2P_{\frac{3}{2}}$ and $^2P_{\frac{1}{2}}$ states of the same configuration through spin–spin and spin–orbit interactions. This is not possible for the $^4P_{\frac{5}{2}}$ state, which can only auto-ionize through the tensor component of the spin–spin interaction and ejection of an electron in an $f_{\frac{5}{2}}$ state. The lifetime as calculated by Pietenpol,[‡] using simple wave functions, is about 1.6×10^{-5} s. This is so long that a lithium atom in this state could travel at thermal velocities a distance of a few cm before decaying by auto-ionization. A comparable situation exists for the corresponding state of He⁻ discussed in Chapter 22.

Because the decay distance for $^4P_{\frac{5}{2}}$ atoms under thermal conditions is comparable with the dimensions of ordinary apparatus, it is possible to investigate experimentally their production by electron impact and to determine their lifetime. Fig. 4.47 illustrates diagrammatically the apparatus used by Feldman and Novick for this purpose.

The general principle is that of the usual crossed-beam method, although it was not necessary to use a.c. background discrimination. $^4P_{\frac{5}{2}}$ metastable atoms, produced by electron bombardment of a lithium

[†] FELDMAN, P. and NOVICK, R., *Atomic collision processes*, ed. McDOWELL, M. R. C., p. 201 (North Holland, Amsterdam, 1964).
[‡] PIETENPOL, J. L., *Phys. Rev. Lett.* **7** (1961) 64.

atom beam, were detected through the Li+ ions resulting from auto-ionization. Referring to Fig. 4.47, the electron gun was mounted on the front of an oven of standard type for the production of atomic beams. The electron beam, which passed through a slit 0·008 in wide, was collimated by a magnetic field of 1000 gauss, which also served to trap ions and electrons in that region.

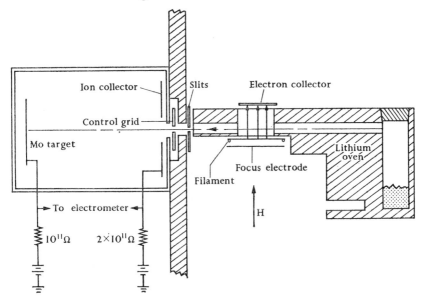

FIG. 4.47. Apparatus of Feldman and Novick for measuring cross-section for production of auto-ionizing metastable states of lithium by electron impact.

The detector for the metastable atoms consisted of an annular ring, biased to collect positive ions, and a control grid placed in front of the entrance and biased negatively so as to reduce any effects due to the presence of charged particles in the beam. It was possible, by reversing the bias on the detector, to collect electrons instead of positive ions so as to provide a partial check that the current to the detector really arose from decay of $^4P_{\frac{5}{2}}$ atoms. This could only be partial because electrons resulting from auto-ionization are emitted with 50 eV energy and are less efficiently collected than the positive ions that have energies less than thermal. It was possible to move the detector parallel to itself so that the distance from the centre of the bombarding region could be varied from 3 to 6 cm.

The decay time for the $^4P_{\frac{5}{2}}$ atoms was determined by observing the variation of the signal at the detector with the distance of the detector from the source. Allowance had to be made for the velocity distribution of the lithium atoms in the beam and for the finite length of the source of the metastable atoms. A mean lifetime of $5 \cdot 1 \pm 1 \times 10^{-6}$ s was obtained, which is of the same order of magnitude as that estimated theoretically by Pietenpol.

A check that the detector signal arose from excitation of lithium atoms and an estimate of the absolute magnitude of the excitation cross-section for the ^4P state was made by using a surface ionization detector to measure the intensity of the lithium atom beam. It was verified that the ratio of the detector signal to the beam intensity for fixed electron bombardment conditions was independent of oven temperature over the range 440–550° C (corresponding to an increase of neutral beam intensity by a factor of 10).

2.1.3. *Methods combining quenching and direct detection of metastable atoms.* A combination of techniques of electric field quenching and electron ejection from surfaces has been used by Lichten and Schultz† to study the excitation of the $2s$ state of H, and by Holt and Krotkov‡ to study the excitation of 2^1S and 2^3S states of He.

Lichten and Schultz adapted a procedure used by Lamb and Retherford§ in the course of their experiments on the Lamb shift in hydrogen. The beam containing a fraction of metastable excited atoms was allowed to impinge on a platinum surface, and the ejected electron current from the surface used to determine the incident flux of metastable atoms. To eliminate background effects the atomic beam could be blocked off from the bombardment region by means of a stop wire (W in Fig. 4.48), while an electrostatic quenching field could be applied across the beam between the bombardment region and the detector, so that ejection from the detector by metastable atoms could be distinguished from that due to photons.

The following measurements were made at each electron energy. First, from the total current, i_1, ejected from the platinum detector and that, i_2, when the quenching field was on, the total quenchable current, $i_1 - i_2$ was obtained. The measurements were repeated with the stop wire in position to interrupt the beam, giving a reduced quenchable current $i'_1 - i'_2$. Both quenchable currents refer to fluxes of metastable atoms incident on the detecting plate. The first included not only the flux due to metastable atoms produced by electron bombardment of the atom

† LICHTEN, W. and SCHULTZ, S., loc. cit., p. 234.
‡ HOLT, H. K. and KROTKOV, R., loc. cit., p. 231.
§ LAMB, W. E. and RETHERFORD, R. C., *Phys. Rev.* **81** (1951) 222.

beam but also that arising from dissociation of the background concentration of molecular hydrogen, whereas the second included the latter only. The difference $i_1-i_2-i'_1+i'_2$ was therefore taken as the wanted signal current.

A number of further check observations were made. It was verified that the background current was unaffected by the quenching field when no hydrogen was admitted to the oven. This showed that no ions

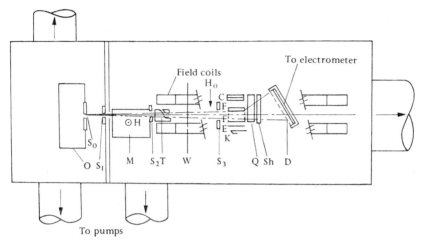

Fig. 4.48. Apparatus of Lichten and Schultz for studying production of metastable hydrogen atoms in direct and exchange collisions with electrons. S_0, S_1, S_2, S_3 are slits, F a field free box, E the electron gun central electrode, Sh an electrostatic shield. The other components are described in the text. (Drawing to scale 1:8.)

or electrons from the electron gun were penetrating to the detector. It was established that the wanted signal was proportional to the electron beam current at currents greater than that, 100 μA, used in the experiment, showing quenching by space charge effects to be unimportant. On the other hand, the signal was not proportional to the pressure in the oven source. This was ascribed to scattering in the beam occurring near the exit slit from the source and was immaterial in relative measurements for which the pressure was held constant.

The general layout of the apparatus is illustrated in Fig. 4.48, with dimensions indicated. The atom source was a tungsten oven O, designed by Hendrie,† operating at a temperature of 3000° K and a gas pressure of 2 torr and giving 91 per cent dissociation. During the experiment a uniform magnetic field H_0 of 575 gauss was applied over the bombarding and detecting regions to collimate the electron beam in the plane of the atomic beam.

† HENDRIE, J. M., J. chem. Phys. 22 (1954) 1503.

The bombarding electron beam from the cathode K was measured on a collector C, made of copper heated to 350° C to prevent the formation of insulating layers that reflect electrons. A positive bias of 45 V was applied to the collector relative to the field-free space traversed by the atom beam and the collector was tilted at an angle of $22\frac{1}{2}°$ to the electron beam.

The detector D was a platinum sheet enclosed in a brass box, the side facing the electron beam being covered with a grid of nichrome wires. It accepted all atoms scattered within an angle of 22° vertically and ±7° horizontally. These angles were sufficient to include 90 per cent of all recoiling metastable atoms when deuterium was used instead of hydrogen. The quencher Q consisted of two parallel plates 1·5 cm apart between which a potential difference of 45 V was applied. Measurements were carried out up to bombarding energies of 40 eV only.

In fact, the experiment was designed not only to measure the relative total cross-section for H(2s) excitation but also to separate out the part that is contributed by exchange of electrons of opposite spin during the collision. This can be accomplished by studying the excitation of polarized atoms. M is a magnet that produced a non-uniform field H and separated out two polarized beams of H(1s) atoms with magnetic quantum numbers $m_s = \frac{1}{2}$ (α-atoms) or $m_s = -\frac{1}{2}$ (β-atoms). After deflexion in the magnet field the beam passed through the field of a specially shaped magnet T that rotated it through 90°, so that the electron beam from the cathode K intersected the atomic beam parallel to the wide side of the section of the latter. The source slit S_0 could be moved so that, after deflexion in the non-uniform field H, either α or β atoms intersected the electron beam. Direct excitation of the 2s state will not change the m_s value of the atomic beam but exchange excitation may do so. In the steady field H_0 of 575 gauss in the excitation and detection region, produced by means of the field coils, the $2p_{\frac{1}{2}}$ ($m_s = -\frac{1}{2}$) and $2s_{\frac{1}{2}}$ ($m_s = -\frac{1}{2}$) states have the same energy, so that mixing occurs, and atoms in this state are quenched with the emission of Ly α-radiation. If, therefore, the H(1s)β beam is bombarded by the electron beam, H(2s) atoms excited directly are immediately quenched, while H(2s) atoms excited by spin exchange may form H(2s) α-atoms, which remain metastable and reach the detector. If Q_{ex} and $Q_T = Q_{\text{ex}} + Q_{\text{direct}}$ are respectively the cross-section for spin exchange and total excitation,

$$\frac{Q_{\text{ex}}}{Q_T} = \frac{S(-)}{S(+)+S(-)}, \tag{25}$$

where $S(-)$ and $S(+)$ are the signals at the detector when β- and α-beams are excited respectively.

To obtain the absolute value of the cross-section, an approximate determination of the electron yield per metastable atom incident on the platinum detector, ζ, was made. A potential was applied between the grid and detector sufficient to quench all the metastable atoms. The resulting electron yield then comes from incidence of Ly α-radiation from the $2p$–$1s$ transitions for which the yield per photon is known. The photoelectron yield for Ly α-photons on an untreated platinum surface has been measured as 0.018 ± 0.005 per photon.† In terms of this the measured value of ζ was 0.065 ± 0.025.

In the experiments of Holt and Krotkov on the excitation of $n = 2$ states in He the 2^1S and 2^3S metastable states could be separated by applying an electric field which quenched the 2^1S atoms by causing mixing with the 2^1P state. An inhomogeneous magnetic field was then used to measure the polarization of the 2^3S atoms by separating the beam into three components, corresponding to magnetic quantum numbers $M_s = \pm 1$ and 0. Direct excitation of the 2^3S state should lead to equal populations of all three magnetic sub-states. Population of the 2^3S state by cascade from higher ^3P states might be expected, however, to populate the $M_s = \pm 1$ substates differently from the $M_s = 0$ substates, so that effects of cascade excitation are manifested by differences in the excitation curves for the respective substates.

Fig. 4.49 shows a schematic diagram of Holt and Krotkov's apparatus. The vacuum system consisted of three parts. The source chamber S contained the electron gun and was filled with helium gas which was excited by an electron beam perpendicular to the plane of the diagram. A beam of metastable and ground state He atoms effused through the slit into the buffer chamber B where an electric field could be applied to quench 2^1S metastable atoms. It then entered the main chamber M where it was split into its three components by the inhomogeneous magentic field. The detector D and its slit could be moved to receive any of the three beam components. The detector consisted of an electron multiplier in which the metastable atoms struck a tungsten surface and the ejected electrons were directly multiplied by the dynode effect.

2.1.4. *Optical methods for measuring metastable atom production rates.* Two optical methods—optical absorption and anomalous dispersion—

† WALKER, W. C., WAINFAN N., and WEISSLER, G. L., *J. appl. Phys.* **26** (1955) 1367; HINTEREGGER, H. E. and WATANABE, K., *J. opt. Soc. Am.* **43** (1953) 604.

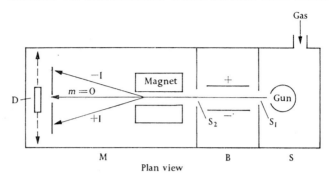

Fig. 4.49. Schematic diagram of apparatus of Holt and Krotkov for studying excitation of He 2^1S and 2^3S metastable states by electron impact. The slits S_1 and S_2 were each of dimensions 0.015×0.635 cm and placed respectively 4·8 cm and 18·8 cm from the centre of the interaction volume (labelled Gun). The slit in front of the detector D was of dimension 0.025×1.59 cm and was 91·1 cm from the centre of the interaction volume.

are available for measurement of the concentration of metastable atoms. Then, if the concentration is proportional to the rate of production, excitation functions for the production of the metastable state may be obtained. Only the absorption method has actually been applied to the determination of excitation functions for metastable states under conditions of excitation of such low current density and pressure that multiple processes are improbable. The anomalous dispersion method has, however, been applied to study the concentration of atoms in metastable states in a glow discharge, from which information about excitation functions can be obtained indirectly.

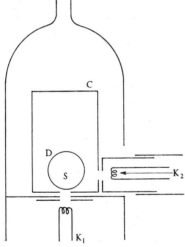

Fig. 4.50. Apparatus of Milatz and Ornstein and of Woudenberg and Milatz for studying excitation of metastable levels of Ne and He.

2.1.4.1. *The optical absorption method*. The optical absorption method has been applied by Milatz and Ornstein† to measure the rate of production of metastable neon atoms by electron impact, and a similar method has been applied to He by Woudenberg and Milatz.‡ Their apparatus is illustrated in Fig. 4.50.

† Milatz, J. M. W. and Ornstein, L. S., *Physica* **2** (1935) 355.
‡ Woudenberg, J. P. M. and Milatz, J. M. W., ibid. **8** (1941) 871.

Electrons from an oxide-coated cathode K_1 were accelerated through a slit system into a cage C where they were collected. Two holes, D, were cut opposite each other in the sides of the cage. Light from the positive column of a neon discharge tube passed through these holes and then fell on the slit of a spectrograph. The plate holder of the spectrograph was replaced by a curved slit in a position to receive light of wavelength 6402 Å. This line corresponds to the transition p_9–s_5 and it is absorbed strongly by neon atoms excited to the s_5 state. Behind the slit was placed a photocell. On varying the potential difference between K_1 and C the intensity of the λ 6402 radiation reaching the photocell varied. The absorption of this radiation on passing through the electron tube was then a measure of the number of metastable atoms present.

Let $\Delta I/I$ be the fractional reduction of intensity of the radiation over the absorption line due to the presence of atoms in an excited state. Then
$$\Delta I/I = l \int \kappa_\nu \, d\nu/\Delta\nu, \tag{26}$$
where $\Delta\nu$ is the width of the absorption line, κ_ν the absorption coefficient of the radiation in the frequency range between ν and $\nu+d\nu$, and l the path length of the beam in the space containing the excited atoms. Then
$$\kappa_\nu \, d\nu = \pi e^2 N_j f_{kj}/mc, \tag{27}$$
where f_{kj} is the oscillator strength associated with the transition giving rise to the absorption and N_j the number of excited atoms per unit volume. Thus
$$N_j = (mc/\pi e^2)(\Delta\nu/l f_{kj})(\Delta I/I). \tag{28}$$

The excitation function for the j state will then be obtained by measuring the variation of $\Delta I/I$ with the energy of the electrons entering C.

Owing to space charge effects the potential at the point where the electron beam passed opposite to the window D was not equal to the potential of C. To determine the true potential of S a subsidiary cathode K_2 was introduced and its potential varied. At a certain stage the beam from this cathode just became visible at the point S. Under these conditions the potential difference between K_2 and S was just equal to the excitation potential of the first level from which visible lines commenced (18·4 V). Thus the energy at S of the electrons emitted from K_1 could be calculated to within about 1 eV.

In the experiments of Milatz and Ornstein the concentration of neon atoms in the s_5 state was about 5×10^8 per cm³ and the ratio $\Delta I/I$ was about 0·015. Hadeishi[†] has used absorption of the λ 6143 line

[†] HADEISHI, T., *UCRL Report* 10477 (1962).

$(2^2P_{\frac{3}{2}} 3^2P_{\frac{1}{2}})_2 \to (2^2P_{\frac{3}{2}} 3^2S_{\frac{1}{2}})_2$ of Ne to measure the excitation function of the $(2^2P_{\frac{3}{2}} 3^2S_{\frac{1}{2}})_2$ metastable atoms in an energy range close to threshold. Unfortunately, in order to get a sufficiently intense absorption signal, gas pressures of the order 0·5–1·5 torr had to be used. Under these conditions the detailed form of the excitation curve was sensitive to pressure, suggesting that multiple processes were playing an important part.

2.1.4.2. *The anomalous dispersion method.* The refractive index μ of a rarefied gas in the neighbourhood of a critical frequency is given by

$$\mu - 1 = \frac{e^2}{2\pi m} \frac{F_{kj}}{\nu_{kj}^2 - \nu^2}, \qquad (29\,\text{a})$$

where

$$F_{kj} = N_j f_{kj}(1 - G_{kj}), \qquad (29\,\text{b})$$

$$G_{kj} = N_k g_j / N_j g_k, \qquad (29\,\text{c})$$

$$f_{kj} = mc^3 A_{kj}(g_k/g_j)/8\pi^2 e^2 \nu_{kj}^2. \qquad (29\,\text{d})$$

In these expressions j, k refer to the lower and upper levels of the absorption line in question, N_j, N_k, g_j, g_k respectively to the corresponding atomic concentrations in atoms cm^{-3} and the statistical weights of the levels. ν_{kj} is the frequency of the line and A_{kj}, f_{kj} are respectively the Einstein A coefficient and the oscillator strength for the transition. In many cases N_k/N_j is small and G_{kj}, the negative dispersion term, can be taken as small compared with unity. If this is the case

$$F_{kj} = N_j f_{kj}. \qquad (30)$$

If excited atoms in the state j are present in a gas a measurement of the anomalous dispersion in the neighbourhood of a critical frequency ν_{kj} will enable F_{kj} to be determined. Then, if the conditions of excitation are altered, the variation in the concentration of the excited atoms can be investigated.

The usual experimental arrangement was developed by Roschdestwensky[†] and has been applied to the study of excited atoms in gas discharges by Ladenburg and his colleagues. Light from an arc lamp S (Fig. 4.51) is divided into two beams by the first plate, P_1, of a Jamin interferometer. One beam passes along the tube T_1 (50–80 cm long and about 1 cm in diameter) containing the gas under investigation at a pressure of about 1 torr. The other beam passes along a similar evacuated tube T_2, the two coherent beams being united by the second plate, P_2, of the interferometer. The interference fringes are focused on to the

[†] ROSCHDESTWENSKY, D., *Annln Phys.* **39** (1912) 307; *Trudỹ gos. vses. issled. prockt. Inst. Giprovostokneft* **2** (1921) no. 13.

slit of a grating spectrograph G of high dispersion. The spectrum produced is traversed by interference fringes, which are almost horizontal owing to the high dispersion of the spectrograph. On both sides of the spectral lines, however, the horizontal fringes are bent in a characteristic way owing to the rapid variation of refractive index with wavelength in the neighbourhood of a critical frequency. Fig. 4.52 (a) shows

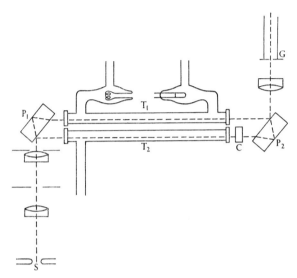

Fig. 4.51. Apparatus for determination of density of excited atoms in gas-discharge tube by measurement of anomalous dispersion.

the appearance of a portion of the spectrum from a neon tube obtained by Ladenburg.†

In most of the work the plane-parallel compensating plate C was placed in the path of the second beam. This has the effect of rotating the fringes relative to the length of the spectrum. The fringes in the neighbourhood of a critical frequency then have a characteristic hooked appearance as shown in Fig. 4.52 (b).

From the positions of the maxima and minima in the fringe system on either side of the critical frequency, F_{kj} can be calculated from the relation
$$F_{kj} = (K\pi m v_{jk}^3 A^2)/e^2 cl, \tag{31}$$
where l is the thickness of the gas layer, A the wavelength separation of the two hooks symmetrically placed with regard to the critical frequency, and K a constant which can be measured for a given apparatus.

† Ladenburg, R., Rev. mod. Phys. 5 (1933) 243.

2.2. Results of the measurements

2.2.1. *Excitation of the 2s metastable states of atomic hydrogen.* Fig. 4.53 shows the excitation functions for the production of the metastable hydrogen $2s_{\frac{1}{2}}$ state as measured by Stebbings *et al.* and by Hils *et al.* The agreement between the form of the curves obtained in the two experiments is very satisfactory.

Absolute values of the excitation cross-section can be obtained by a method suggested by Lichten and Schultz. Allowance has to be made

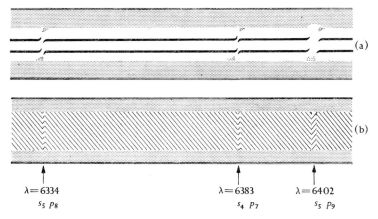

Fig. 4.52. Anomalous dispersion in a neon discharge; (a) without plate C (Fig. 4.51) (b) with plate C.

for an appreciable contribution by radiative transitions from impact-excited $3p$ states. Using tables of radiative transition probabilities it is found that the observed cross-section $Q_p(2s)$ for production of $2s$ atoms is given by

$$Q_p(2s) = Q(2s) + 0 \cdot 23 Q(3p), \qquad (32)$$

where $Q(2s)$, $Q(3p)$ are the cross-sections for direct excitation of the $2s$ and $3p$ states respectively. Calculating these by Born's approximation (see Chap. 7, Table 7.1) gives the theoretical curve shown on the figure that agrees with the measured curve above 200 eV. Subtracting the contribution $0 \cdot 23 Q(3p)$ obtained from the observed form of the $Q(3p)$ excitation curve (Fig. 4.17), the absolute excitation cross-sections given in Fig. 4.54 for direct excitation of the $2s$ state were obtained by Hils *et al.*

Fig. 4.55 (a) shows the measured values obtained in the range up to 40 eV for the total $2s$ excitation cross-section, after correction for cascade from p states using a similar procedure to that described above (eqn (32)), and the cross-section due to the exchange process obtained

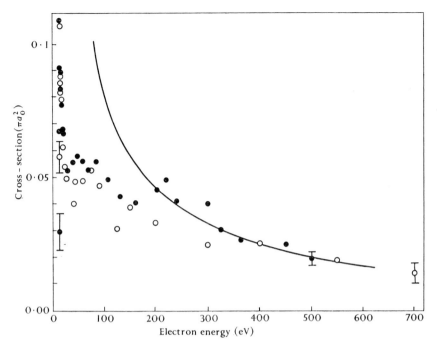

FIG. 4.53. Measured total apparent cross-section for excitation of $2s$ metastable hydrogen atoms. ○ Stebbings, Fite, Hummer, and Brackmann; ● Hils, Kleinpoppen, and Koschmieder; —— cross-section $Q_p(2s) = Q(2s) + 0.23Q(3p)$ calculated by Born's approximation (Chap. 7, § 5.6.1).

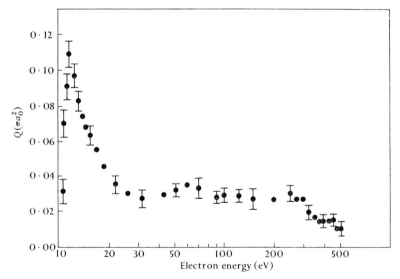

FIG. 4.54. Measured cross-section for excitation of $2s$ metastable state of hydrogen after removal of cascade excitation effects (Hils, Kleinpoppen, and Koschmieder).

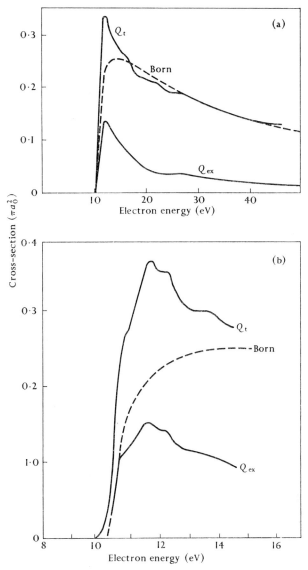

Fig. 4.55. Results of Lichten and Schultz on the measurement of the total cross-section Q_t and spin exchange cross-section Q_{ex} for excitation of H(2s)-metastable states. (b) shows the structure near threshold. The dotted curve is that calculated by Born's approximation (Chap. 7, § 5.6.1).

by Lichten and Schultz. The fine structure near the threshold is shown in Fig. 4.55(b). The absolute value of the cross-section at the maximum is approximately three times as large as that obtained by Hils *et al*. The absolute value obtained by Lichten and Schultz depends on the measured value of the metastable yield, ζ. The value they obtained for H(2s) metastable atoms on platinum was much smaller (less than one-quarter) of the value obtained by Holt and Krotkov for helium metastable atoms on tungsten. An increase in the value of ζ would decrease the value obtained for the absolute cross-section in proportion.

Theoretical discussion of the excitation of the 2s state is given in Chapter 7, § 5.6.1, Chapter 8, § 4.3, and Chapter 9, § 4.

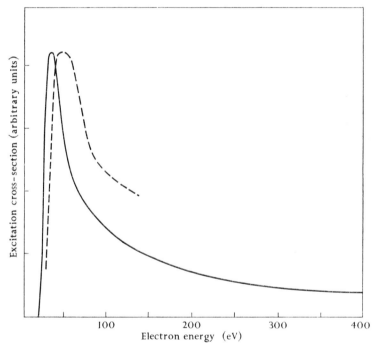

Fig. 4.56. Apparent excitation curve for production of helium metastable atoms. —— Woudenberg and Milatz (2^3S); – – – – Kuprianov ($2^3S + 2^1S$).

2.2.2. *Excitation of the metastable states of* He. It is difficult to obtain reliable excitation functions for particular metastable states over a wide range of energy. The only measurements available are those of Woudenberg and Milatz† for the He 2^3S state, using the absorption of λ 3889 Å as a measure of the metastable concentration, and of Kuprianov,‡ who used surface ionization as a means of detecting metastable intensities but did not separate 2^1S and 2^3S metastable atoms. The curves they obtained are shown in Fig. 4.56, but no attempt

† WOUDENBERG, J. P. M. and MILATZ, J. M. W., *Physica* **8** (1941) 871.
‡ loc. cit. p. 237.

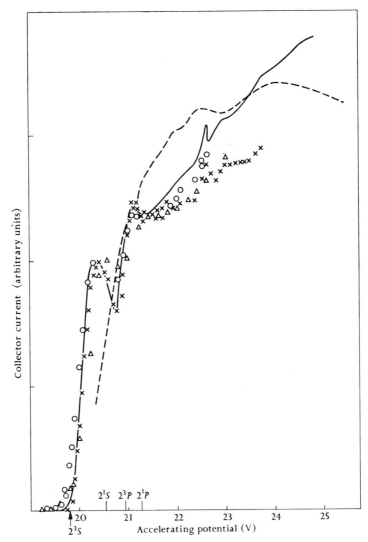

Fig. 4.57. Comparison of structure of He metastable (2^3S+2^1S) excitation curve near threshold obtained by different observers. —— Dowell; △ Holt and Krotkov; × Schulz and Fox; ---- Čermak; ○ Dorrestein.

has been made to allow for cascade processes in order to obtain the true excitation curve.

Much more attention has been paid to the form of the excitation function for He metastable states just above threshold. Fig. 4.57 compares the results obtained by Dorrestein,† Schulz and Fox,‡ Holt

† loc. cit., p. 234.   ‡ loc. cit., p. 235.

and Krotkov,† Čermak,‡ and Dowell.§ The detailed agreement between the different measurements is very good. Difficulty is experienced in measurements of this kind in determining the true electron energy scale, owing to instrumental difficulties such as contact potentials. In Figs. 4.57, 4.61, and 4.63 the accelerating potential scale in volts obtained by Dowell is used. This scale agreed with the known ion onset energies to within ± 0.1 eV for He and Ne and ± 0.2 eV for A, Kr, and Xe. The results of other workers were normalized so that the first maximum in their curves agreed both in energy and in magnitude. The expected onset potentials for the various excited states are shown on the figures.

Three features of the curve of Fig. 4.57 should be noted, viz. the peaks at 20·4 and 21·1 eV and the structure in the 22·3–22·8 eV region. The peak at 20·4 eV lying below the threshold for 2^1S excitation must be attributed to 2^3S excitation.

Information about the origin of the other peaks can be obtained from Fig. 4.58, which shows the excitation functions obtained separately for 2^3S and 2^1S metastable excitation by Holt and Krotkov‖ and by Čermak.†† These two experiments disagree in the relative cross-sections for the 2^3S and 2^1S excitation, but agree in showing the 2^1S excitation to be free of fine structure, so that all the observed structure appears to be associated with 2^3S excitation. Čermak did not obtain the peak at 21·0 eV but found one in the 22·5–23·0 eV region. Holt and Krotkov found a peak in the 2^3S excitation just above 21·0 eV and a shoulder close to 23·0 eV. The most conclusive evidence that this structure is to be associated with 2^3S excitation is to be found, however, in the energy-loss measurements of electrons scattered in helium through a fixed angle, at 72° (Schulz and Philbrick‡‡) and at 0° (Chamberlain and Heideman§§) (see Figs. 5.28 and 5.29).

The absolute value of the cross-section at the 20·4-eV peak was estimated by Schulz and Fox‖‖ to be $(4\pm 1·2)\times 10^{-18}$ cm², although a more accurate determination by Fleming and Higginson††† who obtained $(2·6\pm 0·4)\times 10^{-18}$ cm is described in § 3.2 of Chapter 5.

Fig. 4.59 shows the measurements of Holt and Krotkov of the excitation functions of the $M_s = 0$ and $M_s = 1$ magnetic substates of the 2^3S

---

† loc. cit., p. 231.   ‡ loc. cit., p. 237.   § loc. cit., p. 235.
‖ loc. cit., p. 231.   †† loc. cit., p. 237.
‡‡ SCHULZ, G. J. and PHILBRICK, J. W., *Phys. Rev. Lett.* **13** (1964) 477.
§§ CHAMBERLAIN, G. E. and HEIDEMAN, H. G. M., ibid. **15** (1965) 337.
‖‖ loc. cit., p. 235.
††† FLEMING, R. J. and HIGGINSON, G. S., *Proc. phys. Soc.* **84** (1964) 531.

metastable state. As expected, the excitation curves are the same below the 2^3P threshold where only direct 2^3S excitation is possible. The separation between the two curves above the threshold provides evidence for population of 2^3S states by cascade in the transition 2^3P–2^3S (see § 2.1.3).

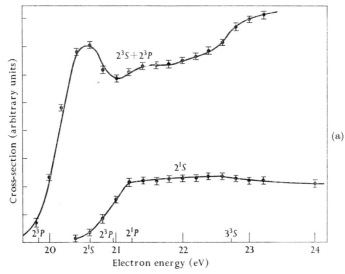

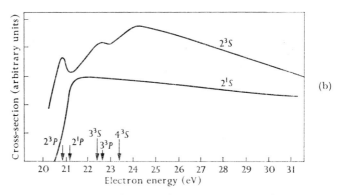

Fig. 4.58. Excitation functions for $n = 2$ states in helium. (a) Measurements of Holt and Krotkov of 2^1S and $2^3S + 2^3P$ excitation. (b) Measurements of Čermak of 2^1S and 2^3S excitation.

The magnitude of the separation is smaller, however, by a factor of 6 than that expected for the predicted 2^3P alignment at threshold. This anomaly is of the same character as that discussed in relation to the polarization of impact radiation near threshold (see § 1).

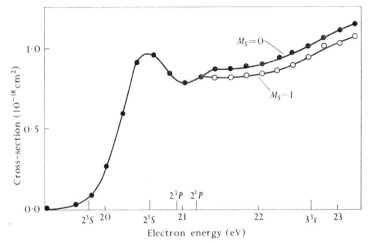

Fig. 4.59. Excitation function for magnetic substates ($M_s = 0, 1$) of the He 2^3S state (measurements of Holt and Krotkov). The peak of the total 2^3S cross-section has been adjusted to 3×10^{-18} cm².

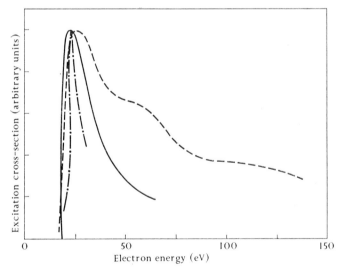

Fig. 4.60. Excitation function for metastable states of neon. ——— Milatz and Ornstein; —·—·— Hadeishi; - - - - Kuprianov.

Theoretical discussion of the excitation of the metastable states of helium is given in Chapter 7, § 5.6.9, Chapter 8, § 6.2, Chapter 9, § 5.

2.2.3. *Excitation of metastable states of other inert gases.* Of the other inert gases most work has been done on neon. Fig. 4.60 shows the results obtained using an optical absorption method by Milatz and Ornstein†

† MILATZ, J. M. W. and ORNSTEIN, L. S., *Physica* **2** (1935) 355.

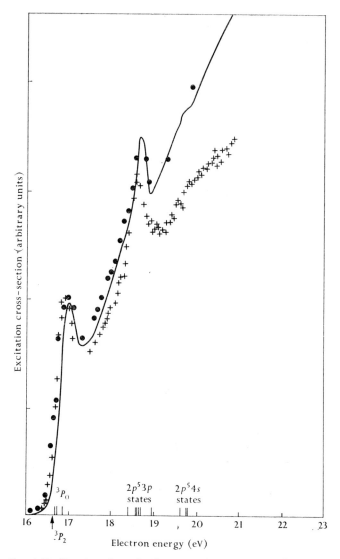

Fig. 4.61. Structure in excitation curve for neon metastable states near threshold. —— Dowell; + Olmsted, Newton, and Street; ● ● ● Dorrestein.

and Hadeishi† and also, using surface ionization detection, by Kuprianov.‡

Fig. 4.61 shows the structure in the metastable excitation curve for Ne just above the threshold. The agreement between the form of the

† HADEISHI, T., *UCRL Report* **10477** (1962).
‡ KUPRIANOV, S. E., loc. cit., p. 237.

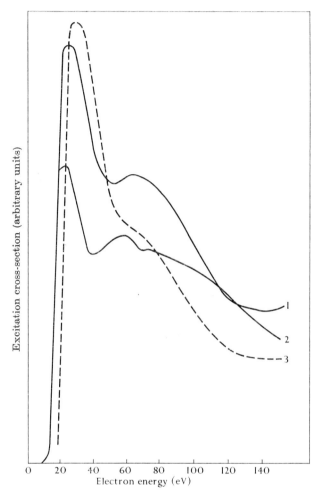

Fig. 4.62. Measurements of Kuprianov of excitation functions for metastable states of (1) Kr, (2) Xe, (3) A.

excitation curves obtained by Dorrestein,† Olmsted et al.,‡ and Dowell§ is once again seen to be good. Experiments that measure excitation to the separate metastable states are not available, so the assignation of the major peaks in the structure to $^3P_2$ or $^3P_0$ excitation cannot be made.

Fig. 4.62 gives the results obtained by Kuprianov for excitation of A, Kr, and Xe metastable atoms over an energy range up to 140 eV, while Fig. 4.63 shows the structure near the threshold in these cases.

† loc. cit., p. 234,
‡ OLMSTED, J., NEWTON, A. S., and STREET, K., J. chem. Phys. **42** (1965) 2321.
§ loc. cit., p. 235.

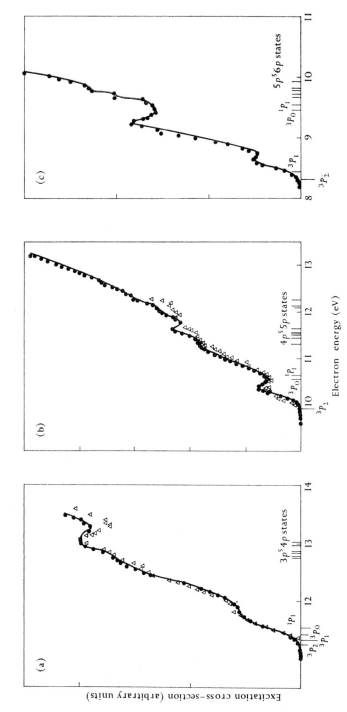

FIG. 4.63. Structure of excitation functions for metastable states near threshold: (a) argon, (b) krypton, (c) xenon. ● Dowell; △ Olmsted, Newton, and Street.

Fig. 4.63 (a) and (b) compares the work of Olmsted *et al.* and Dowell for A and Kr respectively, while measurements for Xe (Fig. 4.63 (c)) have been made only by Dowell.

Of some interest is the form of variation of the metastable excitation cross-section near the threshold. According to Wigner's[†] threshold law, which should apply to neutral atom excitation by electron impact (see Chap. 9, § 10.2), the excitation cross-section should increase like $(E-E_0)^{\frac{1}{2}}$

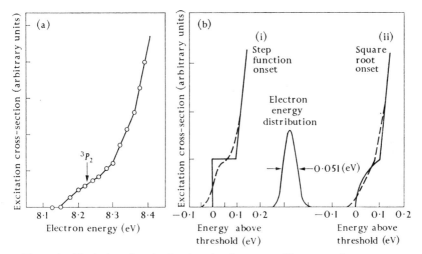

Fig. 4.64. Variation of excitation function for metastable states of xenon near threshold: (a) observed by Dowell; (b) variation expected for (i) step function and (ii) $(E-E_0)^{\frac{1}{2}}$ onset law.

near the threshold energy, E_0. On the other hand, Morrison[‡] has interpreted fine structure in ionization cross-section curves near the threshold to indicate that the cross-section for neutral atom (auto-ionizing) excitation near the threshold should follow a step function.

Fig. 4.64 shows the form of variation near the excitation cross-section obtained for Xe by Dowell (a). It also shows (b) the form of actual variation to be expected for an $(E-E_0)^{\frac{1}{2}}$ and a step-function variation allowing for the known energy distribution of the bombarding electrons. The observed form of variation agrees better with a step function than with the square-root energy excess dependence suggested by Wigner.

2.2.4. *Excitation of the auto-ionizing* $(1s\,2s\,2p)^4P$ *state of lithium.* The results of the measurements of Feldman and Novick§ of the excitation

[†] WIGNER, E. P., *Phys. Rev.* **73** (1948) 1002.
[‡] MORRISON, J. D., *J. appl. Phys.* **28** (1957) 1409; DORMAN, F. H., MORRISON, J. D., and NICHOLSON, A. J. C., *J. chem. Phys.* **32** (1960) 378.
§ loc. cit., p. 240.

of this state are presented in Fig. 4.65, which shows the variation of the detector signal with electron energy. The excitation energy of the ^4P terms has been estimated by Ta-You Wu† as 58 eV. The sharp peak at 58 eV, arising from a threshold at 56 eV, almost certainly arises from excitation of the ^4P term from the ground state. As this is an intercombination transition we would expect it to exhibit a sharp maximum very close to the threshold (compare the excitation of triplet states of

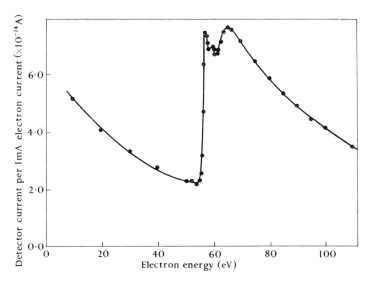

Fig. 4.65. Excitation curve for $(1s\,2s\,2p)^4P_{\frac{5}{2}}$ metastable auto-ionizing state of lithium observed by Feldman and Novick.

helium from the ground singlet state, Fig. 4.19). The small subsidiary maximum at 60 eV and the larger one at 65 eV presumably arise from excitation of other quartet terms associated with core-excited configurations such as $(1s\,2p)^2\;^4P$ and $(1s\,2s\,ns)^4S\,(n \geqslant 3)$. These would populate the $^4P_{\frac{5}{2}}$ state by radiative transitions.

A curve of similar shape was obtained with the detector biased to collect electrons. The nature of the background is uncertain. From these measurements, the geometry of the bombarding system, the electron beam current, and allowance for decay in flight, the absolute cross-section came out to be about 10^{-19} cm^2, which is of the expected order of magnitude.

† Wu, T.-Y., *Phil. Mag.* **22** (1936) 837; Wu, T.-Y. and Shen, S. T., *Chin. J. Phys.* **5** (1940) 150.

3. The measurement of cross-sections for collisions of electrons with excited or ionized atoms

Although the quantitative experimental study of the collisions of electrons with excited or ionized atoms presents special difficulties, the subject has received considerable attention in the last few years because of its importance for the interpretation of the behaviour of high temperature plasmas. These include not only the high density plasmas being generated in the course of research on controlled thermonuclear power but also the solar corona.

So far the only results that have been obtained for collisions with excited atoms rather naturally refer to metastable atoms. Two types of collision processes have been studied—elastic and superelastic collisions. An important technique of the crossed-beam type has been developed successfully for the measurement of the cross-sections for collisions with positive ions (Chap. 3, § 3.1.1). In addition, information about collision rates with excited atoms has been derived from observations on electric discharges.

Ionizing collisions of electrons with ions or excited atoms have already been discussed (Chap. 3, § 3.1.1). In this section we are concerned mainly with electron collisions with ions or excited atoms, leading to further excitation or to de-excitation (superelastic collisions). It is convenient, however, also to discuss here the measurement of the total cross-section for collisions between electrons and excited atoms.

3.1. *Collisions with metastable atoms*

3.1.1. *Superelastic collisions.* The first measurements of cross-sections for superelastic collisions of electrons with metastable Hg atoms were carried out by Latyscheff and Leipunsky.† These were detected from observations of the gain in energy of electrons passing through a region containing metastable Hg atoms. The experiments are described in Chapter 5, § 2.4. In the present section we describe experiments in which superelastic collisions have been studied by measuring directly the concentration of atoms in the initial and final states. In 1953 Phelps and Molnar‡ used the absorption technique (see § 2.1.4) to determine the rate of change of the concentration of metastable 2^1S and 2^3S helium atoms in a discharge afterglow. From such observations, taken in conjunction with measurement of the electron concentration, the cross-section for deactivation of 2^1S atoms to 2^3S by impact of electrons with

---

† LATYSCHEFF, G. D. and LEIPUNSKY, A. I., *Z. Phys.* **65** (1930) 111.
‡ PHELPS, A. V. and MOLNAR, J. P., *Phys. Rev.* **89** (1953) 1202.

a thermal energy distribution was derived. The absorption technique involved was considerably refined by Phelps and Pack.†

The afterglow was produced in the usual way (Chap. 2, § 5 and Chap. 18, § 5) by applying a high-voltage pulse of microwave power. After a suitable interval the pulsed radiation of appropriate wavelength (3889 Å and 5016 Å for selective absorption by 2^3S and 2^1S atoms respectively) was collimated and passed through the afterglow. Transmitted radiation of the desired wavelength was then separated out by an interference filter and focused on the cathode of a photomultiplier. The absorption

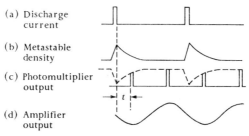

Fig. 4.66. Wave form for the a.c. time-sampling technique for measurement of time-varying optical absorption by afterglow in the measurements of Phelps and collaborators.

was measured in the first experiments by direct observation of the amplified photomultiplier current with and without the afterglow present.

An order of magnitude gain in sensitivity was later achieved in the following way. Referring to Fig. 4.66, (a) illustrates the sequence of breakdown voltage pulses, (b) the time variation of the resulting concentration of metastable atoms and the dotted curves of (c) the variation in transmission of selectively absorbed radiation. The photomultiplier was gated at twice the frequency of the breakdown pulses. In this way the current pulse from the multiplier received soon after breakdown was much more reduced by absorption than the succeeding pulse occurring in a later stage of the afterglow (see the full line curve of Fig. 4.66 (c)).

The amplitude of the fundamental component of the Fourier analysis of the wave form (Fig. 4.66 (d)) of the photomultiplier current is therefore proportional to the difference between the height of the two successive current pulses. By means of a narrow-band amplifier and synchronous detector operating at the same frequency as the breakdown pulse repetition rate, the fundamental component was selected and its amplitude measured. The variation of the absorption with time was then

† Phelps, A. V. and Pack, J. L., *Rev. scient. Instrum.* **26** (1955) 45.

obtained by measuring this amplitude as a function of the time interval between the cessation of the breakdown pulse and the succeeding opening of the photomultiplier gate.

The electron concentration was measured in the same way as in the afterglow experiments described in Chapter 2, § 5, Chapter 18, § 5, and

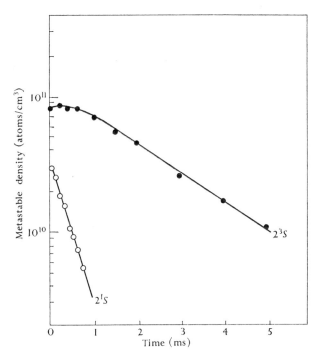

Fig. 4.67. Metastable concentration in helium afterglow as measured by Phelps and Pack ($p = 1\cdot5$ torr).

Chapter 20, § 3 by following the way in which the resonance frequency of a microwave cavity containing the afterglow varied with time in the afterglow phase.

In the experiments of Phelps† using the gated-photomultiplier method the breakdown pulses were of 20-μs duration with a repetition rate of 10, 30, or 90 c/s. The helium pressure used was 1·5 torr and all precautions were taken to ensure its purity. It was obtained by evaporation of the liquid and contained less than two parts in 10^7 of neon. The afterglow cell was evacuated to a pressure less than 10^{-9} torr before introducing the helium. Fig. 4.67 illustrates the observed variation of the concentrations of the 2^1S and 2^3S atoms during the afterglow.

† Phelps, A. V., *Phys. Rev.* **99** (1955) 1307.

Whereas the 2^1S concentration falls rapidly, that of 2^3S at first rises and then falls very much more gradually.

To interpret these results we must allow for loss by diffusion to the walls, by electron impact through superelastic and inelastic collisions, and by impact with neutral atoms in the afterglow. Thus, if N_1, N_3, N_0, N_e are the respective concentrations of 2^1S, 2^3S, 1^1S helium atoms and electrons respectively, we have

$$\frac{\partial N_1}{\partial t} = D_1 \nabla^2 N_1 - \alpha N_0 N_1 - \beta N_e N_1. \tag{33}$$

D_1 is the diffusion coefficient of 2^1S helium atoms in normal helium, while α and β are rate coefficients for destruction of 2^1S helium atoms by neutral atom and electron impact respectively.

If the first two terms on the right-hand side were negligible we would have
$$N_1 = N_1^0 e^{-\nu t}, \tag{34}$$
where
$$\nu = \beta N_e \tag{35}$$
and is independent of the helium gas pressure for a given electron concentration N_e. It is assumed that the change of N_e is very small in a time $1/\nu$ as it is in practice. On the other hand, if either or both of the first two terms on the right-hand side of (33) is important, the value of ν derived from the relation
$$\nu t = \ln(N_1^0/N_1) \tag{36}$$
should depend on the pressure p, since D_1 is proportional to $1/p$ and N_0 to p.

Fig. 4.68 shows the observed variation of ν derived in this way as a function of electron concentration at different gas pressures. It will be seen that ν is both independent of gas pressure and proportional to the electron concentration. This confirms the assumption that only loss by electron impact is important (the only other possibility is that, over the pressure ranges studied, the pressure variations of the two terms combined to produce an accidental pressure independence).

It remains to identify the electron-impact process. Reference to the observed variation of the 2^3S concentration suggests immediately that it is the superelastic collision

$$\text{He}(2^1S) + e \rightarrow \text{He}(2^3S) + e + 0{\cdot}76 \text{ eV}. \tag{37}$$

Thus the initial increase of 2^3S concentration, which persists only over a time of order $1/\nu$, can be ascribed to population from 2^1S through these collisions.

The observed value of β is $3\cdot 5\times 10^{-7}$ cm³ s⁻¹, which gives for the mean cross-section $\bar{Q}$ for the process (37) due to electrons with a thermal energy distribution
$$\bar{Q} = 3\cdot 0 \times 10^{-14}\text{ cm}^2.$$
Comparison with theoretical estimates is discussed in Chapter 8, § 6.2.

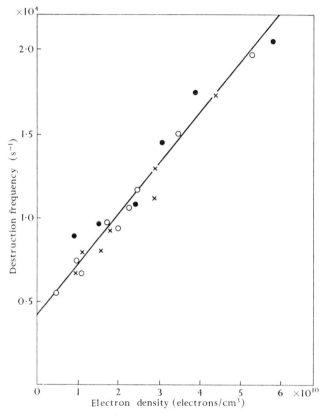

Fig. 4.68. Measured destruction frequency ν for He(2^1S) as a function of central electron density at various pressures ($T = 300°$ K) (Pack and Phelps). ● $p = 17\cdot 2$ torr; ○ $p = 8\cdot 2$ torr; × $p = 2\cdot 7$ torr.

3.1.2. Total cross-sections. Neynaber et al† have applied the technique of Rubin, Perel, and Bederson‡ (see Chap. 1, § 4.2.2) to collisions of slow electrons with a beam of 2^3S metastable helium atoms. These atoms were produced in a hot cathode low-voltage arc discharge. A mixed beam issuing from a hole in the anode (see Fig. 4.69) included

† NEYNABER, R. H., TRUJILLO, S. M., MARINO, L. L., and ROTHE, E. W., *Atomic collision processes*, ed. McDOWELL, M. R. R., p. 1089 (North Holland, Amsterdam, 1964).
‡ RUBIN, K., PEREL, J., and BEDERSON, B., *Phys. Rev.* **117** (1960) 151.

normal, 2^1S, and 2^3S metastable atoms, photons, ions, and electrons. The charged particles were removed by electrostatic deflector plates. To isolate the 2^3S atoms the beam then passed through an inhomogeneous magnetic field. This had no effect on the normal and 2^1S metastable atoms or on the photons. The 2^3S atoms were split because of space quantization into three beams, one undeflected and one each deflected on opposite sides of the undeflected beam as in the experiments of Holt and Krotkov.† One of these deflected beams could then be selected to give a pure 2^3S beam.

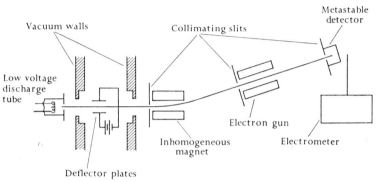

Fig. 4.69. Experimental arrangement used by Neynaber, Trujillo, Marino, and Rothe to study the scattering of electrons by He(2^3S) metastable atoms.

The intensity of the metastable atom beam was monitored by measuring the electron current emitted from a gold target on which the beam impinged normally. This current was about 5×10^{-13} A.

The beam was crossed by an electron current from a gun as shown in Fig. 4.69. Collisions of electrons with the metastable atoms that produce sufficiently large deflections in the paths of the atoms will reduce the intensity of the beam. If Q is the cross-section for such collisions then

$$Q = \frac{\Delta I}{I} \frac{\bar{v}w}{\mathrm{d}n/\mathrm{d}t}, \tag{38}$$

where $\Delta I/I$ is the fractional reduction of the intensity of the metastable atom, as monitored, due to the electron bombardment, $\bar{v}$ is the mean velocity of the metastable atoms at the source, w is the thickness of the metastable atom beam traversed by the electrons, and $\mathrm{d}n/\mathrm{d}t$ is the number of electrons passing per second across the beam.

The ratio $\mathrm{d}n/w\,\mathrm{d}t$ was obtained by dividing the total electron current by the width of the electron beam. The arrangement of the electrodes in

† loc. cit,, p. 231.

the gun and electron collector system used is as indicated in Fig. 4.70. The electrons were accelerated from the grounded cathode by an appropriate positive potential applied to the grid G_1. G_2 was a second grid at the same potential. Both grids were sufficiently thick to reduce field penetration into the interaction region to about 1 per cent of that between G_1 and the cathode or between G_1 and the collector plates P_1 and P_2, which were maintained at a positive potential of 22·5 V. The electron current to P_1 and P_2 was about 0·75 mA at an accelerating voltage of

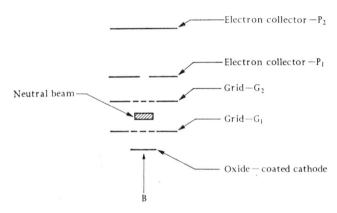

Fig. 4.70. Diagram of low-energy electron gun used by Neynaber et al. in crossed-beam experiment to measure scattering of electrons by metastable helium atoms. Neutral beam dimension 0·205 × 0·016 in; cathode width 0·25 in; slit width in collector P_1 0·005 in.

0·87 V. The energy spread in the electron beam, as measured by retarding potential analysis, varied from 0·32 eV at 0·87 eV mean energy to 0·68 eV at 8·3 eV mean energy.

The effective temperature of the metastable atoms, as measured by a velocity selector, was about 500° K and the angular resolving power of the apparatus, as defined in Chapter 1, § 4, came out to be about 13°.

Fig. 4.71 shows the observed cross-sections Q defined by (38) as a function of electron energy. These cross-sections include contributions from elastic, superelastic, and inelastic collisions, the threshold energy for the latter being 0·76 eV. For all experiments of this kind care must be taken in interpreting Q because a considerable fraction of the collisions may involve only small deflexions. However, for low electron energies, as in these experiments, the angular distribution of the scattered electrons will be nearly isotropic and the observed cross-section will be close to the total cross-section (see Chap. 6, § 3.2).

3.2. Measurement of cross-section for excitation of He+ ions to the 2s state by an electron-ion crossed-beam method

Dance, Harrison, and Smith† have used modulated crossed electron and ion beams to measure the cross-section for the excitation of He+

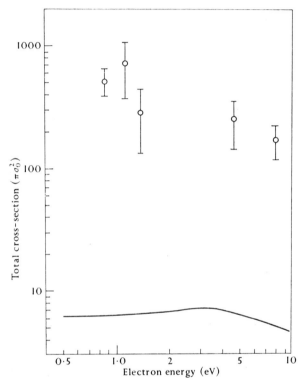

Fig. 4.71. Total collision cross-sections for scattering of electrons by helium 2^3S metastable atoms. The experimental points shown were obtained by Neynaber *et al.* For comparison the full line shows the total cross-section for scattering of electrons by ground state hydrogen atoms (Chap. 1, Fig. 1.11).

ions in the ground state to the 2s state. The principle of the method was similar to that of Dolder, Harrison, and Thonemann‡ described in Chapter 3, § 3.1.1. The metastable ions were detected by quenching them in an electrostatic field leading to Stark mixing of the $2^2s_{\frac{1}{2}}$ and $2^2p_{\frac{1}{2}}$ states followed by the emission of photons of energy 40·8 eV arising from the transitions $2^2p_{\frac{1}{2}} \to 1^2s_{\frac{1}{2}}$.

Fig. 4.72 gives a diagram of their apparatus. The ions were extracted

† Dance, D. F., Harrison, M. F. A., and Smith, A. C. H., *Proc. R. Soc.* A**290** (1966) 74.
‡ Dolder, K. T., Harrison, M. F. A., and Thonemann, P. C., ibid. A**264** (1961) 367.

from an oscillatory-electron type ion source S by an accelerating field, focused by the electrostatic lens L, and bent through 60° in the sector magnet M. A focused beam of He⁺ ions, of energy 5 keV and dimensions 2×1 mm, collimated by the slits S_1 and S_2, traversed the collision chamber X, where it was intersected by an electron beam passing between the gun G and collector E_c. The width of the electron beam, parallel to the direction of the ion beam was 25 mm. Its height, 1·6 mm, was

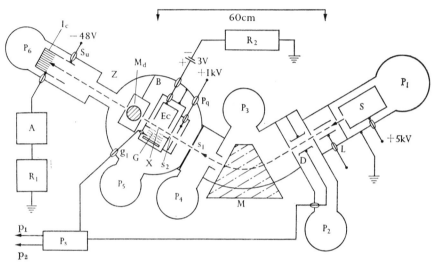

Fig. 4.72. Schematic plan view of apparatus of Dance, Harrison, and Smith for studying excitation of He⁺ ions to $2s$ states. A is a d.c. amplifier, R_1 and R_2 are recorders, and p_1, p_2 are gating pulses supplied to the scalers.

restricted so that only electrons passing through the ion beam were collected. Passing out of the collision region the ion beam entered the metastable He⁺ ion detector M_d. The entrance slit of the metastable detector was 5 cm further along the beam from the exit slit of the collision chamber, so that other excited ions (apart from those in $2s$ metastable states) would have decayed before entering M_d. After passing out of M_d the ion beam was eventually collected by I_c. The deflector D enabled the ion beam to be pulsed, just as in the experiments of Dolder et al. In order to reduce background, great care had to be taken with the pumping. Even with the ion source operating, the pressure near the detector was reduced to 10^{-8} torr. By pulsing the ion beam near the source, the gas formed by neutralization of the He⁺ beam when it was deflected away could be pumped out by the pumps P_2, P_3, and P_4. The narrow tube Z between M_d and I_c served to reduce the back-streaming of neutralized ions from the collector.

Fig. 4.73 shows details of the electron gun and collector. Electrons emitted from the cathode k, after passing through the control grid g_1 and the accelerator grid g_2, were focused by the beam-focusing plates b before passing through the suppressor grid g_3. g_2 and g_3 were always biased positively relative to the anode so as to inhibit secondary electron

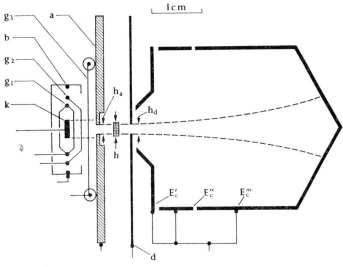

FIG. 4.73. Electron gun and collector in apparatus of Dance, Harrison, and Smith.

emission. The height of the electron beam (1·6 mm) was determined by the dimensions of the anode slit. The hole in the plate d was 1·9 mm wide and the beam was so well collimated that the current to plate d was negligible. The small positive potential between the Faraday cup with electrodes E'_c, E''_c, and E'''_c, and the plate d was sufficient to hold in secondary electrons. The electron beam was pulsed by applying suitable potentials to the control grid g_1.

The He$^+$(2s) ion detector developed by Harrison et al.[†] is illustrated in Fig. 4.74. Ions entered through the tube T_1. The concentric tube T_2 was biased negative relative to T_1, which was earthed. The resulting field quenched more than 90 per cent of the He$^+$(2s) ions. The photons emitted in the quenching process passed through grids in the tubes T_1 and T_2 and a third concentric tube T_3 connected to T_1. The two

[†] HARRISON, M. F. A., DANCE, D. F., DOLDER, K. T., and SMITH, A. C. H., Rev. scient. Instrum. 36 (1965) 1443.

photocathodes Pc_1 and Pc_2 surrounded T_3. Pc_2 was earthed but Pc_1 was maintained at a potential of $+300$ V to attract photoelectrons ejected from Pc_2 into the 17-stage electron multiplier, Mu. The grids in T_2 and T_3 were covered with aluminium films 1000 Å thick, which shielded the photocathodes from charged ions.

Two scalers Sc_1 and Sc_2, connected in parallel, were employed to measure the signal from the photomultiplier. They were gated so that

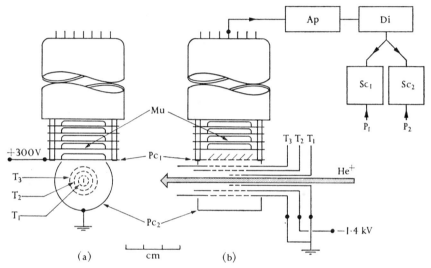

FIG. 4.74. Schematic diagram showing (a) end view, and (b) side view of the $He^+(2s)$ ion detector used by Dance, Harrison, and Smith.

their counting periods were synchronized to the beams pulsing sequence. This sequence was exactly as in the earlier experiments of Dolder, Harrison, and Thonemann (see Chap. 3, § 3.1.1), the four different pulsing modes, $1A, 2A, 1B, 2B$, being again employed and the signal calculated in the same way.

Great care had to be taken to reduce background effects. $He^+(2s)$ ions from the source were eliminated by keeping the operating source potential below 65·4 eV, the threshold energy for the direct excitation of ground state helium atoms to the $2s$ state of He^+. Background due to the excitation of the $He^+(1s)$ beam in collisions with residual gas was reduced by means of the electrode P_q (Fig. 4.72) which was biased to $+1$ kV and served to quench excited ions in the beam immediately before it entered the collision region. A further background was traced to photons produced by collisions of electrons with surfaces in the electron gun and collector, especially with the plate d located in front of the collector. This was reduced by making the opening in this plate larger than that in the anode a.

More troublesome were space charge interference effects between the ion and electron beams, which changed both the ion and electron beam background effects when the beams were crossed. Correction for this effect was made by taking measurements with the biasing potential on the cylinder T_2 of M_d reduced from 1·4 kV to 50 V. Under these circumstances the quenching effect is very small but the background due to electron collisions is almost unaltered. It was established further that the component of the background associated with the ion beam was proportional to $E^{-\frac{1}{2}}$, where E is the energy of the electron beam, so making it possible to estimate the contribution of background effects to the apparent signal.

For a He⁺ ion current of 1 μA and an electron beam current of 4 mA the maximum value of the signal was about 70 counts per second. The background arising from the ion beam alone was about ten times as great and that due to electron collisions with surfaces fifty times as great as the signal. The correction for the apparent signal arising from space charge interference effects was about 20 per cent over a large part of the incident electron energy range (40–800 eV) studied.

As in the case of the excitation of neutral metastables the observed cross-section $Q_\mathrm{p}(E)$ for the production of He⁺(2s) may be expressed as

$$Q_\mathrm{p}(E) = Q_{2s}(E) + Q_\mathrm{c}(E), \qquad (39)$$

where $Q_\mathrm{c}(E)$ is the contribution to the apparent production cross-section due to cascade from higher P states and $Q_{2s}(E)$ the true excitation cross-section for electrons of energy E. Dance et al. wrote

$$\left. \begin{array}{l} Q_\mathrm{c}(E) = \gamma_{3p \to 2s} Q_{1s \to 3p}(E) + \gamma_{4p \to 2s} Q_{1s \to 4p}(E), \quad E < E_5 \\ Q_\mathrm{c}(E) = 0{\cdot}23 Q_{1s \to 3p}(E), \quad E \geqslant E_5 \end{array} \right\}, \qquad (40)$$

and

where E_5 is the excitation potential of the 5p state. $Q_{1s \to np}(E)$ the calculated cross-section for excitation of the np state, and the branching ratios $\gamma_{np \to 2s}$ of transitions from the np state that populate the 2s state are the same as for the corresponding transitions of atomic hydrogen.†

The results obtained were normalized to give absolute values by comparison with calculated values of the cross-section

$$Q_{1s \to 2s}(E) + 0{\cdot}23 Q_{1s \to 3p}(E)$$

† CONDON, E. V. and SHORTLEY, G. H., *The theory of atomic spectra*, p. 136 (Cambridge University Press, 1959).

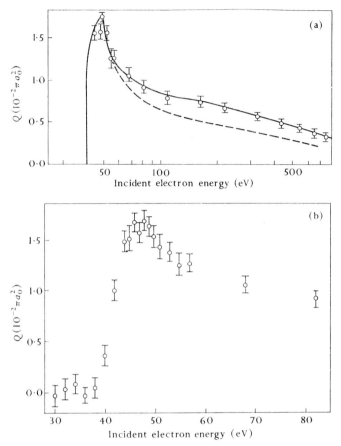

Fig. 4.75. Normalized cross-section Q for production of He$^+$(2s) metastable ions by electron impact excitation of He$^+$(1s) as measured by Dance, Harrison, and Smith. In (a) the full curve gives the experimental results, the error bars corresponding to 90 per cent confidence limits. The dotted curve is the cross-section derived for direct excitation of He$^+$(2s) after excitation by cascade has been deducted. In (b) the structure of the excitation curve near the threshold is shown.

for energies in the range 400–750 eV obtained (Chap. 7 § 5.6.5) using Born's approximation. The absolute values so obtained were used to estimate the efficiency ϵ of the metastable detector, M_d. The value obtained for ϵ was $6 \cdot 77 \times 10^{-3}$, consistent with that, $(8 \cdot 3 \pm 2 \cdot 5) \times 10^{-3}$, estimated by Harrison et al.† from the geometry and other properties of the detector.

Fig. 4.75 (a) shows the experimental values obtained for the He$^+$(2s) excitation cross-section by Dance et al and the excitation function for

† loc. cit., p. 272.

direct excitation of the 2s states of He$^+$ after allowing for the effect of cascade transitions. Fig. 4.75 (b) shows on a larger scale the form of the excitation cross-section variation with electron energy near the threshold when allowance is made for the energy distribution of the electrons. A theoretical discussion is given in Chapter 7, § 5.6.5 and Chapter 8, § 5.

5

THE EXPERIMENTAL ANALYSIS OF THE CROSS-SECTIONS FOR IMPACT OF ELECTRONS WITH ATOMS AND IONS—ANALYSIS OF ENERGY AND ANGULAR DISTRIBUTIONS

1. Introduction

IN the preceding two chapters methods have been described for determining cross-sections for ionization and excitation by direct observation of the ionized and excited atoms or of the radiation emitted from the latter. Information about the cross-sections for excitation and ionization can also be obtained, however, by energy analysis of the electrons after scattering. These methods of investigation are described in the present chapter. They have demonstrated the existence of inelastic processes that proceed through the formation of resonant states similar to those described in Chapter 1 but which are able to break up into electrons and excited atomic states.

In this chapter also are described experimental methods for studying the angular distribution of the scattered electrons from elastic and inelastic collisions. Measurements of the polarization of the scattered electrons are also described.

2. Determination of excitation cross-sections from electron energy loss studies

Most measurements of the energy loss of electrons in gases refer to electron scattering over a limited angular range. Such measurements are described in § 4, where it will be seen that they have given important information about the excitation of resonant, including auto-ionizing, states. It is not possible to obtain from them, however, the total cross-section for the corresponding excitation without a knowledge of the angular distribution of the inelastically scattered electrons. We discuss in this section two electrical methods in which electrons that have lost energy after inelastic scattering through any angle are detected, so that they are free from this limitation and are particularly applicable to the study of the excitation function near the threshold. The first of these methods, developed by Maier-Leibnitz,† is based on an analysis of the energy distribution of electrons moving through a gas, the electrons possessing sufficient energy to excite the gas.

2.1. *Diffusion through a gas of electrons with energy sufficient to produce inelastic collisions*

Consider electrons of mean velocity $\bar{v}$ diffusing through a gas at pressure p in the space between the concentric cylinders A, B of radii ρ_0 and ρ_1 respectively (see Fig. 5.1). It is supposed that there is no electric field between the cylinders. We first calculate the average number of collisions made by an electron reaching the outer cylinder after diffusing from the inner.

Let J be the number of electrons per unit area per second reaching the outer cylinder of radius ρ_1. If D is the diffusion coefficient of the electrons in the gas and $n_e(\rho)$ their concentration at a distance ρ from the axis of the cylinder, then

$$2\pi\rho_1 J = -2\pi\rho D \, dn_e/d\rho. \tag{1}$$

This equation has to be solved with the boundary condition $n_e = 0$ for $\rho = \rho_1$, giving

$$n_e = (J\rho_1/D)\ln(\rho_1/\rho). \tag{2}$$

Of the electrons at radius ρ a fraction α, which is a function of ρ, will eventually be collected by the outer cylinder, while a fraction $1-\alpha$ will rediffuse to the inner cylinder. On the average each electron will experience $n\bar{Q}\bar{v}$ collisions per unit time, where $\bar{Q}$ is the mean total collision cross-section and n is the number of gas atoms per unit volume. The number of collisions per unit time experienced in the space between the cylinders by electrons that will eventually be collected on unit area of the outer cylinder is therefore

$$n\bar{Q}\bar{v} \int_{\rho_0}^{\rho_1} \alpha n_e(\rho)(\rho/\rho_1) \, d\rho. \tag{3}$$

As J electrons reach unit area of the outer cylinder per unit time, the average number ν of collisions they must suffer on passing to the outer cylinder is given by

$$\nu = \frac{n\bar{Q}\bar{v}}{J\rho_1} \int_{\rho_0}^{\rho_1} \alpha\rho n_e(\rho) \, d\rho = \frac{n\bar{Q}\bar{v}}{D} \int_{\rho_0}^{\rho_1} \alpha\rho \ln(\rho_1/\rho) \, d\rho. \tag{4}$$

It remains to determine α. This is done as follows. Suppose at the radius ρ between ρ_0 and ρ_1 there exists a source that emits electrons in

† MAIER-LEIBNITZ, H., *Z. Phys.* **95** (1935) 499.

all directions and is such that a number j_1 reach unit area of the outer cylinder per unit time, and a number j_0 reach unit area of the inner cylinder per unit time.

The equation for the diffusion will be given by (1) with the boundary conditions $n_e = 0$ at $\rho = \rho_0$ and $\rho = \rho_1$ so we obtain two expressions for the density at ρ in the two cases:

$$n_e^1(\rho) = (j_1\rho_1/D)\ln(\rho_1/\rho), \qquad n_e^0(\rho) = (j_0\rho_0/D)\ln(\rho/\rho_0). \tag{5}$$

The problem with the hypothetical source at radius ρ will become identical with the actual problem if we put $n_e^1 = n_e^0$, i.e.

$$j_1\rho_1 \ln(\rho_1/\rho) = j_0\rho_0 \ln(\rho/\rho_0). \tag{6}$$

But $j_1\rho_1$, $j_0\rho_0$ are proportional to the total number of electrons reaching the outer and inner cylinders respectively per unit time, so

$$\alpha = \frac{j_1\rho_1}{j_1\rho_1 + j_0\rho_0} = \frac{\ln(\rho/\rho_0)}{\ln(\rho_1/\rho_0)}. \tag{7}$$

Substituting in (4) we obtain for ν, the average number of collisions made by electrons reaching the outer cylinder,

$$\nu = \frac{n\bar{Q}\bar{v}}{D}\frac{1}{\ln(\rho_1/\rho_0)}\int_{\rho_0}^{\rho_1}\rho\ln(\rho_1/\rho)\ln(\rho/\rho_0)\,d\rho = \frac{n\bar{Q}\bar{v}}{4D}\left\{\rho_0^2 + \rho_1^2 - \frac{\rho_1^2 - \rho_0^2}{\ln(\rho_1/\rho_0)}\right\}. \tag{8}$$

This expression was deduced by Harries and Hertz.†

The diffusion coefficient D is given to a close approximation by

$$D = \bar{v}/3nQ_d, \tag{9}$$

where Q_d is the diffusion cross-section defined in Chapter 2, § 1.1. This gives

$$\nu = \tfrac{3}{4}\bar{Q}Q_d n^2\left\{\rho_1^2 + \rho_0^2 - \frac{\rho_1^2 - \rho_0^2}{\ln(\rho_1/\rho_0)}\right\}. \tag{10}$$

Let η be the proportion of all impacts that are inelastic. Then, if J_1 is the number of electrons that would reach unit area of A per unit time without appreciable loss of energy in the absence of inelastic collisions and J the actual number, $J = J_1 e^{-\nu\eta}$. Thus if J/J_1 can be estimated, $\nu\eta$ can be determined and the total cross-section for excitation calculated from the expression (10) for ν. This expression takes no account of the variation in total diffusion time of electrons traversing the collision chamber. It assumes also that the mean free path of the electrons is short compared with the radial dimensions of the apparatus. Fleming and Higginson‡ have therefore estimated ν by a random-walk procedure

† HARRIES, W. and HERTZ, G., Z. Phys. 46 (1927) 177.
‡ FLEMING, R. J. and HIGGINSON, G. S., Proc. phys. Soc. 84 (1964) 531.

for a collision chamber with inner and outer cylinders of radii 0·4 and 2·0 cm respectively, and a He gas pressure of 0·47 torr. They obtained for ν a value of 26·0±0·8 compared with the value 22·0 given by (10).

2.2. Measurement of excitation cross-sections by the diffusion method

Fig. 5.1 shows the form of apparatus used by Maier-Leibnitz† who applied this method to investigate excitation in He, Ne, and A.

The collisions occurred between two coaxial cylinders A, B of 1·5 and 25 mm radius respectively, and 150 mm long. The cylinder A was

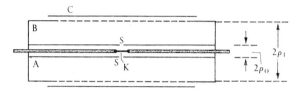

FIG. 5.1. Apparatus of Maier-Leibnitz for the study of excitation cross-sections by the diffusion method.

hollow and a slit S, 0·2 mm wide, was cut in it in the plane of its central section. Inside A was stretched a tungsten filament K, and electrons accelerated from K passed through the slit S into the field-free space between A and B containing the gas under investigation at a pressure of the order of 0·1 torr. Electrons passing between A and B would in general make many collisions with the gas atoms. The greater part of the outer cylinder B consisted of a wire gauze of 0·25 mm mesh. Surrounding this gauze and distant about 1·5 mm from it was a collecting cylinder C. A retarding potential between C and B enabled the energy distribution of the electrons reaching B to be studied.

The current collected by C was measured as a function of the accelerating potential V, between K and A, with the retarding potential U between C and B so adjusted that U/V remained constant ($= \frac{1}{3}$). Apart from the effect of electrons reflected from the cylinder C, the current collected by C under conditions of constant U/V would be expected to remain practically constant, independent of V.

When, however, V becomes large enough to excite one of the levels of the gas the number of electrons reaching C will show a sharp drop. From a study of this decrease in current the absolute yield for the corresponding excitation was estimated.

Fig. 5.2 (a) shows typical results obtained by Maier-Leibnitz in helium at a pressure of 0·326 torr. At about 19 V the current i drops sharply,

† MAIER-LEIBNITZ, H., Z. Phys. **95** (1935) 499.

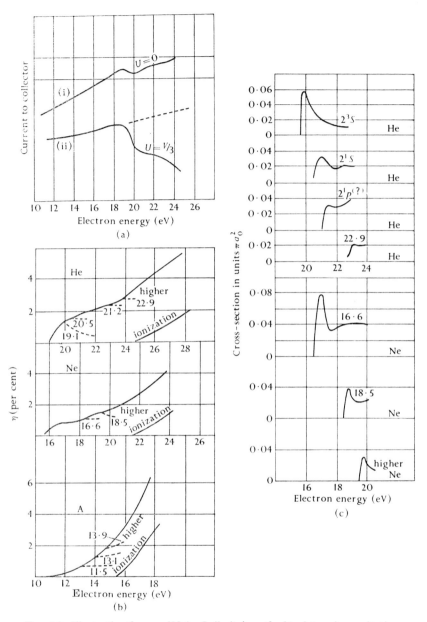

FIG. 5.2. Illustrating the use of Maier-Leibnitz's method to determine excitation cross-sections for He, Ne, and A. (a) Variation with electron energy of the current to the collector in He for (i) $U = 0$, (ii) $U = V/3$. Note the sharp break in the curves at the 19·7-V excitation. (b) Percentage of inelastic impacts as a function of electron energy. The curves for ionization are derived from Smith's measurements (§ 2.3). (c) Excitation functions for individual levels obtained by analysis of (b).

indicating a critical potential. The dotted line shows the estimated course of the curve if there had been no inelastic scattering.

Let i_0 be the value of the current to C if there had been no inelastic collisions, as given by the dotted curve of Fig. 5.2 (a) and i the actual observed current. Then
$$i = i_0 e^{-\nu\eta}, \qquad (11)$$
from which $\eta\bar{Q}$, the effective cross-section for the inelastic impact can be calculated as described above.

Fleming and Higginson† considered more carefully the collection process of electrons that pass through the mesh. Many of these, diffusing through at an angle to the radial direction, cannot reach the collector against the retarding field but re-enter the collision chamber where they make more collisions before again passing through the mesh. Some may re-enter the collision chamber several times before being collected either by the collector C or the mesh B. When allowance is made for these effects equation (11) has to be replaced by
$$i = \frac{i_0(1-S^2\sin\theta_1)e^{-\nu\eta}}{1-S^2\sin\theta_1 e^{-\nu_1\eta}}, \qquad (12)$$
where S is the permeability of the mesh, θ_1 is the maximum angle the direction of motion of an electron through the mesh can make with the radial direction and still be collected on C, and ν_1 the number of collisions an electron re-entering the collision chamber makes before it leaves through the mesh again.

The observed current i must also, however, contain a contribution due to electrons ejected by metastable atoms diffusing from the collision chamber both to the collector and to the mesh. The importance of allowing for this effect was pointed out by Higginson and Kerr.‡ Fleming and Higginson estimated the effect both of re-entry of primary electrons into the collision chamber and of metastable production for their experimental conditions. Calculations were made assuming a value of S in the range 0·10 to 0·40 and of ζ, the electron emission coefficient of the collector and mesh surfaces for metastable atom impact, between 0·2 and 0·3. The range of ν_1/ν chosen for investigation was from 0·4 to 0·5 on the basis of random-walk calculations. The maximum and minimum values obtained for $e^{-\nu\eta}$ on the basis of these ranges of unknown parameters were 0·785 and 0·726 for their experiment and these limits determine the uncertainty of the measurement. Assuming a

† loc. cit., p. 279.
‡ HIGGINSON, G. S. and KERR, L. W., *Proc. phys. Soc.* **77** (1961) 866.

value $2 \cdot 7\pi a_0^2$ for the momentum transfer cross-section, the corresponding value of the 2^3S excitation cross-section at its maximum came to be

$$2 \cdot 6 \times 10^{-18} \text{ cm}^2 \pm 17 \text{ per cent.}$$

Approximately 30 per cent of the collector current was estimated as being due to secondary emission by metastable atoms. For values of the incident electron energy above the ionization potential the presence of the ions disturbed the measurements. An alternative procedure was then adopted.

After ν collisions the number of electrons that have undergone inelastic collisions is proportional to $1-e^{-\nu\eta}$. For a given electron energy a certain fraction of these will be ionizing collisions. Thus the number of ions I collected by C when its potential is made high enough to repel all electrons will be equal to $I_S(1-e^{-\nu\eta})$, where I_S is the saturation value of the positive ion current for a very high pressure. Thus, if I_S is measured and I obtained for a given pressure corresponding to a given ν, η can be estimated.

To separate out the excitation probabilities corresponding to the individual excitations, a retardation analysis was carried out for each initial electron energy and the relative numbers of electrons present corresponding to the various energy losses measured. Smith's[†] results on ionization cross-sections were used to separate ionizing collisions from excitation.

Fig. 5.2 (b) and (c) shows the curves obtained by Maier-Leibnitz and his analysis of them. He estimated the error in the determination as 20 per cent, and his results on ionization are consistent with those of Smith within these limits.

A feature of these curves is the very sharp rise of the excitation cross-section to a maximum when the accelerating potential V is less than a volt above the excitation potential. This might be expected in the case of the 19·7-V excitation in helium corresponding to the transition 1^1S–2^3S involving a change in multiplicity, and possibly for the 20·5-V excitation in helium corresponding to the optically forbidden transition 1^1S–2^1S, but is surprising for the optically allowed transition 1^1S–2^1P corresponding to the energy loss 21·2 V as found by Maier-Leibnitz. It is doubtful whether the results are sufficiently accurate to establish definitely such a feature. Nevertheless, using a similar method Seiler[‡] obtained evidence of a similar maximum in the case of the

[†] SMITH, P. T., *Phys. Rev.* **36** (1930) 1293; **37** (1931) 808.
[‡] SEILER, R., *Z. Phys.* **83** (1933) 789.

6^1S_0–6^1P_1 excitation of mercury vapour. Brattain† using an apparatus in which he collected electrons after inelastic scattering through a large (but uncertain) range of angles (including zero), found a similar effect for this excitation.‡ The matter will be discussed further in § 4.

2.3. *The electron-trap method*

Schulz§ has developed a method for observing excitation functions for electrons with energies within a few eV of the threshold in which,

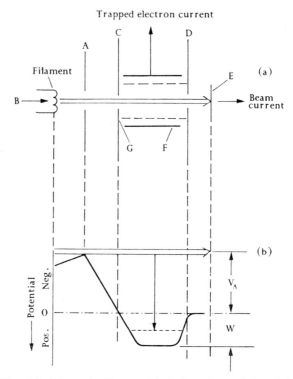

Fig. 5.3. Schematic diagram (a) of the tube, and (b) of the potential distribution along the axis of the tube in the electron-trap experiments of Schulz.

by an ingenious technique, the slow electrons produced by the inelastic collisions in question are collected and measured.

Fig. 5.3 illustrates the electrode system that he used. Electrons from a hot filament are accelerated through a slit in the electrode A into the collision chamber. The electron beam passes along the axis of a

---

† BRATTAIN, W. H., *Phys. Rev.* **34** (1929) 474.
‡ See also Chapter 4, § 1.4.2.2, especially Fig. 4.27.
§ SCHULZ, G. J., *Phys. Rev.* **112** (1958) 150.

cylindrical grid G, which is maintained at the same potential as the entrance and exit electrodes C and D respectively of the chamber. The grid is surrounded by a concentric insulated collector electrode F. Finally, the undeflected electron beam is collected at E.

Taking the potential of the grid G and electrodes C and D as zero, the potential of A is negative and equal to $-V_A$ so that the energy of the electron beam entering the collision chamber is eV_A. The cylindrical collector electrode F is held positive with respect to G by an amount V_F. Some of the field between G and F penetrates through the grid so that, along the axis of the collision chamber, the potential is positive with respect to the grid by an amount W, which depends on the potential V_F and on the grid spacing. This potential well acts as an axial trap for slow electrons that are produced by inelastic collisions in the collision chamber. A magnetic field is applied parallel to the axis of the collision chamber to align the electron beam. The slow electrons trapped in the well will gradually diffuse in a direction perpendicular to this field to be collected eventually on the electrode F.

The energy of the electrons in the beam within the collision chamber is $e(V_A + W)$ so that the electrons collected by F, which are those trapped by the well, must have lost an amount of energy between eV_A and $e(V_A + W)$. Fig. 5.4 illustrates the form of the variation of the current to F with electron accelerating voltage V_A due to excitation of a state with excitation potential V_{ex}, it being assumed that the excitation cross-section increases linearly with energy above the threshold and that the well depth is constant over the length of the collision chamber.

If the electron beam were strictly homogeneous in energy, current could be collected when $V_A > V_{ex} - W$, rising linearly to a maximum at V_{ex}. For higher beam energies the slow electrons produced would possess energies greater than eW and so would not be trapped. No current would therefore be received by F for $V_A > V_{ex}$.

The effect of an energy spread, $e\epsilon$, in the incident beam would be to modify the curve to the form shown by the broken curve in Fig. 5.4. In practice there would also be some blurring of the sharpness of the transition to zero current for $V_A = V_{ex}$ due to scattering of the electrons in the inelastic collisions. Thus electrons will be trapped if they possess a kinetic energy of axial motion less than eW.

The well depth may be determined by potential plotting with an electrolytic model of the electrode system, or directly, using the working equipment in either of the following ways. In both of these the collector F is held at a negative potential with respect to the grid G. The first

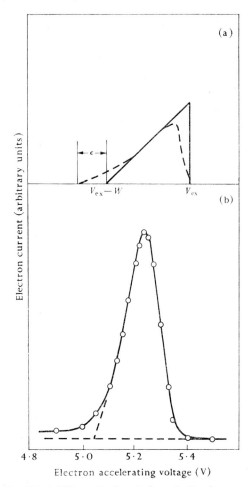

Fig. 5.4. (a) Theoretical peak shape in the electron-trap method for collection of slow electrons due to atomic excitation. ——— for monoenergetic electrons; – – – – for electrons with energy spread $e\epsilon$. (b) Experimental peak shape in the electron-trap method for collection of slow electrons resulting from the excitation of the 6^3P_2 level in mercury.

method depends on observing the threshold for positive ion production for different negative values of V_F, using the linear interpolation technique with the retarding potential difference method as described in Chapter 3, § 2.4.2. The shift of the threshold in each case from the spectroscopic value is just equal to the corresponding well depth.

The second method depends on making a retarding potential analysis of the electron beam. This is retarded as it enters the chamber and the current collected by the application of +3 V to the collector E (see Fig. 5.3). With F and G at the same potential the retarding potential curve will be as shown in the first curve of Fig. 5.5. When F is at a negative potential V_F with respect to G the curve will shift to lower voltages by an amount equal to the well depth W (see Fig. 5.5). For definiteness the shift is measured at the point of inflexion in the retarding potential curve. Values of the well depths per unit potential difference between F and G, $(\Delta V/\Delta V_F)$, are indicated on the figure.

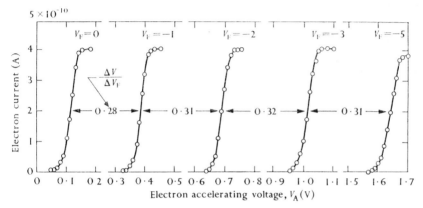

Fig. 5.5. Determination of well depth by the electron-retardation method.

An effective cross-section Q_t can be defined such that

$$NQ_t l = i_F/i, \qquad (13)$$

where i_F is the peak current collected by F for the particular inelastic process, i is the beam current, N the number of gas atoms/cm^3, and l the length of the collision chamber. If V_A is the accelerating potential and W the well depth then we would expect that Q_t is the actual cross-section for the process produced by electrons of energy $e(V_A+W)$ provided $V_A-V_{\text{ex}} < W$.

Care must naturally be taken to work under such conditions that i_F is proportional to gas pressure and to beam current.

In the first experiments of Schulz† a well depth of about 0·2 V was used, but later‡ it was possible to work up to well depths of 3–4 V. The collector F was 10 mm in diameter, while the grid G consisted of ten gold-plated molybdenum wires 0·06 mm in diameter, forming a cylindrical screen 6 mm in diameter, the length of the collision chamber being 152 mm. A solenoid provided the collimating magnetic field, the

† Schulz, G. J., *Phys. Rev.* **112** (1958) 150.
‡ Ibid. **116** (1959) 1141.

magnitude of which could be adjusted between 300 and 1000 gauss. The collector electrodes were plated with platinum black to minimize secondary electron emission. All other electrodes were gold-plated. In both sets of experiments the retarding potential difference method was used so as to reduce effects due to energy spread in the main beam.

The electron-trap method is independent of any secondary electron emission coefficients for metastable atoms. Further, since the trapped electrons cannot be collected by the grid, it should be independent of grid-transmission coefficients. It was found, however, that the measured cross-section at a fixed electron energy increased with increasing well depth even under conditions where the variation of measured electron current with pressure was linear. The cross-section measured by the trapped-electron method for an energy loss of 20·3 eV, corresponding to the peak of the 2^3S excitation, was considerably larger than that measured by Maier-Leibnitz and nearly three times as large as the more recent determination of Fleming and Higginson.

This behaviour is attributed to elastic scattering of electrons through large angles out of the beam direction. Because of the reduction in their kinetic energy in the beam direction these electrons may make many traverses across the tube before another collision reorients their velocity in the axial direction so that they are collected. During this time they may undergo inelastic collisions but it is difficult to estimate their effective path-length and so make the appropriate correction to the cross-section. The fraction of elastically scattered electrons trapped in this way depends on (W/V_A) and on the angular distribution of elastically scattered electrons.

Although the absolute excitation cross-section is not determined reliably its variation with electron energy could be followed out to an energy eW ($\simeq$ 3 to 4 eV) above the threshold. Schulz obtained the excitation curve for helium from the threshold out to 21·8 eV, illustrated in Fig. 5.6, in which it is compared with the corresponding results obtained by the metastable atom collection method (Chap. 4, § 2.1.2). The two sets of results have been normalized to agree at the peak of the 2^3S excitation. The structure is comparable. The departure of the two curves above 21·2 eV is presumably due to the excitation of the 2^1P level which the metastable production method cannot measure.

2.4. *Detection of superelastic collisions*

Changes of energy of a beam of electrons passing through a gas containing atoms in metastable states can be used to demonstrate superelastic collision processes. In this case the electrons gain discrete energies in the collision. Such processes are important in gas discharges and it has been possible to estimate cross-sections for them from studies of metastable atom concentrations in afterglows (see Chap. 4, § 3.1.1).

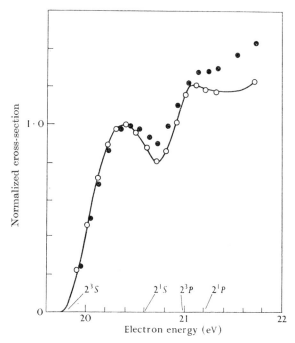

Fig. 5.6. Excitation function for production of metastable atoms in helium. ● data obtained by trapped-electron method; ○ data obtained by metastable production method. Both curves are normalized to unity at the peak of the 2^3S excitation.

It is more difficult, however, to detect such processes from the direct measurement of the energy gain of a beam of electrons passing through a gas containing a concentration of metastable atoms.

Latyscheff and Leipunsky† have demonstrated the existence of superelastic collisions in mercury, in which a metastable mercury atom in the 6^3P_0 state falls to its ground state and gives up energy of 4·7 eV to the incident electron. Their apparatus is illustrated in principle in Fig. 5.7. Electrons from the cathode K were accelerated through the grid G_1 and passed between G_1 and a second grid G_2, at the same potential. A retarding potential between G_2 and A enabled a velocity analysis to

† LATYSCHEFF, G. D. and LEIPUNSKY, A. I., Z. Phys. **65** (1930) 111.

be carried out. The space between G_1 and G_2 contained mercury vapour at a pressure of about 0·01 torr. Light from a Hg arc was now used to illuminate the space between G_1 and G_2 so that the line λ 2537 of the arc excited the mercury vapour and caused a concentration of 6^3P_0 metastable mercury atoms to be produced.

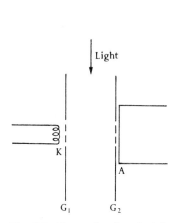

FIG. 5.7. Illustrating the principle of the apparatus of Latyscheff and Leipunsky for studying superelastic collisions of electrons with mercury atoms in the 6^3P_0 metastable state.

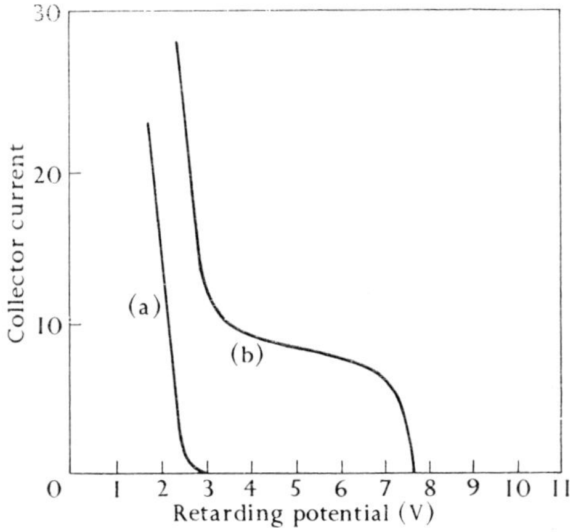

FIG. 5.8. Collector current as a function of retarding potential applied to collector: (a) when the space between G_1 and G_2 is not illuminated; (b) when the space is illuminated. Curve (b) is obtained after making a large correction for photo-electric emission from the collector. The energy of the incident electrons is 2·5 eV. When the space is illuminated the maximum energy of electrons is 7·2 eV, due to electrons having gained 4·7 eV from the 6^3P_0–6^1S_0 transition of the excited atoms.

The velocity analysis of the electron beam was now repeated in the presence of these metastable atoms. Fig. 5.8 shows, for an accelerating potential of 2·5 V between K and G_1, the retardation curves obtained with and without the presence of the metastable Hg atoms. It is seen that when the space between G_1 and G_2 is illuminated the maximum energy of the electrons is 7·2 eV compared with 2·5 eV without illumination. The difference, 4·7 eV, is in very good agreement with the energy of the 6^3P_0 metastable state.

3. Evidence about inelastic cross-sections derived from swarm experiments

3.1. Ionization and excitation coefficient

In Chapter 2, § 1 we have discussed the analysis of observed data on the drift velocity and mean energy of electrons in gases as a function of the ratio F/p of the electric field to the gas pressure. The discussion was restricted to small values of F/p so that only elastic collisions were important and the effect of inelastic encounters could be allowed for by introducing small corrections. To obtain useful information, particularly in the form of checks on assumed values of inelastic cross-sections, it is necessary to work with data at higher values of F/p and take explicit account of the effect of inelastic losses on the electron energy distribution function. Furthermore, it is desirable to work with observable quantities that are themselves more sensitive to inelastic collision rates than the drift velocity and mean energy, although information about inelastic cross-sections has been derived from drift velocity measurement. For this purpose the ionization and excitation coefficients for a drifting swarm of electrons are particularly useful. They are defined as follows.

The ionization coefficient α_i is such that, in diffusing a distance δx in the direction of the electric field, an electron makes $\alpha_i \, \delta x$ ionizing collisions. If $f(\mathbf{v})$ is the velocity distribution function as defined in Chapter 2, § 1.2 and $Q_i(v)$ is the cross-section for ionization by electrons of velocity v, then

$$\alpha_i = \frac{N \int Q_i(v) v f(\mathbf{v}) \, d\mathbf{v}}{\int \xi f(\mathbf{v}) \, d\mathbf{v}}, \tag{14}$$

where ξ is the component of velocity in the direction of the field and N the number of gas atoms/cm³.

In the same way an excitation coefficient may be defined, referring to all non-ionizing excitation processes or to any particular ones.

3.1.1. *Measurement of the ionization coefficient.* The ionization coefficient α_i can be calculated from (14), if the velocity distribution $f(\mathbf{v})$ is known, for different assumptions about $Q_i(v)$ and its variation with electron velocity. A measurement of α_i can then provide evidence about $Q_i(v)$.

The ionization coefficient α_i (first Townsend coefficient) was introduced first by Townsend† in his theory of ionization by collision. According to this theory the gas multiplication factor i/i_0 between two parallel electrodes of separation d is

$$i/i_0 = \frac{e^{(\alpha_i/p)\{p(d-\delta)\}}}{1 - \gamma[e^{(\alpha_i/p)\{p(d-\delta)\}} - 1]}, \tag{15}$$

where i_0 is the electron current in the absence of the gas. The second Townsend coefficient γ is practically the same as the secondary electron emission coefficient for positive ions striking the cathode, while δ is

† TOWNSEND, J. S., *Phil. Mag.* **3** (1902) 557; **5** (1903) 389; **6** (1903) 598.

related to the distance required by the electrons moving from the cathode to reach the equilibrium energy distribution corresponding to the relevant value of F/p.

Breakdown in a steady field will occur when the denominator of (15) vanishes, viz. when

$$(\alpha/p)\{p(d-\delta)\} = \ln\{(1+\gamma)/\gamma\}. \tag{16}$$

In principle, therefore, measurements of either the gas multiplication factor i/i_0, or of the steady field breakdown potential difference may be used to measure α_i/p as a function of F/p.

α_i has usually been obtained from measurements of the gas multiplication factor, using equation (15). An example of a recent determination using modern ultra-high vacuum and gas-purification techniques is provided by the experimental determination of α_i for helium by Davies, Llewellyn Jones, and Morgan.[†] Freedom from contamination is particularly important for helium since, owing to its high ionization potential and small ionization cross-section, small traces of impurity will have a large effect on measured values of the ionization coefficient.

They used a parallel plate ionization chamber, each electrode being of silver, 3 cm diameter, with rounded edges that were thoroughly cleaned and polished before sealing into the glass envelope. The anode was fixed in position while the cathode was attached to a metal bellows. An external screw mechanism enabled the cathode to be moved in order to vary the electrode separation up to a maximum of 70·75 cm while the two electrodes were maintained accurately parallel. The separation, d, could be measured by an external micrometer screw gauge to an accuracy of 0·1 per cent.

Ultra-violet light from a stabilized mercury arc entered the apparatus through a quartz window. It passed through ten holes of diameter 0·5 mm in the central position of the anode before reaching the cathode. An earthed metal cylinder surrounding the electrode system screened the interelectrode space from surface charges on the walls of the glass envelope. After baking for 25 h a residual pressure of 10^{-8} torr was reached in the apparatus. Spectroscopically pure helium gas was further purified before admission to the ionization chamber. The pressure of helium gas in the chamber was measured to within 0·5 per cent by means of a bellows micromanometer used as a null detector. The outside of the bellows was connected to an oil manometer. The null position, when

[†] DAVIES, D. K., LLEWELLYN JONES, F., and MORGAN, C. G., *Proc. phys. Soc.* **80** (1962) 898.

the helium pressure and that recorded by the oil manometer were equal was set by observation using a gas manometer.

Care had to be taken to ensure that the electrode surfaces remained stable throughout the experiment since the ratio i/i_0 depends on the second Townsend coefficient γ whose value is strongly dependent on the nature of the electrode surfaces. These were monitored throughout the experiment by measuring the static breakdown potential difference from time to time. It proved to be reproducible to within ± 0.5 per cent. In measurements of the ionization coefficient of argon Golden and Fisher[†] found difficulty in maintaining stable surfaces, however.

To obtain the initial photoelectric current i_0 a current-voltage characteristic curve was measured for low values of V. If a saturation plateau had been obtained this would have been taken as the value of i_0 in the presence of the gas. Actually some ionization occurs even at low values of V and the characteristic did not show any region where $di/dV = 0$. It was found, however, that di/dV passed through a broad minimum for a wide range of values of F/p around 1 V cm^{-1} torr^{-1}. If i_c is the current when di/dV is a minimum they took $i_0 = ci_c$, where c is constant for given F and p. Eight settings of the electrode separation d were used for a particular value of F/p and the current i measured for each. The current i_c was also measured for each value of d for $F/p = 1$ V cm^{-1} torr^{-1}. Assuming the quantities δ, c, γ remain unchanged throughout the experiment, they can be eliminated from the eight equations of type (15) and α_i determined explicitly for that value of F/p. The measurements of α_i in helium were extended over F/p values ranging from 5 to 100 V cm^{-1} torr^{-1}.

Fig. 5.9 shows the observed variation of i/i_0 with d for $F/p = 50$ V cm^{-1} torr^{-1} and $p = 5.13$ torr. It is seen to agree closely with a curve of form (15), for $\alpha_i = 2.97$ cm^{-1} and $\gamma = 0.11$. The dotted curves show the effect of a ± 10 per cent variation of γ from this value. Careful measurements established that the observed values of α_i/p were functions of F/p only and not of d or p separately. The maximum error of α_i in these measurements was estimated to be ± 8 per cent.

3.1.2. *Study of microwave gas discharge breakdown.* The amplitude of an alternating electric field of angular frequency ω which will produce breakdown in a gas contained within a vessel of specified shape and dimensions can be calculated in terms of related averages over the velocity distribution of the electrons. Breakdown occurs when the rate at which the electrons are produced by ionization becomes equal to the

[†] GOLDEN, D. E. and FISHER, L. H., *Phys. Rev.* **123** (1961) 1079.

rate of loss to the walls by diffusion. Thus if n is the number of electrons per cm³ the continuity equation for electrons is

$$\frac{\partial n}{\partial t} = -D\nabla^2 n + \nu_i n, \tag{17}$$

where ν_i is the number of electrons produced per second by ionization and D is the electron diffusion coefficient. In the absence of any production process the solution of (17) can be written in the form

$$n = \sum_{s=1} e^{-t/\tau_s} n_s(r). \tag{18}$$

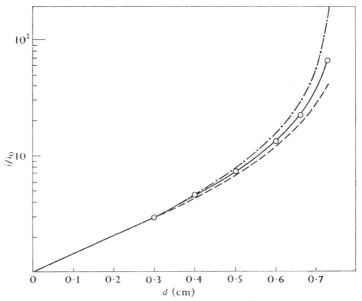

Fig. 5.9. Ionization coefficient of helium—comparison of experimental and calculated current growth curves. i/i_0 is calculated from (15) with $\gamma = 0{\cdot}101$ (—·—·—·), $0{\cdot}112$ (———), and $0{\cdot}123$ (— — — —) respectively and $\alpha_i = 2{\cdot}97$ cm⁻¹. ○ experimental points.

τ_s is such that the solution of

$$\nabla^2 n_s + n_s/(\tau_s D) = 0 \tag{19}$$

is a proper function satisfying the boundary conditions at the walls of the vessel. The length $(\tau_s D)^{\frac{1}{2}}$ is called the *characteristic diffusion length* Λ_s in the sth diffusion mode and will be determined by the geometry of the containing vessel. If we assume that the electron distribution when ionization is present approximates to that in the first diffusion mode for which $s = 1$ then breakdown will occur when

$$D = \Lambda_1^2 \nu_i. \tag{20}$$

In terms of the velocity distribution of the electrons,

$$\nu_i = \int Q_i(v) v f(\mathbf{v}) \, d\mathbf{v}, \tag{21}$$

where $Q_i(v)$ is the ionization cross-section for electrons of velocity v and $f(\mathbf{v})$ is the velocity distribution function. Also (see Chap. 2, § 2)

$$D = 2u\bar{\epsilon}/3eF, \tag{22}$$

where (see Chap. 2, eqns (19a) and (21))

$$u = \int \xi f(\mathbf{v}) \, d\mathbf{v}, \tag{23}$$

$$\bar{\epsilon} = \tfrac{1}{2} m \int v^2 f(\mathbf{v}) \, d\mathbf{v}. \tag{24}$$

This analysis applies to diffusion in a steady field of strength F. We may extend it to apply approximately to the alternating field of r.m.s. strength $\bar{F}$ by taking
$$F = \bar{F}\bar{\nu}_0/(\bar{\nu}_0^2+\omega^2)^{\frac{1}{2}}, \tag{25}$$

where $\bar{\nu}_0$ is the mean elastic collision frequency.†

Substitution of (21)–(25) in the breakdown condition (20) yields a relation involving the parameters p, Λ, ω and the breakdown field $\bar{F}$ through the quantities $\bar{F}\Lambda$, $\bar{F}/p$, p/ω, as well as ν_i and the velocity distribution. For practical application the assumption of a particular diffusion mode can be checked by carrying out observations in vessels of different geometry. If the assumption is correct $\bar{F}\Lambda$ should be a function of $\bar{F}/p$.

As an example of a typical experiment MacDonald and Brown‡ have studied the high-frequency electrical breakdown in helium containing mercury vapour using cylindrical cavities of radius $R = 4\cdot07$ cm and with length $L \ll R$. The dimensions were chosen so that the cavities resonated in the TM_{010} mode in the 10-cm wavelength region. For such cylindrical cavities

$$\frac{1}{\Lambda_1} = \left\{ \left(\frac{n}{L}\right)^2 + \left(\frac{2\cdot405}{R}\right)^2 \right\}^{\frac{1}{2}}. \tag{26}$$

The cavities were constructed of oxygen-free high conductivity copper, baked out for several days at 400° C and the whole vacuum system carefully outgassed before each run.

Fig. 5.10 shows a block diagram of the microwave apparatus used. A continuous wave tunable magnetron supplied up to 150 W of power into a coaxial line leading to the measuring equipment. The power incident on the cavity could be varied by a power divider and was measured by directing, by means of a directional divider, a known fraction of the incident power to a thermistor element whose resistance

† Cf. Chap. 2, § 5, eqn (57).
‡ MacDonald, A. D. and Brown, S. C., *Phys. Rev.* 75 (1949), 411.

could be measured. The slotted section of line enabled the standing wave ratio to be measured and thence the power absorbed in the cavity to be estimated. The unloaded quality factor Q of the cavity could also be calculated from standing wave measurements. The cavities were coupled to the coaxial transmission line by a coupling loop and a second coupling loop provided a transmitted signal to an attenuator, crystal, and meter. The stored energy could be estimated from the absorbed power

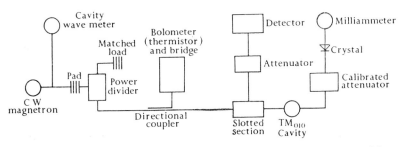

FIG. 5.10. Block diagram of the experimental microwave equipment used by MacDonald and Brown to investigate high-frequency gas discharge breakdown in helium.

and the unloaded Q, while the electric field could be obtained from the stored energy and the known field configuration.†

In the actual experiment, after the cavity had been filled with gas at the desired pressure, the magnetron power was increased until the crystal current reached a maximum and then dropped suddenly to a much smaller value, indicating the onset of breakdown. The breakdown field could then be calculated from the maximum crystal current. A radioactive source near the cavity provided the electrons required to initiate the discharge. The breakdown measurements were reproducible within an experimental error of less than 5 per cent in electric field and 1 per cent in pressure.

For measurements in pure helium Reder and Brown‡ used resonators of length $\frac{1}{4}$, $\frac{1}{8}$, and $\frac{1}{16}$ in. respectively. The relation they found between $\bar{F}\Lambda$ and $\bar{F}/p$ is shown in Fig. 5.11. Their measurements lie on a universal curve, irrespective of the cavity length, thus verifying that the dimensions of the cavity enter into the equations through the combination $\bar{F}\Lambda_1$.

3.1.3. *Measurement of drift velocity at large F/p values.* Measurements of drift velocity of electrons in argon containing a relative concentration of 6.9×10^{-6} of caesium atoms have been used by Nolan and

† Methods of measuring electric field strengths in microwave cavities are discussed by ROSE, D. J. and BROWN, S. C., *J. appl. Phys.* **23** (1952) 719.
‡ REDER, F. H. and BROWN, S. C., *Phys. Rev.* **95** (1954), 885.

Phelps† to obtain information about the excitation function of caesium near the threshold (see Chap. 2, § 7.4). The threshold energy of the first excited state of caesium is low (1·386 eV), so that effects due to inelastic collisions could be seen at quite low F/p values (0·01 to 0·1 V cm^{-1} torr^{-1}). In this energy region electrons in the high-energy tail of the distribution are sufficiently energetic to excite caesium atoms to the first excited state. After such a collision the electron has lost most of

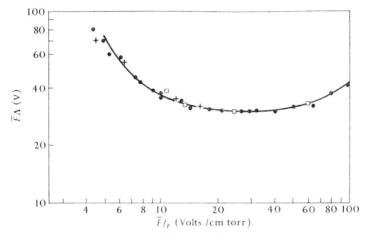

FIG. 5.11. Relation between $\bar{F}\Lambda$ and $\bar{F}/p$ observed by Reder and Brown for high-frequency gas discharge breakdown in pure helium calculated: ———; observed: ● ($L = \frac{1}{4}$ in), + ($L = \frac{1}{2}$ in), □ ($L = \frac{1}{16}$ in).

its energy. These collisions have the effect therefore of enhancing the low-energy region of the distribution function where the cross-section for electron-argon collisions is very small owing to the Ramsauer–Townsend effect (Chap. 1, § 6.2.5).

Nolan and Phelps used an a.c. technique developed originally by Rutherford‡ for the measurement of ion-drift velocities (see Chap. 19, § 2). A square wave alternating potential of frequency f is applied between the two Ni–Cu alloy plates of a parallel plate condenser, with a guard ring round one of the plates. There is a steady source of electrons at one plate but they can only drift across to the other plate during the positive half-cycle. If the drift time $\tau > \frac{1}{2}f$ they will not reach this plate before the field is reversed so that no current flows. By varying f the drift time corresponding to the amplitude of the potential can be obtained.

In the experiment of Nolan and Phelps the whole tube was maintained

† NOLAN, J. F. and PHELPS, A. V., *Phys. Rev.* **140** (1965), A792.
‡ RUTHERFORD, E., *Phil. Mag.* **44** (1897), 422.

at 200° C and under these circumstances caesium deposited on the plates gave rise to thermionic emission from both plates. The two currents i_1, i_2, were not equal, however. The mean resultant current was i_1-i_2. The experiment consisted in measuring this current as a function of f. Fig. 5.12 shows the results plotted against $2f$ for $F/p = 0\cdot032$ and a relative caesium to argon concentration of less than 10^{-7}. Electron

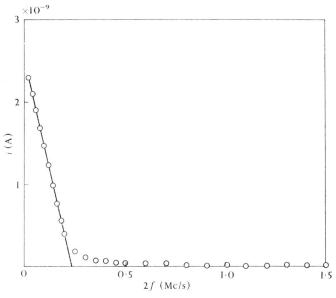

Fig. 5.12. Typical observed variation of average anode current with the square wave frequency (f) in the experiments of Nolan and Phelps on the excitation of caesium.

diffusion leads to the rounding off of the sharp break in the current. The point corresponding to $\tau = \tfrac{1}{2}f$ is obtained by extrapolating the linear portion of the curve to the horizontal axis.

A bellows arrangement in the vacuum wall enabled the separation of the electrodes to be varied. The drift time was found to be accurately proportional to the drift distance for a fixed F/p value.

Fig. 5.13 shows a comparison of the results obtained by Nolan and Phelps for the experiments using caesium-contaminated argon with the drift velocity measurement of Pack and Phelps† for pure argon. The presence of the caesium is seen to cause a marked increase in the drift velocity for F/p values from $0\cdot01$ to $0\cdot1$ V cm^{-1} torr^{-1}. The effect of inelastic collisions on both drift velocity and mean electron energy

† loc. cit. p. 59.

measurements has been analysed by Engelhardt and Phelps† in pure argon at much larger F/p values (up to 35 V cm^{-1} torr^{-1}). In contrast to the drift velocity, inelastic collisions have the effect of reducing the mean electron energy, as is to be expected.

We return later to the discussion of the application of the results of the experiments described in these sections to test the accuracy of assumed inelastic cross-sections. Before doing so, however, it is necessary

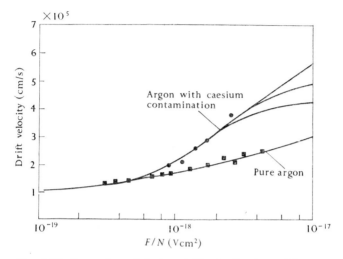

Fig. 5.13. Comparison of electron drift velocity observed in pure argon and in caesium-contaminated argon, for different values of F/N where F is the electric field strength and N the number of argon atoms per cm^3.

to develop methods for calculating the velocity distribution function $f(\mathbf{v})$ when such collisions are important. The ionization and excitation coefficients and frequencies will be very sensitive to the high-energy tail of the distribution, which in turn will be sensitive to the inelastic collision frequency. We consider in the following section how $f(\mathbf{v})$ may be calculated.

3.2. *Calculation of the velocity distribution function when inelastic collisions are important*

When inelastic scattering was ignored the velocity distribution function f was obtained by equating the number $c\,d\gamma$ of representative points leaving the element $d\gamma\,(=d\xi d\eta d\zeta)$ of velocity space, due to the applied field F, to $(b-a)d\gamma$, where a is the number leaving and b the number entering the element per second due to collisions (Chap. 2, § 1.2.1). If the mean energy is small compared with the energy

† ENGELHARDT, A. G. and PHELPS, A. V., *Phys. Rev.* **133** (1964), A375.

of excitation we can ignore the contribution to b from inelastic collisions. The contribution to a will be $NQ_{in}vf$, where N is the number of gas atoms per cm³ and Q_{in} the cross-section for inelastic collisions when the electron velocity is $v = (\xi^2+\eta^2+\zeta^2)^{\frac{1}{2}}$. If this term is included in $(b-a)$ of Chapter 2 the equation (14) of Chapter 2 becomes

$$(eF/m)\frac{df_0}{dv} = -N(Q_d+Q_{in})vf_1, \qquad (27)$$

while (15) is replaced by

$$(eF/3m)\frac{d(v^2f_1)}{dv} = (mN/M)\frac{d}{dv}(v^4Q_df_0) - NQ_{in}v^3f_0, \qquad (28)$$

the notation being as in Chapter 2, § 1.2.1.

Eliminating f_1 we obtain

$$\frac{d^2f_0}{dv^2} + \lambda(v)\frac{df_0}{dv} + \mu(v)f_0 = 0, \qquad (29\,\text{a})$$

where

$$\lambda(v) = \frac{Q_t}{v}\frac{d}{dv}\left(\frac{v}{Q_t}\right) + \frac{3m^3N^2}{Me^2F^2}v^3\,Q_t\,Q_d, \qquad (29\,\text{b})$$

$$\mu(v) = \frac{3m^2N^2}{Me^2F^2v}\left\{mQ_t\frac{d}{dv}(v^4Q_d) - MQ_{in}\,Q_t v^3\right\}. \qquad (29\,\text{c})$$

Here Q_t has been written for Q_d+Q_{in}. In practice Q_d is often $\gg Q_{in}$.

Given Q_{in} and Q_d as functions of electron velocity it is possible to solve (29 a) with the boundary conditions $f_0(0) = 0$, $f_0(\infty) \to 0$. For velocities

$$v > v_a = (2E_a/m)^{\frac{1}{2}},$$

where E_a is the excitation energy, it is usually a good approximation to assume that energy loss due to elastic collisions is negligible compared with that due to inelastic collisions. In that case, omitting the first term on the right side of (28) and substituting in (27), we have, if $Q_{in}/Q_d \ll 1$,

$$\frac{d^2f_0}{dv^2} + \frac{1}{lv}\frac{d}{dv}(lv)\frac{df_0}{dv} = \frac{3m^2}{e^2F^2l^2}\alpha v^2 f_0, \qquad (30)$$

where $\alpha = Q_{in}/Q_d$ is a function of v and the mean free path l, $= 1/NQ_d$, for diffusion. Knowing α and l this equation may be solved to obtain f_0 for $v > v_a$ apart from two arbitrary constants. These may be determined by the requirement that the solution for $v < v_a$, obtained by the method of Chapter 2, § 1.2.1, should join smoothly to that for $v > v_a$ at $v = v_a$, and that

$$4\pi \int_0^\infty f_0(v)v^2\,dv = 1.$$

With the availability of electronic computation there is no difficulty in solving equations such as (29) for a great number of assumed forms for the cross-sections but, up to the present, most calculations have been carried out, either by using the approximation (30) or by representing the cross-sections by analytical expressions so chosen as to make the equation (29) soluble in analytical terms. Some calculations have also been carried out using Jeffreys's method of approximating to the solution of (29) but this is not very satisfactory for present purposes.

We now describe the applications which have been made.

3.3. Interpretation of experimental results

3.3.1. Application to helium. Calculations of distribution functions for helium have been carried out by Smit[†] for F/p ranging from 3 to 10 V cm^{-1} torr^{-1}, by Abdelnabi and Massey[‡] for $F/p = 10$, 20 V cm^{-1} torr^{-1}, by Reder and Brown[§] for $F/p = 6$, 10, 20, and 60 V cm^{-1} torr^{-1}, and by Heylen and Lewis[||] for $F/p = 5$, 20, 50, 100, and 200 V cm^{-1} torr^{-1}. All calculations have been based on the total excitation cross-sections of Maier-Leibnitz (§ 2.2) and the ionization cross-sections of Smith (Chap. 3, § 2.5.1).

The first results were obtained by Smit using the approximation (30). An analytical representation for the cross-sections that enabled equation (29) to be solved in terms of confluent hypergeometric functions was used by Abdelnabi and Massey and by Reder and Brown. Heylen and Lewis solved equation (29) using Jeffreys's approximation. Fig. 5.14 illustrates a number of these distributions.

In this figure the distribution of the symmetrical component is shown as an energy distribution, i.e. the fraction $f_0(E)\,dE$ of electrons with energy between E and $E+dE$. The function plotted, $f_0(E)$, is related to $f_0(v)$ above by

$$f_0(E) = 4\pi\sqrt{2}\,m^{-\frac{3}{2}}E^{\frac{1}{2}}f_0(v). \tag{31}$$

Fig. 5.14 (a) compares three calculations of $f_0(E)$ for $F/p = 20$ V cm^{-1} torr^{-1}. The sharp drop in the distribution function at 20 eV, corresponding to the lowest excitation potential for helium, found by Heylen and Lewis, was not obtained in the other calculations. Fig. 5.14 (b) shows the curves obtained by Reder and Brown out to electron energies of 48 eV.

Fig. 5.15 shows the results of the measurements of the ionization coefficient for helium obtained by Davies et al.[††] in the range of F/p from 5 to 25 V cm^{-1} torr^{-1}. They are somewhat smaller than the values obtained earlier by Townsend and McCallum,[‡‡] the discrepancy ($\simeq 35$ per cent) being most marked at low values of F/p where the effect of small traces of impurity would be most marked.

On the same figure are shown the values of α_i/p calculated by Abdelnabi and Massey,[‡] Heylen and Lewis,[||] by Phelps,[§§] and also by Dunlop,[||||] who

---

[†] SMIT, J. A., *Physica* **3** (1936), 543.
[‡] ABDELNABI, I. and MASSEY, H. S. W., *Proc. phys. Soc.* **A66** (1953) 288.
[§] REDER, F. H. and BROWN, S. C., *Phys. Rev.* **95** (1954) 885.
[||] HEYLEN, A. E. D. and LEWIS, T. J., *Proc. R. Soc.* **A271** (1963) 531.
[††] DAVIES, D. K., LLEWELLYN JONES, F., and MORGAN, C. G., *Proc. phys. Soc.* **80** (1962) 898.
[‡‡] TOWNSEND, J. S. and McCALLUM, S. P., *Phil. Mag.* **17** (1934) 678.
[§§] PHELPS, A. V., *Phys. Rev.* **117** (1960) 619.
[||||] DUNLOP, S. H., *Nature, Lond.* **164** (1949) 452.

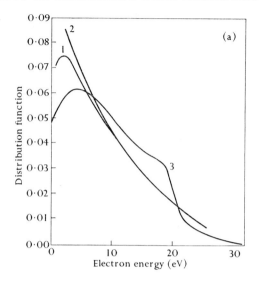
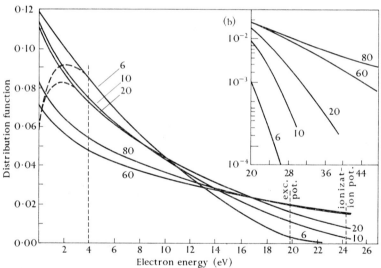

Fig. 5.14. Calculated distribution functions $f_0(E)$ of electrons in helium. (a) For $F/p = 20$ V cm^{-1} torr^{-1}; (1) Abdelnabi and Massey; (2) Reder and Brown; (3) Heylen and Lewis. (b) For various F/p values indicated on curves (Reder and Brown). The full curves are calculated assuming a constant collision frequency, ν, independent of electron energy, E. The dotted curves are calculated assuming a variation of collision frequency like $E^{\frac{1}{2}}$ below 4 eV.

used Smit's distribution. The agreement between the measurements and the calculated ionization coefficients, using the excitation cross-sections of Maier-Leibnitz, is satisfactory. For $F/p = 20$ V cm^{-1} torr^{-1}, for example, Abdelnabi and Massey obtained $\alpha_i/p = 0\cdot 17$ ions cm^{-1} torr^{-1} compared with the observed value of $0\cdot 14$. To examine the sensitivity of the calculated results they also calculated α_i/p assuming excitation cross-sections respectively one-half and twice those found by Maier-Leibnitz, all other cross-sections being unchanged. A roughly proportional effect on α_i/p was obtained, the respective values being $0\cdot 081$ and $0\cdot 31$ ions cm^{-1} torr^{-1}, giving quite strong support to the validity of the Maier-Leibnitz values.

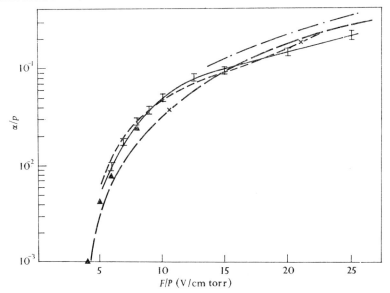

FIG. 5.15. The Townsend first ionization coefficient α_i helium. Calculated: ▲ Dunlop; × Abdelnabi and Massey; – – – – Phelps; —— Heylen and Lewis. Observed: – · – · – · Townsend and McCallum; ⊥ Davies, Llewyllyn-Jones, and Morgan.

The measurements of Reder and Brown of microwave gas discharge breakdown in helium are also in good agreement with those expected on the basis of the Maier-Leibnitz excitation cross-sections and Smith's[†] ionization cross-section. The solid curve of Fig. 5.11, calculated using these cross-sections, is seen to be in good agreement with the measured values.

Frost and Phelps[‡] have taken advantage of earlier observations on microwave breakdown fields in helium containing some mercury vapour[§]

[†] SMITH, P. T., *Phys. Rev.* **36** (1930) 1293.
[‡] FROST, L. S. and PHELPS, A. V., *Westinghouse Research Laboratories Report No. 6-94439–6 R3* (1957).
[§] MACDONALD, A. D. and BROWN, S. C., *Phys. Rev.* **75** (1949) 411.

to compare a metastable atom production coefficient derived from experimental data on excitation cross-sections with one derived from the breakdown data. This is possible because, when mercury vapour is present, almost all of the ionization results from the reaction

$$\text{He}' + \text{Hg} \to \text{He} + \text{Hg}^+ + \text{e}, \tag{32}$$

where He' is a metastable helium atom. In the steady state if n_m is the metastable atom concentration

$$D_m \nabla^2 n_m + \bar{\nu}_m n_e - \bar{\nu}_{\text{Hg}} n_m = 0, \tag{33}$$

where D_m is the coefficient of diffusion of metastable helium atoms in helium at the operating pressure, $\bar{\nu}_m$ is the rate of production of metastable helium atoms per electron per second, and $\bar{\nu}_{\text{Hg}}$ that of destruction of these atoms per metastable atom per second by the process (32).

Similarly, for the electron concentration n_e,

$$D_e \nabla^2 n_e + \bar{\nu}_i n_e + \bar{\nu}_{\text{Hg}} n_m = 0, \tag{34}$$

where D_e is the coefficient of diffusion of electrons and ν_i the rate of ionization of helium atoms per electron per second.

Elimination of n_m from (33) and (34) gives

$$D_e \nabla^4 n_e + (\bar{\nu}_i - D_e \bar{\nu}_{\text{Hg}}/D_m) \nabla^2 n_e - (\bar{\nu}_{\text{Hg}}/D_m)(\bar{\nu}_m + \bar{\nu}_i) n_e = 0. \tag{35}$$

If we now make the usual assumption that $\nabla^2 n_e = -\Lambda_1^{-2} n_e$, where Λ_1 is the fundamental diffusion length, then

$$\bar{\nu}_m = (1 + D_m/\bar{\nu}_{\text{Hg}} \Lambda_1^2)\{(D_e/\Lambda_1^2) - \bar{\nu}_i\}. \tag{36}$$

All the quantities on the right-hand side are known, D_m from the microwave afterglow studies of Phelps (see Chap. 18, § 5), D_e from measurement of drift velocity and mean energy of electrons in helium (see Chap. 2, § 7.1), $\bar{\nu}_{\text{Hg}}$ from the microwave afterglow studies of Phelps, and $\bar{\nu}_i$ from the breakdown studies in pure helium. $\bar{\nu}_m$ can then be calculated for values of $\bar{F}/p$ obtained from measurements of the conditions under which microwave breakdown occurs (§ 3.1.2).

To compare with results derived from observed cross-sections, we must remember that the cross-sections for production of metastable helium atoms by impact of electrons of homogeneous energy, as observed by Dorrestein[†] and Schulz and Fox,[‡] include all contributions from radiation cascade population of the 2^3S and 2^1S states from upper states that have been directly excited. Such data could be taken over immediately for present purposes but for one factor—the imprisonment of resonance radiation. Because of this, cascade contributions involving ^1P states are

[†] DORRESTEIN, R., *Physica* 9 (1942), 433, 447.
[‡] SCHULZ, G. J. and Fox, R. E., *Phys. Rev.* 106 (1957) 1179.

dependent on gas pressure and tube dimensions (see Chap. 4, § 1). It is therefore necessary, before applying the observed data, to analyse it into contributions whose dependence on these quantities can be allowed for (see Chap. 4, § 1.4.2).

Frost and Phelps† therefore write the total metastable atom production rate for electrons of a given velocity, in the form

$$\nu_m = \nu_t + \nu_s + (1-f)\nu_r, \tag{37}$$

where ν_t, ν_r are the respective rates of population of 2^3S and 2^1P states arising either directly or by cascade. ν_s is the rate of population of 2^1S by all processes except radiative transition from 2^1P. f is the fraction of 2^1P excited atoms that undergo a radiative transition to the ground 1^1S state. Because of the imprisonment of resonance radiation, both f and ν_r will depend on pressure and tube dimensions. In fact ν_r is only slightly dependent on these quantities while f may be calculated in terms of them as explained in Chapter 4, § 1.3.1, p. 172.

To derive suitable values for ν_t, ν_s, and ν_r it is necessary to analyse the observed data near the threshold to obtain separate cross-sections for excitation of the various levels. For this purpose most attention was paid to the observations of Maier-Leibnitz (§ 2.2), Dorrestein (Chap. 4, § 2.1.2.1), and Schulz and Fox (Chap. 4, § 2.1.2.1).

The rate of rise of the cross-section for 2^1S just beyond the threshold was taken from the observations of Schulz as 9.6×10^{-8} cm²/V. Given this and the location of the threshold and peak, an excitation cross-section function $Q(2^1S)$ may be drawn in to about 22 eV energy. The observed rate of rise of the 2^3P cross-section (2.6×10^{-18} cm²/V) enables $Q(2^3P)$ to be drawn in from its threshold to about the same energy. Subtraction of $Q(2^1S)+Q(2^3P)$ from the total then gives $Q(2^3S)$ from its threshold to 22 eV. Continuation beyond this is less definite and is based on the general shapes of the cross-sections as discussed in Chapter 4, § 1.4.2.1 together with the value for the cross-sections for excitation of the upper triplet levels derived from analysis of optical excitation data on the same lines as in that section. Fig. 5.16 illustrates the analysis of the metastable production data as carried out by Frost and Phelps. The total production curve was obtained under such low pressure conditions that imprisonment of resonance radiation is unimportant, and at the relatively low electron energies involved cascade population of 2^1S from upper singlet states is very small.

The broken curve shows the rate of metastable production obtained

† FROST, L. S. and PHELPS, A. V., loc. cit., p. 303.

from calculated electron distribution functions and the measured cross-sections as described above. The full curve shows the rate obtained from the microwave breakdown measurements of Macdonald and Brown using cavities of various dimensions. The two curves representing total excitation rate in Fig. 5.16 agree well in the range $3 < F/p < 10$ (V cm^{-1} torr^{-1}).

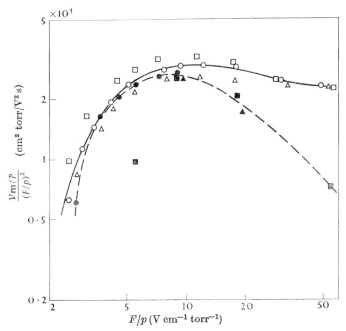

Fig. 5.16. Comparison of total metastable production rate in helium derived from measurements of microwave breakdown in helium contaminated with mercury vapour (full curve) with production rate calculated using measured excitation cross-sections and calculated electron energy distributions (dashed curve).

○ $\Lambda = 0.05055$ cm   ⎫ Measurements of Macdonald and Brown
☐ $\Lambda = 0.1011$ cm   ⎬ for cavities with different
△ $\Lambda = 0.1513$ cm   ⎭ characteristic diffusion length Λ.

Metastable production rate calculated using the distribution functions of Smit ●, Reder and Brown ■, and Abdelnabi and Massey ▲.

The falling off of the calculated curve at higher F/p is attributed by Frost and Phelps to the use in the derivation of the distribution functions of too large a total excitation cross-section above 28 eV.

The total metastable production rate is broken down into its component excitation rates in Fig. 5.17. These curves enable determination of the excitation rate to the 2^3S, 2^1S, or 2^1P levels of helium by electrons for a wide range of pressures and fields.

3.3.2. *Application to other rare gas atoms.* Electron velocity distributions allowing for inelastic collisions have been calculated by Druyvesteyn† and by MacDonald and Betts‡ for neon, by Engelhardt and Phelps§ for argon, and by Heylen and Lewis‖ for neon and argon. All these calculations used inelastic scattering cross-sections obtained by Maier-Leibnitz.

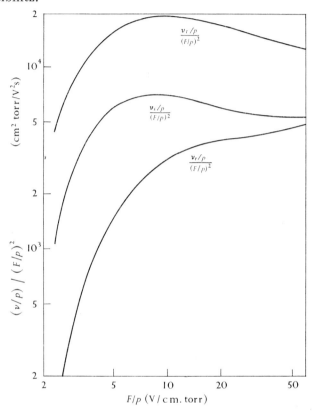

Fig. 5.17. Excitation rates as functions of F/p for the 2^1S (ν_s), 2^3S (ν_t) and 2^1P (ν_r) levels of helium.

Measurements of α_i/p have been made by Townsend and McCallum†† for neon, by Davies and Milne‡‡ for argon, and by Kruithof§§ for neon and argon in the F/p range up to 250 V cm^{-1} torr^{-1}. For neon the results

---

† DRUYVESTEYN, M. J., *Physica* **3** (1936) 65.
‡ MACDONALD, A. D. and BETTS, D. D., *Can. J. Phys.* **30** (1952) 565.
§ ENGELHARDT, A. G. and PHELPS, A. V., *Phys. Rev.* **133** (1964) A375.
‖ HEYLEN, A. E. D. and LEWIS, T. J., *Proc. R. Soc.* **A271** (1963) 531.
†† TOWNSEND, J. S. and MCCALLUM, S. P., *Phil. Mag.* **6** (1928) 857.
‡‡ DAVIES, D. E. and MILNE, J. G. C., *Br. J. appl. Phys.* **10** (1959) 301.
§§ KRUITHOF, A. A., *Physica* **7** (1940) 519.

of the two experiments are in general agreement and they agree with the calculated values of Druyvesteyn up to $F/p = 35$ V cm^{-1} torr^{-1}. Heylen and Lewis extended their calculations up to F/p values of

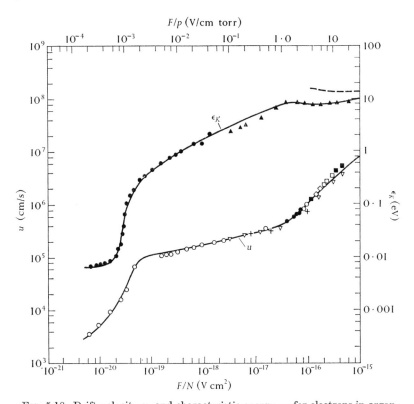

FIG. 5.18. Drift velocity u, and characteristic energy ϵ_K for electrons in argon as functions of F/N. Observed ϵ_K: ● Warren and Parker (77 °K); ▲ Townsend and Bailey (288 °K). Observed u: ○ Pack and Phelps (300 °K); ▽ Errett (293 °K); □ Caren; ◇ Herreng; ■ Riemann; + Bowe; ● Nielsen (293 °K). Calculated: —— Engelhardt and Phelps (77 °K); ---- Heylen and Lewis.

References: WARREN, R. W. and PARKER, J. H., Phys. Rev. 128 (1962), 2661; TOWNSEND, J. S. and BAILEY, V. A., Phil. Mag. 44 (1922) 1033; PACK, J. L. and PHELPS, A. V., Phys. Rev. 121 (1961), 798; ERRETT, D., Ph.D.Thesis, Purdue University 1951; CAREN, R. P., Ph.D. Thesis, Ohio State University, 1961; HERRENG, P., C.r. hebd. Séanc. Acad. Sci., Paris 217 (1943) 75; RIEMANN, H., Ergbn. exakt. Naturw. 22 (1949) 73; BOWE, J. C., Phys. Rev. 117 (1960) 1411 and 1416; NIELSEN, R. A., ibid. 50 (1936) 950; ENGELHARDT, A. G. and PHELPS, A. V., ibid. 133 (1964) A375; Heylen, A. E. D. and LEWIS, T. J., Proc. R. Soc. A271 (1963), 531.

200 V cm^{-1} torr^{-1} but obtained results about 35 per cent less than the experimental values. For argon, on the other hand, their calculated values are in good agreement with the experimental values over the whole range of F/p values up to 200 V cm^{-1} torr^{-1}.

High frequency gas discharge breakdown measurements carried out in neon by MacDonald and Betts† give a relation between breakdown field strength and pressure in good agreement with the calculated values.

Turning now to measurements of transport coefficients, Fig. 5.18 shows the measurements of drift velocity u and characteristic energy ϵ_k for electrons in argon up to F/p values of 35 V cm^{-1} torr^{-1}. They are in very good agreement with the calculated values of Engelhardt and Phelps over the whole range of F/p values, including in particular the region $F/p > 1$ V cm^{-1} torr^{-1} where inelastic collisions have an important influence on the results.

It appears, therefore, that for neon and argon also the excitation cross-sections obtained by Maier-Leibnitz are consistent with the results of all three types of swarm experiment.

3.3.3. *Application to caesium.* The drift velocity measurements of Nolan and Phelps‡ in argon–caesium mixtures (Fig. 5.13) enable a determination of the shape of the caesium excitation curves in the region about 1 eV above the threshold. The curves for argon with caesium contamination (Fig. 5.13) give the expected form of variation of drift velocity with F/N for three different assumptions about the form of the excitation curve. Clearly the present measurements do not enable us to decide between these assumptions. The measured values, however, are consistent with a linear increase of caesium excitation cross-section with a slope of $7 \cdot 1 \times 10^{-15}$ cm^2/eV in the energy range up to 0·5 eV above threshold. It is expected that both the $6P_{\frac{1}{2}}$ and $6P_{\frac{3}{2}}$ states, with thresholds of 1·386 and 1·454 eV respectively, will contribute to the observed inelastic cross-section. The experimental observations are consistent with slopes of $2 \cdot 5 \times 10^{-15}$ cm^2/eV and $5 \cdot 0 \times 10^{-15}$ cm^2/eV for these two excitation cross-sections near threshold.

4. The energy loss of electrons scattered through a fixed angle

The electron energy loss measurements described in § 2 refer to electrons inelastically scattered through all angles. Many measurements have been made of the energy loss of electrons scattered through a fixed angle after collisions in which the target atom is left in an excited or ionized state. Early measurements of this type such as the classical investigations of Lenard§ and of Franck and Hertz‖ were concerned primarily with the measurement of critical potentials and were of the

---

† MACDONALD, A. D. and BETTS, D. D., *Can. J. Phys.* **30** (1952) 565.
‡ NOLAN, J. F. and PHELPS, A. V., *Phys. Rev.* **140** (1965) A792.
§ LENARD, P., *Annln Phys.* **8** (1902) 149.
‖ FRANCK, J. and HERTZ, G., *Verh. der Phys. Ges.* **16** (1914) 457.

utmost importance in providing an essential link between spectroscopic data and inelastic collisions.

4.1. *Methods of velocity analysis of the scattered electrons*

Velocity analyses of the scattered electrons have been made, using retardation methods, magnetic or electrostatic analyses, or combinations of the two last methods.

The retardation method was originally applied to experiments of this type by Franck and Hertz.† Later applications of the method to the

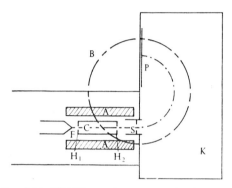

Fig. 5.19. Whiddington's apparatus for the study of the magnetic spectrum of electrons scattered inelastically through small angles.

study of inelastic collisions in mercury were made by Vetterlein,‡ who analysed electrons scattered inelastically through small angles, and by Arnot and Baines§ for electrons scattered inelastically over a large range of angles between 9° and 171°.

The magnetic deflexion method has been employed extensively to study the energy loss of electrons that have undergone inelastic collisions by Whiddington and his associates.∥ Fig. 5.19 illustrates Whiddington's apparatus, which analysed electrons that have been scattered through small angles only.

Electrons from the filament F were accelerated through the hole H_1 into the chamber C containing the gas being investigated. After emerging

---

† Franck, J. and Hertz, G., *Z. Phys.* **17** (1916) 409.
‡ Vetterlein, P., *Annln Phys.* **35** (1939) 251.
§ Arnot, F. L. and Baines, G. O., *Proc. R. Soc.* **A151** (1935) 256.
∥ Jones, H. and Whiddington, R., *Phil. Mag.* **6** (1928) 889; Roberts, J. E. and Whiddington, R., ibid. **12** (1931) 962; Whiddington, R. and Taylor, J. E., *Proc. R. Soc.* **A136** (1932) 651; ibid. **145** (1934) 465; Whiddington, R. and Priestley, H., ibid. 462; Whiddington, R. and Woodroofe, E. G., *Phil. Mag.* **20** (1935) 1109; Lee, A. H., *Proc. R. Soc.* **A173** (1939) 569.

through the hole H_2, 0·145 mm in diameter, the electrons passed through the defining slit S, situated 8·5 mm from H_2, into the (evacuated) camera K. The camera was placed in a magnetic field with lines of force in a direction perpendicular to the plane of the paper, produced by the magnet with pole face B. The electrons were analysed by the magnetic field into a number of beams of different energy corresponding to inelastic scattering after excitation of different states of the molecules of the scattering gas. They were detected by the photographic plate P after deflexion through 180°. The iron cylinder A served to shield the filament F and the collision chamber C from the magnetic field of the velocity analyser. Measurements were carried out on H_2, He, Ne, and A.

Electrostatic analysis has been used by van Atta[†] to study energy losses of electrons scattered inelastically through small angles in He, Ne, and A. Womer[‡] has used a similar method for He. The electrons, after collision, were analysed in energy by a 127° analyser of the Hughes–Rojansky type. An extensive series of measurements using a similar type of analysis has been carried out more recently by Lassettre and colleagues,[§] while Schulz and Philbrick[∥] used cylindrical electrostatic analysers to select a beam of electrons very homogeneous in energy before scattering and to measure their energy loss after scattering through a fixed angle of 72° in helium. Simpson[††] and his collaborators used the apparatus described in Chapter 1, § 5.1 (Fig. 1.6), in which both monochromator and analyser were concentric spherical electrostatic lenses.

In their later work Skerbele and Lassettre[‡‡] used a concentric spherical monochromator similar to those used by Simpson and colleagues to select a narrow band of electron energies. This was used in conjunction with the 127° cylindrical electrostatic analyser used in their earlier work.

Energy loss measurements with electron beams of energy up to 50 keV have been carried out with extremely high energy resolution ($\sim$ 0·02 eV) by Boersch, Geiger, and colleagues,[§§] using a monochromator and analyser of the Wien[∥∥] type with crossed electrostatic and magnetic fields. The

[†] VAN ATTA, L. C., *Phys. Rev.* **38** (1931) 876.
[‡] WOMER, R. L., ibid. **45** (1934) 689.
[§] See, for example, LASSETTRE, E. N. and FRANCIS, S. A., *J. chem. Phys.* **40** (1964) 1208.
[∥] SCHULZ, G. J. and PHILBRICK, J. W., *Phys. Rev. Lett.* **13** (1964) 477.
[††] The apparatus is described by KUYATT, C. E., SIMPSON, J. A., and MIELCZAREK, S. R., *Phys. Rev.* **138** (1965) A385.
[‡‡] SKERBELE, A. M. and LASSETTRE, E. N., *J. chem. Phys.* **40** (1964) 1271.
[§§] BOERSCH, H., GEIGER, J., and HELLWIG, H., *Phys. Lett.* **3** (1962) 64; BOERSCH, H., GEIGER, J., and STICKEL, W., *Z. Phys.* **180** (1964) 415.
[∥∥] WIEN, W., *Verh. dt. phys. Ges.* **16** (1897) 165.

arrangement is shown in Fig. 5.20. The electron beam leaves the gun through a hole H_1 of diameter d_1 with energy E (20–50 keV). Its angular aperture is defined by the diameter d_2 of the hole H_2 in the plate B_2 immediately before the monochromator. The monochromator M selects a very narrow band of electron energies and focuses them on a hole H_5 in the plate B_5 immediately before they enter the collision chamber C. The analyser A focuses electrons of a similar narrow energy band on to the detector D. By varying the energy band passed by the analyser an accurate energy loss analysis can be carried out.

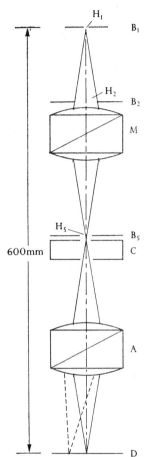

Fig. 5.20. Arrangement of apparatus used by Boersch, Geiger, *et al.* for energy loss measurements with beams of energy up to 50 keV.

The monochromator and analyser (Fig. 5.21) are almost identical in construction. Electrons of the full energy entering the monochromator are reduced in energy to a value E_0 (20–300 eV) by the electrostatic lens L_1 and brought to an intermediate focus to form an image of H_1 at the entrance aperture H_3 of the filter. Electrons of a particular energy determined by the crossed magnetic and electric fields pass through the filter without deflexion and emerge through the exit hole H_4. They are then re-accelerated to their initial energy by the lens L_2, which brings them to a focus on the hole H_5 in the plate B_5. Electrons of any other energy are deflected in the filter and removed from the beam. For operating conditions with $E = 25$ keV, $E_0 = 20$ eV, $d_1 = 0.04$ mm, $d_2 = 0.02$ mm and the separation between B_1 and B_2, $b = 15$ cm, the observed energy width of the transmitted band of electrons at half height was 0.017 eV, giving a resolving power of 1.5×10^6.

4.2. *Results of energy loss measurements*

Fig. 5.22 shows the 0° energy loss spectrum for helium obtained by Skerbele and Lassettre[†] using their high resolution apparatus with incident electrons of energy 250, 300, and 350 eV. Losses corresponding

[†] SKERBELE, A. M. and LASSETTRE, E. N., *J. chem. Phys.* **40** (1964) 1271.

to excitation of the 2^1S, 2^1P, 3^1P, 4^1P, 5^1P, and 6^1P states are clearly seen.

Fig. 5.23 shows the energy loss spectrum at $0°$ obtained by Boersch et al.† for 25-keV electrons in argon. The chief interest lies in the

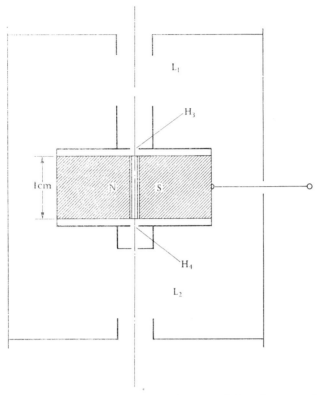

FIG. 5.21. Section through the analyser parallel to the magnetic field, in the apparatus used by Boersch, Geiger et al.

extremely high resolution obtainable for such high energy electrons. The main energy losses observed can be associated with the excitation of known levels of the argon atom as shown in the figure. Boersch et al.‡ have also measured the energy loss spectrum for electrons in atomic hydrogen but with a lower resolution.

Of particular interest was the observation by Whiddington and Priestley§ of an energy loss of 59·25 eV in helium which was interpreted as due to excitation to an auto-ionizing state (see Chap. 9, § 3.2), $2p^2(^1D)$,

† BOERSCH, H., GEIGER, J., and STICKEL, W., Z. Phys. **180** (1964) 415.
‡ BOERSCH, H., GEIGER, J., and REICH, H. J., ibid. **161** (1961) 296.
§ WHIDDINGTON, R. and PRIESTLEY, H., Proc. R. Soc. **A145** (1934) 462.

in which both helium electrons are raised to 2s states. Fig. 5.24 shows a recent energy loss spectrum between 57 and 65 eV energy loss obtained by Simpson, Chamberlain, and Mielczarek† for electrons in helium. Energy loss peaks are visible corresponding to excitation of the following

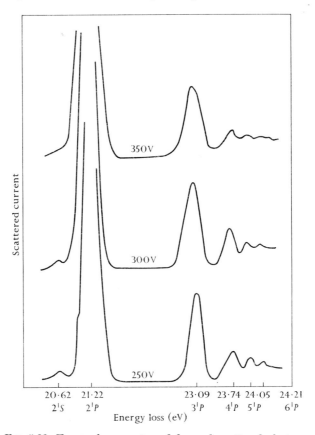

FIG. 5.22. Energy loss spectra of forward scattered electrons in helium as observed by Skerbele and Lassettre (incident electron energies are indicated on each curve).

auto-ionizing states: $(2s^2)^1S$ at 57·9 eV, $(2p^2)^1D$ at 60·0 eV, $(2s2p)^1P$ at 60·1 eV, and $(ns\,mp)^1P$ at 63·6 eV. The first two of these states are most prominent for 90-eV incident electrons and have almost disappeared for the 400-eV energy loss curve where the last two have become very prominent. This is consistent with the shape of the excitation function expected, since the first two transitions are optically forbidden while the

† SIMPSON, J. A., CHAMBERLAIN, G. E., and MIELCZAREK, S. R., Phys. Rev. **139** (1965) A1039.

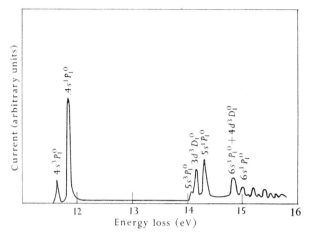

FIG. 5.23. Energy loss spectrum obtained by Boersch et al. for 25-keV electrons scattered at 0° in argon.

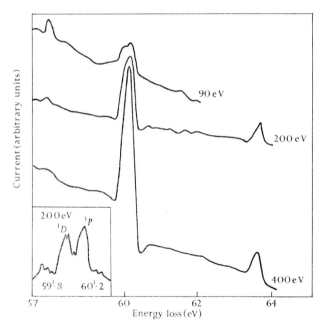

FIG. 5.24. Energy loss spectra of forward scattered electrons in helium in the range 57–65 eV observed by Simpson, Chamberlain, and Mielczarek. The incident energy of the electrons is indicated. The inset shows finer detail of the resonance near 60 eV.

last two are optically allowed. The optical absorption studies of Madden and Codling† (see Chap. 14, § 7.1) do not show the first two auto-ionizing excitations.

The variation with incident electron energy of the relative strength of the energy losses corresponding to optically forbidden excitation of the ^1S and ^1D auto-ionizing states is shown in Fig. 5.25. These curves give effectively values of the quantity $\int_0^{\theta_0} I_n(\theta)\sin\theta\,d\theta$, where $n = {^1S}$ and ^1D respectively, relative to the same quantity for $n = {^1P}$ and θ_0, which

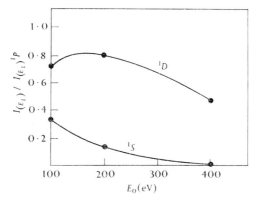

Fig. 5.25. Relative strengths of resonances due to optically forbidden and allowed transitions to auto-ionizing states in He as a function of primary energy E_0 (Simpson et al.).

is determined by the geometry of the experiment, is the minimum scattering angle which will cause an electron to be removed from the beam. The variation of this quantity with incident electron energy is seen to be quite different for the two transitions.

Figs. 5.26 and 5.27 show similar energy-loss spectra obtained by Simpson et al.‡ for excitation of auto-ionizing states of neon and argon.

4.2.1. *Resonances in inelastic scattering.* Systematic studies of the variation with initial electron energy of the strength of energy loss peaks corresponding to inelastic scattering through a fixed angle have revealed resonances in inelastic scattering very similar in nature to those found in elastic scattering (see Chap. 1, § 5). The first observations of such resonances were made by Schulz and Philbrick§ for the inelastic scattering of electrons in helium through an angle of 72°. The solid curve of Fig. 5.28 shows the relative differential cross-sections they observed for

† MADDEN, R. P. and CODLING, K., *Phys. Rev. Lett.* **10** (1963) 516; *J. opt. Soc. Am.* **54** (1964) 268. ‡ loc. cit. p. 314.
§ SCHULZ, G. J. and PHILBRICK, J. W., *Phys. Rev. Lett.* **13** (1964) 477.

electrons of energy between 19·8 and 23·0 eV, scattered in helium in a small angular range round 72° after having lost 19·8 eV energy corresponding to excitation of the 2^3S state. The curve is seen to have a surprisingly complex structure. For comparison the relative cross-

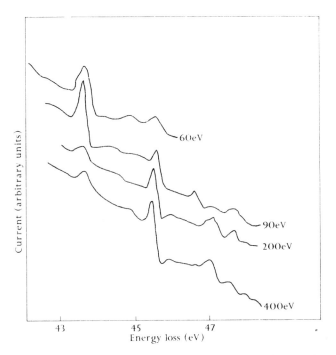

Fig. 5.26. Energy loss spectra of forward scattered electrons in neon as observed by Simpson, Chamberlain, and Mielczarek. The energy of the incident electrons is indicated.

section curve obtained by Schulz and Fox[†] for total metastable production is shown on the same figure. Previously this curve had been interpreted as representing the excitation of the 2^3S state up to 20·6 eV, with the rise commencing at 20·6 eV owing to onset of the 2^1S state. The observations of Schulz and Philbrick indicate that the rise at 20·6 eV is at least partly associated with the excitation of the 2^3S state. This, as well as the peak at 22·4 eV, was interpreted by Schulz and Philbrick to indicate further resonances in helium corresponding to formation of short-lived excited states of the helium negative ion, completely similar to the resonance observed in the elastic scattering of 19·30 eV electrons

† SCHULZ, G. J. and FOX, R. E., *Phys. Rev.* **106** (1957) 1179.

in helium (see Chap. 1, § 6.2.3), which was in fact used to calibrate the energy scale in these experiments.

Resonance structure in forward inelastic scattering in helium has been observed by Chamberlain and Heideman,‡ using the apparatus developed

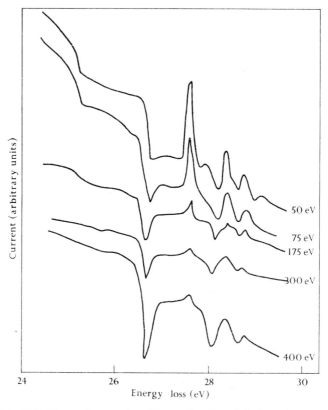

FIG. 5.27. Energy loss spectra of forward scattered electrons in argon between 24- and 30-eV energy loss as observed by Simpson, Chamberlain, and Mielczarek. The energy of the incident electrons is indicated.

by Simpson (Chap. 1, § 5.1) set to detect electrons scattered after having suffered particular energy losses. Fig. 5.29 shows the results obtained for inelastic scattering after excitation of the 2^3S, 2^1S, 2^3P, and 2^1P states of helium. The structure between 20 and 21 eV in the 2^3S excitation is even more complex than that observed by Schulz and Philbrick in electron scattering through 72°. Two peaks 20·4 and 20·8 eV can be

† CHAMBERLAIN, G. E., *Phys. Rev. Lett.* **14** (1965) 581.
‡ CHAMBERLAIN, G. E. and HEIDEMAN, H. G. M., ibid. **15** (1965) 337.

seen, and the cross-section falls almost to zero before the further resonance structure at 22·4–22·6 eV is reached. Practically the whole of the 2^3S excitation appears to be associated with resonance formation. For

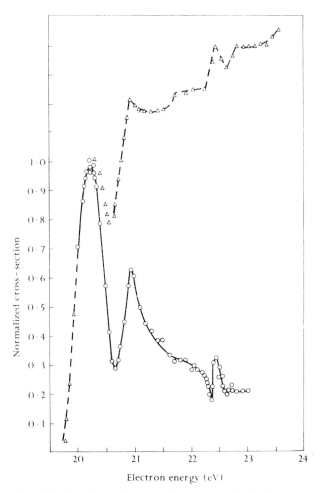

FIG. 5.28. The observed energy dependence of inelastic cross-sections in helium. O-O-O-O- differential cross-section at 72° for excitation of 2^3S (Schulz and Philbrick). -△-△-△-△ total metastable atom production (Schulz and Fox).

both the 2^3S and 2^1S levels there appears to be a peak or step-like rise within 0·06 eV of threshold. Resonant structure appears in all these excitations in the 22·4–22·9 eV region and it can be correlated quite closely with the resonance structure in total scattering observed by

Kuyatt, Simpson, and Mielczarek† in this region (see Chap. 1, § 6.2.3). This lends support to the interpretation of the phenomenon as arising from the formation of resonant states that can then decay through several alternative channels, leading to over-all elastic scattering or to inelastic scattering with excitation of either of the four levels investigated (see Chap. 9, § 5).

The two peaks above 22 eV in the 2^3S channel occur at 22·42 and 22·60 eV. In the 2^3P channel the dominant peak occurs also at 22·60 eV

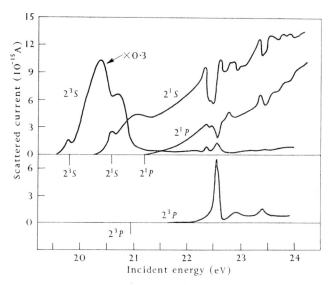

FIG. 5.29. Variation with incident electron energy of the forward scattered electron current after excitation of different 2-quantum states of helium as observed by Chamberlain and Heideman.

but other peaks occur at 22·9 and 23·5 eV that are barely discernible in the 2^3S channel.

The dominant feature in the singlet excitation curves is a rise followed by a sharp dip with minima occurring at 22·53 eV in the 2^1S curve and 22·60 eV in the 2^1P curve. Additional resonances at 22·9 and 23·5 eV are probably associated with those observed in the 2^3P channel.

Above the threshold region the resonances can be associated with He⁻ states formed from the $n = 3, 4, 5$ levels of helium. Chamberlain and Heideman‡ interpret their data as indicating two series of resonant negative ion states (see Chap. 9, § 5).

† KUYATT, C. E., SIMPSON, J. A., and MIELCZAREK, S. R., *Phys. Rev.* **138** (1965) A385.
‡ CHAMBERLAIN, G. E. and HEIDEMAN, H. G. M., *Phys. Rev. Lett.* **15** (1965) 337.

The transmission measurements of Kuyatt, Simpson, and Mielczarek that recorded elastic scattering resonances showed two resonances in helium at 57·1 eV and 58·2 eV. The first of these resonances is 0·8 eV below the lowest doubly excited (auto-ionizing) state of helium reported by Simpson and his collaborators† and has been interpreted by Fano and Cooper (see Chap. 9, § 5) in terms of the formation of a transient He⁻ state in the configuration $(2s^22p)^2P$ with all three electrons in $n = 2$ excited states. Similarly, the 58·2 eV resonance is assigned the configuration $(2s2p^2)^2D$. Chamberlain and Heideman observed these resonant states also in the 2^3S, 2^1S, and 2^1P energy loss curves.

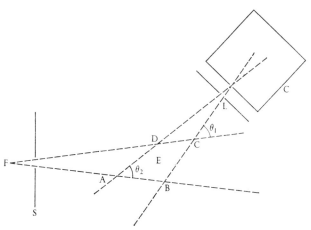

FIG. 5.30. Illustrating the principle of the method of measuring angular distribution of electrons scattered in gases and vapours.

5. Measurement of the angular distributions of the scattered electrons

Very extensive measurements have been made of the angular distribution function $I(\theta)$ both for elastically scattered electrons and electrons scattered in specific inelastic impacts. The general principle of most of the methods used is illustrated in Fig. 5.30. A beam of electrons of definite energy emerges through the slit S into the gas contained in a region free from electrostatic fields. Those electrons scattered from the area $ABCD$ of the beam through angles in the small range between θ_1 and θ_2 ($\simeq \theta_1$) pass through the entrance slit L of a collector. This collector includes some analysing device to ensure that only scattered

† SIMPSON, J. A., MIELCZAREK, S. R., and COOPER, J., *J. opt. Soc. Am.* **54** (1964) 269; SIMPSON, J. A., CHAMBERLAIN, G. E., and MIELCZAREK, S. R., *Phys. Rev.* **139** (1965) A1039.

electrons with energies in the required range are measured. Either the source S or the collector C may be rotated about an axis perpendicular to the plane of the paper at E, so that the variation of the collected current with the angle θ may be measured. The volume of the region from which the scattered current is observed increases as cosec θ so that the observed current at an angle θ must be multiplied by $\sin \theta$ in order to obtain the true relative scattered intensity per unit solid angle.

To ensure that the observations refer only to single scattering it is necessary to work with a sufficiently low gas pressure (checked from the linearity of the relation between pressure and scattered current) and beam current (checked from the linearity of the relation between scattered and beam currents). When working at energies above the ionization energy of the gas the restriction to small beam currents must also be adhered to in order to prevent the setting up of a positive ion space charge large enough to disturb the path of the scattered electrons. It is also of the greatest importance that no stray fields should be present due to charging up of the bounding walls of the scattering chamber. To avoid this the wall is bounded by a metal enclosure, usually cylindrical in form. Finally, to avoid effects due to the presence of impurities that may not only contribute markedly to the scattering but also affect the collecting power of slits, it is necessary to design apparatus so that it can be thoroughly outgassed by baking.

5.1. *Types of apparatus*

5.1.1. *The electron source.* The source used by Bullard and Massey,[†] which is typical of most employed in the earlier work, consisted of a tungsten filament K (Fig. 5.31 (a)) from which the electrons were accelerated through a pair of slits B (4×1.5 mm) and C (2×0.6 mm), 7 mm apart. In some of the later work considerable attention has been paid to the design of the electron source in order to obtain an intense well-collimated electron beam. Lassettre *et al.*[‡] used an electron gun with electrodes and dimensions as shown in Fig. 5.31 (b). The electrodes 1 and 2 were shaped in accordance with a design due to Pierce[§] for space-charge limited operation. The focusing field was provided by electrodes 1, 2, and 3, which contained holes of diameter 2 mm. Electrode 4 which contained a 1-mm diameter hole served to intercept peripheral electrons and to improve beam collimation. Currents of several microamperes of

[†] BULLARD, E. C. and MASSEY, H. S. W., *Proc. R. Soc.* A**130** (1931) 579.
[‡] LASSETTRE, E. N., BERMAN, A. S., SILVERMAN, S. M., and KRASNOW, M. E., *J. chem. Phys.* **40** (1964) 1232.
[§] PIERCE, J. R., *J. appl. Phys.* **11** (1940) 548.

electrons with energy some hundreds of eV were obtained with this source.

Considerably larger currents of up to 0·8 mA in a parallel beam of diameter 1 mm were obtained by Kessler and Lindner† with the gun

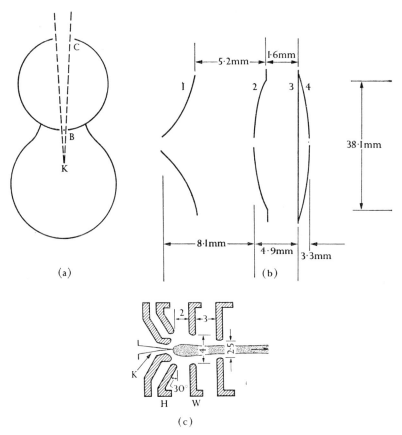

FIG. 5.31. Illustrating (a) the electron source used by Bullard and Massey, (b) the electron gun electrodes as used by Lassettre et al., (c) the electron gun as used by Kessler and Lindner in their experiments on the angular distribution of scattered electrons. The numbers in (c) refer to distances in mm.

shown in Fig. 5.31 (c). The potentials on the cylindrical electrode W and the auxiliary electrode H, relative to cathode potential, depended on the position of the latter and lay between -5 to -50 V and $+50$ to $+150$ V respectively. Electron beams of energy between 200 eV and 4 keV were produced with this gun. Care had to be taken to keep the

† KESSLER, J. and LINDNER, H., Z. Phys. **183** (1965) 1.

pressure in the gun low by means of a separate pumping system in order to avoid electron scattering in the gun itself.

5.1.2. *The monochromator.* For work at high energy resolution it is necessary to use a monochromator to obtain an electron beam of sufficiently small energy spread. Electrostatic deflectors have been employed for this purpose, either of the 127° cylindrical type[†] or the concentric hemispherical type[‡] (see Chap. 1, § 5.1). The theory of the latter type of velocity analyser has been given by Purcell.[§] In the arrangement used by Skerbele and Lassettre the concentric hemispheres were machined from brass and then gold-plated. The radii of the inner and outer hemispheres were 4·2 and 4·8 cm respectively and the entrance and exit apertures were on a circle of 4·5 cm radius. The entrance and exit apertures were grounded, while the inner and outer hemispheres were above and below earth potential by equal amounts. The entrance aperture was a slit 0·005 in wide and 0·04 in long. The exit aperture was a pinhole 0·005 in in diameter. The resolving power R of such an arrangement is given by $R = 4r/S$ where r is the radius of the orbit and S is the slit width or aperture diameter. In the monochromator of Skerbele and Lassettre, $R = 1180$.

5.1.3. *The analyser and collector.* The simplest collector was that used by Bullard and Massey[||] who were concerned with the angular distributions of elastically scattered electrons with energies ranging from 4 to 40 eV. The design of their collector is illustrated in Fig. 5.32 (a). The slits F and G were of width 2 mm and thickness 0·6 mm, were 6 mm apart, and the front was 7 mm from the axis of rotation.

To ensure that elastically scattered electrons alone were collected the outer case E of the collector was kept at such a positive potential with respect to the inner cylinder that only electrons retaining their initial energy would be able to penetrate through the slit G. When working at electron energies above the ionization energy the collection of positive ions was prevented by application of a small positive potential between the outer case E and a surrounding metal enclosure.

With this apparatus the angular range of collection was about 10°. A more elaborate collector of the Faraday cylinder type was used by Arnot[††] in studying the elastic scattering of electrons with energies up to 800 eV. At these higher energies secondary emission from the inner collector and high positive ion currents become serious. Arnot therefore

[†] See, for example, ANDRICK, D. and EHRHARDT, H., ibid. **192** (1966) 99.
[‡] See SKERBELE, A. M. and LASSETTRE, E. N., *J. chem. Phys.* **40** (1964) 1271.
[§] PURCELL, E. M., *Phys. Rev.* **54** (1938) 818.
[||] loc. cit., p. 322.   [††] ARNOT, F. L., *Proc. R. Soc.* A**130** (1931) 655.

used two further insulated enclosing cylinders in his collector, which is illustrated in Fig. 5.32 (b). It consisted of three concentric cylinders containing slits S_1 (8×0·2 mm), S_2 (5×0·2 mm), S_3 (2×2 mm), equally spaced 5 mm apart, and a concentric innermost Faraday cylinder F. The outermost cylinder was kept at the same potential as the nickel cylinder enclosing the scattering space. Taking this as zero, the potentials on the slits S_2, S_3, and the Faraday cylinder were maintained respectively at 6, $-V_0+3$, $-V_0+24$ V, V_0 being equal to the energy of the incident

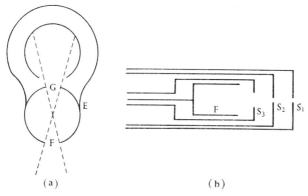

FIG. 5.32. Illustrating design of collectors used (a) by Bullard and Massey, (b) by Arnot, in their experiments on the angular distribution of scattered electrons.

electron in electron volts. The potential between S_1 and S_2 prevented positive ions from entering S_2, that between S_2 and S_1 permitted only elastically scattered electrons to enter S_1, and that between F and S_1 prevented secondary emission from the Faraday cylinder.

With the small dimensions involved, some penetration of fields occurred, so that considerable care had to be taken to check that the applied potentials really did achieve what they were designed for. Owing to the complexity of the collector it was kept fixed and the source rotated, in contrast to the experiments of Bullard and Massey.

In both sets of experiments the electron-beam current employed was of the order of a microampere. The collected currents were of the order 10^{-12} A or less, and were measured by a quadrant or Compton electrometer. The gas pressure used depended on the electron energy, but was usually between 10^{-3} and 10^{-2} torr.

A cylindrical electrostatic analyser has been used by Hughes and McMillen† who were mainly concerned with elastic scattering, and by

† HUGHES, A. L. and MCMILLEN, J. H., *Phys. Rev.* **39** (1932) 585.

Mohr and Nicoll† who have made the most extensive series of measurements of angular distributions for inelastic as well as elastic scattering.

Fig. 5.33 illustrates the arrangement used by Mohr and Nicoll. The electrons from the region O entered the space between two duralumin plates A and B of width 3 cm, curved in concentric circular arcs of radii 2·5 and 3·8 cm, through two slits S_1 (0·27 × 8 mm) and S_2 (0·4 × 5 mm), 7 mm apart. The metal shield P of the scattering chamber was shaped so as to prevent electrons from the scattering chamber from entering

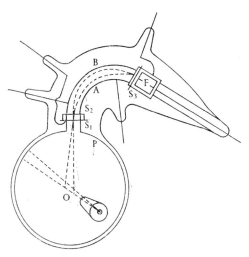

Fig. 5.33. Schematic illustration of apparatus used by Mohr and Nicoll for studying the angular distribution of inelastically scattered electrons.

the analyser in any other way. An adjustable potential difference could be applied to the plates so as to focus electrons of the required energy on the slit S_3 (0·6 × 5 mm), whence they could pass into the Faraday cylinder F. The slits S_2 and S_3 were kept at a potential midway between those of the two plates A and B. A retarding potential was maintained between the Faraday cylinder and the slit to reject all electrons with velocities lower than those it was desired to collect. This was a precaution to remove any background of slow electrons. A potential difference could also be maintained between S_1 and S_2 to accelerate electrons before entering the analyser, if necessary to eliminate disturbance by the earth's magnetic field.

The whole apparatus was enclosed in a Pyrex tube and, except for

† MOHR, C. B. O. and NICOLL, F. H., *Proc. R. Soc.* A**138** (1932) 229, 469; ibid. **142** (1933) 320, 647.

the ground joint that rotated the electron gun, could be thoroughly baked out at 450° C. By suitably enclosing the collision chamber within the metal shield it was possible to maintain a large pressure difference between this chamber and the analyser. With a pressure as high as 10^{-2} torr in the former, the pressure in the latter could be kept down to 10^{-5} torr. Scattering in the angular range 10–155° could be observed with this apparatus.

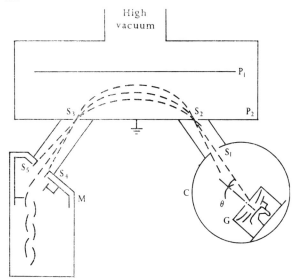

Fig. 5.34. Schematic illustration of the apparatus used by Lassettre et al. for studying the angular distribution of electrons scattered with well-defined energy.

Andrick and Ehrhardt[†] have used a similar type of 127° electrostatic analyser. Reichert[‡] has used a concentric cylindrical sector lens of angle 60° while Lassettre et al.[§] have used a uniform electrostatic field for velocity analysis of the scattered beam.

Fig. 5.34 shows the scattering apparatus of Lassettre et al. The electrostatic field of the velocity analyser was established between the parallel plates P_1, P_2, the latter forming one wall of the rectangular box that holds the analyser. The plate P_1 was supported on steatite insulators, shielded from electrons by sheet metal guards attached to P_2. The entrance slits S_1, S_2 and exit slits S_3, S_4 were mounted in sleeves and

[†] ANDRICK, D. and EHRHARDT, H., Z. Phys. **192** (1966) 99.
[‡] REICHERT, E., ibid. **173** (1963) 392.
[§] LASSETTRE, E. N., BERMAN, A. S., SILVERMAN, S. M., and KRASNOW, M. E., J. chem. Phys. **40** (1964) 1232.

aligned so that electrons enter and leave at an angle of 45° to the plate. The distance between P_1 and P_2 was 6·25 cm and between S_2 and S_3 12·5 cm.

The use of such a uniform field analyser has been studied by Yarnold and Bolton[†] and by Harrower.[‡] An electron image of S_2 of unit magnification is formed at S_3. The resolving power R is given by x/w, the ratio of the distance $S_2 S_3$ to the slit width. In the apparatus of Lassettre et al. $R = 2500$.

In most recent angular distribution experiments, including those of Lassettre et al., an electron multiplier detector has been used for the scattered electrons.

5.2. Apparatus for special conditions

5.2.1. Scattering of very slow electrons.
In order to study the scattering of very slow electrons for which only weak sources are available Ramsauer and Kollath[§] introduced the 'zone' apparatus illustrated in Fig. 5.35. The collectors 1 to 11 were metal plates shaped to the form of the zones of a sphere with centre O. As secondary emission was negligible and no slow electrons arose from inelastic collisions this simple arrangement sufficed and had the advantage of giving comparatively large scattered currents. The geometry of the scattering chamber had to be carefully considered in reducing the observations. With this apparatus Ramsauer and Kollath were able to observe the scattering of electrons with energies as low as 0·6 eV.

5.2.2. Scattering at 180°.
In the conventional type of apparatus described above, scattering cannot be observed out to angles of 180° because of the interference of the source. An apparatus that does enable observations to be made of the scattering at these angles was developed by Gagge.[∥] The principle of his method is illustrated in Fig. 5.36. The whole experiment was carried out in a uniform magnetic field perpendicular to the plane of the paper. Electrons fired from the gun moved in a circular path in the field. After scattering, the electrons moved on another circular path such as $PS_1 S_2$. Corresponding to each such path there is a definite angle of scattering. To select different angles the collector with its slit S_2 could be moved parallel to $S_1 S_2$, the slit S_1 remaining fixed. Allowance had to be made, in calculating the variation of scattering volume with angle of scattering, for the fact that electrons

[†] YARNOLD, G. D. and BOLTON, H. C., J. scient. Instrum. **26** (1949) 38.
[‡] HARROWER, G. A., Rev. scient. Instrum. **26** (1955) 850.
[§] RAMSAUER, C. and KOLLATH, R., Annln Phys. **12** (1932) 529, 837.
[∥] GAGGE, A. P., Phys. Rev. **44** (1933) 808.

pursuing the path PS_1S_2 arose from scattering, through the same angle, at both P and S. Since electrons scattered at 180° did not strike the source in this arrangement, there was no difficulty in carrying out measurements at large angles up to and including 180°.

5.2.3. *Scattering at small angles.* Elastic and inelastic electron scattering of electrons of energy 25 keV through very small angles has been studied by Geiger,† using an apparatus illustrated schematically in

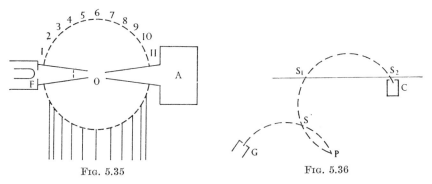

Fig. 5.35.   Fig. 5.36.

Fig. 5.35. 'Zone' apparatus used by Ramsauer and Kollath for studying the angular distribution in the elastic scattering of very slow electrons.

Fig. 5.36. Principle of the apparatus used by Gagge for the study of electrons scattered at large angles.

Fig. 5.37 (a). The electrons, after acceleration from the cathode K, entered the scattering chamber C, containing the scattering gas at a pressure of the order of 1 torr, through a hole H_1. After scattering they emerged through a slit S_1 and passed through a second parallel narrow slit S_2 before entering the electrostatic energy analyser A. After deflexion in the analyser they were focused on to a photographic plate P. Electrons that suffered a definite energy loss, E, were brought to a focus on the plate along a line parallel to the slits S_1 and S_2. Electrons scattered through a particular small angle θ, on the other hand, entered the photographic plate at a definite distance along the line image. By scanning with a microphotometer, therefore, along a line parallel to the slits and corresponding to a particular energy loss, the angular distribution for electrons scattered through small angles for that energy loss could be determined.

The energy analyser used in this work was based on a cylindrical lens of very high dispersion developed by Möllenstedt.‡ The principle is

† GEIGER, J., *Z. Phys.* **175** (1963) 530.
‡ MÖLLENSTEDT, G., *Optik* **5** (1949) 499.

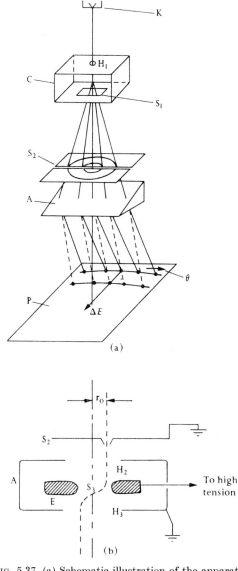

Fig. 5.37. (a) Schematic illustration of the apparatus of Geiger for studying elastic and inelastic scattering through small angles. (b) Principle of the electrostatic analyser used by Geiger.

illustrated in Fig. 5.37 (b). Electrons of energy eV_0 pass through the narrow slit S_2 before entering the analyser A. This consists of a cylindrical metal box maintained at the same potential as S_2 with two circular apertures H_2 and H_3 through which the electron beam passes. Inside the box is an electrode E, maintained at a potential $-V_1$ relative to S_2 ($V_1 < V_0$). The electrode E contains a slit S_3 aligned parallel to S_2 and of length considerably greater than the diameter of the apertures H_2, H_3. The axis of A is displaced relative to the slit S_2 by an amount r_0. For a particular value of V_1 (in Möllenstedt's arrangement about $0 \cdot 85 V_0$) the electron beam passed symmetrically through the analyser and emerged parallel to the incident direction but displaced to the other side of the axis as indicated by the dotted line in Fig. 5.37 (b). For any other ratio V_1/V_0 the electron beam, after passing through the analyser, no longer moves parallel to the axis. For a given V_1 the angle of emergence depends sensitively on V_0 so that Geiger was able easily to detect electrons of energy 25 keV that had lost energies of a few eV in passing through the collision chamber. The theory of the analyser has been discussed by Lippert.† The finite width of the primary beam did not permit measurements of differential cross-sections for elastic scattering using this method to be extended below a scattering angle of approximately 7 mr. For inelastic scattering, however, where the scattered electrons are not masked by the primary beam it was possible to extend the measurements down to angles of 0·23 mr.

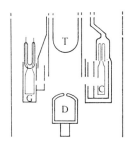

Fig. 5.38. Schematic illustration of the apparatus of Childs and Massey for the study of the angular distribution of electrons scattered by metal vapours.

5.2.4. *Scattering by metal vapours.* Two methods have been used for investigating the scattering from the vapours of metals that do not have sufficient vapour pressure at room temperature. McMillen‡ studied the scattering in potassium vapour by maintaining the whole collision chamber at 150 °C, at which temperature the vapour pressure was sufficient (0·001 torr). Childs and Massey§ adopted, in their experiments on the scattering of cadmium and zinc, an alternative arrangement illustrated in Fig. 5.38. The scattering vapour was provided as a cloud emerging from an oven source D. The vapour was condensed on the liquid air trap T. Electrons from the gun G were fired across the vapour cloud as it emerged from the oven slit and the scattered current measured at various angles by the collector C. Reichert‖ used a similar method in

† Lippert, W., *Optik* **12** (1955) 467.
‡ McMillen, J. H., *Phys. Rev.* **46** (1934) 983.
§ Childs, E. C. and Massey, H. S. W., *Proc. R. Soc.* **A141** (1933) 473; ibid. **142** (1933) 509.
‖ Reichert, E., *Z. Phys.* **173** (1963) 392.

his measurement of the angular distribution of electrons scattered by gold vapour. The gold was melted in an Al_2O_3 crucible mounted inside a molybdenum oven which could be heated to 1700 °C by electron bombardment. The beam of gold atoms was defined by collimating holes in a number of plates and in the collision region it formed a beam of diameter 10 mm with a density of 10^{11} atoms cm^{-3}.

5.2.5. *Crossed-beam method for studying angular distribution of electrons scattered in atomic hydrogen.* The pulsed crossed-beam method using a phase sensitive detector has been applied by Bederson, Malamud, and Hammer[†] and by Gilbody, Stebbings, and Fite[‡] to measure the angular distribution of electrons of energy below 10 eV, scattered elastically in atomic hydrogen. The measurement of the angular distribution of the elastically scattered electrons enables the total elastic scattering cross-section to be deduced (see Chap. 1, § 4.2.1).

The apparatus used was very similar to that used in the earlier experiment of Brackmann et al.[§] to study the cross-section for scattering through a finite angular range round 90° (see Fig. 1.2). The electron collector was replaced by an electron multiplier. It was fixed in position but the electron gun could be rotated about an axis perpendicular to the atomic beam to enable measurements to be made for a range of scattering angles between 30 and 120°. Measurements were carried out for electron energies between 3·8 and 9·4 eV. Both the angular distribution for scattering by atomic hydrogen and the ratio of the differential cross-sections at 90° for scattering by atomic and molecular hydrogen were measured.

5.3. Observed angular distributions

5.3.1. *Elastic scattering.* A thorough survey has been made of the angular distributions of electrons scattered from helium, neon, argon, krypton, xenon, mercury, and gold, and less detailed observations for atomic hydrogen, cadmium, zinc, and potassium. Measurements made of scattering from molecular gases and vapours are described in Chapter 10, §§ 1, 2, and 3.

The results obtained by different observers usually agree very well, so that in giving the observed data the measurements with a particular apparatus are often given instead of a mean of all available observations.

Fig. 5.39 shows the differential cross-sections obtained by Gilbody, Stebbings, and Fite[‡] for the scattering of electrons in atomic hydrogen

[†] BEDERSON, B., MALAMUD, H., and HAMMER, J., *Bull. Am. phys. Soc.* **2** (1957) 172.
[‡] GILBODY, H. B., STEBBINGS, R. F., and FITE, W. L., *Phys. Rev.* **121** (1961) 794.
[§] BRACKMANN, R. T., FITE, W. L., and NEYNABER, R. H., ibid. **112** (1958) 1157.

in the energy range 3·8 to 9·4 eV. Comparison with theoretical expectation is made in Chapter 8, § 2.8. In these experiments the absolute cross-section was obtained by normalizing the measured ratios $Q_A(90°)/Q_M(90°)$ of the differential cross-section to the absolute molecular scattering data at 90° obtained by Ramsauer and Kollath.†

The most conspicuous feature of the angular distributions, at least for

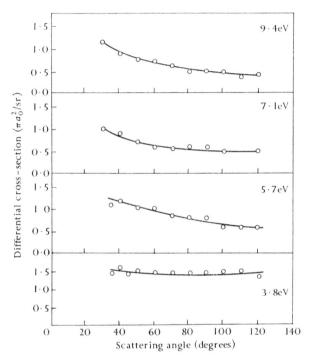

FIG. 5.39. Observed angular distribution of electrons scattered elastically by atomic hydrogen at various energies.

scattering by the heavier atoms, is the appearance of maxima and minima reminiscent of diffraction patterns. This phenomenon was first observed by Bullard and Massey‡ in 1931 for argon and it arises from diffraction of the electron waves by the spherically symmetrical scattering atoms in much the same way as in the scattering of light by spheres of dimensions comparable with the wavelength. This aspect will be discussed in detail in the next chapter. In this section we shall merely call attention to certain salient features.

Referring to Figs. 5.40 and 5.41, in which observed angular

† RAMSAUER, C. and KOLLATH, R., *Annln Phys.* **12** (1932) 529.
‡ BULLARD, E. C. and MASSEY, H. S. W., *Proc. R. Soc.* **A130** (1931) 579.

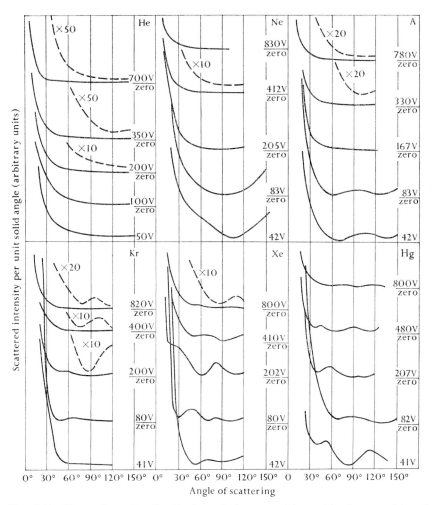

Fig. 5.40. Observed angular distributions of electrons with energies between 30 and 800 eV scattered elastically by the rare gases and by mercury.

distributions for the rare gases and mercury are illustrated, the following features will be noted.

(a) For a given atom the diffraction pattern smooths out at a sufficiently high electron energy leaving a monotonic decrease of scattered intensity with angle.

(b) As the electron energy decreases, the pattern becomes at first more complicated and then begins to smooth out once more at the lowest energies. Thus with mercury the pattern is at its most complex at about 450-eV electron energy.

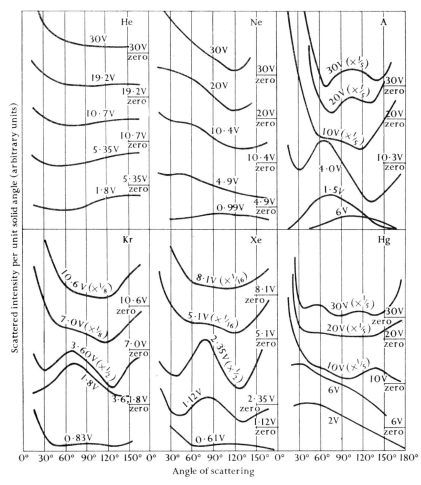

Fig. 5.41. Observed angular distributions of electrons with energies between 0·5 and 30 eV scattered elastically by the rare gases and by mercury.

(c) At low energies, for helium and neon, the variation of scattering with angle becomes very much less marked, so that at the lowest energies it is very nearly uniform.

(d) For the other rare gases the angular distribution remains far from uniform even at the lowest observed energies (0·5 eV). It is to be noted that at these energies the total collision cross-section is very small (Ramsauer–Townsend effect (Chap. 1, § 6.2.5 and Chap. 2, § 7.3)).

(e) The complexity of the diffraction pattern is more marked the greater the atomic number of the scattering atom. Thus helium exhibits very little departure from the monotonic distribution at any energy, neon gives a distribution with a single minimum near 90°, at most, whereas two minima appear in the argon distributions between 4- and 150-eV energy and three or more for the heavier atoms.

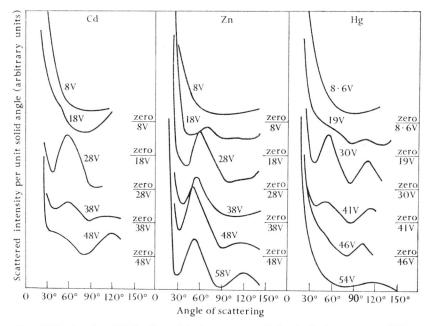

FIG. 5.42. Angular distribution of electrons scattered elastically in Cd, Zn, and Hg.

Fig. 5.42 compares the angular distributions observed for cadmium† and zinc‡ with those for mercury. It will be seen that electrons with energies less than 50 eV or so are scattered in very much the same way by all three atoms. This indicates a great similarity in the scattering field of these atoms at distances from the nucleus at which the energy of the field is of the order 50 eV or less.

Fig. 5.43 compares the angular distributions of elastically scattered electrons from mercury§ and gold‖ in an energy range above 700 eV. Once again the similarity in the two cases is remarkable. The sharp maxima and minima tend to disappear as the energy increases. Absolute values of the differential cross-section are given for mercury.

† CHILDS, E. C. and MASSEY, H. S. W., *Proc. R. Soc.* **A141** (1933) 473.
‡ Ibid. **A142** (1933) 509.
§ KESSLER, J. and LINDNER, H., *Z. Phys.* **183** (1965) 1.
‖ REICHERT, E., ibid. **173** (1963) 392.

5.3.2. *Inelastic scattering.* The first measurements of the angular distribution of inelastically scattered electrons were carried out by Dymond† in He for electrons of 50- to 500-eV energy, scattered after excitation of helium atoms to the 2^1P state. The angular range covered

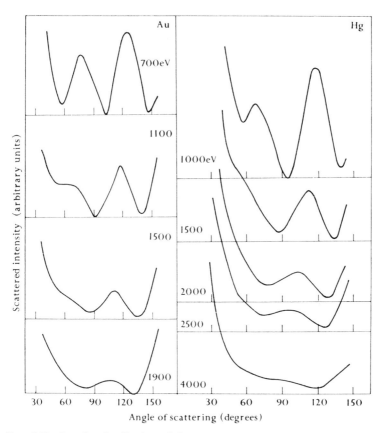

Fig. 5.43. Angular distribution of electrons scattered elastically in gold and mercury as measured by Reichert and by Kessler and Lindner respectively.

was 0–90°. Mohr and Nicoll,‡ using the apparatus shown in Fig. 5.33, have investigated inelastic scattering between 10° and 155° in helium, neon, argon, and mercury, as well as a number of molecular gases including hydrogen and methane. They observed separately not only electrons that had excited the resonance level, but also electrons that had produced ionization and suffered various amounts of energy loss in doing so.

† DYMOND, E. G., *Phys. Rev.* **29** (1927) 433.
‡ MOHR, C. B. O. and NICOLL, F. H., *Proc. R. Soc.* **A138** (1932) 229, 469; ibid. **A142** (1933) 320, 647.

Measurements have also been carried out in mercury vapour by Tate and Palmer.†

More recently an extensive series of measurements of differential cross-sections for inelastic scattering of higher energy electrons (up to 600 eV) through small scattering angles (up to 15°) in helium have been made by Lassettre and his colleagues.‡

The procedure used by Mohr and Nicoll in their comprehensive investigations was as follows. With the electron gun set at a convenient angle the energy distribution of the electrons scattered at that angle was investigated by measuring the variation of the current to the Faraday cylinder with change of plate voltage. A similar procedure was followed by Lassettre and co-workers. Fig. 5.44 shows the energy loss spectrum obtained by Lassettre, Krasnow, and Silverman for the scattering of 511-eV electrons in helium at a number of different angles. Peaks corresponding to 2^1S and 2^1P excitations are clearly seen.§

Fig. 5.45 shows the energy loss spectrum obtained by Mohr and Nicoll for electrons of energy 42 eV scattered through 90° in mercury vapour. The peak due to the 6·7-eV energy loss arising from the excitation of the 6^3P_1 level is clearly

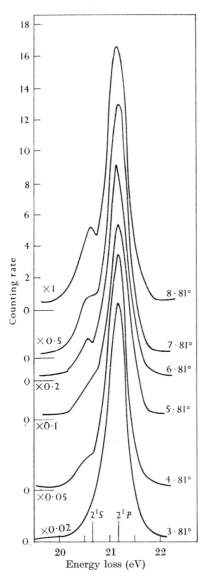

Fig. 5.44. Energy loss spectra in the range 20–22 eV for electrons of incident energy 511 eV scattered in helium through various angles as observed by Lassettre, Krasnow, and Silverman.

† Tate, J. T. and Palmer, R. R., Phys. Rev. **40** (1932) 731.
‡ Lassettre, E. N., Krasnow, M. E., and Silverman, S., J. chem. Phys. **40** (1964) 1242.
§ The 2^1S peak was earlier observed for lower energy electrons by Womer, R. L., Phys. Rev. **45** (1934) 689.

visible. Above 8 eV are unresolved energy losses leading up to the ionization energy at 10·4 eV. At greater energy losses there is a continuous spectrum of electrons scattered by or ejected from atoms in ionizing collisions.

To investigate the angular distribution of electrons scattered after suffering a particular energy loss, the plate voltage was adjusted to focus electrons with the appropriate energy and measurements were taken at different scattering angles by rotating the electron gun. It was verified that the shape of the peaks did not change appreciably throughout the angular range so that the peak height was a true measure of the relative intensity of the scattered electrons. Special care was

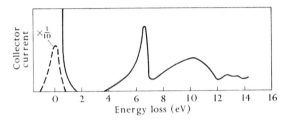

FIG. 5.45. Energy loss spectrum of electrons after scattering through 90° in mercury vapour as observed by Mohr and Nicoll.

taken to check the linearity of the relations between the scattered current and the gas pressure and main beam currents respectively.

The principal features of the observed angular distributions are as follows.

(a) At small angles the scattered intensity falls off very rapidly with angle for inelastic collisions in which the fractional energy loss is not too great. This may be seen from the results of Mohr and Nicoll by reference to Fig. 5.46, in which the angular distributions of electrons scattered in collisions in which the 2^1P level of helium is excited and in which ionization occurs with different energy losses are respectively illustrated. Relative differential cross-section measurements for 2^1P and 3^1P excitation and for ionizing collisions have been made by Silverman and Lassettre for electrons of higher energy (500 eV) and these are shown in Fig. 5.47. Fig. 5.49 gives corresponding curves for argon, and Fig. 5.50 for mercury.

(b) The distributions at small angles of the electrons scattered after exciting the 2^1P and 3^1P levels of helium respectively are nearly the same, the intensity ratio being 2·7 to 1. This is seen from Figs. 5.47 (a) and 5.48. Fig. 5.48 shows the results of Mohr and Nicoll for electrons of 83 eV incident energy scattered in helium.

(c) The differential cross-section for the allowed 2^1P transition falls off much more rapidly with angle than that for the forbidden 2^1S excitation (see Fig. 5.47 (a)).

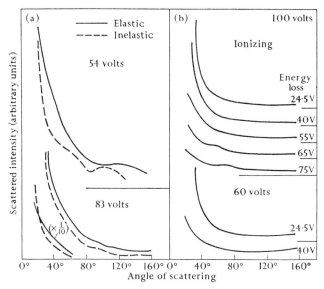

FIG. 5.46. Angular distributions of electrons scattered in helium observed by Mohr and Nicoll. (a) Elastic (———) and inelastic (- - - -) collisions involving excitation of the 2^1P level. (b) Ionizing collisions with different energy losses.

(d) The angular distribution of electrons that have suffered small energy losses in inelastic collisions with heavy atoms exhibit, at large angles, diffraction maxima and minima that closely resemble those appearing in the corresponding distributions of elastically scattered electrons. This may be seen by reference to the observed distributions for argon illustrated in Fig. 5.49 and for mercury in Fig. 5.50. The resemblance between the shape of the distributions at large angles of the elastic and inelastic scattering is found also for the lighter atoms that do not exhibit very complicated diffraction effects.

This remarkable feature of the observations was conclusively shown by Mohr and Nicoll to be associated with single scattering and not to arise from successive collisions, one, elastic, giving rise to the diffraction, and another, inelastic, giving the energy loss. As the intensity of the inelastic scattering at large angles is very much smaller than that of the

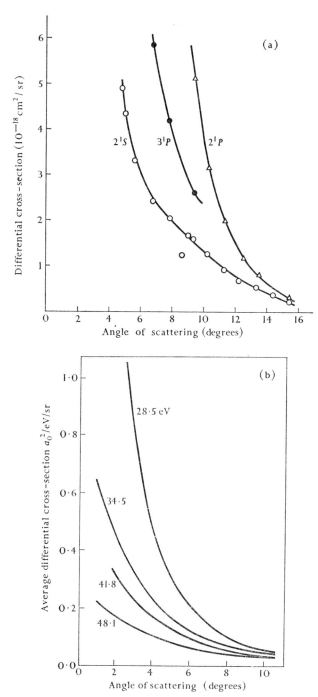

Fig. 5.47. Differential cross-sections for (a) excitation of the 2^1P, 3^1P, and 2^1S states of helium by 500-eV electrons, (b) ionizing collisions in helium with different energy losses of 500-eV electrons derived from the observations of Silverman and Lassettre.

elastic, special care had to be taken to exclude the double scattering explanation. Apart from careful verification that the inelastically scattered current at large angles was proportional to the gas pressure in the experimental range, a study of the relative intensities of the elastic and inelastic scattering showed that at the low gas pressures employed (less than 10^{-3} torr) double scattering would be far too small. Furthermore, Mohr and Nicoll's results for mercury were reproduced by Gagge† with a rather different apparatus.

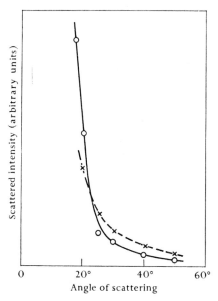

FIG. 5.48. Comparison of angular distributions of electrons of 83 eV incident energy, scattered after exciting the 2^1P and 3^1P levels of helium respectively. $-\bigcirc-\bigcirc-\bigcirc-\ 2^1P$; $\times--\times--\times\ 3^1P$.

(e) The resemblance noted in (d) becomes progressively less marked when the inelastic collision concerned involves a considerable fractional energy loss. Thus in Fig. 5.49 the resemblance between the distribution for electrons scattered elastically in argon and those scattered in a collision involving the 11·6 eV energy loss ceases to be marked when the incident electron energy falls below 40 eV. Further, in Fig. 5.49 a set of distribution curves for electrons that have suffered different losses of energy in producing ionization of argon atoms is shown. At the lowest energy loss the resemblance to the elastic distribution is close but

† loc. cit., p. 328.

becomes progressively less apparent as the energy loss increases. Fig. 5.50 illustrates the corresponding features for mercury.

(*f*) For the lightest atoms, the angular distributions of electrons that have undergone ionizing collisions with large fractional energy loss

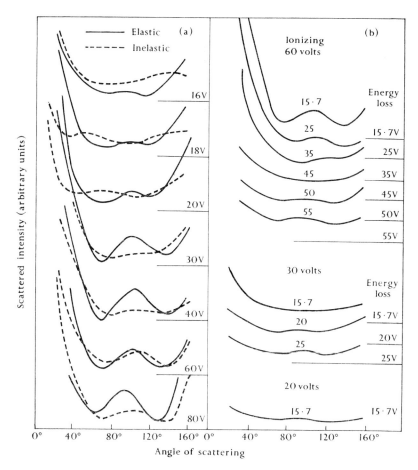

Fig. 5.49. Observed angular distribution of electrons scattered in argon. (a) Elastic (———) and inelastic (– – – –) collisions involving excitation of the $^3P_1$ level in which the energy loss is 11·6 eV. (b) Ionizing collisions with different energy losses.

exhibit at the higher incident energies a maximum that is of a character quite distinct from the diffraction patterns observed with heavy atoms and small fractional energy loss. This is illustrated most clearly in the observed curves for hydrogen in Fig. 5.51. The maximum is most marked at the highest electron energy and moves out to larger angles the greater the fractional energy loss. The same effect appears to a less pronounced extent with helium.

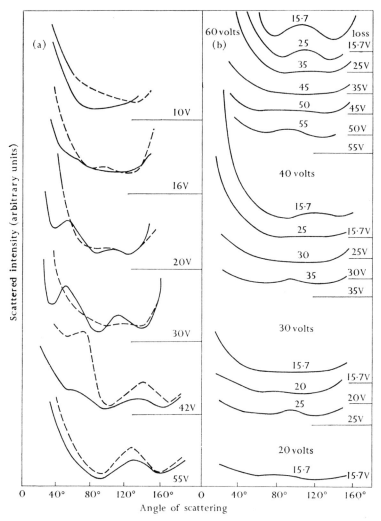

Fig. 5.50. Observed angular distribution of electrons scattered in mercury. (a) Elastic (———) and inelastic (– – – –) collisions involving excitation of the 6^3P_1 level. (b) Ionizing collisions with different energy losses.

The effect is still perceptible for electrons with 200-eV incident energy in methane, nitrogen, and neon, but was not observed for the heavier atoms for electrons with incident energies up to 200 eV. As far as can be judged the position of the maximum for given incident energy and energy loss occurs at nearly the same angle for all the gases in which it was observed. The maximum occurs approximately at the angle expected for an elastic collision of the incident with the atomic electron, giving rise to a scattered electron of the appropriate energy (see Chap. 7, § 5.6.2).

Silverman and Lassettre have measured the differential cross-sections for the excitation of the $(2s2p)^1P$ auto-ionizing state of helium at 60·0 eV energy loss, using incident electrons at energy 500 eV. Fig. 5.52 shows their results. For comparison, mean differential cross-sections for

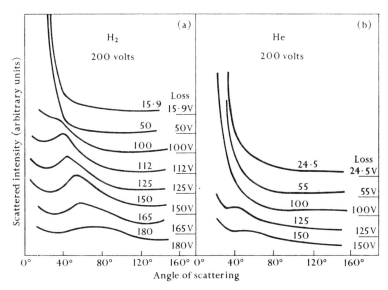

FIG. 5.51. Angular distribution in scattering of fast electrons in ionizing collisions in (a) H_2, (b) He, illustrating a maximum in the distribution for large energy losses.

ionization with losses of 58·9 and 61·3 eV are shown. These correspond to the ionization continuum on either side of the auto-ionizing level. No differences are apparent in the angular distributions in the three cases.

5.3.3. *Excitation of resonant states—differential cross-sections.* Experiments that have established the existence of resonance effects in total, elastic, and inelastic cross-sections for electrons, as functions of electron energy, have been described in Chapter 1, § 6.2 and in § 4.2 of this chapter. As will be discussed in detail in Chapter 9, these effects arise from the resonance formation of short-lived intermediate states by electron capture, which then break up leaving the target atom in its initial state (elastic scattering resonance) or in an excited state (inelastic resonance). The measurement of the angular distribution of scattered electrons of homogeneous energy in the neighbourhood of a resonance is likely to provide valuable information about the nature of the intermediate state (see Chap. 9, § 1). We now discuss measurements of this kind.

(*a*) *Elastic scattering.* We have already described in Chapter 1, § 6.2.2 experiments on the elastic scattering of electrons by atomic

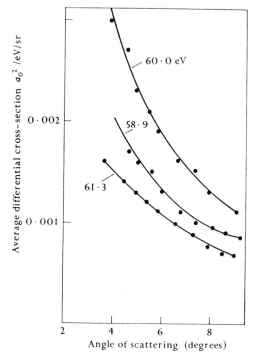

Fig. 5.52. Observed angular distribution of electrons of 500-eV incident energy scattered after suffering energy losses of 58·9, 60·0, and 61·3 eV in helium (Silverman and Lassettre).

hydrogen that gave information about the variation of the resonance effects with angle. These data are discussed in terms of the theoretical interpretation in Chapter 9, § 4.

Fig. 5.53 shows the variation of the intensity of scattering of electrons in helium with energy in the range 18·0–20·5 eV at various angles as observed by Andrick and Ehrhardt.[†] The sharp resonance at 19·3 eV (Chap. 1, § 6.2.3 and Fig. 1.18) is clearly seen. It is convenient in analysing such data to separate approximately the resonant scattering from

[†] ANDRICK, D. and EHRHARDT, H., *Z. Phys.* **192** (1966) 99.

the smoothly varying background scattering. The latter may be extrapolated through the narrow resonance energy region because of its gradual variation so that the intensities $N_r(\theta)$, $N_b(\theta)$ of the resonance and background scattering may be obtained at each electron energy and

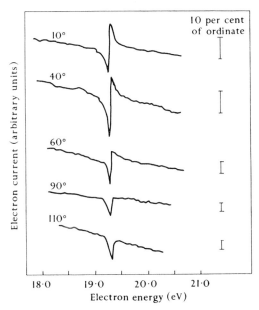

FIG. 5.53. Variation with incident electron energy of the intensity of electrons elastically scattered in helium at different angles of scattering. The intensity variations may be judged from the lines shown on the right-hand side which indicate 10 per cent of the total intensity near the resonance.

scattering angle θ. Ehrhardt and Willmann then define the differential cross-section for resonant scattering $I_r(\theta)$ by

$$I_r(\theta) = \{I_b(\theta)/N_b(\theta)\} \int N_r(\theta)\, dE/\Delta E,$$

the integral being taken over a range ΔE of E large enough to include the resonance effects.

Fig. 5.54 compares $I_r(\theta)$ and $I_b(\theta)$ obtained in this way. It will be seen that, whereas $I_r(\theta)$ is approximately isotropic, $I_b(\theta)$ is of the typical form obtained in experiments in which the electrons are not very homogeneous in energy (cf. Fig. 5.41). The isotropy of $I_r(\theta)$ establishes the character of the intermediate state (see Chap. 9, § 5).

Figs. 5.55 and 5.56 show curves similar to those of Fig. 5.53 obtained by Andrick and Ehrhardt† for neon and argon. In this case a similar analysis to that for helium gives $I_r(\theta)$ closely similar to $\cos^2\theta$ in form, with a nearly zero minimum at 90° about which the distribution is nearly symmetrical. These results are also decisive in establishing the character of the intermediate state.

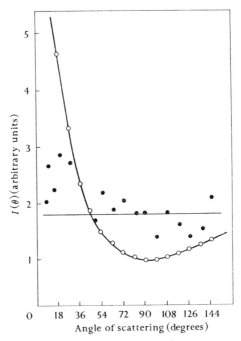

FIG. 5.54. Angular distributions $I_r(\theta)$ and $I_b(\theta)$ of the resonant and background elastic scattering of electrons in helium derived from the observations of Ehrhardt and Meister. $-\bigcirc-\bigcirc-\bigcirc-$ $I_b(\theta)$, ● ● ● $I_r(\theta)$.

(b) *Inelastic scattering.* Ehrhardt and Willmann‡ have studied the angular distribution of electrons scattered inelastically in helium after having lost 19·82 eV corresponding to excitation of the 2^3S state. Incident electron energies in the range 19·82 to 24 eV scattered through an angular range 7° to 110° were studied. Fig. 5.57 shows the scattered intensity obtained as a function of incident electron energy. Resonances were observed at 19·9±0·05, 20·45±0·05, and 21·00±0·05 eV. The last two are evidently to be identified with the two peaks observed in this region by Chamberlain§ but structure in the region above 22·0 eV

† loc. cit., p. 346.
‡ EHRHARDT, H. and WILLMANN, K., *Z. Phys.* **203** (1967) 1.
§ CHAMBERLAIN, G. E., *Phys. Rev. Lett.* **14** (1965) 581.

reported earlier (§ 4.2.1) is seen only in the curves for large angle scattering. Fig. 5.58 shows the angular distribution of the inelastically scattered electrons at the resonance energies and at 22, 23, and 24 eV. These distributions are markedly different for the three resonance energies

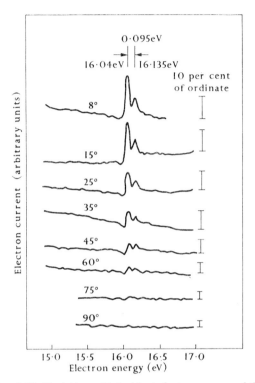

Fig. 5.55. Variation with incident electron energy of the intensity of electrons elastically scattered in neon, at different angles of scattering. The intensity variations may be judged from the lines shown on the right-hand side which indicate 10 per cent of the total intensity near resonance.

but show similarities at the last three energies where no resonance behaviour was observed.

6. Spin polarization of electrons following elastic scattering

Early calculations by Mott† showed that, as a result of the magnetic interaction between the orbital motion and spin, an electron beam, after elastic scattering through an angle θ_1, should be partially polarized in

† See, for example, MOTT, N. F. and MASSEY, H. S. W., *Theory of atomic collisions*, 3rd edn. p. 229 (Clarendon Press, Oxford, 1965).

a direction perpendicular to the plane of scattering. If a scattered beam polarized in this way undergoes a second scatter, the scattered intensity per unit solid angle for scattering through a given angle θ_2 will depend on the azimuth ϕ_2 of the scattering plane relative to the plane of polarization. In particular the scattering into the 'east' direction will be different from that into the 'west' direction. The original calculations

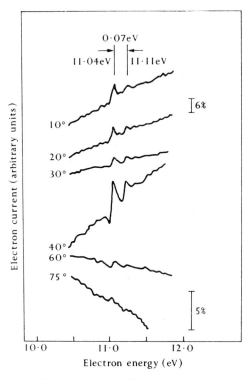

FIG. 5.56. Variation with incident electron energy of the intensity of electrons elastically scattered in argon, at different angles of scattering. The intensity variations may be judged from the lines shown on the right-hand side which indicate a specified percentage of the total intensity near resonance.

were carried out for electrons of energy several hundreds of keV and, after some vicissitudes, the measured asymmetries were found to agree with the theoretical predictions.†

Calculations for slow electrons showed that large polarization effects should be observed in the scattering of low-energy electrons by a heavy

† MOTT, N. F. and MASSEY, H. S. W., *Theory of atomic collisions*, 3rd edn, p. 240 (Clarendon Press, Oxford, 1965).

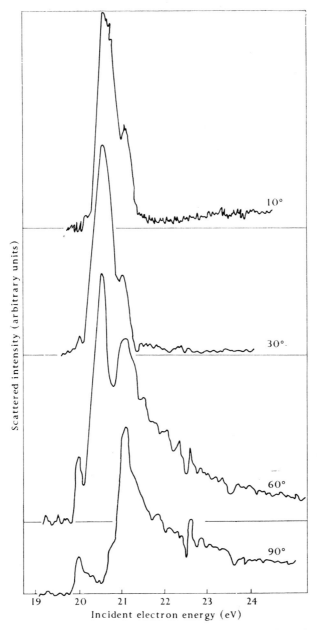

FIG. 5.57. Variation with incident electron energy of the intensity of electrons scattered in helium after exciting the 2^3S state, at different angles of scattering.

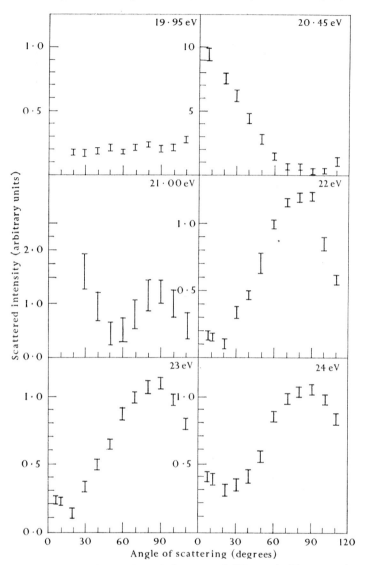

Fig. 5.58. Angular distribution of electrons, of different incident energies as indicated, scattered in helium after exciting the 2^3S state.

nucleus such as gold or mercury (see Chap. 6, § 6). Deichsel† measured the polarization in the scattering of electrons of energy between 1 and 2 keV by mercury, and obtained large values in agreement with the theoretical predictions. Since then a considerable amount of work‡ has

† DEICHSEL, H., *Z. Phys.* **164** (1961) 156.
‡ DEICHSEL, H. and REICHERT, E., *Phys. Lett.* **13** (1964) 125; *Z. Phys.* **185** (1965) 169; STEIDL, H., REICHERT, E., and DEICHSEL, H., *Phys. Lett.* **17** (1965) 31; JOST, K. and

5.6 ENERGY AND ANGULAR DISTRIBUTIONS

been carried out on the scattering of slow electrons in mercury, with the aim both of studying the phenomenon itself and also of developing electron beams with a high degree of polarization.

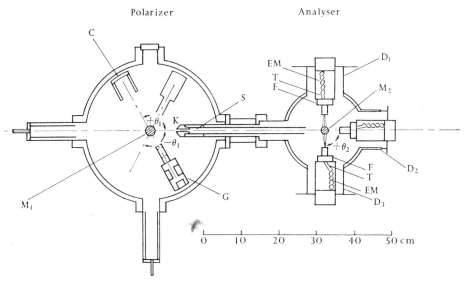

FIG. 5.59. Arrangement of the apparatus used by Deichsel, Reichert, and Steidl to study the double elastic scattering of electrons by mercury.

6.1. Measurement of the spin polarization of the scattered electrons

Fig. 5.59, which shows the apparatus used by Deichsel, Reichert, and Steidl,† is typical of the arrangements used to study the elastic double scattering of electrons in mercury. The two cylindrical scattering chambers were constructed of brass. Electrons from the gun of the first scattering apparatus are incident on a mercury vapour stream M_1 at the centre of the apparatus. The electron current from the gun is collected by the electron collector opposite. Electrons scattered from the mercury through the angle θ_1 enter the channel K where they can be accelerated by the accelerating system S and pass along a connecting tube into the analysing chamber where they are scattered from a second mercury vapour stream M_2 at the centre. Electrons scattered into three fixed detectors are detected. The two detectors (D_1 and D_3) in the 'east' and 'west' positions both detected electrons scattered through the angle $\theta_2 = 90°$. An electrode F in the collector was maintained at a small

KESSLER, J., Phys. Rev. Lett. **15** (1965) 575; Z. Phys. **195** (1966) 1; Phys. Lett. **21** (1966) 524.
† DEICHSEL, H., REICHERT, E., and STEIDL, H., Z. Phys. **189** (1966) 212.

potential relative to the cathode of the electron gun so that electrons that had lost energy greater than 4 eV could not enter the collector (the energy of the lowest excited state of mercury is 4·9 eV). After passing through this electrode the electrons were accelerated again and deflected through 60° in the transverse electrostatic field between the concentric plates T, which served to reject ions and photons, on the first plate of the electron multiplier EM.

The third 'north' detector D_2 measured the total electron current entering the second scattering chamber. The electron gun and its collector opposite could rotate about the first mercury vapour stream M_1 to enable a range of angles θ_1 from 30 to 150° to be covered.

The mercury vapour was obtained from a mercury boiler attached to the apparatus by means of side tubes whose walls were kept heated. The mercury vapour entered the scattering chambers through holes of 1·5 mm diameter a few centimetres above the scattering regions. Water-cooled collectors opposite the holes condensed the mercury streams and fed the mercury back to the boiler by means of side tubes. Liquid-air-cooled cylinders surrounding the tubes through which the mercury vapour stream entered the apparatus prevented the building up of any appreciable mercury vapour pressure in other parts of the apparatus.

If, in the beam scattered out of the first chamber, $N_\uparrow$ and $N_\downarrow$ are respectively the number of electrons with spin 'up' and spin 'down' the polarization $P(\theta_1)$ of the beam scattered through an angle θ_1 is defined by

$$P(\theta_1) = \frac{N_\uparrow - N_\downarrow}{N_\uparrow + N_\downarrow}. \tag{38}$$

After the second scatter the number of electrons, E, scattered into the 'east' detector is proportional to

$$\tfrac{1}{4}\{1+P(\theta_2)\}\{1+P(\theta_1)\}+\tfrac{1}{4}\{1-P(\theta_2)\}\{1-P(\theta_1)\} = \tfrac{1}{2}\{1+P(\theta_1)P(\theta_2)\},$$

while the number W scattered into the 'west' detector is proportional to

$$\tfrac{1}{2}\{1-P(\theta_1)P(\theta_2)\},$$

so that the ratio
$$\left(\frac{E}{W}\right)_+ = \frac{1+P(\theta_1)P(\theta_2)}{1-P(\theta_1)P(\theta_2)}. \tag{39}$$

In measurements of this kind great care has to be taken to correct for the effects of any asymmetries inherent in the apparatus. To eliminate these the experiment was repeated with the gun placed in the symmetrical position relative to the line joining the two scattering volumes. Since this is equivalent to rotation about the $M_1 M_2$ axis it is seen from (38) that $P(-\theta_1) = -P(\theta_1)$. In this case

$$\left(\frac{E}{W}\right)_- = \frac{1+P(-\theta_1)P(\theta_2)}{1-P(-\theta_1)P(\theta_2)}$$

$$= \frac{1-P(\theta_1)P(\theta_2)}{1+P(\theta_1)P(\theta_2)}. \tag{40}$$

Then
$$\frac{(E/W)_+}{(E/W)_-} = \left\{\frac{1+P(\theta_1)P(\theta_2)}{1-P(\theta_1)P(\theta_2)}\right\}^2 \tag{41}$$

and if $\theta_1 = \theta_2$ and the energy of the electrons is the same for the two scatters, $P(\theta_1) = P(\theta_2)$ so that

$$\frac{(E/W)_+}{(E/W)_-} = \left[\frac{1+\{P(\theta_1)\}^2}{1-\{P(\theta_1)\}^2}\right]^2 \tag{42}$$

and $P(\theta_1)$ can be determined.

If $P(\theta_1)$ is known for one energy and scattering angle, the polarization for any other energy and angle can be determined using (41). The apparatus used by Jost and Kessler† was similar to that of Deichsel, Reichert, and Steidl described above. They accelerated the electrons to an energy of 120 keV, however, before entering the analyser, and paid particular attention to the angular resolution of their apparatus, obtaining a resolution of 1°.

6.2. Results of the spin polarization measurements

The measurements of the polarization in the scattering of low-energy electrons made by Deichsel, Reichert, and their colleagues agree in general with those of Jost, Kessler, and their colleagues. The high angular resolution achieved by the latter probably accounts for the higher maximum values of the measured polarization. The polarization is found to be a rapidly varying function of scattering angle and electron energy. Fig. 5.60 shows the results obtained by Jost and Kessler for electrons of a series of energies in the range 180–1700 eV scattered in mercury. The extraordinarily close agreement between the measured and calculated polarizations (given by dashed lines in Fig. 5.60) is discussed in Chapter 6, § 6. The rapid variation of the shape of the curves with electron energy is seen from Fig. 5.61, which shows the detailed structure of the angular distribution of the polarization in the angular range 110–130°, for electrons of energy between 260 and 340 eV.

Deichsel, Reichert, and Steidl have extended their polarization measurements down to incident electron energies of 3·5 eV. Their results

† JOST, K. and KESSLER, J., *Z. Phys.* **195** (1966) 1.

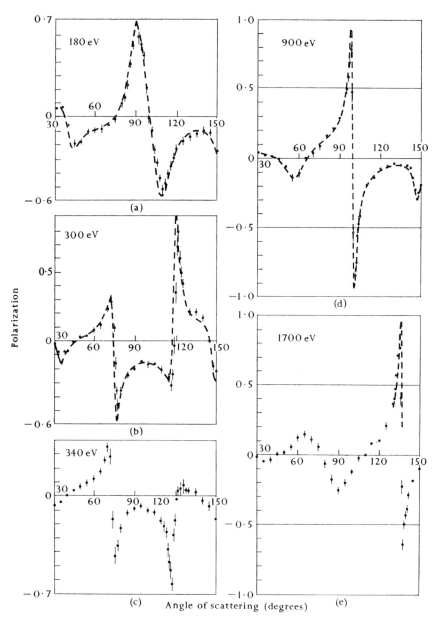

FIG. 5.60. Observed polarization of electrons, scattered elastically by mercury atoms, as a function of angle of scattering for different electron energies in the range 180–1700 eV. ---- theoretical (see Chap. 6, § 6), ⬥ observed.

are shown in Fig. 5.62. In this work the energy of the electrons reaching the analyser was 300 eV.

It is clear that electron beams of a high degree of polarization may be obtained by scattering low-energy electrons from mercury for suitably well-defined electron energies and scattering angles.

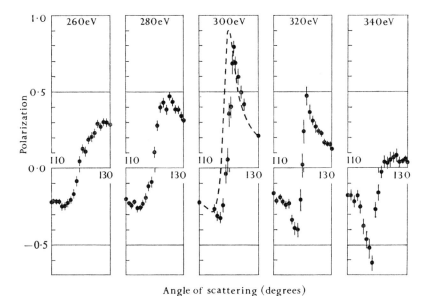

FIG. 5.61. Variation of the polarization of electrons scattered elastically by mercury atoms, with angle of scattering in the range 110–30° at different electron energies.

7. Spin-exchange collisions

The cross-section for scattering of electrons by atoms with exchange of incident and atomic electrons can be measured by studying the polarization of the atom before and after scattering, provided the two electrons that undergo exchange have opposite spin orientations. The method also assumes that spin–orbit interaction is not important, so that an observed change of polarization in scattering is due to electron exchange. In their experiments on electron scattering in atomic hydrogen, described in Chapter 4, § 2.2.1, Lichten and Schultz† measured the cross-section for spin-exchange inelastic collisions leading to the excitation of the $2s_{\frac{1}{2}}$ state of H. We discuss here methods of measuring cross-sections for elastic spin-exchange collisions, especially for electron scattering by alkali atoms.

† LICHTEN, W. and SCHULTZ, S., *Phys. Rev.* **116** (1959) 1132.

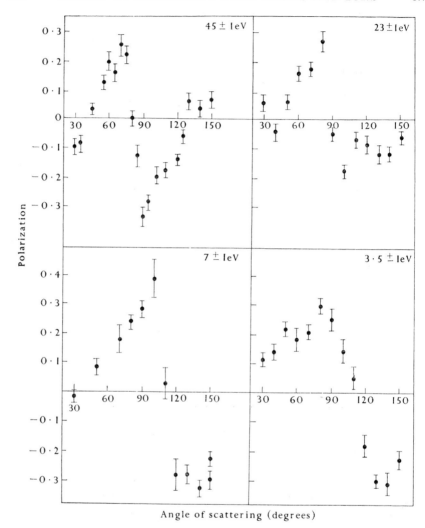

FIG. 5.62. Observed polarization of electrons scattered elastically by mercury atoms, as a function of angle of scattering for different electron energies in the range 3·5 to 45 eV.

7.1. *Atomic-beam method*

The crossed-beam method has been used by Rubin, Perel, and Bederson† to measure the differential cross-section for slow electron scattering in potassium, both for spin-flip and non spin-flip collisions. They used a polarized beam of potassium atoms crossed by an electron beam modulated at a frequency of 30 c/s. Recoil atoms knocked out of the

† RUBIN, K., PEREL, J., and BEDERSON, B., *Phys. Rev.* **117** (1960) 151.

beam were detected by surface ionization on a hot wire connected to a phase-sensitive, narrow-band amplifier locked to the oscillator controlling the electron beam modulation. The polarization of the recoil atoms was analysed by passing them through an inhomogeneous magnetic field. If y is the lateral displacement of the detector perpendicular to the direction of the recoil atom before scattering, θ the corresponding electron scattering angle, and d the distance between the scattering region and the detector,

$$\frac{y}{d} = \frac{mv}{MV}(1-\cos\theta), \tag{43}$$

where mv, MV are respectively the momenta of the electron and the atom. If $p(y)\,dy$ is the probability that the atomic recoil is scattered into the range dy at y, $I(\theta)$ the differential elastic scattering cross-section per unit solid angle through an angle θ, j_e the electron current density, and a the path length of the atomic beam through the scattering region,

$$I(\theta) = \frac{mv}{MV}\frac{vd}{2\pi j_e a}p(y). \tag{44}$$

If f is the fractional polarization of the incident atomic beam and R the ratio of the changed to the original polarization in the recoils, then $I_e(\theta)$, the differential cross-section for elastic scattering with spin exchange, is given by

$$I_e(\theta) = \frac{2I(\theta)(1-R+Rf)}{R+f}. \tag{45}$$

Fig. 5.63 illustrates the apparatus used by Rubin et al. Atoms from the potassium oven are polarized by passing them through an inhomogeneous magnetic field. Atoms of a definite velocity are selected by means of the collimating slits S_1 and S_2. It is necessary to select a definite velocity band $V \to V+dV$ of the atomic beam in order to enable $I(\theta)$ to be calculated using equation (44). The beam is crossed by the modulated electron beam obtained from the electron gun. Recoil atoms, after scattering out of the beam in the direction $\psi\,(=\arctan(y/d))$† relative to the initial direction, are selected by means of the collimating slit and then pass through the analyser, consisting of a second inhomogeneous magnetic field. The surface ionization hot wire detector consists of a length of 10-mil platinum wire mounted in the centre of a brass cylinder containing an opening for the recoil beam. The detector can be swept across the region behind the analyser to collect recoil atoms of both directions of polarization. The assembly of analyser magnet and

† The magnitude of ψ is greatly exaggerated in the figure.

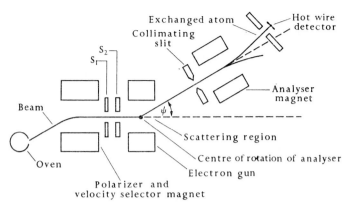

Fig. 5.63. Arrangement of apparatus used by Rubin, Perel, and Bederson to observe separately the differential cross-sections for collisions in which spin flip does and does not occur.

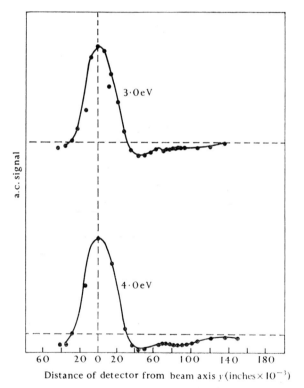

Fig. 5.64. Typical a.c. signals observed using the apparatus shown in Fig. 5.63 (analyser magnet off).

detector can be rotated about the scattering region to cover a range of ψ up to about 10 mrad. The detector is 14 in from the scattering region and very narrow collimating slits have to be used to define the recoil angle ψ. A steady magnetic field of strength about 1000 gauss in the scattering region serves to decouple the nuclear and atomic spin systems so that the change in polarization of the atom may be assumed

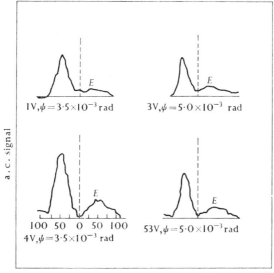

Distance of detector from magnet axis (inches $\times 10^{-3}$)

FIG. 5.65. Typical a.c. signals observed using the apparatus shown in Fig. 5.63 (analyser magnet on), showing signals due to scattered beams with two directions of polarization. E represents the signal due to scattered atoms that have changed their spin states.

to be due solely to a change in the spin orientation of the valence electron. Both the polarizing and analysing magnets are 4 in long with a ratio of field gradient to field of 5 cm^{-1}.

Fig. 5.64 shows the a.c. signal detected when the detector-analyser magnet assembly is swept through the beam direction with the analyser magnet off for electron beam energies of 3 and 4 eV. The positive a.c. signal corresponds to an actual reduction of intensity due to atoms knocked out of the beam, while the negative signal corresponds to recoil atoms moving along different angles ψ. Clearly, if the detector were swept sufficiently far off the axis the area above the zero line should equal the area below for each of these curves. From the magnitude of the detector signal for a given value of y, $I(\theta)$ can be estimated from equation (44).

Fig. 5.65 shows the a.c. signal obtained when the collimator defines a recoil angle ψ corresponding to a scattering angle θ, the analyser field is switched on, and the detector swept across the magnet axis. The two polarized beams due to scattering with and without change of polarization are evident. The ratio of heights of the maxima gives R and,

TABLE 5.1

Electron energy (eV)	Electron scattering angle θ (degrees)	$\dfrac{I_e(\theta)}{I(\theta)}$	$\dfrac{\int_{\theta_{\min}}^{\theta_{\max}} I_e(\theta)\,d\theta}{\int_{\theta_{\min}}^{\theta_{\max}} I(\theta)\,d\theta}$	Q_e (10^{-14} cm²) Lower bound	Upper bound
0·5	47	0·16			
	69	0·46	0·35	0·87	1·6
	84	0·62			
1·0	42	0·14			
	56	0·41	0·27	0·55	1·5
	68	0·44			
2·0	35	0·11			
	38	0·19			
	42	0·24	0·30	0·56	2·0
	47	0·46			
	56	0·63			
3·0	32	0·14			
	38	0·23	0·30	0·45	2·0
	43	0·48			
	51	0·57			
4·0	30	0·12			
	35	0·15	0·25	0·33	1·7
	47	0·24			

knowing f, the cross-section for spin exchange can be calculated from (45).

The results obtained for the spin-exchange cross-sections for potassium are given in Table 5.1.

The upper and lower bounds of the spin-exchange cross-section, Q_e, correspond to different assumptions about the behaviour of $I_e(\theta)$ at larger scattering angles. The lower bound is based on the assumption that $I_e(\theta)/I(\theta)$ will not decrease by more than a factor of 2 on the average at larger angles. The upper bound assumes $I_e(\theta)$ is independent of θ at larger angles. The absolute values of Q_e given are based on Brode's[†] data for Q (see Fig. 1.16).

† BRODE, R. B., *Phys. Rev.* **34** (1929) 673.

Table 5.1 shows that the spin-exchange cross-section $I_e(\theta)$ is much less sharply peaked at small scattering angles than the differential cross-section $I(\theta)$.

7.2. Optical-pumping method

In this method, used first by Dehmelt,† atoms of an alkali metal vapour are first oriented by optical pumping. For example, if light from a sodium arc is circularly polarized (right-handedly) and passed through a flask containing sodium vapour in the direction of a steady magnetic field H_0 the D_2 line λ 5889·4 will be absorbed in the vapour leading to transitions

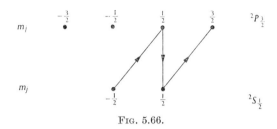

Fig. 5.66.

$^2S_{\frac{1}{2}}$–$^2P_{\frac{3}{2}}$ with $\Delta m_j = 1$. As a result the magnetic substates $m_j = \frac{1}{2}, \frac{3}{2}$ of the $^2P_{\frac{3}{2}}$ state will be populated (see Fig. 5.66). In the de-excitation process transitions may take place from the $^2P_{\frac{3}{2}}$ ($m_j = \frac{1}{2}$) state back to either the $^2S_{\frac{1}{2}}$ ($m_j = \frac{1}{2}$) or $^2S_{\frac{1}{2}}$ ($m_j = -\frac{1}{2}$) states. On the other hand, the formation of the $^2P_{\frac{3}{2}}$ ($m_j = \frac{1}{2}$) state can take place only by transitions from the $^2S_{\frac{1}{2}}$ ($m_j = -\frac{1}{2}$) state. The net result is therefore to 'pump' atoms from the $^2S_{\frac{1}{2}}$ ($m_j = -\frac{1}{2}$) state to the $^2S_{\frac{1}{2}}$ ($m_j = +\frac{1}{2}$) state thus leading to a high degree of polarization of the sodium atoms. A similar effect would be produced by the D_1 line. If free electrons are now introduced into the vessel by means of a subsidiary r.f. discharge or by the proximity of a radioactive source, spin-exchange collisions between the electrons and the sodium atoms will tend to transfer some of the polarization from the sodium atoms to the free electrons. If now an r.f. magnetic field of frequency equal to the free-electron gyromagnetic frequency $\omega_0 = g\mu_0 H_0/\hbar$ (g = gyromagnetic ratio of the electron, μ_0 = Bohr magneton) in a direction perpendicular to H_0 is applied to the vapour the free electrons will be depolarized by gyromagnetic spin resonance. This depolarization in turn will be partially transferred to the sodium atoms by spin-exchange collisions. The transmission of the right-handed circularly polarized D lines will be different for oriented and non-oriented sodium atoms. For example, right-handed circularly

† DEHMELT, H. G., *Phys. Rev.* **109** (1958) 381.

polarized D_1 radiation would not be absorbed at all in passing through sodium vapour in which the atoms were completely oriented ($m_j = \frac{1}{2}$). If the r.f. frequency is kept constant but the steady field intensity varied round H_0, the transmission is changed as the field intensity sweeps through the resonance intensity. The width of the resonance signal of the transmitted radiation depends on the cross-section for spin exchange in collisions between the electrons and the sodium atoms.

In practice an inert gas such as argon is introduced into the flask containing the sodium vapour and acts as a buffer, hindering the diffusion of electrons to the walls. Let N, n be respectively the alkali atom and electron concentrations in the flask, N_+, n_+ having spins up and N_-, n_- having spins down. The sodium atom and electron polarizations, P, p, are given by $P = (N_+ - N_-)/N$; $p = (n_+ - n_-)/n$. The atoms are supposed to be at rest and the electrons to move with velocity $v = (3\kappa T_e/m)^{\frac{1}{2}}$, T_e being the electron temperature, which is close to the ambient temperature. If Q_e is the cross-section for spin-exchange collisions between the free electrons and the sodium atoms, the rates of change of N_+, N_-, n_+ and n_- due to spin exchange are given by

$$\dot{N}_+ = -\dot{n}_+ = -\dot{N}_- = \dot{n}_- = vQ_e(N_+ n_- - N_- n_+). \tag{46}$$

In addition the atoms are continuously polarized by optical pumping and depolarized by relaxation collision effects of characteristic time $2\tau_a$. Similarly the collisions of the electrons with the buffer gas lead to depolarization with characteristic time $2\tau_e$. We then have, if Q_a is the absorption cross-section for the pumping radiation of intensity I,

$$\dot{N}_+ = -(N_+/2\tau_a) + (N_-/2\tau_a) + \tfrac{1}{2}N_- Q_a I + vQ_e(N_- n_+ - N_+ n_-), \tag{47}$$

$$\dot{n}_+ = -(n_+/2\tau_e) + (n_-/2\tau_e) + vQ_e(N_+ n_- - N_- n_+), \tag{48}$$

giving
$$\dot{P} = F(p-P) + (P_I - P)/\tau, \tag{49}$$

$$\dot{p} = f(P-p) - p/\tau_e, \tag{50}$$

where
$$f = vQ_e N \tag{51}$$

is the spin-exchange collision frequency for an electron with sodium atoms,
$$F = vQ_e n \tag{52}$$

is the frequency at which a sodium atom is hit by electrons so that spin exchange takes place,
$$P_I = Q_a I \tau_a \bar{P}/(Q_a I \tau_a + 1) \tag{53}$$

is the equilibrium polarization corresponding to light intensity I that would be obtained in the absence of electron spin exchange, $\bar{P}$ is the saturation polarization obtainable with optical pumping ($\simeq 1$), and

$$\tau = \tau_a/(Q_a I \tau_a + 1). \tag{54}$$

In calculating I allowance must be made for absorption in the pumping cell.

Putting $\dot{p} = \dot{P} = 0$, the equilibrium polarizations p_0, P_0 under the conditions of the experiment can be calculated, viz.

$$p_0 = f\tau_e P_I/(f\tau_e + F\tau + 1), \tag{55}$$

$$P_0 = (f\tau_e + 1)P_I/(f\tau_e + F\tau + 1). \tag{56}$$

The depolarization of the electrons due to magnetic resonance reduces the electron relaxation time, τ_e. It is clear that the width of the electron resonance signal will depend on the spin-exchange collision rate f. Balling, Hanson, and Pipkin† showed that the position of the resonance is also shifted by an amount depending on f. They obtain for the case of an r.f. field $2H_1 \cos \omega t$ perpendicular both to the static field H_0 and to the direction of the beam of circularly polarized light, the reduction δI in the intensity of the transmitted light:

$$\delta I = \text{const} \frac{\omega_1^2 \tau_1 \tau_2}{1 + \omega_1^2 \tau_1 \tau_2 + (\omega_0 - \delta\omega_0 - \omega)^2 \tau_2^2}, \tag{57}$$

where
$$\omega_1 = g\mu_0 H_1/\hbar \tag{58}$$

is the resonance frequency of the electrons in a field of intensity H_1,

$$\tau_1 = \tau_e(F\tau + 1)/(f\tau_e + F\tau + 1) \tag{59}$$

$$\simeq \tau_e/(1 + f\tau_e) \tag{60}$$

(because in practice $F\tau \ll 1$), τ_2 is another similar relaxation time describing physical conditions in the optical pumping cell, and $\delta\omega_0$, the frequency shift due to spin-exchange collisions, can be written

$$\delta\omega_0 = P_0 f K, \tag{61}$$

where K, the scattering shift parameter, can be related to the scattering phases (Chap. 6, § 3) that determine the elastic and spin-exchange cross-sections. Specifically,

$$KQ_e = (\pi/2k^2) \sum (2l+1)\sin 2(\eta_l^- - \eta_l^+), \tag{62}$$

where η_l^-, η_l^+ are respectively the scattering phases for the lth partial wave triplet and singlet electron-alkali atom scattering, k being the electron wave number. In Chapter 8, § 3 it is shown that

$$Q_e = (\pi/k^2) \sum (2l+1)\sin^2(\eta_l^- - \eta_l^+), \tag{63}$$

so that, if the phase difference $(\eta_l^- - \eta_l^+)$ is derived from a measurement of $\delta\omega_0$ using (61) and (62), Q_e can be calculated from (63).

Approximately
$$f \simeq \tau_1^{-1} \simeq \tau_2^{-1}. \tag{64}$$

† BALLING, L. C., HANSON, R. J., and PIPKIN, F. M., *Phys. Rev.* **133** (1964) A607.

If H_1 is small so that $\omega_1 f \ll 1$, the total width of the resonance,
$$\Delta\omega = 2f \tag{65}$$
and thence the ratio
$$2(\delta\omega_0/\Delta\omega) = P_0 K. \tag{66}$$
From a measurement, therefore, of both $\delta\omega_0$ and $\Delta\omega$, K can be determined provided P_0 can be estimated. The phase difference $(\eta_l^- - \eta_l^+)$ can then be calculated from (62) and (63) and Q_e calculated from (63) without an actual measurement of the alkali atom concentration, N, which is subject to much uncertainty. The polarization, P_0, of the alkali atoms during the experiment was estimated by measuring the relative amplitudes of the six Zeeman lines corresponding to the $^2P_{\frac{3}{2}}-^2S_{\frac{1}{2}}$ transition. The relative populations of $^2P_{\frac{3}{2}}$ levels of different m values is determined by P_0.

In his original work on sodium Dehmelt† used the magnitude of the resonance signal to determine f and thence Q_e. For a given light intensity I_0 and thence a given τ (see eqn (54)), let δI_0, δI_1, $\delta \bar{I}$ be respectively the reduction of light intensity without r.f. field, at resonance with an r.f. field of amplitude H_1, and in the absence of free electrons. Then

$$S(H_1) = \frac{\delta I_0 - \delta I_1}{\delta \bar{I}} = \frac{P_0(0) - P_0(H_1)}{P_I}. \tag{67}$$

The maximum possible signal for large H_1 is
$$S(\infty) = f\tau_e F\tau/(F\tau+1)(f\tau_e + F\tau + 1), \tag{68}$$
from (55) and (56).

Franken, Sands, and Hobart‡ used an essentially similar method for determining Q_e for electron-potassium atom collisions. In their case, however, they used a mixture of sodium and potassium vapour in their absorption vessel so that the analysis was correspondingly more complex.

In their work on rubidium,§ caesium,‖ and sodium†† Balling and his collaborators used both the line width $\Delta\omega$ to determine f, and thence Q_e, from equation (65) and the ratio $2\delta\omega_0/\Delta\omega$ of the resonance shift to the line width to determine Q_e from equation (66). For caesium, Balling and Pipkin extended the analysis to allow for the effect of spin-orbit interaction which complicates the relation between Q_e, K and the scattering phases.

Fig. 5.67 shows diagrammatically the apparatus used by Balling for sodium. Light from the source SL (a commercial sodium arc-lamp

† loc. cit., p. 363.
‡ Franken, P., Sands, R., and Hobart, J., *Phys. Rev. Lett.* **1** (1958) 52.
§ Balling, L. C., Hanson, R. J., and Pipkin, F. M., *Phys. Rev.* **133** (1964) A607.
‖ Balling, L. C. and Pipkin, F. M., ibid. **136** (1964) A46.
†† Balling, L. C., ibid. **151** (1966) 1.

operated in an oven) was focused into a parallel beam and made circularly polarized by passing through a polarizing plate P. The circular polarizer could be rapidly changed to give either left or right circularly polarized light.

The light passed through the absorption flask F consisting of a Pyrex cylinder, 8 cm in diameter and 8 cm long. After distilling the sodium into the flask, helium or neon at a pressure of about 40 torr was added as a buffer gas. A continuous r.f. discharge between glass-covered electrodes in a turret T on top of the absorption cylinder provided free electrons which diffused down into the cylinder. After passing through the cylinder the transmitted light intensity was measured by a photocell PC.

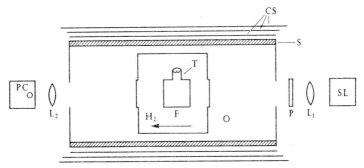

Fig. 5.67. Arrangement of apparatus used by Balling for determining line width and resonance shift in magnetic-resonance optical pumping experiments.

The uniform axial magnetic field of 50 mG was produced by a solenoid S 12 in diameter and 36 in long, enclosed in three concentric cylindrical shields CS to reduce the perturbing effects of external magnetic fields. Since magnetic field inhomogeneities will increase the apparent width of the resonance, care has to be taken to obtain a uniform field. A typical value of the contribution to the apparent width due to such inhomogeneities in Balling's experiment was 100 c/s in a total resonance width of 700 c/s. The r.f. field of frequency about 150 kc/s was produced by a single coil surrounding the flask and was amplitude modulated with a coaxial relay. The demodulated absorption signal was measured using a tuned detector locked to the modulation source.

The absorption flasks were placed in a cylindrical oven O and care taken to reduce temperature gradients across the flask to a minimum.

It was first necessary to determine ω_1. This was done by keeping the r.f. signal generator set at the electron resonance frequency and increasing H_0 until this frequency corresponded to the resonance frequency for sodium atoms. In sweeping through the resonance the sodium resonance signal showed characteristic 'wiggles' due to the nutation of the sodium atomic moment. The frequency, $\omega_1(\text{Na})$ of the sodium 'wiggles', which is proportional to the r.f. field amplitude, H_1, and to the moment of the sodium atom, was measured. The magnetic moment of the electron is approximately four times that of the sodium atomic moment so that the free electron precession frequency ω_1, corresponding to a field of strength H_1, is given by

$$\omega_1 \simeq 4\omega_1(\text{Na}). \tag{69}$$

From equation (57) it is seen that the square of the full width of the

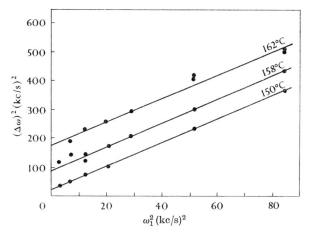

FIG. 5.68. Typical linear plots of the square of the full width $\Delta\omega$ of the electron resonance signal against the square of the free electron precession frequency ω_1, obtained by Balling for sodium at different temperatures.

resonance (after deduction of the contribution to the apparent line width due to the magnetic field inhomogeneity) is given by

$$\frac{(\Delta\omega)^2}{4} = \frac{1}{\tau_2^2} + \omega_1^2 \frac{\tau_1}{\tau_2}, \tag{70}$$

so that a plot of the square of the full width at half-maximum of the electron signal against ω_1^2 should give a straight line of slope $4\tau_1/\tau_2$ and intercept on the $(\Delta\omega)^2$ axis of $4/\tau_2^2$.

Fig. 5.68 shows a typical series of such line plots for various temperatures (i.e. vapour densities). The relation is linear as expected from equation (70) and $\tau_1/\tau_2 = 1\cdot0\pm0\cdot05$. The temperature dependence of $1/\tau_2$ was consistent with that expected from the known dependence of vapour pressure on temperature for sodium vapour, so that the approximation of equation (64), which follows on the assumption that spin-exchange collisions are the dominant spin relaxation mechanism, appears to be justified.

To measure the frequency shift, $\delta\omega_0$, of the electron resonance, the electron signal was observed successively with left- and right-circularly polarized light. The sign of the sodium polarization is opposite in the two cases so that
$$2\delta\omega_0 = \omega_{\text{right}} - \omega_{\text{left}}. \tag{71}$$

Fig. 5.69 shows a plot of $\delta\omega_0/\pi$ against temperature. The dashed curve gives the observed temperature dependence of τ_2^{-1}, demonstrating the agreement to be expected (compare eqns (61) and (64)).

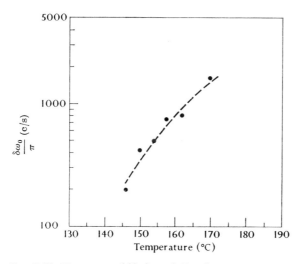

FIG. 5.69. Frequency shift $\delta\omega_0$ of the electron resonance in sodium as a function of temperature. ● observed by Balling, – – – – observed temperature dependence of $1/\tau_2$.

On the assumption (probably not justified) that the phase shift variation with electron energy is not significant, Table 5.2 summarizes the results of measurements of Q_e for low-energy electron collisions with alkali metal atoms. A theoretical discussion is given in Chapter 8, § 8.2.

7.3. *Use of spin-exchange collisions to produce a polarized electron beam*

Byrne and Farago[†] have proposed the use of spin-exchange collisions between a beam of polarized potassium atoms and an electron beam to polarize the electrons. Their proposal has been realized in an experiment carried out by Farago and Siegmann[‡] whose apparatus is shown in

[†] BYRNE, J. and FARAGO, P. S., *Proc. phys. Soc.* **86** (1965) 801.
[‡] FARAGO, P. S. and SIEGMANN, H. C., *Phys. Lett.* **20** (1966) 279.

Fig. 5.70. Potassium atoms leave an oven and pass through an inhomogeneous magnetic field in which they are split into two beams polarized in opposite directions. One of the beams emerges and passes across the

TABLE 5.2

Alkali atom	Temperature	Observer	$Q_e(\times 10^{-14}\ \text{cm}^2)$
Na	400 °C	Dehmelt	2·3
	140 °C	Balling	2·2±0·2
K	200 °C	Franken, Sands, and Hobart	3·0
Rb	20 °C	Balling, Hanson, and Pipkin	2·8±0·3
Cs	20 °C	Balling and Pipkin	3·16±0·25

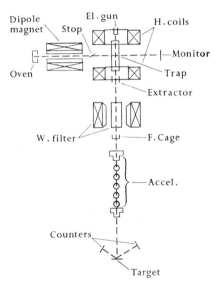

FIG. 5.70. Schematic diagram of the experimental arrangement used by Farago and Siegmann for the production of polarized electron beams through spin-exchange collisions.

diameter of a cylindrical electrode, entering and leaving the inside of the cylinder through two slits.

The cylinder is part of a system of five coaxial cylinders that form an electron trap. Electrons from a gun are fired along the axis of the cylinders. A positive voltage pulse applied to the two cylinders nearest the gun for a time of the order of a few microseconds accelerates electrons moving parallel to the axis into the system. The two cylinders at either end of the system are maintained at a steady negative potential so that at the end of the pulse electrons of energy a few eV are trapped in the

TABLE 5.3

Oven temperature	320 °C
Atomic beam intensity	0.6×10^{12} s^{-1}
Atomic beam polarization	60–80 per cent
Pressure in interaction region	3×10^{-8} torr
Magnetic trapping field	400 gauss
Energy of trapped electrons	3.5 eV
Pulse repetition rate	$f_0 = 2000$–100 c/s
Pulse length	1 μs
Trapping time	$\frac{1}{2}f_0^{-1}$
Mean current of electron beam	10^{-10}–4×10^{-12} A

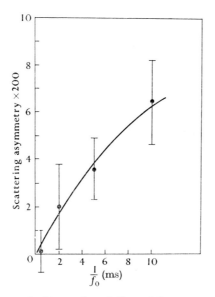

FIG. 5.71. Observed variation of the measured scattering asymmetry with trapping time $1/f_0$ in the experiments of Farago and Siegmann.

cylinder system. An axial magnetic field, produced by a pair of Helmholtz coils, confines the electrons to the axis of the system and they move up and down, crossing and recrossing the polarized potassium atom beam for a time of the order of 1 ms. After this time another positive pulse is applied to the cylinder furthest from the gun and to the extractor electrode, as a result of which electrons are extracted from the trap. As a result of spin-exchange collisions during the repeated transits of the potassium beam the electron beam will be polarized along the direction of the magnetic field (i.e. longitudinally) when it emerges.

In order to change the polarization to a transverse direction the electron beam is passed through a Wien filter consisting of crossed electric and magnetic fields. To measure the polarization the beam is accelerated to an energy of 135 keV before impinging on a thin gold foil. The intensity of electrons scattered through a large angle in a plane perpendicular to the polarization direction is measured by means of two counters set on opposite sides of the electron beam direction.

Typical operating conditions of the apparatus are given in Table 5.3. Fig. 5.71 shows the results obtained for the quantity $200a$, where a is the measured scattering asymmetry, plotted against the trapping time, $1/f_0$, f_0 being the pulse repetition frequency. The experimental measurements are fitted to the curve

$$200a = A(1-\exp(-\lambda/f_0)), \tag{72}$$

with $A = 9\cdot3\pm2\cdot2$; $\lambda = 106\pm46$ s^{-1}.

The expected electron polarization, p, is related to the polarization, P, of the atomic beam by

$$p = P(1+1/f\tau)^{-1}\{1-\exp(-\eta f/f_0)\}, \tag{73}$$

where τ is the relaxation time of electron polarization due to depolarizing collisions with residual gas molecules,

$$f = vQ_e N$$

is the spin-exchange collision frequency of electrons with atoms in the beam (see (51)) and η is the fraction of the trapping time for which the electron is passing through the beam. Comparing λ and f, with Q_e taken as 10^{-14} cm^2, η is about $\frac{1}{3}$.

The highest measured scattering asymmetry corresponds to an electron polarization of about 10 per cent. This was obtained at a mean electron beam intensity of 4×10^{-12} A, equivalent to a peak intensity of about 10^{-8} A.

The prospect of carrying out collision experiments using slow polarized electron beams opens up some interesting possibilities.

6

ELECTRON COLLISIONS WITH ATOMS—THEORETICAL DESCRIPTION—GENERAL AND SEMI-EMPIRICAL THEORY OF ELASTIC SCATTERING

In the previous chapters we have described the information about the various effective cross-sections for collisions of electrons with atoms that has been obtained by a variety of experimental methods. This information is necessarily incomplete and has to be supplemented by theoretical methods. To do this we must first establish a theory which gives a satisfactory description of the observations. If, however, we do not succeed in developing theoretical methods of calculation that provide values of the cross-sections in agreement with all the observed data, it is important to establish under what conditions such a partial theory is valid. Having done this, it will be known when the theory can be used to predict the values of cross-sections which, while difficult to observe, are important in many applications of collision data. For example, the cross-section for excitation of an atom, such as oxygen, by electron impact may be required. This is very difficult to measure experimentally owing to the molecular character of gaseous oxygen.

In general we shall find that the theory of elastic collisions is quite well established. This is also the case for inelastic collisions when the fractional energy loss is small. The theory is still not satisfactory for impacts in which the fractional energy loss is large.

1. Subdivision of the theoretical problem

Even if we regard the nucleus as a fixed centre of force, the collision of an electron with the simplest atom is essentially a three-body problem. Exact solution is impossible and resort must be made to approximations that are not always systematic and between which the relations are sometimes hard to discern.

We shall proceed by first discussing a semi-empirical approach to the calculation of the broad features of the elastic scattering of slow electrons by atoms. This consists in ignoring the inner structure of the atom and treating it as the centre of a force that is a function only of the distance from the nucleus. While the empirical representation of the effective scattering field is based on the field that would be exerted by the atom if unperturbed by the electron, the so-called static field, departures from this field are introduced to obtain the best agreement with observation. These departures are justified by the need to allow for such effects as the disturbance of the atom by the electron and for the possibility of electron exchange occurring during the impact.

An analysis of this sort is very similar to the optical model approach that has been employed in such great detail for the analysis of the scattering of neutrons by nuclei. The major difference is that in the latter analysis the empirical scattering potential includes an imaginary component that allows for the possibility of inelastic collisions and for the formation of a collision complex of appreciable lifetime during the impact. The presence of an imaginary potential in a one-particle Schrödinger equation means that absorption is taking place. When representing a many-body scattering problem by an effective one-body problem, loss of particles of the initial energy, either through inelastic collisions or through temporary capture in a complex, corresponds to absorption and is represented empirically by an imaginary component of effective scattering potential. No allowance for any such component has been made in the semi-empirical analysis of electron scattering by atoms. The fact that a good description of the observed phenomena has nevertheless been obtained is an indication of the relative weakness of the interaction in atomic as distinct from nuclear phenomena. This will be further discussed in Chapter 9.

After discussing the semi-empirical analysis of elastic scattering, we begin in Chapters 7–9 the development of the analytic theory that attempts to take into account the structural details and calculate the various cross-sections directly from the laws of quantum mechanics and the Coulomb interaction between electrons.

The first step in this direction is to deal with relatively high-energy impacts in which the interaction with the atom introduces only a small change in the total energy of the impinging electron. Under these circumstances a systematic perturbation theory, the now famous Born's series of approximations, may be developed. The first approximation in this series, usually called simply Born's approximation, has received widespread application not only in electron collisions with atoms but in all atomic collision phenomena. If the first approximation is not valid it is usually not worth while proceeding to evaluate the higher approximations and resort is made to some less systematic procedure based on the physics of the situation.

We therefore discuss, in Chapter 7, the detailed application of Born's approximation to elastic and then inelastic collisions, and its range of validity. It is possible to include the effect of electron exchange for these 'high'-energy impacts. While normally negligible when the first Born approximation is valid, there are some processes that could not occur at all except through electron exchange and for these the extension of the Born approximation to allow for exchange (the so-called Born–Oppenheimer approximation) is useful.

We proceed further in Chapter 8 to discuss the broad functions of slow collisions for which Born's approximation is invalid. It is best to concentrate attention first on the theory of electron collisions with the simplest atom, hydrogen. For this comparatively simple case, electron scattering may be dealt with to quite high accuracy by using new and powerful variational methods. Application of these methods to more complex atoms is much more difficult and soon becomes out of the question so it is important to develop less laborious methods that depend on isolation of the major effects. This can be done for hydrogen and it is found that three major factors dominate the broad features of the elastic scattering. These are the static field, electron exchange, and the dipole polarization of the atom by the electron. This is found also to be the case for the elastic scattering by the simplest positive ion, that of helium (He^+), which can be treated with the same accuracy as hydrogen.

Extension to elastic scattering by complex atoms is next carried out assuming that the same three factors are dominant as for hydrogen and ionized helium. This proves to be remarkably successful in application to such atoms as neon and argon but less so for others.

Other methods of approximation, the so-called close-coupling methods in which allowance is fully made for the effect of a limited number of excited states, are discussed. These methods are of special interest in connection with the interpretation of resonance phenomena in electron collisions with atoms, the subject of Chapter 9, but before discussing these phenomena explicitly the whole problem of calculating the broad features of inelastic cross-sections for slow electron impacts is considered. At present there is no way of dealing with this except through a close-coupling method of some kind and success has been very patchy.

The problem of calculating ionization cross-sections in low-energy encounters has special features that are next considered.

Finally, in Chapter 9 we discuss fine-structure phenomena in all aspects, concluding with an analysis of the behaviour expected of cross-sections close to the threshold for most of the processes concerned.

2. Elastic scattering—semi-empirical 'optical model' approach

We now seek to describe the elastic scattering of slow electrons by atoms in terms of the scattering of a particle by a structureless centre of force. The choice of this central-force field is based on the static field of an atom but it is understood that modifications may be introduced in this field to improve the agreement with observations. Such modifications are to be expected because an atom is not just a structureless centre and is itself distorted in the encounter. Apart from allowing for this the modifications also allow empirically for electron exchange.

2.1. *The static field of an atom*

The potential energy $V(r)$ of an electron in the field of an undisturbed atom, the static field, when at a distance r from the nucleus is given by

$$V(r) = -\frac{Ze^2}{r} + 4\pi e^2 \left\{ \frac{1}{r} \int_0^r \rho(r') r'^2 \, dr' + \int_r^\infty \rho(r') r' \, dr' \right\}, \tag{1}$$

where $\rho(r)$ is the density of the atomic electrons at r and the nuclear charge is Ze.

The function $\rho(r)$ may be calculated accurately for hydrogen to give

$$V(r) = -e^2 \exp(-2r/a_0)(r^{-1} + a_0^{-1}), \tag{2}$$

where $a_0 \, (= \hbar^2/me^2)$ is the radius of the first Bohr orbit. For other atoms $\rho(r)$ may only be determined approximately, usually by use of the Hartree self-consistent field method or the more refined Hartree–Fock method in which electron exchange effects within the atom are included.

2.2. *The Hartree and Hartree–Fock field of an atom*

The Hartree and Hartree–Fock approximations are associated with the classification of atomic states in terms of configurations. A basic assumption is that the individual electrons move in an effectively central field, which is that of the nucleus partly screened by the presence of the other electrons. The allowed states for an electron in such a field can then be distinguished by the usual quantum numbers n, l or as $1s, 2p$, etc. The totality of atomic electrons can then be assigned to these one-body states or orbitals in any way consistent with the Pauli principle. Any such assignment defines a configuration with a total energy equal to the sum of the energies of the individual orbitals. The ground configuration

will be the one with lowest total energy on this basis. Thus for neon it will be
$$(1s)^2(2s)^2(2p)^6,$$
in which 2 electrons occupy the $1s$ orbital, 2 the $2s$, and 6 the $2p$.

The next step makes allowance for electron exchange and the combination of angular momenta, including that of electron spin. Although it is only the total angular momentum (quantum number J) of all electrons, including spin, that is strictly conserved, the weak interaction between spin and orbital motion means that, for all but heavy atoms, the total orbital and spin angular momentum (quantum numbers L, S, respectively) are effectively also constant. From a given configuration a number of terms arise which we distinguish by the values of L and S. If electron exchange is ignored these terms will have the same energy but otherwise are split. In general, however, the splitting between the terms is considerably less than that between successive configurations.

To proceed further, allowance must be made for interaction between configurations, for fine-structure splitting due to interaction between electron spin and orbital motion, and for hyperfine structure splitting due to interaction between nuclear spin and the electronic motion.

The Hartree approximation neglects exchange and represents the wave function for a given configuration as a simple product of wave functions for each orbital. Thus the wave functions for a $2p$ orbital will have the form
$$r^{-1}P_{21}(r)Y_1(\theta,\phi), \tag{3}$$
where Y_1 is a normalized first-order spherical harmonic. This notation is a general one that is often used, the radial function $P_{21}(r)$ being such that
$$\int_0^\infty \{P_{21}(r)\}^2 \, dr = 1. \tag{4}$$

When this product approximation is used as a trial function in a Ritz variational calculation it is found that the function $P_{nl}(r)$ satisfies a radial wave equation of the form
$$\frac{d^2 P_{nl}}{dr^2} + \left[\frac{2m}{\hbar^2}\{E_{nl}-V_{nl}(r)\} - \frac{l(l+1)}{r^2}\right]P_{nl} = 0, \tag{5}$$
where V_{nl} is of the form (1) with
$$\rho(r) = \sum \alpha_{n'l'}\{P_{n'l'}(r)\}^2. \tag{6}$$
$\alpha_{n'l'}$ is equal to the number of electrons, other than the one concerned, occupying $n'l'$ orbitals.

This means that each electron is considered to move in the combined field of the nucleus and that of each other atomic electron, calculated by averaging its probability density over all orientations. The total energy of the configuration is then given to this approximation by

$$E = \sum E_{nl} - \tfrac{1}{2} \sum \int_0^\infty \{P_{nl}(r)\}^2 \{V_{nl}(r)\}\, dr. \tag{7}$$

In the Hartree–Fock approximation allowance is made for electron exchange by replacing the simple product trial functions by a linear combination of such functions in which the electrons are permuted among the orbitals. The coefficients in the linear combination are then different for each term in order to be consistent with the Pauli principle. The effect is not only to make the individual functions $P_{nl}(r)$ dependent on the term assignment but also to introduce a non-local interaction into the radial equation (5) so it takes the form

$$\frac{d^2 P_{nl}}{dr^2} + \left[\frac{2m}{\hbar^2}\{E_{nl} - V_{nl}(r)\} - \frac{l(l+1)}{r^2}\right] P_{nl} = -\frac{2m}{\hbar^2} \int K_{nl}(r, r')\, dr'. \tag{8}$$

The non-local interaction kernel K_{nl} depends on the orbitals of the remaining electrons and on the term assignment. With this approximation the total energy of the atom is simply given by

$$E = \sum E_{nl}.$$

It is convenient to express the potential $V(r)$ as determined by these approximations in the form

$$V(r) = -Z_p e^2/r,$$

where Z_p, the effective nuclear charge for potential, is a function of r. In general Z_p falls off exponentially for large r so that, according to quantum theory, the scattering cross-section due to the static field of an atom will be finite (see Chap. 1, § 2).

2.3. *The statistical model*

For heavy atoms the Thomas–Fermi[†] statistical atom model may be used. This gives

$$V(r) = -\frac{Ze^2}{r} \phi(r/\mu),$$

where $\mu = 0{\cdot}885 Z^{-\frac{1}{3}} a_0$. Tables of the function ϕ are available.[‡]

[†] See MOTT, N. F. and SNEDDON, I. N., *Wave mechanics and its applications*, Chap. vi (Clarendon Press, Oxford, 1948).
[‡] BUSH, V. and CALDWELL, S. H., *Phys. Rev.* **38** (1931) 1898.

3. Quantum theory of scattering by a centre of force

3.1. *Total cross-section*

Let us consider a particle of mass m and velocity v, which is projected towards a centre of force in such a direction that, if undeviated, it would pass the centre at a distance p. This quantity is called the *impact parameter*. According to classical theory the particle will certainly be deviated unless the force due to the centre vanishes everywhere along its path. In the quantum-theoretical treatment of the problem it is natural to assign a probability $\alpha(p)$ that a particle with impact parameter between p and $p+dp$ will suffer an *observable* deviation. The effective collision area would then be

$$Q_0 = 2\pi \int_0^\infty \alpha(p) p \, dp. \tag{9}$$

To consider other quantum modifications it is convenient to rewrite this in terms of angular momenta. The angular momentum J of a particle about the centre of force is equal to mvp, so that

$$Q_0 = (2\pi/m^2v^2) \int_0^\infty J\beta(J) \, dJ, \tag{10}$$

where β is now the chance that a particle with angular momentum between J and $J+dJ$ should suffer an observable deviation. Now we must take account of the fact that, according to quantum theory, the angular momentum about the centre of force is quantized so that

$$J = \{l(l+1)\}^{\frac{1}{2}}\hbar. \tag{11}$$

Thus, for large l,
$$l \simeq mvp/\hbar. \tag{12}$$

This converts the integral in (10) to the sum

$$Q_0 = (\pi\hbar^2/m^2v^2) \sum_{l=0}^\infty (2l+1)\gamma(l)$$
$$= (\pi/k^2) \sum_{l=0}^\infty (2l+1)\gamma(l), \tag{13}$$

where $k = 2\pi/\lambda$, λ being the wavelength of the incident particle, and
$$\gamma(l) = \beta[\{l(l+1)\}^{\frac{1}{2}}\hbar].$$

We must now consider the probability $\gamma(l)$. It is easy to see how, in a wave theory, $\gamma(l)$ may actually be zero in circumstances in which, on classical theory, it would be unity. For simplicity, consider the scattering by a potential which has the form

$$V(r) = \begin{cases} D & (r < a), \\ 0 & (r > a), \end{cases} \tag{14}$$

where D is a constant. We shall first discuss head-on collisions for which $l = 0$, and for which there is no doubt that $\gamma(0) = 1$ on the classical theory.

According to the quantum theory the motion of the particles, making head-on collisions, will be represented by a train of waves with wavelength $\lambda = h/mv$ outside the obstacle ($r > a$) and $\lambda = h/(m^2v^2 - 2Dm)^{\frac{1}{2}}$ inside. The two trains of different wavelength must join smoothly at the boundary $r = a$. In order to achieve this, and at the same time keep the amplitude finite at the centre $r = 0$, a phase change must be introduced in the train, for $r > a$, relative to that which would exist in the absence of the obstacle. This phase change, due to the obstacle, will be observable at infinity and it alone indicates the presence of the obstacle. However, we must remember now that it is impossible, in principle, to count the waves between the obstacle and the observer so that a phase change that is an integral multiple of 2π will not be observable. In these circumstances the obstacle produces no observable effect on the particles with zero angular momentum, so $\gamma(0) = 0$. This will be so when the obstacle either introduces or eliminates a whole number of complete waves.

On these grounds it is to be anticipated that $\gamma(0)$ will be a periodic function of the phase shift η_0 produced by the scattering potential in the de Broglie waves, which represent the stream of particles of zero angular momentum. It must also vanish with the phase change and never be negative. The simplest function satisfying all of these conditions is $A \sin^2 \eta_0$. In a similar way we take $\gamma(l) = A \sin^2 \eta_l$, where η_l is the phase shift produced in the waves of angular momentum $\{l(l+1)\}^{\frac{1}{2}}\hbar$.

We have now reached the expression

$$Q_0 = (A\pi/k^2) \sum (2l+1)\sin^2 \eta_l.$$

To obtain A we consider the case of scattering by a rigid obstacle of radius a which is such that $ka \gg 1$. Under these conditions, which should certainly be classical, we would expect that $Q_0 = \pi a^2$. Following the relation (12) between l and p we should also expect that $\eta_l \simeq 0$ for $l > l_0$, where
$$l_0 \hbar = mvp,$$
or
$$l_0 = ka.$$

Since l_0 is large
$$Q_0 \simeq (2A\pi/k^2) \int_0^{ka} l \sin^2 \eta_l \, dl. \tag{15}$$

As we expect η_l to be large for $l < ka$ we can replace $\sin^2 \eta_l$ in (15) by

its mean value of $\frac{1}{2}$ to give
$$Q_0 = \tfrac{1}{2}A\pi a^2.$$
This suggests that we should take A as equal to 2 so that Q_0 agrees with the classical value πa^2. In fact, however, we have made no allowance for shadow diffraction, which remains important down to vanishing wavelengths and doubles the value of A—as the wavelength decreases the shadow diffraction is confined to smaller and smaller angles about the forward direction but its integrated effect remains equal to πa^2. We shall discuss this matter further in Chapter 16, § 4.

Meanwhile we have, taking $A = 4$, arrived at the formula
$$Q_0 = (4\pi/k^2) \sum (2l+1)\sin^2\eta_l, \tag{16}$$
which is in fact correct (for proof see § 3.3).

Although the concept of the phase shift, as determining the probability of a deviation, has been introduced in connection with a potential function with a sharp boundary there is no difficulty in applying it to a field, such as the static field of an atom, which falls off gradually to zero at infinity.

3.2. Angular distribution of elastically scattered electrons—differential cross-section

In the classical picture of a collision, a particle with a definite angular momentum about the scattering centre will undergo a definite deviation, but on the quantum theory, as would be expected, this is no longer the case. Associated with each quantized angular momentum there is an amplitude function that gives the distribution in angle of the associated scattered amplitude (not the scattered intensity). The total scattered amplitude is obtained by adding the contributions from the separate angular momenta, each of which contributes at all angles. A suitable weighting factor, related to the probability of any deviation occurring, must be included in each contribution.

Thus the contribution to the amplitude scattered between angles θ and $\theta+d\theta$ from the angular momentum $\{l(l+1)\}^{\frac{1}{2}}\hbar$ is of the form
$$g(\eta_l)F_l(\theta),$$
where $F_l(\theta)$ is the angular function associated with the particular angular momentum and the weight factor $g(\eta_l)$ is a measure of the chance that any deviation will occur. It will vanish when $\eta_l \to 0$ or an integral multiple of π, and can be expected to have a maximum influence when η_l tends to an odd integral multiple of $\tfrac{1}{2}\pi$.

The total scattered amplitude will then be
$$\sum g(\eta_l) F_l(\theta) \tag{17}$$
and the scattered intensity will be
$$\left| \sum_l g(\eta_l) F_l(\theta) \right|^2,$$
allowing for the possibility that $g(\eta_l)$ may be complex, as it is in fact.

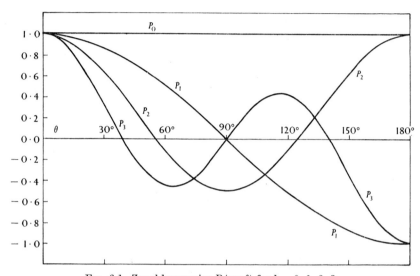

FIG. 6.1. Zonal harmonics $P_l(\cos\theta)$ for $l = 0, 1, 2, 3$.

Since the scattered amplitude is obtained by summing the contributions from different angular momenta, it is to be expected that interference effects, leading to maxima and minima in the angular distribution will arise in some circumstances.

The functions $F_l(\theta)$ are the zonal harmonics $P_l(\cos\theta)$, the forms of which are illustrated in Fig. 6.1 for $l = 0, 1, 2, 3$. It will be noticed that, at $\theta = 0$, all are equal to 1 while, at $\theta = \pi$ they are equal to $(-1)^l$. The number of zeros the functions possess in the range 0 to π is equal to the order l.

The weight factor $g(\eta_l)$ may be calculated by the methods of diffraction theory and is found to be given by
$$g(\eta_l) = (2l+1)\{\exp(2i\eta_l) - 1\}/2ik.$$
This gives for the differential cross-section
$$I_0(\theta)\sin\theta\, d\theta d\phi = \frac{1}{4k^2} \left| \sum (2l+1)\{\exp(2i\eta_l - 1)\} P_l(\cos\theta) \right|^2 \sin\theta\, d\theta d\phi. \tag{18a}$$

THEORY OF ELASTIC SCATTERING

As
$$\int_0^\pi P_l(\cos\theta)P_m(\cos\theta)\sin\theta\,d\theta = \begin{cases} 4\pi/(2l+1) & (l=m), \\ 0 & (l \neq m), \end{cases}$$

it can easily be verified that

$$2\pi \int_0^\pi I_0(\theta)\sin\theta\,d\theta = Q_0.$$

The momentum-transfer cross-section (see Chap. 2, eqn (5)) Q_d is given by

$$Q_d = 2\pi \int_0^\pi (1-\cos\theta)I_0(\theta)\sin\theta\,d\theta$$

$$= (4\pi/k^2)\sum (l+1)\sin^2(\eta_l - \eta_{l+1}). \tag{18 b}$$

3.3. *Proof of quantum formulae*†

The Schrödinger equation for the motion of the electron is

$$\nabla^2 F + \{k^2 - U(r)\}F = 0, \tag{19}$$

where $\quad k^2 = 2mE/\hbar^2, \quad U(r) = 2mV(r)/\hbar^2.$

We require a solution which has the asymptotic form of a plane wave plus an outgoing spherical wave so that

$$F \sim N\{e^{ikr\cos\theta} + f(\theta)r^{-1}e^{ikr}\}, \tag{20}$$

where the axis of the polar coordinates has been taken along the direction of incidence and N is a constant.

Using the formula for the flux corresponding to a wave function ψ,

$$\frac{i\hbar}{4\pi m}(\psi\,\mathrm{grad}\,\psi^* - \psi^*\,\mathrm{grad}\,\psi), \tag{21}$$

we see that (20) corresponds to an incident flux $v|N|^2$, where v, the velocity of the electrons, is given by $k\hbar/m$. The radial flux density corresponding to the outgoing spherical wave is similarly given by

$$v|N|^2|f(\theta)|^2/r^2.$$

Hence the number of scattered electrons incident per second on a small area δS normal to the vector $\mathbf{r}$ is given by

$$v|N|^2|f(\theta)|^2\,\delta S/r^2 = v|N|^2|f(\theta)|^2\,\delta\omega,$$

where $\delta\omega$ is an element of solid angle. The total number scattered per second is then given by

$$v|N|^2 \int |f(\theta)|^2\,d\omega. \tag{22}$$

But if Q is the effective cross-section this number is given by the incident flux which impinges normally on an area Q. Hence

$$v|N|^2 Q = v|N|^2 \int |f(\theta)|^2\,d\omega,$$

so
$$Q = \int |f(\theta)|^2\,d\omega \tag{23}$$

and
$$I(\theta) = |f(\theta)|^2. \tag{24}$$

† See also MOTT, N. F. and MASSEY, H. S. W., *The theory of atomic collisions*, 3rd edn, Chap. ii (Clarendon Press, Oxford, 1965).

To obtain $f(\theta)$ we use the fact that, since F must be a proper function that depends on r and θ only, it may be expanded in a series of Legendre polynomials

$$F = r^{-1} \sum G_l(r) P_l(\cos\theta). \tag{25}$$

On substitution in (19) and use of the fact that the operator ∇^2 can be written

$$\nabla^2 = \frac{1}{r^2}\frac{d}{dr}\left(r^2 \frac{d}{dr}\right) - \frac{L^2}{r^2},$$

where L^2 operates only on θ and is such that

$$L^2 P_l(\cos\theta) = l(l+1) P_l(\cos\theta),$$

we obtain

$$\frac{d^2}{dr^2} G_l + \left\{k^2 - U(r) - \frac{l(l+1)}{r^2}\right\} G_l = 0. \tag{26}$$

We must now find a solution of this equation which vanishes as $r \to 0$ and which behaves asymptotically in such a way that (20) is satisfied. To do this we first note that†

$$e^{ikr\cos\theta} = \sum_l i^l (2l+1) j_l(kr) P_l(\cos\theta), \tag{27}$$

where the $j_l(kr)$ are the spherical Bessel functions‡

$$j_l(x) = (\pi/2x)^{\frac{1}{2}} J_{l+\frac{1}{2}}(x). \tag{28}$$

For large x,

$$j_l(x) \sim x^{-1} \sin(x - \tfrac{1}{2}l\pi). \tag{29}$$

It follows that (25) will have the correct asymptotic form if

$$G_l \sim i^l(2l+1) k^{-1} \sin(kr - \tfrac{1}{2}l\pi) + c_l e^{ikr}, \tag{30}$$

and then

$$f(\theta) = \sum_l c_l P_l(\cos\theta). \tag{31}$$

The c_l may now be determined as follows.

If $r^2 U(r)$ tends to zero for large r the general solution of (26) for large r can be written

$$G_l \sim r\{\alpha j_l(kr) + \beta j_{-l}(kr)\}, \tag{32}$$

where

$$j_{-l}(x) = (\pi/2x)^{\frac{1}{2}} J_{-l-\frac{1}{2}}(x) \sim x^{-1}(-1)^l \cos(x - \tfrac{1}{2}l\pi). \tag{33}$$

We may now write

$$G_l \sim a_l \sin(kr - \tfrac{1}{2}l\pi + \eta_l), \tag{34}$$

where

$$\tan \eta_l = (-1)^l \beta/\alpha. \tag{35}$$

Comparison with (30) shows that

$$a_l e^{-i\eta_l} = i^l(2l+1) k^{-1},$$

$$a_l e^{i\eta_l - \frac{1}{2}i\pi l} = 2ic_l + k^{-1} i^l(2l+1) e^{-\frac{1}{2}l\pi i},$$

so

$$c_l = (e^{2i\eta_l} - 1)(2l+1)/2ik,$$

and

$$f(\theta) = \frac{1}{2ik} \sum_l (2l+1)(e^{2i\eta_l} - 1) P_l(\cos\theta), \tag{36}$$

which, in view of (24), is in agreement with (18a).

† MOTT, N. F. and MASSEY, H. S. W., op. cit., 3rd edn, p. 21 (Clarendon Press, Oxford, 1965).
‡ WHITTAKER, E. T. and WATSON, G. N., *Modern analysis*, 4th edn, Chap. xvii (Cambridge University Press, 1927).

3.4. *The variation of the phase shifts with energy and angular momentum*†

In classical theory, the deviation produced by the scattering potential on the motion of a particle of impact parameter p will be small if

$$V(p) \ll \tfrac{1}{2}mv^2,$$

the kinetic energy of the particle. This may be expressed in a form appropriate for the quantum treatment as follows.

Corresponding to the angular momentum J, the impact parameter $p = J/mv$. We expect then that $\sin^2 \eta_l$ will be small for such values of l that

$$V(J/mv) \ll \tfrac{1}{2}mv^2,$$

where $J = \{l(l+1)\}^{\frac{1}{2}}\hbar$ is the quantized angular momentum corresponding to this value of l, i.e.

$$V(r_0) \ll \tfrac{1}{2}mv^2, \tag{37}$$

where
$$kr_0 \simeq \{l(l+1)\}^{\frac{1}{2}}. \tag{38}$$

This condition is the correct one to apply in the quantum theory, it being understood that, because of the uncertainty of position characteristic of the wave treatment, r_0 is to be interpreted as meaning all values of r in the neighbourhood of $\{l(l+1)\}^{\frac{1}{2}}/k$.

Before examining the consequences of this condition for the calculation of Q it is of interest to see how it arises in terms of the form of the de Broglie waves. In the classical theory, if the angular momentum about the centre of force is J, then the impact parameter is certainly J/mv. On the quantum theory the impact parameter is most likely to be J/mv, but there is a finite chance that it has some other value. This is exhibited by the form of the radial probability amplitude functions when the scattering potential is negligible. These are given by $g_l(r)/r$, where

$$\frac{d^2 g_l}{dr^2} + \left\{k^2 - \frac{l(l+1)}{r^2}\right\} g_l = 0 \tag{39}$$

and $g_l = 0$ at the centre $r = 0$. The term $l(l+1)r^2$ arises in the following way. In a Schrödinger equation there appears in its place $2mV/\hbar^2$, where V is the potential energy. If we substitute for V the energy of the centrifugal force $J^2/2mr^2$ with $J = \{l(l+1)\}^{\frac{1}{2}}\hbar$ then this term becomes $l(l+1)/r^2$ as in (39).

The forms of the functions g_l/r, which according to the analysis of § 3.3 are the spherical Bessel functions $j_l(kr)$, are illustrated in Fig. 6.2 for $l = 0, 1, 2, 3$. These exhibit the main feature that the first maximum appears close to $\{l(l+1)^{\frac{1}{2}}/k$ as l increases.

† See also MOTT, N. F. and MASSEY, H. S. W., op. cit., 3rd edn, Chap. ii (Clarendon Press, Oxford, 1965).

We can now see how the condition (37) for a small phase shift arises. Considering a particular value of l, it is clear that to introduce an extra half-wave, i.e. produce a phase shift of $\frac{1}{2}\pi$, the scattering potential must be large enough to affect the wavelength very markedly, at least as far as r_0, the first maximum, where $r_0 \simeq \{l(l+1)\}^{\frac{1}{2}}/k$. To produce such an effect at any point it must be comparable there with the kinetic energy

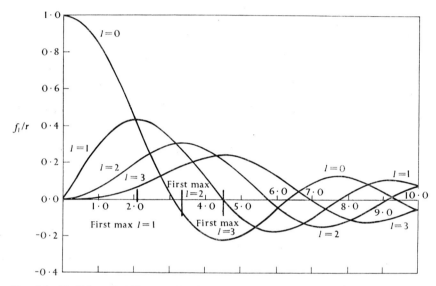

FIG. 6.2. Radial probability amplitude functions, $g_l(r)/r = (\pi/2kr)^{\frac{1}{2}} J_{l+\frac{1}{2}}(kr) = j_l(kr)$, when the scattering potential is negligible ($l = 0, 1, 2, 3$). The position of the first maximum is indicated in each case.

$\frac{1}{2}mv^2$, i.e. $V(r_0)$ must be comparable with, or greater than, $\frac{1}{2}mv^2$. With the condition (37) satisfied this is not so and no large phase shift is introduced.

We may now obtain a qualitative picture of the behaviour of η_l as the energy and angular momentum vary. Consider first a scattering potential of the form $V(r) = C/r^s$. Then the condition (37) for a small phase η_l is that $C/r_0^s \ll E$, where E is the kinetic energy. This requires that $r_0 \gg (C/E)^{1/s}$,

i.e.
$$l \gg E^{\frac{1}{2}-1/s} C^{1/s} (2m/\hbar^2)^{\frac{1}{2}} = l_0, \quad \text{say}. \tag{40}$$

From this result it follows that, if $s > 2$, all phases, except perhaps the zero-order one, must tend to zero as the kinetic energy tends to zero. It also follows that, at a given kinetic energy, all phases for l much greater than l_0 will be small and that the greater the value of E the larger the number of important phases in the series (16) and (18). For very low kinetic energies, only the zero-order phase will be important and the scattering cross-section will reduce simply to

$$Q_0 = \frac{4\pi}{k^2} \sin^2 \eta_0.$$

These conclusions remain valid for a scattering potential that falls off exponentially, as for an atomic field.

There is one further point which we must discuss, which is important in dealing with scattering by attractive potentials such as those of atoms. A particular angular momentum makes only a small contribution to the scattering if $\sin \eta_l$ is small. This is so, not only if η_l is nearly zero, but also if it is nearly equal to an integral multiple of π, say $s\pi$. It is largely conventional what we choose as the low-velocity limit of a phase, zero or $s\pi$, but there is one special choice that has real theoretical advantages. Although when the condition (37) is satisfied the amplitude of the wave function for the particles of lth order angular momentum will be small within the scattering field, except in very special circumstances that will be discussed shortly, the strength of the field may be so large that additional zeros of the wave function are introduced within it. This is illustrated in Fig. 6.3, which compares the plane wave functions for $l = 0$, 1, and 2 with the corresponding

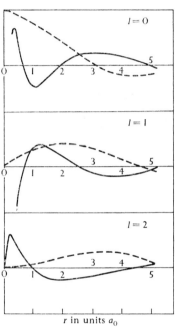

FIG. 6.3. Radial wave functions ($l = 0, 1, 2$) for krypton for $ka_0 = 1$ (13·54 eV) showing how the scattering field introduces additional zeros. ——— wave modified by field; - - - - plane wave.

ones for motion in the field of a krypton atom. It may occur even in the low-velocity limit. In view of this it is natural to define the phase η_l so that it tends to an integral multiple s of π, where s is the number of zeros introduced by the field in the wave function for zero-velocity particles. This has the great advantage that, in the limit of very fast collisions, the phase so defined will tend to zero. This will be so because, as the energy increases, the number of additional zeros will gradually decrease. A further advantage is that, with this convention, η_l is a steadily decreasing function of l for a fixed electron velocity and a given atomic field. This is to be expected because, with increase of l, the effective penetration into the field is reduced and hence also its effect in distorting the wave. Use will be made of this in § 4.6.

Fig. 6.4 (a) illustrates typical forms of variation of the phase η_0 with electron energy, following the above convention. Similar behaviour is characteristic of phases for other values of l.

Reference may be made also to Figs. 6.6 and 6.8, which illustrate the behaviour of the phases for certain atomic fields.

3.5. *Effective range expansions for* $k^{2l+1} \cot \eta_l$ †‡

We can specify the behaviour of the phase shifts at low energies more closely in terms of effective range expansions.

First we consider the case in which the scattering potential V falls off faster than any power of r, at large r. We then have, for all l,

$$k^{2l+1} \cot \eta_l = -\frac{1}{a_l} + \tfrac{1}{2}k^2 r_l + \text{terms of higher order,} \tag{41}$$

where a_l and r_l are constants known respectively as the *scattering length* and the *effective range* for angular momentum $l(l+1)^{\frac{1}{2}}\hbar$. It follows that

$$\lim_{k \to 0} Q = q_0 = 4\pi a^2, \tag{42}$$

where a is the scattering length for zero angular momentum.

As an indication of the mode of derivation of (41) we obtain the result for $l = 0$. The wave equation for this radial motion with wave number k_1 is

$$\frac{d^2 G_1}{dr^2} + (k_1^2 - U)G_1 = 0, \tag{43}$$

where $U = 2mV/\hbar^2$. Similarly, when the wave number is k_2

$$\frac{d^2 G_2}{dr^2} + (k_2^2 - U)G_2 = 0. \tag{44}$$

Hence multiplying (43) by G_2 and (44) by G_1, subtracting and integrating over r, gives

$$\left[G_2 \frac{dG_1}{dr} - G_1 \frac{dG_2}{dr} \right]_a^b = (k_2^2 - k_1^2) \int_a^b G_1 G_2 \, dr. \tag{45}$$

Similarly, if H_1, H_2 are the respective solutions of (43) and (44) when $U = 0$

$$\left[H_2 \frac{dH_1}{dr} - H_1 \frac{dH_2}{dr} \right]_a^b = (k_2^2 - k_1^2) \int_a^b H_1 H_2 \, dr. \tag{46}$$

† BETHE, H. A., *Phys. Rev.* **76** (1949) 38.
‡ BLATT, J. M. and JACKSON, J. D., ibid. **76** (1949) 18.

We now choose G_1, G_2 to satisfy

$$G_1(0) = 0, \qquad G_2(0) = 0,$$
$$G_1 \sim \sin k_1 r \cot \eta_0(1) + \cos k_1 r = H_1,$$
$$G_2 \sim \sin k_2 r \cot \eta_0(2) + \cos k_2 r = H_2. \tag{47}$$

Thus H_1 and H_2 are equal for all r to the asymptotic forms of the corresponding functions G_1, G_2.

Noting that

$$H_1(0) = H_2(0) = 1, \quad H_1'(0) = k_1 \cot \eta_0(1), \quad H_2'(0) = k_2 \cot \eta_0(2)$$

we have, subtracting (46) from (45) and making $a \to 0$, $b \to \infty$,

$$k_1 \cot \eta_0(1) - k_2 \cot \eta_0(2) = (k_2^2 - k_1^2) \int_0^\infty (G_1 G_2 - H_1 H_2)\,dr.$$

In the limit when $k_2 \to k_1$,

$$\frac{d}{dk^2}(k \cot \eta_0) = \int_0^\infty (H_1^2 - G_1^2)\,dr, \tag{48}$$

giving

$$k \cot \eta_0 = -\frac{1}{a} + \tfrac{1}{2} k^2 r_0 + O(k^4), \tag{49}$$

where

$$\tfrac{1}{2} r_0 = \int_0^\infty (H_1^2 - G_1^2)\,dr. \tag{50}$$

Next we consider the case in which the potential V falls off as r^{-s} where $s > 2$. We then find that, if $l < \tfrac{1}{2}(s-3)$,

$$\lim_{k \to 0} k^{2l+1} \cot \eta_l = -\frac{1}{a_l}, \tag{51}$$

but the effective range does not exist unless $l < \tfrac{1}{2}(s-5)$. For $l > \tfrac{1}{2}(s-3)$

$$\lim_{k \to 0} k^{s-2} \cot \eta_l = \text{const.} \tag{52}$$

An important case arises in practice owing to polarization of an atom by an electron during impact (see Chap. 8, §§ 2.3, 2.4). In this case $s = 4$ so for all $l > 0$

$$\lim_{k \to 0} k^2 \cot \eta_l = \text{const.} \tag{53}$$

For $l = 0$, O'Malley, Spruch, and Rosenberg† have shown that

$$k \cot \eta_0 = -\frac{1}{a} + \frac{\pi \alpha k}{3a^2 a_0} + \frac{2\alpha k^2}{3a a_0} \ln\left(\frac{\alpha k^2}{16 a_0}\right) + O(k^2) \tag{54a}$$

† O'Malley, T. F., Spruch, L., and Rosenberg, L., J. Math. Phys. 2 (1961) 491.

for a potential having the asymptotic form

$$V \sim -\frac{1}{2}\frac{\alpha e^2}{r^4}.$$

Thus the scattering length exists but not the effective range.

For such a potential,

$$k^2 \cot \eta_1 = \frac{15a_0}{\pi\alpha}\{1+a_1 k+O(k^2)\} \tag{54 b}$$

in agreement with (53), and in general

$$k^2 \cot \eta_l = \{(2l+3)(2l+1)(2l-1)a_0/\pi\alpha\}(1+O(k^3)), \quad l > 1. \tag{54 c}$$

It follows that[†] for the scattering of low-energy electrons by a polarized atomic field the total cross-section may be expanded in the form

$$Q_0^t = 4\pi\{a^2+(2\pi/3a_0)\alpha ak+(8/3a_0)\alpha a^2 k^2 \ln(ka_0)+bk^2+...\}, \tag{55 a}$$

while the momentum-loss cross-section (18 b) is

$$Q_d = 4\pi\{a^2+(4\pi/5a_0)\alpha ak+(8/3a_0)\alpha a^2 k^2 \ln(ka_0)+ck^2+...\}, \tag{55 b}$$

and the differential cross-section (18 a) is

$$I(\theta)\,d\omega = \{a^2+(\pi/a_0)\alpha ak \sin \tfrac{1}{2}\theta+(8/3a_0)\alpha a^2 k^2 \ln(ka_0)+O(k^2)\}\,d\omega. \tag{55 c}$$

3.6. *Relation of the phase shifts to number of bound states*

It is of interest to note the connection here between the scattering power of the field and the bound energy levels that can exist within it.[‡] If, with the convention described above, the phase $\eta_l \to s\pi$ as the velocity tends to 0, then s bound energy levels exist with angular momentum $\{l(l+1)\}^{\frac{1}{2}}\hbar$. A special case arises when the uppermost energy level falls at the energy zero. In that case the wave function is not small within the range of the scattering potential but very large, and the corresponding value of $\sin^2 \eta_l$ falls off more slowly with velocity. Indeed, for $l = 0$ it tends in the low-velocity limit to unity, giving an infinite cross-section. This resonance effect is quite a sharp one and can be ignored in most of the applications to atoms.

A repulsive scattering potential, as it increases the local wavelength, cannot introduce any new waves but can only cancel some that would otherwise be present within its range of action. In the low-velocity limit, as the unperturbed wavelength is infinitely great, there is no wave to eliminate and all the phases tend to zero according to our convention.

[†] O'Malley, T. F., *Phys. Rev.* **130** (1963) 1020.

[‡] See Mott, N. F. and Massey, H. S. W., *The theory of atomic collisions*, 3rd edn, pp. 156–9 (Clarendon Press, Oxford, 1965).

This corresponds to the fact that in a repulsive field no bound energy levels can exist.

The close relationship between the phase shifts and the existence and energies of bound states can be expressed in terms of formulae of the type (49). Thus suppose that a bound state of energy $-\kappa^2\hbar^2/2m$ and zero angular momentum exists. Then (49) may be expressed in the form

$$k\cot\eta_0 = -\kappa + \tfrac{1}{2}(k^2+\kappa^2)r_b + \text{higher order terms}, \qquad (56)$$

involving an expansion in powers of $k^2+\kappa^2$ instead of k^2. If κ is small the convergence will still be good. In this way the phase shift may be determined from the known energy of a bound state which lies close to the threshold.

In terms of the analysis above (56) is obtained by taking in (49) $k \to -i\kappa$ so, since $G_1 \sim e^{-\kappa r}$, $\cot\eta_0(1) = -i$ and $H_1 = e^{-\kappa r}$.

It also follows from (49) that

$$\kappa = \frac{1}{a} + \tfrac{1}{2}\kappa^2 r_b + \text{higher order terms}, \qquad (57)$$

so that, from knowledge of the scattering length and effective range, the energy of a bound state close below the threshold can be determined.

Although, if the interaction falls off as r^{-4}, the effective range r_0 does not exist, the range r_b does exist and (57) remains valid.

3.7. 'Classical' approximation for the phase η_l

In order to test whether the theory we have discussed is capable of describing any of the main features of the observed cross-sections, it is necessary to have some convenient method for estimating the phase shifts that are produced by a particular atomic field. This may be done by taking account of the fact that the variation of the potential $V(r)$ within a wavelength is small, except for very low electron energies.

The general form of the radial amplitude function for a given angular momentum, in the absence of a scattering potential, is illustrated in Fig. 6.2. The number of wavelengths between the scatterer and any great distance R from it would be given by R/λ if λ were a constant. When λ varies we may define a local wavelength λ_a, valid between r and $r+dr$, such that

$$\lambda_a = h \bigg/ \left[2m\left\{E - \frac{\hbar^2}{2m}\frac{l(l+1)}{r^2}\right\}\right]^{\tfrac{1}{2}}.$$

Under the conditions stated above the number of wavelengths will become

$$\int_{r_1}^{R} dr/\lambda_a.$$

The lower limit r_1 may be taken as the greatest distance at which $1/\lambda_a$ vanishes. This amounts to ignoring the small contribution from that part of the wave that exists within the classical closest distance of approach.

When the scattering potential is present, the number of wavelengths, to the same approximation, will be changed to

$$\int_{r_2}^{R} \mathrm{d}r/\lambda'_a$$

where
$$\lambda'_a = h \bigg/ \left[2m \left\{ E - V - \frac{\hbar^2}{2m} \frac{l(l+1)}{r^2} \right\} \right]^{\frac{1}{2}}$$

and r_2 is the greatest distance at which $1/\lambda'_a$ vanishes.

The phase shift is equal to 2π times the difference in the number of waves between the scatterer and infinity so that, approximately,

$$\eta_l = \int_{r_2}^{\infty} \left\{ k^2 - \frac{2mV}{\hbar^2} - \frac{l(l+1)}{r^2} \right\}^{\frac{1}{2}} \mathrm{d}r - \int_{r_1}^{\infty} \left\{ k^2 - \frac{l(l+1)}{r^2} \right\}^{\frac{1}{2}} \mathrm{d}r. \quad (58)$$

With this approximation the phase will be given with the appropriate multiple of π, according to our convention. This is because the formula is derived by counting the actual number of waves present with and without the field.

This approximation may be derived more rigorously from Jeffreys's approximate solution[†] of a second-order differential equation such as (26). It is very useful for the study of the scattering of electrons by atoms, as in most circumstances it gives at least a fair estimate of the phase considered. For $l = 0$ the formula (58) clearly breaks down as no zero r_1 or r_2 exists. A more accurate approximation, given by Langer,[‡] in which $l(l+1)$ in (58) is replaced by $(l+\frac{1}{2})^2$ may be used, however, even for $l = 0$.

When the phase is small a more accurate approximation, due to Mott,[§] may be used. It gives

$$\eta_l = -\tfrac{1}{2}\pi \frac{2m}{\hbar^2} \int_0^{\infty} V(r)\{J_{l+\frac{1}{2}}(kr)\}^2 r \, \mathrm{d}r, \quad (59)$$

where $J_{l+\frac{1}{2}}(kr)$ is the usual Bessel function.

[†] JEFFREYS, H., *Proc. Lond. math. Soc.* (2) **23** (1923) 428; MOTT, N. F. and MASSEY, H. S. W., *The theory of atomic collisions*, 3rd edn, Chap. i, § 6 (Clarendon Press, Oxford, 1965).
[‡] LANGER, R. E., *Phys. Rev.* **51** (1937) 669.
[§] MOTT, N. F., *Proc. Camb. phil. Soc. math. phys. Sci.* **25** (1929) 304.

THEORY OF ELASTIC SCATTERING

When greater accuracy is required, the phases may be calculated by numerical solution of the appropriate differential equation.

3.8. *Variational methods for determining the phase η_l*†

In § 3.4 we have discussed the close relationship between the phase shifts and the energies of the bound states for a given angular momentum. As one of the most effective means of obtaining good approximate values of binding energies has come from applications of the calculus of variations it is natural to seek similar methods for obtaining approximate phase shifts.

One of the most valuable features of the variational method for determining the energy of the lowest state is that it always gives an upper bound to this energy. The problem is less definite in this respect for dealing with excited states. It seems at least likely that a variational method can be found to give a lower bound to the scattering length when there are no bound states because the lowest state then is the one with zero energy.

These expectations are largely justified and variational methods have proved to be very powerful in dealing with collision phenomena. We shall indicate here how the method operates in the simplest case, that of approximately determining phase shifts, but we shall have occasion to refer to applications in much more complicated circumstances (Chap. 8, § 4.1).

We consider the approximate determination of the phase η_0. The wave function G_0 from which η_0 is obtained will satisfy

$$\frac{d^2 G_0}{dr^2} + \{k^2 - U(r)\} G_0 = 0, \tag{60}$$

where $U = 2mV/\hbar^2$. In addition

$$G_0(0) = 0, \qquad G_0 \sim A \sin(kr + \eta_0). \tag{61}$$

As usual we assume that $U(r)$ vanishes faster than r^{-2} as $r \to \infty$. The variational relation on which the method depends is based on the integral

$$I = \int_0^\infty G\left\{\frac{d^2}{dr^2} + k^2 - U(r)\right\} G \, dr, \tag{62}$$

which obviously vanishes when $G = G_0$. We now ask what will be the value δI of I if a trial function G_t is substituted for G_0, such that

$$G_t(0) = 0, \qquad G_t \sim A' \sin(kr + \eta_t). \tag{63}$$

† See MOTT, N. F. and MASSEY, H. S. W., op. cit., 3rd edn, Chap. vi (Clarendon Press, Oxford, 1965).

If $\delta G\ (= G_t - G_0)$ is small then
$$\delta G \sim \delta A \sin(kr+\eta_0) + A\cos(kr+\eta_0)\,\delta\eta, \tag{64}$$
where $\qquad \delta A = A' - A, \qquad \delta\eta = \eta_t - \eta_0.$

Substituting $G_t = G_0 + \delta G$ in (62) we find

$$\begin{aligned}\delta I &= \int_0^\infty \delta G\left\{\frac{d^2}{dr^2}+k^2-U(r)\right\}G_0\,dr + \int_0^\infty G\left\{\frac{d^2}{dr^2}+k^2-U(r)\right\}\delta G\,dr \\ &= \int_0^\infty \left\{G\frac{d^2}{dr^2}(\delta G) - \delta G\frac{d^2 G}{dr^2}\right\}dr \\ &= \left[G\frac{d}{dr}(\delta G) - \delta G\frac{dG}{dr}\right]_0^\infty \\ &= -Ak^2\,\delta\eta. \end{aligned} \tag{65}$$

The variational principle which we use depends on the choice of the normalizing factor A. One of the most useful choices is to take $A = \sec\eta_0$ so
$$G \sim \sin kr + \tan\eta_0 \cos kr, \tag{66}$$
and (65) becomes $\qquad \delta I = -k\,\delta(\tan\eta_0). \tag{67}$

This gives Kohn's variational method.† To apply it a trial function is chosen which satisfies
$$G_t(0) = 0, \qquad G_t \sim \sin kr + \lambda \cos kr$$
and contains n adjustable parameters $c_1,\ldots, c_n$. Using this function the integral I_t is calculated where

$$I_t = \int_0^\infty G_t\left\{\frac{d^2}{dr^2}+k^2-U(r)\right\}G_t\,dr. \tag{68}$$

The condition (67) will be satisfied to the first order if $c_1,\ldots, c_n, \lambda$, are chosen so that
$$\frac{\partial I_t}{\partial c_r} = 0 \qquad (r = 1,\ldots, n), \tag{69 a}$$

$$\frac{\partial I_t}{\partial \lambda} = -k. \tag{69 b}$$

In addition we should also have $\partial I_t/\partial A = 0$ which requires
$$I_t = 0. \tag{70}$$

As there are only $n+1$ variables and $n+2$ equations it is not possible to satisfy all the conditions unless G_t is an exact solution. In Kohn's

† KOHN, W., *Phys. Rev.* **74** (1948) 1763.

method no attempt is made to satisfy $I_t = 0$, the $n+1$ variables being determined from the linear equations (69). Having determined these the best approximation for the phase shift is to use (67) once more so that
$$\tan \eta_0 \approx k^{-1} I_t + \lambda, \tag{71}$$
correct again to the first order.

An alternative introduced by Hulthén† is to reject the equation (69 b) and satisfy (70) instead. In that case the resulting value of λ gives the desired approximation to $\tan \eta_0$. This method suffers from the practical complication of requiring solution of a quadratic equation and n linear equations instead of $(n+1)$ linear equations.

In the limit of zero energy ($k \to 0$) the Kohn method gives a variational approximation to the scattering length such that
$$a = a_t - I_t, \tag{72}$$
with
$$G_t(0) = 0, \quad G_t \sim a_t - r. \tag{73}$$
It may be shown then‡ that if there are no bound states the approximation found for a is an upper bound. Introduction of more flexibility in the trial function, through introduction of more parameters, will improve the approximation but the convergence will always be from above. The situation, as might be expected, is exactly similar to that which appears in variational approximation to the binding energy of the lowest state.

If one or more bound states exist in the field of the scattering centre the situation is less simple. Considering first the case of one bound state, suppose that ψ_t is a trial function which, when used to give an approximation to the energy of this state, leads to a value ϵ_t which is < 0. Then an upper bound to the scattering length is given by§
$$a_t = I_t + \epsilon_t^{-1} \left\{ \int_0^\infty \psi_t \left(\frac{d^2}{dr^2} - U(r) \right) G_t \, dr \right\}^2. \tag{74}$$
Alternatively, if the trial function G_t can be written in the form
$$G_t = G_{t'} + b\psi_t, \tag{75}$$
where b is an adjustable parameter and $G_{t'}$ satisfies the same boundary conditions as G_t, then the scattering length calculated by Kohn's method is again an upper bound. Generalization of these results to cases of more than one bound state is not difficult.

† Hulthén, L., *K. fysiogr. Sällsk. Lund Förh.* **14** (1944) 1.
‡ Spruch, L. and Rosenberg, L., *Phys. Rev.* **116** (1959) 1034.
§ Rosenberg, L., Spruch, L., and O'Malley, T. F., ibid. **118** (1960) 184.

It is important to remember that the upper bound refers only to the scattering length and not to the approximate phase shifts determined at finite energies. Much more elaborate procedures must be introduced if bounded approximations to such phase shifts are required.

Hulthèn's method gives upper bounds under the same conditions as does that of Kohn but unlike the latter it cannot be assumed that use of a more flexible trial function will improve the approximation.

Another method that has received considerable application is that of Rubinow,† which is based on choice of A as $\operatorname{cosec} \eta_0$ in (61), so that

$$G \sim \cot \eta_0 \sin kr + \cos kr, \qquad (76)$$

and
$$\delta(I - k \cot \eta_0) = 0. \qquad (77)$$

With a trial function
$$G_t \sim \mu \sin kr + \cos kr, \qquad (78)$$

the approximation to the phase shift is given by
$$\cot \eta_0 \approx -k^{-1} I_t + \mu. \qquad (79)$$

In the low-energy limit it gives
$$\frac{1}{a} \simeq \frac{1}{a_t} + I_t. \qquad (80)$$

Unlike Kohn's method it does not give an upper bound to the scattering length.

Many other principles may be introduced depending on the choice of A but most emphasis is placed on Kohn's method because of its minimal properties and the ease of generalization to more complex, many-body, problems. For the calculation of phase shifts for scattering by central fields it is so readily possible to use rapid means of accurate numerical integration of the appropriate differential equation that a variational approximation is rarely needed. We have introduced it here so that the procedure can be followed for the simplest case—generalization to quite complex cases can usually be made without difficulty (see Chap. 8, § 4.1).

3.9. *Scattering by a Coulomb field*‡

We have explicitly excluded from consideration cases in which the scattering potential $V(r)$ does not fall faster than r^{-2} at large distances r. However, if we are to discuss the collisions of electrons with positive

---

† Rubinow, S. I., ibid. **98** (1955) 183.
‡ See Mott, N. F. and Massey, H. S. W., op. cit., 3rd edn, Chap. iii (Clarendon Press, Oxford, 1965).

6.3 THEORY OF ELASTIC SCATTERING

ions we must be prepared to deal with cases in which

$$V(r) \sim -Ze^2/r, \qquad (81)$$

where Ze is the effective nuclear charge.

We begin by discussing first the case in which $V(r)$ is of the form (81) for all r, the case of a pure Coulomb field. Consider the scattering by this field of an electron incident along the z-direction with wave number k. The wave function describing the collision can no longer be described in terms of an incident plane wave e^{ikz} plus an outgoing spherical wave with asymptotic form $r^{-1}e^{ikr}f(\theta)$. It is found that asymptotically the function must have the form

$$\exp\{ikz + i\alpha \ln k(r-z)\} + r^{-1}\exp\{ikr - i\alpha \ln 2kr\}f_c(\theta), \qquad (82)$$

where $\alpha = -Ze^2 m/k\hbar^2 = -Ze^2/\hbar v$, v being the electron velocity.

The first term, which corresponds to the incident plane wave in the cases we have considered hitherto, shows that, even at infinity, the incident wave is distorted by the field it is to encounter. This can be understood in relation to classical theory. At infinity we would expect the wave-fronts to be normal to the classical trajectories which, in the pure Coulomb case, are the asymptotes of hyperbolae pointing along the z-direction. It may be shown that in fact the wave-fronts lie on surfaces defined by

$$z - \frac{Ze^2}{mv^2}\ln\{k(r-z)\} = \text{const.} \qquad (83)$$

and not on the surfaces $z = \text{const.}$ Similar arguments apply to the asymptotic form of the outgoing waves.

Having obtained a solution which is finite at the origin and satisfies (82) the differential scattering cross-section is given as usual by

$$I(\theta)\,d\omega = |f_c(\theta)|^2\,d\omega.$$

Exact solution of the wave equation

$$\nabla^2\psi + (k^2 + 2mZe^2/\hbar^2 r)\psi = 0 \qquad (84)$$

is possible and gives

$$f_c(\theta) = -\frac{Ze^2}{2mv^2}\operatorname{cosec}^2 \tfrac{1}{2}\theta \exp[-i\alpha \ln\{\tfrac{1}{2}(1-\cos\theta)\} + i\pi + 2i\eta_0], \qquad (85)$$

where

$$e^{2i\eta_0} = \Gamma(1+i\alpha)/\Gamma(1-i\alpha). \qquad (86)$$

With the form (85) for $f_c(\theta)$ the differential cross-section gives the familiar Rutherford scattering formula

$$I(\theta)\,d\omega = \frac{Z^2 e^4}{4m^2 v^4}\operatorname{cosec}^4 \tfrac{1}{2}\theta\,d\omega. \qquad (87)$$

We shall denote the solution of the equation (84) which is finite at the origin and satisfies (82) by $\psi_c^Z(k; r, \theta)$. The corresponding function that represents a wave incident in the direction (λ, μ) will be given by

$$\psi_c^Z(k; r, \Theta), \tag{88}$$

where $\qquad \cos\Theta = \cos\theta\cos\lambda + \sin\theta\sin\lambda\cos(\phi-\mu).$

When $r = 0$, $\qquad |\psi_c^Z|^2 = 2\pi\alpha/(e^{2\pi\alpha}-1) = C_0^2, \quad \text{say}. \tag{89}$

If $\qquad |\alpha| = |Z|e^2/\hbar v \ll 1 \tag{90}$

then $|\psi_c^Z|^2 \to 1$ as for an unmodified plane wave. Under these circumstances the scattering by the Coulomb field may be treated by Born's approximation (Chap. 7) both as regards the amplitude and phase of the scattered wave. Even when $|\alpha|$ is not small Born's approximation gives the amplitude correctly. When $|\alpha|$ is large classical theory would be expected to be valid but it also gives the amplitude correctly for all $|\alpha|$. This fortuitous coincidence of classical and exact quantum-theoretical results for the differential cross-section for scattering by a pure Coulomb field is very fortunate in view of the major role played by the Rutherford scattering formula in supporting the nuclear model of the atom.

When α is large and positive (repulsive Coulomb field) $|\psi_c^Z|^2 \ll 1$ for small r, while when α is large in magnitude and negative, $|\psi_c^Z|^2$ is of order $|\alpha|$ and is much larger than for a plane wave.

The amplitude $f_c(\theta)$ appearing in (85) may be expressed in terms of phase shifts in the same form as (36), viz.

$$f_c(\theta) = \frac{1}{2ik}\sum_l (e^{2i\eta_l}-1)(2l+1)P_l(\cos\theta), \tag{91}$$

with $\qquad e^{2i\eta_l} = \Gamma(1+l+i\alpha)/\Gamma(1+l-i\alpha). \tag{92}$

These phase shifts are such that the solution of the equation

$$\frac{d^2 G_l}{dr^2} + \left\{k^2 + \frac{2mZe^2}{\hbar^2 r} - \frac{l(l+1)}{r^2}\right\}G_l = 0$$

which vanishes at $r = 0$ has the asymptotic form

$$G_l^c \sim \sin(kr - \tfrac{1}{2}l\pi - \alpha\ln 2kr + \eta_l). \tag{93a}$$

We distinguish also a second solution H_l^c chosen so as to have the asymptotic form

$$H_l^c \sim \cos(kr - \tfrac{1}{2}l\pi - \alpha\ln 2kr + \eta_l). \tag{93b}$$

3.10. Scattering by a modified Coulomb field†

We now consider the scattering by a potential $V(r)$, which has the asymptotic form $-Ze^2/r$ but which departs from this form at smaller distances. In that case the solution of the equation

$$\frac{d^2 G_l}{dr^2} + \left\{ k^2 - \frac{2mV(r)}{\hbar^2} - \frac{l(l+1)}{r^2} \right\} G_l = 0, \qquad (94)$$

which vanishes at $r = 0$, will have the asymptotic form

$$G_l \sim \sin(kr - \tfrac{1}{2}l\pi - \alpha \ln 2kr + \eta_l + \sigma_l), \qquad (95)$$

where σ_l is an *additional* phase shift due to departure of the potential from the Coulomb form at small distances.

The differential cross-section for the scattering now takes the form

$$I(\theta) \, d\omega = |f_c(\theta) + f_m(\theta)|^2 \, d\omega, \qquad (96)$$

where $f_c(\theta)$ is the amplitude (91) arising for a pure Coulomb field and $f_m(\theta)$ is the additional contribution due to departure from the Coulomb force at small distances. It is given by

$$f_m(\theta) = \frac{1}{2ik} \sum_l (2l+1) e^{2i\eta_l}(e^{2i\sigma_l}-1) P_l(\cos\theta). \qquad (97)$$

The ratio of the scattered intensity to that for a pure Coulomb field is given by

$$R = |1+N|^2, \qquad (98)$$

where

$$N = (2\hbar v/Ze^2) \sin^2 \tfrac{1}{2}\theta \exp(i\alpha \ln \sin^2 \tfrac{1}{2}\theta) \times$$
$$\times \sum (2l+1)(e^{2i\sigma_l} \sin \sigma_l) e^{2i(\eta_l - \eta_0)} P_l(\cos\theta). \qquad (99)$$

Effective range expansions may be obtained in terms of the phases σ_l but because of the long range of the Coulomb field they do not apply to $k^{2l+1} \cot \sigma_l$. Instead we take‡§

$$M = k^{2l+1}\{(2l+1)!!\}^2 C_l^2 \left[\cot \sigma_l + \frac{2\alpha}{C_0^2} \{g(\alpha) - \ln|\alpha|\} \right], \qquad (100)$$

where

$$C_l = \frac{2^l C_0}{(2l+1)!} \prod_{s=1}^{l} (s^2 + \alpha^2)^{\frac{1}{2}}, \qquad (101)$$

$$g(\alpha) = -\gamma + \sum_s \frac{\alpha^2}{s(s^2+\alpha^2)}, \qquad (102)$$

and γ is Euler's constant.

---

† See MOTT, N. F. and MASSEY, H. S. W., op. cit., 3rd edn, Chap. iii, § 6 (Clarendon Press, Oxford, 1965).
‡ BETHE, H., *Phys. Rev.* **76** (1949) 38.
§ GAILITIS, M., *Zh. éksp. teor. Fiz.* **44** (1963) 1974; *Soviet Phys. JETP* **17** (1963) 1328.

It may then be shown that M, which tends to $k^{2l+1}\cot\sigma_l$ when $\alpha \to 0$, may be expanded about $k = 0$ in the form

$$M = M_0 + M_1 k^2 + \ldots. \tag{103}$$

Moreover, this may be continued to negative values of k^2 so as to relate the phase shifts σ_l to the energies of the bound states and obtain very useful results. In a field which has the asymptotic form of a Coulomb field there will be an infinite discrete set of bound states for each value of l. These states can be classified in terms of the limit to which they tend when the field becomes a pure Coulomb one at all distances. Thus we may write for the energy of a state

$$E_{nl} = -\frac{2\pi^2 m e^4}{\nu_{nl}^2 h^2}, \tag{104}$$

where $\nu_{nl} = n - \mu_{nl}$. μ_{nl} tends to zero in the limit of the pure Coulomb field and is known as the *quantum defect*.

It may then be shown,† by taking the limit of M from above and below $k^2 = 0$, that

$$\lim_{n\to\infty} \cot \pi\mu_{nl} = \lim_{k\to 0} \cot \sigma_l. \tag{105}$$

Furthermore it is possible to regard μ_{nl}, by analytical continuation, as a continuous function μ_l of κ^2 where $-2m\kappa^2/\hbar^2$ is the energy of a bound state. If this function is extrapolated to small positive energy $k^2\hbar^2/2m$ then (105) may be extended to give

$$\frac{\cot\sigma_l(k^2)}{1-e^{2\pi\alpha}} = \cot\pi\mu_l(k^2). \tag{106}$$

Seaton‡ has shown that this extrapolation may be carried out most effectively in terms of a function $Y_l(k^2)$ which is such that

$$A_l(k^2)Y_l(k^2) = \tan\{\pi\mu(k^2)\}, \tag{107 a}$$

where
$$A_l(k^2) = \prod_{s=0}^{l}(1+s^2/\alpha^2). \tag{107 b}$$

The bound state energies (104) can be regarded as arising from solutions of the equation

$$\tan\{\pi\nu(k^2)\} = -\tan\{\pi\mu(k^2)\} = -A_l(k^2)Y_l(k^2). \tag{108}$$

It may be proved that $Y_l(k^2)$ can be expanded in the form

$$Y_l(k^2) = \frac{\sum a_m k^{2m}}{\sum b_s k^{2s}} \tag{109}$$

both for $k^2 \gtrless 0$. Having obtained $\mu(-\kappa_n^2)$ from the spectroscopic data,

† SEATON, M. J., *Mon. Not. R. astr. Soc.* **118** (1958) 504.
‡ SEATON, M. J., *Proc. phys. Soc.* **88** (1966) 815.

for different values of the total quantum number n, $Y_l(-\kappa_n^2)$ is obtained from (108). Using (109) values of a_m, b_s with m and s small, are obtained which give the best fit to Y_l. The form (109) is then used for extrapolation to $k^2 > 0$.

4. Application to calculation of the broad features of elastic cross-sections of atoms

In Chapter 1, § 6.1, the measured total cross-sections of different atoms towards electrons are described. We can now see how the broad features of these observations may be understood. This discussion will be based on the formula for the elastic scattering cross-section

$$Q_0 = \sum q_l,$$

where
$$q_l = (4\pi/k^2)(2l+1)\sin^2 \eta_l$$

and is known as the lth-order partial cross-section.

4.1. *The Ramsauer–Townsend effect*

Let us consider first the behaviour of very slow electrons. If the energy is low enough the only important phase will be η_0 and

$$Q_0 = (4\pi/k^2)\sin^2 \eta_0.$$

As $k \to 0$, $\eta_0 \to s\pi$ where s is, in general, a whole number determined by the strength of the atomic field. For a weak field $s = 0$ and the variation of η_0 with k is of the form of curve I in Fig. 6.4 (a). As the energy of the electron increases the phase η_0 rises to a maximum that will be less than π; the field is not strong enough to introduce an additional wave even at zero energy. The corresponding form for q_0 that can occur in this case is illustrated in Fig. 6.4 (b). In the low-velocity limit q_0 will be finite as $\sin \eta_0$, in general, tends to zero as k. As $\sin \eta_0$ is never zero at any higher energy the cross-section is also never zero. Clearly the Ramsauer–Townsend effect does not arise in these cases.

For a stronger field, for which $s = 1$, the variation of η_0 with k can have the form either of curve II or curve III of Fig. 6.4 (a). With curve II q_0 is never zero except in the special case in which it vanishes as $k \to 0$. On the other hand, with curve III q_0 returns to the value π at a finite electron energy and the variation of q_0 with k is of the form typical of the Ramsauer–Townsend effect. Similar possibilities arise when $s = 2$, or more. The vanishing cross-section occurs because, at a particular energy, the scattering potential is just strong enough to introduce a whole number of additional waves within its range at this energy.

One further condition for the appearance of the Ramsauer–Townsend effect, which must not be overlooked, is that at the zero of q_0 the contributions to Q from the partial cross-sections q_1, q_2,..., etc., must be negligible.

A detailed confirmation of the explanation has been given by Holtsmark,† who calculated the phases, by accurate numerical solution of the differential equations, for scattering of electrons by argon atoms. He

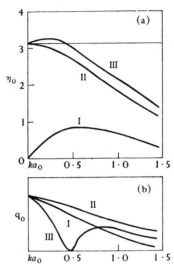

FIG. 6.4. (a) Typical forms of variation of the phase η_0 with electron energy. Curve I refers to a field that is not strong enough to introduce an additional zero, so that there is no stationary s-state in this field. The phase η_0 therefore tends to zero in the low velocity limit. Curves II and III refer to a field strong enough to introduce one additional zero so that one stationary s-state exists in the field and the phase η_0 tends to π in each case, from below in Curve II and from above in Curve III. (b) Variation of the corresponding cross-sections $q_0 = (4\pi/k^2)\sin^2\eta_0$ with electron energy. Curves I, II, and III refer to the same conditions as the corresponding curves of (a).

found that, whereas the scattering potential given by the Hartree field did not lead to the Ramsauer–Townsend effect, a modification to this field, which was arrived at by introducing a polarization correction, did give very good agreement with observation. Fig. 6.5 illustrates the modification that was introduced. This modification essentially represents a correction to the approximation that regards the atomic field as unperturbed by the incident electron. It was obtained by assuming that the polarization effectively introduced an additional attraction that fell off, at large distances r from the atom, as $-\frac{1}{2}\alpha e^2/r^4$, where α is the

† HOLTSMARK, J., *Z. Phys.* **55** (1929) 437.

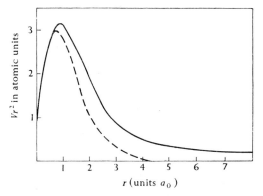

FIG. 6.5. Field used by Holtsmark in the calculation of electron scattering in argon. Quantities are expressed in atomic units. ---- Hartree field; —— Hartree field with polarization.

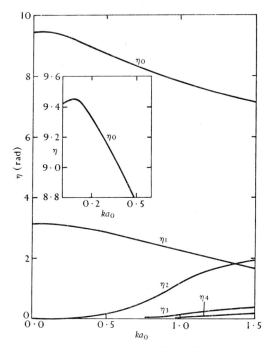

FIG. 6.6. Phases for argon using the field of Fig. 6.5 'with polarization'. The variation of η_0 with electron energy is clearly similar to case III of Fig. 6.4 (a). η_0 tends to 3π at zero energy.

polarizability of the argon atom ($= 10\cdot819$ atomic units). This asymptotic form was smoothly joined to the unmodified field so that the polarization was negligible at distances of the order $0\cdot6a_0$ from the nucleus.

The phases calculated by Holtsmark for this field are illustrated in Fig. 6.6, and Fig. 6.7 shows how the Ramsauer–Townsend effect is reproduced by substitution in the formula (16). Reference to the phases shows that, for argon, $s = 3$.

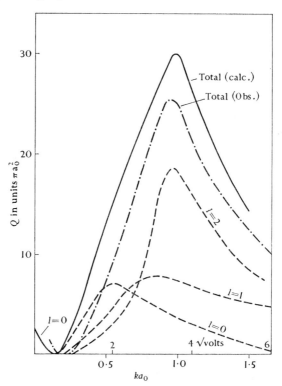

FIG. 6.7. Partial and total cross-sections for elastic scattering in argon calculated from the phases of Fig. 6.6. The Ramsauer–Townsend effect follows from the variation of η_0 and hence of q_0 with electron energy. The strong maximum in the total elastic cross-section for electrons of about 13-eV energy is due to η_2 passing through $\tfrac{1}{2}\pi$ at this energy. – – – – partial cross-section for $l = 0, 1, 2$; ——— total calculated cross-section; — · — · — measured results.

It is now of interest to consider the contribution from higher-order cross-sections. These are very small at the Ramsauer–Townsend minimum, but reference to Fig. 6.7 shows that at higher energies first q_1 and then q_2 becomes important. The maximum observed in the cross-section for electrons of about 13-eV energy is largely due to q_2. Higher-order phases never produce a marked individual effect, for the atomic field of argon is not strong enough to produce, at any energy, a large phase shift in the waves associated with more than two units of angular momentum. It will be noted that the second-order cross-section is dominant when the corresponding phase η_2 is nearly equal to $\tfrac{1}{2}\pi$, as would be expected.

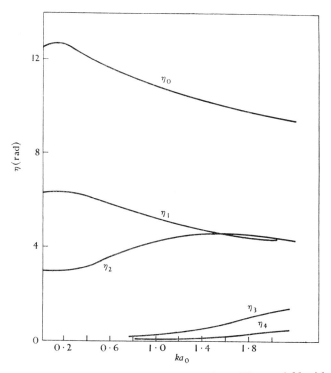

FIG. 6.8. Phases calculated for krypton using a Hartree field with polarization. η_0 tends to 4π at zero energy. Note that the η_0, η_1, η_2 phases are approximately equal to those for argon (Fig. 6.6) with the addition of π. (To convert from ka_0 to $\sqrt{(\text{volt})}$ multiply by 3·68.)

4.2. *Similar behaviour of the heavier rare gases*

The reason why the heavier rare gases also produce a Ramsauer–Townsend effect can be traced to the quasi-periodic behaviour of the cross-section q_0 as the atomic number of the scattering atom changes. In going from argon to krypton the atomic field becomes just so much stronger that, for low-energy collisions, one whole additional wavelength is added within the range of the field. This increases the zero-order phase by π but does not change q_0. Similarly, in going from krypton to xenon, the zero-order phase increases again by nearly π. Confirmation of this is provided by the detailed calculations of Holtsmark† for krypton, using

† HOLTSMARK, J., *Z. Phys.* **66** (1930) 49.

again a Hartree field modified by a polarization correction. The phases which he obtained are illustrated in Fig. 6.8 and the corresponding cross-sections in Fig. 6.9.

Argon, krypton, and xenon give similar cross-sections, not only for low-velocity electrons, but also over a wide energy range. Thus for all three a maximum appears at an electron energy near 13 eV. This is due

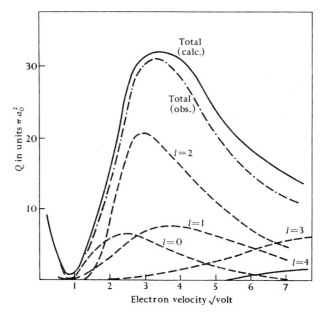

Fig. 6.9. Partial and total cross-sections for elastic scattering in krypton calculated from the phases of Fig. 6.8. – – – – partial cross-sections for $l = 0, 1, 2, 3, 4$; ——— total calculated cross-section; —·—·— total measured cross-section.

to periodic behaviour of $\sin \eta_2$ in proceeding from one rare gas to the other. Thus, referring to Fig. 6.7, it will be seen that, for argon, the maximum arises from the maximum of the second-order partial cross-section when $\eta_2 = \frac{1}{2}\pi$. For krypton, and electrons of the same energy, the same phase is nearly equal to $\frac{3}{2}\pi$ (see Fig. 6.8) and presumably for xenon it is nearly $\frac{5}{2}\pi$.

One further point to be noticed concerns the highest-order phase which reaches a maximum in excess of $\frac{1}{2}\pi$. For argon it is η_2 (see Fig. 6.6), but for krypton it is η_3 (see Fig. 6.8) and for xenon, presumably η_4. This has important consequences when the angular distribution of scattered electrons is considered (see § 4.6).

4.3. *Behaviour of neon and helium*

The marked similarity in the behaviour of xenon, krypton, and argon does not extend to the lighter rare gases, neon and helium. This is because, in passing say from argon to neon, the phase η_0, for low electron energies, does not pass through a complete period. It appears, in fact, that the molecule methane gives a mean scattering field which provides a phase η_0 differing from that for argon by nearly π and thus becomes the fourth member of the series, preceding argon (see Chap. 10, § 3.4 and Fig. 10.24).

The neon field is never strong enough to produce a phase shift of $\tfrac{1}{2}\pi$ in any order except η_0 and η_1, while the weaker helium field cannot even produce such a phase shift in the first-order wave.

TABLE 6.1

In the limit of very low velocities	$\eta_0 \to 3\pi, 4\pi, 5\pi$ from above for A, Kr, Xe respectively, the slope $d\eta_0/dk$ increasing in that order. Ramsauer–Townsend effect.
	$\eta_0 \to \pi, 2\pi$ from below for He, Ne respectively. No Ramsauer–Townsend effect.
At cross-section maximum	$\eta_2 \to \tfrac{1}{2}\pi, \tfrac{3}{2}\pi, \tfrac{5}{2}\pi$ for A, Kr, Xe respectively. No such effect in He or Ne.
Highest-order phase reaching $\tfrac{1}{2}\pi$	η_2, η_3, η_4 for A, Kr, Xe respectively.
	η_0, η_1 for He, Ne respectively.

The effectiveness of the fields of the rare-gas atoms in producing phase shifts is summarized in Table 6.1.

4.4. *Large cross-sections for alkali metals*

A conspicuous feature of the observed cross-sections is the very large magnitude attained for the heavier alkali metal atoms at low electron energies. Thus referring to Chapter 1, Fig. 1.16, the cross-section for sodium will be seen to reach a value of $400\pi a_0^2$ for an electron energy of 3 eV. At this energy the wavelength λ is $44a_0$. The maximum value of the partial cross-section q_l, $(2l+1)\lambda^2/\pi$, is therefore $(2l+1)(44/\pi)^2\pi a_0^2$. To explain the observed cross-section for sodium it is only necessary to suppose that the cross-section has practically its maximum value for the energy concerned, i.e. that it is nearly equal to an odd multiple of $\tfrac{1}{2}\pi$ at that energy. Reference to Fig. 6.6 shows that, for argon, η_1 is equal to $\pi - 0.25$ and is making very little contribution to the cross-section. For the alkali metals, however, the atomic field extends over much greater distances than for argon so that, for low-energy electrons, the first-order phase shifts are much bigger.

4.5. *Similarity of behaviour of chemically similar atoms*

We have already discussed the similarity of the cross-section velocity curves for the heavier rare-gas atoms. This similarity between elements with similar chemical properties, i.e. between elements occupying similar positions in the periodic table, extends also to the alkali metals and the zinc, cadmium, mercury triad. In terms of the quantum collision theory this similarity, for low-velocity electrons, can be explained if the phase shifts important at these low velocities (η_0 and η_1) change by nearly a whole multiple of π in passing through a complete period of the periodic table. This we found to be the case for the heavier rare gases. It is also true for the alkali metals and the zinc, cadmium, mercury triad.

In order to show the generality of this behaviour, Allis and Morse[†] calculated the cross-sections to be expected for a schematic atomic field given by

$$V = \begin{cases} -Ze^2\left(\dfrac{1}{r} - \dfrac{1}{r_0}\right) & (r \leqslant r_0), \\ 0 & (r \geqslant r_0), \end{cases} \quad (110)$$

for which the accurate calculation of the phases could be carried out without too much labour for a variety of values of the constants Z and r_0. Defining a quantity $\beta = (Zr_0/2a_0)^{\frac{1}{2}}$, Allis and Morse showed that, for a fixed value of kr_0 ($= x$) the partial cross-sections are quasi-periodic in β with a period unity. Examples of this are given in Fig. 6.10. Allis and Morse were able to show that a change of β by unity did correspond to passage through one complete period of the periodic table. To do this the best choice of β for different atoms was made in the following way.

By using the rules given by Slater[‡] for determining atomic radii the constant r_0 was obtained directly, since Slater considers the electrons in any shell as equivalent to a spherical shell of negative charge of radius equal to the mean radius for the shell. These shells screen the positive charge of the nucleus according to definite rules depending on the quantum numbers of the shell electrons. From the potential energy V due to this distribution Z was also determined as follows.

Fig. 6.11 shows the radial variation of rV as given by Slater's rules for an atom. Allis and Morse approximated to this form by means of the broken line in the figure, drawn so as to enclose with the axes the same area as the Slater curve. The intercept of this curve at the $r = 0$ axis is equal to Ze^2 in the expression (110). The values of β found in this way for the different elements are given in Table 6.2.

[†] ALLIS, W. P. and MORSE, P. M., *Z. Phys.* **70** (1931) 567.
[‡] SLATER, J. C., *Phys. Rev.* **36** (1930) 57.

This shows that, between the heavier atoms in a column of the periodic table, β does change by very nearly unity. With the lighter atoms the change is not so nearly unity, as would be expected since the cross-sections for helium and neon do not resemble those for the heavier rare

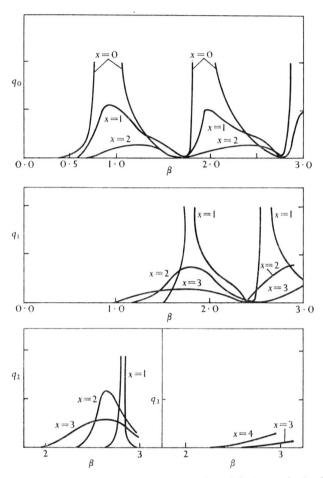

FIG. 6.10. Partial cross-sections for scattering of electrons of a fixed energy by different atomic fields. x ($= kr_0$) is proportional to the electron velocity. The cross-sections are quasi-periodic in the quantity β ($= (Zr_0/2a_0)^{\frac{1}{2}}$) with period unity.

gases. Lithium also falls out of place in the alkali metal series and the experimental data (Fig. 1.16) support this.

Morse† carried out a similar investigation with a schematic potential
$$V = -(Ze^2/r)\exp(-2r/r_0),$$

† MORSE, P. M., *Rev. mod. Phys.* **4** (1932) 577.

defined in terms of a similar pair of parameters Z and r_0 but which has no definite boundary. The results obtained with this potential agreed closely with those from (110).

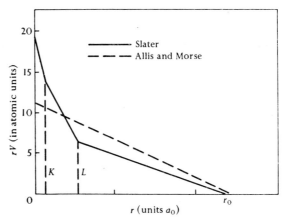

Fig. 6.11. Radial variation of rV assumed by Allis and Morse, compared with that given by Slater's rules.

TABLE 6.2

Lithium	1·36	Zinc	3·77	Helium	0·77
Sodium	2·54	Cadmium	4·87†	Neon	1·73
Potassium	3·51			Argon	2·68
				Krypton	3·66

† The value of β for Cd was not given by Allis and Morse but has been estimated by the method described in their papers.

4.6. Angular distribution for scattering of low-energy electrons by rare gas atoms

We first point out some general features of the way the angular distribution will vary with energy. At very low energies only η_0 is important and

$$I_0(\theta) = \frac{1}{4k^2} |\{\exp(2i\eta_0)-1\}P_0|^2.$$

Reference to Fig. 6.1 shows that $P_0(\cos\theta)$ is equal to 1 for all θ, so that the angular distribution should be uniform under these conditions.

With increasing energy the phase η_1 will become appreciable and the distribution will be modified by the introduction of the harmonic $P_1(\cos\theta) = \cos\theta$. When $\eta_1 = \frac{1}{2}\pi$ we can expect the angular distribution to be largely determined by this function, possessing a low minimum at 90°. If the atomic field is strong enough the phase η_2 will approach $\frac{1}{2}\pi$

as the energy increases still further so, in this energy range, we expect the distribution to be strongly influenced by the harmonic $P_2(\cos\theta)$ giving minima near 60° and 120° and a maximum at 90° (see Fig. 6.1), and so on. It is not to be expected, of course, that there will be many circumstances in which the contributions from the other angular momenta will be negligible so that the distribution, while similar in these general features to that given by a single harmonic, will usually be more complicated in detail.

Referring now to Table 6.1, which summarizes the general scattering properties of the rare-gas atoms, it will be seen that the following general forms of the angular distributions are to be expected for the scattering of slow electrons:

Helium. Very flat, because η_0 is the only phase which passes through a value of $\tfrac{1}{2}\pi$ whereas η_1, and all higher phases, are always small.

Neon. The influence of η_1 should be apparent at not too low velocities in producing a minimum at 90°. No important influence of higher order phases is to be expected.

Argon. In the neighbourhood of the cross-section maximum the influence of η_2 should be strong (see Table 6.1 which shows that $\eta_2 \simeq \tfrac{1}{2}\pi$ there). The angular distribution should therefore show minima near 60° and 120° and a maximum at 90°.

Krypton. The influence of the harmonic $P_3(\cos\theta)$ is likely to be important at energies near 50 eV, and beyond, for which the phase η_3 is near $\tfrac{1}{2}\pi$ (see Fig. 6.8). As the contribution from η_2 is also large, at least at the lower energies in this range, the interference between the two contributions may rather modify the distribution. Near the cross-section maximum (10 eV) the distribution should be dominated by $P_2(\cos\theta)$.

Xenon. The harmonic $P_4(\cos\theta)$ will play an important part here, though significant contributions from second- and third-order phases will complicate the picture.

In Fig. 6.12 a number of observed angular distributions† are illustrated that show that these predictions are substantially correct. The sequence of curves representing the angular distributions for 80 eV electrons scattered by He, Ne, A, Kr, and Xe atoms are particularly interesting in this respect. It is quite clear by comparing these curves with those given respectively by the single harmonics $(P_0)^2$, $(P_1)^2$, $(P_2)^2$, $(P_3)^2$, $(P_4)^2$,

† For references see Chapter 5, § 5.3.

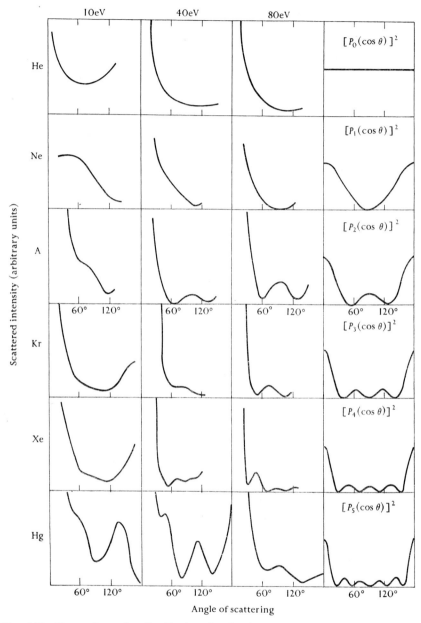

FIG. 6.12. Observed angular distributions for elastic scattering of 10-, 40-, and 80-eV electrons in helium, neon, argon, krypton, xenon, and mercury together with distributions $P_l(\cos\theta)^2$ corresponding to the dominant phase η_l for the 80-V case in each gas.

that the predominant phase angle changes successively from η_0 to η_4 in going from helium to xenon, as predicted.

It is possible to trace the gradual change in the atomic field in proceeding from phosphorus to potassium with reference to the scattering of 80-eV electrons. Mohr and Nicoll† measured the angular distribution

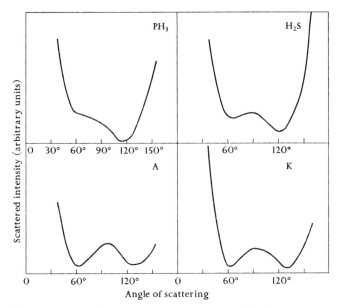

Fig. 6.13. Comparison of observed angular distributions for elastic scattering of 80-eV electrons by phosphine, hydrogen sulphide, argon, and potassium, showing that η_2 is the most important phase for all four at these energies.

for scattering of such electrons by phosphine PH_3 and hydrogen sulphide H_2S. Owing to the comparatively small effect of the hydrogen atom these distributions are essentially those characteristic of phosphorus and sulphur atoms (see Chap. 10, § 2.1). McMillen‡ has measured the distribution for scattering by potassium. The three distributions as well as one for argon are illustrated in Fig. 6.13. Although it is clear that η_2 is the dominant phase for all four, it is clearly less important for sulphur and phosphorus, as would be expected from their somewhat weaker fields. Similar relations are found between nitrogen and neon (η_1 dominant), between zinc, krypton, and bromine (η_3 dominant), and between iodine and xenon (η_4 dominant).

† Mohr, C. B. O. and Nicoll, F. H., *Proc. R. Soc.* A**138** (1932) 469.
‡ McMillen, J. H., *Phys. Rev.* **46** (1934) 983.

For 40-eV electrons the distributions for xenon and krypton have already become less variable, showing that the phases η_3 and η_4 respectively are becoming less important. For 10-eV electrons it is clear that these phases have become unimportant, but the form of the distribution shows that η_2 is still effective. At this energy the distribution for helium has already come close towards uniformity, indicating that η_1 and higher-order phases are becoming unimportant.

It was noted in Chapter 5, § 5.3 that even for very low-energy electrons (0·5 eV) in argon, krypton, and xenon the angular distribution is far from uniform (see Fig. 5.41). This is rather surprising at first sight because for these electrons η_1, η_2 and all higher phases are very small (Figs. 6.6 and 6.8). It must be remembered, however, that in this energy range the Ramsauer–Townsend effect occurs. As explained in § 4.1, this effect arises because η_0 varies with energy according to curve III of Fig. 6.4 and is equal to π at the cross-section minimum. Hence, although η_1 and η_2 are very small at 0·5 eV, their contributions to the scattering are by no means necessarily small compared with that from η_0. The resultant angular distribution may therefore be characteristic of first- or second-order scattering, and reference to Figs. 6.6 and 6.8 shows that η_2 must indeed be relatively important. The variability of the angular distributions for electrons of very low energy in argon, krypton, and xenon may thus be regarded as additional confirmation of the general correctness of the explanation of the Ramsauer–Townsend effect given in § 4.1.

To understand the behaviour of the angular distributions at higher electron energies it is best to start from the limiting case of very high electron energies for which a simplified approximation may be used. We therefore defer discussion of this aspect to Chapter 7, which deals with Born's approximation. Before concluding this chapter, however, a short account will be given (§ 6) of certain relativistic modifications of the theory which are significant for scattering by heavy atoms and which lead to the possibility of polarization of electron spin after scattering.

5. Pressure shift of the high series terms of the alkali metals—relation to the elastic cross-section of the perturbing atoms for very low-energy electrons

We describe here a method for obtaining the elastic cross-sections from various atoms for collisions with very low-energy electrons which supplements the data obtained by methods discussed in Chapter 1.

If the valence electron of an alkali metal atom is excited to a state with principal quantum number $n \simeq 30$, the radius of its Bohr orbit is about $1000a_0$. A sphere of this radius, at atmospheric pressure, encloses about 10^4 atoms. The valence electron moves very slowly through these atoms in a manner essentially similar to that of a very slow free electron through a gas. The associated wavelength is of the order $200a_0$. The presence of the foreign atoms will modify the energy of the excited state concerned for two reasons—the mean potential energy in which the valency electron moves will be modified, and the foreign atoms will be polarized by the charged core of the alkali atom.

To calculate† the change in the energy due to the modification of the mean field acting on the valence electron we start from the wave equation for the motion of that electron in the form

$$\nabla^2\psi + \frac{2m}{\hbar^2}\left(E - V - \sum_s v_s\right)\psi = 0. \tag{111}$$

V is the potential energy due to the alkali atom core and v_s that due to the sth perturbing atom. To obtain the average energy shift we consider the equation for the wave-amplitude ψ averaged over a region which, while containing many foreign atoms, is of dimensions small compared with the wavelength of the electron. Throughout such a region V will not vary appreciably and we have

$$\overline{\nabla^2\psi} + \frac{2m}{\hbar^2}(E - V)\overline{\psi} - \frac{2m}{\hbar^2}\sum \overline{v_s\psi} = 0, \tag{112}$$

where the bars denote the averaging process described.

We consider now the evaluation of $\overline{v_s\psi}$. For this purpose it is convenient to take the origin of coordinates at the centre of the sth atom. We consider a region S enclosing this atom, which is of dimensions large compared with the range of v_s but small compared with the distance between neighbouring atoms and hence with the electron wavelength. Throughout this region the negative potential energy V due to the alkali atom core may be regarded as constant and equal to V_s. The wave equation in the new coordinate system takes the form

$$\nabla^2\psi + \left(k_s^2 - \frac{2m}{\hbar^2}v_s\right)\psi = 0 \tag{113}$$

within the region S, k_s^2 being given by $2m(E-V_s)/\hbar^2$. As v_s is negligible outside S it is only necessary to obtain ψ from (113).

The equation (113) is of exactly the same form as that concerned in any elastic scattering problem (see § 3.3). The potential energy v_s will produce distortion of the otherwise plane waves whose amplitude satisfies

$$\nabla^2\psi + k^2\psi = 0.$$

Employing the usual resolution into partial waves of different angular momenta it is clear that, for the long wavelengths concerned, only the waves with zero angular momenta about the centre of the sth atom will be appreciably perturbed by v_s (see § 3.4). The others have such small amplitude near the centre of the sth atom that their contribution to $\overline{v_s\psi}$ is negligible.

† The following treatment is due to FERMI, E., *Nuovo Cim.* **11** (1934) 157. See also MARGENAU, H. and WATSON, W. W., *Rev. mod. Phys.* **8** (1936) 22.

The waves with zero angular momentum about the centre ($r = 0$) of the sth atom are spherically symmetrical about that centre so that, for our purposes, we may write
$$\psi = u(r)/r,$$
where
$$u \sim A \sin(k_s r + \eta_0),$$
η_0 being the phase shift produced by the field v_s of the perturbing atom (see § 3.4).

Now k_s is very small and so also will be η_0, apart from an integral multiple of π that is immaterial, so we may write
$$u \sim A k_s (r + \eta_0/k_s).$$
At these values of r, however, u must be nearly equal to $r\psi$ so that $A = \psi/k_s$.

The low-velocity limit of the elastic cross-section of a perturbing atom is $4\pi a^2$, where a is the scattering length such that
$$a^2 = \eta_0^2/k_s^2.$$
We may therefore write
$$u \sim (r + a)\psi, \tag{114}$$
where a is likely to be of the order of the gas kinetic radius of the perturbing atom.

To evaluate $\overline{v_s \psi}$ we have now
$$\overline{v_s \psi} = \frac{4\pi}{W} \int_0^R v_s u r \, dr,$$
where the upper limit R is large compared with the range of v_s but small compared with k_s^{-1}, and $W = \tfrac{4}{3}\pi R^3$. Using (113),
$$\frac{4\pi}{W} \int_0^R v_s u r \, dr = \frac{h^2}{2\pi m W} \int_0^R \left(\frac{d^2 u}{dr^2} + k_s^2 u\right) r \, dr$$
$$= \frac{h^2}{2\pi m W}\left[r \frac{du}{dr} - u\right]_0^R + \frac{k_s^2 h^2}{2m}\psi.$$

As $u(0) = 0$ and at the upper limit (114) is valid we have
$$\overline{v_s \psi} = -\frac{a h^2}{2\pi m W}\psi + \frac{k_s^2 h^2}{2m}\psi.$$

The second term on the right-hand side is negligible because $k_s R$ and $k_s a$ are both $\ll 1$.

If there are N perturbing atoms per unit volume we may obtain $\sum \overline{v_s \psi}$ by replacing W by $1/N$. This gives, on substitution in (112),
$$\nabla^2 \psi + \frac{2m}{\hbar^2}\left(E + \frac{a h^2 N}{2\pi m} - V\right)\psi = 0. \tag{115}$$

The effect of the foreign atoms therefore gives an average decrease of the energy of the valence electron equal to $a h^2 N/2\pi m$, i.e. a frequency shift
$$\Delta \nu = -a \hbar N/m,$$
where $4\pi a^2$ is the low-velocity limit of the elastic collision cross-section of a perturbing atom.

The sense of this effect depends on the sign of a. This is not fixed by the size of the elastic cross-section for the given perturbing atom. However, if the phase η_0 approaches its low-velocity limit from above,

a will be positive; if from below, negative. In the former case the cross-section will exhibit a minimum at a low electron energy so that a frequency shift towards the red should only occur from this effect if the perturbing atoms show a Ramsauer–Townsend effect with slow electrons (see § 4.1). Otherwise the shift is towards the violet.

The polarization energy is given by

$$\Delta E_p = -\tfrac{1}{2}\alpha e^2 \sum_s \frac{1}{R_s^4},$$

where R_s is the distance of the sth perturbing atom from the nucleus of the alkali metal atom. The sum over all perturbers is approximately given by

$$\sum \frac{1}{R_s^4} = 4\pi N \int_{R_0}^{\infty} \frac{R^2\,dR}{R^4},$$

where the lower limit R_0 is the mean distance $(3/4\pi N)^{\frac{1}{3}}$ of a perturbing atom from the nucleus of the alkali metal atom. This gives a frequency shift

$$\Delta\nu_p = -10(e^2/h)\alpha N^{\frac{4}{3}}$$

towards the red, which is usually small compared with that due to the first effect considered.

Measurements of the frequency shifts of the highest terms of the principal series of the spectra of the alkali metal atoms due to perturbation by various foreign gases have been carried out by a number of investigators.† The results are in good agreement with the theory and provide the only values at present available for the low-velocity limits of the elastic cross-sections of different atoms and molecules.

It is found that the shift of the highest terms is independent of the nature of the alkali atom and of the particular term concerned. The rare gases A, Kr, and Xe, which exhibit a Ramsauer–Townsend effect, all produce a broadening that is strongly asymmetric towards the red, while He and Ne and Hg, which show no Ramsauer–Townsend effect, all produce a broadening asymmetric towards the violet. Fig. 6.14 illustrates these results.

Table 6.3. gives the values of the low-velocity limit of the elastic cross-section for different gases derived from the measured shifts.

The relation of these results to values obtained by other methods and

---

† AMALDI, E. and SEGRÈ, E., *Nuovo Cim.* **11** (1934) 145; FUCHTBAUER, CHR., SCHULZ, P., and BRANDT, A. F., *Z. Phys.* **90** (1934) 403 (Na and K); FUCHTBAUER, CHR. and GÖSSLER, F., ibid. **93** (1935) 648; FUCHTBAUER, CHR. and REIMERS, H. J., ibid. **95** (1935) 1 (Cs); TSI-ZÉ, N. and SHANG-YI, C., *Phys. Rev.* **51** (1937) 567.

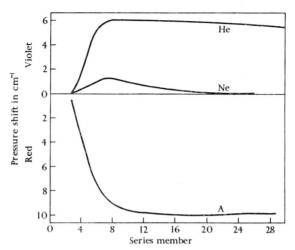

Fig. 6.14. Pressure shift of the principal lines of caesium in helium, neon, and argon. The shift is towards the red or the violet according as the foreign gas does or does not exhibit a Ramsauer–Townsend effect.

TABLE 6.3

Effective cross-sections for elastic scattering of very low-velocity electrons by gas atoms derived from observations of the pressure shift of high series terms in alkali metal spectra

Gas	He		Ne		A		Kr	Xe	Hg	N_2	H_2
	†	‡	†	‡	†	‡	§	‖	‖	†	††
Cross-section (πa_0^2)	4·94,	5·03	0·073,	0·078	8·01,	7·70	38·7	132	11·9	1·6	3·92

† FÜCHTBAUER, CHR., SCHULZ, P., and BRANDT, A. F., *Z. Phys.* **90** (1934) 403.
‡ TSI-ZÉ, N. and SHANG-YI, C., *Phys. Rev.* **51** (1937) 567.
§ FÜCHTBAUER, CHR. and REIMERS, H. J., *Z. Phys.* **95** (1935) 1.
‖ FÜCHTBAUER, CHR. and GÖSSLER, F., ibid. **93** (1935) 648.
†† AMALDI, E. and SEGRÈ, E., *Nature, Lond.* **133** (1934) 141; *Nuovo Cim.* **11** (1934) 145.

by semi-empirical and other theory is discussed in Chapter 8, §§ 6.1 and 7.1.1.

6. Relativistic effects† including spin polarization

It would not seem important to consider relativistic effects in discussing the scattering of electrons with incident energies up to some keV. After all, the rest energy of an electron is 546 keV. However, an electron is

† See MOTT, N. F. and MASSEY, H. S. W., *The theory of atomic collisions*, 3rd edn, Chap. ix (Clarendon Press, Oxford, 1965).

accelerated as it approaches an atom and, if the atomic field is powerful enough, there may well be sufficient probability of the electron attaining a velocity comparable with that of light at some stage in the collision for relativistic modifications to be significant. In Chapter 5, § 6 experimental evidence of the importance of one relativistic effect, the polarization of slow electrons on scattering by mercury atoms, was described. We now discuss the relativistic theory of scattering by a centre of force and its application to the scattering of electrons with incident energies of a few eV up to several keV, particularly by heavy atoms.

According to Dirac's theory of the electron the differential cross-section $I(\theta)\,d\omega$ and the polarization $P(\theta)$ for scattering of initially unpolarized electrons of incident energy E (including the rest energy mc^2) and velocity v by a centre of force of potential energy $V(r)$ are given in terms of two sets of phase shifts η_l, η_{-l-1} by

$$I(\theta)\,d\omega = \{|f|^2 + |g|^2\}\,d\omega, \tag{116}$$

$$P(\theta) = i(fg^* - f^*g)/(|f|^2 + |g|^2), \tag{117}$$

where
$$f(\theta) = \frac{i}{2k}\sum_l \{(l+1)(1-e^{2i\eta_l}) + l(1-e^{2i\eta_{-l-1}})\}P_l(\cos\theta), \tag{118}$$

$$g(\theta) = \frac{i}{2k}\sum_l \{e^{2i\eta_l} - e^{2i\eta_{-l-1}}\}P_l(\cos\theta). \tag{119}$$

Here
$$k^2 = \frac{1}{\hbar^2}\left(\frac{E^2}{c^2} - m^2c^2\right)^{\frac{1}{2}}. \tag{120}$$

The phase shift η_l is such that the asymptotic form of the solution of the equation

$$\frac{d^2 G_l}{dr^2} + \left\{k^2 - \frac{l(l+1)}{r^2} - \frac{2\gamma m}{\hbar^2}V(r) - U(r) + (l+1)W(r)\right\}G_l = 0, \tag{121}$$

with
$$\gamma = (1-v^2/c^2)^{-\frac{1}{2}}, \qquad U(r) = \frac{V^2}{\hbar^2 c^2} - \frac{3}{4}\frac{\zeta'^2}{\zeta^2} + \frac{1}{2}\frac{\zeta''}{\zeta},$$

$$W(r) = \frac{\zeta'}{r\zeta}, \qquad \zeta = \frac{mc}{\hbar}(1+\gamma - V/mc^2), \qquad \zeta' = d\zeta/dr, \qquad \zeta'' = d^2\zeta/dr^2,$$

is given by
$$G_l \sim \sin(kr - \tfrac{1}{2}l\pi + \eta_l). \tag{122}$$

η_{-l-1} is similarly given from the corresponding equation for G_{-l-1}, which differs only in that $(l+1)W(r)$ is replaced by $-lW(r)$.

The non-relativistic case is regained when $c \to \infty$, which causes $U(r)$ and $W(r)$ to vanish and $\gamma \to 1$.

In atomic units (121) becomes

$$\frac{d^2 G_l}{dr^2} + \left[k^2 - \frac{l(l+1)}{r^2} - 2\gamma V(r) + \alpha^2 \left\{ V^2(r) + \frac{1}{2} \frac{V''(r)}{\gamma + 1 - \alpha^2 V(r)} \right. \right.$$
$$\left. \left. - \frac{3}{4} \frac{\alpha^2 V'^2(r)}{(\gamma + 1 - \alpha^2 V(r))^2} \right\} - \frac{l+1}{r} \frac{\alpha^2 V'(r)}{\gamma + 1 - \alpha^2 V(r)} \right] G_l = 0, \quad (123)$$

where $\alpha = e^2/\hbar c$ is the fine structure constant $1/137\cdot 037$.

Of the terms in (123) the factor before $V(r)$ arises directly from the Lorentz–Fitzgerald contraction. The last term, depending on the scattering force rather than the scattering potential, arises from spin-orbit interaction and depends on whether the electron spin is parallel or antiparallel to the orbital angular momentum.

For scattering by an atom we write $V(r) = -Z_p(r)/r$, where Z_p is the effective nuclear charge for potential. At low electron energies and small r, for which $Z_p \simeq Z$ the full nuclear charge, (123) can be simplified to

$$\frac{d^2 G_l}{dr^2} + \left[k^2 - \frac{l(l+1)}{r^2} + \frac{2Z}{r} + \frac{Z^2 \alpha^2}{r^2} - \frac{ly + \frac{3}{4}y^2}{r^2} \right] G_l = 0 \quad (124)$$

with
$$y = \frac{Z\alpha^2/r}{2 + Z\alpha^2/r}.$$

This shows that the relativistic modifications will be more important the larger Z and the smaller l. In particular if αZ is not small the modification is likely to be significant even at very low energies for small l. As the main contribution to the differential cross-section at low velocities come from such values of l it may be appreciably influenced by the additional terms in (124).

Browne and Bauer† have carried out detailed calculations involving numerical solution by electronic computation of equations of the form (123) for selected cases. Their results strongly support the general conclusions we have outlined (Table 6.4).

Thus, referring to Table 6.4, which gives their results for total cross-sections calculated with and without relativistic modifications for scattering by the Hartree–Fock fields of He, Kr, Cs, and Hg, it will be seen that the effects are most marked for 2-eV electrons and relatively unimportant at 200 eV. For the lightest atom, helium, they are negligible at all energies but for the heaviest atom the effect is very large at 2 eV and amounts to 20 per cent at 20 eV.

The sensitivity of the cross-sections, total and differential, for mercury at 2 eV to the assumed potential is very great both in relativistic and

† BROWNE, H. N. and BAUER, E., *Phys. Rev. Lett.* **16** (1966) 495.

TABLE 6.4

Comparison of total cross-sections (in units πa_0^2) for scattering of slow electrons by the Hartree–Fock (non-relativistic) fields of various atoms calculated with and without relativistic modifications

Atom	Electron energy					
	2 eV		20 eV		200 eV	
	rel.	non-rel.	rel.	non-rel.	rel.	non-rel.
He	76·0	76·1	8·20	8·20	0·752	0·753
Kr	110	108	31·7	29·8	13·2	13·2
Cs	228	286	65·8	61·2	20·7	21·0
Hg	12·8	1·55	50·9	59·3	21·0	20·3

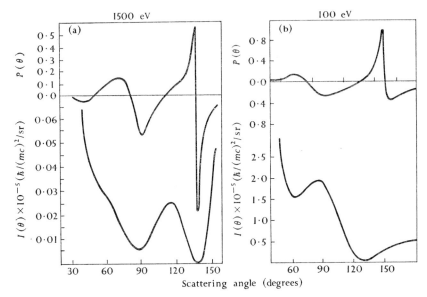

FIG. 6.15. The scattered intensity $I(\theta)$ and polarization $P(\theta)$ for scattering of electrons by mercury atoms. (a) Electrons of energy 1·5 keV calculated by Holzwarth and Meister. (b) Electrons of energy 100 eV calculated by Schonfelder.

non-relativistic theory, so the large relativistic modification cannot be disentangled from other effects that cause the effective scattering field to depart from the Hartree–Fock form. Differential cross-sections are sensitive in much the same way though to a more marked extent.

Much more interest attaches to the polarization produced by the scattering of slow electrons. The formula (117–19) for the polarization

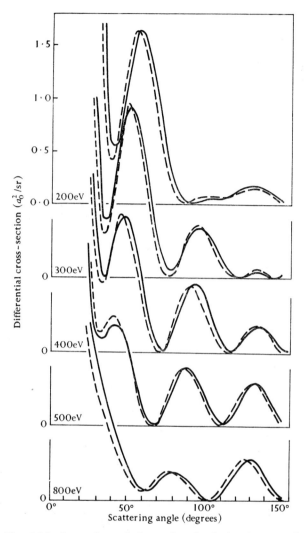

Fig. 6.16. Comparison of observed and calculated scattered intensities for scattering of electrons by mercury atoms. —— calculated; – – – – observed.

was first derived by Mott† and applied to the case of scattering by a pure Coulomb field. He showed that for scattering by gold nuclei the polarization due to scattering through 90° has a maximum of 0·26 for electron kinetic energies of 165 keV. Several years later Massey and Mohr‡ investigated the effect of screening by the atomic electrons.

† MOTT, N. F., *Proc. R. Soc.* A**135** (1932) 429.
‡ MASSEY, H. S. W. and MOHR, C. B. O., ibid. A**177** (1941) 341.

Using an approximate Hartree field for gold and calculating the phase shifts by the classical approximation described in § 3.7 they found no appreciable modification of Mott's results at energies of 50 keV or higher but they did show that, owing to diffraction effects occurring, larger polarizations could be expected in the scattering of even quite slow electrons by atoms such as gold or mercury.

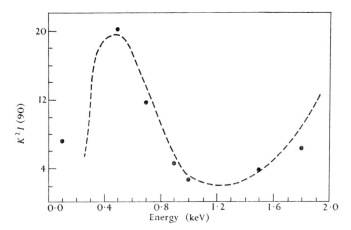

Fig. 6.17. Comparison of observed and calculated values of the intensity $I(90°)$ of scattering of electrons by gold atoms. ● calculated; – – – – observed.

The expression $fg^* - f^*g$ appearing in the numerator of the expression (117) for the polarization depends on the differences between the phase shifts η_l, η_{-l-1}. In the energy range of interest, even for such heavy atoms as gold, the only important contribution comes from $\eta_1 - \eta_{-2}$, which remains roughly constant and $\simeq 0{\cdot}3$ rad over a wide energy range. The denominator of (117) will vary widely owing to diffraction effects so that, near diffraction minima, the polarization will be quite large.

Examples to illustrate this were worked out by Massey and Mohr but they pointed out that much more accurate calculations needed to be carried out before reliable predictions could be made. Such calculations were first performed by Holzwarth and Meister[†] using a relativistic Hartree potential for mercury computed by Mayers.[‡] Fig. 6.15 (a) illustrates the calculated variation of $I(\theta)$ and $P(\theta)$ with θ for electrons of kinetic energy 1·5 keV scattered by mercury atoms. It will be seen that large magnitudes for P occur near the minima of $I(\theta)$ as expected.

† HOLZWARTH, G. and MEISTER, H. J., *Nucl. Phys.* **59** (1964) 56.
‡ MAYERS, D. F., *Proc. R. Soc.* **A241** (1957) 93.

Finally, an extensive series of calculations of both $I(\theta)$ and $P(\theta)$ have been carried out by Bunyan and Schonfelder† also using a relativistic Hartree field. Fig. 6.15 (b) shows Schonfelder's‡ results for $I(\theta)$ and $P(\theta)$ for 100-eV electrons scattered by mercury. Even at this energy polarizations as high as 0·8 or more are expected.

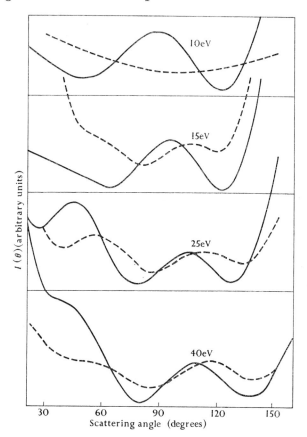

FIG. 6.18. Comparison of observed and calculated values of the intensity $I(\theta)$ of scattering of slow electrons by mercury atoms. ——— calculated; - - - - observed.

The experimental techniques which have been used so successfully to detect and measure $P(\theta)$ for quite slow electrons have been described in Chapter 5, § 6. One of the most remarkable features of this work is the extraordinarily good agreement obtained between theory and observation down to energies as low as 180 eV. Thus in Fig. 6.16 comparison

† BUNYAN, P. J. and SCHONFELDER, J. L., *Proc. phys. Soc.* **85** (1965) 455.
‡ SCHONFELDER, J. L., ibid. **87** (1966) 163.

is made with the differential cross-sections observed by Kessler and Lindner† and in Fig. 5.60 (Chap. 5) with the polarizations observed by Jost and Kessler.‡ In each case the scattering atom is mercury. Over the entire energy range 180–1700 eV there is no evidence of any significant discrepancy. A further check is shown in Fig. 6.17 where comparison is made between the calculated values of $I(90°)$ for gold with the observations of Reichert§ (see Chap. 5, § 6) over the energy range 250 to 1800 eV.

On the other hand, at electron energies of 40 eV and below the agreement is lost, as may be seen from Fig. 6.18, which compares theoretical values of $I(\theta)$ as a function of θ for electrons of these energies with observations made by Deichsel and Reichert.‖ Failure of the simple theory at these energies is not surprising because electron exchange and atom distortion effects can no longer be neglected (see Chap. 8, § 2).

† KESSLER, J. and LINDNER, H., *Z. Phys.* **183** (1965) 1.
‡ JOST, K. and KESSLER, J., ibid. **195** (1966) 1.
§ REICHERT, E., ibid. **173** (1963) 392.
‖ Private communication to Schonfelder (1965).

7

ELECTRON COLLISIONS WITH ATOMS—THEORETICAL DESCRIPTION— BORN'S APPROXIMATION

In Chapter 6 an account is given of the theory of the scattering of electrons by a centre of force and its application to a semi-empirical description and interpretation of the broad features of the elastic scattering of electrons by atoms. We now begin the discussion of the scattering problem in terms of the detailed atomic structure, including the existence of atomic states. Provided the energy of the incident electron is large compared with the binding energy of the atomic electrons an approximate method of wide applicability, Born's approximation, may be used. In this chapter we discuss this method, its application to the elastic and inelastic collisions of electrons with atoms and its range of validity. To illustrate the method we begin by discussing the scattering of electrons by hydrogen atoms.

1. Born's approximation and the scattering of electrons by hydrogen atoms

If $\mathbf{r}_1$, $\mathbf{r}_2$ are the respective coordinates of the incident and atomic electron the wave equation for the system is

$$\left[\nabla_1^2 + \nabla_2^2 + \frac{2m}{\hbar^2}\left(E + \frac{e^2}{r_1} + \frac{e^2}{r_2} - \frac{e^2}{r_{12}}\right)\right]\Psi = 0 \qquad (1)$$

in which the total energy $E = E_0 + E_t$ where E_0 is the energy of the ground state of the atom and E_t is the kinetic energy of the incident electron.

We write now $\qquad \Psi = e^{i\mathbf{k}_0 \cdot \mathbf{r}_1}\psi_0(r_2) + \phi(\mathbf{r}_1, \mathbf{r}_2),$ (2)

where $e^{i\mathbf{k}_0 \cdot \mathbf{r}_1}$ represents the incident wave moving in the direction of the unit vector $\hat{\mathbf{k}}_0$ with wave number $k_0 = (2mE_t/\hbar^2)^{\frac{1}{2}}$ and ψ_0 is the wave function of the ground state of the atom. In the absence of any interaction between the electron and the atom the first term of (2) would be the complete solution. The existence of the interaction, however, introduces further terms denoted by $\phi(\mathbf{r}_1, \mathbf{r}_2)$. When $E_t \gg E_0$ we can assume that these terms will be relatively small.

7.1 BORN'S APPROXIMATION

To determine ϕ approximately we expand it in the form

$$\phi(\mathbf{r}_1, \mathbf{r}_2) = \left(\sum_n + \int\right) F_n(\mathbf{r}_1)\psi_n(\mathbf{r}_2), \tag{3}$$

where the ψ_n are the wave functions of the various atomic states, the integral being included to allow for the existence of the continuum. On substitution in (2) and thence in (1) we obtain

$$\sum (\nabla_1^2 + k_n^2) F_n(\mathbf{r}_1)\psi_n(\mathbf{r}_2)$$
$$= \frac{2me^2}{\hbar^2}\left\{\left(\frac{1}{r_{12}} - \frac{1}{r_1}\right)e^{i\mathbf{k}_0 \cdot \mathbf{r}_1}\psi_0(r_2) + \sum_n \left(\frac{1}{r_{12}} - \frac{1}{r_1}\right) F_n(\mathbf{r}_1)\psi_n(\mathbf{r}_2)\right\}, \tag{4}$$

where
$$k_n^2 = \frac{2m}{\hbar^2}(E_t - E_n + E_0), \tag{5}$$

E_n being the energy of the nth atomic state.

In deriving (4) use has been made of the fact that

$$\left\{\nabla_2^2 + \frac{2m}{\hbar^2}(E_n + e^2/r_2)\right\}\psi_n = 0. \tag{6}$$

Multiplying both sides of (4) by ψ_n^* and integrating over $\mathbf{r}_2$ we have, since
$$\int \psi_n^* \psi_j \, d\mathbf{r}_2 = 0 \quad (n \neq j),$$

$$(\nabla_1^2 + k_n^2) F_n = U_{0n} e^{i\mathbf{k}_0 \cdot \mathbf{r}_1} + \sum U_{nm}(\mathbf{r}_1) F_m(\mathbf{r}_1), \tag{7}$$

where
$$U_{nm} = \frac{2me^2}{\hbar^2} \int \psi_n^*(\mathbf{r}_2)\left(\frac{1}{r_{12}} - \frac{1}{r_1}\right)\psi_m(\mathbf{r}_2) \, d\mathbf{r}_2 \tag{8a}$$

$$= \frac{2m}{\hbar^2} V_{nm}. \tag{8b}$$

So far we have made no approximation but now we take account of the fact that, on the right-hand side of the equations (7), the first term involving the incident wave is much larger than the other terms under the conditions we have assumed. We then have approximately

$$(\nabla_1^2 + k_n^2) F_n = U_{0n} e^{i\mathbf{k}_0 \cdot \mathbf{r}_1}. \tag{9}$$

In particular, for elastic scattering we have

$$(\nabla^2 + k_0^2) F_0 = U_{00} e^{i\mathbf{k}_0 \cdot \mathbf{r}_1}, \tag{10}$$

where
$$\frac{\hbar^2}{2me^2} U_{00} = \int |\psi_0(r_2)|^2 \left(\frac{1}{r_{12}} - \frac{1}{r_1}\right) d\mathbf{r}_2, \tag{11}$$

and is the static field of the atom in its ground state.

At a great distance r from the nucleus the elastically scattered electrons will appear as an outgoing spherical wave

$$r^{-1} e^{ik_0 r} f_0(\theta), \tag{12}$$

so the differential cross-section for elastic collisions is given by
$$I_0(\theta)\sin\theta\,d\theta d\phi = |f_0(\theta)|^2\sin\theta\,d\theta d\phi. \tag{13}$$
To determine $f_0(\theta)$ we need to obtain a solution of (10) which, while a proper function throughout space, has the asymptotic form (12). This may readily be obtained by the method of Green's function† and gives
$$f_0(\theta) = -\frac{1}{4\pi}\int U_{00}\,e^{i(\mathbf{k}_0-\mathbf{k}_0').\mathbf{r}}\,d\mathbf{r}, \tag{14}$$
where $\mathbf{k}_0'$ is a vector in the direction of scattering of magnitude k_0.

We may deal with the inelastic scattering in the same way. If the excitation of the nth state is energetically possible $k_n^2 > 0$ and we may obtain a solution of (9) with the asymptotic form
$$r^{-1}e^{ik_n r}f_{0n}(\theta,\phi). \tag{15}$$
The scattered flux into the solid angle $\sin\theta\,d\theta d\phi$ of electrons that have excited the nth state is $v_n|f_{0n}|^2\sin\theta\,d\theta d\phi$, where v_n ($= k_n\hbar/m$), is the velocity of an electron after exciting the nth state, so the differential cross-section is given by
$$I_{0n}(\theta,\phi)\sin\theta\,d\theta d\phi = (v_n/v_0)|f_{0n}(\theta,\phi)|^2\sin\theta\,d\theta d\phi. \tag{16}$$
As for $f_0(\theta)$ we have
$$f_{0n}(\theta,\phi) = -\frac{1}{4\pi}\int U_{0n}\,e^{i(\mathbf{k}_0-\mathbf{k}_n).\mathbf{r}}\,d\mathbf{r}. \tag{17}$$
If excitation of the nth state is not energetically possible $k_n^2 < 0$ and the functions F_n will have asymptotic form
$$r^{-1}e^{-|k_n|r}f_{0n}(\theta,\phi) \tag{18}$$
giving no scattered flux.

2. Scattering by a helium ion—the Coulomb–Born approximation

If the target is a He$^+$ ion instead of a neutral H atom, allowance must be made for the long-range effect of the Coulomb field. Thus for collisions with a helium ion the formula (17) for the scattered amplitude f_{0n} becomes
$$f_{0n}(\theta,\phi) = -\frac{1}{4\pi}\int U_{0n}(\mathbf{r}')\psi_c^1(k_0;r',\theta')\psi_c^1(k_n;r',\pi-\Theta')\,d\mathbf{r}', \tag{19}$$
where the Coulomb functions $\psi_c^1(k_0;r',\theta')$, $\psi_c^1(k_n;r',\Theta')$, defined and discussed in Chapter 6, § 3.9 (88), replace the plane waves $\exp(i\mathbf{k}_0.\mathbf{r}')$, $\exp(i\mathbf{k}_n.\mathbf{r}')$ respectively. Θ is such that
$$\cos\Theta = \cos\theta\cos\theta' + \sin\theta\sin\theta'\cos(\phi-\phi').$$

† MOTT, N. F. and MASSEY, H. S. W., *The theory of atomic collisions*, 3rd edn, Chap. v (Clarendon Press, Oxford, 1965).

3. Generalization to scattering by complex atoms

Let $\mathbf{r}_1, ..., \mathbf{r}_N$ be the coordinates of the N atomic electrons. The differential cross-section for excitation of the nth state from the ground state is then given by (16) with

$$U_{0n}(\mathbf{r}) = \frac{2me^2}{\hbar^2} \int \sum_{s=1}^{N} \frac{1}{|\mathbf{r}-\mathbf{r}_s|} \psi_0(\mathbf{r}_1,...,\mathbf{r}_N)\psi_n^*(\mathbf{r}_1,...,\mathbf{r}_N) \, d\mathbf{r}_1...d\mathbf{r}_N, \quad (20)$$

ψ_0, ψ_n being the respective wave functions of the ground and nth state of the atom.

Similarly for excitation of an ion (19) applies with U_{0n} as in (20).

4. Application to elastic scattering—angular distributions at high energy

We note first that the elastic scattering by the actual atom according to Born's approximation is the same as that which would be obtained, making the same approximation, for the scattering of the electron by a centre of force exerting the same field as the static atomic field $\hbar^2 U_{00}/2m = V_{00}$. The Schrödinger equation for the latter problem is

$$(\nabla^2 + k_0^2 - U_{00})\psi = 0. \quad (21)$$

Writing
$$\psi = e^{i\mathbf{k}_0 \cdot \mathbf{r}} + F_0(\mathbf{r}), \quad (22)$$

where F_0 represents the scattered wave, considered of small amplitude, we have
$$(\nabla^2 + k_0^2)F_0 = U_{00}(e^{i\mathbf{k}_0 \cdot \mathbf{r}} + F_0). \quad (23)$$

Neglecting the relatively small second term on the right-hand side of (23) we obtain exactly the same equation as (10) for F_0.

This result has been derived for scattering by a hydrogen atom but it obviously applies to any atom. It is important to note, however, that we cannot assume that, when the kinetic energy of the incident particle is not large enough for Born's approximation to be valid the elastic scattering can still be calculated from (21) without assuming F_0 to be small. Although this seems to give useful results at intermediate energies it is well to remember that the exact system of equations is the infinite coupled set (7) so that the behaviour of F_0 depends to some degree on that of all the F_n. We shall return to this matter in Chapter 8.

In terms of spherical polar coordinates
$$|\mathbf{k}_0 - \mathbf{k}_0'| = 2k_0 \sin \tfrac{1}{2}\theta$$

and
$$f_0(\theta) = -\frac{1}{4\pi} \int U_{00}(r') \exp(2ik_0 r' \sin \tfrac{1}{2}\theta \cos\theta') \, d\mathbf{r}'$$
$$= -\frac{2m}{\hbar^2} \int V_{00}(r') \frac{\sin(2k_0 r' \sin \tfrac{1}{2}\theta)}{2k_0 \sin \tfrac{1}{2}\theta} r' \, dr'. \quad (24)$$

The function $f_0(\theta)$ may be expanded in a series of zonal harmonics for comparison with the exact expression (Chap. 6 (36)) and it is found that, in the approximate formula, the phase η_l in the exact formula is replaced by

$$\zeta_l = -\tfrac{1}{2}\pi \frac{2m}{\hbar^2} \int_0^\infty V_{00}(r)\{J_{l+\frac{1}{2}}(k_0 r)\}^2 \, dr. \tag{25}$$

It may be shown that,† when η_l is small it is given, to a good approximation by ζ_l (see Chap. 6 (59)).

The function $f_0(\theta)$ falls off steadily as θ increases, exhibiting no maxima and minima. The higher the energy of impact, the steeper the angular distribution becomes. To see the reason for this more clearly it is convenient to transform the expression for $f_0(\theta)$ by writing the potential $V_{00}(r)$ in the form

$$V_{00}(r) = -\frac{Ze^2}{r} + e^2 \int \frac{\rho(r')}{|\mathbf{r}-\mathbf{r}'|} \, d\mathbf{r}', \tag{26}$$

where the first term represents the potential due to the nucleus, of charge Ze, and the second that due to the atomic electrons whose density at distance r from the nucleus is $\rho(r)$. It may be shown‡ that

$$f_0(\theta) = \frac{e^2}{2mv_0^2}\{Z - F(\theta)\}\operatorname{cosec}^2 \tfrac{1}{2}\theta, \tag{27}$$

where

$$F(\theta) = 4\pi \int_0^\infty \rho(r') \frac{\sin(2k_0 r' \sin\tfrac{1}{2}\theta)}{2k_0 \sin\tfrac{1}{2}\theta} r' \, dr' \tag{28}$$

and is known as the atom form factor.

The function $r^2\rho(r)$ vanishes at the origin and possesses a maximum at $r = r_0$, say. For electron energies such that $k_0 r_0 \sin\tfrac{1}{2}\theta$ is large $F(\theta)$ will be small owing to the rapid oscillations of $\sin(2kr'\sin\tfrac{1}{2}\theta)$ for $r' \simeq r_0$. In particular $F(\theta) \to 0$ as $k_0 \to \infty$. In that case

$$f_0(\theta) \to (Ze^2/2mv_0^2)\operatorname{cosec}^2 \tfrac{1}{2}\theta$$

and the differential cross-section $I_0(\theta)\,d\omega$ has the familiar form of the Rutherford formula for scattering by a bare nucleus:

$$I_0(\theta)\,d\omega = (Z^2 e^4 / 4m^2 v_0^4)\operatorname{cosec}^4 \tfrac{1}{2}\theta \, d\omega. \tag{29}$$

However, as long as k_0 is finite, $I_0(\theta)$ will tend to a finite value for $\theta = 0$ given by

$$I_0(0) = (4m^2/\hbar^4)\left\{\int V_{00}(r')r'^2 \, dr'\right\}^2. \tag{30}$$

This limit is independent of the electron energy.

† See MOTT, N. F. and MASSEY, H. S. W., op. cit., 3rd edn, Chap. v, § 2 (Clarendon Press, Oxford, 1965).   ‡ op. cit. Chap. v, § 1.

We therefore have, at high energies, a distribution falling from the constant value at $\theta = 0$ to a value, at a moderate to large angle θ, which decreases as $(v_0 \sin \tfrac{1}{2}\theta)^{-4}$. The distribution thus gets rapidly steeper as the electron energy increases.

If the Fermi–Thomas statistical model for the atomic field is used (Chap. 6, § 2.3), $I_0(\theta)Z^{-\frac{2}{3}}$ may be calculated as a function of $Z^{-\frac{1}{3}}v_0 \sin \tfrac{1}{2}\theta$, giving in one table the distribution function for all atoms for which the model is valid. Such a table is given in *The theory of atomic collisions* by Mott and Massey, 3rd edition, Chapter 16, p. 462.

For the lighter atoms the statistical model is not very accurate and it is then necessary to use the Hartree self-consistent field or, if available, the Hartree–Fock field. Tables of $I_0(\theta)$ calculated in this way are also given in *The theory of atomic collisions*, Chapter 16, p. 460.

The validity of Born's approximation calculated from the Hartree–Fock field may be checked for helium from the observations of Hughes, McMillen, and Webb[†] (Chap. 5, Fig. 5.40). The comparison is given in Fig. 7.1 (a). It will be seen that the agreement is very good for the highest electron energies, but becomes progressively worse at the lower energies, at which two distinct types of disagreement may be distinguished. At large angles the distribution is flatter, and at small, steeper, than the calculated.

Fig. 7.1 (b) exhibits a similar comparison for argon, krypton, and xenon for 800-V electrons, the theoretical curves being calculated from the Hartree fields. The most interesting feature here is that, while good agreement is obtained over a considerable part of the angular range, at angles greater than 60° there is a considerable divergence. For argon it takes the same form as for helium, the observed distribution being very flat, but, for krypton and xenon, the observed curve exhibits maxima and minima.

It is not difficult to trace the way in which deviations from Born's approximation will begin to set in at large angles, as the electron energy decreases. We may write for the exact scattered amplitude $f(\theta)$

$$f(\theta) = f_b(\theta) + \{f(\theta) - f_b(\theta)\}, \tag{31}$$

where

$$f(\theta) - f_b(\theta) = (2ik)^{-1} \sum (2l+1)\{\exp(2i\eta_l) - 1 - 2i\zeta_l\}P_l(\cos\theta), \tag{32}$$

η_l is here the exact value for the lth-order phase, ζ_l that, (25), given from Born's approximation $f_b(\theta)$ for the scattered amplitude (see (27)). At high energies η_l is small and nearly equal to ζ_l for all l. As the energy

[†] Hughes, A. L., McMillen, J. H., and Webb, G. M., *Phys. Rev.* **41** (1932) 154.

decreases, the first phase that becomes too large for it to be given closely by the expression (25) is η_0, as this is the largest phase. Under these conditions the amplitude will take the form

$$f(\theta) = f_\mathrm{b}(\theta) + \{\exp(2i\eta_0) - 1 - 2i\zeta_0\}/2ik. \tag{33}$$

$f_\mathrm{b}(\theta)$ is very small at large angles, so the effect of the additional constant term will be to make the scattering very flat at these angles. This is just

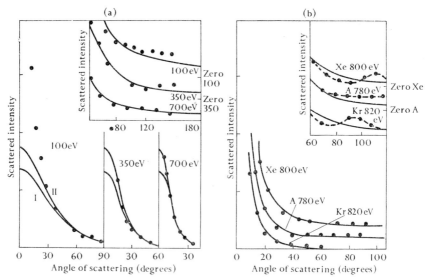

FIG. 7.1. Observed angular distribution of elastically scattered electrons compared with theoretical calculations based on Born's approximation and the Hartree atomic fields for (a) He for 100-, 350- and 700-eV electrons. (b) A, Kr, Xe for electrons of about 800 eV energy.

what is observed in helium (see Fig. 7.1 (a)) for 100-V and in argon (Fig. 7.1 (b)) for 780-V electrons.

As the energy decreases η_1 will become large, if the scattering field is strong enough, giving rise to an additional correcting term in the amplitude of the form $\{\exp(2i\eta_1) - 1 - 2i\zeta_1\}\cos\theta$. This will predominate over the zero-order correction when η_1 approaches $\tfrac{1}{2}\pi$ and the deviation from Born's approximation at large angles will have a form characteristic of $P_1(\cos\theta)$ with a minimum near 90°. This is observed with argon for 500-V, and neon for 200-V, electrons.

If the atomic field is large enough η_2 will next approach $\tfrac{1}{2}\pi$ and dominate the distribution. This is noticeable with 84-V electrons in argon (see Fig. 7.2) for which the distribution exhibits the minimum at 60° and maximum at 90° characteristic of $P_2(\cos\theta)$. No important influence

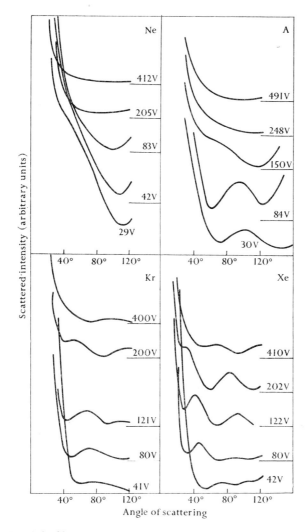

Fig. 7.2. Observed angular distribution of the electrons of energy 30–400 eV scattered elastically in Ne, A, Kr, Xe, showing how the shape of the distribution changes as the energy decreases and higher order harmonics become important.

of this term appears for neon at any energy (see Fig. 6.12 of Chap. 6 or Figs. 5.40 and 5.41 of Chap. 5) because the phase shift η_2 is never large for that atom. Again no effect due to phases of higher order than η_2 is apparent in argon, whereas, in krypton, η_3 is clearly important for 200-V electrons, for example, and in xenon for 400-V electrons. This is illustrated in Fig. 7.2. At lower energies (120 eV) in xenon, η_4 is clearly effective, but no sign of its influence is detectable in the results for krypton.

We see then that the different harmonics become successively effective in the same order both as the energy increases from the lowest value and as it decreases from very high values. In the second case, of course, the effect is the correction to be superposed on the smoothly falling distribution given by Born's approximation. This is the explanation of the phenomenon noted in Chapter 6, § 4.6 that the diffraction pattern becomes at first more complicated as the energy increases and then smoothes out again with further increase of energy, after going through the same sequence in the reverse order. The maximum complexity is determined by the highest order harmonic for which the phase attains a value $\frac{1}{2}\pi$. The heavier the atom, the greater this order will be, and hence the more complex the pattern.

So far we have been concerned with deviations from Born's approximation at large angles of scattering. It was pointed out, referring to Fig. 7.1 (a), that for helium the observed scattering at small angles exceeds that given by the approximation to an extent that decreases as the energy increases. The nature of this effect, which arises from interaction between elastic and inelastic scattering, will be discussed in Chapter 8, § 6.1.

For electrons of energy as high as 20 keV the elastic scattering at very small angles agrees quite well, at least in its variation with angle, with that predicted by Born's approximation. This is shown in Fig. 7.3, in which a comparison is made between Geiger's observations (see Chap. 5, § 5.2.3) for helium, neon, argon, and krypton[†] at angles of 5×10^{-2} rad and below, and calculated values obtained using the Hartree–Fock fields. In helium the observed values were normalized to agree with the calculated value at $\theta = 9 \cdot 4 \times 10^{-3}$ rad and for krypton at $\theta = 4 \times 10^{-2}$ rad. Instead of normalizing to agree at a definite angle for neon and argon the results were adjusted to give the best over-all fit with the theory. In all cases the agreement as to shape is very satisfactory.

It would be of considerable interest to trace the small angle scattering down to gradually lower energies to determine, for a number of atoms, when departures from Born's approximation become significant.

5. Inelastic collisions

5.1. *The cross-sections for excitation and ionization*

The expression (16) for I_{0n} with U_{0n} given by (8 a) may be simplified

[†] GEIGER, J., *Z. Phys.* **175** (1963) 530; ibid. **177** (1964) 138.

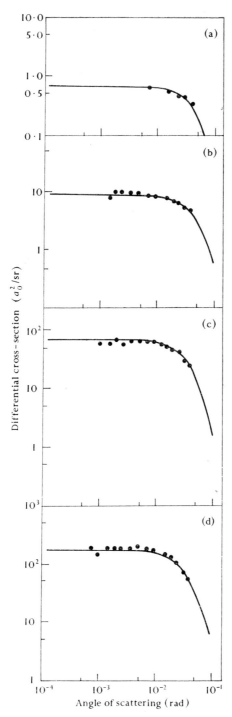

FIG. 7.3. Comparison of observed ● and calculated ——— angular distribution of electrons of energy 20 keV scattered elastically through small angles in (a) helium, (b) neon, (c) argon, and (d) krypton.

by carrying out the integration over the coordinates of the incident electron.† This gives

$$I_{0n} = \frac{4\pi^2 m^2 e^4}{h^4} \frac{k_n}{k_0} \left| \frac{4\pi}{K^2} \int \psi_0 \left(\sum_1^N \exp\{i(\mathbf{k}_0 - \mathbf{k}_n) \cdot \mathbf{r}_j\} \right) \psi_n^* \, d\mathbf{r}_1 \ldots d\mathbf{r}_N \right|^2 \quad (34)$$

where $K^2 = k_0^2 + k_n^2 - 2k_0 k_n \cos\theta$. In many cases the transition concerned can be regarded as one involving a single electron only so that, henceforward, we shall drop the summation over the atomic electrons.

We have for the cross-section Q_{0n} for excitation of the nth state when I_{0n} is independent of ϕ

$$Q_{0n} = 2\pi \int_0^\pi I_{0n}(\theta) \sin\theta \, d\theta.$$

It is convenient to change the variable of integration from θ to K so that

$$Q_{0n} = \frac{2\pi}{k_0 k_n} \int_{K_{\min}}^{K_{\max}} I_{0n}(K) K \, dK, \quad (35)$$

where $K_{\min} = k_0 - k_n$, $K_{\max} = k_0 + k_n$.

Since $k_0^2 = k_n^2 + (2m/\hbar^2)(E_n - E_0)$, we have, for collisions in which the energy transfer is small,

$$K_{\min} = k_0 - k_n \simeq (k_0^2 - k_n^2)/2k_0$$
$$= (E_n - E_0) m / k_0 \hbar^2, \quad (36)$$

since $k_0 + k_n \simeq 2k_0$.

An approximate expression may be found for Q_{0n} for these collisions.‡ The function I_{0n} falls off very steeply as K increases. This may be seen as follows. We have

$$I_{0n} = \frac{4m^2}{\hbar^4} \frac{k_n}{k_0} \frac{e^4}{K^4} |\epsilon_{0n}(K)|^2, \quad (37)$$

where

$$\epsilon_{0n}(K) = \int \psi_0 \exp(iKz) \psi_n^* \, d\mathbf{r}. \quad (38)$$

The integral will obviously be very small when many oscillations of $\exp(iKz)$ occur within the range of the functions ψ_0, ψ_n, i.e. when

$$K^2 \gg Z^2/a_0^2 = K_0^2, \quad \text{say}, \quad (39)$$

Z being the effective nuclear change in the ground state so that the radius of the ground-state orbit is a_0/Z. Alternatively, since $Z^2 e^2/2a_0$ is the ionization energy V_i of the normal atom, I_0 will be negligible when

$$K^2 \gg 2mV_i/\hbar^2 = K_0^2. \quad (40)$$

† BETHE, H., *Annln Phys.* 5 (1930) 325.
‡ See MOTT, N. F. and MASSEY, H. S. W., *The theory of atomic collisions*, 3rd edn, Chap. xvi, § 9 (Clarendon Press, Oxford, 1965).

To evaluate $\epsilon_{0n}(K)$ approximately we expand
$$e^{iKz} = 1 + iKz + \ldots . \tag{41}$$
Since $\int \psi_0 \psi_n^* \, d\mathbf{r} = 0$ we have
$$\epsilon_{0n}(K) \simeq iK \int \psi_0 z \psi_n^* \, d\mathbf{r}$$
$$= iKz_{0n}, \quad \text{say.} \tag{42}$$
This gives
$$I_{0n} = \frac{4m^2}{\hbar^4} \frac{k_n}{k_0} \frac{e^4}{K^2} |z_{0n}|^2, \tag{43}$$
an approximation certainly valid when $K < K_0$. For greater values of K, $I_{0n}(K)$ may be neglected so that we have, on substitution in (35),
$$Q_{0n} \simeq \frac{8\pi m^2 e^4}{k_0^2 \hbar^4} |z_{0n}|^2 \int_{K_{\min}}^{K_0} \frac{dK}{K} \tag{44a}$$
$$\simeq \frac{4\pi m^2 e^4}{k_0^2 \hbar^4} |z_{0n}|^2 \ln \frac{2mv_0^2}{E_n - E_0}, \tag{44b}$$
on insertion of the expressions (36) and (40) for $K_{\min}$ and K_0 respectively.

This shows that the cross-section falls off as $v_0^{-2} \log v_0$ at high velocities v_0 of impact provided z_{0n} does not vanish. If the transition from the ground to the nth state is optically allowed this will be so but, if it is not, z_{0n} will vanish and it is necessary to proceed to the next term in (41), to give
$$\epsilon_{0n}(K) \simeq K^2 (z^2)_{0n}$$
and
$$Q_{0n} \simeq \frac{4\pi m^3 e^4}{k_0^2 \hbar^6} |(z^2)_{0n}|^2 |E_0|. \tag{45}$$
This differs from the result for the optically allowed transition in that the rate of decrease at high velocities is slightly faster, as v_0^{-2} instead of $v_0^{-2} \ln v_0$.

A corresponding expression may be found for the cross-section for ionization. We first consider ionization in which the wave number κ of the ejected electron lies between κ and $\kappa + d\kappa$, i.e. so that its energy lies between $\kappa^2 \hbar^2 / 2m$ and $(\kappa^2 + 2\kappa \, d\kappa)\hbar^2 / 2m$. The corresponding cross-section $Q_{0\kappa} d\kappa$ is then found, by the same method as used above, to be given approximately by
$$Q_{0\kappa} \, d\kappa = \frac{4\pi m^2 e^4}{k_0^2 \hbar^4} |z_{0\kappa}|^2 \left(\ln \frac{2mv_0^2}{E_\kappa - E_0} \right) d\kappa. \tag{46}$$
The total cross-section for ionization, Q_i, is given by
$$Q_i = \int_0^{K_{\max}} Q_{0\kappa} \, d\kappa.$$

This may be evaluated approximately, using (39), to give
$$Q_i = \frac{2\pi e^4}{mv_0^2} \frac{a}{|E_0|} \ln\left(\frac{2mv_0^2}{C}\right), \tag{47}$$
where
$$a = \int |z_{0\kappa}|^2 \, d\kappa$$
and C is a certain mean of $E_\kappa - E_0$ which is of the order E_0.

The ionization cross-section thus decreases as the velocity of the incident electron increases at the same rate as that for an optically allowed excitation.

These approximations are very convenient for impacts with small fractional energy loss but they are not correct when a considerable fraction of the incident energy is lost. Exact evaluation of the expressions for I_{0n} and Q_n must then be carried out.

5.2. Relation to optical transition probabilities—generalized oscillator strengths

5.2.1. *Optical transition probabilities and oscillator strengths and cross-sections.* We begin by summarizing a number of formulae relating to optical transitions in atoms. These will also be found convenient in the discussions in Chapters 14 and 15.

The optical transition probability A_{ji} from a non-degenerate state j to a non-degenerate state i of a one-electron atom is such that the intensity of radiation of frequency ν_{ij}, given by
$$h\nu_{ij} = E_j - E_i, \tag{48}$$
radiated per second from N_j atoms in the excited state j is
$$N_j A_{ji} h\nu_{ij}. \tag{49}$$
In terms of the wave functions ψ_i, ψ_j of the initial and final states†
$$A_{ji} = (64\pi^4 \nu_{ij}^3 e^2/3hc^3)|\mathbf{r}_{ij}|^2, \tag{50}$$
where
$$\mathbf{r}_{ij} = \int \psi_i \mathbf{r} \psi_j^* \, d\mathbf{r},$$
ψ_i and ψ_j being the respective wave functions of the ith and jth states.

Consider now the rate at which energy is absorbed by N_i atoms in the state i from radiation of frequency ν_{ij} and density $\rho(\nu)$. This we write
$$N_i B_{ij} h\nu_{ij} \rho(\nu_{ij}). \tag{51}$$
It may then be shown that‡
$$B_{ij} = (c^3/8\pi h\nu_{ij}^3) A_{ji}. \tag{52}$$

† SLATER, J. C. and FRANK, N. H., *Introduction to theoretical physics*, 1st edn, p. 395 (McGraw-Hill, New York, 1933).
‡ EINSTEIN, A., *Phys. Z.* **18** (1917) 121.

A_{ji} and B_{ij} are often expressed in terms of the oscillator strength f_{ij}†
for the transition defined by

$$\mathrm{f}_{ij} = (8\pi^2 m \nu_{ij}/3h)|\mathbf{r}_{ij}|^2, \tag{53}$$

so that
$$A_{ji} = (8\pi^2 \nu_{ij}^2 e^2/mc^3)\mathrm{f}_{ij}, \tag{54}$$

$$B_{ij} = (\pi e^2/mh\nu_{ij})\mathrm{f}_{ij}. \tag{55}$$

These formulae may be generalized to apply to transitions in which one of the states lies in the continuum. Thus using (55) we see that (51) becomes
$$\frac{\pi e^2}{m} N_i \mathrm{f}_{ij} \rho(\nu_{ij}). \tag{56}$$

If now we consider the rate at which energy is absorbed by the N_i atoms from radiation of frequency between ν and $\nu+d\nu$, through transitions to the continuum, we may write this as

$$(\pi e^2/m) N_i \Delta \mathrm{f} \rho(\nu), \tag{57}$$

where $\Delta \mathrm{f}$ is the sum of the oscillator strength over the continuum states whose energies lie between E and $E+dE$ where

$$E = h\nu - E_i. \tag{58}$$

We may calculate $\Delta \mathrm{f}$ from (53) and (50) if we normalize the continuum wave function to the number of states per unit frequency range per unit volume, which is $(m^2 v/h^2)\,d\omega$ where $d\omega$ is the element of solid angle around the asymptotic direction (θ, ϕ) of motion of the electron in the continuum state. Thus we have

$$\Delta \mathrm{f} = (8\pi^2 m^3 v\nu/3h^3)\,d\nu \int |\mathbf{r}_{i\nu}|^2\,d\omega,$$

where the continuum wave function used in calculating $\mathbf{r}_{i\nu}$ has the asymptotic form of a plane wave of unit amplitude (modified by the Coulomb field, see Chap. 6, § 3.9) and the corresponding outgoing scattered wave.

It is usual to write $\Delta \mathrm{f}$ in terms of a differential oscillator strength $d\mathrm{f}/d\nu$ so that
$$\frac{d\mathrm{f}}{d\nu} = (8\pi^2 m^3 v\nu/3h^3) \int |\mathbf{r}_{i\nu}|^2\,d\omega. \tag{59}$$

We may also write the rate (57) in terms of the cross-section $Q_i^E(\nu)$ for photo-ionization of the atom by radiation of frequency ν. Thus we find for the rate
$$N_i c Q_i^E(\nu)\rho(\nu)\,d\nu. \tag{60}$$

† We use f_{ij} to distinguish the oscillator strength from the scattered amplitude, f_{ij}, for a transition between the two states excited by electron impact.

Equating this to (57) gives

$$Q_i^E(\nu) = (\pi e^2/mc)\,\mathrm{d}f/\mathrm{d}\nu \qquad (61)$$

$$= (8\pi^3 m^2 e^2 v\nu/3ch^3)\int |\mathbf{r}_{i\nu}|^2\,\mathrm{d}\omega. \qquad (62)$$

In a similar way we may consider the intensity of transitions from a continuous state of energy E to a discrete state i. The number of transitions per second that radiate quanta with energy between ν and $\nu+\mathrm{d}\nu$ is given according to (49) and (54) by

$$N_E(8\pi^2 e^2 \nu^2/mc^3)\frac{\mathrm{d}f}{\mathrm{d}\nu}\,\mathrm{d}\nu. \qquad (63)$$

Also if Q_E^i is the cross-section for a transition from the continuum state to the discrete state i the number of transitions per second is

$$N_E v Q_E^i g(\nu)\,\mathrm{d}\nu, \qquad (64)$$

where $g(\nu)\,\mathrm{d}\nu$ is the number of electron continuum states per unit volume in the energy range from $E-E_i$ to $E-E_i+h\,\mathrm{d}\nu$. Hence, equating (63) and (64),

$$Q_E^i = \{8\pi^2 e^2 \nu^2/mc^3 v g(\nu)\}\,\mathrm{d}f/\mathrm{d}\nu \qquad (65)$$

$$= \{8\pi\nu^2/c^2 v g(\nu)\}Q_i^E. \qquad (66)$$

Since
$$g(\nu) = 4\pi m^2 v/h^2, \qquad (67)$$

$$Q_E^i = (2h^2\nu^2/m^2v^2c^2)Q_i^E \qquad (68)$$

$$= (16\pi^3 \nu^3 e^2/3hc^3 v)\int |\mathbf{r}_{i\nu}|^2\,\mathrm{d}\omega. \qquad (69)$$

Finally, if both states lie in the continuum, we have for the cross-section $Q_E^{E'}\,\mathrm{d}\nu$ for a process in which the electron radiates a quantum of energy in the range $h\nu$ to $h(\nu+\mathrm{d}\nu)$ in changing its energy from E to between E' and $E'+\mathrm{d}E'$,

$$Q_E^{E'}\,\mathrm{d}\nu = \frac{64\pi^4 m^2 e^2 v' \nu^3}{3h^3 c^3 v}\iint |\mathbf{r}_{EE'}|^2\,\mathrm{d}\omega\mathrm{d}\nu. \qquad (70)$$

This is obtained simply from (69) by noting that the final wave function replacing ψ_i must be normalized to the number of energy levels per unit frequency range per unit volume.†

The corresponding absorption cross-section $Q_{E'}^E\,\mathrm{d}E'$ when the number of electrons per cm³ with energy between E' and $E'+\mathrm{d}E'$ is $n(E')\,\mathrm{d}E'$ is given by (62) with the initial wave function normalized to this electron

† It is important to note the normalization of the continuum wave functions is such that, whereas ψ_E has the asymptotic form of a Coulomb-distorted plane wave of unit amplitude plus an outgoing spherical wave, that of $\psi_{E'}$ differs in that the plane wave is associated with an incoming spherical wave. (See SOMMERFELD, A., *Annln Phys.* **11** (1931) 257.)

density, i.e.
$$Q^E_{E'} \, dE' = (8\pi^3 m^2 e^2 v v' / 3 c h^3) n(E') \int |\mathbf{r}_{EE'}|^2 \, d\omega dE'. \tag{71}$$

There is no difficulty in generalizing these results to transitions between degenerate states. Thus if g_i, g_j are the degrees of degeneracy of the two states

$$A_{ji} = \frac{1}{g_j} \sum_{\alpha=1}^{g_j} A^\alpha_{ij}, \qquad B_{ij} = \frac{1}{g_i} \sum_{\beta=1}^{g_i} B^\beta_{ij},$$

so that
$$B_{ij} = (c^3 / 8\pi h \nu^3_{ij})(g_j / g_i) A_{ji}. \tag{72}$$

Generalization to transitions involving many-electron atoms and ions presents no difficulty. All that is necessary is to interpret $\mathbf{r}$ in the matrix element as the sum of the position vectors of all the electrons, relative to the nucleus as origin.

The relation between the photo-ionization and electron capture cross-sections Q^E_i and Q^i_E now involves the degeneracies, if any, of the ith state of the atom, of the ion that is left after photo-ionization, and of the free electron. Denoting these by g_a, g_c, g_e respectively for the whole atom and for the core left after ionization we have in place of (68)

$$Q^i_E = (2 v^2 h^2 / m^2 c^2 v^2)(g_a / g_c g_e) Q^E_i. \tag{73}$$

5.2.2. Sum rules for oscillator strengths.
The optical oscillator strengths obey a number of important sum rules of which the simplest one is, for a one-electron atom,

$$\sum_j f_{ij} + \int \frac{df}{d\nu} d\nu = 1. \tag{74}$$

In terms of the cross-section Q^E_i this becomes

$$(mc/\pi e^2) \int Q^E_i(\nu) \, d\nu + \sum_j f_{ij} = 1, \tag{75}$$

which is often useful for checking observed or calculated photo-ionization or photodetachment cross-sections (see Chaps. 14 and 15).

The proof of (74) is simple. We have
$$f_{ij} = (2m/3\hbar^2)(E_i - E_j) \left| \int \psi_i \mathbf{r} \psi^*_j \, d\mathbf{r} \right|^2. \tag{76}$$

Now
$$\nabla^2 \psi_i + (2m/\hbar^2)(E_i - V)\psi_i = 0, \tag{77}$$
$$\nabla^2 \psi^*_j + (2m/\hbar^2)(E_j - V)\psi_j = 0. \tag{78}$$

Multiplying (77) by $x \psi^*_j$, (78) by $x \psi_i$, subtracting, and integrating gives

$$(2m/\hbar^2)(E_i - E_j) \int \psi_i x \psi^*_j \, d\mathbf{r} = \int (\psi_i \nabla^2 \psi^*_j - \psi^*_j \nabla^2 \psi_i) x \, d\mathbf{r}. \tag{79}$$

By use of Green's theorem the right-hand integral reduces to

$$2 \int \psi_j^* \frac{\partial \psi_i}{\partial x} \, d\mathbf{r}. \tag{80}$$

Multiplying both sides by $\int \psi_i^* x \psi_j \, d\mathbf{r}$ and summing over j† gives

$$(2m/\hbar^2) \sum_j (E_i - E_j) \left| \int \psi_i x \psi_j^* \, d\mathbf{r} \right|^2 = 2 \sum_i \int \psi_i^* x \psi_j \, d\mathbf{r} \int \psi_j^* \frac{\partial \psi_i}{\partial x} \, d\mathbf{r}$$

$$= 2 \int \psi_i^* x \frac{\partial \psi_i}{\partial x} \, d\mathbf{r}, \tag{81}$$

since

$$\sum_j \left(\int \psi_i^* x \psi_j \, d\mathbf{r} \right) \psi_j = x \psi_i^*.$$

But

$$2 \int \psi_i^* x \frac{\partial \psi_i}{\partial x} \, d\mathbf{r} = \int x \frac{\partial}{\partial x} |\psi_i|^2 \, d\mathbf{r}$$

$$= \int |\psi_i|^2 \, d\mathbf{r}$$

$$= 1,$$

so the sum on the left-hand side of (81) is equal to unity. Since the same results are obtained with y and z substituted for x the sum rule is established.

For an atom containing N electrons the sum rule becomes

$$\sum_j f_{ij} + \int \frac{df_{iE}}{dE} \, dE = N. \tag{82}$$

A second important sum is related to the theory of the refractive index μ of a gaseous medium composed of free atoms. This is given by

$$\frac{\mu^2 - 1}{\mu^2 + 2} = \frac{4\pi N \alpha}{3V}, \tag{83}$$

where N and V are Avogadro's number and the molar volume respectively and α is expressed in terms of the oscillator strengths by

$$\alpha = (e^2/4\pi^2 m) \sum_j \{f_{ij}/(\nu_{ij}^2 - \nu^2)\}. \tag{84}$$

When $\nu \to 0$, α is the static polarizability α_0 of the atoms of the medium so that we have the sum rule‡

$$\sum_j \frac{f_{ij}}{(E_i - E_j)^2} + \int (E_i - E)^{-2} \frac{df}{dE} \, dE = m\alpha_0/e^2\hbar^2. \tag{85}$$

The two sums involved in (82) and (85) are particular examples of the general sum

$$S_n = \sum_j (E_i - E_j)^n f_{ij} + \int (E_i - E)^n \frac{df}{dE} \, dE \tag{86}$$

for $n = 0$ and -2 respectively.

† This summation includes integration over the continuum states.
‡ VAN VLECK, J. H., *The theory of electric and magnetic susceptibilities* (Clarendon Press, Oxford, 1932).

For other values of n we have†

$$S_{-1} = (2m/3\hbar^2) \int \psi_0^* \Big(\sum_{s=1}^{N} \mathbf{r}_s\Big)^2 \psi_0 \, d\mathbf{r}_1 \ldots d\mathbf{r}_N, \tag{87}$$

$$S_1 = \tfrac{4}{3}\Big\{E_T - (\hbar^2/2m) \int \psi_0^* \Big(\sum_{s,t} \mathrm{grad}_s . \mathrm{grad}_t\Big)\psi_0 \, d\mathbf{r}_1 \ldots d\mathbf{r}_N\Big\}, \tag{88}$$

where E_T is the total energy of the atom, and

$$S_2 = (4\pi Z e^2 \hbar^2/3m) \int \psi_0^* \Big(\sum_{s} \delta(\mathbf{r}_s)\Big)\psi_0 \, d\mathbf{r}_1 \ldots d\mathbf{r}_N, \tag{89}$$

where $\delta(\mathbf{r}_s)$ is the three-dimensional Dirac δ-function and Z is the nuclear charge.

These further sum rules are of value for checking observed and calculated transition probabilities and cross-sections although, apart from S_0 and S_{-2}, it is necessary in using them to calculate certain average values over the ground state electron probability distribution. Applications are described in Chapter 14, § 7.1 and Chapter 15, § 5.1.

5.2.3. *Alternative forms for the transition matrix element.* So far we have expressed the formulae in terms of the matrix element $\mathbf{r}_{ij}$, which is usually referred to as the dipole length matrix element. It is possible to express $\mathbf{r}_{ij}$ in alternative forms and, of these, particular interest attaches to the so-called dipole velocity and acceleration matrix elements.

Thus we have shown in (79) and (80) that, for a one-electron atom,

$$x_{ij} = (\hbar^2/m)(E_i - E_j)^{-1} \int \psi_j^* \frac{\partial \psi_i}{\partial x} d\mathbf{r} \tag{90}$$

and hence
$$\mathbf{r}_{ij} = i\hbar(E_i - E_j)^{-1} \mathbf{v}_{ij}, \tag{91}$$

where $\mathbf{v}$ refers to the velocity operator $(-i\hbar/m)\mathrm{grad}$.

Carrying out a similar transformation with $\mathbf{v}_{ij}$ in place of $\mathbf{r}_{ij}$ gives

$$\mathbf{v}_{ij} = i\hbar(E_i - E_j)^{-1} \mathbf{a}_{ij}, \tag{92}$$

where $\mathbf{a}$ is now the acceleration operator $e^2\mathbf{r}/r^3$.‡ Thus

$$\mathbf{r}_{ij} = -\hbar^2(E_i - E_j)^{-2} \mathbf{a}_{ij}. \tag{93}$$

For an atom containing N electrons, the matrix elements are summed in all cases over the appropriate operators for the individual electrons.

Substituting (93) and (91) for $\mathbf{r}_{ij}$ gives three alternative expressions for the matrix elements that determine the transition probabilities. If *exact* atomic wave functions ψ_i, ψ_j are used all three will give exactly

---

† VINTI, J. P., *Phys. Rev.* **41** (1932) 432; see also DALGARNO, A. and LEWIS, J. T., *Proc. R. Soc.* A**233** (1955) 70.
‡ In terms of a potential energy of interaction V, $\mathbf{a} = -m^{-1} \mathrm{grad}\, V$.

the same results. This will not be so if only approximate forms are used for ψ_i and ψ_j. In this case a guide to the accuracy of the approximations will be provided by the extent to which the three formulae give approximately equal results. It is even more important to note that, whereas $\mathbf{a}_{ij}$ depends mainly on the wave functions at small distances from the nucleus, $\mathbf{r}_{ij}$ depends mainly on their values at large distances and $\mathbf{v}_{ij}$ on those at intermediate distances. Usually, approximate wave functions are obtained by variational methods which depend on accurate determination of the energy. For this purpose the main contribution comes from the wave functions near their maxima. These are at intermediate distances and so should give good values for $\mathbf{v}_{ij}$. They will be less accurate for $\mathbf{r}_{ij}$ and $\mathbf{a}_{ij}$. Hence, if a variational approximation is used for ψ_i and ψ_j, $\mathbf{v}_{ij}$ is often the most reliable matrix element to use. On the other hand, there may be circumstances in which ψ_i and ψ_j are known well asymptotically and then it will be best to use $\mathbf{r}_{ij}$. Applications on these lines will be discussed in §§ 5.1, 7.1, 7.2 of Chapter 15.

5.2.4. *Relation to Born's collision theory.* We return now to the formula (37) for the differential cross-section $(k_0 k_n)^{-1} I_{0n}(K) K \, dK d\phi$ for excitation of the nth state from the ground state by electron impact, namely

$$(k_0 k_n)^{-1} I_{0n}(K) K \, dK d\phi = (4m^2 e^4/\hbar^4 k_0^2 K^3)|\epsilon_{0n}(K)|^2 \, dK d\phi, \quad (94)$$

where

$$\epsilon_{0n}(K) = \int \psi_0 \, e^{iKz} \psi_n^* \, d\mathbf{r} \quad (95)$$

and $K\hbar$ is the momentum change suffered by the incident electron in the collision.

This formula shows that $(k_0/k_n)I_{0n}(K)$ is a function of K only and we now investigate the relation of the limiting form when $K \to 0$ to the optical oscillator strengths defined in § 5.2.1.

For this purpose we define a generalized oscillator strength

$$f_{0n}(K) = (2m/K^2\hbar^2)(E_n - E_0)|\epsilon_{0n}(K)|^2. \quad (96)$$

When $K \to 0$ this reduces to $f_{0n}(0)$ the usual optical oscillator strength defined in (53). We have then

$$I_{0n}(K) = (2me^4 k_n/k_0 K^2\hbar^2) f_{0n}(K)(E_n - E_0)^{-1}. \quad (97)$$

Hence if Born's first approximation is valid we may derive $f_{0n}(K)$ from observations of $I_{0n}(K)$ using (97). This derivation should be independent of the electron energy and angle of scattering, $f_{0n}(K)$ being a function of K only. Verification of this is in itself an important test of the validity of the approximation. Extrapolation to $K = 0$ should then yield the optical oscillator strength $f_{0n}(0)$ which can be checked against values derived from optical observations and, if determined for a sufficient number of upper states including continuum states, against the sum rules given in § 5.2.2. If the transition is not optically allowed $f_{0n}(0) = 0$ so that $f_{0n}(K)$ should extrapolate to zero. Applications of these considerations will be made in § 5.6.6 of this chapter and in §§ 1.3, 3.2, 4.2, 5.1, and 7.1 of Chapter 13.

The generalized oscillator strengths also satisfy sum rules for fixed values of K. Thus it is not difficult to show that

$$\sum_n f_{0n}(K) + \int \frac{df_{0E}}{dE}(K) \, dE = N, \tag{98}$$

where N is the number of atomic electrons, the method of proof following closely on that of (81).

5.2.5. *Alternative forms for Born matrix elements.*† It is possible to transform the matrix element $\epsilon_{0n}(K)$ to obtain alternative forms in much the same way as the transformation of the dipole matrix element $\mathbf{r}_{ij}$ discussed in § 5.2.3. Thus corresponding to the relation (91) we have

$$\epsilon_{0n}/iK = \int \psi_n^* \psi_0 (e^{iKz}/iK) \, d\mathbf{r}$$
$$= -(\hbar^2/m)(E_0 - E_n + K^2\hbar^2/2m)^{-1} \int \psi_n^* \frac{\partial \psi_0}{\partial z} e^{iKz} \, d\mathbf{r}. \tag{99}$$

Just as for the dipole matrix element this transformation may be useful in carrying out calculations with approximate wave functions. Thus the integral in (99) may be less sensitive to imperfections in these functions than the integral which appears in the usual formula for ϵ_{0n}.

5.3. *Impact parameter formulation*

We may approach the problem of calculating cross-sections for excitation of discrete states in an alternative way, making use of time-dependent perturbation theory. Although the results obtained are the same as given from Born's first approximation when it is applicable, the impact parameter formulation is often useful as a starting point for further approximations, particularly in dealing with collisions between heavy particles (see, for example, § 5.3.5 as well as Chap. 8, § 4.4, Chap. 17, §§ 4 and 5, and Chap. 18, § 8). For this reason we shall describe the method and its relation to Born's approximation as well as to semi-classical theory.

5.3.1. *Excitation by a time-dependent perturbation.*‡ It is convenient to begin by considering the excitation of an atom by a perturbing

† ALTSCHULER, S., *Phys. Rev.* **87** (1952) 992.
‡ See MOTT, N. F. and MASSEY, H. S. W., *The theory of atomic collisions*, 3rd edn, Chap. xxi, § 2 (Clarendon Press, Oxford, 1965).

potential that is a function of the time. Initially, at time $t = t_0$, the atom is in the ground state of energy E_0 so that the wave function is of the form

$$\Psi_0(\mathbf{r}, t) = \psi_0(\mathbf{r})\exp(-iE_0 t/\hbar), \tag{100}$$

where
$$(H-E_0)\psi_0 = 0, \tag{101}$$

H being the Hamiltonian operator appropriate to the internal motion within the atom. Due to the perturbation, at any subsequent time t there is a finite probability of finding the atom in the sth excited state, say. The full wave function for the perturbed system will satisfy

$$H\Psi - i\hbar \frac{\partial \Psi}{\partial t} = -V(r,t)\Psi. \tag{102}$$

We may expand Ψ in terms of the orthogonal functions ψ_s of the unperturbed atom so that

$$\Psi(\mathbf{r},t) = \sum_s a_s(t)\psi_s(\mathbf{r})\exp(-iE_s t/\hbar). \tag{103}$$

If at $t = t_0$ this solution reduces to $\Psi_0(r, t_0)$ then we see that, at time t the chance that the atom is now in the sth excited state is simply given by $|a_s(t)|^2$.

To determine $a_s(t)$ approximately we proceed as follows. Substituting (103) in (102) and using the fact that

$$(H-E_s)\psi_s = 0, \tag{104}$$

we have

$$i\hbar \sum_s \left[\left\{\frac{\mathrm{d}}{\mathrm{d}t}a_s(t)\right\}\psi_s(\mathbf{r})\exp(-iE_s t/\hbar)\right] = V(r,t)\Psi(\mathbf{r},t). \tag{105}$$

Multiplying both sides by $\psi_s^*(\mathbf{r})\exp(iE_s t/\hbar)$ and integrating over all space we have

$$\frac{\mathrm{d}}{\mathrm{d}t}(a_s(t)) = -\frac{i}{\hbar}\exp(iE_s t/\hbar)\int \psi_s^*(\mathbf{r})V(r,t)\Psi(\mathbf{r},t)\,\mathrm{d}\mathbf{r}. \tag{106}$$

Integrating and noting that $a_s(t_0) = 0$, $s \neq 0$ we have

$$a_s(t) = -\frac{i}{\hbar}\int_{t_0}^{t}\left\{\exp(iE_s t/\hbar)\int \psi_s^*(\mathbf{r})V(r,t)\Psi(\mathbf{r},t)\,\mathrm{d}\mathbf{r}\right\}\mathrm{d}t. \tag{107}$$

So far we have made no approximation but we must now do so in order to proceed further, Ψ being the unknown exact solution of (102). If the perturbation is small we may assume that all the a_s remain small when $s \neq 0$. Under these circumstances $\Psi(\mathbf{r}, t)$ will never depart very much from the initial form $\Psi_0'(\mathbf{r},t)$. Substituting Ψ_0' for Ψ' we obtain

$$a_s(t) = -\frac{i}{\hbar}\int_{t_0}^{t} \exp\{i(E_s-E_0)t/\hbar\}V_{s0}(t)\,dt, \qquad (108)$$

where
$$V_{s0}(t) = \int \psi_s^* V(r,t)\psi_0\,d\mathbf{r}.$$

We note for later reference (see Chap. 18, § 8) that, on substituting (103) for Ψ' on the right-hand side of (106) we have the infinite set of equations

$$\frac{d}{dt}a_s(t) = -\frac{i}{\hbar}\sum_q a_q(t)V_{sq}(t)\exp\{i(E_s-E_q)t/\hbar\}. \qquad (109)$$

5.3.2. *Excitation by a moving centre of force.* To apply this to the collision problem we suppose that the perturbation arises from the passage of a charged particle past the atom, it being assumed that the motion of the particle can be treated classically. We shall return to the examination of the validity of this assumption at a later stage.

For simplicity we take the atom as one of hydrogen, denoting the position vector of the atomic electron relative to the nucleus as $\mathbf{r}_2$ and that of the passing charged particle as $\mathbf{r}_1$ which will be a function of t. We then have

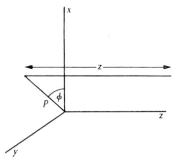

Fig. 7.4.

$$V = -Ze^2\left(\frac{1}{r_1} - \frac{1}{|\mathbf{r}_1-\mathbf{r}_2|}\right) \qquad (110)$$

and
$$V_{s0}(t) = Ze^2\int \psi_s^*(\mathbf{r}_2)\frac{1}{|\mathbf{r}_2-\mathbf{r}_1(t)|}\psi_0(\mathbf{r}_2)\,d\mathbf{r}_2, \qquad (111)$$

Ze being the charge on the particle.

In general $\mathbf{r}_1$ must be calculated as a function of t from classical orbit theory, but for present purposes we consider the simple case in which the deflecting force experienced by the charged particle throughout its motion is so small that its path is effectively a straight line and its velocity remains at the constant value $\mathbf{v}$. We now define the position of the particle in terms of cylindrical coordinates (z,p,ϕ) with the nucleus of the atom as origin and z-axis along the direction of motion of the charged particle. p is the impact parameter. Referring to

Fig. 7.4 we see that, if X, Y, Z are the rectangular coordinates of the particle
$$X = p\cos\phi, \qquad Y = p\sin\phi, \qquad Z = vt, \tag{112}$$
the origin of time being taken at the instant of closest approach to the nucleus. We now have
$$a_s(p,\infty) = -\frac{i}{\hbar v}\int_{-\infty}^{\infty} V_{s0}(X,Y,Z)\exp\{i(E_s-E_0)Z/\hbar v\}\,\mathrm{d}Z. \tag{113}$$

If the atom is bombarded by a beam of N charged particles per cm² per second the number incident per second on the areal element $p\,\mathrm{d}p\mathrm{d}\phi$ will be
$$Np\,\mathrm{d}p\mathrm{d}\phi.$$

It follows that the number of particles scattered per second after excitation of the sth state of the atom will be
$$N\int_0^\infty\!\!\int_0^{2\pi} |a_s(p,\infty)|^2 p\,\mathrm{d}p\mathrm{d}\phi = NQ_{0s}, \tag{114}$$
where Q_{0s} is the total effective cross-section for the inelastic collision.

5.3.3. *Relation to Born's approximation*

Before proceeding further it is of interest to show the relationship of the formula (114) for Q_{0s} to the corresponding formula given by Born's approximation. At the moment they appear very different. This analysis is of special interest in connection with heavy particle collisions (see Vols. III and IV).

We shall not attempt to prove that they are equivalent in the general case but will show how they can be related by dealing in detail with the special case in which V_{s0} is spherically symmetrical so that
$$Q_{0s} = 2\pi\int_0^\infty |a_s(p,\infty)|^2 p\,\mathrm{d}p, \tag{115}$$
where
$$a_s(p,\infty) = -\frac{i}{\hbar v}\int_{-\infty}^\infty V_{s0}\{\sqrt{(p^2+Z^2)}\}\exp(iZ\Delta E/\hbar v)\,\mathrm{d}Z \tag{116}$$
and we have written $E_s - E_0 = \Delta E$.

According to Born's approximation
$$Q_{0s} = 2\pi(k_s/k_0)\int_0^\infty |f_{0s}(\vartheta)|^2 \sin\vartheta\,\mathrm{d}\vartheta, \tag{117}$$
where (see (8b) and (17)),
$$f_{0s} = (2\pi m/h^2)\int V_{s0}(r)e^{i(\mathbf{k}_0-\mathbf{k}_s)\cdot\mathbf{r}}\,\mathrm{d}\mathbf{r}, \tag{118}$$
$\hat{\mathbf{k}}_0\cdot\hat{\mathbf{k}}_s = \cos\vartheta$, $k_0^2 - k_s^2 = 2m\Delta E/\hbar^2$.

Expanding the plane waves in the form†
$$e^{i\mathbf{k}_0\cdot\mathbf{r}} = \sum i^l(2l+1)j_l(k_0 r)P_l(\cos\theta), \tag{119}$$
$$e^{-i\mathbf{k}_s\cdot\mathbf{r}} = \sum i^{-l}(2l+1)j_l(k_s r)P_l(\cos\Theta), \tag{120}$$

† MOTT, N. F. and MASSEY, H. S. W., *The theory of atomic collisions*, 3rd edn, p. 21 (Clarendon Press, Oxford, 1965).

7.5 BORN'S APPROXIMATION

where
$$\cos\Theta = \cos\theta\cos\vartheta + \sin\theta\sin\vartheta\cos(\phi-\varphi), \tag{121}$$

and using the integral properties of the spherical harmonics we have

$$f_{0s} = (2m/\hbar^2)\sum(2l+1)P_l(\cos\vartheta)\int_0^\infty j_l(k_0 r)V_{s0}(r)j_l(k_s r)\,dr. \tag{122}$$

Hence
$$Q_{0s} = (16\pi m^2/\hbar^4)(k_s/k_0)\sum_l(2l+1)|\beta_l|^2, \tag{123}$$

where
$$\beta_l = \int_0^\infty j_l(k_0 r)V_{s0}(r)j_l(k_s r)\,dr. \tag{124}$$

If r_0 is the effective range of the atomic field the convergence of the series (123) will be very slow when $k_0 r_0$, $k_s r_0$ are large. Under these conditions

$$Q_{0s} \simeq 2(16\pi m^2/\hbar^4)(k_s/k_0)\int|\beta_l|^2 l\,dl. \tag{125}$$

Also for large l,[†] the spherical Bessel functions take the approximate form

$$j_l(kr) \simeq \begin{cases} (k\mu_l)^{-\frac{1}{2}}\sin\left(\tfrac{1}{4}\pi + \int_{r_l}^r \mu_l\,dr\right) & (r > r_l), \\ (k\nu_l)^{-\frac{1}{2}}\exp\left(\int_{r_l}^r \nu_l\,dr\right) & (r < r_l), \end{cases} \tag{126}$$

$$\mu_l^2 = k^2 - (l+\tfrac{1}{2})^2/r^2, \tag{127}$$

$$\nu_l^2 = -k^2 + (l+\tfrac{1}{2})^2/r^2, \tag{128}$$

and $r_l = (l+\tfrac{1}{2})/k$.

We may ignore the exponentially decreasing part for $r < r_l$ as its contribution to the integral will be small (cf. Fig. 6.3). This gives

$$\beta_l \simeq \tfrac{1}{2}(k_0 k_s)^{-\frac{1}{2}}\int_0^\infty [\mu_l(k_0 r)\mu_l(k_s r)]^{-\frac{1}{2}}V_{s0}(r)\times$$

$$\times\left[\cos\int_{r_l}^r\{\mu_l(k_0 r')-\mu_l(k_s r')\}\,dr' + \sin\int_{r_l}^r\{\mu_l(k_0 r')+\mu_l(k_s r')\}\,dr'\right]dr. \tag{129}$$

Under the conditions we have assumed the sine term oscillates very rapidly so that its contribution may be ignored. For the argument of the cosine we have

$$\mu_l(k_0 r')-\mu_l(k_s r') = \{\mu_l^2(k_0 r')-\mu_l^2(k_s r')\}/\{\mu_l(k_0 r')+\mu_l(k_s r')\}.$$

Now
$$\mu_l^2(k_0 r')-\mu_l^2(k_s r') = k_0^2 - k_s^2$$
$$= 2m\,\Delta E/\hbar^2,$$

and since $k_0^2 - k_s^2 \ll k_0^2$,

$$\mu_l(k_0 r')+\mu_l(k_s r') \simeq 2k_0\{1-(l+\tfrac{1}{2})^2/k_0^2 r'^2\}^{\frac{1}{2}}$$
$$= (2mv_0/\hbar)\{1-(l+\tfrac{1}{2})^2\hbar^2/m^2 v_0^2 r'^2\}^{\frac{1}{2}}.$$

Hence
$$\mu_l(k_0 r')-\mu_l(k_s r') \simeq (\Delta E/\hbar v_0)[1-\{(l+\tfrac{1}{2})^2\hbar/mv_0 r'^2\}^2]^{-\frac{1}{2}}. \tag{130}$$

Furthermore we may write in (129)

$$(k_0 k_s)^{\frac{1}{2}} \simeq k_0, \qquad \{\mu_l(k_0 r)\mu_l(k_s r)\}^{-\frac{1}{2}} \simeq \{\mu_l(k_0 r)\}^{-1},$$

[†] JEFFREYS, H. and JEFFREYS, B., *Methods of mathematical physics*, Chap. 21 (Cambridge University Press, 1946).

so
$$\beta_l \simeq \frac{1}{2k_0} \int_0^\infty V_{s0}(r)\{\mu_l(k_0 r)\}^{-1} \cos\left\{\frac{\Delta E}{\hbar v_0} \int_p^r (1-p^2/r'^2)^{-\frac{1}{2}} dr'\right\} dr, \qquad (131)$$

where $p = (l+\tfrac{1}{2})\hbar/mv_0$.

The integral in the argument of the cosine is easily evaluated to give

$$\int_p^r r'(r'^2-p^2)^{-\frac{1}{2}} dr' = (r^2-p^2)^{\frac{1}{2}}. \qquad (132)$$

Hence, since $\mu_l(k_0 r) = k_0(1-p^2/r^2)^{\frac{1}{2}}$,

$$\beta_l = \frac{1}{2k_0^2} \int_0^\infty V_{s0}(r) r(r^2-p^2)^{-\frac{1}{2}} \cos\left\{\frac{\Delta E}{\hbar v_0}(r^2-p^2)^{\frac{1}{2}}\right\} dr$$

$$= \frac{1}{2k_0^2} \int_0^\infty V_{s0}\{\sqrt{(z^2+p^2)}\} \cos(z\Delta E/\hbar v_0) \, dz$$

$$= \frac{1}{4k_0^2} \int_{-\infty}^\infty V_{s0}\{\sqrt{(z^2+p^2)}\} \exp(iz\Delta E/\hbar v_0) \, dz$$

$$= \frac{i\hbar^2}{4mk_0} a_s(p, \infty).$$

To recover (114) we merely change the variable in (125) from l to p and take $k_0 \simeq k_s$.

The correlation between the angular momentum quantum number l and the impact parameter p is just what would be expected in a semi-classical theory, for the classical angular momentum is mvp which is quantized as $(l+\tfrac{1}{2})\hbar$ when l is large.

It is clear that we can only expect the formula (114) to be valid when the energy of the charged particle is great compared with that required to produce the excitation. In addition it is necessary for the classical description of its motion to be valid so that the particle can be represented by a wave packet which is of dimensions small compared with the impact parameter p and that p should be large compared with the spread of the atomic field.

5.3.4. *Excitation of optically allowed transitions*†

Although we have discussed the case in which V_{s0} is spherically symmetrical there is no difficulty in establishing the equivalence of the two formulae under the same conditions when V_{s0} is not spherically symmetrical. One of the most important cases, that of excitation of optically allowed transitions, is one example of this kind.

To evaluate V_{s0} we have, since

$$r_{12}^{-1} = \sum_s \gamma_s(r_1, r_2) P_s(\cos\Theta_{12})$$

where
$$\cos\Theta_{12} = \cos\theta_1 \cos\theta_2 + \sin\theta_1 \sin\theta_2 \cos(\phi_1-\phi_2)$$

† Seaton, M. J., *Proc. phys. Soc.* **79** (1962) 1105.

and
$$\gamma_s(r_1, r_2) = \begin{cases} r_1^s/r_2^{s+1} & (r_2 > r_1) \\ r_2^s/r_1^{s+1} & (r_2 < r_1), \end{cases}$$
$$V_{s0} = Ze^2 \int \psi_s^*(\mathbf{r}_2) \sum_s \gamma_s(r_1, r_2) P_s(\cos\Theta_{12}) \psi_0(r_2) \, d\mathbf{r}_2. \tag{133}$$

For optically allowed transitions the only term that contributes is that with $s = 1$ so that
$$V_{s0} = Ze^2 \int \psi_s^*(\mathbf{r}_2) \gamma_1(r_1, r_2) \cos\Theta_{12} \psi_0(r_2) \, d\mathbf{r}_2. \tag{134}$$

Ignoring the contribution to this integral from large r_2 we have
$$V_{s0} \simeq (Ze^2/r_1^3) \int \psi_s^*(\mathbf{r}_2) r_1 r_2 \cos\Theta_{12} \psi_0(r_2) \, d\mathbf{r}_2$$
$$= (Ze^2/r_1^3) \int \psi_s^*(\mathbf{r}_2)(x_1 x_2 + y_1 y_2 + z_1 z_2) \psi_0(r_2) \, d\mathbf{r}_2$$
$$= (Ze^2/r_1^3)\mathbf{r}_1 \cdot \int \psi_s^*(\mathbf{r}_2) \mathbf{r}_2 \psi_0(r_2) \, d\mathbf{r}_2$$
$$= (Ze^2/r_1^3)\mathbf{r}_1 \cdot \mathbf{r}_{0s}. \tag{135}$$

Hence, from (113),
$$|a_{0s}(p, \infty)|^2 = (Z^2 e^4/\hbar^2 v_0^2)|\mathbf{r}_{0s}|^2 \left| \int_{-\infty}^{\infty} \exp(i\Delta E z/\hbar v)(\mathbf{r}/r^3) \, dz \right|^2. \tag{136}$$

In terms of cylindrical coordinates
$$\mathbf{r} = p\cos\phi\,\mathbf{i} + p\sin\phi\,\mathbf{j} + z\mathbf{k}.$$

Also†
$$\int_{-\infty}^{\infty} \frac{e^{i\alpha z}}{(z^2+p^2)^{3/2}} \, dz = 2\int_0^{\infty} \frac{\cos\alpha z}{(z^2+p^2)^{3/2}} \, dz = 2\alpha^2 \int_0^{\infty} \frac{\cos r}{(r^2+\alpha^2 p^2)^{3/2}} \, dr$$
$$= -2(\alpha/p) K_1(\alpha p), \tag{137a}$$
where K_1 is the Bessel function of the second kind.

Further†
$$\int_{-\infty}^{\infty} \frac{z e^{i\alpha z}}{(z^2+p^2)^{3/2}} \, dz = -i\frac{\partial}{\partial\alpha} \int_{-\infty}^{\infty} \frac{e^{i\alpha z}}{(z^2+p^2)^{3/2}} \, dz = 2i\alpha K_0(\alpha p). \tag{137b}$$

Hence we have
$$|a_{0s}(p, \infty)|^2 = (4Z^2 e^4/\hbar^2 v_0^2)|\mathbf{r}_{0s}|^2 \alpha^2 \{K_0^2(\alpha p) + K_1^2(\alpha p)\}, \tag{138}$$
where
$$\alpha = \Delta E/\hbar v_0.$$

In terms of the oscillator strength f_{0s} for the transition we have, from (53),
$$|a_{0s}(p, \infty)|^2 = (2Z^2 e^4 \Delta E/mv_0^4 \hbar^2) f_{0s}\{K_0^2(\alpha p) + K_1^2(\alpha p)\}. \tag{139}$$

We are now in a position to relate the impact parameter method to Born's approximation. The total cross-section for all collisions leading to excitation of the sth state, in which the impact parameter exceeds p_0, is given by
$$Q_{0s}^{p_0} = 2\pi \int_{p_0}^{\infty} |a_{0s}(p, \infty)|^2 p \, dp$$
$$= (4\pi Z^2 e^4 \Delta E f_{0s}/mv_0^4 \hbar^2) \int_{p_0}^{\infty} p\{K_0^2(\alpha p) + K_1^2(\alpha p)\} \, dp$$
$$= -(4\pi Z^2 e^4 f_{0s}/mv_0^2 \Delta E)\alpha p_0 K_0(\alpha p_0) K_1(\alpha p_0). \tag{140}$$

† JEFFREYS, H. and JEFFREYS, B., *Methods of mathematical physics*, Chap. 21, p. 558 (Cambridge University Press, 1946).

For high velocity impacts $\alpha p_0 \ll 1$ and we use the series expansions†

$$K_0(z) = -\ln(\tfrac{1}{2}z) - \gamma + \text{higher order terms},$$
$$K_1(z) = -1/z + \text{higher order terms},$$

γ, which is Euler's constant, equals 0·5772 or $\ln 1\cdot 781$.

We have then, for these impacts,

$$Q_{0s}^{p_0} \simeq (4\pi Z^2 e^4 \mathrm{f}_{0s}/mv_0^2 \Delta E)\ln(1\cdot 1229 \hbar v_0/p_0 \Delta E). \tag{141}$$

Turning to (44a) and using the definition (53) of f_{0s},

$$Q_{0s} = (4\pi Z^2 e^4 \mathrm{f}_{0s}/mv_0^2 \Delta E)\ln(K_0 \hbar v/\Delta E), \tag{142}$$

so that the high energy (Bethe) approximation agrees with the impact parameter formula provided we take

$$p_0 = 1\cdot 1229/K_0. \tag{143}$$

However, whereas with the Bethe approximation with constant momentum transfer cut-off K_0 the formula (142) is obtained at all impact energies, formula (141) is replaced by the more accurate formula (140) when we use a fixed impact parameter cut-off. This is equivalent to neglecting the contribution from incident angular momenta

$$(l+\tfrac{1}{2})\hbar > mv_0 p_0.$$

It has been suggested that (140) with a suitable value for p_0 is capable of giving better results at low electron energies than the full Born approximation. This is usually true in the sense that, taking p_0 of the order of the atomic radius, (140) gives a smaller and hence better result at these energies. How much theoretical significance attaches to this when, at all values of p, $|a_s(p,\infty)|^2 \ll 1$ is doubtful.

5.3.5. *Application to transitions between closely coupled states.* However, many cases arise in which the oscillator strength for the transition is so high that $|a_s(p,\infty)|^2$ becomes > 1 already for $p > p_1$ where $p_1 > p_0$. We refer to such a case as one of strong coupling. It is reasonable to assume that, for $p < p_1$, a_{0s} oscillates quite rapidly as a function of p so that the mean value of $|a_s(p,\infty)|^2$ for $p < p_1$ is $\tfrac{1}{2}$. Seaton‡ has provided evidence in support of this by considering a schematic model and has suggested the following procedure.

We choose a value p_m of p such that

$$|a_s(p_m,\infty)|^2 = \tfrac{1}{2}$$

† WHITTAKER, E. T. and WATSON, G. N., *Modern analysis*, 4th edn, p. 374 (Cambridge University Press, 1927).
‡ SEATON, M. J., *Proc. phys. Soc.* A**68** (1955) 457; ibid. **79** (1962) 1165.

and then
$$Q_{0s} = 2\pi \int_0^\infty |a_s(p_m, \infty)|^2 p \, dp$$
$$\simeq 2\pi \int_0^{p_m} \tfrac{1}{2} p \, dp + \int_{p_m}^\infty |a_s(p, \infty)|^2 p \, dp$$
$$= \tfrac{1}{2}\pi p_m^2 + Q_{0s}^{p_m}, \tag{144}$$

where $Q_{0s}^{p_m}$ is given by (140).

Applications of this method will be discussed in Chapter 8, §§ 4.4 and 8.1.

5.4. Electron exchange†

According to the formula (34) the chance of exciting a level with a multiplicity different from that of the ground state would be zero for an atom in which spin-orbit coupling is negligible. Physically, this is due to the impossibility of changing the total electron spin in the impact; mathematically, to the different symmetry properties possessed by the functions if they correspond to states of different multiplicity so that the integral in (34) always vanishes. Thus, for the helium singlet states, ψ is symmetrical in the coordinates of the two electrons whereas for the triplet states it is antisymmetrical.

It must be remembered, however, that it is the total spin of the atom plus the incident electron which must be conserved, not necessarily that of the atom alone. The formula (34) requires conservation of the spin of the atomic levels because it ignores the possibility of electron exchange occurring in the collision.

Suppose that the total electron spin quantum number in the initial state of the atom is s so that the multiplicity is $2s+1$. The spin quantum number of the incident electron is $\tfrac{1}{2}$, so that the total spin quantum number of the atom+incident electron is either $s \pm \tfrac{1}{2}$. If a state of the atom with spin s' is excited the total spin will be $s' \pm \tfrac{1}{2}$. Hence, in order for this excitation to be possible without change of the total spin, $s' \pm \tfrac{1}{2}$ must fall within $s \pm \tfrac{1}{2}$, i.e. $s' = s-1$, s, or $s+1$. The multiplicity of the state can therefore change by ± 2, but this may occur only if the spin direction of the outgoing electron is opposite to that of the incident— electron exchange must have taken place in the collision, for we are assuming that spin-orbit interaction is negligible so the spin of the incident electron cannot be reversed by the impact. Exchange can also occur between electrons with the same spin but in this case it will not be associated with a change of multiplicity.

† OPPENHEIMER, J. R., *Phys. Rev.* 32 (1928) 361; MOTT, N. F. and MASSEY, H. S. W., *The theory of atomic collisions*, 3rd edn, Chap. xv, § 1, Chap. xvi, § 10 (Clarendon Press, Oxford, 1965).

These considerations need modification if spin-orbit coupling is not negligible. This will be so for heavy atoms such as mercury. Under these conditions the multiplet classification in terms of the total spin (L–S coupling) is no longer a good approximation. For a two-electron system such as the outer shell of mercury this means that certain triplet states include a singlet admixture and vice versa. The formula (34) will not give a vanishing cross-section for a singlet–triplet transition when this is so—either or both of the functions ψ_0, ψ_n will not be completely symmetric and antisymmetric respectively in the space coordinates of the two electrons.

5.4.1. *The Born–Oppenheimer approximation.* It is possible to allow for electron exchange to a degree of accuracy comparable with Born's approximation. In the formula (17), $f_{0n}(\theta)$ can be regarded as the amplitude contributed by direct scattering. It represents a cross mean of the interaction energy between the electron and the atom, averaged over the initial and final wave functions of the combined system. The wave functions are $\psi_0(r_2)e^{i\mathbf{k}_0 \cdot \mathbf{r}_1}$, $\psi_n(r_2)e^{i\mathbf{k}_n \cdot \mathbf{r}_1}$ respectively. We would expect that the corresponding amplitude $g_{0n}(\theta)$ due to exchange should be given by a similar expression in which the initial wave function is as above but, in the final one, the electrons 1 and 2 are interchanged so that it becomes $\psi_n(\mathbf{r}_1)e^{i\mathbf{k}_n \cdot \mathbf{r}_2}$.

This is indeed the case. For collisions of electrons with hydrogen atoms the exchange amplitude is given by†

$$g_{0n}(\theta) = -(2\pi m e^2/h^2) \int\int \left(\frac{1}{r_1} - \frac{1}{r_{12}}\right) \psi_0(r_2) \psi_n^*(\mathbf{r}_1) \times$$
$$\times \exp i(\mathbf{k}_0 \cdot \mathbf{r}_1 - \mathbf{k}_n \cdot \mathbf{r}_2)\, d\mathbf{r}_1\, d\mathbf{r}_2. \quad (145)$$

In this case, however, it is to be noted that the contribution from the nuclear interaction $1/r_1$ does not vanish. This must be regarded as a defect in the approximation. Nevertheless, neglect of this term is not justifiable, as modifications in the formula that eliminate it would also modify the contribution from the interelectronic interaction. Until a more satisfactory formula is available it is best to retain the nuclear interaction term. One further question occurs here. It is not clear whether this term should be taken as $1/r_1$ or $1/r_2$. This difficulty does not arise if the wave functions ψ_0, ψ_n are exact solutions of the wave equation for the atom as both terms then give the same result.‡ If

† Mott, N. F. and Massey, H. S. W., *The theory of atomic collisions*, 3rd edn, Chap. viii, p. 414 (Clarendon Press, Oxford, 1965).
‡ Bates, D. R., Fundaminsky, A., and Massey, H. S. W., *Phil. Trans. R. Soc.* A243 (1950) 93,

approximate solutions only are available for ψ_0 and ψ_n, this is no longer true in general. In such a case there exist no means of deciding whether the prior, $1/r_1$, or post, $1/r_2$, interaction, or some mean between them, is the best to choose.

The total scattered amplitude is obtained by a linear combination of f and g determined by the number of electrons and multiplicities of the states involved. Thus for atomic hydrogen, in which there can be no change of multiplicity, $|f_{0n}|^2$ in the expression for I_{0n} is replaced by

$$\tfrac{1}{4}|f_{0n}+g_{0n}|^2+\tfrac{3}{4}|f_{0n}-g_{0n}|^2. \tag{146}$$

For helium, on the other hand, for which the ground state is a singlet, excitation can occur either to a singlet or triplet state. In the former case, in which no multiplicity change occurs $|f_{0n}|^2$ is replaced by $|f_{0n}-g_{0n}|^2$ and in the latter, which can only arise from electron exchange, by $3|g_{0n}|^2$. g_{0n} is now given by

$$g_{0n}(\theta) = -(2\pi me^2/h^2) \int\int\int \left(\frac{2}{r_1}-\frac{1}{r_{12}}-\frac{1}{r_{13}}\right)\psi_0(r_2,r_3)\psi_n^*(\mathbf{r}_1,\mathbf{r}_3)\times$$
$$\times \exp i(\mathbf{k}_0.\mathbf{r}_1-\mathbf{k}_n.\mathbf{r}_2)\,d\mathbf{r}_1\,d\mathbf{r}_2\,d\mathbf{r}_3. \tag{147}$$

Again, if ψ_0 and ψ_n are exact solutions of the helium wave equation, it is immaterial whether the interaction is taken as in (147) or replaced by one in which 1 and 2 are interchanged. This holds also if the Hartree self-consistent approximations (see Chap. 6, § 2) to ψ_0 and ψ_n are used, provided the energy difference between the states is taken to be that given by the same approximation. For other approximations, such as those obtained by use of the variation method, or by the Hartree–Fock method, it is no longer true. This presents a serious difficulty in applying these formulae to atoms with many electrons and already leads to uncertainty for helium. This and other aspects of the approximate theory of electron exchange are further discussed in § 5.6.9.

For convenience of reference we shall refer to the approximation in which $g_{0n}(\theta)$ is included in forms such as (147) as the Born–Oppenheimer approximation, to distinguish it from the Born approximation in which $g_{0n}(\theta)$ is ignored.

5.4.2. *Ochkur's approximation.* Ochkur† has obtained a simple means of approximating to the exchange amplitudes at not too low energies of impact. At such energies the contribution to (145) from the term in $1/r_1$ is very small. We now use the Fourier transform of r_{12}^{-1}

$$r_{12}^{-1} = (1/2\pi^2) \int \exp\{i\mathbf{q}.(\mathbf{r}_1-\mathbf{r}_2)\}q^{-2}\,d\mathbf{q}, \tag{148}$$

† OCHKUR, V. I., *Soviet Phys. JETP* **18** (1964) 503.

so that

$$g_{0n}(\theta) = (-me^2/\pi h^2) \iiint \exp\{i\mathbf{q}.(\mathbf{r}_1-\mathbf{r}_2)\} q^{-2} \psi_0(r_2) \psi_n^*(\mathbf{r}_1) \times$$
$$\times \exp\{i(\mathbf{k}_0.\mathbf{r}_1 - \mathbf{k}_n.\mathbf{r}_2)\} \, d\mathbf{r}_1 d\mathbf{r}_2 d\mathbf{q}$$
$$= (-me^2/\pi h^2) \iiint \exp[i\{(\mathbf{q}+\mathbf{k}_0).\mathbf{r}_1 - (\mathbf{q}+\mathbf{k}_n).\mathbf{r}_2\}] \times$$
$$\times q^{-2} \psi_0(r_2) \psi_n^*(\mathbf{r}_1) \, d\mathbf{r}_1 d\mathbf{r}_2 d\mathbf{q}$$
$$= (-me^2/\pi h^2) \iiint \exp[i\{\mathbf{p}.(\mathbf{r}_1-\mathbf{r}_2)+\mathbf{K}.\mathbf{r}_2\}] \times$$
$$\times |\mathbf{p}-\mathbf{k}_0|^{-2} \psi_0(r_2) \psi_n^*(\mathbf{r}_1) \, d\mathbf{r}_1 d\mathbf{r}_2 d\mathbf{p}, \quad (149)$$

where $\mathbf{k}_0 - \mathbf{k}_n = \mathbf{K}$.

At sufficiently high energies we neglect p in comparison with k_0 in $|\mathbf{p}-\mathbf{k}_0|^{-2}$. This leaves

$$\int \exp\{i\mathbf{p}.(\mathbf{r}_1-\mathbf{r}_2)\} \, d\mathbf{p} = 8\pi^3 \delta(\mathbf{r}_1-\mathbf{r}_2), \quad (150)$$

giving

$$g_{0n}(\theta) = -(8\pi^2 me^2/k_0^2 h^2) \int \exp(i\mathbf{K}.\mathbf{r}_1) \psi_0(r_1) \psi_n^*(\mathbf{r}_1) \, d\mathbf{r}_1$$
$$= (K^2/k_0^2) f_{0n}(\theta). \quad (151)$$

At high energies $f_{0n}(\theta)$ is concentrated at very small angles for which

$$K \simeq k_0 - k_n$$
$$= (E_n - E_0) m / k_0 h^2, \quad (152)$$

giving $\qquad g_{0n}(\theta) \simeq \{(E_n - E_0)/E\}^2 f_{0n}(\theta). \quad (153)$

Thus for electrons of 150 eV energy $g_{0n}(\theta)$ is only about 10^{-2} of $f_{0n}(\theta)$, while at 400 eV it will have fallen to $10^{-3} f_{0n}(\theta)$.

It follows that the cross-section for excitation of a state that involves change of multiplicity will fall very rapidly with electron energy as it increases above 100 eV or so. Equally well it follows that, when no change of multiplicity is involved, the effect of electron exchange on the cross-section is insignificant under the same energy conditions. In fact, under these circumstances, the Born–Oppenheimer approximation adds little as compared with the simpler Born approximation. When it is valid the difference between the two is very small.

The Ochkur approximation to the Born–Oppenheimer exchange integral is only a good representation of that integral when the electron energy is high enough for it to be small. There is no theoretical justification for supposing that it remains a good approximation to the true exchange amplitude at low electron energies, even though it is then much smaller than the full Born–Oppenheimer approximation. Any apparent success that the extrapolated use of the Ochkur approximation may have at low energy in improving agreement with experiment must be regarded as empirical.

5.5. *Theoretical limit to the magnitude of collision cross-sections*

As we are about to discuss the range of validity of the Born approximation it is important to have available any general formulae that enable an upper limit to be placed on the size of the cross-section. Although it is not possible to do this when the angular momentum $\{l(l+1)\}^{\frac{1}{2}}\hbar$ about the centre of force is not known to lie within fairly narrow limits, a definite limit may be obtained for the partial cross-section due to electrons with a definite angular momentum. Correspondence with classical theory indicates that such a limit is likely to exist.

The contribution dQ to the cross-section due to particles with angular momentum about the centre of force between J and $J+dJ$ is, as in Chapter 6 (10),
$$dQ = 2\pi J \, dJ/m^2v^2, \tag{154}$$
where m is the mass and v the velocity of the electron. Allowing for quantization of angular momentum so that $J \, dJ \simeq (l+\tfrac{1}{2})\hbar$, we would expect this to be replaced in quantum theory by
$$q_l = \pi(2l+1)p(l)/k^2, \tag{155}$$
where $k = mv/\hbar$ and $p(l)$ is a probability factor $\not> 1$. This suggests that the maximum cross-section for electrons of angular momentum $\{l(l+1)\}^{\frac{1}{2}}\hbar$ should be $\pi(2l+1)/k^2$, but care must be taken to allow for the effect of edge diffraction (see Chap. 6, p. 381 and Chap. 16, § 3).

A detailed formulation† of the problem in terms of the conservation of particles shows that the limits are as follows:

(a) The maximum possible partial cross-section for all inelastic collisions is $\pi(2l+1)/k^2$.

(b) When this maximum is obtained the partial cross-section for elastic collisions must also be $\pi(2l+1)/k^2$.

(c) The maximum possible partial cross-section for all collisions is $4\pi(2l+1)/k^2$, in which case there are no inelastic collisions.

These limits are not useful when many partial cross-sections contribute. This is the case for direct excitation of atoms by electron impact. On the other hand, as will appear below, the exchange contribution, when it is at all large, usually arises from incident electrons with one, or at most two, particular angular momenta. The limit $\pi(2l+1)/k^2$ provides an effective check in these cases.

† See MOTT, N. F. and MASSEY, H. S. W., *The theory of atomic collisions*, 3rd edn, Chap. xii, § 1 (Clarendon Press, Oxford, 1965).

5.6. Application of Born's first approximation and its range of validity

5.6.1. *Excitation of atomic hydrogen.* The calculation of differential and total cross-sections for excitation of atomic hydrogen from the ground state according to the formulae (16) and (17) of Born's first approximation offers no special difficulty and may be carried out analytically.

A general formula for the matrix element $\epsilon_{0,nlm}(K)$ of (94) for excitation of a state with quantum numbers n, l, m has been given by Massey and Mohr.† For the different states with $n = 2$ and 3 their formulae give the following expressions for $I_{0,nl}(K)$ and for the generalized oscillator strengths $f_{0,nl}(K)$. Dropping for brevity the suffix 0, we have

$n = 2, l = 0$
$$f_{20}(K) = \tfrac{3}{16} K^2(k_0/k_2) I_{20}(K) = 3 \times 2^{15} K^2 a_0^2 (9+4K^2 a_0^2)^{-6}; \quad (156)$$

$n = 2, l = 1$
$$f_{21}(K) = \tfrac{3}{16} K^2(k_0/k_2) I_{21}(K) = 3^3 \times 2^{13} (9+4K^2 a_0^2)^{-6}; \quad (157)$$

$n = 3, l = 0$
$$f_{30}(K) = \tfrac{2}{9} K^2(k_0/k_3) I_{30}(K) = 3^5 2^{11} K^2 a_0^2 (16+27K^2 a_0^2)^2 \times$$
$$\times (16+9K^2 a_0^2)^{-8}; \quad (158)$$

$n = 3, l = 1$
$$f_{31}(K) = \tfrac{2}{9} K^2(k_0/k_3) I_{31}(K) = 3^4 2^{14} (16+27K^2 a_0^2)^2 \times$$
$$\times (16+9K^2 a_0^2)^{-8}; \quad (159)$$

$n = 3, l = 2$
$$f_{32}(K) = \tfrac{2}{9} K^2(k_0/k_3) I_{32}(K) = 3^5 2^{20} K^2 a_0^2 (16+9K^2 a_0^2)^{-8}. \quad (160)$$

In Table 7.1 differential cross-sections $I_{0,n,l}(\theta)\,d\omega$ for excitation of these levels are given as a function of angle of scattering while in Fig. 7.5 the generalized oscillator strengths are illustrated as functions of K.

For small values of K, and hence small angles of scattering, ϵ_{nl} behaves like K^l except for s states for which it behaves like K^2. It follows that the differential cross-section rises most rapidly at small angles for p states for which it behaves like K^{-2}. For s and d states it is independent of K at small K and for states of higher angular momenta it behaves like some positive power of K.

This behaviour is clearly seen by reference to Table 7.1, which shows how steeply the angular distribution of the scattered electrons falls with angle for excitation of p states, i.e. optically allowed transitions, compared with that for s and d states.

† MASSEY, H. S. W. and MOHR, C. B. O., *Proc. R. Soc.* A**132** (1931) 605.

The concentration at small angles increases as the electron energy increases so that in the forward direction the scattered intensity rises quite rapidly as the electron energy increases.

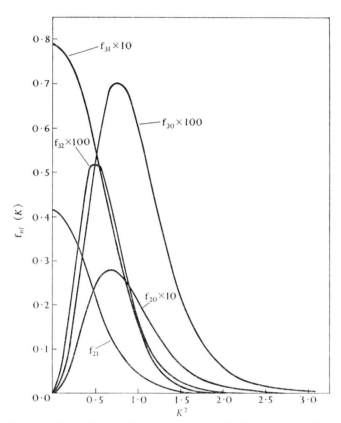

FIG. 7.5. Generalized oscillator strengths $f_{nl}(K)$ for transitions from the ground state to the different 2- and 3-quantum excited states of atomic hydrogen.

In terms of a semi-classical picture optically allowed transitions are readily excited in distant collisions, a result which is associated with the fact that these transitions arise from excitation of an oscillatory dipole moment.

The total cross-sections $Q_{0,nl}$ may be calculated analytically† for particular states by use of (35). They may be expressed in the form

$$Q_{0,nl} = 4\pi(|E_0|/E)|z_{0n}|^2[G_{nl}\{1+(n/n+1)^2(k_0+k_n)^2a_0^2\} - G_{nl}\{1+(n/n+1)^2(k_0-k_n)^2a_0^2\}], \quad (161)$$

† ELSASSER, W., Z. Phys. **45** (1927) 522; GOLDSTEIN, L., Annls Phys. **19** (1933) 305.

TABLE 7.1

Differential cross-sections (in units a_0^2 per sterad) and total cross-sections (in units πa_0^2) for excitation of various levels in hydrogen

State excited	Electron energy (eV)	Angle of scattering						Total cross-section (πa_0^2)
		0°	5°	10°	20°	30°	40°	
2s	100	0·887	0·771	0·518	0·133	0·024	0·0042	0·057
	200	0·936	0·705	0·324	0·031	0·0024	0·0$_3$24	0·029
	400	0·961	0·547	0·013	0·0034	0·0$_3$12	0·0$_5$78	0·015
2p	100	99·4	23·7	5·01	0·346	0·028	0·0028	0·751
	200	215·6	13·3	1·63	0·039	0·0014	0·0$_4$79	0·470
	400	448·1	5·43	0·33	0·0022	0·0$_4$35	0·0$_5$13	0·246
3s	100	0·121	0·113	0·090	0·032	0·0067	0·0013	0·011
	200	0·127	0·110	0·066	0·0085	0·0$_3$72	0·0$_4$73	0·0058
	400	0·130	0·096	0·032	0·0010	0·0$_3$37	0·0$_5$24	0·0030
3p	100	11·33	3·72	1·00	0·098	0·0095	0·0010	0·128
	200	24·63	2·40	0·39	0·013	0·0$_3$50	0·0$_4$29	0·081
	400	51·22	1·12	0·096	0·0$_3$77	0·0$_4$13	0·0$_6$47	0·044
3d	100	0·220	0·175	0·091	0·0105	0·0$_3$77	0·0$_4$59	0·0082
	200	0·243	0·152	0·043	0·0011	0·0$_4$26	0·0$_6$95	0·0043
	400	0·255	0·100	0·010	0·0$_4$43	0·0$_6$37	0·0$_7$85	0·0022

where E is the energy of the incident electron. z_{0n} is the dipole moment matrix element for the transition from the ground state to the state with total quantum n for which $l = 1$ and is given by[†]

$$|z_{0n}|^2 = \frac{2^8}{3} \frac{n^7 (n-1)^{2n-8}}{(n+1)^{2n+5}} a_0^2. \tag{162}$$

For the states with $n = 2$ and 3 the functions G_{nl} are as follows:

$$n = 2, l = 0 \quad G_{20}(x) = -\tfrac{1}{5}x^{-5}, \tag{163}$$

$$n = 2, l = 1 \quad G_{21}(x) = \ln(1-1/x) + x^{-1} + \tfrac{1}{2}x^{-2} + \tfrac{1}{3}x^{-3} + \tfrac{1}{4}x^{-4} + \tfrac{1}{5}x^{-5}, \tag{164}$$

$$n = 3, l = 0 \quad G_{30}(x) = -\tfrac{5}{3}x^{-5} + \tfrac{4}{3}x^{-6} - \tfrac{8}{21}x^{-7}, \tag{165}$$

$$n = 3, l = 1 \quad G_{31}(x) = \ln(1-1/x) + x^{-1} + \tfrac{1}{2}x^{-2} + \tfrac{1}{3}x^{-3} + \tfrac{1}{4}x^{-4} + \tfrac{1}{5}x^{-5} - \tfrac{4}{3}x^{-6}, \tag{166}$$

$$n = 3, l = 2 \quad G_{32}(x) = -\tfrac{4}{21}x^{-7}. \tag{167}$$

In Table 7.1 cross-sections calculated from these formulae for these states are given for electron energies of 100, 200, and 400 eV. The most obvious feature of these results is that, for a given value of n, the

[†] SCHRÖDINGER, E., *Annln Phys.* **80** (1926) 437.

cross-section for the optically allowed transition ($l = 1$) is by far the largest. Next in order of magnitude is that with $l = 0$ (s–s transition), then $l = 2$, $l = 3$ and so on.

For high energy encounters for which

$$(k_0+k_n)a_0 \gg 1, \qquad (k_0-k_n)a_0 \ll 1,$$

$$G_{n1}\{1+(n/n+1)^2(k_0-k_n)^2 a_0^2\} \to \ln\{(n/n+1)^2(k_0-k_n)^2 a_0^2\}, \qquad (168)$$

$$G_{n1}\{1+(n/n+1)^2(k_0+k_n)^2 a_0^2\} \to 0, \qquad (169)$$

so
$$Q_{0,n1} \simeq 4\pi(|E_0|/E)|z_{0,n1}|^2 \ln\left\{\frac{n^2-1}{(n+1)^2}\frac{4E}{E_n-E_0}\right\} \qquad (170)$$

which for not too small n does not differ greatly from Bethe's approximation (44 b).

We have ignored the possibility of electron exchange in deriving the differential and total cross-sections given in Table 7.1. This is because, for electron energies of 100 eV and higher, the contribution from exchange is negligible as discussed in § 5.4.

Although Born's first approximation becomes increasingly inaccurate at electron energies below 100 eV, as we shall see below, it is sometimes possible, knowing the nature of the deviation to be expected from comparison with experiment for some states, empirically to correct results obtained with its use. It is also possible to improve the results in a more formal way as described in Chapter 8, § 4.4. We therefore show in Fig. 7.6 the total cross-sections calculated by Born's first approximation right down to threshold energies, for a number of excitation processes from the ground state.

5.6.2. *Ionization of atomic hydrogen.* When we turn to discuss ionization we note first that to specify a particular ionization process due to incidence of an electron of energy $k_0^2 \hbar^2/2m$ we need to give the energy $\kappa^2 \hbar^2/2m$ of the ejected electron as well as the polar angles (θ, ϕ), (λ, μ) respectively of the directions of motion of the scattered and ejected electrons relative to some fixed direction such as that of incidence. Corresponding to such a process we have a differential cross-section $I_{0\kappa}(\theta, \phi; \lambda, \mu)\,d\kappa\,d\omega\,d\Omega$, where $d\omega$, $d\Omega$ are elements of solid angle about the respective directions of scattering and of ejection.

In terms of the usual formulation of Born's first approximation

$$I_{0\kappa} = |f_b(k', \kappa)|^2, \qquad (171)$$

where

$$f_b(k', \kappa) = (2\pi m e^2/h^2) \iint \psi_0(r_1)\psi_c^1(\mathbf{r}_1, \kappa) r_{12}^{-1} \exp\{i(\mathbf{k}_0-\mathbf{k}').\mathbf{r}_2\}\,d\mathbf{r}_1\,d\mathbf{r}_2. \qquad (172)$$

$\psi_c^Z(r,\kappa)$ is the properly normalized wave function for motion of an electron with energy $\kappa^2\hbar^2/2m$ in the asymptotic direction $\boldsymbol{\kappa}$ in the field of a nucleus of charge Ze (see Chap. 6, § 3.9). $k\hbar$, $\mathbf{k}'\hbar$ are the respective momenta of the incident and scattered electron.

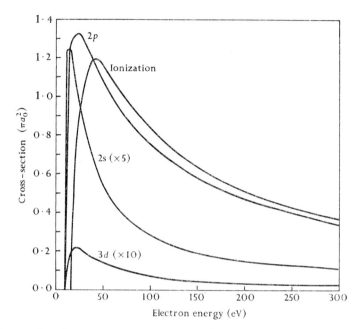

FIG. 7.6. Total cross-sections for excitation of 2- and 3-quantum excited states of atomic hydrogen by electron impact, calculated by Born's approximation. The ionization cross-section is also included for comparison.

In terms of the energy E_0 of the ground state
$$k^2 = k'^2 + \kappa^2 - 2mE_0/\hbar^2.$$
The total ionization cross-section is then taken to be
$$Q_i^b = \int_0^{\kappa_m}\!\!\int\!\!\int I_{0\kappa}(\theta,\phi;\lambda,\mu)\,d\kappa d\omega d\Omega, \tag{173}$$
where
$$\kappa_m^2 = k^2 + 2mE_0/\hbar^2.$$

Before proceeding further we must consider carefully what we mean by scattered and ejected electrons. Electrons are in fact indistinguishable so we cannot distinguish between a collision in which the ejected electron has wave number κ and the scattered k' and one in which the ejected electron has wave number k' and the scattered κ. The formula (172) does, however, make a distinction in terms of the forms taken for the wave functions representing the motion of the two electrons after the collision. Thus for the ejected electron the wave function is that for a state in which the electron moves with positive energy $\kappa^2\hbar^2/2m$ in the unscreened field of the nucleus, but for the scattered electron it is simply a plane wave undisturbed by any Coulomb interaction. In other words, it is assumed that, whereas the ejected electron screens off the nuclear field completely from the scattered electron, the latter exerts no screening effect at all on the former. This would be expected to hold to a good approximation only when, at large distances, the scattered electron is moving faster than the ejected.

Thus we have the following classical argument.† Suppose the electrons, with wave vectors $\mathbf{k}_1$, $\mathbf{k}_2$ are at the points $\mathbf{r}_1$, $\mathbf{r}_2$ relative to the nucleus where r_1 and r_2 are both large. Let ζe be the effective charge acting on electron 1. Then classically

$$\frac{\zeta}{r_1} = \frac{1}{r_1} - \frac{1}{|\mathbf{r}_1 - \mathbf{r}_2|}. \tag{174}$$

Also, since both electrons left the origin simultaneously at a certain time, their positions at a time t later will be given by

$$r_1 = (k_1\hbar/m)t, \qquad r_2 = (k_2\hbar/m)t, \tag{175}$$

so
$$\zeta = 1 - k_1/|\mathbf{k}_1 - \mathbf{k}_2|$$
$$= 1 - \left(1 + \frac{\mathbf{k}_1 \cdot \mathbf{k}_2}{k_1^2} + \ldots\right) \qquad (k_1 > k_2),$$
$$= 1 - \left(\frac{k_1}{k_2} + \mathbf{k}_1 \cdot \mathbf{k}_2 \frac{k_1}{k_2^3} + \ldots\right) \qquad (k_1 < k_2). \tag{176}$$

If we ignore directional terms it follows that $\zeta = 0$ for $k_1 > k_2$, indicating no screening by the electron of lower energy.

It would seem on this basis that a more correct procedure in deriving the equivalent of Born's approximation for excitation of discrete states would be to assume that (173) applies only to collisions for which $\kappa \leqslant k'$, that is, collisions in which electron 2 emerges with the higher energy. We would then have

$$Q_i^{b'} = \int_0^{\kappa'_m} I_{0\kappa}(\theta, \phi; \lambda, \mu)\, d\kappa d\omega d\Omega, \tag{177}$$

where
$$\kappa_m'^2 = \tfrac{1}{2}(k_0^2 + 2mE_0/\hbar^2). \tag{178}$$

Collisions in which $\kappa \geqslant k'$ would then be considered to arise from

† PETERKOP, R. K., *Izv. Akad. Nauk SSSR* Ser. Fiz. **27** (1963) 1012; GELTMAN, S., RUDGE, M. R. H., and SEATON, M. J., *Proc. phys. Soc.* **81** (1963) 373.

electron exchange. Thus corresponding to (172) we have the exchange amplitude

$$g_{bo}(k',\kappa) = -(2\pi me^2/\hbar^2) \int\int \psi_0(r_1)\left(\frac{1}{r_1}-\frac{1}{r_{12}}\right)\psi_c^1(\mathbf{r}_2,\kappa)\exp\{i(\mathbf{k}_0\cdot\mathbf{r}_2-\mathbf{k}'\cdot\mathbf{r}_1)\}\,d\mathbf{r}_1 d\mathbf{r}_2 \quad (179)$$

which, according to the above considerations, should be a good representation only when $k' > \kappa$, i.e. when electron 1 is moving with the greater energy.

We would then have

$$I_{0\kappa} = \begin{cases} |f_b(k',\kappa)|^2 & (k' > \kappa), \\ |g_{bo}(\kappa,k')|^2 & (\kappa > k'), \end{cases} \quad (180)$$

so

$$Q_i^{bo} = \int_0^{\kappa_m} |f_b(k',\kappa)|^2\,d\kappa d\omega d\Omega + \int_{\kappa_m'}^{\kappa_m} |g_{bo}(\kappa,k')|^2\,d\kappa d\omega d\Omega. \quad (181)$$

A detailed analysis of the asymptotic form of the wave function† representing the motion of the two outgoing electrons shows that a more logically consistent approximation is to take

$$f(k',\kappa) = (-2\pi me^2/\hbar^2)\exp\{i\Delta(k',\kappa)\}\int\int \psi_0(r_1)\psi_c^{Z'}(\mathbf{r}_2,k')\times$$

$$\times\left\{\frac{1-Z}{r_1}+\frac{1-Z'}{r_2}-\frac{1}{r_{12}}\right\}\exp(i\mathbf{k}_0\cdot\mathbf{r}_2)\psi_c^Z(\mathbf{r}_1,\kappa)\,dr_1 dr_2, \quad (182)$$

where

$$\frac{Z}{\kappa}+\frac{Z'}{k'} = \frac{1}{\kappa}+\frac{1}{k'}-\frac{1}{|\mathbf{\kappa}-\mathbf{k}'|}, \quad (183)$$

$$\Delta(k',\kappa) = \frac{2Z}{\kappa}\ln\{\kappa/(\kappa^2+k'^2)^{\frac{1}{2}}\}+\frac{2Z'}{k'}\ln\{k'/(\kappa^2+k'^2)^{\frac{1}{2}}\}. \quad (184)$$

This formula, apart from the presence of the phase $\Delta(k',\kappa)$, is not surprising in view of (176). If we take $Z = 1$, $Z' = 0$ we regain Born's approximation because the functions ψ_0, ψ_c^Z are orthogonal to each other. However, according to (176) if we take $Z = 1$ then

$$Z' = 1-k'/|\mathbf{\kappa}-\mathbf{k}'|. \quad (185)$$

With this formulation the exchange amplitude $g(k',\kappa)$ is given by

$$g(k',\kappa) = f(\kappa,k'), \quad (186)$$

and (see (146))

$$I_{0\kappa} = \tfrac{1}{4}\{|f(k',\kappa)-f(\kappa,k')|^2+3|f(k',\kappa)+f(\kappa,k')|^2\}. \quad (187)$$

† PETERKOP, R. K., *Latv. PSR Zināt Akad. Vest.* **9** (1960) 79; *Proc. phys. Soc.* **77** (1961) 1220; *Zh. éksp. teor. Fiz.* **41** (1962) 1938; **43** (1962) 616; *Optika Spektrosk.* **13** (1962) 153; *Izv. Akad. Nauk SSSR Ser. Fiz.* **27** (1963) 1012; RUDGE, M. R. H. and SEATON, M. J., *Proc. R. Soc.* **A283** (1965) 262.

A plethora of further approximations† follow from this formulation. We shall indicate some of these that have received application:

(a) Born approximation, see (173).

(b) Truncated Born approximation, see (177).

(c) Born–Oppenheimer approximation, see (181).

(d) Spherical-average Born approximation. For this we take in (182) $Z = 1$; $Z' = 0$, $k' > \kappa$, $Z' = 1-k'/\kappa$, $k' < \kappa$ and evaluate the total cross-section by integration of the resulting $\iint I_{0\kappa} \, d\omega d\Omega$ over all κ from 0 to κ_m.

(e) Truncated spherical-average Born approximation. As for (d) but with the final integration over κ from 0 to κ'_m only.

(f) Born-exchange approximations. These depend on use of (186). The difficulty arises, however, that the phase difference $\Delta(k',\kappa) - \Delta(\kappa,k')$ is only given by (184) if (183) holds accurately. If some approximate relation between Z and Z' is substituted for (183) it is difficult to determine what is the best assumption to make about the phase difference. One set of approximations is therefore based on taking in (187)‡

$$g(k',\kappa) = f(\kappa,k') = \exp\{i\,\delta(\kappa,k')\} f_\text{b}(\kappa,k'),$$

$$f(k',\kappa) = f_\text{b}(k',\kappa), \tag{188}$$

where f_b is the usual Born approximation to the scattered amplitude, and various assumptions are made about δ. A further possibility is to use (188) but with f_b replaced by the appropriate spherical-average Born approximation.

When the energy E of the incident electron is large so $E/|E_0| \gg 1$ the different approximations all give much the same result for the total ionization cross-section. This is because, under these conditions, $f(k',\kappa)$ falls off rapidly as κ increases and is unimportant for $\kappa > k'$. The various approximations are concerned essentially only with the way in which $f(k',\kappa)$ is to be calculated when $\kappa > k'$. This will be clearer when we discuss the actual results for atomic hydrogen in more detail.

Before doing this there is one further general point. If the atomic electron were ejected in a free collision, the effect of the nucleus being ignored, the conservation of momentum would ensure that, if an electron is observed to be moving after the collision with wave number k' in a

---

† RUDGE, M. R. H. and SEATON, M. J., loc. cit., p. 464.

‡ PETERKOP, R. K., loc. cit., p. 464; GELTMAN, S., RUDGE, M. R. H., and SEATON, M. J., loc. cit., p. 463.

direction making an angle θ with the incident direction then
$$k\cos\theta = k'. \tag{189}$$
This applies to both the 'ejected' and 'scattered' electrons. Thus
$$k\cos\lambda \simeq \kappa \tag{190}$$
or, since $k^2 = \kappa^2 + k'^2 + \kappa_0^2$ where $\kappa_0^2 \hbar^2/2m$ is the ionization energy
$$k^2 + \kappa^2 - 2k\kappa\cos\lambda \simeq k'^2 + \kappa_0^2. \tag{191}$$

Although this relationship will be blurred out by the effect of the nuclear interaction we would still expect that the angular distribution of the 'ejected' electrons for fixed κ would be a maximum when (191) is approximately satisfied. Since $\kappa \ll k$ this will be for angles λ not far from $\tfrac{1}{2}\pi$.

A similar argument applies to the angular distribution of the scattered electrons, a maximum or at least an inflexion being expected where
$$k\cos\theta \simeq k'. \tag{192}$$
Since $k' \simeq k$ this will be close to the forward direction.

We shall now describe the detailed application to atomic hydrogen. In the first instance, in discussing the differential cross-sections we shall work only in terms of Born's first approximation (173).

Integration of the differential cross-section $I_{0\kappa}(\theta,\phi;\lambda,\mu)\,d\kappa d\omega d\Omega$ over Ω gives the angular distribution of the scattered electrons with energy $2mk'^2/\hbar^2$. The results may be expressed analytically in the form†

$$\begin{aligned}J_{0\kappa}(\theta,\phi)\,d\kappa d\omega &= \int I_{0\kappa}(\theta,\phi;\lambda,\mu)\,d\Omega d\kappa d\omega \\ &= (2^{10}\kappa k'/a_0^8\, kK^2)\{K^2 + \tfrac{1}{3}(\mu^2+\kappa^2)\} \times \\ &\quad \times \{\mu^4 + 2\mu^2(K^2+\kappa^2) + (K^2-\kappa^2)^2\}^3 \{1-\exp(-2\pi\mu/\kappa)\}^{-1} \times \\ &\quad \times \exp\!\left\{-\frac{2\pi}{\kappa}\arctan\frac{2\mu\kappa}{K^2-\kappa^2+\mu^2}\right\} d\kappa d\omega,\end{aligned} \tag{193}$$

where $\mu = 1/a_0, \quad K = |\mathbf{k}-\mathbf{k}'|.$

For $\kappa^2 \ll k^2$ the form of the function $J_{0\kappa}(\theta,\phi)$ is generally very similar to that for excitation of optically allowed discrete transitions. Fig. 7.7 illustrates some typical examples. In some cases in which the energy loss suffered by the incident electron is considerable an inflexion appears at small angles. This is a manifestation of the blurred-out conservation of momentum in the electron–electron collision involved. It must be remembered, of course, that too much reliance should not be placed on

† MASSEY, H. S. W. and MOHR, C. B. O., *Proc. R. Soc.* A**140** (1933) 613.

the results for the scattered electrons when the energy loss is a considerable fraction of the total incident energy.

Integration of $I_{0\kappa}(\theta, \phi; \lambda, \mu)$ over $d\omega$ gives the angular distribution of electrons ejected with energy $\kappa^2\hbar^2/2m$. This integration must be carried out numerically. Some typical results are shown in Fig. 7.8.

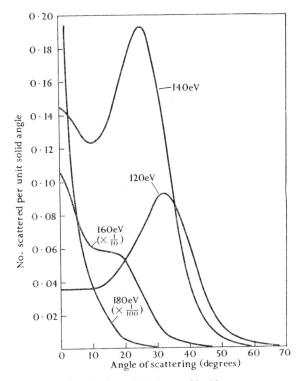

FIG. 7.7. Angular distribution of electrons of incident energy 200 eV scattered after ionizing collisions with hydrogen atoms, calculated by Born's first approximation (193) for various final electron energies as indicated.

It will be seen that many of these curves exhibit maxima at angles comparable with $\frac{1}{2}\pi$, the reasons being as discussed above in terms of momentum relations.

By integrating $J_{0\kappa}(\theta, \phi)$ over $d\omega$ we obtain the energy distribution of the ejected electrons in the form

$$Q_{0\kappa}\, d\kappa = \int J_{0\kappa}(\theta, \phi)\, d\omega d\kappa. \tag{194}$$

This integration can be carried out numerically to give energy (or velocity) distributions, a sample of which is shown in Fig. 7.9. These

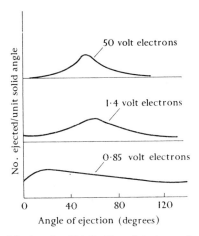

Fig. 7.8. Angular distribution of electrons ejected in ionizing collisions of atomic hydrogen with electrons of incident energy 200 eV, calculated by Born's first approximation.

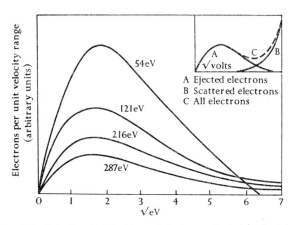

Fig. 7.9. Velocity distribution of electrons ejected in ionizing collisions of atomic hydrogen with electrons, as calculated by Born's first approximation. The incident electron energy is indicated on each curve.

distributions follow the general form we have described earlier. Thus for a given incident energy the probability is a maximum for ejection of electrons with a few eV energy only and is clearly very small when $\kappa^2 \simeq \tfrac{1}{2}k^2$.

In practice both ejected and scattered electrons would be observed and the total velocity distribution would have the general form shown dotted in Fig. 7.9. To a large extent this can be obtained merely by adding the distributions for the two sets of electrons, but this will not be accurate where they overlap appreciably—in this region interference will occur between the amplitudes of the two electron waves. For high impact energies the number of electrons ejected or scattered with energies in this region will be small.

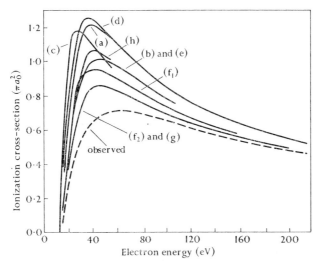

FIG. 7.10. Comparison of cross-sections for ionization of atomic hydrogen by electron impact, calculated using various approximations, with those observed. (a) Born approximation. (b) and (e) Truncated Born approximation. (c) Born–Oppenheimer approximation. (d) Spherical average Born approximation. (f_1) Born-exchange approximation $\delta = 0$; (f_2) Born-exchange approximation

$$\delta(k, k') = \arg f_b(k', k) - \arg f_b(\kappa, k').$$

(g) Ochkur approximation. (h) Distorted wave approximation (see Chap. 8, § 4.6).

Finally, by integration over κ we obtain the total ionization cross-section. At this stage we introduce the further approximations discussed above in (a) to (f). Fig. 7.10 illustrates the total ionization cross-sections calculated according to the various prescriptions (a)–(f).† At electron energies above about 200 eV all give effectively the same cross-sections. Considering first the so-called Born approximations, it will be seen that truncation of the range of integration over κ (approximation (b)) reduces the maximum cross-section by about 20 per cent. On the other hand, use of the spherical average interaction (d) has little effect, giving merely a small increase. Truncation in this case leads to nothing new as it yields merely approximation (b) once more.

† RUDGE, M. R. H. and SEATON, M. J., *Proc. R. Soc.* A**283** (1965) 262.

The Born–Oppenheimer approximation (c) increases the cross-section very substantially near the threshold so the maximum falls at a considerably lower energy. We also include the result obtained† if Ochkur's approximate form (151) for the exchange amplitude is used. Because Ochkur ignores terms that grow large near the threshold, results obtained with his formula show a maximum at much the same energy as do the Born results but the magnitude of the maximum is reduced to about 70 per cent of that given by approximation (a).

TABLE 7.2

Relative probabilities of different types of collision of electrons in atomic hydrogen

Type of collision	Energy of incident electrons (eV)				
	100	200	400	1000	10 000
	Percentage of all collisions				
Elastic	12·2	10·2	9·8	8·7	6·5
Excitation of 2-quantum levels	33·5	33·6	39·0	42·8	45·3
,, ,, 3- ,, ,,	5·9	5·8	6·8	6·3	7·0
,, ,, 4- ,, ,,	2·2	2·0	2·2	2·4	2·6
,, ,, 5- ,, ,,	1·0	0·9	1·0	1·2	1·2
Excitation of higher quantum levels	1·7	1·7	2·0	2·2	2·3
All discrete levels	44·3	44·0	51·0	54·8	58·4
Ionization	43·5	45·8	39·2	36·5	35·1
Total cross-section (πa_0^2)	2·45	1·59	0·79	0·37	0·049

Finally, results are given for the Born-exchange approximations assuming for the phase difference $\delta(k',\kappa)$ in (188) either $\delta = 0$ or $\arg f_b(k',\kappa) - \arg f_b(\kappa,k')$, the latter assumption yielding the smallest cross-section given by this type of approximation. Once again the maximum appears at nearly the same energy as for the Born approximations but is reduced in magnitude by as much as 30 per cent or so in the extreme case.

5.6.3. *Relative probabilities of different types of collision.* In Table 7.2 the calculated relative probabilities of different types of collision of electrons with hydrogen atoms are given over an energy range from 100 to 10^4 eV, throughout most of which Born's approximation is valid (see § 5.6.4). For heavier atoms the relative importance of elastic scattering will be greater as this tends to increase as Z^2, where Z is the atomic number, whereas the inelastic collision cross-sections increase more nearly as Z.

† PRASAD, S. S., *Proc. phys. Soc.* **85** (1965) 57.

5.6.4. *Atomic hydrogen—comparison with observation.* The most direct comparison with experiment can be made for the ionization cross-sections. As described in Chapter 3, § 2.3 the observed cross-sections† were normalized by relation to the absolute values for molecular hydrogen and did not make any assumptions about the range of validity of Born's first approximation. The cross-sections observed up to

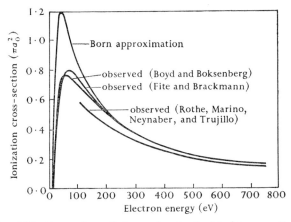

FIG. 7.11. Comparison of observed and calculated cross-sections for ionization of atomic hydrogen by electrons with energies up to 800 eV.

energies of nearly 800 eV are compared with the calculated cross-section according to the full Born approximation (a) in Fig. 7.11. It will be seen that, for energies above about 200 eV, the agreement is good but as the energy falls below this the calculated values begin to rise above the observed to an increasing extent. It is also true that at these energies the various approximations discussed in § 5.6.2 begin to yield different results.

Reference to Fig. 7.10 shows on an expanded energy scale how the different approximations compare with the observations. It will be seen that all give too large maximum cross-sections. The best results are given by the Born-exchange approximation (f_2) and by the Ochkur form of the Born–Oppenheimer approximation. It is difficult to judge whether the improvement is empirical or whether it means that the modifications introduced by these approximations are really the most important

† FITE, W. L. and BRACKMANN, R. T., *Phys. Rev.* **112** (1958) 1141; BOYD, R. L. F. and BOKSENBERG, A., *Proc. 4th Int. Conf. Ioniz. Phenom. Gases*, vol. 1, p. 529 (North Holland, 1960); ROTHE, E. W., MARINO, L. L., NEYNABER, R. H., and TRUJILLO, S. M., *Phys. Rev.* **125** (1962) 582; MCGOWAN, J. W. and FINEMAN, M. A., *4th Int. Conf. Phys. Electron. Atom. Collisions*, Quebec, 1965, p. 429 (Science Bookcrafters, New York, 1965).

at these energies. So many other factors may be involved (see Chap. 8) that the improvement may be more apparent than real. However, in any case it may well be that use of these approximations provides a means of improving the results of the Born approximation, irrespective of their theoretical significance. Their value must then be judged by the ease with which they may be evaluated.

We shall return in Chapter 9, § 10.8 to the discussion of the behaviour of the cross-section close to the threshold.

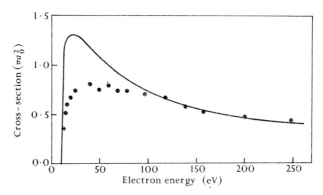

FIG. 7.12. Comparison of observed ●●● and calculated ——— cross-sections for excitation of the $2p$ state of atomic hydrogen by electrons with energies up to 250 eV.

For the excitation of the $2s$ and $2p$ states comparison with observation suggests a similar type of divergence from the predictions of Born's first approximation, but the evidence is less definite because the observed cross-sections are normalized by assuming the approximation to be valid at sufficiently high energies. Thus Fig. 7.12 compares the shape of the calculated and observed† cross-section velocity curves for the $2p$ excitation and this strongly suggests a similar situation to that for ionization.

For the $2s$ excitation the experimental situation is still less definite (Chap. 4, § 2.2.1) but even so it seems likely that the same position prevails.

5.6.5. *Excitation and ionization of* He^+ *and other hydrogen-like ions.* The only additional feature that must be introduced when calculating the cross-sections for excitation and ionization of one-electron ions, such as He^+, is that the Coulomb–Born approximation (see § 2) must be used.

Burgess‡ has applied this approximation to calculate cross-sections

† FITE, W. L. and BRACKMANN, R. T., *Phys. Rev.* **112** (1958) 1141; FITE, W. L., STEBBINGS, R. F., and BRACKMANN, R. T., ibid. **116** (1959) 356.
‡ BURGESS, A., *Mem. Soc. R. Sci. Liége* **4** (1961) 299.

for excitation of the 2p state. To present results in a form that permits comparison between the shapes of the excitation functions for ions of different charge, including in the limit the neutral hydrogen atom, we show in Fig. 7.13 the scaled cross-sections $(E_i/E_{iH})^2 Q$, where E is the excitation energy for the ion and E_{iH} that for atomic hydrogen, as functions of the reduced energy E/E_i. Results are given for He$^+$ ($Z = 2$) and for the limiting case $Z \to \infty$. As would be expected for sufficiently high energies, somewhat greater than $5E_i$, all these reduced excitation

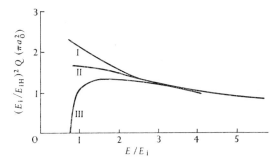

FIG. 7.13. Calculated cross-sections $(E_i/E_{iH})^2 Q$ for excitation of the 2p states of hydrogen-like ions as functions of electron energy E. I, $Z \to \infty$; II, $Z = 2$; III, atomic hydrogen.

functions tend to equality. This is to be expected because, at such energies, the effect of the Coulomb interaction is to distort the incident waves only very slightly (see Chap. 6 (90)).

For the ions the cross-section tends to a finite limit at the threshold whereas it vanishes for the neutral atom. Behaviour of cross-sections near the threshold will be further discussed in Chapter 9, § 10.

Analysis of the results for ionization is complicated by the fact that the accuracy of the first calculations† carried out using the Coulomb–Born (CB), Coulomb–Oppenheimer (CBO), and Coulomb–Born-exchange approximations (CBE) corresponding to (a), (c), and (f) of § 5.6.2, p. 465, was not adequate. Further more accurate calculations‡ have dealt only with the truncated Coulomb–Born approximation (TCB) (b) of § 5.6.2, p. 465, and the CBE approximation. It is often difficult to understand the significance of the results, particularly for inner-shell ionization.

Fig. 7.14 illustrates the comparison between the results obtained for the last two approximations and assumed nuclear charges of 2 and 128

† BURGESS, A. and RUDGE, M. R. H., *Proc. R. Soc.* A**273** (1963) 372.
‡ RUDGE, M. R. H. and SCHWARTZ, S. B., *Proc. phys. Soc.* **86** (1965) 773; **88** (1966) 563.

respectively. It is convenient in presenting these results to illustrate the reduced cross-section

$$Q_i^r = (E_i/E_{iH})^2 Q_i,$$

where Q_i is the actual total ionization cross-section, E_i is the minimum energy required to remove the electron from the ion, and E_{iH} that from hydrogen. If we plot the energy scale in terms of E/E_i then, at sufficiently high energies, all the Q_i^r tend to the same curve, which is that

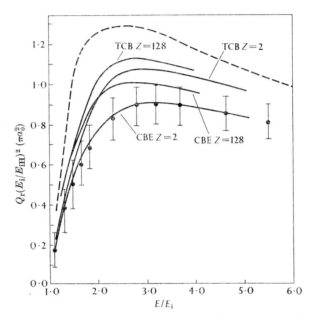

FIG. 7.14. Comparison of observed reduced cross-section for further ionization of He$^+$ by electron impact with the results of various theoretical approximation. ●, observed. Full line curves theoretical approximations as indicated. The dotted curve is a less accurate theoretical curve for the CB approximation.

for atomic hydrogen calculated by the simple Born approximation. A tentative curve is also sketched in for He$^+$ ($Z = 2$) for the full CB approximation. This is obtained by making empirical corrections to the earlier less accurate calculations.

Finally, in Fig. 7.14 we also show the observed results of Dolder, Harrison, and Thonemann† (see Chap. 3, § 3.1.1). It will be seen that the agreement with the results of the CBE approximation is good but the other approximations, and particularly the full CB, overestimate

† DOLDER, K. T., HARRISON, M. F. A., and THONEMANN, P. C., *Proc. R. Soc.* A**264** (1961) 367.

the cross-section. On the whole the agreement is much better than for ionization of hydrogen.

We would expect, following the same trend, that for K-shell ionization of considerably heavier atoms such as nickel and silver the agreement should be well-nigh perfect with the CBE approximation, which for these atoms hardly differs from the CBO.

Evidence about the cross-sections for ionization of the K-shells of nickel ($Z = 28$) and of silver ($Z = 47$) is forthcoming from measurements of X-ray fluorescent yields as described in Chapter 3, § 5. There are two difficulties in comparing these observed cross-sections with theory. One is that no theoretical results other than those carried out by the simple plane-wave Born approximation are available for $Z = 28$, and while Burgess and Rudge[†] carried out calculations by the Coulomb equivalents of (a), (c), and (f) of § 5.6.2, p. 465, their results are not accurate to 15 per cent, at least at some reduced energies. The second difficulty is that relativistic effects must be taken into account for these atoms even at reduced energies as low as 4. However, Fig. 7.15 (a) and (b) illustrate the position, as far as it can be pieced together, for the K-shell ionization of nickel and of silver respectively.

For nickel, in Fig. 7.15 (a), the results of the simple plane wave, non-relativistic and relativistic,[‡] Born[§] approximations are shown. It will be seen that relativistic effects are appreciable for $E/E_i > 4$. The experimental results for the relativistic curve agree very well for $E/E_i >$ about 7 and only fall about 10 per cent below the calculated maximum that occurs for $E/E_i \simeq 3\cdot 8$. Results are also included for the truncated CB and CBE approximations for $Z = 2$ and $Z = 128$. Since the results for nickel will be bracketed between these it seems likely that the CBE approximation will, in fact, give good agreement with observation, though probably a little too large.

The situation for silver is rather similar. For this case no relativistic calculations have been made. Referring to Fig. 7.15 (b) the simple plane-wave Born approximation§ is seen to give quite good agreement near the rather weak maximum. With a relativistic correction included the agreement might well extend to higher energies. The same considerations as for nickel suggests that the CBE approximation again gives good results, while the TCB approximation results are probably too large.

[†] BURGESS, A. and RUDGE, M. R. H., *Proc. R. Soc.* A**273** (1963) 372.
[‡] ARTHURS, A. M. and MOISEIWITSCH, B. L., ibid. A**247** (1958) 550.
[§] BURHOP, E. H. S., *Proc. Camb. phil. Soc. math. phys. Sci.* **36** (1940) 43.

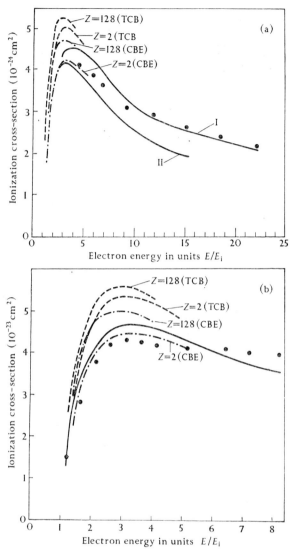

Fig. 7.15. (a) Comparison of observed and calculated cross-sections for K-shell ionization of nickel. ● observed. ———, —·—· theoretical approximations as indicated. —— plane wave Born approximation: I relativistic, II non-relativistic. (b) Comparison of observed and calculated cross-sections for K-shell ionization of silver. ● observed. ———, —·—· theoretical approximations as indicated. —— plane wave non-relativistic Born approximation.

It may be objected that the absolute cross-sections for inner-shell ionization are insufficiently accurate to apply these tests but, for nickel at least, the observed values at the higher energies fit very well to the plane-wave relativistic Born approximation.

Burgess, Hummer, and Tully† have also calculated the cross-sections for excitation of the 2s states of hydrogen-like ions with the results shown in Fig. 7.16, which are generally similar to those for the 2p state. Comparison may be made with the observations of Dance, Harrison, and Smith‡ (see Chap. 4, § 3.2) on the 2s excitation of He⁺. Allowance must be made for population of the 2s state in these experiments through cascade transitions from upper states also excited by electron impact.

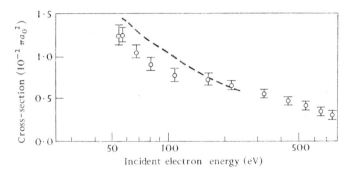

Fig. 7.16. Comparison of observed and calculated cross-sections for excitation of the 2s state of He⁺ by electron impact. ---- calculated. (Burgess, Hummer, and Tully). $\bar{\underline{\phi}}$ observed (Dance, Harrison, and Smith).

This has been done by Burgess et al. who calculated the 2s excitation function to be expected in the experiments. These results are compared with the observations in Fig. 7.16. The two results begin to diverge at low energies in the usual way but the rise of the theoretical as compared with the observed results is not so marked as for ionization of atomic hydrogen.

5.6.6. *Excitation of helium—calculated cross-sections.* Because of the extensive experimental information about cross-sections for excitation and ionization of helium we can hope to obtain much more information about the range of applicability of the approximations of Born and of Born–Oppenheimer from helium than from any other atom. The only adverse factor is the need to use approximate forms for the atomic wave functions, thereby introducing some additional uncertainty.

We give in Table 7.3 the differential and total cross-sections for excitation of a number of singlet states of helium by electrons with energies of 100, 200, and 400 eV, calculated using Born's approximation

---

† BURGESS, A., HUMMER, D. G., and TULLY, J. A., *2nd Int. Conf. Phys. electron. atom. Collisions*, p. 173 (Benjamin, New York, 1961).
‡ DANCE, D. F., HARRISON, M. F. A., and SMITH, A. C. H., *Proc. R. Soc.* A290 (1966) 74.

and the following approximate atomic wave functions. For the ground state the Hartree–Fock wave function† is used. It may be represented analytically by

$$\psi_0(r_1, r_2) = u(r_1)u(r_2),$$

where
$$u(r_1) = N(\pi a_0^3)^{-\frac{1}{2}}(e^{-Zr/a_0} + ce^{-2Zr/a_0}), \quad (195)$$

with $N = 1\cdot 48423$, $Z = 1\cdot 4558$, $c = 0\cdot 6$.

This function, while simple, is quite a good approximation because it not only gives a good value for the total energy of the atom but also for

TABLE 7.3

Differential cross-sections per unit solid angle I_{0n} (in units a_0^2 per steradian) and total cross-sections (in units πa_0^2) for excitation of various singlet levels in helium

State excited	Electron energy (eV)	Angle of scattering						Total cross-section (πa_0^2)
		0°	5°	10°	20°	30°	40°	
2^1S	100	0·162	0·150	0·124	0·059	0·021	0·006	0·023
	200	0·175	0·148	0·095	0·022	0·004	0·0006	0·0125
	400	0·181	0·124	0·057	0·005			
2^1P	100	5·92	3·58	1·413	0·214	0·035	0·0062	0·155
	200	14·42$_5$	3·60	0·737	0·045	0·0034	0·0$_3$35	0·111
	400	31·60	2·03$_5$	0·229	0·0047	0·0$_3$15		0·072
3^1P	100	1·21$_5$	0·795	0·348	0·062	0·011	0·0022	0·0384
	200	2·99	0·872	0·200	0·014	0·0011	0·0$_3$1	0·0275
	400	6·50	0·527$_5$	0·067	0·0017	0·0$_4$5		0·0180
3^1D	100	0·012$_5$	0·011	0·0080	0·0023$_5$	0·0$_3$45	0·0$_4$7	0·0$_3$93
	200	0·015$_5$	0·012	0·0059	0·0$_3$59	0·0$_3$39	0·0$_5$3	0·0$_3$52$_5$
	400	0·017	0·010	0·0026	0·0$_4$55	0·0$_5$1		0·0$_3$28
4^1P	100	0·452	0·303	0·137	0·025$_5$	0·0048	0·0010	0·0152
	200	1·116	0·342	0·081$_5$	0·0062	0·0$_3$5	0·0$_4$5	0·0109
	400	2·456	0·211	0·028	0·0$_3$7	0·0$_4$5		0·0071

the diamagnetic susceptibility (calculated $-1\cdot 92\times 10^{-6}$ units as compared with $-1\cdot 9\times 10^{-6}$ units observed) which depends on the wave function at larger distances.

For excited states other than S states the wave functions were taken to be of the form first proposed by Eckart,‡ namely

$$\psi(r_1, r_2) = 2^{-\frac{1}{2}}\{\psi_{nlm}(\alpha|r_1)\psi_{100}(\beta|r_2) + \psi_{nlm}(\alpha|r_2)\psi_{100}(\beta|r_1)\}, \quad (196)$$

where $\psi_{nlm}(Z|r)$ denotes the normalized wave function for an electron moving in the field of a bare nucleus of charge Ze, with quantum numbers

† GREEN, L. C., MULDER, M. M., LEWIS, M. N., and WOLL, J. W., *Phys. Rev.* **93** (1954) 757.
‡ ECKART, C., ibid. **36** (1930) 878.

n, l, m, while α and β may be determined by a variational procedure. For many purposes $\alpha \simeq 2$, $\beta \simeq 1$.

For excited S states it is necessary to allow for the departure of ψ_{nlm} from the hydrogenic form. Thus Marriott and Seaton† took for the 2^1S state

$$\psi_{200} = N_{200}(4\pi a_0^3)^{-\frac{1}{2}}\{e^{-\beta r/a_0} - S(r/a_0)e^{-\gamma r/a_0}\}, \qquad (197)$$

where $N_{200} = 0.568$, $\beta = 1.136$, $\gamma = 0.464$, $S = 0.317$. The parameters were chosen by a variational method, it being required that the 2^1S function be orthogonal to the simple Hylleraas variational wave function for the ground state

$$\psi_{100} = N_{100}(\pi a_0^3)\exp\{-1.6875(r_1+r_2)/a_0\}. \qquad (198)$$

We have used for this state a linear combination of the function (197) and the Hartree–Fock function (195) so chosen that the combination is orthogonal to the latter function.

The results shown in Table 7.3 exhibit the same general features as for atomic hydrogen. Thus the optically allowed transitions are, for a given n, much the most strongly excited. While for all cases the angular distribution of the scattered electrons falls rapidly from a maximum at $\theta = 0$, the descent is steepest for the optically allowed transitions.

Turning now to the excitation of triplet states that can only take place through electron exchange we give in Table 7.4 differential and total cross-sections for electron energies ranging from 100 to 400 eV for a number of cases. These cross-sections have been calculated in most cases using Ochkur's 'high-energy' approximation to the exchange amplitude (see § 5.4.2). The same ground state wave function (195) was used as for excitation of the singlet states. For excited states other than S states the antisymmetrical version of (196) was used, while for 2^3S the function taken was

$$\psi(2^3S) = 2^{\frac{1}{2}}\{v_1(r_1)v_2(r_2) - v_1(r_2)v_2(r_1)\},$$

where

$$v_1(r) = (\mu^3 a^3/\pi)^{\frac{1}{2}}e^{-\mu ar}, \quad v_2(r) = (\mu^5/3\pi N)^{\frac{1}{2}}\{re^{-\mu r} - (A/\mu)e^{-\mu br}\}, \qquad (199)$$

with
$$a = 3.28, \quad b = 2.57, \quad \mu a_0 = 0.61,$$
$$A = 0.539, \quad N = 0.546.$$

This is the variational approximation first proposed by Morse, Young, and Haurwitz.‡

† MARRIOTT, R. and SEATON, M. J., *Proc. phys. Soc.* A**70** (1957) 296.
‡ MORSE, P. M., YOUNG, L. A., and HAURWITZ, E. S., *Phys. Rev.* **48** (1935) 948.

The striking feature of the results for these intercombination transitions in the energy range concerned is the very rapid fall of the total cross-section with electron energy in all cases. This is as expected from the general discussion given in § 5.4. Thus for the 2^3P excitation, for example, the cross-section falls by a factor of 4 while the electron energy E increases from 122 to 166 eV. This follows the approximate

TABLE 7.4

Differential cross-sections per unit solid angle I_{0n} (in units a_0^2 per steradian) and total cross-sections (in units πa_0^2) for excitation of various triplet levels in helium

State excited	Electron energy (eV)	Angle of scattering						Total cross-section (πa_0^2)
		0°	5°	10°	20°	30°	40°	
2^3S	100	0·0$_4$7	0·00016	0·00057	0·0025	0·0040	0·0033	0·0025
	200	0·0$_5$4	0·0$_4$45	0·00029	0·00099	0·00079	0·00060	
	400	0·0$_6$4	0·0$_4$25	0·00016	0·00022			
2^3P	100	0·0028	0·0040	0·0067	0·0092	0·0065	0·0034	0·0043
	200	0·0004	0·0011	0·0022	0·0018	0·0007	0·0002	0·00054
	400	0·0$_4$5	0·0004	0·0006	0·0002	0·0$_4$3	0·0$_5$6	0·0$_4$7
3^3P	100	0·0008	0·0011	0·0018	0·0027	0·0021	0·0012	0·0014
	200	0·0001	0·0003	0·0006	0·0006	0·0002	0·0$_4$7	0·00017
	400	0·0$_4$1	0·00011	0·00019	0·0$_4$7	0·0$_4$1	0·0$_5$2	0·0$_4$2
3^3D	100	0·0$_4$1	0·0$_4$15	0·0001	0·0$_4$8	0·0$_4$4	0·0$_4$46	0·0$_4$46
	200	0·0$_6$55	0·0$_5$41	0·0$_4$18	0·0$_4$24	0·0$_5$75	0·0$_5$18	0·0$_5$6
	400	0·0$_7$37	0·0$_5$21	0·0$_5$7	0·0$_5$23	0·0$_6$26	0·0$_7$33	0·0$_6$75

dependence of the cross-section on E^{-5}. It is also noteworthy that the angular distribution of the scattered electrons is somewhat less steep than for the corresponding singlet excitations, a result which again follows directly from Ochkur's approximation (see § 5.4.2).

The effect of electron exchange on the excitation of the singlet states is small at energies of 100 eV or higher.

5.6.7. *Excitation of helium—comparison with observation—optically allowed transitions.* We shall discuss first the comparison for optically allowed transitions—the excitation of the ^1P states and ionization.

The only measured absolute values for the total cross-section for the 2^1P excitation are those of Jobe and St. John† (Chap. 4, § 1.4.2.1). Fig. 7.17 illustrates the comparison of these values with the calculated values given in Table 7.3. It will be seen that, whereas there is quite

† loc. cit., p. 191.

good agreement at an electron energy of 400 eV, the theoretical values become progressively too large as the electron energy falls.

Further evidence about the validity of Born's approximation in the same energy range comes from the experiments of Lassettre and his collaborators (Chap. 5, § 4). According to Born's approximation, if the

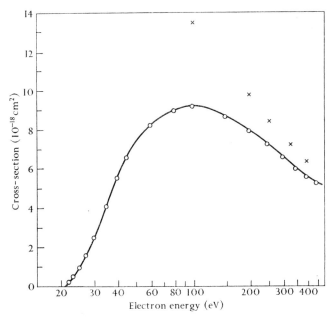

FIG. 7.17. Comparison of observed and calculated cross-sections for excitation of the 2^1P state of helium by electron impact. × calculated (Born's approximation). –O–O–O– observed (Jobe and St. John).

differential cross-section for excitation of a discrete state is expressed in the form
$$I_{0n}(K)\,d\omega,$$
where $K\hbar$ is the momentum change suffered by the electrons in the impact, then $k_n I_{0n}/k_0$ is a function of K only, k_0, k_n being the initial and final electron wave numbers respectively. This conclusion may be checked from observations of the relative differential cross-sections $I_{0,2P}$ for excitation of the 2^1P state by electrons of three different kinetic energies plotted as a function of K^2 (Fig. 7.18). It will be seen that all points lie on a smooth curve that would be expected since, at the electron energies concerned $k_n/k_0 \simeq 1$ so I_{0n} should be a function of K^2 only.

Using the formula (97), the generalized oscillator strength $f_{2p}(K)$ for the transition can be derived apart from a normalizing factor. To fix this factor Lassettre and Jones† calculated $f_{2p}(K)$ from the formulae (43) and (97) using the following wave functions (expressed in atomic units). In the ground state‡

$$\psi_0 = (8\pi^2)^{-\frac{1}{2}}\exp\{-S(r_1+r_2)\}\{A+B(r_1+r_2)+C(r_1^2+r_2^2)+$$
$$+Dr_1r_2+Er_1r_2\cos\theta_{12}+Fr_{12}\},$$

with

$A = 0\cdot1922,\qquad B = -0\cdot0194,\qquad C = 0\cdot0250,\qquad D = -0\cdot0366,$
$\qquad\qquad E = 0\cdot0122,\qquad F = 0\cdot0680,\qquad S = 1\cdot82,\qquad\qquad(200\,\text{a})$

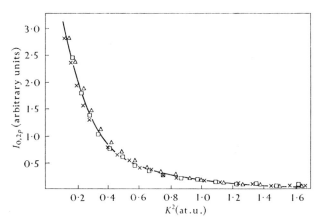

Fig. 7.18. Observed differential cross-sections $I_{0,2P}$ for scattering of electrons after excitation of the 2^1P state showing that, for incident electron energies in the range 400–600 eV, $I_{0,2P}$ is a function of the electron momentum change $K\hbar$. × 417 eV, △ 511 eV, □ 604 eV, incident electron energy.

and in the excited state§

$$\psi_{2p} = (8\pi^2)^{-\frac{1}{2}}\{G(r_1,r_2)\cos\theta_1+G(r_2,r_1)\cos\theta_2\},$$
where $\qquad G(r_1,r_2) = p(r_1-H)\exp\{-2(ar_1+r_2)\},$
with $\qquad p = 0\cdot0159,\qquad H = 0\cdot0595,\qquad a = 0\cdot25.\qquad(200\,\text{b})$

The observed cross-section was normalized‖ so as to agree on the average with the value calculated for $K^2 = 0\cdot1968, 0\cdot2446, 0\cdot2980$, and $0\cdot3570$ a.u. corresponding to scattering of 511 eV electrons through angles of $4\cdot0, 4\cdot5, 5\cdot0$, and $5\cdot5°$ respectively. Over this range the ratio of observed to calculated cross-sections varied by 3·2 per cent. It is clear by reference

† Lassettre, E. N. and Jones, E. A., J. chem. Phys. **40** (1964) 1218.
‡ Hylleraas, E., Z. Phys. **54** (1928) 347.
§ Wheeler, J. A., Phys. Rev. **43** (1933) 258.
‖ Lassettre, E. N., Berman, A. S., Silverman, S. M., and Krasnow, M. E., J. chem. Phys. **40** (1964) 1232.

to Fig. 7.19 that the fit is then very good over the entire range of the observations from $K^2 = 0 \cdot 2$ to $K^2 = 2 \cdot 0$ a.u. In the limit $K = 0$ the calculated optical oscillator strength is 0·268. This is to be compared with the calculated value 0·275 obtained by Dalgarno and Stewart† using the same six-parameter Hylleraas variational wave function for the ground state, and for the 2^1P state a function of the form (196) with

$$\alpha = 2 \cdot 003, \quad \beta = 0 \cdot 965. \tag{201}$$

Use of more elaborate wave functions will have little effect on the form of the generalized oscillator strength as a function of K. Indeed,

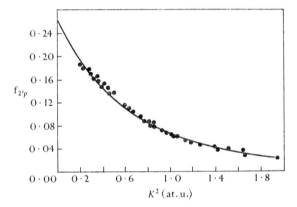

FIG. 7.19. Generalized oscillator strength f(K) for the 1^1S–2^1P transition in helium. —— calculated using the wave functions (200) and (201). ● observed by Silverman and Lassettre the scale being adjusted to give agreement with the full line curve at $K^2 = 0 \cdot 3$ a.u.

using the wave functions (195) and (196), from which the values of Table 7.3 were obtained, the optical oscillator strength comes out to be 0·278 in fortuitously close agreement with the best available value, while the form of f(K) is indistinguishable from that shown in Fig. 7.19.

The evidence clearly suggests that Born's approximation gives good results for the excitation of the 2^1P states, at least for electron energies of 400 eV or higher. Having normalized the differential cross-section for the excitation of the 2^1P state it is possible to derive absolute values of oscillator strengths for other transitions. Thus by extrapolation of the generalized oscillator strengths derived from the observed differential cross-sections for excitation of 3^1P a value is obtained for the optical oscillator strength which is given in Table 7.5.

† DALGARNO, A. and STEWART, A. L., *Proc. phys. Soc.* **76** (1960) 49.

Values for 4^1P and 5^1P excitations were obtained in later experiments† (see Chap. 5, § 5.3.2) by comparison of the intensity of scattering per unit solid angle, in the forward direction, of electrons that have excited the respective states. From these observations the generalized oscillator strengths may be derived from (97) for values of K^2 corresponding to zero scattering angle. Using (36) it may be seen that, for electrons of 500-eV incident energy these values are small, of order 0·015 a.u. For such low values of K^2, $f(K) \simeq f(0)$ the optical oscillator strength.

TABLE 7.5

Optical oscillator strengths for 1^1S–m^1P transitions in helium

			2^1P	3^1P	4^1P	5^1P
From electron impact observations	Lassettre et al.	Zero scattering angle method	0·275	0·075	0·031	0·0149
		Extrapolation	0·275	(0·067)		
	Geiger		0·312	0·090		
Calculated	Wave functions (200 a) and (200 b)		0·275	0·0746	0·0304	0·0153
	Wave functions (195) and (196)		0·278	0·074	0·0297	

The oscillator strengths obtained in this way—not only for transitions to 4^1P and 5^1P but also to the 3^1P states—are given in Table 7.5. For the latter state the agreement with results obtained from extrapolation is not unsatisfactory. Comparison with values calculated‡ using the ground-state wave function (200 a) and the appropriate modification of (196) for the excited states, reveals very good agreement. It is of interest to note also that the calculated values using the simpler functions (195) and (196), which provide the data of Table 7.3, also agree closely.

A further test of the range of validity of Born's approximation may be carried out as follows. From the observed values of the generalized oscillator strength as a function of K^2 for the 3^1P excitation the total cross-section may be calculated as a function of energy right down to the threshold, assuming Born's approximation to be valid at all energies. The resulting cross-section may then be compared as a function of electron energy with the optical excitation function for the 3^1P states obtained from observations of electron impact spectra§ (see Chap. 4, § 1.4.2.1). Fig. 7.20 illustrates this comparison, the curves being normalized

† SKERBELE, A. M. and LASSETTRE, E. N., *J. chem. Phys.* **40** (1964) 1271.
‡ DALGARNO, A. and STEWART, A. L., *Proc. phys. Soc.* **76** (1960) 49.
§ THIEME, O., *Z. Phys.* **78** (1932) 412; LEES, J. H., *Proc. R. Soc.* A**137** (1932) 173; GABRIEL, A. H. and HEDDLE, D. W. O., ibid. A**258** (1960) 124; ST. JOHN, R. M., MILLER, F. L., and LIN, C. C., *Phys. Rev.* **134** (1964) A888.

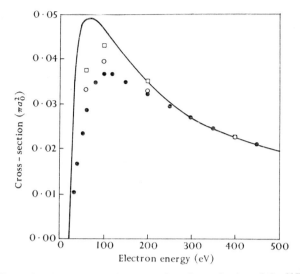

Fig. 7.20. Comparison of cross-sections for excitation of the 3^1P state of helium by electron impact. —— calculated from observations of Silverman and Lassettre. □ observed by Thieme. ○ observed by Lees. ● observed by St. John, Miller, and Lin. The scales are adjusted so all agree with the absolute values given by St. John et al. at 400 eV.

TABLE 7.6

Cross-sections for excitation of 3^1P and 4^1P levels of helium by electron impact

Electron energy (eV)	Cross-sections in 10^{-20} cm²			
	3^1P		4^1P	
	(calc.) (Born's approx.)	(obs.) (St. John et al.)	(calc.) (Born's approx.)	(obs.) (St. John et al.)
100	338	320	134	152
200	242	290	96	132
400	158	200	62·5	95

so as to agree at 400 eV electron energy. The same feature appears as for optically allowed transitions in H discussed in § 5.6.4. As the electron energy falls Born's approximation overestimates the cross-section by an increasing factor. Further comparison may be made with absolute values of the cross-sections observed from electron impact spectra (see Chap. 4, § 1.4.2.1). In Table 7.6 such a comparison is shown with the cross-sections for 3^1P and 4^1P excitation measured by St. John, Miller, and Lin. It will be seen that, for both states, the cross-section at 400 eV, calculated by Born's approximation and the helium wave functions (195) and (196), is smaller than the observed. This may be partly due to the difficulty of obtaining very accurate absolute values from optical measurements of the intensity of impact excitation (see Chap. 4, § 1.3)—the evidence from the energy-loss measurements of Lassettre et al. supports the calculated values at these energies. As the electron energy falls the ratio of the calculated to the observed cross-section increases so that the relative shapes of calculated and observed excitation functions is similar to that shown in Figs. 7.17 and 7.20.

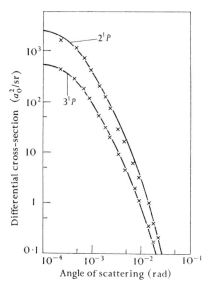

Fig. 7.21. Comparison of observed and calculated angular distributions of electrons of 25 keV incident energy scattered through small angles after exciting the 2^1P and 3^1P states of helium. × observed. —— calculated.

Finally, we have evidence at much higher electron energies (25 keV) from the experiments of Geiger using the technique described in Chapter 5, § 5.2.3. His results for elastic scattering in helium have been discussed in § 4. He also measured the variation of intensity with scattering angle θ in the range $2 \cdot 3 \times 10^{-4} \leqslant \theta \leqslant 4 \times 10^{-2}$ rad, of electrons that have excited the 2^1P and 3^1P states. These were normalized by comparison with the observed elastic scattering, which in turn was normalized as described in § 4, p. 434.

Fig. 7.21 shows a comparison between the observed and calculated differential cross-sections. The agreement is remarkably good. Extrapolation to obtain the optical oscillator strengths yields the values given in Table 7.5. For 2^1P there is agreement within the experimental error with the results obtained at much lower electron energies by Lassettre et al., but for 3^1P Geiger's result is appreciably higher.

5.6.8. *Ionization of helium.* We now have the problem of choosing an appropriate wave function for a continuum state. The most elaborate function used for this state is the exchange-polarization approximation used in the calculation of the elastic scattering of electrons by He+ ions and discussed in Chapter 8, § 5. With this function for the continuum and the Hartree–Fock function (195) for the ground state, Sloan† obtained with the Born approximation the ionization cross-section shown in Fig. 7.25. As an indication of the extent to which the result depends on the assumed wave function for the continuum, the cross-section obtained by Peach,‡ using an undistorted Coulomb wave in the field of a charge e, for this function is also shown. The difference, though appreciable, is not of major importance.

Comparison with the observed results of Smith§ (see Chap. 3, § 2.5) shows that, as for hydrogen and ionized helium, the Born approximation gives results that become increasingly too large as the electron energy falls. The calculated maximum cross-section exceeds the observed by a factor of about 1·5.

Further evidence about the validity of Born's approximation, which does not depend on accurate knowledge of helium wave functions, is provided from the observations of Silverman and Lassettre|| and of Kuyatt and Simpson.†† Both pairs of investigators observed the energy-loss spectrum of electrons with a few hundreds of eV energy as described in Chapter 5, § 4. Their measurements extend well into the continuum. For the excitation of continuum states the differential cross-section may be expressed in terms of differential generalized oscillator strengths so that, corresponding to (97), we have in the notation of § 5.6.2, for an energy loss ΔE within the continuum

$$I_{0\kappa}(K)\,\mathrm{d}\kappa = (2me^4k'/kK^2\hbar^2)\frac{\partial f_{0\kappa}}{\partial \kappa}\frac{\mathrm{d}\kappa}{\Delta E}$$

$$= 2(e^4k'\kappa/kK^2)\frac{\partial f_{0\kappa}}{\partial \epsilon}\frac{\mathrm{d}\kappa}{\Delta E}, \tag{202}$$

where ϵ is the energy of the ejected electron. Just as for excitation of discrete states we have for a given value of ΔE, and hence of K, that $(k/k')I_{0\kappa}$ is a function of K only, and from observations of this function $\partial f_{0\kappa}/\partial \epsilon$ may be obtained. Extrapolation of this function to zero K gives the differential optical oscillator strength $\partial f_{0\kappa}(0)/\partial \epsilon = h^{-1}\partial f_{0\kappa}(0)/\partial \nu$ and hence the photo-ionization cross-section from (61).

In Fig. 7.22 the photo-ionization cross-section derived from the

† SLOAN, I. H., *Proc. phys. Soc.* **85** (1965) 435.
‡ PEACH, G., ibid. **87** (1966) 381.
§ SMITH, P. T., *Phys. Rev.* **36** (1930) 1293.
|| SILVERMAN, S. M. and LASSETTRE, E. N., *J. chem. Phys.* **40** (1964) 1265.
†† KUYATT, C. E. and SIMPSON, J. A., *Atomic collision processes*, p. 191 (North Holland, Amsterdam, 1964).

electron impact data is compared with that directly measured by methods described in Chapter 14, § 7.1. The agreement is not unsatisfactory.

It is also possible to obtain directly from the measurements made by Silverman and Lassettre† the angular distribution of the scattered electrons for a fixed energy loss and the energy distribution of the ejected

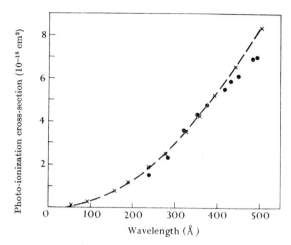

FIG. 7.22. Comparison of the photo-ionization cross-section of helium derived from analysis of electron impact data with that derived from direct observation. ●●● from electron impact data. ×——× from direct observation.

electrons for fixed incident energy. Fig. 7.23 illustrates a comparison between observed and calculated angular distributions for two incident electron energies and a number of energy losses. The calculated values are those of Peach,‡ which do not differ very much from the more elaborate calculations of Sloan. The agreement is seen to be very good.

Comparison of observed and calculated energy distributions of ejected electrons is shown in Fig. 7.24 and again the agreement is good.

From this evidence it seems that Born's approximation gives very good results for the differential and total cross-sections for ionization of helium by electrons of energy greater than about 200 eV.

We now consider results given by some of the modified approximations discussed in § 5.6.2 for atomic hydrogen. In general the conclusions are similar.

Sloan§ evaluated a Born-exchange approximation (see § 5.6.2 (f)). In this case, when exchange is included, we must allow for transitions

---

† SILVERMAN, S. M. and LASSETTRE, E. N., loc. cit.
‡ PEACH, G., unpublished.
§ SLOAN, I. H., Proc. phys. Soc. 85 (1965) 435.

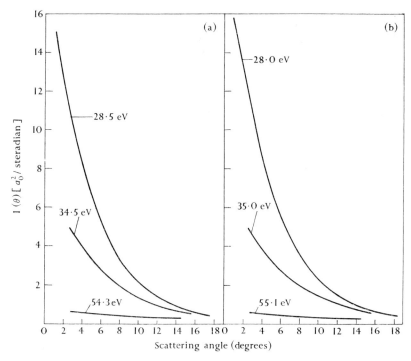

Fig. 7.23. Angular distribution of electrons scattered in helium after suffering ionizing collisions and a definite energy loss. (a) Derived from electron impact observations of Silverman and Lassettre. (b) Calculated by Peach.

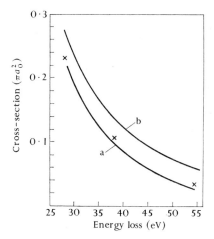

Fig. 7.24. Energy distribution of ejected electrons in ionizing collisions with helium atom. (a) Derived from electron impact observations of Silverman and Lassettre. (b) Calculated by Peach. × Calculated by restricting the range of angular integration to that observed.

to triplet as well as singlet states of the continuum. The differential cross-section $I_{0\kappa}\,d\kappa d\omega d\Omega$ is now given by

$$I_{0\kappa} = |f(k',\kappa) - g^+(k',\kappa)|^2 + 3|g^-(k',\kappa)|^2, \qquad (203)$$

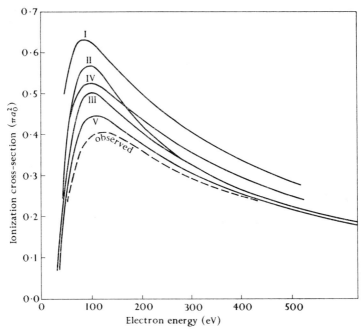

Fig. 7.25. Cross-sections for ionization of helium by electron impact. I. Calculated by Born's approximation (Sloan). II. Calculated using a Coulomb wave for the ejected electron (Peach). III. Calculated by the truncated Born approximation (Peach). IV. Calculated by a Born-exchange approximation (Sloan). V. Calculated by the Ochkur form of the Born–Oppenheimer approximation (Peach). – – – – observed (Smith).

where g^+, g^- are the amplitudes for singlet and triplet exchange transitions respectively. Corresponding to (188) for atomic hydrogen we now have

$$f(k',\kappa) = e^{i\delta(k',\kappa)} g^+(\kappa, k') \qquad (204)$$

$$|g^-(k',\kappa)| = |g^+(k',\kappa)|, \qquad (205)$$

where, if approximate expressions are used for f and g, the phase $\delta(k',\kappa)$ in (204) is indeterminate. Sloan assumed that $\delta(k',\kappa) = \arg\Gamma(1-i/\kappa a_0)$ and obtained the cross-sections shown in Fig. 7.25.

The truncated Born approximation ((b) of § 5.6.2) and the Ochkur form of the Born–Oppenheimer approximation (see p. 456) have been calculated by Peach† using a pure Coulomb wave for the ejected electron.

† Peach, G., Proc. phys. Soc. 87 (1966) 381.

Her results are also included in Fig. 7.25. It will be seen that the latter of these two approximations gives the best agreement with observation, as is also the case for atomic hydrogen and ionized helium. At the least the Ochkur procedure gives a good empirical estimate. It will be noted that the proportional difference between the full Born approximation and the other approximations is somewhat less than for hydrogen.

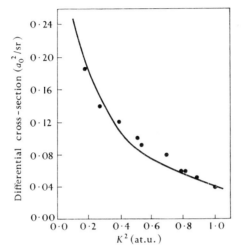

Fig. 7.26. Comparison of observed and calculated differential cross-sections for excitation of the 2^1S state of helium by electron impact. ● observed points. —— calculated curve.

5.6.9. *Excitation of optically forbidden transitions in helium.* We next consider the excitation of optically forbidden transitions. The strongest of these for a given total quantum number of the excited one-electron orbital are the ^1S–^1S transitions. Evidence concerning the validity of Born's first approximation is forthcoming for the two strongest of these transitions, 1^1S–2^1S and 1^1S–3^1S, from the observations of Lassettre and his collaborators of the energy-loss spectrum of electrons of a few hundred eV incident energy, as described in Chapter 5, § 4.

Fig. 7.26 is a comparison of the observed† and calculated differential cross-section for excitation of the 2^1S state by electrons of 500-eV incident energy, the calculations being those of Fox‡ using the wave functions (195) and (197). The agreement is very good indeed,

---

† LASSETTRE, E. N., KRASNOW, M. E., and SILVERMAN, S. M., *J. chem. Phys.* **40** (1964) 1242.
‡ Fox, M. A., *Proc. phys. Soc.* **86** (1965) 789.

particularly when it is remembered that the observed values are normalized with reference to the observed scattered intensity at one angle and energy for electrons which have excited the 2^1P states.

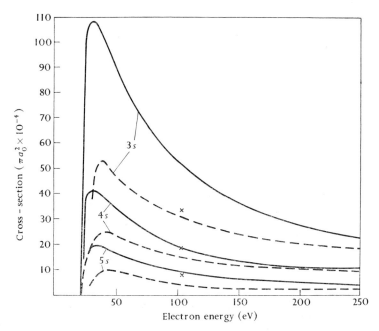

Fig. 7.27. Cross-sections for excitation of the n^1S states of helium for $n = 2$ to 6. —— calculated. - - - - observed by St. John, Miller, and Lin, × absolute values observed by Gabriel and Heddle.

Fox has extended his calculations to the excitation of ^1S states with n ranging from 3 to 10. For this work he chose for ψ_{n00} a function of the form
$$\psi_{n00} = \psi'_{n00} + a\psi_{100}, \tag{206}$$
where the constant a is chosen so that ψ_{n00} is orthogonal to the ground state function ψ_{100} taken to be of the form (195). ψ'_{n00} was taken to have the form described in Chapter 14, § 4.1, the constants being determined from the quantum defects of the corresponding S states. Fox finds, as for the excitation of the ^1P states, that the shape of the excitation function is the same for all states of the series, just as is found from observations of excitation functions (Chap. 4, Fig. 4.19). This may be seen by reference to Fig. 7.27, which gives the calculated excitation functions for $n = 3$ to 5. In the same figure the observations of St. John, Miller, and Lin (see Chap. 4, § 1.4.2.1) are also shown, as well as the absolute

values taken from the observations of Gabriel and Heddle† (see Chap. 4, § 1.4.2.1) at 100 eV. It will be seen that the agreement is only moderately satisfactory. This may be partly due to difficulty in obtaining good ^1S wave functions and partly to uncertainties in the observed values. A further check is obtained from the observations of Lassettre, Meyer, and Longmire‡ who find that the ratio of the cross-section for excitation of the 3^1S to that of the 2^1S state is 0·22, based on comparison of the differential cross-section at zero scattering angle for electrons of 200-eV energy. The ratio calculated by Fox§ is 0·211.

We next consider the excitation of ^1D states. For these states, in which the peak electron density occurs at an even greater distance from the nucleus than for the corresponding ^1P states, it is particularly important that the ground-state wave function used in the

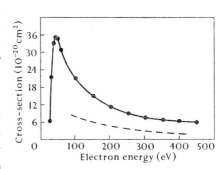

FIG. 7.28. Comparison of observed and calculated cross-sections for excitation of the 3^1D state of helium. – – – – calculated, Born's approximation. ●–●–● observed by St. John, Miller, and Lin.

calculations should be a good approximation at large distances. Fig. 7.28 shows a comparison between the absolute ^1D excitation function observed by St. John, Miller, and Lin‖ and that calculated using the wave functions (195) and (196). The agreement is not satisfactory. Thus, although the shape of the two functions is very similar, at least at electron energies greater than 100 eV, the absolute values disagree by a factor of 2.†† However, if the simpler approximation (198) to the ground-state wave function is used, the calculated value comes out to be nearly five times too small. This is because this approximation gives too compact a charge distribution, so there is insufficient overlap with that for the excited 3^1D state. Extension of the comparison to energies below 100 eV suggests that Born's approximation gives good results as far as the shape of the cross-section is concerned to quite low energies, in contrast to the excitation of the ^1P states.

It remains to consider the excitation of the triplet states. According to Table 7.4 the cross-sections for these intercombination transitions fall

---

† GABRIEL, A. H. and HEDDLE, D. W. O., *Proc. R. Soc.* A**258** (1960) 124.
‡ LASSETTRE, E. N., MEYER, V. D., and LONGMIRE, M. S., *J. chem. Phys.* **41** (1964) 2952.   § loc. cit., p. 491.   ‖ loc. cit., p. 190.
†† Recent measurements by MOUSSA, H. R. (Thesis, Leiden 1967) agree with the calculated to 30 per cent.

very rapidly with electron energy in the range 100–400 eV. Experimental confirmation of these very small values has proved difficult, because it is hard to reduce secondary effects that populate the triplet states to such small proportions that they contribute less than the very weak direct excitation processes. However, Jobe and St. John have been able to do this for the excitation of the 2^3P state and they find an excitation function (see Chap 4, § 1.4.2.1) that falls off as fast as

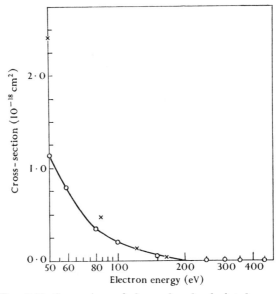

FIG. 7.29. Comparison of observed and calculated cross-sections for excitation of the 2^3P state of helium. ○–○–○ calculated, Born–Oppenheimer approximation. × observed by Jobe and St. John.

the calculated values given in Table 7.4. In fact there is quite good numerical agreement between the calculated and observed values for electron energies greater than 120 eV, as may be seen from the comparison shown in Fig. 7.29.

5.6.10. *Summarizing remarks—the applicability of Born's approximation to inelastic collisions in helium.* The evidence, although still incomplete and rather imprecise, suggests that Born's approximation gives quite good values for inelastic cross-sections in helium when the energy of the incident electrons exceeds 200 eV. For transitions associated with a quadrupole moment (^1S–^1D transitions) the range of validity may extend to considerably lower energies.†

For intercombination transitions it is likely that the Born–

† See footnote †† on p. 493.

Oppenheimer approximation also gives good results for energies above 200 eV but, because of the experimental difficulties in measuring the very small cross-sections at these energies and making adequate allowance for even very weak secondary effects, this is still largely unconfirmed except for the 2^3P excitation.

5.6.11. *Application to other atoms and ions.* A number of other calculations of cross-sections for excitation and ionization of more complex atoms and ions have been carried out using the Born or Coulomb–Born approximation as appropriate. For ionization the other related approximations described in § 5.6.2 have also been used in many cases. The main aim of these calculations has been to provide data for application to plasma physics or in astrophysics. These data are usually very difficult to obtain experimentally so that there are few cases in which comparison between theory and experiment is possible. We shall discuss these first.

Geiger[†] has measured the differential cross-sections for inelastic collisions of 20-keV electrons in neon, argon, and krypton, which have led to $(np)^6 {}^1S_0 \to (np)^5(\overline{n+1}s)$ $^3P_1$ and $^1P_1$ transitions. As for helium, this was done by comparison with the intensity of elastic scattering normalized as in Fig. 7.3. From these results the optical oscillator strengths for the transitions were deduced in the usual way and were found to agree with theoretical expectation within the rather large uncertainty of the calculations. Thus for the transition in neon, argon, and krypton respectively, Geiger derived f values (summed over $^3P_1$ and $^1P_1$ final states) of 0·140, 0·233, and 0·346. Corresponding theoretical values are 0·121,[‡] 0·133,[§] 0·163[||] for neon, 0·22,[‡] 0·25,[§] and 0·33[||] for argon and 0·40$_5$[||] for krypton.

The cross-section for excitation of the 3^1P state of sodium from the ground state has been calculated by Fundaminsky[††] using Born's approximation. His results are shown in Fig. 8.28 in comparison with the observations of Haft[‡‡] and Christoph.[§§] It will be seen that the calculated values begin to rise above the observed at an electron energy of about 20 eV, which is about eight times the threshold value. Further discussion of this case in terms of improved approximations is given in Chapter 8, § 8.3.

[†] GEIGER, J., *Z. Phys.* **177** (1964) 138.
[‡] GOLD, A. and KNOX, R. S., *Phys. Rev.* **113** (1959) 834.
[§] KNOX, R. S., ibid. **110** (1958) 375.
[||] COOPER, J. W., ibid. **128** (1962) 681.
[††] BATES, D. R., FUNDAMINSKY, A., LEECH, J. W., and MASSEY, H. S. W., *Phil. Trans. R. Soc.* A**243** (1950) 117.   [‡‡] HAFT, G., *Z. Phys.* **82** (1933) 73.
[§§] CHRISTOPH, W., *Annln Phys.* **23** (1935) 51.

Cross-sections for ionization of lithium,† beryllium,† sodium,‡ magnesium,‡ and neon§ have been calculated using Born's approximation. For the first two atoms Hartree–Fock wave functions were used for the initial and final states of atom and ion respectively and for neon Hartree wave functions. Distorted Coulomb waves were used for the ejected electron. Calculations have also been carried out‖ for lithium, beryllium, sodium, and magnesium, using the Born-exchange and Ochkur approximations. Comparison with observed data is shown in Fig. 7.30.

Turning now to theoretical results for which no comparable experimental data exist we note first the calculations, using Born's approximation, of cross-sections for excitation of already excited hydrogen atoms to higher states†† including states of the continuum. An interesting case is the excitation‡‡ of the $2s$–$2p$ transition discussed further in Chapter 8, § 4.4.

Cross-sections for excitation of the resonance terms of lithium and sodium and of the isoelectronic positive ions have been calculated§§ by the Born and Born–Oppenheimer approximation and their appropriate Coulomb–Born modifications. This work has also been extended‖‖ to the transitions $2s$–np, $2 \leqslant n \leqslant 7$ for Be II, N V, and Ne VIII.

Excitation cross-sections have also been calculated using the Coulomb–Born approximation for other cases of astrophysical interest such as Mg II $3s$–$3p$,††† Ca II $4s$–$3d$,††† O VI $2s$–$2p$,§§ Fe VIII $3d$–$4s$, $4p$, $4f$, $5s$, $5f$, $6s$, $7f$,‡‡‡ Fe XIV $3s^2 3p\ ^2P$–$3s\ 3p^2\ ^2P$, ^2D, ^2S§§§ and $3s^2 3p\ ^2P_{\frac{1}{2}}$–$3s^2 3p\ ^2P_{\frac{3}{2}}$,‖‖‖ Fe XV $3s^2\ ^1S$–$3s\ 3d\ ^1D$, $3s\ 3p\ ^2P$–$3s\ 3d\ ^3D$†††† and Fe XVI $3s$–$3p$.‡‡‡‡

† Peach, G., *Proc. phys. Soc.* **85** (1965) 709.
‡ Ibid. **87** (1966) 375.
§ Ledsham, F., Thesis, London, 1951.
‖ Peach, G., *Proc. phys. Soc.* **87** (1961) 381.
†† Boyd, T. J. M., ibid. **72** (1958) 523; McCrea, D. and McKirgan, T. V. M., ibid. **75** (1960) 235; McCoyd, G. C., Milford, S. N., and Wahl, J. J., *Phys. Rev.* **119** (1960) 149; Fisher, L., Milford, S. N., and Pomilla, F. R., ibid. **119** (1960) 153; McCoyd, G. C. and Milford, S. N., ibid. **130** (1963) 206; Omidvar, K., ibid. **140** (1965) A38; Kingston, A. E. and Lauer, J. E., *Proc. phys. Soc.* **87** (1966) 399; **88** (1966) 597.
‡‡ Seaton, M. J., ibid. **A68** (1955) 457.
§§ Bely, O., *C. r. hebd. Séanc. Acad. Sci., Paris* **254** (1962) 3075; Bely, O., Tully, J., and van Regemorter, H., *Annls Phys.* **8** (1963) 303.
‖‖ Bely, O., *Proc. phys. Soc.* **88** (1966) 587.
††† Van Regemorter, H., *C. r. hebd. Séanc. Acad. Sci., Paris* **252** (1961) 2514, 2667.
‡‡‡ Krueger, T. K. and Czyzak, S. J., *Ap. J.* **144** (1966) 381.
§§§ Petrini, D. *C. r. hebd. Séanc. Acad. Sci., Paris* **264** (1967) 41.
‖‖‖ Blaha, M., *Bull. astr. Insts. Csl.* **13** (1962) 81.
†††† Bely, O., and Blaha, M., in course of publication.
‡‡‡‡ Krueger, T. K. and Czyzak, S. J., *Mem. R. astr. Soc.* **69** (1965) 145.

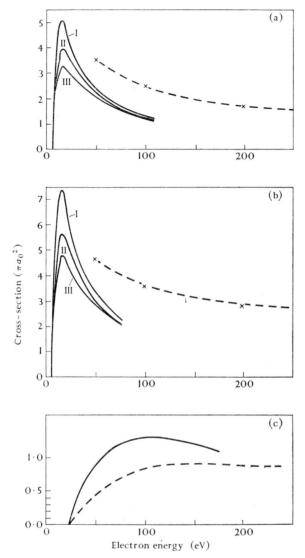

Fig. 7.30. Comparison of observed and calculated cross-sections for ionization of various atoms by electron impact. (a) Lithium and (b) sodium ×----× observed,† —— I. calculated by Born's approximation, —— II. calculated by Born-exchange approximation, —— III. calculated by Ochkur approximation, (c) Neon —— calculated by Born's approximation, ---- observed.‡

Burke, Tait, and Lewis§ have calculated cross-sections for excitation of the transitions between the 2- and 3-quantum states of the outer

† McFarland, R. H. and Kinney, J. D., loc. cit. p. 130.
‡ Smith, P. T., loc. cit. p. 129.
§ Burke, P. G., Tait, J. H., and Lewis, B. A., Proc. phys. Soc. 87 (1966) 209.

electron in N V using the Coulomb–Born approximation, by electrons with energy extending from the threshold to 455 eV. These are of interest in connection with the determination of the electron temperature in a high-temperature plasma by measuring the intensity ratio of the $2s$–$2p$ and $2s$–$3p$ transitions in three-electron ions. Except for the $2s$–$3p$ and $2p$–$3s$ transitions the Coulomb–Born approximation is likely to give results correct to within 20–30 per cent even close to the threshold and more accurately at higher energies. Burke, Tait, and Lewis, from a study of selected more accurate approximations (see Chap. 8) estimated correction factors for the $2s$–$3p$ and $2p$–$3s$ cases.

As for excitation, a number of ionization cross-sections of astrophysical interest have been calculated. These include O IV,[†] using the Coulomb–Born–Oppenheimer method, and Fe XV and Fe XVI,[‡] using the truncated Coulomb–Born and Coulomb–Born exchange approximations. In all cases the bound-state wave functions involved were those given by the Hartree–Fock field. For O IV the continuum wave function (for the 'ejected' electron) was taken to be an undistorted Coulomb wave, while for Fe XV and Fe XVI allowance for distortion of this wave was made by determining an effective local distorting potential which yielded the Hartree–Fock bound state functions.

The likely validity of these various calculations in different energy ranges must be judged against the evidence from those cases in which experimental check is possible. Because of the doubtful physical basis for using the approximations at electron energies which are not large compared with the threshold energy, it is not always easy to weigh the evidence. This is particularly true when the case under consideration is not very similar to any for which the applicability of the approximation is well known.

Calculations made using more elaborate approximations are discussed in Chapter 8.

5.6.12. *Angular distribution of the totality of inelastically scattered electrons.* It is sometimes of importance to know the intensity of all inelastic scattering in a given angular range (see Chap. 10, § 1.3). Morse[§] has shown that, if Born's approximation is valid, the differential cross-section for all inelastic collisions of electrons with an atom of atomic number Z is given by

$$\left(\sum_n + \int\right) I_{0n}(\theta)\, d\omega = (Ze^4 m^2/4k^4\hbar^4 \sin^4 \tfrac{1}{2}\theta) S(k\sin \tfrac{1}{2}\theta)\, d\omega, \quad (207)$$

[†] Trefftz, E., *Proc. R. Soc.* A**271** (1963) 379.
[‡] Rudge, M. R. H. and Schwartz, S. B., *Proc. phys. Soc.* **88** (1966) 579.
[§] Morse, P. M., *Phys. Z.* **33** (1932) 443.

where S is a certain function determined by the ground-state wave function of the atom, k is the wave number of the incident electron, and θ must be such that
$$\theta \gg |E_0|/E,$$
where E_0 is the energy of the ground state of the atom and E the electron energy. Heisenberg† showed how S may be calculated in terms of the Thomas–Fermi statistical model of the atom. Bewilogua,‡ using Heisenberg's formula, carried out the calculations giving values of S shown in Table 7.7.

TABLE 7.7

Differential cross-sections $I_{0n}(\theta)$ for total inelastic scattering of fast electrons by atoms

($\sqrt{\text{volts}}$) $\sin \tfrac{1}{2}\theta / Z^{\frac{2}{3}}$	S	$4Z^{5/3} I_{\text{in}}(\theta)$ in a_0^2/sr
0·278	0·319	9920
0·556	0·486	924
1·112	0·674	81·7
1·668	0·776	18·6
2·224	0·839	6·35
2·784	0·880	2·72
3·337	0·909	1·36
3·893	0·929	0·75
4·449	0·944	0·45
5·005	0·954	0·28
5·561	0·963	0·19

Under the same angular limitations, for atomic hydrogen§

$$S = \frac{\cos\theta}{\cos^4 \tfrac{1}{2}\theta}\left\{1 - \frac{1}{(1+\tfrac{1}{2}k^2 a_0^2 \sin^2 \tfrac{1}{2}\theta)^4}\right\}. \tag{208}$$

5.6.13. *Detachment of electrons from H^- ions by electron impact.* The cross-sections for detachment of electrons from H^- ions by electron impact has been calculated using the truncated Born approximation (177) of § 5.6.2 by McDowell and Williamson.‖ By ignoring contributions other than from s–p transitions of the attached electron, and using the expansion (41), the cross-section may be expressed in the form

$$Q^{\text{d}}(E) = \frac{e^2}{2\pi\alpha_0 a_0 E} \int_0^{\frac{1}{2}(E-A)} \frac{Q_{\text{p}}^{\text{d}}(W)}{W+A} \ln\left(\frac{4E}{W+A}\right) dW,$$

where α_0 is the fine structure constant, A the electron affinity, and

---

† HEISENBERG, W., *Phys. Z.* **32** (1931) 737.   ‡ BEWILOGUA, L., ibid. 740.
§ BETHE, H., *Annls Phys.* **5** (1930) 325.
‖ McDOWELL, M. R. C. and WILLIAMSON, J. H., *Phys. Lett.* **4** (1963) 159.

$Q_\text{p}^\text{d}(W)$ is the cross-section for photodetachment by photons of energy $A+W$. Using accurately calculated values for $Q_\text{p}^\text{d}(W)$ (see Chap. 15, § 5.1) McDowell and Williamson obtained the detachment cross-section shown in Fig. 7.31.

This calculation ignores the repulsive effect of the ionic charge on the incident electron. To allow for this McDowell and Williamson followed a procedure originally due to Geltman.† They multiplied the cross-sections by a factor

$$P(E) = 2\pi\alpha/(\text{e}^{2\pi\alpha}-1),$$

where $\alpha = e^2/\hbar v$, v being the incident electron velocity. $P(E)$ is the value of $|\psi_\text{c}|^2$ at $r = 0$, ψ_c being the wave function for motion in a Coulomb

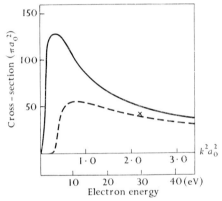

Fig. 7.31. Comparison of observed and calculated cross-sections for detachment of electrons from H⁻ ions by electron impact. —— calculated without allowance for repulsion of the incident electron, - - - - calculated with such allowance. × observed by Tisone and Branscomb.

field e^2/r, which has the asymptotic form of a modified plane wave of unit amplitude and the corresponding scattered wave (see Chap. 6, § 3.9). Inclusion of this factor reduces the cross-section, as seen from Fig. 7.31. Comparison with the observed results of Tisone and Branscomb (Chap. 3, § 4) shows reasonable agreement, particularly when allowance is made for the inaccuracy of the experimental results. Other calculations using Born's approximation directly but with rather simplified wave functions for the H⁻ ion have been carried out by Smernov and Chibisov‡ and they give rather similar results, not markedly distinct on a log–log plot from those of McDowell and Williamson.

† GELTMAN, S., *Proc. phys. Soc.* **75** (1960) 67.
‡ SMERNOV, B. M. and CHIBISOV, M. J., *Zh. éksp. teor. Fiz.* **49** (1965) 841; *Soviet Phys. JETP* **22** (1966) 585.

8

ELECTRON COLLISIONS WITH ATOMS— THEORETICAL DESCRIPTION—ANALYTICAL THEORY FOR SLOW COLLISIONS

1. Introduction

IN Chapter 7 we described Born's approximate method for calculating cross-sections, elastic and inelastic, for collisions of electrons with atoms in terms of the atomic structure. At low impact energies, less than say ten times the excitation energy, the first approximation is no longer valid in general. The problem of calculating cross-sections under these conditions is a difficult one and reliable results can only be obtained, if at all, by employing much more elaborate analytical and computational procedures. It is not possible to replace the clear-cut procedure of Born's approximation by one of comparable generality and definiteness. Nevertheless, since the first edition of this book was published, very considerable progress has been made particularly, but not exclusively, in dealing with elastic scattering. This has been largely due to the availability of high-speed computers and to the great proportional increase in research effort. In this chapter we shall describe this work in so far as it refers to the broad features, as distinct from the fine structure, of collision cross-sections as functions of impact energy. The theoretical interpretation of the fine structure will be discussed in Chapter 9.

We shall begin by considering the various methods that may be used for calculating cross-sections for slow collisions with atomic hydrogen. At first we shall concentrate on the elastic scattering at energies below the excitation threshold. Although there are still few experimental results of sufficient accuracy to check the theory it is possible, because of the comparative simplicity of the two-electron problem, to determine the elastic cross-section at these energies to a very reliable accuracy. Although the methods required to do this are very elaborate and not readily applicable to more complex atoms they provide a reliable background against which to check the suitability of less elaborate approximations which can be so applied.

After dealing with elastic scattering we shall then proceed to discuss inelastic scattering as well as elastic beyond the excitation threshold. Although the situation here is less satisfactory we can nevertheless introduce procedures that are likely to be useful for other atoms and ions.

The analysis for atomic hydrogen can be applied, with allowance for the net Coulomb field, to collisions with He$^+$ and other ions with only one residual electron.

After discussing such applications, the next problem in order of simplicity, the collisions of electrons with helium atoms, will be considered. This will be followed by an account of progress made in dealing with more complex atoms.

We deal next with the theory of the polarization of impact radiation the experimental investigation of which has been described in Chapter 4, § 1.3.2.

Finally, we give an account of the attempts to use classical theory to obtain inelastic cross-sections for collisions of slow electrons with atoms.

2. Collisions with hydrogen atoms—elastic scattering at energies below the excitation threshold

2.1. *The generalized variational method*[†]

It is convenient to base the analysis on an extension of the variational method for the calculation of phase shifts for elastic scattering of particles by a centre of force described in Chapter 6, § 3.8. We consider in place of (62) of that chapter, the integral

$$I_{11} = \iint \Psi_2^t(\mathbf{r}_1, \mathbf{r}_2) L \Psi_1^t(\mathbf{r}_1, \mathbf{r}_2) \, d\mathbf{r}_1 \, d\mathbf{r}_2, \qquad (1)$$

where
$$L = \nabla_1^2 + \nabla_2^2 + (2me/\hbar^2)\left\{\left(\frac{1}{r_1} + \frac{1}{r_2} - \frac{1}{r_{12}}\right) + E\right\}, \qquad (2)$$

E being the total energy.

If Ψ_1^t is an exact solution of the wave equation for the combined system of hydrogen atom plus electron then $I_{11} = 0$. Furthermore, if it describes the scattering of electrons with energy $k^2\hbar^2/2m$ below the excitation threshold, from incident direction $\mathbf{n}_0$ to final direction $\mathbf{n}_1$ by hydrogen atoms in the ground state then

$$\Psi_1^t \sim \{e^{ik\mathbf{n}_0\cdot\mathbf{r}_1} + f(\mathbf{n}_0, \mathbf{n}_1)e^{ikr_1}r_1^{-1}\}\psi_0(r_2), \qquad (3)$$

where $\psi_0(r_2)$ is the wave function of the ground state of the hydrogen atom. The differential electron scattering cross-section is then given by

$$|f(\mathbf{n}_0, \mathbf{n}_1)|^2 \, d\omega. \qquad (4)$$

[†] KOHN, W., *Phys. Rev.* **74** (1948) 1763; MOTT, N. F. and MASSEY, H. S. W., *The theory of atomic collisions*, 3rd edn, Chap. xiv, § 2 (Clarendon Press, Oxford, 1965).

If instead of exact solutions of $L\Psi = 0$ we substitute for Ψ_1, Ψ_2 trial functions Ψ_1^t, Ψ_2^t which have the asymptotic form

$$\Psi_1^t \sim \{e^{ik\mathbf{n}_0\cdot\mathbf{r}_1} + f^t(\mathbf{n}_0, \mathbf{n})e^{ikr_1}r_1^{-1}\}\psi_0(r_2), \tag{5a}$$

$$\Psi_2^t \sim \{e^{-ik\mathbf{n}_1\cdot\mathbf{r}_1} + f^t(-\mathbf{n}_1, \mathbf{n})e^{ikr_1}r_1^{-1}\}\psi_0^*(r_2), \tag{5b}$$

then it is easy to show that, to the second order,

$$\delta I_{11} = -4\pi\,\delta f(\mathbf{n}_0, \mathbf{n}_1) \tag{6}$$

so that, to the same order, the true scattered amplitude is given by

$$f(\mathbf{n}_0, \mathbf{n}_1) = f^t(\mathbf{n}_0, \mathbf{n}_1) + I_{11}^t/4\pi. \tag{7}$$

If we simply use plane wave trial functions then (7) reduces to Born's first approximation.

We note also that there is no difficulty in extending this analysis to allow for electron exchange. The trial functions now not only must have the asymptotic forms (5a, b) for large r_1, but for large r_2,

$$\Psi_1^t \sim g^t(\mathbf{n}_0, \mathbf{n})e^{ikr_2}r_2^{-1}\psi_0(r_1), \tag{8a}$$

$$\Psi_2^t \sim g^t(-\mathbf{n}_1, \mathbf{n})e^{ikr_2}r_2^{-1}\psi_0^*(r_1). \tag{8b}$$

(7) remains valid as before. To obtain a comparable relation for $g(\mathbf{n}_0, \mathbf{n}_1)$ we replace I_{11}^t by the integral

$$I_{12}^t = \iint \Psi_2^t(\mathbf{r}_2, \mathbf{r}_1) L\Psi_1^t(\mathbf{r}_1, \mathbf{r}_2)\,\mathrm{d}\mathbf{r}_1\,\mathrm{d}\mathbf{r}_2, \tag{9}$$

and then
$$\delta I_{12} = -4\pi\,\delta g(\mathbf{n}_0, \mathbf{n}_1); \tag{10}$$

so, to the second order,

$$g(\mathbf{n}_0, \mathbf{n}_1) = g^t(\mathbf{n}_0, \mathbf{n}_1) + I_{12}^t/4\pi. \tag{11}$$

With plane wave trial functions (11) gives the Born–Oppenheimer approximation for g.

The problem now is to choose trial functions that are not only sufficiently good representations of the exact function but are not so complicated as to make the calculation of I_{11}^t and I_{12}^t impracticable. In fact, with such a comparatively simple system and modern methods of automatic computation it is possible to go so far that reliable results are obtained. Nevertheless, it is very important to obtain as much guidance as possible in choosing the trial functions. To obtain some help in this respect we shall return to the consideration of the problem in terms of eigenfunction expansions.

2.2. *The eigenfunction expansion*

In Chapter 7, § 1 we represented the wave function describing the collision of an electron with a hydrogen atom in the form

$$\Psi(\mathbf{r}_1, \mathbf{r}_2) = e^{ik\mathbf{n}_0\cdot\mathbf{r}_1}\psi_0(r_2) + \phi(\mathbf{r}_1, \mathbf{r}_2) \tag{12}$$

and then expanded $\phi(\mathbf{r}_1, \mathbf{r}_2)$ in the form

$$\phi(\mathbf{r}_1, \mathbf{r}_2) = \left(\sum_n + \int\right) F_n(\mathbf{r}_1)\psi_n(\mathbf{r}_2), \tag{13}$$

where ψ_n is the wave function of the nth excited state of the hydrogen atom. The assumption of the form (12) is convenient when $\phi(\mathbf{r}_1, \mathbf{r}_2)$ is small but when it is not necessarily so it is more convenient not to distinguish between the unperturbed plane-wave term and the remainder. If we write instead of (12) simply

$$\Psi(\mathbf{r}_1, \mathbf{r}_2) = \left(\sum_n + \int\right) F_n(\mathbf{r}_1)\psi_n(\mathbf{r}_2), \tag{14}$$

then we must have

$$F_0(\mathbf{r}_1) \sim e^{i k \mathbf{n}_0 \cdot \mathbf{r}_1} + r_1^{-1} e^{i k r_1} f_0(\theta_1, \phi_1), \tag{15}$$

and

$$(\nabla^2 + k_n^2) F_n(\mathbf{r}_1) = \sum U_{nm} F_m(\mathbf{r}_1), \tag{16}$$

where U_{nm} is defined in Chapter 7 (8). In particular, for the elastic scattering

$$(\nabla^2 + k^2) F_0(\mathbf{r}_1) = \sum U_{0m} F_m(\mathbf{r}_1), \tag{17}$$

or

$$(\nabla^2 + k^2 - U_{00}) F_0(\mathbf{r}_1) = \sum_{m \neq 0} U_{0m} F_m(\mathbf{r}_1). \tag{18}$$

The asymptotic form of the functions $F_n(\mathbf{r})$ ($n \neq 0$) will depend on the sign of k_n^2. In general $k_n^2 > 0$ only for a finite number of states n. As the integral in (14) includes the possibility of ionization there will always be an unlimited range of values of $k_n^2 < 0$.

Then, if $k_n^2 > 0$,

$$F_n(\mathbf{r}_1) \sim r_1^{-1} e^{i k_n r_1} f_n(\theta_1, \phi_1), \tag{19}$$

while, if $k_n^2 < 0 = -\kappa_n^2$,

$$F_n(\mathbf{r}_1) \sim r_1^{-1} e^{-\kappa_n r_1} f_n(\theta_1, \phi_1). \tag{20}$$

In terms of these asymptotic forms for large r_1 we can think of the six-dimensional $(\mathbf{r}_1, \mathbf{r}_2)$ space as divided into asymptotic regions corresponding to the terms in the expansion (14) in each of which the atom is in a definite stationary state and the interaction between the F_n is negligible. Such regions are known as *channels* and a channel is said to be *open* when $k_n^2 > 0$ so that there is a finite outgoing current in the channel at large r_2. Otherwise it is said to be *closed*. These concepts may be generalized to more complex collisions and are frequently useful (see in particular Chap. 9, § 10).

2.2.1. *Analysis of elastic scattering in terms of phase shifts.*[†] In Chapter 6, § 3 we discussed the solution of the equation of the form (18) in which the right-hand side vanishes—the case of scattering by a centre

---

[†] MOTT, N. F. and MASSEY, H. S. W., *The theory of atomic collisions*, 3rd edn, Chap. xii, § 1.4 (Clarendon Press, Oxford, 1965).

of force of potential $\hbar^2 U_{00}/2m$. It was shown that $f_0(\theta, \phi)$ could be expressed in the form

$$f_0(\theta, \phi) = \frac{1}{2ik} \sum (2l+1)(e^{2i\eta_l}-1) P_l(\cos\theta), \qquad (21)$$

where the η_l are *real* phase shifts defined in terms of the asymptotic form of the wave function for motion of electrons, with orbital angular momentum quantum number l, under the potential $\hbar^2 U_{00}/2m$.

If no open inelastic channels exist so the F_n all have the asymptotic form (20) it may be shown that $f_0(\theta, \phi)$ can still be expressed in the form (21) involving *real* phase shifts η_l even though the η_l can now only be determined in principle in terms of the solutions of an infinite set of coupled equations.

On the other hand, if any inelastic channel is open, it is no longer possible to represent $f_0(\theta, \phi)$ in the form (21) with real phase shifts η_l. Instead they will be complex with a positive imaginary part. This is because the outgoing flux of elastically scattered electrons for given l will no longer be equal to the incoming as some of the latter electrons will have suffered inelastic collisions. From the point of view of the flux of electrons with the initial energy such collisions cause a reduction of the net flux as if absorption had occurred, and this adds a positive imaginary component to the phase shift.

2.2.2. *The forward intensity of elastic scattering—dispersion relations.*
An important general theorem relates the imaginary component of the amplitude of the elastic scattering in the forward direction to the total scattering cross-section. Thus, when only elastic scattering is possible,

$$f_0(0) = \frac{1}{2ik} \sum (2l+1)(e^{2i\eta_l}-1)$$

and

$$\text{im } f_0(0) = -\tfrac{1}{2}k^{-1} \sum (2l+1)(\cos 2\eta_l - 1)$$
$$= k^{-1} \sum (2l+1) \sin^2 \eta_l$$
$$= (k/4\pi) Q_0 \quad \text{from Chap. 6 (16)}. \qquad (22)$$

This result is actually of much greater generality and may be shown to apply even when inelastic channels are open, provided the total elastic cross-section Q_0 on the right hand is replaced by Q_t the total cross-section for all collisions, elastic and inelastic.

It is also possible[†] to relate the real component of the forward scattered amplitude $f_0(0, E)$ at a particular energy E to an integral over the

† GERJUOY, E. and KRALL, N. A., *Phys. Rev.* **119** (1960) 705.

imaginary component, taken over all energies, in the form

$$\operatorname{re} f_0(0, E) = f_\mathrm{b}(0, E) - R(E) + \frac{1}{\pi} \int \frac{\operatorname{im} f_0(0, E')}{E' - E} \, dE', \qquad (23\,\mathrm{a})$$

where $f_\mathrm{b}(0, E)$ is the forward elastically scattered amplitude according to Born's approximation and $R(E)$ is determined by the properties of the bound states for the particle in the field of the scatterer. If the wave functions of these states are known then $R(E)$ may be calculated.

A relation of the form (23 a) is known as a dispersion relation.

Since
$$\operatorname{im} f_0(0, E) = (k/4\pi) Q_\mathrm{t}(E), \qquad (23\,\mathrm{b})$$

it follows that (23 a) and (23 b) together make it possible to determine $f_0(0, E)$ and hence $I_0(0, E)$, if the total cross-section Q_t is known at all energies that contribute significantly to the right-hand integral in (23 a), and the wave functions for the bound states are also known. The forms taken by (23 a) and (23 b) for scattering by hydrogen and by helium atoms including electron exchange are given in §§ 2.9 and 6.1, pp. 519 and 542 respectively.

2.2.3. *The truncated eigenfunction expansion approximation.* It is to be noted that the equations (16) may be written in the form

$$(\nabla^2 + k_n^2) F_n = -(2me^2/\hbar^2) \int \left(\frac{1}{r_1} - \frac{1}{r_{12}} \right) \psi_n^*(\mathbf{r}_2) \Psi(\mathbf{r}_1, \mathbf{r}_2) \, d\mathbf{r}_2. \qquad (24)$$

A possible new method of approximation now suggests itself. This is by truncating the sum in (14). We find then that if we choose as a trial function Ψ^t in (5) the form

$$\Psi^t(\mathbf{r}_1, \mathbf{r}_2) = \sum_{n=0}^{s} F_n(\mathbf{r}_1) \psi_n(\mathbf{r}_2), \qquad (25)$$

then the variational method shows that the best choice for the F_n are the solutions of the coupled equations

$$(\nabla^2 + k_n^2) F_n(\mathbf{r}_1) = \sum_{n=0}^{s} U_{nm} F_m(\mathbf{r}_1),$$

which are also obtained by substituting (25) for Ψ in (24).

In this way we have a series of trial functions that are conveniently referred to as *one-state*, *two-state*, etc., approximations. For the one-state approximation
$$\Psi^t(\mathbf{r}_1, \mathbf{r}_2) = F_0(\mathbf{r}_1) \psi_0(\mathbf{r}_2), \qquad (26)$$

we have
$$(\nabla^2 + k^2 - U_{00}) F_0 = 0, \qquad (27)$$

and the problem is the same as that of scattering by the central field, i.e. the static field of the atom. The fact that the optical model potentials described in Chapter 6 are based on the static field shows that this is probably already not a bad approximation.

However, before proceeding further, it is convenient to show how electron exchange may be brought into the picture. To satisfy the Pauli principle the over-all wave function, including the spin, must be antisymmetric in the coordinates of the two electrons. Hence in place of (26) we now take, for the one-state exchange case

$$\Psi^t(\mathbf{r}_1, \mathbf{r}_2) = F_0^{\pm}(\mathbf{r}_1)\psi_0(r_2) \pm F_0^{\pm}(\mathbf{r}_2)\psi_0(r_1), \tag{28}$$

the $\pm$ signs being taken according as the electron spins are antiparallel or parallel respectively. On using this trial function in conjunction with the variation method discussed above we find† that $F_0^{\pm}$ should satisfy an integro-differential equation of the form

$$(\nabla^2 + k^2 - U_{00})F_0^{\pm}(\mathbf{r}) = \pm \int K_{00}(\mathbf{r}, \mathbf{r}')F_0^{\pm}(\mathbf{r}')\,d\mathbf{r}', \tag{29}$$

where
$$K_{00}(\mathbf{r}, \mathbf{r}') = -\psi_0(r)\psi_0(r')\{k^2 + \lambda_0^2 + U(|\mathbf{r}-\mathbf{r}'|)\}, \tag{30}$$

$-\lambda_0^2 \hbar^2/2m$ being the energy of the ground state of hydrogen. The effect of electron exchange and the Pauli principle is to introduce the additional interaction term on the right-hand side of (29). This is an example of a non-local interaction because its magnitude at a point r depends not only on the value of the amplitude function F_0 at r but at all points of space.

If we obtain solutions of (29) that have the asymptotic form (15) then the differential cross-section averaged over spin is given by

$$\tfrac{1}{4}|f^+|^2 + \tfrac{3}{4}|f^-|^2. \tag{31}$$

This we call the one-state exchange approximation. In fact it will be found that inclusion of exchange in this way is important so that we shall henceforward include it from the outset.

A two-state exchange approximation that it is natural to choose next is to take

$$\Psi^t(\mathbf{r}_1, \mathbf{r}_2) = F_0^{\pm}(\mathbf{r}_1)\psi_0(r_2) \pm F_0^{\pm}(\mathbf{r}_2)\psi_0(r_1) + F_1^{\pm}(\mathbf{r}_1)\psi_1(r_2) \pm F_1^{\pm}(\mathbf{r}_2)\psi_1(r_1), \tag{32}$$

where ψ_1 is the wave function for the 2s state of hydrogen. This gives rise, through the variational method, to two coupled integro-differential

† MOTT, N. F. and MASSEY, H. S. W., *The theory of atomic collisions*, 3rd edn, Chap. xv, § 1.3 (Clarendon Press, Oxford, 1965).

equations
$$(\nabla^2+k^2-U_{00})F_0^{\pm}(\mathbf{r})\mp \int K_{00}(\mathbf{r},\mathbf{r}')F_0^{\pm}(\mathbf{r}')\,\mathrm{d}\mathbf{r}'$$
$$= U_{01}F_1^{\pm}(\mathbf{r})\pm \int K_{01}(\mathbf{r},\mathbf{r}')F_1^{\pm}(\mathbf{r}')\,\mathrm{d}\mathbf{r}', \quad (33\,\mathrm{a})$$
$$(\nabla^2+k_1-U_{11})F_1^{\pm}(\mathbf{r})\mp \int K_{11}(\mathbf{r},\mathbf{r}')F_1^{\pm}(\mathbf{r}')\,\mathrm{d}\mathbf{r}'$$
$$= U_{10}F_0^{\pm}(\mathbf{r})\pm \int K_{10}(\mathbf{r},\mathbf{r}')F_0^{\pm}(\mathbf{r}')\,\mathrm{d}\mathbf{r}', \quad (33\,\mathrm{b})$$

where K_{01}, K_{11} are further non-local interaction kernels arising from exchange. We refer to the solution obtained without approximation from these equations as the 1s–2s close-coupled exchange approximation—the designation close-coupled is included because no assumption has been made about the magnitude of the coupling terms U_{01}, K_{01}.

It is obviously possible to continue this indefinitely but the law of diminishing returns begins to operate quite quickly. In practice it seems hardly worth while to go beyond the 1s–2s–2p three-state close-coupling exchange approximation because the influence of the continuum states, which contribute to the integral in (14) and which are neglected in these approximations, is substantially more important than that of the higher discrete states.

The 1s–2s–2p close-coupling exchange approximation has proved to be so fruitful in many ways (see §§ 2.6, 4, 5 of this chapter and § 3 of Chap. 9) that it is worth discussing the analysis in terms of angular momentum. Neglecting spin-orbit interaction, the total orbital and spin angular momenta will be constants of the motion for the combined system of atom plus electron and will be specified by the respective quantum numbers L, S.

Since $S = 0, 1$ there will be separate sets of singlet and triplet states of the combined system. We now determine the coupling possibilities in each of these sets in terms of the orbital angular momenta. If l_1, l_2 refer to the orbital angular-momentum quantum-numbers of the atomic and colliding electrons we have, from the rules for combination of angular momenta for $L = 0$; $l_1 = 0$, $l_2 = 0$; $l_1 = 1$, $l_2 = 1$. Corresponding to each of these we have the terms in the eigenfunction expansion

$$r_1^{-1}\psi_{1s}(r_2)\,G_{1s,0}(r_1); \quad r_1^{-1}\psi_{2s}(r_2)\,G_{2s,0}(r_1); \quad r_1^{-1}\psi_{2p}(r_2)\,G_{2p,1}(r_1),\dagger \quad (34)$$

where $r_1^{-1}G_{2p,1}$, for example, denotes the wave function for motion of the electron in the 2p channel with angular-momentum quantum-number 1. We have therefore three coupled equations linking the three terms (34).

For $L = 1$ we have the possibilities $l_1 = 0, l_2 = 1; l_1 = 1, l_2 = 0; l_1 = 1, l_2 = 1$. Of these, the third possibility is distinguished by possessing even parity whereas the others have odd parity. This means that the corresponding term is not coupled to the other three and we are left with coupling between the terms

$$r_1^{-1}\psi_{1s}(r_2)\,G_{1s,1}(r_1); \quad r_1^{-1}\psi_{2s}(r_2)\,G_{2s,1}(r_1); \quad r_1^{-1}\psi_{2p}(r_2)\,G_{2p,0}(r_1),\dagger \quad (35)$$

and so on for larger L.

Below the excitation threshold we may describe the elastic scattering

† $\psi_{2p}(r_2)$ is here the radial part of the 2p wave function.

in terms of the real phase shifts η_l^+, η_l^- corresponding to the singlet and triplet state of the combined system. Because we are considering an electron incident on a hydrogen atom in its ground state $l = L$.

The position is very similar to that encountered in calculating the energies of bound states of complex atoms. The assumption that the state belongs to a definite configuration (see Chap. 6, § 2.2) gives a good initial approximation but the improvement achieved by allowing for interaction with further configurations is very gradual.

Two alternative courses are open. The first possibility is to fashion an approximate method to include the major physical factors that are likely to influence the low-energy elastic scattering, and the second is to use more elaborate trial functions in the variational method of § 2.1, just as has been done so successfully for calculating the energy of the ground state of helium.

To begin the discussion of the first possibility we note that while, in the 1s exchange approximation, the effects of the undisturbed (static) atomic field and of electron exchange are included there is no explicit allowance for disturbance of the atomic field by the incident electron. In terms of the eigenfunction expansion (14) the effect of truncation to include only the first term is to suppose that the atomic wave function is undisturbed. To allow for such disturbance higher terms in the expansion must be included. When we allow for exchange we go some way because the function (28) when expanded in the form (14) certainly introduces contributions from functions $\psi_n(\mathbf{r}_2)$ other than $\psi_0(r_2)$. This allowance is incomplete and in a sense haphazard because it comes in only because of allowance for the indistinguishability of electrons. The most important disturbance is likely to be dipole polarization which, it may be shown, introduces a further effective interaction, which has the asymptotic form

$$V_{\mathrm{p}} = -\tfrac{1}{2}\alpha e^2/r^4, \tag{36}$$

where α is the electric polarizability of the atom.

Before discussing this it is useful to consider a method which, while only applicable to the calculation of the phase shifts $\eta_0^\pm$ for elastic scattering by hydrogen atoms, leads to additional insight into the form of the atomic distortion terms.

2.3. *Allowance for distortion of the atom by the incident electron—Temkin's analysis of s-scattering*†

For the special case of the elastic scattering of slow electrons by atomic hydrogen which involves only continuum states of the two-electron

† TEMKIN, A., *Phys. Rev. Lett.* **4** (1960) 566; *Phys. Rev.* **126** (1962) 130.

system with zero total angular momentum, the wave function for the total system will depend only on the respective magnitudes r_1, r_2 of the vectors joining the proton to each electron and the angle ϑ between them. The wave equation in terms of these coordinates is

$$\left[\frac{\partial^2}{\partial r_1^2}+\frac{2}{r_1}\frac{\partial}{\partial r_1}+\frac{\partial^2}{\partial r_2^2}+\frac{2}{r_2}\frac{\partial}{\partial r_2}+\left(\frac{1}{r_1^2}+\frac{1}{r_2^2}\right)\frac{1}{\sin\vartheta}\frac{\partial}{\partial\vartheta}\left(\sin\vartheta\frac{\partial}{\partial\vartheta}\right)+\right.$$
$$\left.+(2me^2/\hbar^2)\left(\frac{1}{r_1}+\frac{1}{r_2}-\frac{1}{r_{12}}\right)+2mE/\hbar^2\right]\Psi = 0, \quad (37)$$

where $E = E_0 + 2mk^2/\hbar^2$, E_0, being the energy of the ground state of hydrogen.

The solution of (37) may be expanded in zonal harmonics in the form

$$\Psi(r_1, r_2, \vartheta) = (r_1 r_2)^{-1}\sum_{l=0}^{\infty}(2l+1)^{\frac{1}{2}}\Phi_l(r_1, r_2)P_l(\cos\vartheta). \quad (38)$$

Since
$$\Psi(r_1, r_2, \vartheta) = \pm\Psi(r_2, r_1, \vartheta)$$

according as the electron spins are antiparallel or parallel respectively, we must have
$$\Phi_l(r_1, r_2) = \pm\Phi_l(r_2, r_1),$$

and in addition
$$\Phi_l(r_1, 0) = 0,$$

$$\left.\begin{array}{l}\Phi_0(r_1, r_2) \sim \sin(kr_1+\eta_0^{\pm})r_2\psi_0(r_2) \\ \Phi_l(r_1, r_2) \to 0, \quad l > 0\end{array}\right\} \text{ as } r_1 \to \infty.$$

It is possible, using these equations, to obtain very good values for the phase shifts $\eta_0^{\pm}$ as has been shown by Temkin. If all terms in (38) are ignored except that for which $l = 0$, a first approximation $\eta_0^{(0)\pm}$ is obtained that is equivalent to the results that would be obtained if, in the expansion (14), all s states were included but no others. Inclusion of $\Phi_1, \Phi_2,...$ allows successively for dipole, quadrupole,..., etc., distortion of the atom. Temkin has shown that

$$\Phi_l(r_1, r_2) \sim -\frac{1}{(2l+1)^{\frac{1}{2}}}\frac{\sin(kr_1+\eta_0^{\pm})}{r_1^{l+1}}\left(\frac{r_2^{l+2}}{l+1}+\frac{r_2^{l+1}}{l}\right)\psi_0(r_2). \quad (39)$$

In Table 8.1 the relative importance of the dipole and quadrupole terms is given. It will be seen that, particularly for the η_0^+ phases, the former terms introduce important contributions but the quadrupole are very much less significant.

2.4. *Allowance for distortion of the atom by the incident electron—the polarized-orbital and exchange-adiabatic approximations*

It seems from this analysis that an approximation that included the effect of the static field, of exchange, and of dipole polarization should

TABLE 8.1

Successive approximations to zero order phase shifts for elastic scattering of electrons in atomic hydrogen, calculated by Temkin's method

Electron wave number k (at.u.)	Singlet phase shift (η_0^+) (rad)				Triplet phase shift (η_0^-) (rad)				
	$\eta_0^{(0)}$	$\delta_1\eta_0$	$\delta_2^{(1)}\eta_0$	$\delta_2^{(2)}\eta_0$	η_0	$\eta_0^{(0)}$	$\delta_1\eta_0$	$\delta_2^{(1)}\eta_0$	$\delta_2^{(2)}\eta_0$	η_0
0·01	3·064	0·026	−0·008	0·004	3·086	3·118	0·0050	0·0003	0·0001₅	3·124
0·05	2·759	0·117	−0·035	0·019	2·86	3·025	0·0193	0·0013	0·0008	3·046
0·10	2·420	0·187	−0·045	0·030	2·59	2·909	0·0303	0·0021	0·0014	2·942
0·20	1·895	0·215	−0·030	0·034	2·11	2·681	0·0379	0·0021	0·0023	2·723
0·40	1·269	0·165	−0·009	0·026	1·45	2·259	0·0379	0·0013	0·0028	2·301
0·80	0·728	0·126	−0·002	0·020	0·87	1·617	0·0282	0·0005	0·0018	1·647

$\delta_1\eta_0$ is the first order correction to $\eta_0^{(0)}$ due to dipole distortion.
$\delta_2^{(1)}\eta_0$ is the second order correction to $\eta_0^{(0)}$ due to dipole distortion.
$\delta_2^{(2)}\eta_0$ is the first order correction to $\eta_0^{(0)}$ due to quadrupole distortion.

give very good results. One semi-empirical method of proceeding would be to include the latter effect by adding an effective scattering potential (see (36))

$$V_p = -\frac{1}{2}\frac{\alpha e^2}{r^4}q(r), \tag{40a}$$

where $q(r)$ is a convergence factor that satisfies

$$q(r) \to 1 \qquad (r \to \infty),$$
$$r^{-4}q(r) \to \text{const} \quad (r \to 0). \tag{40b}$$

One popular choice† is to take

$$q(r) = r^4/(r^2+l^2)^2, \tag{41}$$

where l is an adjustable parameter of the order of atomic dimensions. Temkin and Lamkin‡ have introduced a less empirical method, which has proved to be very effective in application, not only to elastic scattering by atomic hydrogen, but by many other atoms and positive ions.

They take a trial function of the form

$$\Psi^t = (1 \pm P_{12})\{\psi_0(r_2) + \phi_0(\mathbf{r}_1, \mathbf{r}_2)\}F_0(\mathbf{r}_1), \tag{42}$$

in which ϕ_0 represents the distortion of the atomic wave function during the collision and P_{12} is the operator which interchanges 1 and 2. ϕ_0 is determined in the following way. Provided $r_1 > r_2$, so that the incident electron is at a greater distance from the nucleus than the atomic electron, the root mean square velocity of the latter will be greater than that of the incident electron, the kinetic energy of which is initially less than the excitation threshold. As an approximation for ϕ_0 under these conditions we can regard the atomic electron wave function as following nearly adiabatically the changing position of the incident electron—in other words, when $r_1 > r_2$ we take for $\phi(\mathbf{r}_1, \mathbf{r}_2)$ the change in the atomic wave function when the incident electron is fixed at $\mathbf{r}_1$. For $r_1 < r_2$ this viewpoint is no longer valid and it is assumed that, under these conditions, $\phi_0 \simeq 0$. This is a reasonable approximation in view of the symmetry conditions satisfied by Ψ^t.

With these assumptions the perturbation giving rise to $\phi_0(\mathbf{r}_1, \mathbf{r}_2)$ is given by

$$v(\mathbf{r}_1, \mathbf{r}_2) = \begin{cases} -e^2\left(\dfrac{1}{r_2} - \dfrac{1}{r_{12}}\right) & (r_1 > r_2), \\ 0 & (r_1 < r_2). \end{cases} \tag{43}$$

The further approximation is made of expanding $v(\mathbf{r}_1, \mathbf{r}_2)$ in a multipole

† BATES, D. R. and MASSEY, H. S. W., *Phil. Trans. R. Soc.* A**239** (1946) 269.
‡ TEMKIN, A. and LAMKIN, J. C., *Phys. Rev.* **121** (1961) 788.

expansion and retaining only the first, dipole term, viz.
$$v(\mathbf{r}_1, \mathbf{r}_2) \simeq 2(r_1/r_2^2)\cos\vartheta, \tag{44}$$
where ϑ is the angle between $\mathbf{r}_1$ and $\mathbf{r}_2$.

Neglect of higher terms is justified by the results of Temkin's analysis of the exact problem given in Table 8.1.

$\phi_0(\mathbf{r}_1, \mathbf{r}_2)$ may now be evaluated by standard methods and it is found that
$$\phi_0(\mathbf{r}_1, \mathbf{r}_2) = -\epsilon(r_1, r_2) r_2^{-2}(r_1 + \tfrac{1}{2}r_1^2)\cos\vartheta\, \psi_0(r_2), \tag{45}$$
with
$$\epsilon(r_1, r_2) = \begin{cases} 0 & (r_2 < r_1), \\ 1 & (r_2 > r_1). \end{cases}$$

The asymptotic form of the distortion term $\phi_0(\mathbf{r}_1, \mathbf{r}_2)F_0(\mathbf{r}_1)$ in (42) is therefore given by
$$-(r_1 + \tfrac{1}{2}r_1^2) r_2^{-2} \cos\vartheta \sin(kr_1 + \eta_0) \psi_0(r_2). \tag{46}$$

This agrees with the asymptotic form of the contribution
$$3^{\frac{1}{2}}(r_1 r_2)^{-1} \cos\vartheta\, \Phi_1(\mathbf{r}_1, \mathbf{r}_2)$$
due to dipole distortion in Temkin's analysis of the exact problem, as may be seen from (39). Since Temkin finds that this distortion term is the most important it seems that there is some prospect that (42) and (45) may give good results.

The formulation using the trial function (42) was not carried through by Temkin and Lamkin in terms of the full variational procedure. They required simply that F_0 should satisfy
$$\iint \psi_0(r_2) L\Psi^t\, d\mathbf{r}_1\, d\mathbf{r}_2 = 0, \tag{47}$$
where L is as in (2). Writing F_0 in the form of the usual expansion
$$F_0(\mathbf{r}) = r^{-1} \sum G_l(r) P_l(\cos\theta), \tag{48}$$
we find that G_l must satisfy
$$\left[\frac{d^2}{dr^2} + k^2 - \frac{l(l+1)}{r^2} - \frac{2m}{\hbar^2}\left\{V_{00} - \frac{1}{2}\frac{e^2\alpha(r)}{r^4}\right\}\right] G_l$$
$$= \int \{K_l(r, r') + K_l^{(1)}(r, r')\} G_l(r')\, dr'. \tag{49}$$

V_{00} is the usual mean static potential, $K_l(r, r')$ the kernel representing the non-local interaction due to electron exchange, as in (30), and $-\tfrac{1}{2}\alpha(r)e^2/r^4$ is the additional local interaction due to dipole distortion of the atom. $\alpha(r)$ is given by
$$a_0^{-3}\alpha(r) = \tfrac{9}{2} - \tfrac{2}{3}e^{-r/a_0}\left\{\left(\frac{r}{a_0}\right)^5 + \frac{9}{2}\left(\frac{r}{a_0}\right)^4 + 9\left(\frac{r}{a_0}\right)^3 + \frac{27}{2}\frac{r}{a_0} + \frac{27}{4}\right\}. \tag{50}$$

$9a_0^3/2$ is the static polarizability of the hydrogen atom so $\alpha(r)/r^4$ has the asymptotic form (36).

In addition there are further non-local terms, represented by $K_l^{(1)}(r, r')$, which arise from the presence of ϕ_0 in (42).

The approximation based on exact solution of the complete equation (49) is known as the *polarized-orbital* approximation. If the non-local terms involving $K_l^{(1)}$ are neglected it is referred to as the *exchange-adiabatic* approximation. For many purposes the latter is sufficiently satisfactory. We describe the results obtained by these approximations in § 2.6 and Table 8.2.

2.5. *Use of many-parameter variational trial functions*

The second possibility is to use more elaborate trial functions in the variational method of § 2.1 just as is done so successfully for calculating the energy of the ground state of helium. For the calculation of the zero order phase shifts $\eta_0^\pm$ a suitable trial function is of the form

$$\Psi^t = \Phi^t(r_1, r_2, r_{12}) \pm \Phi^t(r_2, r_1, r_{12}), \tag{51}$$

where

$$\Phi^t = r_1^{-1}\{\sin kr_1 + \lambda s(r_1, r_2, r_{12})\cos kr_1\}\psi_0(r_2) + \chi(r_1, r_2, r_{12}). \tag{52}$$

$s(r_1, r_2, r_{12})$ is a function chosen to satisfy

$$s(r_1, r_2, r_{12}) \to 1 \quad \text{as } r_1 \to \infty; \to 0 \quad \text{as } r_1 \to 0,$$

while $\chi(r_1, r_2, r_{12})$ is bounded everywhere and $\to 0$ as $r_2 \to 0$. A procedure of this kind may also be extended to the calculation of higher order phase shifts. The most accurate and reliable results for the phase shifts have been obtained in this way. A detailed description of the choice of trial functions and of the results will be given in §§ 2.6 and 2.7.

Burke and Taylor[†] have combined the truncated eigenfunction expansion approach with a more usual variational trial function by taking

$$\Psi_t = \sum_{n=0}^{s} \psi_n(\mathbf{r}_1) F_n(\mathbf{r}_2) + \chi(r_1, r_2, r_{12}).$$

The variational problem then reduces to the solution of a set of coupled differential and algebraic equations.

2.6. *Results obtained by various methods—the zero order phase shifts $\eta_0^\pm$*

We give in Table 8.2 values of the zero order phase shifts calculated according to the following methods.

(a) Static field approximation. This involves numerical solution of (27).

(b) The one-state 1s (exchange) approximation. To obtain this the single integrodifferential equations (29) are solved numerically.

[†] BURKE, P. G. and TAYLOR, A. J., *Proc. phys. Soc.* **88** (1966) 549.

TABLE 8.2

Zero order phase shifts ($\eta_0^\pm$) for elastic scattering of electrons by hydrogen atoms calculated by various methods

ka_0	Electron energy (eV)	Static potential	Exchange	$1s$–$2s$	$1s$–$2s$–$2p$	Exchange adiabatic	Polarized orbital	Variational M. & M. (53)	Variational Schwartz (54), (55)	Temkin $\eta_0^{(0)}$	η_0
η_0^+											
0	0	(−9·45)	(8·095)	(8·05)	(6·74)	(6·5)	(5·7)	(7·4)	(5·95)	(7·8)	(5·6)
0·1	0·135	0·721	2·396	2·404	2·491	2·522	2·583	2·484	2·553	2·42	2·59
0·2	0·54	0·973	1·871	1·878	1·974	2·025	2·114	2·003	2·067	1·895	2·11
0·3	1·22	1·046	1·508	1·519		1·654	1·750	1·649	1·696	1·53	1·74
0·4	2·17	1·056	1·239	1·256		1·374	1·469	1·425	1·415	1·269	1·45
0·5	3·39	1·045	1·031	1·046	1·082	1·157	1·251	1·250	1·202	1·065	1·23
0·6	4·87	1·021	0·869	0·89	0·93				1·041		
0·7	6·63	0·993	0·744	0·77	0·82				0·930		
0·8	8·67	0·963	0·651	0·70				0·857	0·886	0·728	0·87
η_0^-											
0	0	(−9·45)	(2·350)	(2·33)	(1·89)	(1·9)	(1·9)	(2·33)	(1·77)	(2·34)	(1·76)
0·1	0·135	0·721	2·908	2·901	2·935$_5$	2·949	2·945	2·909	2·939	2·91	2·942
0·2	0·54	0·973	2·679	2·680	2·715$_3$	2·737	2·732	2·680	2·717	2·68	2·723
0·3	1·22	1·046	2·461	2·461	2·50	2·528	2·519	2·447	2·500	2·46	2·516
0·4	2·17	1·057	2·257	2·257$_5$	2·28	2·329	2·320	2·248	2·294	2·26	2·301
0·5	3·39	1·045	2·070	2·070	2·096	2·146	2·133	2·029	2·105	2·07	2·112
0·6	4·87	1·022	1·901	1·90	1·92$_5$			1·909	1·933		
0·7	6·63	0·993	1·749	1·753	1·77				1·780		
0·8	8·67	0·963	1·614	1·616				1·621	1·643	1·62	1·647

Bracketed numbers refer to scattering lengths.

(c) The two-state 1s–2s (exchange) approximation. This requires numerical solution of the coupled integrodifferential equations arising from use of the trial function (32).

(d) The three-state 1s–2s–2p (exchange) approximation (see p. 508). The 2p states are now also included in the trial function and two sets of coupled integrodifferential equations must be solved, one set involving three coupled equations.

(e) Temkin's analysis, in terms of the coordinates r_1, r_2, ϑ, described in § 2.3.

(f) The polarized-orbital approximation.

(g) The exchange-adiabatic approximation.

(h) Variational methods using functions involving adjustable parameters.

Results are given for two such methods all using trial functions of the form (51) (52). The first† is a comparatively simple form which, in atomic units, is given by

$$s(r_1, r_2, r_{12}) = \{1+(b+cr_{12})e^{-r_1}\}(1-e^{-r_1}); \quad \chi = 0. \tag{53}$$

The second, due to Schwartz,‡ is much more elaborate. It takes the form

$$s(r_1, r_2, r_{12}) = 1 - e^{-\kappa r_1} \tag{54 a}$$

and

$$\chi(r_1, r_2, r_{12}) = \sum_{l,m,n>0} a_{lmn} e^{-\kappa(r_1+r_2)} r_{12}^l \{r_1^m r_2^n \pm r_1^n r_2^m\}. \tag{54 b}$$

In addition, in order to introduce directly the effect of dipole distortion, Schwartz replaced $\psi_0(r_2)$ in (52) by

$$\psi_0\left\{1 - \frac{a}{r_2}f_1 - \frac{\alpha}{r_2^2}f_2 - \frac{r_1 + \frac{1}{2}r_1^2}{r_2}\cos\vartheta f_3\right\}, \tag{55}$$

in which $f_1, f_2,$ and f_3 are factors that remove the singularities at $r_2 = 0$, and α is the polarizability of atomic hydrogen. The introduction of this modification materially improves the convergence at very low energies.

In addition to the phase shifts, in Table 8.2 the scattering lengths are given in the zero energy limit. All methods except the Temkin method (e), the polarized-orbital and the exchange-adiabatic approximation (f) and (g) are based on the Kohn variational method and give upper bounds to the scattering length.

It will be seen that Temkin's method gives results that agree very closely with those obtained by Schwartz with his elaborate variational calculations and it is very probable that these results are close to the

† MASSEY, H. S. W. and MOISEIWITSCH, B. L., *Proc. R. Soc.* **A205** (1950) 483.
‡ SCHWARTZ, C., *Phys. Rev.* **124** (1961) 1468.

exact values. On this basis the polarized-orbital method also gives remarkably good results. The simpler exchange-adiabatic approximation is also quite good and for η_0^+ is a little more satisfactory than the three-state ($1s$–$2s$–$2p$) exchange approximation. For η_0^- it is slightly less satisfactory. As might be expected the $1s$–$2s$ exchange approximation, which does not allow for dipole polarization at all (this comes from the p-state terms in the eigenfunction expansion), is markedly less accurate and indeed is only a small improvement on the one-state exchange approximation.

The complete inadequacy of the static field approximation at these energies is manifest. Introduction of exchange gives a marked improvement which for the η_0^- is almost adequate for most purposes. This is because, when the space wave function is antisymmetric in the electrons, the chance of finding the electrons close together is reduced. This in turn reduces the distortion of the atomic wave function through interaction with the impinging electron. Even for η_0^+ inclusion of the dipole polarization through the polarized-orbital or even the exchange-adiabatic method yields a good approximation.

Comparison of the results obtained using different variational trial functions of the form discussed under (b) shows that very elaborate functions must be assumed to obtain results better than those given by the exchange polarization and *a fortiori* by the polarized-orbital method.

2.7. *The first and second order phase shifts* $\eta_1^\pm, \eta_2^\pm$

All of the methods with the exception of (c) that are listed in § 2.6 for calculating $\eta_0^\pm$ are also applicable to the calculation of higher order phase shifts. We give in Fig. 8.1 results obtained for $\eta_1^\pm$ and in Fig. 8.2 for $\eta_2^\pm$. The trial functions used in methods (h)† for $\eta_2^\pm$ are the natural generalizations of (54a) and (54b) for states of angular momentum $6^{\frac{1}{2}}\hbar$ involving no terms in r_{12}. For $\eta_1^\pm$ † one trial function is of this form and the other is the generalization of the form used by Schwartz,‡ without inclusion of any factor corresponding to (55).

The conclusions are very much the same as for the $\eta_0^\pm$ phases. There is no doubt that the polarized-orbital and exchange-adiabatic methods give good results. It may even be that they give better results than the variational methods (h). The trial functions used are not as elaborate as those used by Schwartz‡ for $\eta_0^\pm$, which include the dipole distortion factor (55).

† TAYLOR, A. J., unpublished.   ‡ loc. cit., p. 516.

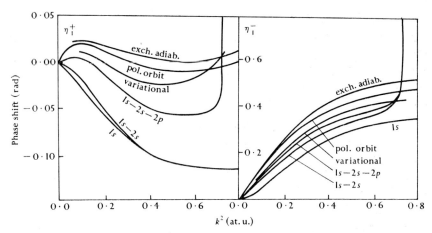

Fig. 8.1. Calculated phase shifts $\eta_1^{\pm}$ for elastic scattering of electrons by atomic hydrogen. The labels $1s$, $1s-2s$, $1s-2s-2p$ refer to the one, two, and three-state exchange approximations respectively. The variational results are those of Taylor.

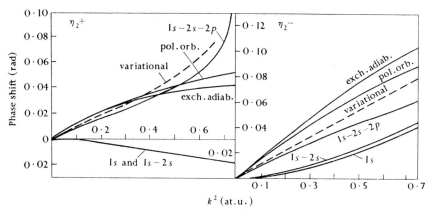

Fig. 8.2. Calculated phase shifts $\eta_2^{\pm}$ for elastic scattering of electrons by atomic hydrogen. The labels $1s$, $1s-2s$, $1s-2s-2p$ refer to the one, two, and three-state exchange approximations respectively. The variational results are those of Taylor.

2.8. Comparison with observation

Fig. 8.3 compares total cross-sections calculated using the various approximations to the phase shifts with observed values† (see Chap. 1, § 6.1). The accuracy of the latter is not high enough to distinguish between calculated values derived from the more elaborate approximations. Within experimental error they all agree very well.

† BRACKMANN, R. T., FITE, W. L., and NEYNABER, R. H., *Phys. Rev.* **112** (1958) 1157; NEYNABER, R. H., MARINO, L. L., ROTHE, E. W., and TRUJILLO, S. M., ibid. **124** (1961) 135.

A similar situation applies to comparison of observed† (Chap. 5, § 5.3.1) and calculated differential cross-sections shown in Fig. 8.4. No experimental results are available which could discriminate between various calculated values for the small phase shifts $\eta_1^\pm$, to say nothing of $\eta_2^\pm$.

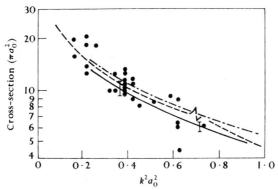

FIG. 8.3. Comparison of calculated and observed cross-sections for elastic scattering of electrons by atomic hydrogen. — — — — calculated by $1s$–$2s$–$2p$ close-coupling approximation. — · — · — calculated by polarized-orbital method. ——— observed by Neynaber, Marino, Rothe, and Trujillo. ••• observed by Brackmann, Fite, and Neynaber.

We are nevertheless left with the feeling that the exchange-adiabatic method is likely to be very useful for relatively simple calculation of elastic cross-sections below the threshold for atoms more complex than hydrogen. This will be reinforced when we proceed to consider cases in which experimental checks, either direct or indirect, are available.

2.9. *Dispersion relations and the elastic scattering of electrons by atomic hydrogen*

In § 2.2.2 it was pointed out that the real component of the forward elastic scattered amplitude $f(0, E)$ at energy E can be related to an energy integral over the imaginary component in the form

$$\mathrm{re}\,f(0, E) = f_\mathrm{b}(0, E) - R(E) + \frac{1}{\pi}\int \frac{\mathrm{im}\,f(0, E')}{E' - E}\,\mathrm{d}E'.$$

f_b is the amplitude calculated by Born's approximation and $R(E)$ is determined by the wave functions of the bound states of the particle in the field of the scatterer.

For collisions of electrons with hydrogen atoms we must deal with two scattered amplitudes $f \pm g$ according as the electron spins are respectively antiparallel and parallel. We have then

$$\mathrm{re}\{f(0, E) \pm g(0, E)\}$$
$$= f_\mathrm{b}(0, E) \pm g_\mathrm{b}(0, E) - R^\pm(E) + \frac{1}{\pi}\int \frac{\mathrm{im}\{f(0, E') \pm g(0, E')\}}{E' - E}\,\mathrm{d}E'.$$

† GILBODY, H. B., STEBBINGS, R. F., and FITE, W. L., *Phys. Rev.* **121** (1961) 794.

The conservation theorem (22) now applies in the form
$$\text{im}\{f(0, E) \pm g(0, E)\} = (k/4\pi)Q^\pm(E),$$

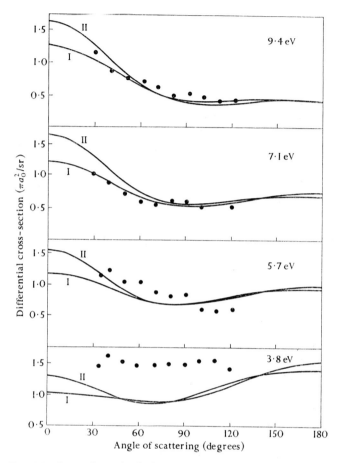

Fig. 8.4. Comparison of calculated and observed differential cross-sections for elastic scattering of electrons by atomic hydrogen. Curves I and II, calculated by $1s$–$2s$–$2p$ three-state exchange approximation and by polarized-orbital method respectively. ●●● observed by Gilbody, Stebbings, and Fite.

where k is the electron wave number. The measured total cross-section is given by
$$Q = \tfrac{1}{4}Q^+ + \tfrac{3}{4}Q^-,$$
so that
$$\text{im}\{f(0, E) - \tfrac{1}{2}g(0, E)\} = (k/4\pi)Q.$$
We therefore have
$$\text{re}\{f(0, E) - \tfrac{1}{2}g(0, E)\} = f_\text{b}(0, E) - \tfrac{1}{2}g_\text{b}(0, E) - \tfrac{1}{4}R^+ - \tfrac{3}{4}R^- + \frac{1}{4\pi^2}\int \frac{Q(E')k'}{E' - E}\,\text{d}E'.$$

For an electron interacting with a hydrogen atom there is only one bound state and that is for the antisymmetric case so R^+ vanishes and we are left with the relation

$$\text{re}\{f(0, E) - \tfrac{1}{2}g(0, E)\} = f_{\text{b}}(0, E) - \tfrac{1}{2}g_{\text{b}}(0, E) - \tfrac{3}{4}R^-(E) + \frac{1}{4\pi^2}\int \frac{Q(E')k'}{E' - E}\,\mathrm{d}E'. \quad (56)$$

Although f_{b}, g_{b}, and R^- may be calculated, the usefulness of this relation is reduced by the fact that the observed forward scattered intensity is not proportional to $|f(0, E) - \tfrac{1}{2}g(0, E)|^2$ but $\tfrac{1}{4}|f(0, E) + g(0, E)|^2 + \tfrac{3}{4}|f(0, E) - g(0, E)|^2$ so that a direct consistency check cannot be made between measured quantities on the right- and left-hand sides. However, a useful limitation on observed cross-sections $Q(E)$ was derived by Krall and Gerjuoy[†] as follows.

At zero incident electron energy (56) becomes

$$\text{re}\{f(0, 0) - \tfrac{1}{2}g(0, 0)\} = f_{\text{b}}(0, 0) - \tfrac{1}{2}g_{\text{b}}(0, 0) - \tfrac{3}{4}R^-(0) + \frac{1}{2\pi^2}\int_0^\infty Q(k')\,\mathrm{d}k'.$$

But
$$\text{re}\{f(0, 0) \pm g(0, 0)\} = -a^\pm$$

where $a^\pm$ are the zero-energy scattering lengths. Also
$$f_{\text{b}}(0, 0) - \tfrac{1}{2}g_{\text{b}}(0, 0) = -2a_0$$

so
$$\text{re}\{f(0, 0) - \tfrac{1}{2}g(0, 0)\} = -\tfrac{3}{4}a^- - \tfrac{1}{4}a^+$$
$$= -\tfrac{3}{4}R^- - 2a_0 + \frac{1}{2\pi^2}\int_0^\infty Q(k')\,\mathrm{d}k'.$$

There seems to be little doubt that a^- and a^+ are both > 0 (see Table 8.2) so that

$$\int_0^\infty Q(k')\,\mathrm{d}k' \leqslant 2\pi^2(\tfrac{3}{4}R^- + 2a_0).$$

Using wave functions for the bound state of H$^-$ given by Geltman, $R^- = 7.36a_0$ so
$$\frac{1}{2\pi^2}\int_0^\infty Q(k')\,\mathrm{d}k' \leqslant 7.52a_0.$$

Taking the most accurate calculated values of $a^\pm$ (see Table 8.2) with the same value of R^- gives
$$\frac{1}{2\pi^2}\int_0^\infty Q(k')\,\mathrm{d}k' = 4.5a_0.$$

The left-hand integral evaluated using the total cross-section observations of Brackman, Fite, and Neynaber (Chap. 1, § 6.1) gives $(5.1 \pm 0.5)a_0$, which is consistent.

3. Spin-exchange collisions

For some applications, particularly in astrophysics, the cross-section for a collision between a slow electron and a hydrogen atom that results in reversal of the spin of the atomic electron is required. Thus one of the few strong lines radiated in the radio emission from the galaxy[‡] is the

[†] KRALL, N. A. and GERJUOY, E., *Phys. Rev.* **120** (1960) 143.
[‡] PAWSEY, J. L. and BRACEWELL, R., *Radio astronomy*, Chap. vi (Clarendon Press, Oxford, 1954).

21-cm line of hydrogen that arises from a transition between the two hyperfine structure levels associated with the ground state. In the upper of these levels the electron and nuclear spins are parallel, in the lower antiparallel. It is therefore clear that, in order to understand the observed emission intensities, a knowledge of the rate of population of the upper level from the lower by collision is required. Such collisions require the reversal of the electron spin in the atom. This will occur if electron exchange takes place between the atomic electron and an incident electron of opposite spin. The cross-section for such a collision may be calculated in terms of the phase shifts $\eta_l^\pm$ as follows.

Let α, β be the spin wave functions corresponding to the z-components of spin $\pm\tfrac{1}{2}\hbar$ respectively.

The spin wave function for the singlet state of the two-electron system associated with a symmetric function of the space coordinates is then
$$2^{-\frac{1}{2}}\{\alpha(1)\beta(2)-\alpha(2)\beta(1)\}, \tag{57}$$
while the three functions for the triplet state that are associated with an antisymmetric function of the space coordinates are
$$\alpha(1)\alpha(2),\quad 2^{-\frac{1}{2}}\{\alpha(1)\beta(2)+\alpha(2)\beta(1)\},\quad \beta(1)\beta(2), \tag{58}$$
the two electrons being distinguished as 1, 2.

Solutions for the two-electron wave functions describing a collision between an electron and a hydrogen atom have already been obtained. That associated with a symmetric orbital function has the asymptotic form for large r_1
$$2^{-\frac{1}{2}}\{\alpha(1)\beta(2)-\alpha(2)\beta(1)\}(1+P_{12})[\psi_0(r_2)\{e^{ikr_1\cos\theta_1}+r_1^{-1}e^{ikr_1}f^+(\theta_1)\}],$$
where
$$f^+(\theta) = \frac{1}{2ik}\sum i^l(2l+1)(e^{2i\eta_l^+}-1)P_l(\cos\theta). \tag{59}$$

Similarly for the solutions associated with an antisymmetric orbital function the asymptotic forms are
$$\left.\begin{array}{l}\alpha(1)\alpha(2)\\ 2^{-\frac{1}{2}}\{\alpha(1)\beta(2)+\alpha(2)\beta(1)\}\\ \beta(1)\beta(2)\end{array}\right\}(1-P_{12})[\psi_0(r_2)\{e^{ikr_1\cos\theta_1}+r_1^{-1}e^{ikr_1}f^-(\theta_1)\}], \tag{60}$$
where f^- differs from f^+ in the replacement of η_l^+ by η_l^-.

By addition of (59) and (60) we obtain a solution with asymptotic form
$$\psi_0(r_2)[e^{ikr_1\cos\theta_1}\alpha(2)\beta(1)+r_1^{-1}e^{ikr_1}\{\tfrac{1}{2}(f^++f^-)\alpha(2)\beta(1)+$$
$$+\tfrac{1}{2}(f^+-f^-)\alpha(1)\beta(2)\}]. \tag{61}$$

This solution corresponds to an electron with spin wave function β incident on an atom in which the electron spin wave function is α and two outgoing waves, one with the same spin wave function and one with the spins exchanged.

TABLE 8.3

Cross-section $Q(1, 0)$ for excitation of the hyperfine structure transition $(F, 1 \to 0)$ in atomic hydrogen by electron impact

Electron wave number (a.u.)	0·00	0·10	0·20	0·30	0·40	0·50
Electron energy (eV)	0·00	0·135	0·54	1·22	2·17	3·39
$Q(1, 0)$ (πa_0^2)	4·4	3·5	2·3	1·44	0·93	0·62

The differential cross-section for spin exchange is therefore given by

$$I_{se}(\theta)\, d\omega = \tfrac{1}{4}|f^+ - f^-|^2\, d\omega \tag{62}$$

and the total cross-section by

$$Q_{se} = \tfrac{1}{4} \int |f^+ - f^-|^2\, d\omega \tag{63}$$

$$= (\pi/k^2) \sum_l (2l+1)\sin^2(\eta_l^+ - \eta_l^-). \tag{64}$$

With unpolarized electrons incident on unpolarized atoms the effective cross-section for excitation of one hyperfine structure state from the other must take into account unfavourable orientation factors. Thus exchange of electrons with the same spin will have no effect.

When account is taken of these factors the effective cross-section for the hyperfine transition $(1 \to 0)$, ignoring the very small energy difference between the two states, is given by $\tfrac{1}{4}Q_{se}$.

In general, the cross-section for excitation of a hyperfine structure transition in which the total (nuclear+electronic) spin quantum number changes from F to F' is given by†

$$Q(F, F') = \frac{2F'+1}{2(2I+1)}\, Q_{se}, \tag{65}$$

where I is the nuclear spin quantum number. For hydrogen $F = 1$, $F' = 0$, $I = \tfrac{1}{2}$ so giving the factor $\tfrac{1}{4}$. The factor multiplying Q_{se} is simply the ratio of the number of possible orientations of the final total spin to the number of orientations of the separate electron and nuclear spins.

Table 8.3 gives $Q(1, 0)$ for the deactivation of the upper state of the 21-cm emission line of hydrogen calculated from the variationally determined phase shifts.

† DALGARNO, A., *Proc. R. Soc.* **A262** (1961) 132; DALGARNO, A. and RUDGE, M. H. R., ibid. **286** (1965) 519.

4. Collisions with hydrogen atoms—elastic and inelastic collisions with electrons at energies above the excitation threshold

4.1. *The generalized variational method*†

We now turn our attention to collisions of hydrogen atoms with electrons which, while slow, possess energy sufficient to excite the atoms. Under these conditions Temkin's analysis can no longer be carried through and the polarized-orbital method is not valid.

There is no difficulty in principle in further generalizing the variational method of § 2.1 to include inelastic scattering. Suppose that we wish to determine approximately the differential cross-section for an inelastic collision in which the sth state is excited, the electron wave number changing from k_0 to k_s and its direction of motion from $\mathbf{n}_0$ to $\mathbf{n}_s$.

We then work from an integral I_{0s}, which is an immediate generalization of the integral I_{11} of (1). The function Ψ_1^* in I_{11} must be generalized so

$$\Psi_1^* \sim e^{ik_0 \mathbf{n}_0 \cdot \mathbf{r}_1} \psi_0(r_2) + \sum_{p=0}^{p_m} f_{0p}(\mathbf{n}_0, \mathbf{n}) e^{ik_p r_1} r_1^{-1} \psi_p(\mathbf{r}_2). \tag{66}$$

This allows for the fact that inelastic collisions are now energetically possible for states such that $p < p_m$. Ψ_2^* in I_{11} is replaced by Ψ_s^* where

$$\Psi_s^* \sim e^{-ik_s \mathbf{n}_s \cdot \mathbf{r}_1} \psi_s^*(\mathbf{r}_2) + \sum_{q=0}^{p_m} f_{sq}(-\mathbf{n}_s, \mathbf{n}) e^{ik_q r_1} r_1^{-1} \psi_q(\mathbf{r}_2). \tag{67}$$

Ψ_s^* is thus the wave function which describes a collision in which the electron is incident in the direction $\mathbf{n}_s$ with wave number k_s.

Using trial functions which have the asymptotic forms (66), (67) we find

$$f_{0s}(\mathbf{n}_0, \mathbf{n}_s) = f_{0s}^t(\mathbf{n}_0, \mathbf{n}_s) + I_{0s}^t/4\pi \quad (s = 0,...,p_m). \tag{68}$$

Extension to include exchange may be carried out as in § 2.1.

It follows that we may use the eigenfunction expansion method in the same way as before. Thus with a two-state approximation (32) we obtain the same equations (33) as before. The only difference is that the second state may be excited so that the boundary conditions for $F_1^{\pm}$ now become

$$F_1^{\pm}(r) \sim e^{ik_1 r} r^{-1} f_1^{\pm}(\theta, \phi), \tag{69}$$

in place of (20). Solution of these coupled equations then gives approximate differential cross-sections for both the elastic scattering and inelastic scattering in which the state 1 is excited. Similar considerations apply when more states are taken into account.

† KOHN, W., *Phys. Rev.* **74** (1948) 1763; MOTT, N. F. and MASSEY, H. S. W., *The theory of atomic collisions*, 3rd edn, Chap. xiv, § 2 (Clarendon Press, Oxford, 1965).

4.2. Distorted-wave method†

It is clear that practical limitations soon restrict the number of states which can be included in an eigenfunction expansion. Thus it is certainly practicable to take into account the $1s$, $2s$, and $2p$ states of atomic hydrogen but extension to 3-quantum states already taxes the capacity of large computers. An alternative procedure, which is much less limited from the computational point of view, is known as the method of distorted waves.

We return to the exact infinite set of coupled equations (16)

$$(\nabla^2+k_n^2)F_n(\mathbf{r}) = \sum_m U_{nm} F_m(\mathbf{r}). \tag{70}$$

In general we can expect that the non-diagonal matrix elements U_{nm} ($n \neq m$) will be much smaller than the diagonal elements U_{nn}. Thus in the expression (Chap. 7 (8)) for U_{nm} the product $\psi_n^* \psi_m$ will in general change sign one or more times within the range of integration when $n \neq m$, so that some cancellation will occur. This will be absent when $n = m$.

If this is so it should be a good approximation when calculating the cross-section for excitation of the sth state to reduce (70) to the two equations

$$(\nabla^2+k_0^2-U_{00})F_0 = 0, \tag{71 a}$$

$$(\nabla^2+k_s^2-U_{ss})F_s = U_{s0} F_0, \tag{71 b}$$

F_0, F_s satisfying the usual boundary conditions (15), (69) respectively. F_0 is just the function which describes electron scattering by the potential $\hbar^2 U_{00}/2m$. We can similarly obtain a function $\mathscr{F}_s(r, \theta)$, which describes the elastic scattering by the potential $\hbar^2 U_{ss}/2m$ so that

$$\mathscr{F}_s(r, \theta) \sim e^{ik_s r \cos\theta} + r^{-1}f_s(\theta)e^{ik_s r}. \tag{72}$$

The solution of the inhomogeneous equation (71 b) may now be written down in terms of a suitable Green's function‡ and gives for $f_s(\theta, \phi)$,

$$f_s(\theta, \phi) = -(1/4\pi) \int U_{s0}(\mathbf{r}')F_0(r', \theta')\mathscr{F}_s(r', \pi-\Theta) \, d\mathbf{r}', \tag{73}$$

where $$\cos\Theta = \cos\theta \cos\theta' + \sin\theta \sin\theta' \cos(\phi-\phi'). \tag{74}$$

Comparing this formula with the corresponding Born approximation (Chap. 7 (17)) it will be seen that the plane waves $e^{i\mathbf{k}_0 \cdot \mathbf{r}}$, $e^{i\mathbf{k}_s \cdot \mathbf{r}}$ are replaced by the functions F_0, $\mathscr{F}_s$ which are distorted from the plane wave form by the respective interactions V_{00}, V_{ss} $\{= (\hbar^2/2m)U_{00}, U_{ss}\}$. These are the

---
† MOTT, N. F. and MASSEY, H. S. W., *The theory of atomic collisions*, 3rd edn, Chap. xiii, § 2 (Clarendon Press, Oxford, 1965).
‡ MOTT, N. F. and MASSEY, H. S. W., ibid. Chap iv, § 4 (Clarendon Press, Oxford, 1965).

mean static fields of the atom in the initial and final states respectively. For this reason (73) is referred to as the distorted-wave approximation.

The validity of the approximation depends on the non-diagonal matrix elements U_{sm} all being small. It is *not* sufficient that U_{s0} should be small. This is because there may be matrix elements U_{sp} that are not small so it is not possible to isolate the excitation of the sth state from that of the pth (even if this be only virtual). However, under many circumstances the method is a very useful one. It is a weak coupling approximation that does not require in addition that the interaction should be so weak that the diagonal terms are also small, as in Born's approximation.

At low energies of impact, not far above the excitation threshold, the distortion of the plane waves by the atomic fields U_{00}, U_{ss}, will be large and the deviation from Born's approximation will be very marked.

There is no difficulty in generalizing the distorted-wave method to include electron exchange. Thus, in the case of excitation of the first excited state for which the full two-state approximation is obtained from solution of the coupled equations (33), we take as the equations of the distorted wave approximation

$$(\nabla^2 + k_0^2 - U_{00})F_0^\pm(\mathbf{r}) \mp \int K_{00}(\mathbf{r}, \mathbf{r}')F_0^\pm(\mathbf{r}')\,d\mathbf{r}' = 0, \tag{75}$$

$$(\nabla^2 + k_1^2 - U_{11})F_1^\pm(\mathbf{r}) \mp \int K_{11}^\pm(\mathbf{r}, \mathbf{r}')F_1^\pm(\mathbf{r}')\,d\mathbf{r}'$$
$$= U_{10}F_0^\pm(\mathbf{r}) \pm \int K_{10}(\mathbf{r}, \mathbf{r}')F_0^\pm(\mathbf{r}')\,d\mathbf{r}'. \tag{76}$$

These equations give

$$f_1^\pm(\theta, \phi) = -(1/4\pi) \int \left\{ U_{10}(r')F_0^\pm(\mathbf{r}') \pm \int K_{10}(\mathbf{r}', \mathbf{r}'')F_0^\pm(\mathbf{r}'')\,d\mathbf{r}'' \right\} \times$$
$$\times \mathscr{F}_1^\pm(r', r - \Theta) \} \,d\mathbf{r}', \tag{77}$$

in which $F_0^\pm$, $\mathscr{F}_1^\pm$ are now waves distorted not only by the local interactions U_{00}, U_{11} but also by the corresponding non-local interactions determined by the respective kernels K_{00}, K_{11}, i.e. $\mathscr{F}_1^\pm$ satisfies

$$(\nabla_1^2 + k_1^2 - U_{11})\mathscr{F}_1^\pm \mp \int K_{11}(\mathbf{r}, \mathbf{r}')F_1^\pm(\mathbf{r}')\,d\mathbf{r}' = 0. \tag{78}$$

The application of the distorted-wave method to calculation of ionization cross-sections is not straightforward as it is difficult to isolate a single state of the continuum and it is not at all obvious what to take for U_{ss} in such a case. This is especially difficult for hydrogen because the only contribution to U_{ss}, apart from that arising from the attraction of the proton, comes from the ejected electron. For complex atoms the contribution from the latter source is so small compared with that arising from interaction with the atomic core that it may be ignored and there is no difficulty in writing down the form of the distorted-wave approximation.

It is possible in principle to extend the distorted-wave formulation to allow for some degree of close coupling. This may be conveniently based on the exact equations (18). The usual form of the distorted-wave approximation consists in taking for Ψ on the right-hand side of (24),

$$\Psi = F_0(\mathbf{r}_1)\psi_0(r_2), \tag{79}$$

where
$$(\nabla_1^2 + k_0^2 - U_{00})F_0 = 0. \tag{80}$$

This may be extended, for example, when $n \neq 0$, by substituting for $F_0(\mathbf{r}_1)\psi_0(r_2)$ the corresponding solution for the two-state approximation, $F_0(\mathbf{r}_1)\psi_0(r_2) + F_1(\mathbf{r}_1)\psi_1(r_2)$, where F_0, F_1 satisfy the coupled equations

$$(\nabla^2 + k_0^2 - U_{00})F_0 = U_{01}F_1, \tag{81 a}$$

$$(\nabla^2 + k_1^2 - U_{11})F_1 = U_{10}F_0. \tag{81 b}$$

In this way it is possible to deal with a problem in which there is weak coupling between the nth state and all other states but strong coupling between the ground and first excited states.

As far as atomic hydrogen is concerned the three-state $1s$–$2s$–$2p$ close coupling approximation $\Psi_{(3)}(\mathbf{r}_1, \mathbf{r}_2)$ should be used (see p. 508) so that F_n satisfies the equation

$$(\nabla^2 + k_n^2 - U_{nn})F_n = -(2me^2/\hbar^2) \int \left(\frac{1}{r_1} - \frac{1}{r_{12}}\right)\psi_n^*(\mathbf{r}_2)\Psi_{(3)}(\mathbf{r}_1, \mathbf{r}_2)\,d\mathbf{r}_2 \quad (n > 2), \tag{82}$$

which may be solved in the usual way by the Green's function method. There is no difficulty in extending this approximation to include exchange.

4.3. *The excitation of the 2s and 2p states of hydrogen*

The first application in detail of the distorted wave method (§ 4.2) in which distortion by the non-local exchange interaction as well as the mean static field was included, was that of Erskine and Massey[†] for the excitation of the 2s level of H. They determined the distorted wave functions by a variational method. Although the cross-section

$$Q = \tfrac{1}{4}Q^+ + \tfrac{3}{4}Q^-,$$

where Q^+, Q^- are the separate contributions from singlet and triplet scattering respectively, did not come out to be very different to that given by the Born–Oppenheimer approximation, this was clearly

[†] ERSKINE, G. A. and MASSEY, H. S. W., *Proc. R. Soc.* A**212** (1952) 521.

coincidental. Thus, whereas the latter formulation gave $Q^+ \ll Q^-$, the reverse behaviour was found with the distorted-wave method. Calculations were later carried out by Ochkur† in which the distorted-wave functions were calculated by electronic computation from the appropriate integrodifferential equations. They confirmed the general conclusions of the earlier work but differed considerably in numerical detail.

Similar calculations were carried out for the excitation of the $2p$ state by Khashaba and Massey‡ using variationally determined distorted wave functions and by Ochkur† who calculated them numerically. For this case the effect of distortion was not so marked, partly because many more states of relative angular momentum, for which the wave functions are little distorted, contribute than for $2s$.

Burke and Schey§ applied the $1s–2s–2p$ truncated eigenfunction expansion to the problem. Comparison of their results with observation and with the results of other theoretical approximations is shown in Fig. 8.5 (a) for $2s$ and 8.5 (b) for $2p$. For $2s$ the experimental position is somewhat unclear as discussed in Chapter 4, § 2.2.1. However, for $2p$ even the $1s–2s–2p$ approximation appears to overestimate the cross-section by only a little less than the Born approximation. To investigate thesed iscrepancies further Burke, Ormonde, Taylor, and Whitaker‖ carried out even more elaborate calculations in the energy range quite close to the threshold, between 10·4 and 12·7 eV. In addition to the $1s–2s–2p$ expansion method calculated at close energy intervals they also carried out two other calculations. One of these involved the solution of the coupled equations for the six-state $1s–2s–2p–3s–3p–3d$ truncated expansion and the other the use of a trial function in which a linear combination of terms of the same form as the correlation terms in the Schwartz trial function for elastic scattering by H (see (54 (a) and (b))) was added to a $1s–2s–2p$ eigenfunction expansion. In the latter case the application of the Kohn variational method leads to a set of coupled integrodifferential equations for the 'free' wave function and a set of linear equations for the coefficients in the correlation terms.

Fig. 8.6 (a) and (b) shows the results obtained by these three methods for the excitation of the $2s$ and $2p$ states respectively. At electron energies below about 11·5 eV the three methods do not give very different results. Some structure is apparent very close to the threshold, a sharp

† OCHKUR, V. I., *Vest. Leningr. gos. Univ.* **4** (1958) 53.
‡ KHASHABA, S. and MASSEY, H. S. W., *Proc. phys. Soc.* **71** (1958) 574.
§ BURKE, P. G. and SCHEY, H. M., *Phys. Rev.* **126** (1962) 147.
‖ BURKE, P. G., ORMONDE, S., and WHITAKER, W., *Proc. Phys. Soc.* **92** (1967), 319; TAYLOR, A. J. and BURKE, P. G., ibid. **336**.

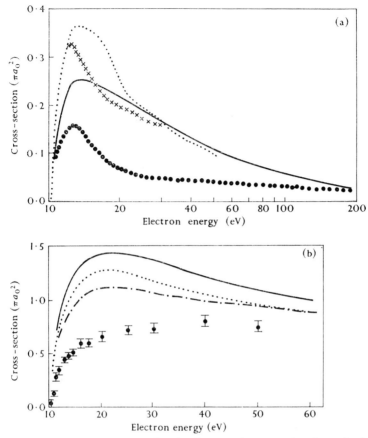

Fig. 8.5. (a) Comparison of calculated and observed cross-sections for excitation of the $2s$ state of hydrogen by electron impact. —— calculated by Born's approximation. .... calculated by $1s$–$2s$–$2p$ close-coupling method. ●●● observed (Stebbings et al.). ××× observed (Lichten and Schulz). (b) Comparison of calculated and observed cross-sections for excitation of the $2p$ state of hydrogen by electron impact. —— calculated by Born's approximation. —·—·— calculated by distorted-wave method. .... calculated by $1s$–$2s$–$2p$ close-coupling method. ●●● observed (Fite et al.).

peak being present for the excitation of both states, according to the calculations carried out with the three-state expansion+20 correlation terms. The peak is present even when the correlation terms are omitted but it is blunter. Direct experimental evidence for the existence of the peak for the $2p$ excitation is provided from the observations of Chamberlain, Smith, and Heddle† (see Chap. 4, § 1.4.1). Between 11·5 eV and the 3-quantum excitation threshold (12·1 eV) the six-state approximation

† CHAMBERLAIN, G. E., SMITH, S. J., and HEDDLE, D. W. O., Phys. Rev. Lett. 12 (1964) 647.

gives rise to resonance effects of the type considered in Chapter 9 and we shall defer further discussion of them until § 4 of that chapter. It is to be noted, however, that beyond the 3-quantum threshold the six-state approximation gives results for both $2s$ and $2p$ excitation, which are

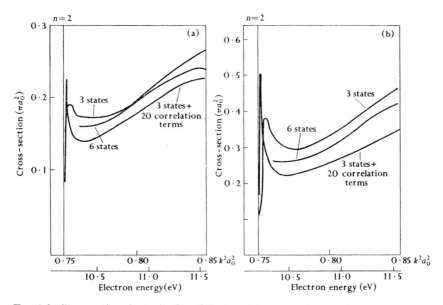

FIG. 8.6. Cross-sections for excitation of the $2s$ and $2p$ states of atomic hydrogen by electron impact at energies close to the threshold calculated by three different theoretical approximations as indicated. (a) $2s$, (b) $2p$.

very different from the three-state, showing that the convergence of the truncated expansion is still poor. This may explain the discrepancies with experiment still apparent in Fig. 8.5.

A quite different approach to the calculation of excitation cross-sections has been introduced by Presnyakov, Sobelman, and Vainshtein[†] that attempts to take into account more completely the interaction between the atomic and incident electrons. To do this we return to (24), which is an exact equation for the function F_n but contains the complete collision wave function $\Psi(\mathbf{r}_1, \mathbf{r}_2)$. In terms of this equation the scattered amplitude $f_n(\theta, \phi)$ associated with excitation of the nth state from the ground state is given by

$$f_n(\theta, \phi) = -(2\pi m e^2/h^2) \int \left(\frac{1}{r_1} - \frac{1}{r_{12}}\right) e^{-i\mathbf{k}_n \cdot \mathbf{r}_1} \psi_n^*(\mathbf{r}_2) \Psi(\mathbf{r}_1, \mathbf{r}_2) \, d\mathbf{r}_2. \tag{83}$$

To obtain the Born approximation we approximate to Ψ by writing

$$\Psi = \psi_0(r_1) e^{i\mathbf{k}_0 \cdot \mathbf{r}_2}. \tag{84}$$

† VAINSHTEIN, L., PRESNYAKOV, L., and SOBELMAN, I., Soviet Phys. JETP **18** (1964) 1383.

Presnyakov, Sobelman, and Vainshtein† propose as an alternative to write

$$\Psi = \psi_0(r_1)\chi(\mathbf{r}_1, \mathbf{r}_2), \tag{85}$$

where $\chi(\mathbf{r}_1, \mathbf{r}_2)$ must have the correct asymptotic form for large r_2

$$\chi(\mathbf{r}_1, \mathbf{r}_2) \sim e^{i\mathbf{k}_0 \cdot \mathbf{r}_2}. \tag{86}$$

On substitution into the full wave equation satisfied by Ψ, namely

$$\left\{\nabla_1^2 + \nabla_2^2 + \frac{2m}{\hbar^2}\left(\frac{e^2}{r_1} + \frac{e^2}{r_2} - \frac{e^2}{r_{12}} + E\right)\right\}\Psi = 0, \tag{87}$$

we find that χ must satisfy

$$\left\{\nabla_1^2 + \nabla_2^2 + \frac{2me^2}{\hbar^2}\left(\frac{1}{|\mathbf{r}_1+\mathbf{r}_2|} - \frac{1}{|\mathbf{r}_1-\mathbf{r}_2|}\right) + k_0^2\right\}\chi$$
$$= \frac{2me^2}{\hbar^2}\left(\frac{1}{|\mathbf{r}_1+\mathbf{r}_2|} - \frac{1}{r_2}\right)\chi - 2\nabla_1\ln\psi_0 \cdot \nabla_1\ln\chi. \tag{88}$$

At this stage it is convenient to introduce new variables

$$\boldsymbol{\rho} = \tfrac{1}{2}(\mathbf{r}_2-\mathbf{r}_1), \quad \mathbf{R} = \tfrac{1}{2}(\mathbf{r}_1+\mathbf{r}_2), \tag{89}$$

which are the respective coordinates of the relative position of the two electrons and of their centre of mass. We then have

$$\left\{\tfrac{1}{2}(\nabla_R^2+\nabla_\rho^2) + (me^2/\hbar^2)\left(\frac{1}{R}-\frac{1}{\rho}\right)+k_0^2\right\}\chi = (me^2/\hbar^2)\left(\frac{1}{R}-\frac{2}{|\mathbf{R}+\boldsymbol{\rho}|}\right)\chi - 2\nabla_1\ln\psi_0 \cdot \nabla_1\ln\chi. \tag{90}$$

So far the equations are still exact but are much too complicated to use.

The most obvious simplification to try at first is to neglect the right-hand side of (90) so that χ would then represent the scattering of two free electrons by their mutual repulsion while their centre of mass moves in the nuclear field as if it possessed a charge e and mass $2m$. However, this is open to the objections that the neglected terms are of order ρ^{-1} for large ρ and tend to infinity as $k_0 \to 0$. To reduce this difficulty Presnyakov, Sobelman, and Vainshtein introduced an effective screening constant ζ, so rewriting (90) in the form

$$\left\{\tfrac{1}{2}(\nabla_R^2+\nabla_\rho^2) + (me^2/\hbar^2)\left(\frac{\zeta}{R}-\frac{\zeta}{\rho}\right)+k_0^2\right\}\chi$$
$$= (me^2/\hbar^2)\left(\frac{\zeta}{R}-\frac{2}{|\mathbf{R}+\boldsymbol{\rho}|}+\frac{1-\zeta}{\rho}\right)\chi - 2\nabla_1\ln\psi_0 \cdot \nabla_1\ln\chi. \tag{91}$$

If ζ is taken as $k_0/(k_0+\lambda_0)$, where $\lambda_0^2\hbar^2/2m$ is the ionization energy, then the terms on the right-hand side which for large ρ tend to ρ^{-1} and diverge as $k_0 \to 0$ are reduced in importance. There is then some justification for neglecting the right-hand side of (91), giving an equation for χ that is separable and soluble in terms of the wave functions for motion in a Coulomb field described in Chapter 6, § 3.9. Having obtained χ (85) may be substituted in (83) to give an explicit integral for the scattering amplitude. Unfortunately, the correct evaluation of this integral involves serious mathematical difficulties and it is not possible at present to attain other than semi-empirical results, at least in the interesting region close to the threshold.

4.4. *The excitation of the 2s–2p transition*

Interest in the determination of the cross-section for impact excitation of the 2s–2p transition in atomic hydrogen was first aroused because

† loc. cit., p. 530.

of the importance of metastable H $2s$ atoms in certain astrophysical problems. The first of these concerned the possible contribution of radiative transitions to the ground state, with emission of two quanta, to the continuous spectrum of gaseous nebulae,† while the second came from the possibility of detecting microwave radiation from the solar chromosphere due to radiative transitions between $2s$ and $2p$ states.‡ To estimate the concentration of metastable atoms in any environment a knowledge of the rate of $2s$–$2p$ transitions due to electron collisions, among others, is certainly required.

Breit and Teller§ estimated the cross-section by Born's approximation, which is of doubtful validity at the low electron energies ($\simeq 1$ eV) of interest. Purcell‖ therefore carried out a calculation by an impact parameter method essentially similar to that outlined in Chapter 7, § 5.3.5 (144). The critical impact parameter p_m for 1-eV electrons came out to be about $30a_0$ but as the total excitation cross-section is around $5000\pi a_0^2$ it is clear that the effect of close coupling is not very large. A similar conclusion was arrived at by Seaton†† who applied a partial wave analysis to the Born cross-section in terms of the angular momentum quantum number l of the incident electron. He found that for $l < l_0$, where l_0 is almost 8 for 1-eV electrons, the partial cross-sections exceeded the maximum given in § 5.5 of Chapter 7. Corresponding to the assumptions made in obtaining the impact parameter formula (144) of Chapter 7, he assumed that for $l < l_0$ the mean partial cross-section is one-half the maximum possible $(2l+1)\pi/k^2$. For $l > l_0$ the Born partial cross-sections were taken. His results agree well with those of Purcell. Calculations carried out by Omidvar‡‡ by the $1s$–$2s$–$2p$ close-coupling approximation for values of $l < l_0$ show that Seaton's assumption is sufficiently accurate.

From the astrophysical point of view the rate coefficient W for the process in cm³ s⁻¹ averaged over a Maxwellian distribution of electron velocities at a temperature of order 10^4 °K is required. Results obtained by the different methods are given in Table 8.4. Separate values are given in the table for transitions to the two fine structure components $2p_{\frac{1}{2}}$, $2p_{\frac{3}{2}}$ of the $2p$ state.

† SPITZER, L. and GREENSTEIN, J. L., *Astrophys. J.* **114** (1951) 407; KIPPER, A. Y., *Tartu astr. Obs. Bull.* **32** (1952) 63.
‡ WILD, J. P., *Astrophys. J.* **115** (1952) 206.
§ BREIT, G. and TELLER, E., ibid. **91** (1940) 215.
‖ PURCELL, E. M., ibid. **116** (1952) 457.
†† SEATON, M. J., *Proc. phys. Soc.* **A68** (1955) 457.
‡‡ OMIDVAR, K., *Phys. Rev.* **133** (1964) A970.

It is of interest to note that, in fact, for astrophysical applications excitation of the transitions by proton impact occurs at a considerably faster rate and for this the effect of strong coupling is much more important (see Vol. IV).

TABLE 8.4

Calculated rate coefficients for impact excitation of the 2s–2p transition in atomic hydrogen

Electron temperature (°K)	Transition	Rate coefficient ($cm^3 s^{-1}$)		
		Born	Purcell (impact parameter)	Seaton (partial wave analysis)
10^4	$2s$–$2p_{\frac{1}{2}}$	0.25×10^4	0.24×10^4	0.22×10^4
	$2s$–$2p_{\frac{3}{2}}$	0.42×10^4	0.38×10^4	0.35×10^4
2×10^4	$2s$–$2p_{\frac{1}{2}}$	0.20×10^4	0.17×10^4	0.17×10^4
	$2s$–$2p_{\frac{3}{2}}$	0.34×10^4	0.29×10^4	0.27×10^4

4.5. *The excitation of higher discrete states*

Very few calculations have been carried out by any of the more refined methods for the excitation of the higher discrete states of atomic hydrogen. Ochkur[†] has applied the distorted-wave method to calculate the partial cross-sections for the excitation of the $3s$, $3p$, $3d$, $4s$, and $4d$ states in which the scattered electron possesses zero angular momentum, while a number of calculations have been carried out by Ormonde and Smith[‡] and by Ormonde[§] using the truncated eigenfunction expansion method. The slowness of the convergence of such expansions in dealing with the upper states makes it difficult to apply the method effectively within the capacity of available computers.

4.6. *Ionization*

Taylor[∥] has calculated the ionization cross-section using the extended form of the distorted-wave approximation discussed in § 4.2, exchange being included. The distortion of the scattered electron wave is neglected but the $1s$–$2s$–$2p$ exchange approximation is used to describe the initial state of the system. Results obtained in this way are shown in Fig. 7.10. It will be seen that they do not differ greatly from those obtained by the

[†] OCHKUR, V. I., loc. cit., p. 528.
[‡] ORMONDE, S. and SMITH, K., *Atomic collision processes*, ed. McDOWELL, M. R. C. (North Holland, Amsterdam, 1964).
[§] ORMONDE, S., *Proc. 4th Int. Conf. Phys. Electron. Atom. collisions*, p. 20 (Quebec, 1965).
[∥] TAYLOR, A. J., unpublished.

Born-exchange approximation (Chap. 7, § 5.6.2 (*f*)), lying somewhat above the observed values.

Similar results have been obtained by Veldre, Vinkalns, and Karule.† They calculated the distortion of the incident wave by the mean static field of the atom and also by this field supplemented by a polarization potential. The results are not sensitive to the distorting field and agree quite well with those of Taylor.

5. Collisions with He$^+$ ions

The calculation of cross-sections for elastic and inelastic scattering of slow electrons by helium ions presents a two-electron problem which is very similar to the corresponding one discussed above for hydrogen atoms. Certain differences arise, however, because of the long-range Coulomb field of the ion. In some respects this is a complication. It renders impracticable an analysis on similar lines to that of Temkin for hydrogen (see § 2.3) and makes all numerical evaluation more complex. Furthermore, experimental measurement of cross-sections is also difficult though progress is being made (see Chap. 3, § 3 and Chap. 4, § 3.2). There is, however, a credit side. Although it is not yet practicable to measure differential cross-sections for elastic scattering of slow electrons by He$^+$, it is possible to derive the associated phase shifts from the wavelengths of lines of the various Rydberg series of the helium atomic spectrum. This may be done through the relationship between quantum defects and phase shifts for the motion of electrons in a modified Coulomb field as described in Chapter 6, § 3.10.

Although the analysis of that section referred to a single electron problem it is plausible to assume that it may be generalized to cases in which more than one electron is concerned. We begin by discussing elastic scattering below the excitation threshold using the observed quantum defects as checks on results calculated by different approximate methods.

5.1. *Elastic scattering below the excitation threshold*

As discussed in Chapter 6, § 3.10 the elastic scattering by an ion is best described in terms of the phase shifts σ_l produced in the incident Coulomb wave by the modified field of the ionic core. As for scattering by hydrogen we must distinguish two sets of phase shifts $\sigma_l^{\pm}$ for the singlet and triplet cases. Table 8.5 gives a comparison of values for

---

† VELDRE, V. J., VINKALNS, I., and KARULE, E., *Atomic collision processes*, ed. MCDOWELL, M. R. C., p. 253 (North Holland, Amsterdam, 1964).

He⁺ obtained by the following methods, all of which have been discussed above for scattering by atomic hydrogen.

  (i) Static field approximation.
  (ii) $1s$-exchange approximation.†
  (iii) $1s$–$2s$–$2p$ exchange approximation.‡
  (iv) Exchange-adiabatic approximation.†
  (v) Polarized-orbital approximation.†

TABLE 8.5

Zero-order phase shifts $\sigma_0^{\pm}$ (rad) for scattering of electrons by He^+ ions calculated by different approximations

k (a.u.)	static potential (exact solution)	$1s$-exchange (exact solution)	Exchange-adiabatic	Polarized-orbital	$1s$–$2s$–$2p$ (exchange)	Quantum defect
σ_0^+						
0		0·387	0·420	0·444		0·438
0·491		0·366	0·399	0·424	0·385	0·417
0·779		0·341	0·375	0·400	0·364	0·389
1·076		0·318	0·352	0·377	0·346	0·357
1·353		0·302	0·235	0·361	0·342	0·327
1·897	0·468	0·290	0·321	0·344		0·276
2·198	0·443	0·283				0·254
σ_0^-						
0		0·920	0·945	0·942		0·932
0·491		0·893	0·918	0·915	0·909	0·903
0·779		0·855	0·879	0·875	0·865	0·861
1·076		0·802	0·827	0·832	0·808	0·803
1·353		0·748	0·772	0·775	0·752	0·740
1·897	0·468	0·645	0·669	0·675		0·609
2·198	0·440	0·604				0·539

As an indirect but effective experimental check on these results we include values for the phase shifts derived from extrapolation of the observed quantum defects of the $ns\,^1S$ and $ns\,^3S$ Rydberg series terms in helium.

In a detailed analysis Seaton§ used the formula (107) of Chapter 6 to extrapolate the quantum defects. According to this formula the phase shifts $\sigma_0^{\pm}$ for incident electrons of wave number k are given in

† Sloan, I. H., *Proc. R. Soc.* **A281** (1964) 151.
‡ Burke, P. G., McVicar, D. D., and Smith, K., *Phys. Lett.* **12** (1964) 215; Burke, P. G. and McVicar, D. D., *Proc. phys. Soc.* **86** (1965) 989.
§ Seaton, M. J., ibid. **88** (1966) 815.

terms of functions $Y_0^\pm(k^2)$ by

$$\tan \sigma_0^\pm = Y_0^\pm(k^2),$$

where $Y_0^+(k^2)$ can be expanded in the form

$$Y_0^\pm(k^2) = \frac{\sum_m a_m k^{2m}}{\sum_s b_s k^{2s}}, \qquad (92)$$

which has been continued analytically to negative values of k^2. The bound states for such values of k^2 are given by the solutions of

$$\tan \pi\mu^\pm(k^2) = Y_0^\pm(k^2),$$

where $\mu^\pm$ are the quantum defects.

From the observed values of the quantum defects the first few coefficients in the expression (92) for $Y_0^\pm$ can be determined. Thus Seaton finds, in atomic units,

$$Y_0^+(k^2) \simeq (0\cdot46873 + 0\cdot04737 k^2)/\{1 + 0\cdot34937 k^2\},$$

$$Y_0^-(k^2) \simeq 1\cdot34559/\{1 + 0\cdot25870 k^2\}.$$

These expansions are only valid for small k^2, but they should give very accurate values for $\sigma_0^\pm$ at the limit $k \to 0$.

It will be seen from these values, given in Table 8.5, that the polarized orbital method gives very good results both for σ_0^+ and σ_0^-, while the exchange-adiabatic method is not much less satisfactory. Both give better results, particularly σ_0^+, than the 1s-exchange approximation, although the latter is quite good. In general all the methods that include exchange are more effective than for scattering by atomic hydrogen—the residual Coulomb field is already beginning to dominate the situation.

Very similar conclusions follow from corresponding comparisons for the $\sigma_1^\pm$ and $\sigma_2^\pm$ phases which are given in Figs. 8.7 and 8.8. The importance of the 1s-2s-2p exchange approximation for dealing with resonance effects will be discussed in Chapter 9, § 3.

5.2. *Excitation of 2s and 2p states of* He^+

Fig. 8.9 gives a comparison of the cross-sections for excitation of the 2s state of He^+ by electrons with energies ranging from threshold (40·5 eV) to 200 eV, calculated according to the 1s-2s-2p close-coupling approximation and observed by Harrison, Smith, and Dance (see Chap. 4, § 3.2).

The evidence suggests that the values calculated from the 1s-2s-2p close-coupling approximation are likely to be considerably too large

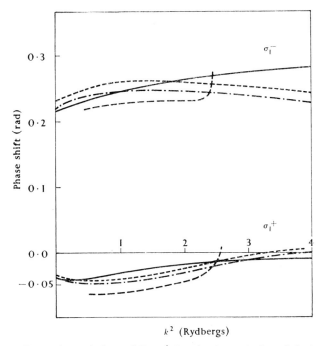

FIG. 8.7. Comparison of phase shifts $\sigma_1^\pm$ for elastic scattering of electrons by helium ions, calculated by different approximations with those derived by quantum defect extrapolation. – – – – calculated by $1s$–$2s$–$2p$ exchange approximation. — · — · — calculated by polarized-orbital method. - - - - calculated by exchange-adiabatic approximation. ——— derived by quantum defect extrapolation.

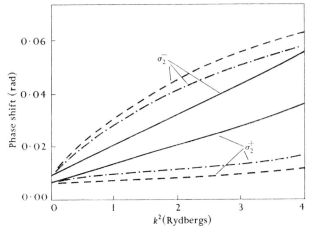

FIG. 8.8. Comparison of phase shifts $\sigma_2^\pm$ for elastic scattering of electrons by helium ions, calculated by different approximations, with those derived by quantum defect extrapolation. — · — · — calculated by polarized-orbital method. - - - - calculated by exchange-adiabatic approximation. ——— derived by quantum defect extrapolation.

near the threshold. This is somewhat surprising in view of the success of the approximation in other applications to He$^+$ (see Chap. 9, § 3.1).

To examine the convergence of the close-coupling series Burke, McVicar, and Smith included also closely coupled $3s$ and $3p$ states in a test calculation for 68-eV electrons. They found only a few per cent change in the calculated excitation cross-section.

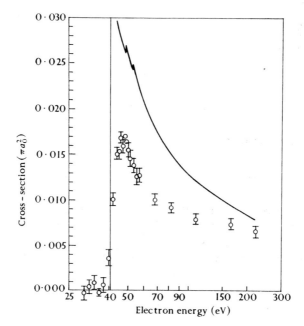

FIG. 8.9. Comparison of observed and calculated cross-sections for excitation of the $2s$ state of He$^+$ by electron impact. ——— calculated by $1s$–$2s$–$2p$ exchange approximation. ⌀ observed.

No observed data are available for excitation of the $2p$ state but Fig. 8.10 gives the cross-section calculated by the $1s$–$2s$–$2p$ approximation. Inclusion of $3s$ and $3p$ states has little effect.

6. Collisions with helium atoms

The fact that we are now dealing with a three-electron problem not only means that the different symmetry properties are required of the wave function but also precludes the use of exact atomic wave functions in the calculations. Despite the additional uncertainty introduced because of this a great deal of attention has been paid to the study of slow electron collisions with helium atoms. This is because of the much greater wealth and accuracy of experimental data available than for the only less complex atom, hydrogen.

We begin as usual by discussing elastic scattering.

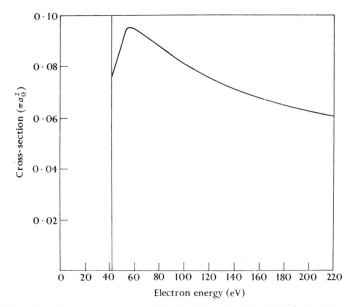

FIG. 8.10. Cross-sections for excitation of the 2p state of He⁺ by electron impact calculated by the 1s–2s–2p exchange approximation.

6.1. *Elastic scattering*

Of the various methods which are discussed in § 2 for calculating elastic cross-sections for electron collisions with hydrogen atoms the ones which are also applicable to helium are the polarized-orbital method, the truncated eigenfunction expansion method, and variation methods of the Kohn and similar types. In fact the only extensive calculations carried out up to the time of writing have used the exchange-adiabatic approximation to the polarized-orbital method. It also happens, because of certain special symmetry conditions, that, in contrast to the case of hydrogen (§ 2.9), a dispersion relation may be used to determine directly the forward scattered amplitude in terms of a correction to that given by the Born–Oppenheimer approximation. This correction is given in terms of an integral over all positive energies of a quantity involving the total collision cross-section (elastic+inelastic). We now discuss the application of these methods.

Since the ground state is of zero total spin, the over-all state of the three-electron system of atom plus incident electron must be a doublet.

The trial wave function, in the polarized-orbital method, including the spin coordinates, now takes the form

$$\Psi(\mathbf{r}_1, \mathbf{r}_2, \mathbf{r}_3) = 3^{-\frac{1}{2}} \sum_{\text{cyclic}} \{\psi_0(r_1, r_2) + \phi_0(\mathbf{r}_1, \mathbf{r}_2, \mathbf{r}_3)\} F_0(\mathbf{r}_3) \chi(1, 2, 3). \quad (93)$$

$\psi_0(r_1, r_2)$ is the ground state wave function of the helium atom and $\chi(1, 2, 3)$ is the doublet spin function

$$\chi(1, 2, 3) = \{\alpha(1)\beta(2) - \alpha(2)\beta(1)\}\alpha(3), \quad (94)$$

where α, β are single electron spin functions corresponding to axial spin quantum numbers $\pm\frac{1}{2}$ respectively. F_0 has the usual asymptotic form (15).

The function ϕ_0, which represents the effect of the dipole distortion, is calculated in the same way as for hydrogen. The appropriate integro-differential equation for F_0 is now determined by requiring

$$\sum_{\text{spin}} \iint \psi_0(r_1, r_2) \chi(1, 2, 3)(H - E) \Psi(\mathbf{r}_1, \mathbf{r}_2, \mathbf{r}_3) \, d\mathbf{r}_1 \, d\mathbf{r}_2 = 0, \quad (95)$$

where H denotes the operator

$$-\frac{\hbar^2}{2m}(\nabla_1^2 + \nabla_2^2 + \nabla_3^2) - e^2\left(\frac{2}{r_1} + \frac{2}{r_2} + \frac{2}{r_3} - \frac{1}{r_{12}} - \frac{1}{r_{23}} - \frac{1}{r_{31}}\right).$$

F_0 is then found to satisfy an equation of exactly the same form as (49). The effective polarization potential will have the asymptotic form

$$V_p \sim -\tfrac{1}{2}\alpha e^2/r^4 \quad (96)$$

but in practice, since it is necessary to use approximate atomic wave functions ψ_0, α will not be equal to the observed polarizability.

In all applications up to the present the exchange-adiabatic approximation in which the polarization kernel $K^{(1)}$ appearing on the right-hand side of (49) is ignored, has been used. Apart from direct use of the method just outlined other almost equivalent calculations have been carried out that have differed only in the means adopted for determining the polarization potential. In particular the potential derived by Bethe[†] has been used. This is given by

$$V_p(r) = -\frac{9}{2}\frac{e^2}{a_0}(\zeta r)^{-4}[1 - \tfrac{1}{3}e^{-2\zeta r}\{1 + 2\zeta r + 6\zeta^2 r^2 + \tfrac{20}{3}\zeta^3 r^3 + \tfrac{4}{3}\zeta^4 r^4\} - \tfrac{2}{3}e^{-4\zeta r}(1 + \zeta^4 r^4)], \quad (97)$$

where ζ is chosen so that $V_p(r)$ has the asymptotic form (96) with α the observed polarizability. This fixes ζ as $(9/\alpha a_0)^{\frac{1}{4}}$. This potential differs from that derived by the usual polarized-orbital method in that it has the correct asymptotic form and magnitude.

[†] BETHE, H., *Handb. Phys.*, Vol. 24, Pt. I, p. 339 (1943).

Calculations of this kind have been carried out with three slightly different assumptions as follows (see (195) of Chap. 7).

(i)† ψ_0 given by the Hartree–Fock approximation. V_p calculated by the polarized orbital procedure. The value of α comes out to be 1·559 a.u.

(ii)‡ ψ_0 given by the Hartree–Fock approximation, V_p calculated from (97) with the observed value of α, 1·376 a.u.

TABLE 8.6

Phase shifts η_0, η_1, and η_2 for elastic scattering of electrons by helium atoms calculated by exchange-adiabatic approximations, according to the assumptions (i), (ii), (iii)

Electron wave number k (a.u.)	Phase shifts (rad)								
	η_0			η_1			η_2		
	(i)	(ii)	(iii)	(i)	(ii)	(iii)	(i)	(ii)	(iii)
0·5	2·505	2·493	2·493	0·0093	0·1016	0·106	0·012	0·011	0·011
1·0	1·972	1·963	1·963	0·300	0·318	0·323	0·050	0·049	0·050
1·5	1·607	1·604	1·604	0·408	0·423$_5$	0·424	0·097$_5$	0·102	0·104
2·0	1·360	1·360	1·360	0·436$_5$	0·466	0·447	0·143	0·146	0·148

(iii)§ ψ_0 given by the variational approximation of Shull and Löwdin∥

$$\psi_0(r_1, r_2) = N[\exp\{-(ar_1+br_2)\}+\exp\{-(ar_2+br_1)\}], \qquad (98)$$

where $a = 2\cdot1832/a_0$, $b = 1\cdot1886/a_0$, V_p calculated from (97) but with a theoretical value of the polarizability, 1·32 a.u.

The results obtained in these three calculations do not differ very much, as may be seen from the calculated phase shifts given in Table 8.6 and the calculated zero energy scattering lengths given in Table 8.8. It will therefore be unnecessary to distinguish between them henceforward.

It is of interest also to compare the calculated phase shifts and zero energy scattering lengths calculated with and without inclusion of exchange and polarization. Reference to Table 8.7 shows that both the latter effects are clearly important, particularly for the first-order phase shifts.

Before discussing the comparison with experiment we consider the application of the dispersion relation to the determination of the forward

† LAWSON, J., MASSEY, H. S. W., WALLACE, J., and WILKINSON, D., *Proc. R. Soc.* A**294** (1966) 149.
‡ LA BAHN, R. W. and CALLAWAY, J., *Phys. Rev.* **135**A (1964) 1539.
§ WILLIAMSON, J. H. and MCDOWELL, M. R. C., *Proc. phys. Soc.* **85** (1965) 719.
∥ SHULL, H. and LÖWDIN, P. O., *J. chem. Phys.* **25** (1956) 1035.

scattered amplitude as a function of electron energy. Helium forms no stable negative ions and the dispersion relation (see § 2.2.2) takes the form†

$$\mathrm{re}\,f(0, E) = f_\mathrm{b}(0, E) - g_\mathrm{b}(0, E) + \frac{1}{\pi} \int \frac{\mathrm{im}\,f(0, E')}{E'-E}\, dE', \quad (99)$$

where $f(0, E)$ is the forward scattered amplitude for electrons of energy E. $f_\mathrm{b}(0, E)$ and $g_\mathrm{b}(0, E)$ are the direct and exchange forward elastically scattered amplitudes according to the Born and Born–Oppenheimer approximations discussed in Chapter 7, § 5.4.1.

TABLE 8.7

Comparison of phase shifts η_0, η_1, and η_2 calculated with and without inclusion of exchange and polarization

Electron wave number k (a.u.)	Phase shifts (rad)								
	η_0			η_1			η_2		
	(a)	(b)	(c)	(a)	(b)	(c)	(a)	(b)	(c)
0.5		2.47	2.505			0.099		—	0.0122
1.0	1.41	1.89	1.972	0.060	0.185	0.300	0.0052	0.0130	0.050
1.5		1.52	1.607		0.287	0.408		0.0431	0.097

(a) Calculated without inclusion of exchange or polarization.
(b) Calculated with inclusion of exchange but not polarization.
(c) Calculated with inclusion of exchange and polarization.

Through the conservation relation (§ 2.2.2)

$$\mathrm{im}\,f(0, E) = \frac{k}{4\pi} Q_\mathrm{t}(0, E), \quad (100)$$

where Q_t is the total collision cross-section (elastic+inelastic) we may write (99) in the form

$$\mathrm{re}\,f(0, E) = f_\mathrm{b}(0, E) - g_\mathrm{b}(0, E) + \frac{1}{4\pi^2} \int k' \frac{Q_\mathrm{t}(0, E')}{E'-E}\, dE'. \quad (101)$$

By combining (100) and (101) the forward scattered intensity may be obtained as a function of electron energy provided $Q_\mathrm{t}(0, E')$ is known over a sufficiently wide range of energy for the right-hand integral in (101) to converge. To obtain Q_t the total cross-section observed by Normand‡ (Chap. 1, § 6.1) may be used up to an energy of 144 eV. Between 144 and 7400 eV the ionization cross-section observed by Smith§ may be used in conjunction with cross-sections for elastic scattering

† GERJUOY, E. and KRALL, N. A., *Phys. Rev.* **127** (1962) 2105.
‡ NORMAND, C. E., ibid. **35** (1930) 1217.
§ SMITH, P. T., ibid. **36** (1930) 1293.

and for excitation of the 2^1P, 3^1P, and 4^1P states calculated by Born's approximation. Allowance for excitation of higher ^1P states can be made by assuming that the cross-section for a fixed energy falls off approximately as n^{-3} where n is the principal quantum number (see Chap. 4, § 1.4.2.1). Excitation of other states may be neglected. Extrapolation to still higher energies may be carried out by fitting an analytical formula to the form of Q_t already calculated.

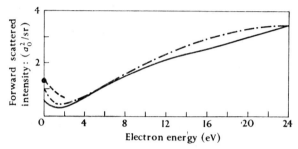

FIG. 8.11. The forward scattered intensity for electrons in helium as a function of electron energy from 0 to 24 eV. ——— calculated from the dispersion relation Chap. 8 (101) with the Hartree–Fock ground state wave function for helium. ● calculated in the zero energy limit from the dispersion relation Chap. 8 (101) with the 6-parameter Hylleraas ground state wave function Chap. 7 (200 a) for helium. —·—·— calculated from the exchange-adiabatic approximation. ---- calculated from the effective range expansion.

This procedure was carried out by Lawson, Massey, Wallace, and Wilkinson.† They evaluated f_b and g_b using the Hartree–Fock approximation to the helium wave function. The resulting forward scattering intensity as a function of electron energy is shown in Fig. 8.11 for electron energies between 0 and 24 eV and in Fig. 8.15 for energies between 24 and 400 eV.

In Fig. 8.11 comparison is made with the corresponding results obtained using the exchange-adiabatic approximation. The agreement is generally good but is improved very much at the low-energy limit by using a more accurate helium ground state wave function to evaluate f_b and g_b. If the Hylleraas 6-parameter variational function‡ (see Chap. 7 (200 a)) is used, then the zero energy limit for the forward scattered intensity, which is equal to $Q_t(0)/4\pi$, comes out to be $1\cdot32a_0^2$, which is in close agreement with the value $1\cdot28a_0^2$ obtained from the exchange-adiabatic approximation using (ii) of p. 541.

† LAWSON, J., MASSEY, H. S. W., WALLACE, J., and WILKINSON, D., Proc. R. Soc. A**294** (1966) 149.
‡ HYLLERAAS, E. A., Z. Phys. **54** (1929) 347.

O'Malley[†] has analysed the scattering to be expected at low energies in terms of effective range expansions for the phase shifts. Because of the polarizability of the atoms we may use the formulae (55 a) and (55 c) of Chapter 6. Ignoring terms of order k^2 or smaller, these involve one single free parameter a that is equal to the zero energy scattering length.

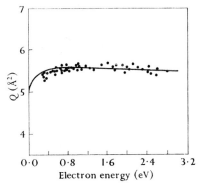

FIG. 8.12. Illustrating the fitting of experimental data (●) on the total cross-section Q for scattering of electron atoms by the effective range formula (the smooth curve).

TABLE 8.8

Zero energy scattering lengths for helium

Method of derivation	Scattering lengths in a_0
Exchange-adiabatic method	1·13–5
Dispersion relation	1·15
Effective range fit[†] to observed total cross-sections[‡]	1·19
Drift velocity and characteristic energy measurements	
(a) Frost and Phelps[§]	1·18
(b) Crompton, Elford, and Jory[‖]	1·20
Pressure shift in alkali metal spectra[††]	1·06

Fig. 8.12 shows the fit obtained with the experimental observations of Golden and Bandel[‡] (see Chap. 1, § 5) when the experimental value $1·36a_0^3$ is taken for the polarizability and a is chosen to have the value $1·19a_0$. This compares very well both with values determined from exchange-adiabatic theory and the dispersion relation, and with values derived from the measurement of electron transport properties (see

[†] O'MALLEY, T. F., *Phys. Rev.* **130** (1963) 1020.   [‡] loc. cit., p. 19.
[§] FROST, L. S. and PHELPS, A. V., *Phys. Rev.* **136** (1964) A1538.
[‖] CROMPTON, R. W., ELFORD, M. T., and JORY, R. L., *Aust. J. Phys.* **20** (1967) 369.
[††] See Table 6.3.

Chap. 2) and from the pressure shift in alkali metal spectra (Chap. 6, § 5) as may be seen by reference to Table 8.8.

It seems from this comparison that the zero energy scattering length for helium is $1·18a_0$ to within 5 per cent.

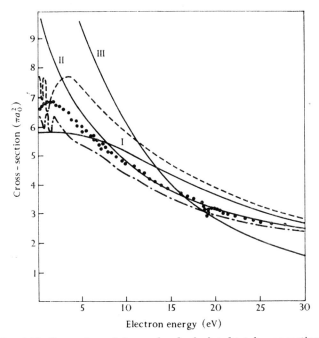

FIG. 8.13. Comparison of observed and calculated total cross-sections for scattering of electrons by helium atoms. I, calculated by exchange-adiabatic approximation including exchange and polarization. II, calculated including exchange but not polarization. III, calculated as scattering by the static field only. $----$ observed (Ramsauer and Kollath). $—·—·—$ observed (Normand). ●●● observed (Golden and Bandel).

We now extend the comparison to the total and differential elastic cross-sections. Fig. 8.13 illustrates the comparison of observed total cross-sections† with the results of three theoretical calculations. Curve III is obtained ignoring both exchange and polarization, curve II includes exchange but not polarization, while curve I includes all three. It can be seen that the last gives the closest fit to the observations although curve II is not very different. Curve III deviates very strongly at low energies.

Fig. 8.14 shows the comparison between observed and calculated

† RAMSAUER, C. and KOLLATH, R., *Annln Phys.* **3** (1929) 336; NORMAND, C. E., *Phys. Rev.* **35** (1930) 1217; GOLDEN, D. E. and BANDEL, H. W., ibid. **138** (1965) A14.

differential cross-sections for three electron energies below the excitation threshold. As in Fig. 8.13, three sets of calculated curves are shown. The observed results† are normalized so as to agree with the prediction of the exchange-adiabatic method at 90°.

It is immediately clear that the differential cross-sections are much more variable as functions of angle than given by the simple static approximation in which exchange and polarization are ignored. For

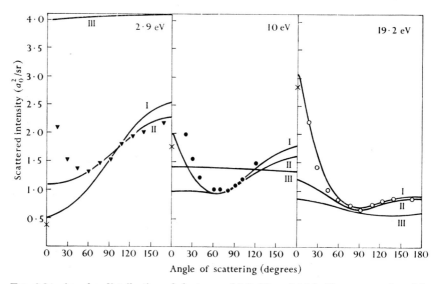

FIG. 8.14. Angular distribution of electrons of 2·9, 10, and 19·2 eV energy scattered in helium. I, calculated by exchange-adiabatic approximation including exchange and polarization. II, calculated including exchange but no polarization. III, calculated as scattering by the static field only. ×, forward scattered intensity calculated from the dispersion relation. ○ observed 19·2 eV, ● observed 10 eV, ▼ observed 2·9 eV, normalized in each case so as to agree with I at 90°.

electrons of 19·2-eV energy the exchange-adiabatic approximation gives very good agreement with the observations. Furthermore, the observed curve extrapolates very smoothly to the limit at zero angle given by the dispersion relation. Inclusion of exchange but not polarization gives quite good agreement at large angles but gives too little scattering at small angles.

At the lowest electron energy shown (2·9 eV) the situation is less satisfactory. The exchange-adiabatic approximation gives an angular distribution in which the intensity increases steadily with increasing angle of scattering. The zero angle limit is not very different from

† BULLARD, E. C. and MASSEY, H. S. W., *Proc. R. Soc.* A**133** (1931) 637; RAMSAUER, C. and KOLLATH, R., *Annln Phys.* **12** (1932) 529.

that given by the dispersion relation. In addition, according to the effective range formula† (see Chap. 6 (55 c)), the angular distribution is given by

$$I(\theta) = a^2 + (\pi/a_0)\alpha ak\sin\tfrac{1}{2}\theta + (8/3a_0)\alpha a^2 k^2 \ln(ka_0) + O(k^2 a_0^2), \quad (102)$$

where a is the zero energy scattering length, increasing steadily as θ goes from 0 to π, at least for sufficiently small k. The observed distribution, however, has a minimum at about 50° and rises quite rapidly at smaller angles.

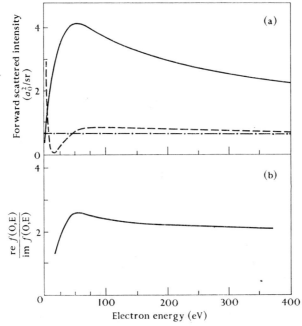

Fig. 8.15. (a) The forward scattered intensity for elastic collisions of electrons in helium as a function of electron energy from 25 to 400 eV. ——— calculated from the dispersion relation Chap. 8 (101) with the Hartree–Fock ground state wave function for helium. ---- calculated by Born–Oppenheimer approximation. —·—·— calculated by Born approximation. (b) The ratio of the real to the imaginary part of the forward scattered amplitude for elastic collisions of electrons in helium calculated from the dispersion relation Chap. 8 (101) with the Hartree–Fock ground-state wave function for helium.

The intermediate energy (10 eV) is also intermediate as far as comparison between theory and experiment is concerned.

At electron energies greater than about 25 eV the exchange-adiabatic approximation is no longer valid but the forward scattered intensity may still be obtained from (101) as a function of electron energy and is given in Fig. 8.15 (a) over the energy range from 25 to 400 eV. The

† O'Malley, loc. cit. p. 544.

ratio of the real to the imaginary component of the scattered amplitude is given in Fig. 8.15 (b). Comparison with the constant value given by Born's approximation shows that the latter is quite inadequate for predicting the forward scattered intensity at electron energies as high as 400 eV. The Born–Oppenheimer approximation is also quite unsatisfactory.

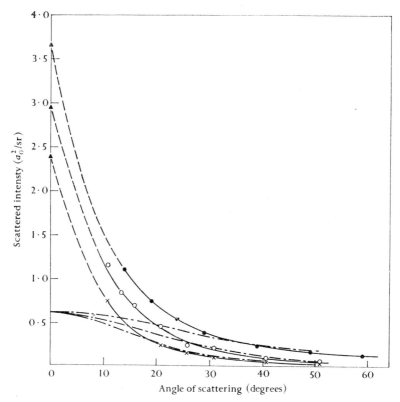

Fig. 8.16. Angular distribution of electrons of 100, 200, and 350 eV scattered elastically in helium. Observed points: ● 100 eV, ○ 200 eV, × 350 eV. ▲ Zero angle limits determined from the dispersion relation (101) —·—·— calculated by Born's approximation.

Direct comparison with observation is not possible. However, in Fig. 8.16 angular distributions for electrons of energy 100, 200, and 350 eV observed by Hughes, McMillen, and Webb (Fig. 5.40),† which extend down to angles of 15°, are shown. Normalization is such that the observed intensity at from 10° to 147° for 700-eV electron energy agrees with that

† HUGHES, A. L., MCMILLEN, J. H., and WEBB, G. M., *Phys. Rev.* **41** (1932) 154.

given by Born's approximation (see Chap. 7, § 4). It will be seen that the observed distributions, which show increasing departure from the results of Born's approximation at small scattering angles as the electron energy falls, all extrapolate smoothly to the zero angle limit given from the results shown in Fig. 8.15 (a).

At 25 eV the experimental results of Hughes, McMillen, and Webb cannot be relied upon as far as absolute magnitude is concerned. This is certainly true at 25 eV at which energy direct comparison with observed total cross-sections may be made. The uncertainty probably extends to the observations at 50 and 75 eV for which any check by comparison with Born's approximation near 90° is not applicable. At 100 eV and above, this comparison may be made because Born's approximation is expected to hold at such energies near 90°.

To summarize, we see that below the excitation threshold we have a nearly self-consistent detailed interpretation of the electron scattering. The exchange-adiabatic approximation, the dispersion relation, and the effective range expansion all give results which agree well with experimental values of the zero energy scattering length. Discrepancies still remain between the calculated and observed differential cross-sections at energies of 100 eV and less, but these probably arise from inaccuracy in the observations at small scattering angles. As for elastic scattering by H and He^+ the exchange-adiabatic approximation gives good results below the excitation threshold.

At energies of 100–400 eV the dispersion and conservation relations show that the forward scattered intensity considerably exceeds that given by Born's approximation and this is consistent with the experimental results. There is need for further observations of differential cross-sections in which the emphasis is placed on determination of accurate absolute values and on measurements at small scattering angles.

6.2. *Excitation of the 2-quantum states—elastic collisions with metastable atoms*

The first application of the distorted wave method to excitation of a 2-quantum state of helium was made by Massey and Moiseiwitsch.† They calculated cross-sections for excitation of the 2^3S and 2^1S states, ignoring any interaction between them. For the ground-state wave function they used the simple Hylleraas form (198) of Chapter 7 and for the excited states variational approximations due to Morse, Young,

---

† MASSEY, H. S. W. and MOISEIWITSCH, B. L., *Proc. R. Soc.* A**227** (1955) 38.

and Haurwitz.† The distorted wave functions were calculated using the Kohn variational method.

The striking feature of their results was the sharp peak in the excitation function for the 2^3S state close to the threshold. This arose because the effective field (non-local exchange and mean static) which a 2^3S atom exerted on the electron was such as to provide a bound state only a little below this energy. This has the effect of concentrating the wave function for the free motion just above the threshold energy very much in the neighbourhood of the atom, so enhancing the cross-section. In a sense the peak represents the tail-end of a virtual resonance in the excitation cross-section occurring just below the threshold. It is closely related to the resonance effects that are discussed in Chapter 9, which arise essentially from the existence of bound states for the motion of an electron in the field of an excited atom or molecule.

The existence of a sharp peak in the 2^3S excitation function had been known from the work of Maier-Leibnitz (Chap. 5, § 2.2) but a similar peak, not given by the distorted wave calculations, was also known to exist in the 2^1S excitation function. The calculations did not allow for the fact that the 2^1S and 2^3S states are closely coupled at low electron energies, a fact that became apparent from the observations of Phelps‡ on the rate coefficient for deactivation of 2^1S to 2^3S metastable atoms in a discharge afterglow (see Chap. 4, § 3.1.1). These rates were so high as to leave no doubt of the strength of the coupling. Marriott§ then carried out a close-coupling calculation of the rate coefficient using essentially a truncated expansion involving the 2^1S and 2^3S states only. This calculation, carried out using desk machines, was the first in which two coupled integrodifferential equations were solved with boundary conditions appropriate to a collision problem. Disappointing results were obtained, the calculated values being about six times too small. A likely source of the discrepancy was the neglect of coupling with the 2^3P and 2^1P states, some further evidence for which was provided by the lack of success of a distorted wave calculation of the 2^3P excitation function, the calculated value at 25 eV being about seven times too large.∥

Burke, Cooper, Ormonde, and Taylor†† have solved by electronic computation the coupled equations arising from an eigenfunction expansion including all four 2-quantum states but not the ground state. The

† loc. cit., p. 479.   ‡ PHELPS, A. V., *Phys. Rev.* **99** (1955) 1307.
§ MARRIOTT, R., *Proc. phys. Soc.* A**70** (1957) 288.
∥ MASSEY, H. S. W. and MOISEIWITSCH, B. L., *Proc. R. Soc.* A**258** (1960) 147.
†† BURKE, P. G., COOPER, J. W., ORMONDE, S., and TAYLOR, A. J., *Abst. 5th Int. Conf. Phys. Electron. Atom. Collisions*, Leningrad, p. 376, (Publishing House Nauka, 1967).

wave functions for the excited states were represented by a properly symmetrical combination of a 1s wave function for an electron in the field of an He$^+$ ion and a 2-quantum state wave function in the form of a sum of exponentials that approximated closely to the numerical solutions of the Hartree–Fock equations under these conditions. Fig. 8.17

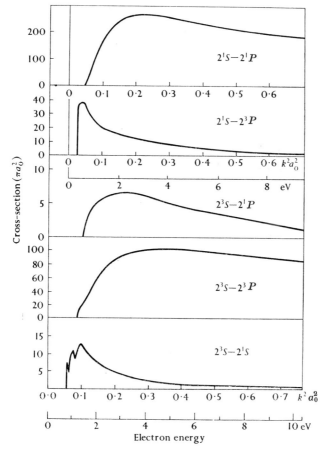

Fig. 8.17. Cross-sections for excitation of various transitions between 2-quantum states of helium calculated by use of four-state eigenfunction expansion with close coupling.

shows the results that they obtained for the excitation of various transitions within the 2-quantum states. These are little affected by any allowance for coupling through the ground state as this coupling is weak. In fact, the solutions available for the 2-quantum coupled system may be used as distorted waves in a calculation of the cross-section for transition from the ground state. The failure of the simple distorted wave calculations referred to above is not because the method is inapplicable but because the distorted waves in the field of the excited atom must include allowance for the strong coupling between all four of the 2-quantum states.

The cross-section for the superelastic transition 2^1S–2^3S at thermal velocities comes out to be $360\pi a_0^2$, which agrees very well with that observed by Phelps,† $340\pi a_0^2$.

In view of these considerations we would expect that some of the structure apparent in the cross-sections shown in Fig. 8.17, which arise from particular features of the wave functions for the composite system of 2-quantum states, will appear in the cross-sections for excitation from the ground state. Burke, Cooper, and Ormonde‡ in an earlier calculation included the ground state in a five-state expansion, but made the assumption that the approximate bound-state wave functions were exact solutions of the Schrödinger equation. This led to unsatisfactory results for weak transitions, particularly those involving S and P states of the total system. Nevertheless they obtained promising results in the energy range between the 2^3S threshold, at 19·84 eV, and 22·0 eV. It seems likely that, when these calculations are repeated without the unsatisfactory assumption, the cross-sections for excitation of the 2-quantum states will be obtained in good agreement with observation (see Chap. 9, § 5).

Further interesting evidence is provided by the experiments of Ehrhardt and Willmann§ (Chap. 5, § 5.3.3) who measured the angular distributions of electrons of closely defined energy after exciting the 2^3S state. Their results (Fig. 5.58) show that for 19·95-eV electrons the distribution is nearly uniform characteristic of s scattering, but it is markedly different at 20·45 eV with a minimum at 90° characteristic of p scattering, while at 21·00 eV it has changed again, having a minimum near 60° and maximum near 90° as expected for d-scattering. These energies correspond to peaks in the excitation function for the 2^3S state. The calculations of Burke, Cooper, Ormonde, and Taylor that determine the distorted wave functions in the field of 2-quantum excited helium atoms (allowing for interaction between all four sets of states) would predict just the effects observed. They find that, close to the observed energies, the 'free' wave functions for the ^2S, ^2P, and ^2D states of the overall system respectively exhibit resonance behaviour. In the field of a 2^3S helium atom these correspond to outgoing electrons in s, p,

† loc. cit., p. 550.
‡ BURKE, P. G., COOPER, J. W., and ORMONDE, S., *Phys. Rev. Lett.* **17** (1966) 345.
§ EHRHARDT, H. and WILLMANN, K., *Z. Phys.* **203** (1967) 1.

and d states respectively in giving rise to angular distributions of the form observed.

7. Collisions with Ne, Ar, Kr, Xe atoms
7.1. Elastic scattering

7.1.1. *Effective range analysis.* The behaviour of the total, momentum-loss, and differential cross-sections for elastic scattering of very slow electrons by atoms of the heavier rare gases have been analysed by O'Malley[†] in terms of effective range expansions in much the same way as for helium. This involves the use of the formulae (54 a, b, c) and (55 a, b, c) of Chapter 6 with the observed values of the respective polarizabilities[‡] (2·65, 11·0, 16·6, and 27·0a_0^3 for Ne, Ar, Kr, and Xe respectively) taken for α.

As explained in Chapter 6, § 4.1 the Ramsauer–Townsend effect arises from the behaviour of the phase η_0. According to (54 a) at very low energy, apart from an integral multiple of π,

$$\eta_0 \simeq -ak - (\pi/3a_0)\alpha k^2, \qquad (103)$$

where a is the zero-energy scattering-length. For neon as for helium $a > 0$ and η_0 decreases monotonically from its value at zero energy. On the other hand, for argon, krypton, and xenon $a < 0$ and η_0 varies as in Figs. 6.6 and 6.8. The minimum in the total cross-section, the Ramsauer–Townsend effect, will occur where η_0, according to (103), vanishes. If k_t is the electron wave number at the minimum then $a \simeq -(\pi/3a_0)\alpha k_t$. This gives

$$a = -1·9, \quad -3·7, \quad -6·2 a_0 \qquad (104)$$

for argon, krypton, and xenon respectively.

According to Chapter 6 (55 b) the momentum loss cross-section Q_d will also exhibit a minimum for the same rare gases. This will occur for an electron wave number $k_m \simeq 5k_t/6$. For argon k_t corresponds to an electron energy 0·37 eV so that the minimum in Q_d should occur at 0·25 eV in very good agreement with its location as determined by Frost and Phelps[§] from an analysis of drift velocities of slow electrons in argon (see Chap. 2, § 7.3 and Fig. 2.22).

Referring to the observed angular distributions (Fig. 5.41) of the electrons from each of the four gases at very low energies we note that there is evidence that the p-phase η_1 vanishes at some electron energy

[†] loc. cit., p. 544.
[‡] VAN VLECK, J. H., *The theory of electric and magnetic susceptibilities* (Clarendon Press, Oxford, 1932).
[§] FROST, L. S. and PHELPS, A. V., *Phys. Rev.* **136** (1964) A1538.

between 1 and 1·5 eV. If the higher order phases η_3, η_5,... are negligible, vanishing of η_1 will be manifest in an angular distribution symmetrical about 90°. Allowing for the small contributions from the higher odd phases O'Malley finds that the observed data indicate the vanishing of η_1 at electron energies of 1·5, 1·1, 1·0, and 0·8 eV for neon, argon, krypton, and xenon respectively. Substitution in (54 b) of Chapter 6 then gives for these respective atoms

$$a_1 \simeq 1\cdot66,\ 8\cdot0,\ 12\cdot8,\ \text{and}\ 23\cdot2 a_0^3. \tag{105}$$

Used in connection with (54 b) this gives the p-phases up to energies of several eV. The higher, small, phases will be given well in this energy range by (54 c). To extend the expansion for η_0 to cover this range it is necessary to take into account terms in k^2. Thus we have

$$k^{-1}\sin\eta_0 = -a - (\pi/3a_0)\alpha k - (4/3a_0)\alpha a k^2 \ln(ka_0) + fk^2 + \ldots. \tag{106}$$

To determine the further parameter f it is chosen to provide the best fit with the observed total cross-section up to electron energies of the order 1·0 eV, assuming the other phases to be determined as described above. In this analysis the parameter a was also regarded as adjustable to yield

$$a = 0\cdot24,\ -1\cdot70,\ -3\cdot7,\ -6\cdot5 a_0, \tag{107}$$

for Ne, Ar, Kr, and Xe respectively. For the latter three gases these may be compared with the rougher values (104) derived from the observed energy at which the Ramsauer–Townsend minimum occurs.

Fig. 8.18 shows the agreement that is obtained in this for all four cases as well as the extrapolation to zero energy for neon and argon that follows through use of the effective range expansion.

Comparison may now be made with results of observations by microwave and drift velocity methods of the cross-sections at very low energies. This is best done in terms of the zero energy scattering length. The microwave observations of Phelps et al.† (see Chap. 2, § 7.3) give the momentum loss cross-section for a mean electron energy of about 0·039 eV. Using the relation

$$\pm(Q_\mathrm{d}/4\pi)^{\frac{1}{2}} = a + (2\pi/5a_0)\alpha k + \ldots,$$

assuming that k corresponds to the mean energy and that a has the same sign as in (107), a may be obtained. Analysis of the drift velocity data (Chap. 2, § 7.3) is complicated by the existence of three sets of solutions for the Q_d-energy variation derived from the data for each of the three gases, Ar, Kr, and Xe, investigated. However, at energies of 0·013 eV

† Phelps, A. V., Fundingsland, O. T., and Brown, S. C., Phys. Rev. 84 (1951) 559.

for argon and 0·023 eV for Kr and Xe the three sets are coincident. a may then be determined from the values of Q_d at these energies.

Finally, a may also be obtained from observations of the pressure shift of the lines of alkali metal spectra (Chap. 6, § 5) due to the presence of the appropriate rare gas.

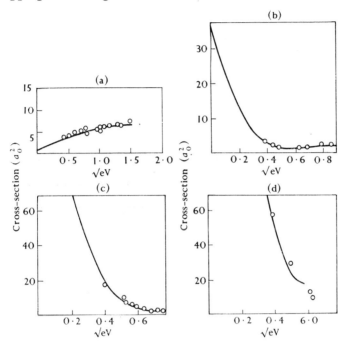

Fig. 8.18. Illustrating the fitting, by an effective range expansion, of the variation with electron velocity of the total cross-section for collision of slow electrons with neon, argon, krypton, and xenon. (a) neon, (b) argon, (c) krypton, (d) xenon. —— from effective range expansion. ○ experimental points.

In Table 8.9 the values of a derived in these ways are compared with those (107) derived from the effective range analysis. The agreement is seen to be remarkably good and indicates that the zero energy scattering length is known certainly within 10 per cent for Ar, Kr, and Xe. The position is less satisfactory for neon.

7.1.2. *Application of the exchange-adiabatic approximation.* The exchange-adiabatic method has been applied by Thompson† to calculate cross-sections for elastic scattering of electrons by neon and argon. In principle the method is essentially the same as for helium but involves more extensive computation. Once again the wave function describing

† THOMPSON, D. G., *Proc. R. Soc.* A**294** (1966) 160.

the collision satisfies an integrodifferential equation of the form (49) with $K_l^{(1)}(r, r')$ put equal to zero, so that there are three main contributing interactions—the mean static potential V of the atom calculated from the Hartree–Fock field, the non-local exchange interaction determined by the kernel $K_l(r, r')$, and the polarization potential V_p, which has the asymptotic form

$$V_p = -\tfrac{1}{2}\alpha e^2/r^4, \qquad (108)$$

where α is a calculated approximation to the atomic polarizability following the method of Sternheimer.† The calculation of α also requires the use of the Hartree–Fock approximation to the atomic orbitals.

TABLE 8.9

Zero energy scattering lengths (in units a_0) for neon, argon, krypton, and xenon

	Scattering length			
	Ne	A	Kr	Xe
I. Effective range expansion and total cross-section observations	0·24	−1·70	−3·7	−6·5
II. Microwave observations	0·39‡ 0·18§		−3·2‡	−5·6‡
III. Drift velocity observations‖		−1·169	−3·2	−6·0
IV. Pressure shift in alkali†† metal spectra	0·03	−1·86	−3·8	−6·9
V. Exchange-adiabatic approximation	0·35	−1·60		
VI. Approximation including exchange but ignoring polarization	1·05	+1·50		

In Thompson's calculation α was found to have the values $2\cdot20$ and $14\cdot2a_0^3$ for neon and argon respectively to be compared with $2\cdot65$ and $11\cdot0a_0^3$ as observed. For neon the calculated values of α and of V_p were used but for argon α as calculated was scaled down in the ratio $11/14\cdot2$ so that α in (108) had the observed value.

The values obtained for the zero-order scattering lengths are given in Table 8.9. Corresponding values obtained by neglecting polarization are also included. In view of the basic character of the calculations, depending as they do only on knowledge of the Hartree–Fock wave functions of the atoms concerned (apart from the downward scaling

† STERNHEIMER, R. M., *Phys. Rev.* **96** (1954) 951. The chief contribution to the polarizability comes from interaction between p- and d-orbitals. That from s–p and p–s interactions is small and was neglected in the calculations.
‡ PHELPS, A. V., FUNDINGSLAND, O. T., and BROWN, S. C., ibid. **84** (1951) 559.
§ GILARDINI, A. L. and BROWN, S. C., ibid. **105** (1957) 31.
‖ See Chapter 2, § 7.3.   †† See Chapter 6, § 5.

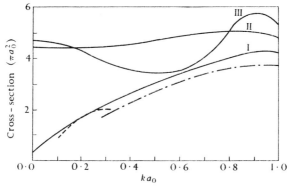

FIG. 8.19. Comparison of observed and calculated total cross-sections for scattering of electrons by neon. Full line curves: I, calculated by exchange-adiabatic method. II, calculated with allowance for exchange but not polarization. III, calculated neglecting exchange and polarization (scattering by static field only). — · — · — observed (Brüche).† — — — — observed (Ramsauer and Kollath).‡

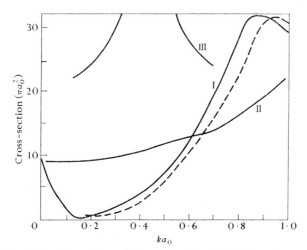

FIG. 8.20. Comparison of observed and calculated total cross-sections for scattering of electrons by argon. Full line curves: I, calculated by exchange-adiabatic method. II, calculated allowing for exchange but not polarization. III, calculated neglecting exchange and polarization (scattering by static field only). — — — — observed (Ramsauer and Kollath).‡

of α to 80 per cent of its calculated value for argon) the results obtained from the full approximation are quite good. It is clear, however, that, if polarization is ignored, the agreement is lost, and that allowance for electron exchange is also important.

† loc. cit. p. 25.   ‡ loc. cit. p. 24.

Figs. 8.19 and 8.20 illustrate the comparison between observed and calculated total cross-sections. Once again the full approximation gives remarkably good results but both exchange and polarization must be allowed for to achieve this.

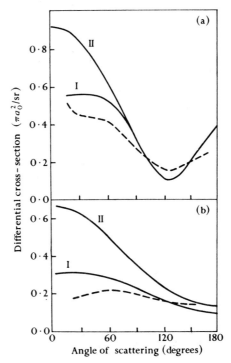

FIG. 8.21. Comparison of calculated and observed differential cross-sections for elastic scattering of electrons by neon. (a) 13·6 eV electrons. (b) 2·2 eV electrons. Full-line curves: I, calculated by exchange-adiabatic method. II, calculated allowing for exchange but not polarization. – – – – observed (Ramsauer and Kollath).†

Finally, in Figs. 8.21 and 8.22 observed and calculated angular distributions of the scattered electrons are compared. The full approximation passes even this sensitive test, for the predicted distributions are very similar to the observed at each energy, particularly when it is noted how rapidly the form of the distribution changes with electron energy in this low energy region.

7.2. *Angular distribution of inelastically scattered electrons—diffraction effects*

In Chapter 5, § 5.3.2 the experiments of Mohr and Nicoll‡ on the

† loc. cit. p. 328.
‡ NICOLL, F. H. and MOHR, C. B. O., *Proc. R. Soc.* **A142** (1933) 320, 647.

angular distribution at large angles of the intensity of scattering of electrons after exciting a particular state of an atom are described. The most prominent feature of their results was the observation of diffraction maxima and minima in these angular distributions, closely resembling those appearing in the distributions of elastic scattering at the same incident energy. This close similarity (see Figs. 5.49 and 5.50) tends to disappear as the electron energy is reduced to such an extent that the wave numbers of the inelastic and elastically scattered electrons are markedly different.

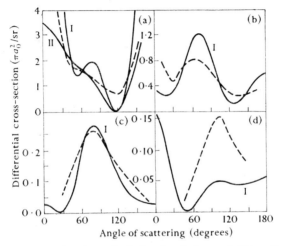

FIG. 8.22. Comparison of calculated and observed differential cross-sections for elastic scattering of electrons by argon. (a) 9 eV, (b) 3·3 eV, (c) 1·2 eV, (d) 0·60 eV electrons. Full-line curves: I, calculated by exchange-adiabatic method. II, calculated allowing for exchange but not polarization. − − − − observed (Ramsauer and Kollath).†

The existence of diffraction effects in the inelastic scattering of electrons that have suffered a definite energy loss is not difficult to understand. An electron can be thought of as first losing energy in exciting the atom, a process resulting in general in little deviation, and then undergoing diffraction by the field of the excited atom. Equally it might first be diffracted without energy loss and then suffer the inelastic transition. If the field of the excited atom does not differ very much from that of the normal atom, and the energy loss suffered in the excitation process is small compared with the incident energy, we would furthermore expect the diffraction pattern to be very similar to that for elastic scattering—the inelastic transition does not affect the angular

† loc. cit. p. 328.

distribution to any marked extent. It is true that the field of the excited atom will be of much longer range than that of the normal atom but, at the electron energies involved and for large-angle scattering, the effective scattering field is that at quite short distances where the two fields are nearly equal.

Massey and Mohr† carried out calculations for the inelastic scattering of electrons by neon and argon by the distorted-wave method to check the validity of these ideas. They were not particularly concerned with reproducing closely the actual observed distributions for neon and argon, but with comparing the inelastic scattering distribution at large angles for electrons that had excited an outer p electron to the lowest unoccupied s orbital with that for elastic scattering, calculated with the same assumptions about the atomic field. The wave functions for the orbitals concerned were obtained using Slater's rules for effective screening constants. It was found that the main contribution to the distorted wave matrix elements came from such large values of r that the distorted wave functions could be replaced by their asymptotic forms. Because of this it was only necessary to determine the phase shifts produced in the incident and final wave functions by the initial and final atomic fields respectively. This was done using the semi-classical approximation (Chap. 6, § 3.7).

The results obtained are shown in Fig. 8.23. Not only do they exhibit the close similarity between the elastic and inelastic diffraction patterns but they also agree quite well with the observations.

8. Collisions with alkali metal atoms

Considerable attention has been paid to the calculation of elastic cross-sections for electron scattering by alkali metal atoms. Despite its complexity, caesium has attracted theoretical attention for a special reason. As mentioned in Chapter 2, § 7.4 it is the most suitable working substance in the proposed magnetohydrodynamic (MHD) method for direct conversion of heat to electricity. Knowledge of the scattering cross-section for low-energy electrons is an important design requirement.

It is unwise to assume that methods that apply well to the calculation of elastic scattering by atoms with closed shells will be equally effective in dealing with other atoms. The alkali metal atoms represent the opposite extreme, with a single outer electron loosely bound in an s-orbital of large radius. Because of the small energy separation from the neighbouring p-orbital, a very large fraction (over 98 per cent) of the

† MASSEY, H. S. W. and MOHR, C. B. O., *Proc. R. Soc.* A**146** (1934) 880.

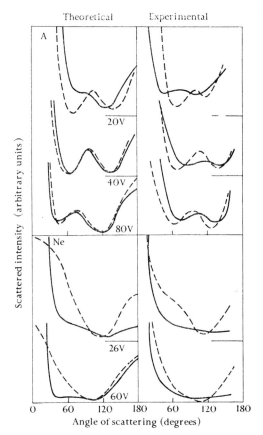

Fig. 8.23. Comparison of observed and calculated angular distributions for electrons of different incident energies scattered elastically and inelastically (after excitation of the resonance transition) in neon and argon. – – – – elastic, ——— inelastic.

polarizability arises from interaction between these orbitals. We would, therefore, expect that dipole distortion of an alkali atom in an encounter with a slow electron could be taken into account very completely by using a truncated eigenfunction expansion involving only two states— the ground state and one in which the outer electron occupies the neighbouring p-orbital.

On the other hand, because of the large polarizability, difficulties might be anticipated in applying the polarized-orbital or exchange-adiabatic methods. In fact, as we shall see, the s–p truncated eigenfunction expansion (TEE), does not give as good a fit with the experimental data as might be expected and there is no doubt of the difficulty of dealing with polarization effects by methods which have been very successful for other atoms. The situation for all the alkali metals seems to be a particularly sensitive one so that small differences in the approximate wave functions for the atoms may produce quite large modifications even in the s–p TEE method. It is also important to remember that the experimental observations are not as reliable as for the rare gases, particularly at very low electron energies.

8.1. *Application of the s–p close-coupling approximation to elastic scattering*

The s–p close-coupling equations have been solved for all the alkali metal atoms, for electron energies below the excitation threshold by Karule,† and for energies above the threshold up to 5 eV by Karule and Peterkop.‡ For lithium, Hartree–Fock wave functions were used for the valence electrons and to determine the potential due to the atomic core, while for the remaining atoms a semi-empirical procedure§ was used. With these wave functions the polarizabilities of Li, Na, and K come out to be 167, 162, and $303a_0^3$, respectively. These are to be compared with experimental values 148 ± 14, 144 ± 18, and $342\pm40a_0^3$, showing that the wave functions are not yet as accurate as would be desired.

The appropriate coupled equations were solved for incident electrons with total angular momentum quantum numbers l from 0 to 7. Exchange was taken fully into account for $l \leqslant 4$.

Fig. 8.24 (a)–(d) illustrates the comparison between the calculated total cross-sections and those observed by beam techniques for Li, Na, K, and Cs atoms.‖ The calculated elastic cross-section is shown, as distinct from the total cross-section, at electron energies above the excitation threshold. It will be seen that the agreement is far from perfect. For Na, K, and Cs the calculated cross-section does not exhibit the maximum observed at electron energies between 1 and 3 eV. However, a sharp maximum appears in the calculated cross-section for sodium at a somewhat lower energy (0·2 eV). It appears that this is largely due to the η_1^- phase shift passing through $\tfrac{1}{2}\pi$ near this energy. If this were to occur at a somewhat higher energy of around 1 eV the peak would be displaced to around this energy and would be somewhat smaller,

---

† KARULE, E. M., *Proc. 4th Int. Conf. Phys. electron. atom. Collisions*, p. 139 (Quebec, 1965).

‡ KARULE, E. M. and PETERKOP, R. K., *Optika Spektrosk.* **16** (1964) 958; *Optics Spectrosc., N.Y.* **16** (1964) 519.

§ GASPAR, R., *Acta Phys. Hung.* **2** (1952) 151; ANDERSON, E. M. and SMOKINA, E. S., *Rept. LVU-SC Riga* (1963). ‖ See Chap. 1, pp. 29–30.

giving closer agreement with the shape of the observed curve. The sensitivity of the phase shifts to the assumed atomic wave functions is quite high so that a small change in these functions would bring about the desired result. An example of the sensitivity is provided by comparison of Karule's results for the zero-energy scattering lengths a^+, a^- with those obtained by Salmona and Seaton† who also used the $3s$–$3p$ close-

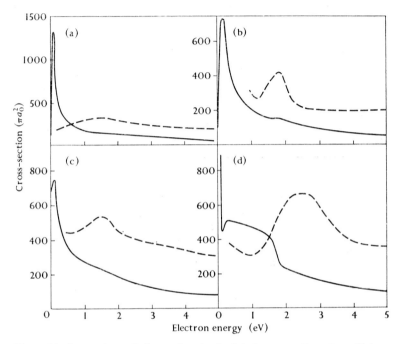

Fig. 8.24. Comparison of observed and calculated cross-sections for collisions of electrons with alkali metal atoms. —— calculated elastic cross-section by s–p close-coupling approximation. - - - - observed total cross-section. (a) lithium, (b) sodium, (c) potassium, (d) caesium.

coupling approximation with Hartree–Fock wave functions for both core and valence electrons. This comparison is given in Table 8.10. Included also in this table are results obtained by Salmona and Seaton in which exchange and polarization were successively ignored. The effect of exchange with the core electrons is also revealed.

For potassium there is no sharp peak in the calculated cross-section at very low energies, but a weak maximum does exist that could again be transformed to a sharper one at a higher energy by a comparatively small change of phase shifts.

† Salmona, A. and Seaton, M. J., *Proc. phys. Soc.* **77** (1961) 617.

The situation is less clear for caesium and is not helped by reference to data about the momentum loss cross-section obtained by various swarm methods (see Chap. 2, § 7.4). These vary over such a wide range as to be of little value, but it is worth noting that two observations, those of Flavin and Meyerand[†] and of Chen and Raether[‡] both give very high cross-sections between 800 and $1000\pi a_0^2$ at electron energies of about 0·07 eV, which is not inconsistent with the theoretical result. It

TABLE 8.10

Zero-energy scattering lengths for elastic collisions of electrons with sodium atoms, calculated by various approximations

	Scattering length in units a_0						
	(1) I	(2) I	(3) I	(3) II	(4) I	(4) II	(5) I
a^+	−2·4	36	−60	−18	13·9	4·2	12
a^-	−2·4	36	29	5·7	−8	−5·9	−9

I calculated by Salmona and Seaton; II calculated by Karule.

(1) Static field only (no 3p coupling, no exchange).
(2) Static field plus polarization but no exchange (3p coupling but not using antisymmetrized wave functions).
(3) Static field plus exchange with outer electrons but no polarization (antisymmetrized wave functions but no 3p coupling).
(4) Static field plus exchange with outer electrons plus polarization.
(5) As for (4) but including exchange with core electrons (full 3s–3p close-coupling approximation).

is also to be noted that the calculated caesium cross-section rises very sharply as the energy falls below 1·55 eV, as in the observed beam data, but does not then fall again as observed. A small change in the assumed wave functions may again be all that is necessary to improve the agreement with experiment as to the shape of the cross-section.

For lithium, unfortunately, experimental results are more difficult to obtain and those available are considerably less accurate than for the heavier atoms. No experimental evidence exists for the presence of a sharp peak in the cross-section, although this is certainly a feature of the calculated cross-section at 0·1 eV. This is below the lowest energy investigated but the trend of the present observations is against the existence of a peak. In this case the close-coupling *s–p* calculations have

[†] FLAVIN, R. K. and MEYERAND, R. G., *Atomic collision processes*, ed. McDOWELL, M. R. C., p. 59 (North Holland, Amsterdam, 1964).
[‡] CHEN, C. L. and RAETHER, M., *Phys. Rev.* **128** (1962) 2679.

also been carried out by Burke† and his results agree very well with those of Karule and Peterkop and of Karule.

Angular distributions of scattered electrons have been measured only for potassium.‡ Fig. 8.25 illustrates a comparison between calculated and observed distributions for 5-eV electrons, the curves being normalized so as to agree at 130°. The shapes of the two curves are not dissimilar, showing that at this energy the important calculated phase shifts are roughly of the right relative magnitude.

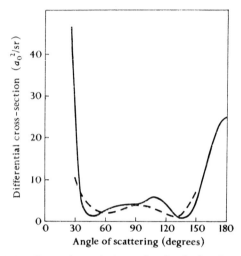

FIG. 8.25. Comparison of observed and calculated angular distribution for scattering of 5-eV electrons by potassium atoms. ——— calculated by $3s$–$3p$ close-coupling method. – – – – observed (McMillen) normalized to agree at 130°.

Many attempts have been made to calculate elastic cross-sections by some version of the exchange-adiabatic method, but the sensitivity of the results to the means adopted to cut-off the polarization term is so great that little or no success has been achieved.

As an example we show in Fig. 8.26 a comparison of results obtained for lithium, which is in principle the simplest case. Three sets of cross-section–electron energy curves are given, all calculated by an exchange-adiabatic method with only slightly different assumptions. Lawson and Massey§ followed closely the method of Temkin, using simple analytical expressions for the Hartree–Fock wave functions, their calculated polarizability being $135a_0^3$ as compared with the observed

† BURKE, P. G., unpublished.
‡ McMILLEN, J. H., *Phys. Rev.* **46** (1934) 983.   § Unpublished.

value $148 \pm 14 a_0^3$. Vinkalns, Karule, and Obed'kov† calculated the polarization potential by a less elaborate perturbation procedure, but used more accurate variational wave functions giving a calculated

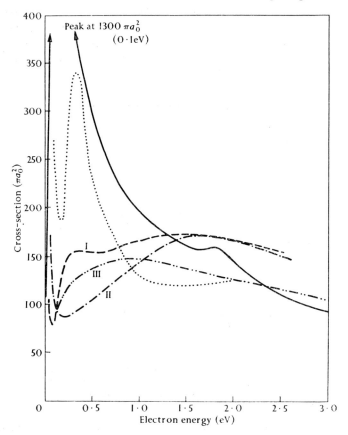

FIG. 8.26. Comparison of calculated elastic cross-sections for collisions of electrons with lithium. ——— $2s$–$2p$ close coupling. Exchange-adiabatic approximations: – – – – I. Vinkalns. —·—·— II. Massey and Lawson. —··—··— III. Stone. Polarized orbital approximation . . . . Stone.

polarizability of $165 a_0^3$. Stone‡ also used these wave functions, obtaining the same polarizability, but otherwise followed more nearly the Temkin method. It will be seen from Fig. 8.26 that there is a general resemblance between the results obtained in these three calculations. However, Stone§ has also carried out calculations by the full polarized-orbital method and it will be seen that, in contrast with the cases of H

---

† VINKALNS, I. ZH., KARULE, E. M., and OBED'KOV, V. D. *Optika Spektrosk.* **17** (1964) 197; *Optics Spectrosc.*, N.Y. **17** (1964) 105.
‡ STONE, P. M., *Phys. Rev.* **141** (1966) 137. § Unpublished.

and He$^+$ (Chap. 8, §§ 2.4 and 5.1 respectively) this gives very different results, an example of the high sensitivity. This is further reflected in the difference between all of these results and those of the 2s–2p close coupling.

A similar situation exists for sodium, as shown in Fig. 8.27. For this atom the calculations were carried out by El-Wakeil,† both for the

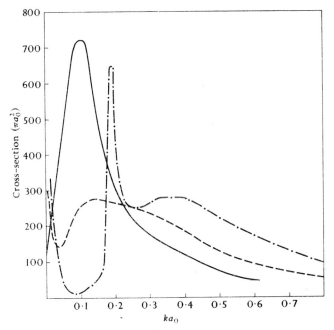

Fig. 8.27. Comparison of calculated elastic cross-sections for collisions of electrons with sodium. ——— 3s–3p close coupling. —·—·— polarized-orbital method. — — — exchange-adiabatic approximation.

exchange-adiabatic and polarized-orbital approximations. He used the Temkin method and Hartree–Fock wave functions, giving a calculated polarizability of $172a_0^3$ that is somewhat high (the observed value is $144 \pm 18a_0^3$). As for sodium, the two approximations give quite different results, which are also very different from the results of the 3s–3p close-coupling calculation. There is a remarkable resemblance in each case to the form of the result for lithium.

The sensitivity is no less for caesium.‡

† El-Wakeil, A. N., Thesis, London, 1966.
‡ Crown, J. C. and Russek, A., *Phys. Rev.* **138** (1965) A669.

8.2. Spin-exchange cross-sections

According to the $3s$–$3p$ close-coupling calculations of Karule the singlet (a^+) and triplet (a^-) zero-energy scattering lengths for Li, Na, K, and Cs are as given in Table 8.11. The results of Salmona and Seaton† for Na are also given.

TABLE 8.11

Singlet (a^+) and triplet (a^-) scattering lengths for alkali metal atoms, calculated by ns–np close-coupling method, and the corresponding exchange cross-section Q_{se}

	Li	Na	K	Cs
a^+ (in a_0)	3·7	4·2(12)	0·5	−4·0
a^- (in a_0)	5·7	−5·9(−9)	−15	−25
Q_{se} calc (πa_0^2)	4	102(441)	240	441
Q_{se} obs. (πa_0^2)		230‡ 220§	340‖	365††

Bracketed values for sodium are those of Salmona and Seaton.

From these the cross-section Q_{se} for electron spin exchange collisions in their zero energy limit may be obtained (see Chap. 8, § 2) using

$$Q_{se} = \pi(a^+ - a^-)^2.$$

Comparison may be made with the spin-exchange cross-sections for thermal electrons in Na and K from optical pumping experiments (see Chap. 5, § 7.2). From Table 8.11 it is seen that the agreement is moderately satisfactory, certainly much better than is obtained for other collision properties of the alkali metal atoms.

8.3. Excitation of ns–np (resonance) transitions

The cross-section for the excitation of the resonance p state for each alkali metal atom may be derived from the close-coupling calculations of Karule and Peterkop.‡‡ The values obtained for sodium are illustrated in Fig. 8.28 in comparison with observed values (see Chap. 4, § 1.4.4.2). The agreement is good but is only checked up to electron energies of about twice the threshold value.

The close-coupling calculations can be extended to higher energies but good results may be obtained by following the same procedure as used by Purcell§§ and by Seaton‖‖ for the $2s$–$2p$ transition in atomic hydrogen (see Chap. 8, § 4.4).

---

† loc. cit., p. 563.   ‡ DEHMELT, H. G., *Phys. Rev.* **109** (1958) 381.
§ BALLING, L. C., ibid. **151** (1966) 1.
‖ FRANKEN, P., SANDS, R., and HOBART, J., *Phys. Rev. Lett.* **1** (1958) 118.
†† BALLING, L. C., HANSON, R. J., and PIPKIN, F. M., *Phys. Rev.* **133** (1964) A607.
‡‡ loc. cit. p. 562.   §§ loc. cit., p. 532.   ‖‖ loc. cit., p. 532.

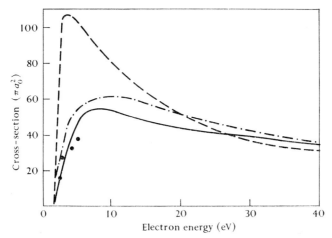

Fig. 8.28. Comparison of observed and calculated cross-sections for excitation of the $3p$ state of sodium by electron impact. —— observed. – – – – calculated by Born approximation. — · — · — calculated by Seaton. ● ● ● calculated by $3s$–$3p$ close-coupling method.

TABLE 8.12

Cross-sections (in πa_0^2) for excitation of the resonance levels of Li, Na, K, and Cs by electron impact

Electron energy (eV)	Li	Na	K	Cs
1·6				40
1·8			15	29
2·0	10		31	27
2·5		16		
3·0	53	27	50	53
4·0	44	33	61	68
5·0	44	38	66	73

Thus in Fig. 8.28 the results obtained in this way by Seaton are shown and are seen to agree well with the observed data.

In Table 8.12 cross-sections for excitation of the resonance levels of Li, Na, K, and Cs atoms calculated by Karule and Peterkop† using the close-coupling method are given. The only remarkable feature of these results is the maximum close to the threshold for caesium. Apart from Na there are few observed data to compare with, but for caesium Nolan and Phelps‡ (Chap. 5, § 3.3.3) have derived, from analysis of

† loc. cit., p. 562.
‡ NOLAN, J. F. and PHELPS, A. V., *Phys. Rev.* **140** (1965) A792.

electron drift velocities in caesium vapour, the slope of the cross-section, assumed to rise linearly from the threshold (1·386 and 1·454 eV respectively for excitation of the $6p_{\frac{1}{2}}$ and $6p_{\frac{3}{2}}$ states). Assuming a single excitation threshold at 1·386 eV they found a slope of

$$7·1 \times 10^{-15} \text{ cm}^2 (80\pi a_0^2)/\text{eV}.$$

This is compatible with the calculated results.

9. Collisions with atoms and ions with incomplete outer p shells

9.1. *Introductory*

We now consider a further type of scattering problem. So far we have considered cases in which the target atom either has completely closed shells or a single outer electron. In all these cases the ground configuration consists of a single term only, ^1S for the rare gas atoms and ^2S for hydrogen and the alkali metals. However, whereas for the rare gases and hydrogen the energy separation of the ground and first excited configuration is large, it is very small for the alkali metals. A close-coupling treatment involving both these configurations is therefore necessary for the latter atoms.

If the target atom or ion has an outer open p shell with between 2 and 4 electrons included, more than one term is associated with the ground configuration. Thus for neutral oxygen the ground configuration $(1s)^2(2s)^2(2p)^4$ gives rise to ^3P, ^1D, and ^1S terms of which the ^3P is the lowest, the ^1D and ^1S terms lying respectively 1·9 and 4·2 eV higher. These separations are so small that we must expect that, in discussing low-energy collisions of electrons with oxygen atoms, strong coupling between the ^3P, ^1D, and ^1S terms must be allowed for. Once this is done it can be assumed that coupling of any of these terms with those arising from excited configurations will be relatively small.

Considerable attention has been devoted to the theoretical study of collisions of slow electrons with neutral atoms and ions of this type with incomplete p shells, because they are of much importance in upper atmospheric physics, astrophysics, and plasma physics. Furthermore, it is always difficult to study these collisions experimentally as even the neutral atoms cannot be studied in the bulk gaseous form.

For the discussion of collisions that involve transitions between terms of the same configuration it is sometimes convenient to introduce the dimensionless quantities known as collision strengths as defined by Seaton.† We distinguish the different terms by numbers n (or n') and

† SEATON, M. J., *Phil. Trans. R. Soc.* A**245** (1953) 469.

write the cross-section $Q(n, n')$, for excitation of the n'th term from the nth, in the form
$$Q(n, n') = (\pi/g_n k_n^2)\Omega(n, n'), \quad (109)$$
where k_n is the wave number of the incident electron and g_n the statistical weight of the initial term. Since from the principle of detailed balancing
$$Q(n, n') = (k_{n'}^2/k_n^2)(g_{n'}/g_n)Q(n', n),$$
we have
$$\Omega(n, n') = \Omega(n', n). \quad (110)$$

We may analyse a collision strength into contributions from different states of initial and final angular momentum of the colliding electron. If l and l' are the respective angular-momentum quantum numbers of these states then
$$\Omega(n, n') = \sum_l \sum_{l'} \Omega^{ll'}(n, n'). \quad (111)$$
It follows from the conservation theorem (Chap. 7, § 5.5) that
$$\sum_l \Omega^{ll'}(n, n') \leqslant (2l+1)g_<, \quad (112)$$
where $g_<$ is the smaller of g_n, $g_{n'}$.

Still further analysis is possible if LS coupling is a good approximation for the total system of atom plus electron. For this system, in which the atom is in its initial state and the electron initial angular-momentum quantum number is l, there will exist a set of possible values L^T, S^T for the total orbital and spin angular momentum quantum numbers. We distinguish each of these pairs by a suffix and may then write
$$\Omega^{ll'}(n, n') = \sum_i \Omega_i^{ll'}(n, n').$$
The chance that L^T, S^T, which are conserved in a collision, will have the values L_i^T, S_i^T will be
$$(2L_i^T+1)(2S_i^T+1)/2(2l+1)g_n,$$
so that the conservation theorem requires
$$\Omega_i^{ll'}(n, n') \leqslant \tfrac{1}{2}(2L_i^T+1)(2S_i^T+1). \quad (113)$$
For transitions within a configuration there is no change of parity so that $l' \sim l$ must be even. Close coupling arises mainly when $l = l'$.

The formulation of the problem in terms of truncated eigenfunction expansions was first carried out by Seaton.† For fixed values L^T, S^T, M_L^T, M_S^T of the total orbital and spin quantum numbers for the combined system of atom and electron, and their z-components, the wave function for this system may be written
$$\Psi(L^T, S^T, M_L^T, M_S^T) = \sum_{L^c, S^c} \sum_p \sum_l \sum_{m_l, m_s} \epsilon_p r_p^{-1} G_l(L^T, S^T, L^C, S^C, r_p) \times$$
$$\times B_l(L^T, S^T, m_l, m_s)\chi_l(m_l, m_s|\vartheta_p, \varphi_p, \sigma_p) \times$$
$$\times \psi(L^C, S^C, M_L^C, M_S^C | \mathbf{r}_1, ..., \mathbf{r}_{p-1}, \mathbf{r}_{p+1}, ..., \mathbf{r}_t). \quad (114)$$

† loc. cit., p. 570.

Here $\chi_l(m_l, m_s|\vartheta_p, \varphi_p, \sigma_p)$ is an eigenfunction of the angular and spin coordinates of the pth electron corresponding to angular momentum quantum numbers l, s, m_l, m_s. $\psi(L^C, S^C, M_L^C, M_S^C|\mathbf{r}_1,...,\mathbf{r}_{p-1},\mathbf{r}_{p+1},...,\mathbf{r}_t)$ is a properly antisymmetrized eigenfunction for an atomic term with angular momentum quantum numbers L^C, S^C, M_L^C, M_S^C respectively where $M_L^C + m_l = M_L^T$, $M_S^C + m_s = M_S^T$. The coefficients B_l, which may be determined by vector coupling formulae, are such that summation over m_l and m_s gives an eigenfunction for an over-all state with angular momentum quantum numbers L^T, S^T, M_L^T, M_S^T. The allowed values of l over which the sum $\sum_l$ is taken are given by

$$L^T - L^C, \quad ..., \quad L^T + L^C.$$

Finally, the ϵ_p are chosen so that the sum over p is antisymmetric for interchange of every pair of electrons.

Confining the expansion to the three terms of the ground configuration which are distinguished by n and distinguishing different pairs of values of L^T, S^T by i, we may rewrite (114) in the form

$$\Psi^i(M_L^T, M_S^T) = \sum_{n=1}^{3} \sum_l \left\{ \sum_{p=1}^{t} \epsilon_p\, G_{nl}^i(r_p) r_p^{-1} \phi_{nl}^i(M_L^T, M_S^T|\mathbf{r}_1,...,\mathbf{r}_{p-1},\mathbf{r}_{p+1},...,\mathbf{r}_t) \right\} \quad (115)$$

where $\phi_{nl}^i = \sum_{m_l, m_s} B_{i,l}(m_l, m_s) \chi(m_l, m_s|\vartheta_p, \varphi_p, \sigma_p) \times$

$$\times \psi_n(M_L^T - m_l, M_S^T - m_s|\mathbf{r}_1,...,\mathbf{r}_{p-1},\mathbf{r}_{p+1},...,\mathbf{r}_t). \quad (116)$$

The cross-sections are independent of M_L^T, M_S^T so only one function Ψ^i need be considered.

Coupled integrodifferential equations for the functions $G_{nl}^i(r)$ may be obtained in the usual way from

$$\int \phi_{nl}^{i*}(H-E)\Psi^i\, d\mathbf{r}_1 ... d\mathbf{r}_{p-1}\, d\mathbf{r}_{p+1} ... d\mathbf{r}_t = 0. \quad (117)$$

As an example, for $L^T = 1$, $S^T = \tfrac{1}{2}$ for oxygen the possible values of l associated with each term are

$$^3P, 0, 1, 2; \quad ^1D\ 1, 2, 3; \quad ^1S\ 1.$$

At first sight this would suggest that seven coupled equations would arise, but no coupling occurs between the even and odd states so that there are two uncoupled sets, one of three equations $^3P, l = 0, 2$ and $^1D, l = 2$ and one of four equations $^3P, l = 1$, $^1D, l = 1, 3$ and $^1S, l = 1$.

The solution of sets of coupled equations of this kind is a formidable task even with modern computers that were not available at the time when Seaton first attacked the problem. Close coupling is confined to cases in which $l = 1$. For oxygen this applies for the $^2D(^3P + ^1D)$ and

$^2P(^3P+^1D+^1S)$ cases. Seaton therefore used the distorted-wave method to calculate the inelastic cross-sections for all other significant cases. Nevertheless he was still faced by one pair and one triad of coupled equations to deal with by more accurate methods. He proceeded to do this in the following way.

The energies of the terms are determined by certain averages over the electrostatic interactions e^2/r_{ij} between pairs of electrons. Expanding this interaction in zonal harmonics in the form

$$e^2/r_{ij} = e^2 \sum \gamma_s P_s(\cos\theta_{ij}),$$

where
$$\gamma_s = \begin{cases} r_i^s/r_j^{s+1} & (r_j > r_i), \\ r_j^s/r_i^{s+1} & (r_j < r_i), \end{cases}$$

it is found that only terms with $s = 0, 1, 2$ contribute and of these it is only the quadrupole term $s = 2$ that leads to an energy separation of the atomic terms. As these separations are relatively small Seaton neglected them consistently in a first approximation. With the quadrupole terms removed and the colliding electron wave numbers associated with each atomic term now the same, the equations may be uncoupled and solved as linear combinations of solutions of single integrodifferential equations. Thus, if we have two coupled equations

$$\left(\frac{d^2}{dr^2} + k^2 - U\right) G_0 = U_{01} G_1, \tag{118 a}$$

$$\left(\frac{d^2}{dr^2} + k^2 - U\right) G_1 = U_{01} G_0, \tag{118 b}$$

and require a solution such that

$$G_0 \sim \sin kr + \alpha e^{ikr}, \quad G_0(0) = 0,$$
$$G_1 \sim \beta e^{ikr}, \quad G_1(0) = 0,$$

we note that if

$$\mathcal{G}^+ = G_0 + G_1, \quad \mathcal{G}^- = G_0 - G_1,$$

$$\left\{\frac{d^2}{dr^2} + k^2 - (U \pm U_{01})\right\} \mathcal{G}^\pm = 0. \tag{119}$$

These two uncoupled equations may be solved to obtain solutions with asymptotic form

$$\mathcal{G}^\pm \sim A^\pm \sin(kr + \eta^\pm). \tag{120}$$

From these by suitable choice of $A^\pm$ we find

$$\alpha = (1/4i)(e^{2i\eta^+} + e^{2i\eta^-} - 2), \tag{121 a}$$

$$\beta = (1/4i)(e^{2i\eta^+} - e^{2i\eta^-}), \tag{121 b}$$

and
$$|\beta|^2 = \tfrac{1}{4}\sin^2(\eta^+ - \eta^-). \tag{122}$$

There is no difficulty in extending this to more than two coupled equations.

Having determined these so-called exact-resonance solutions they are then used to obtain a better approximation as follows. If the exact equations are

$$\left\{\frac{d^2}{dr^2}+k^2-(U+\delta U_0)\right\}G_0 = (U_{01}+\delta U_{01})G_1, \quad (123\text{ a})$$

$$\left\{\frac{d^2}{dr^2}+k^2-(U+\delta U_1)\right\}G_1 = (U_{01}+\delta U_{01})G_0, \quad (123\text{ b})$$

where δU_0, δU_1, and δU_{01} are all small, we have still, without approximation,

$$\left\{\frac{d^2}{dr^2}+k^2-(U\pm U_{01})\right\}\mathscr{G}^\pm$$
$$= \pm\delta U_{01}\mathscr{G}^\pm + \tfrac{1}{2}(\delta U_0\pm\delta U_1)\mathscr{G}^+ + \tfrac{1}{2}(\delta U_0\mp\delta U_1)\mathscr{G}^-. \quad (124)$$

If $\mathscr{G}_0^\pm$ are the solutions of the equation (119) satisfying (120) then a second approximation is obtained by solving (124) with $\mathscr{G}_0^\pm$ substituted for $\mathscr{G}^\pm$ on the right-hand side.

TABLE 8.13

Oxygen $(1s)^2(2s)^2(2p)^4$		Nitrogen $(1s)^2(2s)^2(2p)^3$		Carbon $(1s)^2(2s)^2(2p)^2$	
^3P		^4S		^3P	
^1D	1·96 eV	^2D	2·37 eV	^1D	1·26 eV
^1S	4·17 eV	^2P	3·56 eV	^1S	2·67 eV

Improved methods of obtaining better approximations based on use of a variational principle have been derived by Seaton and his collaborators, but we first describe the results of a full solution of the sets of coupled equations for collisions of electrons with neutral oxygen, nitrogen, and carbon.

9.2. Collisions with neutral oxygen, carbon, and nitrogen atoms

The collisions of electrons with neutral oxygen, carbon, and nitrogen atoms resulting either in elastic scattering or in transition between terms of the ground configuration have been investigated in thorough detail by Smith, Henry, and Burke,[†] who solved the relevant coupled equations directly by non-iterative methods using electronic computation.

Table 8.13 gives the designations of the terms involved for each atom, together with their energy separations.

Fig. 8.29 illustrates the calculated cross-sections for elastic scattering

[†] SMITH, K., HENRY, R. J. W., and BURKE, P. G., *Phys. Rev.* **157** (1967) 51.

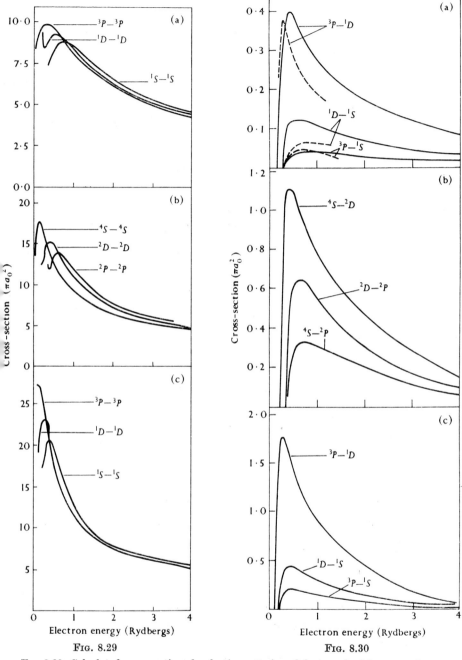

Fig. 8.29. Calculated cross-sections for elastic scattering of electrons by (a) oxygen atoms in the ^3P ground state and in the ^1D and ^1S excited states of the ground configuration; (b) nitrogen atoms in the ^4S ground state and in the ^2D and ^2P excited states of the ground configuration; (c) carbon atoms in the ^3P ground state and in the ^1D and ^1S excited states of the ground configuration.

Fig. 8.30. Calculated cross-sections for excitation of transitions within the ground configurations of (a) oxygen, (b) nitrogen, (c) carbon. In (a) the broken curves are the results of earlier calculations by Seaton.

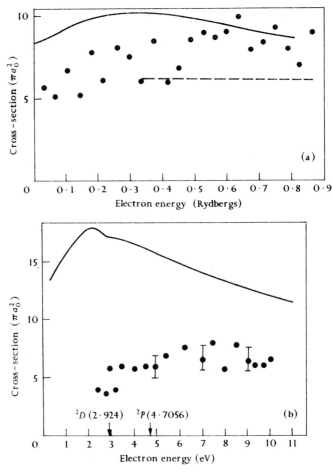

Fig. 8.31. Comparison of observed and calculated total cross-sections for collisions of electrons with (a) atomic oxygen in the ground ^3P state, —— calculated (Smith, Henry, and Burke), ●●● observed (Aubrey, Bedersen, and Sunshine), – – – – observed (Trujillo et al.). (b) atomic nitrogen in the ground ^4S state, —— calculated (Smith, Henry, and Burke), ●●● observed (Trujillo et al.).

of electrons by the respective atoms in their ground states and in each of the excited states associated with the ground configuration, while Fig. 8.30 gives the different inelastic cross-sections for transitions between terms of the ground configuration.

In Fig. 8.30 (a) comparison is made between the full solutions for the inelastic cross-section and those obtained earlier by Seaton using the method outlined in § 9.1. Except for the ^1D–^1S case the agreement is quite good.

Comparison with experiments for the total cross-section (elastic and inelastic) for scattering by oxygen and nitrogen in the ground ^3P and ^4S states respectively is shown in Fig. 8.31.

For oxygen there is good agreement with the observations of Sunshine et al. (see Chap. 1, § 6.1) at energies above 7 eV but at lower energies the calculated cross-section is larger. This may be due to neglect of polarization effects arising from interaction between the ground and excited configuration.

The situation is much less satisfactory for nitrogen, the calculated cross-section being well above the observed (Chap. 1, § 6.1) over the whole energy range.

9.3. Collisions with ionized atoms

We now give results obtained for excitation of transitions within the ground configurations of ionized atoms obtained by the methods developed by Seaton and his colleagues which have been described in § 9.1 above.

For convenience we shall distinguish the three terms arising from the ground configuration, in order of increasing energy, by the numbers 1, 2, 3. Thus $\Omega(1, 2)$ is the collision strength for a transition from the ground term to the one of next highest energy and so on.

In the first instance we consider collisions in which the incident electron has energy sufficient to excite the $1 \to 3$ transition. We denote the initial wave number of the electron by k_1, that after exciting the terms 2 and 3 by k_2, k_3 respectively. Since we are dealing with excitation of positive ions the collision strength $\Omega(1, 3)$ is finite at the threshold, $k_3^2 = 0$ (see Chap. 9, § 10.4)

In a given configuration the coupling between the terms decreases as the ionic charge increases. This is seen clearly from Table 8.14,† which compare collisions strengths Ω^{pp} for excitation of transitions between the $2p^2$ configuration of various ions by incident p electrons as calculated by the exact-resonance close-coupling method and by the distorted-wave method which assumes the coupling to be small. As pointed out in § 9.1 it is only the p–p transitions of the incident electron which lead to close coupling. Whereas for N+ the distorted-wave method gives seriously incorrect results it is much better for O++ and very good for Ne4+.

In Table 8.15† values of the collision strengths for ions with $2p^2$ configuration are given.

† SARAPH, H. E., SEATON, M. J., and SHEMMING, J., Proc. phys. Soc. 89 (1966) 27.

As the variation of the collision strengths with k_3^2 is slow we give results for other configurations† in Table 8.16 for $k_3^2 = 0$ only.

To illustrate the effects of a change of the total quantum number of the outer configuration we give in Table 8.17 the collision strengths for

TABLE 8.14

Collision strengths for excitation of terms of the ground $2p^2$ configuration of various ions for $k_3^2 = 0$, calculated with full allowance for close coupling (ER method) and by the distorted-wave method

Ion	Close-coupling method			Distorted-wave method		
	$\Omega^{pp}(1, 2)$	$\Omega^{pp}(1, 3)$	$\Omega^{pp}(2, 3)$	$\Omega^{pp}(1, 2)$	$\Omega^{pp}(1, 3)$	$\Omega^{pp}(2, 3)$
N^+	2·685	0·304	0·030	1·237	0·092	0·004
O^{++}	1·664	0·209	0·012	1·396	0·180	0·008
Ne^{4+}	0·690	0·088	0·003	0·649	0·084	0·003

TABLE 8.15

Collision strengths for excitation of transitions between the ground terms of ions with outer $2p^2$ configurations

Ion	$k_3^2 a_0^2$	$\Omega(1, 2)$	$\Omega(1, 3)$	$\Omega(2, 3)$
N^+	0·0	3·136	0·342	0·376
	0·2	3·203	0·391	0·424
	0·4	3·289	0·428	0·456
O^{++}	0·0	2·391	0·335	0·310
	0·2	2·398	0·345	0·319
	0·4	2·388	0·351	0·326
Ne^{4+}	0·0	1·376	0·218	0·185
	0·2	1·352	0·216	0·186
	0·4	1·328	0·212	0·187
Mg^{6+}	0·0	0·800	0·128	0·123
S^{10+}	0·0	0·353	0·055	0·065
Zn^{24+}	0·0	0·070	0·010	0·017
$\lim_{Z \to \infty} (Z^2\Omega)$		42·8	5·83	12·5

$3p^2$ ions calculated for $k_3^2 = 0·0_35$.† These are to be compared with the corresponding results for the $2p^2$ ions in Table 8.15. It will be seen that for ions with charge $+3e$ or less the collision strengths for the $3p^2$ configurations are higher than for $2p^2$ but for ions with greater charge the reverse is the case. The same result applies to the p^3 and p^4 configurations—collision strengths are available for the $3p^3$ cases,‡ S^+

† CZYZAK, S. J., KRUEGER, T. K., SARAPH, H. E. and SHEMMING, J., *Proc. phys. Soc.* **92** (1967) 1146.
‡ CZYZAK, S. J. and KRUEGER, T. K., ibid. **90** (1967) 623.

Cl^{++}, Ar^{3+}, K^{4+}, Ca^{5+}, V^{8+}, and Fe^{11+} and for the $3p^4$ cases† Cl^+, Ar^{++}, K^{3+}, Ca^{4+}, V^{7+}, Cr^{8+}, Mn^{9+}, Fe^{10+}, and Ni^{12+}.

When the incident electron energy falls below the threshold for excitation of the highest term so that $k_3^2 < 0$ but $k_2^2 > 0$, resonance effects

TABLE 8.16

Collision strengths for excitation of transitions between the ground terms of ions with outer $2p^3$ and $2p^4$ configuration, for $k_3^2 = 0$

Ion	$2p^3$			Ion	$2p^4$		
	$\Omega(1,2)$	$\Omega(1,3)$	$\Omega(2,3)$		$\Omega(1,2)$	$\Omega(1,3)$	$\Omega(2,3)$
O^+	1·43	0·428	1·70	F^+	1·34	0·147	0·193
F^{++}	1·25	0·461	1·67	Ne^{++}	1·27	0·164	0·188
Ne^{3+}	1·04	0·427	1·42	Na^{3+}	1·13	0·163	0·157
Na^{4+}	0·837	0·359	1·22	Mg^{4+}	0·973	0·146	0·129
Mg^{5+}	0·652	0·289	0·942	Al^{5+}	0·792	0·123	0·107
Ar^{11+}	0·208	0·089	0·340	Ar^{10+}	0·314	0·050	0·051
Kr^{29+}	0·035	0·140	0·069	Br^{27+}	0·054	0·008	0·012
$\lim_{Z\to\infty}(Z^2\Omega)$	31·6	11·7	67·2	$\lim_{Z\to\infty}(Z^2\Omega)$	42·8	5·83	12·5

TABLE 8.17

Collision strengths for excitation of transitions between the ground terms of ions with outer $3p^2$ configuration, for $k_3^2 = 0 \cdot 0_3 5$

Ion	$\Omega(1,2)$	$\Omega(1,3)$	$\Omega(2,3)$
P^+	6·312	1·124	1·110
S^{++}	4·966	1·068	0·961
Cl^{3+}	1·993	0·328	1·030
Ar^{4+}	1·192	0·141	0·945
K^{5+}	0·742	0·090	0·787
Ca^{6+}	0·564	0·068	0·667
V^{9+}	0·286	0·035	0·406
Cr^{10+}	0·233	0·029	0·345

will occur (see Chap. 9, § 1) but, for most applications, the collision strengths $\Omega(1,2)$ averaged over the resonances are sufficient.

A valuable check, and indeed a means of further improving the accuracy of the calculations, is to use the technique of extrapolation of quantum defects of spectral terms. By an extension of that method already outlined in Chapter 6, § 3.10 to a many-channel system it is possible to extrapolate the results obtained for the collision problem in which an electron excites terms of the ground configuration of an ion of charge ze, to obtain the quantum defects of terms of the ion of

† CZYZAK, S. J., KRUEGER, T. K., SARAPH, H. E. and SHEMMING, J., loc. cit., p. 578.

charge $(z-1)e$. These may be checked against values derived from spectroscopic observations. If they do not agree satisfactorily it is possible to readjust the collision strengths so that extrapolation gives better results and in this way improve the collision data. This technique will be referred to in more detail in Chapter 9, § 10.5.

It must be remembered that in all these calculations no direct allowance is made for interaction with other configurations, so that polarization effects are not included. Because of this the extrapolated quantum defects are all too small. The empirical method of correction of the collision strengths by modifying them so as to lead to improved extrapolated quantum defects does tend to allow for polarization.

9.4 Transitions between fine structure levels

Excitation of transitions between fine structure levels of the ground terms of ions such as N^+, O^{++}, N^{++}, etc., is an important cooling mechanism in H II interstellar regions and in planetary nebulae. Similar excitation in neutral atomic oxygen plays a role in determining electron temperatures in certain levels in the ionosphere.

As might be expected the cross-sections for excitation of transitions between these levels can be represented by formulae very similar to that for spin exchange collisions, involving differences between phase shifts for certain over-all states including the ion or atom in question and a free electron.† For excitation of ions these phase shift differences may be expressed in terms of suitable extrapolated quantum defects.

Thus consider the collision strength $\Omega(^2P_{\frac{1}{2}}, {}^2P_{\frac{3}{2}})$ for excitation of an ion with an outer p or p^5 shell. If σ_1, σ_3; π_1, π_3; δ_1, δ_3 are the extrapolated quantum defects for the (^2P) ns^1S, ^3S, ^1P, ^3P, ^1D, and ^3D series the partial strengths Ω^{ss} and Ω^{pp}, corresponding to incident and scattered s and p electrons respectively, are given by

$$\Omega^{ss} = 1\cdot 333 \sin^2\pi(\sigma_1-\sigma_3),$$

$$\Omega^{pp} = 1\cdot 667 \sin^2\pi(\delta_1-\delta_3) + 1\cdot 111 \sin^2\pi(\pi_3-\delta_3) +$$
$$+ 0\cdot 889 \sin^2\pi(\sigma_3-\pi_3) + 0\cdot 556 \sin^2\pi(\pi_1-\delta_3) +$$
$$+ 0\cdot 556 \sin^2\pi(\pi_3-\delta_1) + 0\cdot 444 \sin^2\pi(\sigma_1-\pi_3) +$$
$$+ 0\cdot 444 \sin^2\pi(\sigma_3-\pi_1) + 0\cdot 333 \sin^2\pi(\pi_1-\pi_3).$$

The coefficients are determined from the formulae for coupling of angular momenta.

† SEATON, M. J., Proc. R. Soc. A218 (1953) 400.

8.9 ANALYTICAL THEORY FOR SLOW COLLISIONS

Table 8.18 gives values of the collision strengths for fine structure transitions in O^+, O^{++}, and O^{+++} obtained in this way. A number of other cases of astrophysical interest have been worked out by the same method.

TABLE 8.18

Collision strengths for excitation of transitions between fine structure levels

Ion	Shell	Transition	Collision strength
†O^+	$2p^3$	$^2P_{\frac{1}{2}}-^2P_{\frac{3}{2}}$	0·33
		$^2D_{\frac{1}{2}}-^2D_{\frac{5}{2}}$	0·85
‡O^{++}	$2p^2$	$^3P_0-^3P_1$	0·31
		$^3P_0-^3P_2$	0·25
		$^3P_1-^3P_2$	0·96
§O^{+++}	$2p$	$^2P_{\frac{1}{2}}-^2P_{\frac{3}{2}}$	0·92

10. The theory of the polarization of impact radiation

In Chapter 4, §§ 1.3.2, 1.4.2.3 we described the experimental methods used and the results obtained in the measurement of the polarization of radiation emitted when an electron beam of nearly homogeneous energy passes through a gas at low pressure.

In terms of reference axes in which Oz is taken in the direction of the beam and Ox in that of observation the polarization of the emitted light in a given wavelength region is given by

$$P = 100\frac{I^{\|}-I^{\perp}}{I^{\|}+I^{\perp}},$$

where $I^{\|}$, $I^{\perp}$ are the intensities of the radiation emitted with electric vector along Oz, Oy respectively.

To discuss the calculation of P in terms of collision cross-sections and optical transition probabilities we examine first the excitation of atoms which in the ground state possess neither electron nor nuclear spin and no orbital angular momentum. A state excited by electron impact must then also be one without nuclear spin but it may possess electron spin and orbital angular momentum. We further restrict the discussion at this stage to cases in which the impact-excited state is without electron spin. If it possesses a total orbital angular momentum $\{L(L+1)\}^{\frac{1}{2}}\hbar$ then it will possess $2L+1$ degenerate substates corresponding to the

† SEATON, M. J. and OSTERBROCK, D. E., *Astrophys. J.* **125** (1957) 66.
‡ SEATON, M. J., *Proc. R. Soc.* A**218** (1953) 400.
§ OSTERBROCK, D. E., *Astrophys. J.* **142** (1965) 1423.

allowed values $-L, -L+1, ..., L$ of the magnetic quantum number M_L. We denote the impact-excited state by α and write Q_M^α for the cross-section for impact excitation of a particular substate.

Apart from any cascade effects due to radiative transitions from upper states, the population of a particular substate will be proportional to Q_M^α. To calculate the polarization of radiation emitted in transitions from the state α to a lower state β we introduce the radiative transition probabilities $A_M^{\alpha\beta}(y)$, $A_M^{\alpha\beta}(z)$ for emission of radiation from the M substate of α, in a transition to state β, with electric vector in the direction of Oy and Oz respectively (see Chap. 7. § 5.2.1). We then have

$$P = 100 \frac{\sum_M \{A_M^{\alpha\beta}(z) - A_M^{\alpha\beta}(y)\} Q_M^\alpha}{\sum_M \{A_M^{\alpha\beta}(z) + A_M^{\alpha\beta}(y)\} Q_M^\alpha}. \tag{125}$$

With symmetry about the beam direction Oz,

$$A_M^{\alpha\beta}(x) = A_M^{\alpha\beta}(y) = \tfrac{1}{2}\{A_M^{\alpha\beta} - A_M^{\alpha\beta}(z)\},$$

giving
$$P = 100 \frac{\sum_M \{3A_M^{\alpha\beta}(z) - A_M^{\alpha\beta}\} Q_M^\alpha}{\sum_M \{A_M^{\alpha\beta}(z) + A_M^{\alpha\beta}\} Q_M^\alpha}. \tag{126}$$

Furthermore, the total transition probability $A_M^{\alpha\beta}$ must be independent of M so that we may write

$$P = 100 \frac{3K_z - K}{K_z + K}, \tag{127}$$

where
$$K_z = (1/A^{\alpha\beta}) \sum_M A_M^{\alpha\beta}(z) Q_M^\alpha, \tag{128}$$
$$K = \sum Q_M^\alpha. \tag{129}$$

Thus, as for the radiation emitted in the $n^1D\text{-}n'^1P$ transition in helium

$$A_{\pm 2}^{\alpha\beta}(z) : A_{\pm 1}^{\alpha\beta}(z) : A_0^{\alpha\beta}(z) = 0 : 3 : 4,$$
$$A_0^{\alpha\beta}(z) : A_0^{\alpha\beta}(y) = 4 : 1,$$

and $Q_{+M}^\alpha = Q_{-M}^\alpha$, we have

$$P = 300 \frac{(Q_0 + Q_1 - 2Q_2)}{5Q_0 + 9Q_1 + 6Q_2}. \tag{130}$$

Similarly we have the formulae for other singlet transitions:

$^1P\text{-}^1S$: $100(Q_0 - Q_1)/(Q_0 + Q_1)$, (131 a)

$^1P\text{-}^1D$: $100(Q_0 - Q_1)/(7Q_0 + 13Q_1)$, (131 b)

$^1D\text{-}^1P$: as in (130), (131 c)

$^1D\text{-}^1F$: $300(Q_0 + Q_1 - 2Q_2)/(15Q_0 + 29Q_1 + 26Q_2)$. (131 d)

At the excitation threshold all cross-sections vanish except Q_0. This is because, under these conditions, the scattered electron then leaves with vanishing energy and hence vanishing orbital angular momentum.

As the z-component of angular momentum of the incident electron is zero there can be no change in this component for the electron due to the collision. But the total orbital angular momentum of electron plus atom is conserved in the collision. Hence since $M = 0$ initially for the atom it must remain so.

Referring to (131a–d) we see that at the excitation threshold the percentage polarizations of ^1P–^1S, ^1P–^1D, ^1D–^1P, and ^1D–^1F radiations is 100, 14·3, 60, and 6·7 per cent respectively, quite independent of any knowledge of the cross-sections.

Considerable difficulty has been experienced in checking experimentally these conclusions about the threshold polarization (see Chap. 4, § 1.4.2.3). For a long time the observations suggested that the polarization vanishes at the threshold. However, some supporting evidence is now available for the 4^1D–2^1P and 5^1D–2^1P lines in helium. Fig. 4.30 shows the observed results of McFarland[†] and of Heddle and Keesing,[‡] which tend to the theoretical value 60 per cent at the threshold.

10.1. *Inclusion of electron spin*

Much more definite evidence has been obtained for lines arising from impact excitation of alkali metal atoms but these involve states α and β possessing both nuclear and electron spin. We must therefore show how the theory may be extended to such cases.[§]

Again proceeding in stages we first consider cases in which the nuclear spin remains zero, as, for example, with the lines of the triplet series in ^{4}He. We suppose that the fine structure levels of the upper state may be well resolved so that the separation is large compared with the natural line width. A particular quantum sub-level of the upper states may be designated, within the accuracy of LS coupling, by the angular momentum quantum numbers $SLJM_J$, where J is that for the total (spin+orbital) angular momentum and M_J for its z-component. The polarization of the radiation emitted in the transition $SLJ \rightarrow SL'J'$ is now given by (127) with

$$K_z = (1/A^{\alpha\beta}) \sum_{M_J} A^{\alpha\beta}_{M_J} Q^{\alpha}_{M_J}, \qquad (132\,\text{a})$$

$$K = \sum_{M_J} Q^{\alpha}_{M_J}. \qquad (132\,\text{b})$$

It is possible to transform these expressions into sums over the cross-

[†] McFarland, R. H., *Phys. Rev.* **136** (1964) A1240.
[‡] Heddle, D. W. O. and Keesing, R. G. W., *Proc. R. Soc.* A**299** (1967) 212.
[§] Oppenheimer, J. R., *Z. Phys.* **43** (1927) 27; *Proc. natn. Acad. Sci. U.S.A.* **13** (1927) 800; *Phys. Rev.* **32** (1929) 361; Penney, W. G., *Proc. natn. Acad. Sci. U.S.A.* **18** (1932) 231; Percival, I. C. and Seaton, M. J., *Phil. Trans. R. Soc.* A**251** (1958) 113.

sections $Q^\alpha_{M_L}$ by using vector-coupling angular-momentum transformation theory. The details have been worked out by Percival and Seaton who find

$$K_z = \frac{A(SLJ \to SL'J')}{(2S+1)A(SLJ)} \sum_{M_S M_L M_J} [C^{SLJ}_{M_S M_L M_J} C^{J'1J}_{M'_J 0 M_J}]^2 Q_{|M_L|}, \quad (133\,\mathrm{a})$$

$$K = \frac{A(SLJ \to SL'J')}{(2S+1)A(SLJ)} \frac{2J+1}{2L+1} \sum_{M_L} Q_{|M_L|}, \quad (133\,\mathrm{b})$$

where $C^{SLJ}_{M_S M_L M_J}$ and $C^{J'1J}_{M'_J 0 M_J}$ are Clebsch–Gordan coefficients.†

$A(SLJ \to SL'J')$ is the radiative transition probability from the upper state to the lower state with $J = J'$, while $A(SLJ)$ is the sum of these probabilities over the final value of J. We have

$$A(SLJ \to SL'J') = (2L+1)(2J'+1)W^2(LJL'J';S1)A(SL \to SL'), \quad (134)$$

where W is a Racah coefficient.†

Table 8.19‡ gives formulae for the polarizations for a number of transitions.

An important difficulty in current theory was pointed out and resolved by Percival and Seaton.‡ Consider a $^2P \to {}^2S$ transition. According to the values given in Table 8.19 the polarization of the multiplet is given by

$$P = 300 \frac{(Q_0 - Q_1)}{7Q_0 + 11Q_1}, \quad (135)$$

and this is independent of the doublet separation. In the limit, however, in which this separation tends to zero we have essentially the same situation as for the $^1P \to {}^1S$ transition, and we would expect the polarization to be given by

$$P = 100 \frac{(Q_0 - Q_1)}{Q_0 + Q_1}. \quad (136)$$

Why does this differ from (135)? Percival and Seaton showed that in fact (135) and (136) are two limiting cases depending on the ratio

$$\epsilon = 2\pi \delta\nu / A \quad (137)$$

of the fine structure separation $h\,\delta\nu$ to the line width $A\hbar$ where A is the total radiation transition probability from the state α. If $K_z^{(0)}$, $K^{(0)}$ are the values of K_z, K calculated on the assumption of zero electron spin and $K_z^{(\infty)}$, $K^{(\infty)}$ the corresponding values when the spin is included

† RACAH, G., *Phys. Rev.* **62** (1942) 438.
‡ PERCIVAL, I. C. and SEATON, M. J., *Phil. Trans. R. Soc.* A**251** (1958) 113.

TABLE 8.19
Polarization formulae for transitions involving fine structure components (no nuclear spin)

Transitions from ^3P states
$P = 100G(Q_0-Q_1)/(h_0Q_0+h_1Q_1)$

Transition	G	h_0	h_1	Threshold value
$^3P_0 \to {}^3D_1$	0			0
$^3P_1 \to {}^3D_1, {}^3S_1$	1	3	5	33·3
$\to {}^3D_2$	-1	13	27	$-7\cdot7$
$^3P_2 \to {}^3D_1, {}^3S_1$	21	47	73	44·7
$\to {}^3D_2$	-7	11	29	$-63\cdot6$
$\to {}^3D_3$	1	7	13	14·3
$^3P \to {}^3D$	3	73	143	4·1
$\to {}^3S$	15	41	67	36·6

Transitions from ^3D states
$P = 100G(Q_0+Q_1-2Q_2)/(h_0Q_0+h_1Q_1+h_2Q_2)$

Transition	G	h_0	h_1	h_2	Threshold value
$^3D_1 \to {}^3P_0$	3	5	9	6	60
$\to {}^3P_1$	-3	7	15	18	$-42\cdot9$
$\to {}^3P_2, {}^3F_2$	3	41	81	78	7·3
$^3D_2 \to {}^3P_1$	3	9	17	14	33·3
$\to {}^3P_2, {}^3F_2$	-3	7	15	18	$-42\cdot9$
$\to {}^3F_3$	3	29	57	54	10·3
$^3D_3 \to {}^3P_2, {}^3F_2$	18	41	76	58	43·9
$\to {}^3F_3$	-9	11	25	34	$-81\cdot8$
$\to {}^3F_4$	3	15	29	26	20
$^3D \to {}^3P$	213	671	1271	1058	31·7
$\to {}^3F$	213	2171	4271	4058	8·1

$^2P \to {}^2S$ transitions
$P = 100G(Q_0-Q_1)/(h_0Q_0+h_1Q_1)$

Transition	G	h_0	h_1	Threshold value
$^2P_{\frac{1}{2}} \to {}^2S$	0			0
$^2P_{\frac{3}{2}} \to {}^2S$	3	5	7	60
$^2P \to {}^2S$	3	7	11	42·9

completely as in (135) then the correct polarization is given by

$$P_\epsilon = 100 \frac{3K_z^{(\epsilon)}-K^{(\epsilon)}}{K_z^{(\epsilon)}+K^{(\epsilon)}}, \tag{138}$$

where $\quad K_z^{(\epsilon)} = \dfrac{K_z^{(0)}+\epsilon^2 K_2^{(\infty)}}{1+\epsilon^2}, \quad K^{(\epsilon)} = \dfrac{K^{(0)}+\epsilon^2 K^{(\infty)}}{1+\epsilon^2}. \tag{139}$

In the limit $\epsilon \to 0$ (138) tends to (127) and, in the limit $\epsilon \to \infty$, to (132).

10.2. *Effects of nuclear spin*

Very similar considerations apply to nuclear spin and hyperfine structure. For example,† for the ^2P–^2S transitions in atomic hydrogen,

$$A = 6\cdot 25 \times 10^{-8} \text{ s}^{-1},$$

so
$$A/2\pi c = 3\cdot 3 \times 10^{-3} \text{ cm}^{-1}.$$

The fine structure separation $\delta \nu/c$ is $0\cdot 36$ cm^{-1} so that the factor ϵ for the fine structure is of order 100 and (132) applies if no nuclear spin effects are taken into account. This gives

$$P(^2P_{\frac{3}{2}}-^2S) = 100(Q_0-Q_1)/(1\cdot 667Q_0 + 2\cdot 333Q_1). \tag{140}$$

On the other hand, the hyperfine structure separation for the ^2P levels is $0\cdot 79 \times 10^{-3}$ cm^{-1}, which gives $\epsilon \simeq 0\cdot 3$. The polarization must therefore be calculated from (138), where $K_z^{(0)}$, $K^{(0)}$ are now the values that yield (132) while $K_z^{(\infty)}$, $K^{(\infty)}$ refer to the formula that would be obtained with maximum influence ($\epsilon \to \infty$) of the nuclear spin for which the quantum number is $\frac{1}{2}$. This latter formula is

$$P(^2P_{\frac{3}{2}}-^2S) = 100(Q_0-Q_1)/(2\cdot 467Q_0 + 3\cdot 933Q_1), \tag{141}$$

whereas the correct formula taking $\epsilon = 0\cdot 3$ should be

$$P(^2P_{\frac{3}{2}}-^2S) = 100(Q_0-Q_1)/(1\cdot 694Q_0 + 2\cdot 388Q_1). \tag{142}$$

This is not far from (140) because ϵ is small.

For the total radiation ^2P–^2S the appropriate values are

$100(Q_0-Q_1)/(2\cdot 333Q_0 + 3\cdot 667Q_1)$, ignoring nuclear spin;

$100(Q_0-Q_1)/(3\cdot 533Q_0 + 6\cdot 067Q_1)$, with maximum influence of nuclear spin;

$100(Q_0-Q_1)/(2\cdot 375Q_0 + 3\cdot 749Q_1)$, with correct allowance for nuclear spin.

A remarkable confirmation of the theory of Percival and Seaton has been provided by the beautiful experiments of Hafner, Kleinpoppen, and Krüger‡ described in Chapter 4, § 1.3.4. They measured the polarization of the resonance (^2P–^2S) lines of ^{6}Li, ^{7}Li, and ^{23}Na excited by electron impact. According to the theory§ the expected threshold polarizations are as given in Table 8.20. It will be seen that the three cases cover an interesting range of values of the discriminating parameter ϵ which is determined by the ratio of hyperfine structure separation to line width. For ^{6}Li ϵ is small, for ^{23}Na larger, while for ^{7}Li it is

---

† PERCIVAL, I. C. and SEATON, M. J., loc. cit., p. 584.
‡ HAFNER, H., KLEINPOPPEN, H., and KRÜGER, H., *Phys. Lett.* **18** (1965) 270.
§ FLOWER, D. R. and SEATON, M. J., *Proc. phys. Soc.* **91** (1967) 59.

intermediate. The observed threshold polarizations in all these cases agree well with that calculated for the measured value of ϵ.

TABLE 8.20

Threshold polarizations of resonance radiation excited in lithium and sodium by electron impact

Atom	Nuclear spin	Calculated			Observed
		(a) $\epsilon \ll 1$	(b) $\epsilon \gg 1$	(c) from measured ϵ	
^{6}Li	1	42·9	14·0	37·5	39·7±3·8
^{7}Li	$\frac{3}{2}$	42·9	12·9	21·6	20·6±3·0
^{23}Na	$\frac{3}{2}$	42·9	12·9	14·1	14·8±1·8

11. Classical theory and electron–atom collisions

We have discussed in Chapter 6, § 3.7 how a semi-classical approximation may be used for the evaluation of the phase shifts that determine the cross-section for scattering of particles by a centre of force. In Chapter 16 the conditions under which classical dynamics may be used to describe such scattering are analysed in detail. The problem of determining when classical methods are likely to give useful results when dealing with three or more body problems is more difficult. Thus, consider a collision between a particle A and an atom B. In order that a classical description should be satisfactory it is not only necessary that the relative motion of A and B should follow classical dynamics but so also should the internal motion within B. There will, of course, be circumstances in which it is a good approximation to treat the relative motion classically but not the internal motion. Examples of such semi-classical methods are discussed in Chapters 17, 18, and 19. Our immediate interest here is in the application of fully classical methods to discuss electron collisions with atoms. According to the correspondence principle a classical description in such cases should be valid when the initial and final quantum numbers of the atomic state are large and large changes occur between them during the collision. In particular, ionization of a highly excited atom by electron impact should be well represented except for the contribution from distant encounters in which the energy transfer is relatively small.

Until the Monte Carlo calculations carried out by Abrines, Percival, and Valentine† all attempts to apply classical methods to electron–atom

† ABRINES, R., PERCIVAL, I. C., and VALENTINE, N. A., *Proc. phys. Soc.* **89** (1966) 515.

collisions made the further approximation of treating the encounter essentially as a binary one between two electrons. Thus in 1912 J. J. Thomson† treated ionization in this way, assuming the atomic electron to be initially at rest. In an encounter in which the impact parameter of the relative motion of the incident and atomic electron is p' the energy ϵ transferred to the atomic electron is given, according to classical orbit theory, by

$$p'^2 = \frac{e^4}{E_1}\left(\frac{1}{\epsilon} - \frac{1}{E_1}\right), \qquad (143)$$

E_1 being the initial kinetic energy of the incident electron.

The cross-section $dQ(\epsilon)$ for an energy transfer between ϵ and $\epsilon+d\epsilon$ is therefore given by

$$\frac{dQ(\epsilon)}{d\epsilon} = \frac{\pi e^4}{E_1 \epsilon^2}. \qquad (144)$$

For the total ionization cross-section Q_i we require $\epsilon \geqslant E_i$, the ionization energy, in (143), or we integrate (144) from E_i to E_1 to give

$$Q_i = \frac{\pi e^4}{E_1}\left(\frac{1}{E_i} - \frac{1}{E_1}\right). \qquad (145)$$

Several refinements have been introduced in this analysis. Thus Thomas‡ and Williams§ allowed for initial motion of the atomic electron by supposing it to possess a spherically symmetrical distribution of velocity with energy E_2. It is then found that (144) is replaced by

$$\frac{dQ(\epsilon)}{d\epsilon} = \frac{\pi e^4}{E_1}\left(\frac{1}{\epsilon^2} + \frac{4}{3}\frac{E_2}{\epsilon^3}\right). \qquad (146)$$

Further explicit account was taken|| of the atomic nucleus in accelerating the incident electron so that its kinetic energy, when it makes a close encounter with the atomic electron, is enhanced approximately by an amount equal to the initial potential energy of the struck electron. This means that, in (146), E_1 is replaced by $E_1+E_2+E_i$. However, it was further pointed out by Webster, Hansen, and Duveneck†† that allowance should also be made for the focusing effect of the nuclear field on the incident electron beam. They found that this introduced a further factor $(E_1+E_2+E_i)/E_i$, so restoring the original formula (146).

A different type of improvement was first introduced by Burgess‡‡

---

† THOMSON, J. J., *Phil. Mag.* **23** (1912) 449.
‡ THOMAS, L. H., *Proc. Camb. phil. Soc. math. phys. Sci.* **23** (1927) 714.
§ WILLIAMS, E. J., *Nature, Lond.* **119** (1927) 489.
|| THOMAS, L. H., *Proc. Camb. phil. Soc. math. phys. Sci.* **23** (1927) 829.
†† WEBSTER, D. L., HANSEN, W. W., and DUVENECK, F. B., *Phys. Rev.* **43** (1933) 839.
‡‡ BURGESS, A., *Atomic collision processes*, ed. MCDOWELL, M. R. C., p. 237 (North Holland, Amsterdam, 1964).

and developed in further detail by Vriens.† Burgess realized that, as long as the theory was based essentially on binary encounters, there is no reason why the dynamics of the encounter should not allow for the Pauli principle. In other words, the differential cross-section for the two-electron encounter could be taken as that given by Mott‡ in which the Pauli principle is allowed for, rather than the appropriate version of classical Rutherford scattering. It is then found that

$$\frac{\mathrm{d}Q(\epsilon)}{\mathrm{d}\epsilon} = \left(\frac{\mathrm{d}Q(\epsilon)}{\mathrm{d}\epsilon}\right)_\mathrm{d} + \left(\frac{\mathrm{d}Q(\epsilon)}{\mathrm{d}\epsilon}\right)_\mathrm{e} + \left(\frac{\mathrm{d}Q(\epsilon)}{\mathrm{d}\epsilon}\right)_\mathrm{i},$$

where the first term arises from a direct collision in which the atomic electron gains energy ϵ, the second from an exchange collision in which the atomic and incident electrons change places, and the third from interference effects. We then find for $E_1 > \epsilon$, in comparison with (146),

$$\left(\frac{\mathrm{d}Q(\epsilon)}{\mathrm{d}\epsilon}\right)_\mathrm{d} = \frac{\pi e^4}{E_1}\left(\frac{1}{\epsilon^2} + \frac{4}{3}\frac{E_2}{\epsilon^3}\right),$$

$$\left(\frac{\mathrm{d}Q(\epsilon)}{\mathrm{d}\epsilon}\right)_\mathrm{e} = \frac{\pi e^4}{E_1}\left\{\frac{1}{(E_1+E_\mathrm{i}-\epsilon)^2} + \frac{4}{3}\frac{E_2}{(E_1+E_\mathrm{i}-\epsilon)^3}\right\},$$

$$\left(\frac{\mathrm{d}Q(\epsilon)}{\mathrm{d}\epsilon}\right)_\mathrm{i} = -\frac{\pi e^4}{E_1}\frac{1}{E_1+E_\mathrm{i}}\left\{\frac{1}{\epsilon} + \frac{1}{E_1+E_\mathrm{i}-\epsilon}\right\}. \qquad (147)$$

It is here assumed, in calculating the interference term, that the electron spins are oriented at random.

To calculate the ionization cross-section we must now be careful to avoid counting events twice so that integration over ϵ must be carried out between limits E_i, $\tfrac{1}{2}(E_1+E_\mathrm{i})$. This gives

$$Q_\mathrm{i} = \frac{\pi e^4}{E_1}\left\{\left(\frac{1}{E_\mathrm{i}} - \frac{1}{E_1}\right) + \tfrac{2}{3}E_2\left(\frac{1}{E_\mathrm{i}^2} - \frac{1}{E_1^2}\right) - \frac{\ln(E_1/E_\mathrm{i})}{E_1+E_\mathrm{i}}\right\}. \qquad (148)$$

So far no attention has been paid to the choice of the initial energy E_2 of the atomic electron. A simple approximation to take for this is $\tfrac{1}{2}m\overline{v^2}$, where $\overline{v^2}$ is the mean square velocity of the electron in its initial state. We shall discuss a more elaborate approximation after consideration of the Monte Carlo method.

The application of the Monte Carlo method§ to the calculation of classical cross-sections for ionization of hydrogen atoms by electrons has been carried out by Abrines, Percival, and Valentine.∥ In classical

† VRIENS, L., *Phys. Rev.* **141** (1966) 88; *Proc. phys. Soc.* **89** (1966) 13.
‡ MOTT, N. F., *Proc. R. Soc.* **A126** (1930) 259.
§ WALL, F. T., HILLER, L. A., and MAZUR, J., *J. chem. Phys.* **35** (1961) 1284.
∥ ABRINES, R., PERCIVAL, I. C., and VALENTINE, N. A., loc. cit., p. 587.

theory an electron incident with initial energy E_1 may transfer an amount ϵ of energy to the bound electron. Three possibilities arise:

(a) if $\epsilon < E_i$ we have direct scattering,

(b) if $E_i < \epsilon < E_1$ we have ionization,

(c) if $E_1 < \epsilon$ we have exchange scattering.

The problem is essentially a statistical one because of the distribution in phase space of the initial state of fixed total energy. This distribution per unit of phase space for a single electron revolving with total energy E_i about a proton is given by

$$\rho(\mathbf{r}, \mathbf{p}) = N\delta\left(E_i - \frac{e^2}{r} - \frac{p^2}{2m}\right).$$

Integrating this over the spherically symmetrical distribution in ordinary space gives the momentum distribution function

$$\rho(p) = (8p_0^5/\pi^2)(p^2+p_0^2)^{-4}, \tag{149}$$

where
$$p_0^2 = 2|E_i|/m.$$

It is of particular interest that the momentum distribution for any state of a hydrogen atom with total quantum number n in which the states of different angular momentum are uniformly filled, is of the same form.† This tends to encourage the belief that, even for small n, the classical treatment may not be too unsatisfactory.

The Monte Carlo method consists in carrying out a statistically significant sample of classical trajectory calculations chosen consistently to satisfy the distribution (149) and the conditions defining the incident beam. If $Q_i^{cl}(E_A, E_1)$ is the classical ionization cross-section when the energy of the atom is E_A and that of the incident electron is E_1, then simple dimensional arguments show that

$$Q_i^{cl}(\alpha E_A, \alpha E_1) = \alpha^{-2} Q_i^{cl}(E_A, E_1).$$

If, therefore, $Q_i^{cl}(E_A, E_1)$ is calculated as a function of E_1 for fixed E_A, the cross-sections for other values of E_A may be derived. This simple scaling law does not apply in quantum mechanics because of the occurrence of the dimensional constant h.

Fig. 8.32 shows the results obtained by Abrines, Percival, and Valentine‡ for the ionization of the ground state of hydrogen. Comparison is made with observed cross-sections (see Chap. 3, § 2.5.1) and with those calculated by Born's approximation (see Chap. 7, § 5.6.2). At low energies the classical result is considerably closer to the experimental

† Fock, V., *Z. Phys.* **98** (1935) 145.   ‡ loc. cit., p. 587.

results but at higher energies it falls off too rapidly, as E^{-1} instead of as $E^{-1}\ln E$ given by Born's approximation.

In Fig. 8.33 the Monte Carlo cross-sections are compared with those calculated by different binary approximations. It will be seen that the formula (146) taking E_2 as given by $\tfrac{1}{2}\overline{mv^2}$, gives much too large a cross-section at low energies. If in the outside factor E_1 is replaced by

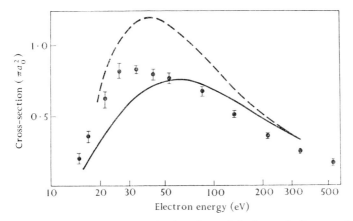

FIG. 8.32. Cross-sections for ionization of atomic hydrogen in the ground state. $\bar{\phi}$ calculated classically by the Monte Carlo method. —— observed. – – – calculated by Born's approximation.

$E_1+E_2+E_i$, that is to say, the focusing effect of the nuclear field is ignored, much closer agreement with the Monte Carlo cross-section is obtained. The effect of symmetry, involving the use of (148) does not make a great deal of difference. If the initial kinetic energy of the atomic electron is taken to be distributed according to (149) instead of taken at the constant value $\tfrac{1}{2}\overline{mv^2}$ the binary cross-section is a little reduced but the effect is not very great.

Some further insight into the usefulness of the binary approximations is obtained by comparison of the differential cross-sections $dQ/d\epsilon$ calculated by these approximations with those obtained by the Monte Carlo method. Such a comparison is shown in Fig. 8.34 for incident electrons of 22 and 54 eV respectively. It will be seen that the binary approximations fail when the energy transfer ϵ becomes small. Apart from difficulties of identification of the range of classical energy loss corresponding to excitation of a discrete quantum state, it appears that a classical binary approximation will not be satisfactory for the evaluation of excitation cross-sections, for which the energy transfer, in the sense we have been considering it, is always small.

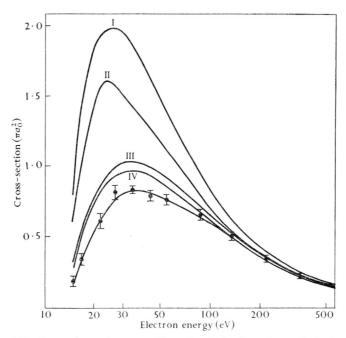

FIG. 8.33. Comparison of cross-sections for ionization of atomic hydrogen calculated classically by the Monte Carlo method with those calculated by different binary approximations. ⊕ calculated classically by the Monte Carlo method: —— I. binary approximation (146) with struck electron possessing the kinetic energy $\frac{1}{2}m\overline{v^2}$, —— II. binary approximation (148) allowing for the Pauli principle, with the struck electron possessing the kinetic energy $\frac{1}{2}m\overline{v^2}$, —— III. binary approximation (148) allowing for the Pauli principle and ignoring the focusing effect of the nuclear field, the struck electron possessing the kinetic energy $\frac{1}{2}m\overline{v^2}$, —— IV. As for III but assuming the struck electron to possess the initial momentum distribution (149).

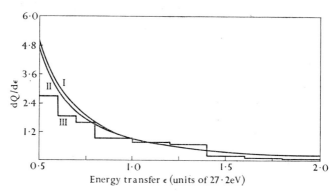

FIG. 8.34. Differential cross-section $dQ/d\epsilon$ for ionization of atomic hydrogen by electrons (a) of 22-eV and (b) of 54-eV incident energy. I. binary approximation (147) allowing for the Pauli principle and ignoring the focusing effect of the nuclear field, the struck electron possessing the kinetic energy $\frac{1}{2}m\overline{v^2}$. II. as for I but assuming the electron to possess the initial momentum distribution (149). III. Monte Carlo calculation.

As discussed in Volume IV the classical Monte Carlo calculations for ionization of hydrogen atoms by protons† give better agreement with observation than is the case for electrons shown in Fig. 8.32. This may be because of the impossibility in the Monte Carlo method of allowing for the Pauli principle.

As the validity of the classical method will improve as the quantum number n of the initial atomic state increases it seems likely from the already moderately good results obtained for ionization of the ground state of hydrogen that, for $n > 3$, the Monte Carlo cross-section should be quite good, at least up to energies below $4E_i^{(n)}$, where $E_i^{(n)}$ is the ionization energy for a particular value of n. It must be remembered, however, that it is assumed that all angular momentum states associated with a particular n are equally populated.

The binary approximation in the semi-empirical form in which E_1 in (148) is replaced by $E_1+E_2+E_i$ is likely to give good results but, if focusing by the nuclear field is included, so that (148) is regained, the approximation is much less satisfactory.

Application of classical theory to excitation is difficult because of the ambiguity associated with the relation between a range of classical final energy states and a discrete quantum state. In any case a binary approximation, unless modified empirically, will not give good results for excitation.

† ABRINES, R. and PERCIVAL, I. C., *Proc. phys. Soc.* 88 (1966) 861.

9

RESONANCE PHENOMENA—THRESHOLD BEHAVIOUR

So far in the theoretical discussion of Chapters 6, 7, and 8 we have been concerned with the broad features of the energy variation of the cross-sections for electron impact with atoms. While it is true that such features as the Ramsauer–Townsend effect have been considered and a theoretical description provided, the peaks present in the elastic cross-section as a function of electron energy in these cases are of the order of some eV wide. The much sharper peaks observed with high resolution equipment (see Chaps. 1, § 6.2 and 5, § 5.3.3) cannot be understood theoretically in the same way and we now consider how they arise. We shall introduce these resonance effects in more than one way as it is an advantage from the point of view of physical interpretation and for the analysis of experimental data to have a many-sided approach to the problem.

We shall begin by discussing the method developed particularly by Fano† that is aimed especially at a description of auto-ionization. After showing how the experimental results may be qualitatively accounted for, the relation of this method to the truncated eigenfunction method of Chapter 8, § 2.2.3 is next discussed and detailed application made to interpretation of experimental data. We then proceed to introduce the basic scattering matrices and extend the effective range formulae derived in Chapter 6, § 3.5 for the one-body case to a many-body system. One of the advantages of the effective range expansion is that it may be used by a process of analytic continuation to relate the energies of bound states to the scattering parameters. In its generalized form the corresponding continuation makes it possible to derive the form of the variation with energy of cross-sections near the threshold for onset of a new process as, for example, the variation of the elastic cross-section near the onset for excitation of some state. We shall conclude by discussing this threshold behaviour and its relation to resonance effects.

† FANO, U., *Nuovo Cim.* **12** (1935) 156; *Phys. Rev.* **124** (1961) 1866; see also earlier work of RICE, O. K., *J. chem. Phys.* **1** (1933) 375.

1. Resonance phenomena and auto-ionization

We consider a two-electron system such as a helium atom or negative hydrogen ion. It is often convenient to order the energy states in terms of a zero order approximation in which we imagine the interelectronic interaction V_{12} to be switched off, so that the states are specified in terms of electron assignments to one-particle hydrogen-like states or orbitals just as explained in Chapter 6, § 2.2. These assignments define a set of configurations. Any actual state of the system will be a superposition of such configurations although in many cases one, or at most a few, configurations will be dominant.

For a two-electron system there will be a set of configurations

$$(1s)^2, \quad (1s)(2s), \quad (1s)(3s), \quad ..., \quad (1s)(E_c s)$$
$$(1s)(2p), \quad (1s)(3p), \quad ..., \quad (1s)(E_c p) \qquad \text{I}$$

in which one electron is always in the ground $1s$ orbital. $(E_c s), (E_c p),...$ denote continuum orbitals of positive energy E_c and orbital angular momentum quantum numbers $0, 1,...$.

In addition to this set there will be others in which both electrons are excited. We will have, based on $2s$,

$$2s^2, \quad (2s)(3s), \quad ..., \quad (2s)(E_c s)$$
$$(2s)(2p), \quad (2s)(3p), \quad ..., \quad (2s)(E_c p) \qquad \text{II}s$$

and on $2p$

$$(2p)^2, \quad (2p)(3p), \quad ..., \quad (2p)(E_c p)$$
$$(2p)(3s), \quad ..., \quad (2p)(E_c s) \qquad \text{II}p$$

and so on.

The actual situation in helium is illustrated in Fig. 9.1, distinction being made between singlet and triplet configurations. Corresponding to each single electron orbital (nl) there exists a Rydberg series of states converging to a limit at an energy $E_i(\text{He}) + 4Rh/n^2$ above the ground state, R being the Rydberg constant and $E_i(\text{He})$ the ionization energy of neutral helium (24·57 eV). Above this limit the series becomes a continuum. The important point for present purposes is that the energy of any of these doubly excited configurations will coincide with that of a member of the singly excited set (I) in which the excited orbital (El) is one of the continuum. This follows because

$$2E(n, l) > E(1s),$$

where $E(n, l)$ is the energy of the nth hydrogenic orbital.

An energy degeneracy therefore exists between each bound doubly-excited configuration and the singly-excited continuum. In the neighbourhood of the energy of such a doubly-excited configuration the true wave function will involve a superposition of these two configurations.

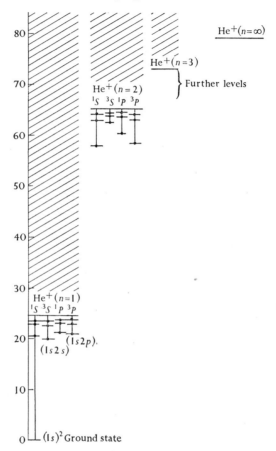

FIG. 9.1. Energy diagram for neutral He atoms, showing the different series and associated continua.

It follows, because of the admixture of the singly excited continuum, that the corresponding state will not be a bound one. Thought of in dynamic terms the situation is then as follows. At some instant the state may be well represented by the bound doubly-excited configuration but this cannot persist because a finite admixture of the continuum configuration will lead to break-up. The doubly-excited configuration has a finite lifetime determined by the strength of the interaction with the singly-excited continuum with overlapping total energy.

9.1 RESONANCE PHENOMENA—THRESHOLD BEHAVIOUR

On the other hand, some effects will arise in elastic scattering, for example, because of the existence of the doubly-excited configurations. The incident electron moving in the field of the single electron atom or ion represents a configuration of the singly-excited continuum. If the total energy of this configuration is close to that of a doubly-excited configuration, i.e. an energy resonance exists, then in the course of the interaction with the atom the dominant configuration may change and the electron spend some time captured in the doubly-excited configuration before once more reverting to the continuum and escaping. The delay produced by the capture will affect the phase shift of the incident electron (see § 9, p. 643) and hence the elastic scattering. We can expect these effects to appear only in narrow ranges of energy as they depend on the existence of energy degeneracy between the continuum and a doubly-excited configuration. It is to be noted that the doubly-excited configurations based on 3-quantum orbitals such as $3s$ will interact not only with the singly-excited continuum ($1s, E_c$) but also with the doubly-excited ($2s, E_c$ and $2p, E_c$). In general doubly-excited configurations based on the nth quantum orbital will interact with all continua based on lower quantum orbitals.

1.1. *Perturbation theory involving the interaction of a discrete state and a continuum*

To describe this situation we need to develop the appropriate perturbation theory where the perturbation has the effect of raising degeneracy between discrete and continuum states. We begin by considering the simplest case in which one discrete state and a single continuum are involved. This already includes a number of the resonance features with which we are concerned and the analysis permits of ready generalization to more complex situations.

One of the first questions that arises is the choice of the approximate wave functions for the discrete and continuum configurations, the so-called basis functions. This choice may clearly be made in many different ways and the aim is to select the functions so as to make the calculation both practical and accurate. For this purpose physical insight is important. In order to avoid inessential complications we shall make some rather drastic approximations in the treatment in this section. A more accurate treatment by a different, less physically descriptive, method is given in § 3. Consider first the special case of the scattering of s-wave electrons by hydrogen atoms, ignoring exchange effects. The full wave equation for the system is

$$H\Psi \equiv (H_0 + H_1)\Psi = E\Psi, \tag{1}$$

where, in terms of the electron coordinates relative to the proton,

$$H_0 = -(\hbar^2/2m)(\nabla_1^2+\nabla_2^2)-e^2\left(\frac{1}{r_1}+\frac{1}{r_2}\right), \qquad (2)$$

$$H_1 = e^2/r_{12}. \qquad (3)$$

As wave functions for the doubly-excited configuration we take the product

$$\phi(\mathbf{r}_1,\mathbf{r}_2) = \chi_a(\mathbf{r}_1)\chi_b(\mathbf{r}_2), \qquad (4)$$

where χ_a, χ_b are wave functions for excited states of hydrogen atoms corresponding to energies ϵ_a, ϵ_b.† Thus

$$H_0\phi = (\epsilon_a+\epsilon_b)\phi, \qquad (5)$$

so that $\epsilon_a+\epsilon_b$ is the zero-order approximation to the energy of the configuration. A better, first-order, approximation will be E_d where

$$E_d = \epsilon_a+\epsilon_b+\langle\phi|H_1|\phi\rangle. \qquad (6)$$

We then have
$$\langle\phi|H|\phi\rangle = E_d. \qquad (7)$$

For the continuum configuration with energy E

$$\psi_E(\mathbf{r}_1,\mathbf{r}_2) = \chi_0(r_2)F_0(E-\epsilon_0,\mathbf{r}_1), \qquad (8)$$

where χ_0 is the wave function for the ground state of hydrogen of energy ϵ_0 and $F_0(\mathbf{r})$ is the wave function for motion of an electron in the mean static potential field V_{00} of the hydrogen atom with energy $E-\epsilon_0$. Because χ_0 is orthogonal to χ_a, $a \neq 0$, ϕ and ψ_E are also orthogonal. Since

$$\left(\frac{\hbar^2}{2m}\nabla_1^2+\frac{e^2}{r_2}\right)\chi_0 = -\epsilon_0\chi_0 \qquad (9)$$

and
$$\left\{\frac{\hbar^2}{2m}\nabla_2^2+V_{00}(r_1)\right\}F_0 = -(E-\epsilon_0)F_0, \qquad (10)$$

$$\int \chi_0(r_2)H\chi_0(r_2)F_0(\mathbf{r}_1)\,d\mathbf{r}_2 = EF_0(\mathbf{r}_1). \qquad (11)$$

The wave functions F_0 may be chosen as real and form an orthonormal set so that, from (11),

$$\langle\psi_{E''}|H|\psi_{E'}\rangle = E'\int F_0(E''-\epsilon_0,\mathbf{r}_1)F_0(E'-\epsilon_0,\mathbf{r}_1)\,d\mathbf{r}_1$$
$$= E'\delta(E''-E'). \qquad (12)$$

Finally
$$\langle\psi_E|H|\phi\rangle = \langle\psi_E|H_0|\phi\rangle+\langle\psi_E|H_1|\phi\rangle,$$
$$= (\epsilon_a+\epsilon_b)\langle\psi_E|\phi\rangle+\langle\psi_E|H_1|\phi\rangle, \quad \text{from (5)},$$
$$= \langle\psi_E|H_1|\phi\rangle, \qquad (13)$$

† This is a crude approximation in general and is improved in the treatment of § 3.

because ψ_E and ϕ are orthogonal. We write, for brevity,
$$\langle\psi_E|H|\phi\rangle = V_{E\phi} \tag{14}$$
and note that $V_{E\phi}$ vanishes with the interaction term H_1. Furthermore, because the functions ψ_E are normalized per unit energy range, $V_{E\phi}^2$ has the dimensions of energy.

Given the three relations (7), (12), and (13) we may now proceed in a manner essentially similar to that used in dealing with perturbation theory applied to discrete degenerate states. The analysis applies to any system for which relations of the form (7), (12), and (13) exist.

The true wave function Ψ_E for the complete system for energy E is now expanded in terms of ϕ and $\psi_{E'}$ in the form
$$\Psi_E = a\phi + \int b_{E'}\psi_{E'}\,dE', \tag{15}$$
contributions from configurations with greatly differing energies being ignored. In particular, it is assumed that the doubly-excited configuration does not interact with any other continuum than the one we have considered. This will be so if the energy of the incident electron is below the energy necessary to produce excitation. The modifications necessary when interaction can occur with more than one continuum will be discussed in § 1.2.

Since Ψ_E is to be an eigenfunction of the full Hamiltonian with energy E
$$\langle\phi|H|\Psi_E\rangle = E\langle\phi|\Psi_E\rangle,$$
$$= Ea, \quad \text{from (15)}. \tag{16}$$
Also from (15), $\quad H\Psi_E = aH\phi + \int b_{E'}H\psi_{E'}\,dE',$
so
$$\langle\phi|H|\Psi_E\rangle = a\langle\phi|H|\phi\rangle + \int b_{E'}\langle\phi|H|\psi_{E'}\rangle\,dE'$$
$$= E_d a + \int b_{E'}V_{E'\phi}\,dE'. \tag{17}$$
Hence
$$E_d a + \int b_{E'}V_{E'\phi}\,dE' = Ea. \tag{18}$$
Similarly, from (14),
$$V_{E'\phi}a + E'b_{E'} = Eb_{E'}. \tag{19}$$

The only difficulty in relating $b_{E'}$ to a through (19) is that $E-E'$ may be zero. To circumvent this we adopt a procedure introduced by Dirac† in similar circumstances. The formal solution of (19) for $b_{E'}$ is written
$$b_{E'} = \left\{\frac{1}{E-E'} + z(E)\delta(E-E')\right\}V_{E'\phi}a, \tag{20}$$
and it is understood that, on substitution in (15) or (18), the principal

† DIRAC, P. A. M., *Z. Phys.* **44** (1927) 585.

part of the integral over $E-E'$ is to be taken and $z(E)$ determined as follows.

Substituting (20) in (18) we have

$$E_d + \Delta E + z(E) V_{E\phi}^2 = E, \tag{21}$$

where

$$\Delta E = P \int \frac{V_{E'\phi}^2}{E-E'} \, dE', \tag{22}$$

P denoting the 'principal part of'. Hence

$$z(E) = (E - E_d - \Delta E)/V_{E\phi}^2. \tag{23}$$

We are especially concerned with the determination of the asymptotic form of Ψ_E for large r_1. The contribution from $a\phi$ in (15) will vanish exponentially and can be ignored. Using (20) and (23) and the asymptotic form

$$\psi_E \sim \chi_0(r_2)\{n(E)\}^{\frac{1}{2}} r_1^{-1} \sin\{k(E)r_1 + \eta(E)\}, \tag{24}$$

where $k(E) = (2mE/\hbar^2)^{\frac{1}{2}}$ and

$$n(E) = (1/8\pi^3) \, dk/dE = m/2\pi k h^2, \dagger \tag{25}$$

we find

$$\Psi_E \sim a\chi_0(r_2) r_1^{-1} \bigg[\int n^{\frac{1}{2}}(E') \sin\{k(E')r_1 + \eta(E')\} V_{E'\phi}(E-E')^{-1} \, dE' + \\ + z(E) n^{\frac{1}{2}}(E) \sin\{k(E)r_1 + \eta(E)\} V_{E\phi} \bigg]. \tag{26}$$

The integral in (26) reduces to $-\pi n^{\frac{1}{2}}(E)\cos\{k(E)r_1 + \eta(E)\} V_{E\phi}$, so that

$$\Psi_E \sim a\chi_0(r_2) n^{\frac{1}{2}}(E)(\pi^2 + z^2)^{\frac{1}{2}} V_{E\phi} r_1^{-1} \sin\{k(E)r_1 + \eta(E) + \sigma(E)\}, \tag{27}$$

where

$$\sigma(E) = -\arctan\{\pi/z(E)\}$$
$$= -\arctan\{\pi(V_{E\phi})^2/(E - E_d - \Delta E)\}, \quad \text{from (23)}. \tag{28}$$

The presence of the interaction of the continuum configuration with the discrete one produces an additional phase shift σ given by (28). This is very small when E differs greatly from E_d but changes rapidly by π in going through an energy range of order $(V_{E\phi})^2$ about the resonance value $E_d + \Delta E$. In other words, we have an elastic scattering resonance of width $(V_{E\phi})^2$. It is important to notice that the existence of the interaction between the configurations produces a small shift of the resonance from E_d.

The partial elastic cross-section is now given by

$$Q_0^{\text{el}} = (\pi/k^2) |e^{2i(\eta+\sigma)} - 1|^2$$
$$= (\pi/k^2) |-e^{2i\sigma} + 1 + e^{-2i\eta} - 1|^2 \tag{29a}$$
$$= (\pi/k^2) \left| \frac{-2i\pi V_{E\phi}^2}{E - E_d - \Delta E + i\pi V_{E\phi}^2} + e^{-2i\eta} - 1 \right|^2. \tag{29b}$$

† This is the normalizing factor consistent with (12).

The first term in the modulus sign is the contribution from the resonance while $e^{-2i\eta}-1$ arises from the scattering by the mean potential. If the resonance term dominates

$$Q_0^{\text{el}} = \frac{4\pi}{k^2}\frac{\tfrac{1}{4}\Gamma^2}{(E-E_r)^2+\tfrac{1}{4}\Gamma^2}, \tag{30}$$

where
$$\Gamma = 2\pi V_{E\phi}^2, \tag{31}$$
$$E_r = E_d + \Delta E. \tag{32}$$

This is usually referred to as the Breit–Wigner† one-level formula as it was first derived by these authors, using a somewhat different method. It is distinguished further as 'one-level' because it applies to a case in which one discrete resonance state alone is involved. It has the shape of a resonance peak centred about E_r of half-thickness Γ. In practice the form of the cross-section in passing through the resonance region will depend on the interference between the resonance and mean potential scattering terms in (29).

The partial cross-section (29) may be written in the form

$$Q_0^{\text{el}} = (4\pi/k^2)\sin^2(\eta+\sigma) = (4\pi/k^2)\sin^2\eta\,\frac{(q+\epsilon)^2}{1+\epsilon^2}, \tag{33}$$

where $q = -\cot\eta$, $\epsilon = (E-E_r)/\tfrac{1}{2}\Gamma$. The factor $(q+\epsilon)/(1+\epsilon^2)$, which gives the shape of the resonance, is of the same form as that which arises in other examples of resonance effects (see § 1.2.1). Fig. 9.2 shows how the resonance shape varies with q which is often called the *shape profile parameter* or *profile index*.

To obtain the coefficient a of the discrete wave function in (15) we use the condition that the Ψ_E determined from (15) should form an orthonormal set, i.e.

$$\int \Psi_{E''}^*\Psi_{E'}\,\mathrm{d}\mathbf{r}_1\,\mathrm{d}\mathbf{r}_2 = \delta(E'-E''). \tag{34}$$

By comparison of (27) with (24) we see that we must have

$$aV_{E\phi}(\pi^2+z^2)^{\tfrac{1}{2}} = 1,$$

so that
$$a = -\sin\sigma/\pi V_{E\phi}, \tag{35}$$

and thence, from (20),

$$b_{E'} = -\frac{V_{E'\phi}}{\pi V_{E\phi}}\frac{\sin\sigma}{E-E'} + \cos\sigma\,\delta(E-E'). \tag{36}$$

The significance of the form (35) for a is interesting. Since, from (28),

$$|a(E)|^2 = V_{E\phi}^2/\{(E-E_r)^2+\pi^2 V_{E\phi}^4\}, \tag{37}$$

we see that the contribution from the discrete wave function ϕ is no longer limited to the single energy E_d but is spread over a continuous

† BREIT, G. and WIGNER, E., *Phys. Rev.* **51** (1937) 593.

range of values of E with a maximum at E_r^* close to E_d of half-width $2\pi V_{E\phi}^2$ (see Fig. 9.3). This is because the interaction with the continuum leads to a finite mean lifetime Δt for the discrete state. According to

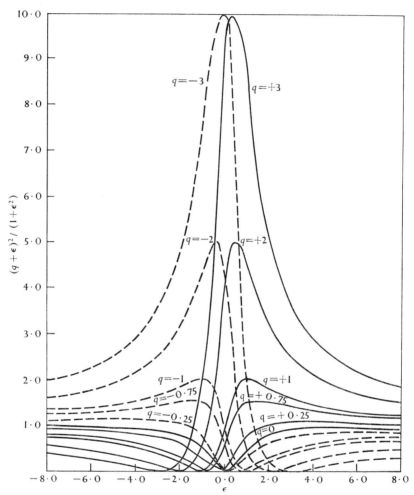

Fig. 9.2. The resonance profile function for different values of the profile index q.

the uncertainty principle this will lead to an uncertainty of the energy of the state of order $\hbar/\Delta t$. Accordingly we expect $\Delta t \simeq \hbar/2\pi V_{E\phi}^2$.

1.1.1. *Cross-sections for excitation of unbound states from the ground state.* We now consider the effect of an interaction of the type we have been considering on the behaviour of the cross-section for excitation of an unbound state from the ground state, as a function of the impact energy. The source of excitation may either be particle or photon

9.1 RESONANCE PHENOMENA—THRESHOLD BEHAVIOUR

impact. In either case the excitation cross-section will depend on the final wave function Ψ_E through the square of some matrix element

$$\langle \Psi_E | T | \Psi_0 \rangle \tag{38}$$

between the unbound state and the ground state with wave function Ψ_0.

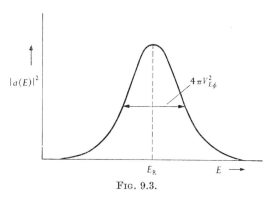

Fig. 9.3.

Using the form (15) for Ψ_E with (35) and (36) for a, b_E, respectively, we have
$$\langle \Psi_E | T | \Psi_0 \rangle = -A \sin \sigma + B \cos \sigma, \tag{39}$$
where
$$A = (1/\pi V_{E\phi}) \langle \Phi | T | \Psi_0 \rangle, \tag{40a}$$
$$B = \langle \psi_E | T | \Psi_0 \rangle, \tag{40b}$$
and σ is as given in (28).

B is thus the matrix element which would apply if there were no interaction of the unperturbed continuum with a discrete state. In A, Φ is given by

$$\Phi = \phi + P \int V_{E'\phi} \psi_{E'} (E-E')^{-1} \, dE' \tag{41}$$

and is the wave function for the discrete state including an admixture of continuum states due to its interaction with the continuum.

Since σ varies very rapidly as the energy E of the continuum state passes through E_r, (39) shows that the matrix element, and hence the cross-section, for the process will also vary rapidly in the same energy range. We may measure the resonance effect by the ratio R of the cross-section to that which would arise in the absence of the interaction with the discrete state. Although the latter cannot be measured directly it may be determined by smooth extrapolation from energies outside the resonance region. We have

$$R = \langle \Psi_E | T | \Psi_0 \rangle^2 / \langle \psi_E | T | \Psi_0 \rangle^2 \tag{42}$$
$$= \{(A/B)\sin \sigma - \cos \sigma\}^2 \tag{43}$$
$$= (q+\epsilon)^2/(1+\epsilon^2), \tag{44}$$

where
$$\epsilon = -\cot\sigma$$
$$= (E-E_r)/\tfrac{1}{2}\Gamma, \quad \text{as in (33),} \tag{45}$$
and now
$$q = \langle\Phi|T|\Psi_0\rangle/\pi V_{E\phi}\langle\psi_E|T|\Psi_0\rangle. \tag{46}$$
We note that
$$\tfrac{1}{2}\pi q^2 = \langle\Phi|T|\Psi_0\rangle^2/\Gamma\langle\psi_E|T|\Psi_0\rangle^2, \tag{47}$$
which is the ratio of the cross-section for excitation to the modified discrete state to that for excitation into a band of continuum states of width Γ.

Just as for elastic scattering (see (33)) the shape of the resonance is determined by two parameters ϵ and q. The ratio R has a zero minimum value when $\epsilon = -q$ and a maximum of $1+q^2$ when $\epsilon = 1/q$. Fig. 9.2 illustrates a number of different line shapes calculated for different values of q.

1.2. *Interaction between one discrete state and two or more continua* .

In § 1.1 we discussed effects arising from the interaction of a single discrete state with one continuum only. This analysis would be approximately applicable, for example, to resonance of a doubly-excited level based on $2s$ with a singly-excited continuum in which one electron remains in the $1s$ state. In fact, as may be seen from Fig. 9.1, levels based on higher excited states will interact with more than one continuum. We now discuss this situation which, in general, includes many important features. Thus it raises the possibility of resonance effects in inelastic cross-sections at energies close to the excitation threshold. If a doubly-excited level of helium, say $(3s)^2$, interacts both with the $(1s, E_c)$ and $(2s, E_c)$ continuum states we would expect an effect both on the cross-section for elastic scattering of electrons of energy E_c by normal helium ions and also for inelastic scattering resulting in excitation of the ion to the $2s$ level. Furthermore, the formulae we obtain are of wide general application in the familiar Breit–Wigner form.

Consider first for simplicity the case of interaction of a discrete state ϕ with two continua, the wave functions for which are ψ_E, ω_E respectively. As before (see (7), (12), and (13)) we have
$$\langle\phi|H|\phi\rangle = E_d, \quad \langle\psi_{E'}|H|\psi_{E''}\rangle = E'\delta(E''-E'),$$
$$\langle\psi_E|H|\phi\rangle = V_{E\phi}, \tag{48}$$
but in addition we must now add
$$\langle\omega_{E'}|H|\omega_{E''}\rangle = E'\delta(E''-E'), \quad \langle\omega_E|H|\phi\rangle = W_{E\phi}. \tag{49}$$
We also, for simplicity, take $(\omega_{E'}|H|\psi_{E''}) = 0$ which amounts to neglecting the contribution to inelastic scattering that comes from 'direct'

9.1 RESONANCE PHENOMENA—THRESHOLD BEHAVIOUR

transitions between continua without proceeding through the formation of an unstable doubly-excited complex.

In place of (15) we have

$$\Psi_E = a\phi + \int b_{E'}\psi_{E'}\,dE' + \int c_{E'}\omega_{E'}\,dE'. \tag{50}$$

Following a similar procedure to that used in § 1.1 we obtain, in place of (18) and (19), the three equations

$$E_d a + \int V_{E'\phi} b_{E'}\,dE' + \int W_{E'\phi} c_{E'}\,dE' = Ea, \tag{51}$$

$$V_{E'\phi} a + E' b_{E'} = E b_{E'}, \tag{52}$$

$$W_{E'\phi} a + E' c_{E'} = E c_{E'}. \tag{53}$$

From these equations two orthogonal solutions may be obtained, one of which is closely similar in form to the single solution obtained for the case of interaction with a single continuum. This solution, which we call Ψ_{E1}, is given by

$$a_1 = -\sin\sigma/\pi U_{E\phi}, \qquad b_{E'1} = -\frac{V_{E'\phi}}{U_{E\phi}}\frac{1}{\pi}\frac{\sin\sigma}{E-E'} + \cos\sigma\,\delta(E-E'),$$

$$c_{E'1} = (W_{E'\phi}/V_{E'\phi}) b_{E'1}, \qquad \sigma = -\arctan\{\pi U_{E\phi}^2/(E-E_d-G(E))\},$$

$$G(E) = P\int U_{E\phi}^2 (E-E')^{-1}\,dE', \qquad U_{E\phi}^2 = V_{E\phi}^2 + W_{E\phi}^2. \tag{54}$$

The second solution Ψ_{E2}, which is orthogonal to Ψ_{E1}, is given by

$$a_2 = 0, \qquad b_{E'2} = (W_{E'\phi}/U_{E\phi})\,\delta(E-E'), \qquad c_{E'2} = -(V_{E'\phi}/W_{E'\phi}) b_{E'2}. \tag{55}$$

This solution does not depend on σ, and shows no peculiarities near the resonance energy.

Although any linear combinations of these two solutions is also a solution the choice (54) and (55) is the suitable one for our purpose. Thus, consider any normalized linear combination $\Psi_\lambda = \alpha \Psi_{E1} + \beta \Psi_{E2}$. In Ψ_λ the contribution from the unperturbed continuum is just that which would arise if the unperturbed continuum function were

$$\psi_c = (\alpha b_1 + \beta b_2)\psi_E + (\alpha c_1 + \beta c_2)\omega_E. \tag{56}$$

It may be shown that the matrix element

$$\langle \phi | H | \psi_c \rangle$$

is a maximum when $\beta = 0$, i.e. when $\Psi_\lambda = \Psi_{E1}$. The selection of (54) and (55) ensures the greatest degree of separation into one function which is affected by the interaction with the discrete state and one that is not.

By following a similar procedure to that described in § 1.1, (24), (25), and (26), we find

$$r_1\Psi_{E1} \sim -(V_{E\phi}/U_{E\phi})\chi_0(r_2)n^{\frac{1}{2}}(E_0)\sin(k_0 r_1+\eta_0+\sigma)-$$
$$-(W_{E\phi}/U_{E\phi})\chi_1(r_2)n^{\frac{1}{2}}(E_1)\sin(k_1 r_1+\eta_1+\sigma), \quad (57\,\text{a})$$

$$r_1\Psi_{E2} \sim (W_{E\phi}/U_{E\phi})\chi_0(r_2)n^{\frac{1}{2}}(E_0)\sin(k_0 r_1+\eta_0)-$$
$$-(V_{E\phi}/U_{E\phi})\chi_1(r_2)n^{\frac{1}{2}}(E_1)\sin(k_1 r_1+\eta_1). \quad (57\,\text{b})$$

Here $\chi_0(r_2)$, $\chi_1(r_2)$ are wave functions for the ground state and first excited state of the residual one-electron system while k_0 and k_1 are the wave numbers for the motion of the electron in the continuum state of total energy E associated with the one-electron system in these respective states. Thus $k_{0,1}^2 = (2m/\hbar^2)(E-\epsilon_{0,1}) = (2m/\hbar^2)E_{0,1}$.

The wave function $\Psi_{\text{coll}}(E)$ describing the collision of an electron with the one-electron system in its ground state will have the asymptotic form

$$r_1\Psi_{\text{coll}}(E) \sim \chi_0(r_2)(\sin k_0 r_1+f_0 e^{ik_0 r_1})+\chi_1(r_2)f_1 e^{ik_1 r_1}. \quad (58)$$

The partial elastic cross-section is then given by

$$Q_0^{\text{el}} = 4\pi|f_0|^2/k_0^2 \quad (59\,\text{a})$$

and the corresponding inelastic cross-section by

$$Q_0^{\text{in}} = 4\pi(k_0/k_1^3)|f_1|^2. \quad (59\,\text{b})$$

To obtain the correct asymptotic form we must take for $\Psi_{\text{coll}}(E)$ an appropriate linear combination of Ψ_{E1} and Ψ_{E2}. Writing then

$$\Psi_{\text{coll}}(E) = c\Psi_{E1}+d\Psi_{E2}, \quad (60)$$

we see that, in the asymptotic form of Ψ_{coll}, the coefficient of $e^{-ik_1 r_1}$ is proportional to

$$cW_{E\phi}\,e^{-i\sigma}+dV_{E\phi}. \quad (61)$$

Since according to (58) this must vanish we have

$$\Psi_{\text{coll}}(E) = c\{\Psi_{E1}-(W_{E\phi}/V_{E\phi})\,e^{-i\sigma}\Psi_{E2}\}. \quad (62)$$

Using (57) and (59) we then find

$$Q_0^{\text{el}} = \frac{\pi}{k_0^2}\left|\frac{i\Gamma_0}{E-E_r+\frac{1}{2}i\Gamma}+e^{-2i\eta_0}-1\right|^2, \quad (63\,\text{a})$$

$$(k_1/k_0)^{\frac{1}{2}}f_1 = \frac{\frac{1}{2}\Gamma_0^{\frac{1}{2}}\Gamma_1^{\frac{1}{2}}}{E-E_r+\frac{1}{2}i\Gamma}e^{i(\eta_0+\eta_1)}, \quad (63\,\text{b})$$

$$Q_0^{\text{in}} = \frac{\pi}{k_0^2}\frac{\Gamma_0\Gamma_1}{(E-E_r)^2+\frac{1}{4}\Gamma^2}, \quad (63\,\text{c})$$

where

$$\Gamma_0 = 2\pi V_{E\phi}^2, \quad \Gamma_1 = 2\pi W_{E\phi}^2, \quad \Gamma = \Gamma_0+\Gamma_1, \quad E_r = E_d+G. \quad (64)$$

9.1 RESONANCE PHENOMENA—THRESHOLD BEHAVIOUR

If the background elastic scattering is ignored,

$$Q_0^{el} = \frac{4\pi}{k_0^2} \frac{\frac{1}{4}\Gamma_0^2}{(E-E_r)^2+\frac{1}{4}\Gamma^2}. \tag{65}$$

The width of the resonance is now increased to $\Gamma = \Gamma_0+\Gamma_1$ owing to the possibility of interaction with the second continuum. By providing an alternative mode for break-up of the two-electron system in the doubly-excited discrete state the lifetime of the state is reduced to $\hbar/\Gamma$. Furthermore, (63 a) and (65) may be interpreted in the following way. We may regard

$$\frac{\pi}{k_0^2} \frac{\Gamma_0 \Gamma}{(E-E_r)^2+\frac{1}{4}\Gamma^2} \tag{66}$$

as the cross-section for capture of the incident electron by the one-electron system in its ground state to form a two-electron system in the discrete state ϕ. Γ_0/Γ is then the chance that this complex will break up by departure of the incident electron with its initial energy. Γ_1/Γ on the other hand, is the chance that the break-up occurs by ejection of the electron with reduced energy leaving the one-electron system in the first excited state.

There is no difficulty in extending this analysis to interaction of a discrete state with any number of continua. In the general case, in which collision processes occur through the formation of a complex of finite lifetime as an intermediate stage, we may write for the cross-section for the inelastic process involving a particular kind of break-up of the complex

$$Q_0^n = \frac{\pi}{k_0^2} \frac{\Gamma_0 \Gamma_n}{(E-E_r)^2+\frac{1}{4}\Gamma^2}, \tag{67}$$

where $\Gamma = \Gamma_0+\Gamma_1+....$ Γ_0, Γ_1, etc., referred to as the *partial resonance widths*, arise from break-up leaving the initial target atom in its ground state and successive excited states respectively. For the elastic scattering the formula (63 a) applies with the new interpretation of Γ. These formulae in their general form were first derived by Breit and Wigner[†] and have received much application in nuclear physics.[‡]

1.2.1. *Cross-sections for excitation of continuum states, of energy E, from the ground state.* We now extend the analysis of § 1.2 to apply when the discrete state interacts with more than one continuum. As usual we deal first with the simplest case, that of two continua.

The matrix element $\langle \Psi_E|T|\Psi_0\rangle^2$ of that section is now replaced by

$$\langle \Psi_{E1}|T|\Psi_0\rangle^2+\langle \Psi_{E2}|T|\Psi_0\rangle^2. \tag{68}$$

[†] BREIT, G. and WIGNER, E., *Phys. Rev.* **51** (1937) 593.
[‡] See BLATT, J. M. and WEISSKOPF, V. P., *Theoretical nuclear physics*, Chap. ix (Wiley, New York; Chapman and Hall, London, 1952).

Of these terms it is only the first that includes any resonance effect—
the second varies smoothly with the energy.

We may cast the first term in a form similar to (44). Thus, referring
to (39),
$$\langle \Psi_{E1}|T|\Psi_0^*\rangle = -A_1 \sin\sigma + B_1 \cos\sigma, \tag{69}$$
where
$$A_1 = (1/\pi U_{E\phi})\langle\Phi|T|\Psi_0^*\rangle, \tag{70a}$$
$$B_1 = \langle\psi_E^m|T|\Psi_0^*\rangle. \tag{70b}$$

Here, by an obvious extension of (41),
$$\Phi = \phi + P\int U_{E'\phi}\psi_{E'}^m(E-E')^{-1}\,dE', \tag{71}$$
with
$$\psi_E^m = (V_{E\phi}/U_{E\phi})\psi_E + (W_{E\phi}/U_{E\phi})\omega_E. \tag{72}$$

As explained in § 1.2 above, ψ_E^m is the linear combination of the two
continuum functions that gives the maximum interaction with the
discrete state.

We may now write
$$\langle\Psi_{E1}|T|\Psi_0^*\rangle^2 = \frac{(q+\epsilon)^2}{1+\epsilon^2}\langle\psi_E^m|T|\Psi_0^*\rangle^2, \tag{73}$$
where
$$q = \langle\Phi|T|\Psi_0^*\rangle/\pi U_{E\phi}\langle\psi_E^m|T|\Psi_0^*\rangle, \qquad \epsilon = -\cot\sigma = (E-E_r)/\tfrac{1}{2}\Gamma. \tag{74}$$

The cross-section for the process concerned, which will be proportional
to $\langle\Psi_E|T|\Psi_0^*\rangle^2$, can therefore be written in the form
$$Q = \frac{(q+\epsilon)^2}{1+\epsilon^2}Q_a + Q_b, \tag{75}$$
where Q_a and Q_b vary slowly with the energy E. At energies far from
the resonance energy E_r
$$Q = Q_a + Q_b. \tag{76}$$

The most obvious effect of the existence of two interacting continua
is to provide a gradually varying background so that when $\epsilon = -q$ the
cross-section no longer vanishes.

We may write (75) in the form
$$Q = Q_a + Q_b + \frac{q^2 + 2q\epsilon - 1}{1+\epsilon^2}Q_a. \tag{77}$$

The maximum of $(q+\epsilon)^2/(1+\epsilon^2)$ occurs at $\epsilon = 1/q$ so at this maximum
$$Q = Q_{\max} = (Q_a + Q_b) + q^2 Q_a. \tag{78}$$
In the same way we have at the zero minimum of $(q+\epsilon)^2/(1+\epsilon^2)$,
$$Q = Q_{\min} = (Q_a + Q_b) - Q_a. \tag{79}$$
Hence, if $|q| < 1$ the resonance produces a greater reduction from the
background on one side than it does an increase above the background

9.1 RESONANCE PHENOMENA—THRESHOLD BEHAVIOUR

on the other as illustrated in Fig. 9.4. This net reduction in the cross-section over the resonance region must be made up by a widely distributed net increase in other energy regions.

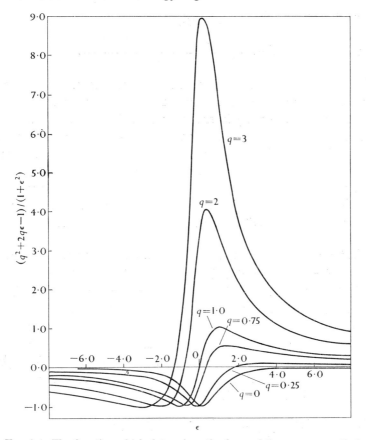

Fig. 9.4. The function which determines the form of the resonance effect relative to the smooth background.

When the excitation process is due to photon absorption the case shows up in the appearance of narrow windows in the absorption spectrum at the energies concerned. This remarkable effect is described in detail in Chapter 14, § 7.3.

Fano and Cooper† find it useful to introduce a third parameter ρ in such formulae as (75). This is defined by

$$\rho^2 = Q_a/(Q_a+Q_b)$$
$$= \langle \psi_E^m|T|\Psi_0^{\circ}\rangle^2/\{\langle \psi_E^m|T|\Psi_0^{\circ}\rangle^2+\langle \omega_E|T|\Psi_0^{\circ}\rangle^2\}, \tag{80}$$

† Fano, U. and Cooper, J. W., *Phys. Rev.* **137** (1964) A1364.

because, at a great distance from the resonance, the effect of the discrete state is negligible and the transition probability is the sum of the contributions from transitions to each of the two unperturbed continuum states. We may write

$$\rho = \langle \psi_E^m | \psi_E^T \rangle, \tag{81}$$

where

$$\psi_E^T = \frac{\langle \Psi_0 | T | \psi_E \rangle \psi_E + \langle \Psi_0 | T | \omega_E \rangle \omega_E}{\langle \Psi_0 | T | \psi_E \rangle^2 + \langle \Psi_0 | T | \omega_E \rangle^2}. \tag{82}$$

ψ_E^T is the linear combination of the two continuum functions that maximizes the continuum contribution to the matrix element of T from the ground state. As ψ_E^m is similarly that linear combination which maximizes the interaction that produces auto-ionization, ρ is essentially a correlation coefficient between matrix elements for auto-ionization and for the excitation transition concerned.

The generalization of these results to apply to cases in which interaction occurs with more than two continua is obvious. Thus (77) remains valid with q as in (74), ψ_E^m being the obvious extension of (72).

1.3. Extension to scattering of electrons with non-zero angular momentum

Although we have restricted the analysis of §§ 1.1, 1.2 to the scattering of s-electrons there is no difficulty in extending it to cases in which the electron possesses any quantized angular momentum.

Thus, if we are considering the scattering of electrons, with insufficient energy to produce excitation, by an atom in a ^1S state and there exists a doublet resonance state of orbital angular momentum $\{L(L+1)\}^{\frac{1}{2}}\hbar$, the differential cross-section for scattering through an angle θ into the solid angle $d\omega$ will be given by (see Chap. 6, § 3.2)

$$I(\theta)\,d\omega = (\pi/k^2)\Big|\sum_{l \neq L}(2l+1)(e^{2i\eta_l}-1)P_l(\cos\theta) + \\ +(2L+1)\Big\{\frac{\Gamma e^{2i\eta_L}}{E-E_r+\tfrac{1}{2}i\Gamma}+e^{2i\eta_L}-1\Big\}P_L(\cos\theta)\Big|^2 d\omega, \tag{83}$$

where Γ and E_r are the level-width and energy associated with the resonant state.

It is clear from this formula that the relative magnitude and shape of the resonance effect will vary with angle of scattering. From observation of this variation the angular momentum of the resonance state can be obtained.

The total elastic cross-section can be written

$$Q^{el} = Q_a\frac{(q_L+\epsilon)^2}{1+\epsilon^2}+Q_b, \tag{84}$$

where

$$Q_a = (4\pi/k^2)(2L+1)\sin^2\eta_L, \qquad Q_b = (4\pi/k^2)\sum_{l \neq L}(2l+1)\sin^2\eta_l, \qquad (85)$$

$$q_L = -\cot\eta_l, \qquad \epsilon = (E-E_r)/\tfrac{1}{2}\Gamma. \qquad (86)$$

This is of the same form as the total cross-section (75) for excitation in general, Q_b being the background cross-section that does not include resonance. The same considerations apply as in § 1.2.1. If $|q_L| < 1$, i.e. $2s\pi+\tfrac{1}{4}\pi < \eta_l < 2s\pi+\tfrac{3}{4}\pi$, where $s = 0, 1,...$, the resonance effect results in a reduction from the background on one side of the resonance greater than the increase above the background on the other side. The reverse is the case if $|q_L| > 1$.

In many experiments to measure Q^{el} as a function of E the energy resolving power will not be great enough to follow the details of the variation across the resonance. Under these circumstances the quantity measured is the integral over an energy band large compared with the level width Γ around the resonance energy. As $\epsilon \to \infty$, on the extreme wings of the resonance,

$$Q = Q_a + Q_b,$$

so we have

$$\int_{\text{res}} Q\,\mathrm{d}E = \int_{\text{res}} (Q_a+Q_b)\,\mathrm{d}E + \tfrac{1}{2}\Gamma \int_{\text{res}} Q_a \frac{q^2+2q\epsilon-1}{1+\epsilon^2}\,\mathrm{d}\epsilon. \qquad (87)$$

For the second integral on the right-hand side of (87) the limits may be taken as $\pm\infty$, without serious error. We then have

$$\int_{\text{res}} Q\,\mathrm{d}E = \int_{\text{res}} (Q_a+Q_b)\,\mathrm{d}E + (2\pi/k^2)\Gamma(2L+1)\pi(q^2-1)\sin^2\eta_L \qquad (88)$$

$$= \int_{\text{res}} (Q_a+Q_b)\,\mathrm{d}E + (2\pi^2/k^2)\Gamma(2L+1)\cos 2\eta_L. \qquad (89)$$

The observed effect of the resonance will therefore be either to produce a peak or a depression of the cross-section at the resonance energy according as $\cos 2\eta_L$ is $>$ or < 0, respectively.

1.4. *Variation of resonance parameters along a Rydberg series*

We consider the various resonance states in an atom forming a Rydberg series converging to the energy E_j of the jth excited state of the singly ionized atom. The energy of any particular member of the series may be written

$$E_{jn} = E_j - E_\text{H}/n^{*2}, \qquad (90)$$

where E_H is the ionization energy of hydrogen and $n*$ is an effective quantum number given by

$$n^* = n-\mu, \qquad (91)$$

where μ is the quantum defect (see Chap. 6, § 3.10). Corresponding to each of these states there will be a line width Γ_{jn}, and parameters q_{jn}, ρ_{jn}. We now examine how these parameters vary with n. This depends on the variation of the corresponding discrete wave functions ϕ_{jn}.

Two such functions, for fixed j, will not differ appreciably in form in the interior of the atom because $E_{jn}-E_{jn'}$ is small compared with the kinetic energy of the electrons there. Hence ϕ_{jn}, $\phi_{jn'}$ will, in this region, differ only by a normalizing factor so we may write

$$\phi_{jn} = N_{jn}\bar{\phi}_j, \tag{92}$$

where $\bar{\phi}_j$ is independent of n.

If the matrix elements which appear in the formula (74) for q and (81) for ρ and which involve ϕ, are determined mainly by the values of the functions within the atom, it follows that

$$q_{jn} \simeq \bar{q}_j, \qquad \rho_{jn} = \bar{\rho}_j, \tag{93}$$

where $\bar{q}_j$, $\bar{\rho}_j$ are constant for the given Rydberg series.

On the other hand, it follows from (64) that

$$\Gamma_{jn} = N_{jn}^2 \bar{\Gamma}_j, \tag{94}$$

where $$\bar{\Gamma}_j = 2\pi\{\langle\bar{\phi}|H|\psi_E\rangle^2 + \langle\bar{\phi}|H|\omega_E\rangle^2 + ...\}. \tag{95}$$

For large n, $N_n \simeq \text{const}/n^3$.

In accord with the normalization of the continuum wave functions to unit energy range, Cooper and Fano† extended the formula to small n by taking N_{jn}^2 as the mean separation of energy levels close to E_{jn}, i.e.

$$\begin{aligned} N_{jn}^2 &= \tfrac{1}{2}(E_{j,n+1}-E_{j,n-1}) \\ &= E_H 2n^*/(n^{*2}-1)^2 \\ &\simeq 2E_H/n^3, \quad \text{for large } n. \end{aligned} \tag{96}$$

It then follows that

$$\Gamma_{jn}/\tfrac{1}{2}(E_{j,n+1}-E_{j,n-1}), = \bar{\bar{\Gamma}}_j, \tag{97}$$

should be effectively constant over the Rydberg series.

The accuracy of these approximations will be compared with detailed theoretical determinations in § 3.1 and will be applied to discuss resonance effects in absorption spectra in Chapter 14, § 4.3.

2. Summary of resonance effects and qualitative comparison with experimental results in electron scattering by atoms

Before describing means for calculating the resonance energies, widths, and other characteristics, which involves considering the relation of

† COOPER, J. W. and FANO, U., *Phys. Rev.* **137** (1964) A1364.

resonance effects to the truncated eigenfunction expansion method (Chap. 8, § 2.2.3) for calculating collision cross-sections, we pause to summarize the nature of the effects which may occur. These are as follows.

(a) *Elastic scattering of electrons*

(i) *By neutral atoms.* The doubly-excited discrete states that produce the resonances are states of the negative atomic ion based on an excited state of the neutral atom. Because the effective attractive field acting on the temporarily attached electron will be small, as the core is neutral, the energies of the doubly-excited, autodetaching states will be only a little below that of the respective excited states of the neutral atom. In other words, the resonance energies will be closely below the thresholds for excitation of the successive neutral atomic states. Associated with each such state there would in general only be a finite number of resonance states because of the short range of the effective interaction between neutral atom and electron.

At each resonance energy the cross-section will exhibit a variation of the typical form shown in Fig. 9.4.

The resonance effects observed by Schulz† and by Kleinpoppen and Raible‡ (see Chap. 1, § 6.2.2) in the intensity of electron scattering by atomic hydrogen as a function of energy at a fixed angle are of this type, occurring close below the threshold for excitation of the 2-quantum states. Similar considerations apply to the effects observed in elastic scattering by the rare gases (see Chap. 1, §§ 6.2.3, 6.2.4, 6.2.5).

(ii) *By positive ions.* Once again the resonance states can be related to excited states of the ion but there will be two marked differences from the neutral atom case. Thus there will be an unlimited number of resonance states associated with each ionic state and the lowest of these resonance states falls substantially lower in energy than the associated ionic state. Thus for scattering by He^+ the $2s^2\ ^1S$ resonance state falls 7·2 eV below the $2s$ state of He^+.

Resonance effects of this kind have not yet been observed in elastic scattering of electrons by ions because of obvious experimental difficulties. However, the presence of resonance states, as expected, may be verified by experiments on inelastic scattering of electrons by the corresponding neutral atom which lead to ionization (see (b) (i)).

† Schulz, G. J., *Phys. Rev. Lett.* 13 (1964) 583.
‡ Kleinpoppen, H. and Raible, V., *Phys. Lett.* 18 (1965) 24.

(b) *Inelastic scattering of electrons*

(i) *Effects on the ionization cross-section.* According to the theory we have discussed above, the cross-section of ionization of an atom by electron impact will be modified over certain limited energy ranges through interaction of the continuum states of the atom with doubly excited states of finite lifetime. Superposed on the contribution from direct ionization there will be peaks due to excitation of discrete but auto-ionizing states (see Chap. 3, § 2.5.3). These are the same states that lead to the occurrence of resonance effects in elastic scattering of electrons by the singly ionized atom as discussed in (a) (ii).

The occurrence of a resonance effect will be manifest in the energy loss spectrum of the electrons scattered through a fixed angle after ionizing the atom. The observed effects described in Chapter 5, § 4.2.1 are of this type.

(ii) *Effects on excitation cross-sections.* Resonance effects will appear in the cross-sections for excitation of discrete atomic states in relatively restricted circumstances.

The resonance states that influence elastic scattering lie close below the thresholds, of energy E_2, E_3,..., for excitation of the second, third,... excited states of the atom. Suppose that one such resonance state occurs at an energy $E_3 - \delta E_3$. This will not only produce a resonance effect of the type discussed under (a) (i) in the elastic scattering but also in inelastic scattering which leads to excitation of the second excited state. Whether this will show up against the background due to direct excitation depends on the strength of the coupling parameters which may often be very small.

There remains one further possibility that must be remembered. Between certain atomic states with closely similar energies the coupling may be so close that additional resonances arise which are not allowed for in the perturbation treatment we have described. This may be so, for example, between the 2^1S, 2^3S, 2^1P, and 2^3P states of helium. Further mention of these possibilities will be made in § 5, in relation to the experimental results described in Chapter 5, § 5.3.3.

(c) *Photo-ionization*

Similar effects are to be expected in ionization of atoms or ions by photons to those described for electron impact in (b) (i). In these cases the absorption cross-section of the atom is observed as a function of photon energy above the ionization threshold. Such experiments are

9.2 RESONANCE PHENOMENA—THRESHOLD BEHAVIOUR 615

described in Chapter 14, § 5.5 and their interpretation discussed in terms of resonance theory in Chapter 14, §§ 4.3 and 7.1–3.

(d) Ionization by impact of heavy particles

The energy loss spectrum, above the ionization threshold, for heavy particles, such as atomic or molecular ions, incident on particular atoms will exhibit similar resonance effects to that for electron impact as described in (b) (i). Observations of this kind and their interpretation are discussed in Volume IV.

3. Accurate calculation of resonance parameters—relation to coupled equations

The theory we have outlined in § 1 is essentially a perturbation treatment in which the interaction between the discrete and continuum states is assumed to be small. For many purposes this is a valid approximation for electron impacts with atoms and in any case it seems to bring out clearly the physical principles involved. The most accurate calculations of resonance parameters to date have been made by working from a somewhat more direct approach based on the usual expansion of the collision wave function in terms of the unperturbed atomic eigenfunctions† (see Chap. 8, § 2.2).

Consider, for example, the scattering of electrons by hydrogen atoms. It would seem reasonable to suppose that, in order to obtain a collision wave-function that includes the effects of resonances due to H⁻ states based on the 2s and 2p states of the neutral atom, we need only extend the eigenfunction expansion to include the 2s and 2p atomic functions. Is it possible then to see how the effects arise if this is done? To demonstrate this we simplify the conditions still further by ignoring the energy degeneracy between the 2s and 2p states and seeking only resonance effects due to H⁻ states based on the 2s state. Furthermore, we suppose that the electron energy is so low that only s-scattering is involved. We also neglect exchange effects.

The collision wave function is now approximated by

$$\Psi = \psi_0(r_2)F_0(r_1) + \psi_1(r_2)F_1(r_1), \qquad (98)$$

where ψ_0, ψ_1 are the respective wave functions of the 1s and 2s states of atomic hydrogen. If k is the incident electron wave number, which is below the threshold for excitation of the 2s state, then

$$r_1 F_0(r_1) \equiv G_0(r_1) \sim \sin kr_1 + \alpha e^{ikr_1}, \qquad (99\,\text{a})$$

$$r_1 F_1(r_1) \equiv G_1(r_1) \sim \beta e^{-\kappa r_1}, \qquad (99\,\text{b})$$

† FESHBACH, H., *Ann. Phys.* **5** (1958) 357; **19** (1962) 287.

where $k^2-k_0^2 = -\kappa^2$, k_0^2 being written for $2m|E_0-E_1|/\hbar^2$ with E_0, E_1 the respective energies of the 1s and 2s states. The partial elastic cross-section is then $4\pi|\alpha|^2/k^2$. As in the analysis of Chapter 8, § 2.2.3 the best approximation for the trial functions

$$F_0(r_1) = r_1^{-1}G_0(r_1), \qquad F_1(r_1) = r_1^{-1}G_1(r_1)$$

is by taking them to be the solutions of the coupled equations

$$\frac{d^2 G_0}{dr^2} + (k^2 - U_{00})G_0 = U_{01}G_1, \qquad (100\,\text{a})$$

$$\frac{d^2 G_1}{dr^2} - (\kappa^2 + U_{11})G_1 = U_{10}G_0, \qquad (100\,\text{b})$$

which satisfy the asymptotic conditions (99 a) (99 b) as well as the conditions $G_0(0) = 0$, $G_1(0) = 0$. The functions U_{00}, U_{01}, U_{11} are given by (Chap. 7 (8 a))

$$U_{ij} = (2me^2/\hbar^2) \int \left(-\frac{1}{r_1} + \frac{1}{r_{12}}\right) \psi_i(\mathbf{r}_2)\psi_j^*(\mathbf{r}_2)\,d\mathbf{r}_2. \qquad (101)$$

As both ψ_0 and ψ_1 are real, $U_{ij} = U_{ji}$.

$\hbar^2 U_{11}/2m$ is the mean interaction of an electron with the atom in a 2s state, so the equation which is the homogeneous equivalent of (100 b), namely

$$\frac{d^2 \mathscr{G}_1}{dr^2} - (\kappa^2 + U_{11})\mathscr{G}_1 = 0, \qquad (102)$$

is the Schrödinger equation for motion of an electron under the influence of this interaction. In general U_{11}, which in any case is attractive, will be large enough to provide one or more bound states of energy

$$-\hbar^2\kappa_1^2/2m, \quad -\hbar^2\kappa_2^2/2m, \quad \ldots.$$

Corresponding to each of these states there will be bound state wave functions X_1, X_2,.... If the coupling $U_{10} = U_{01}$ were vanishingly small we would have when $k^2 - k_0^2 = \kappa_1^2$, κ_2^2,... two degenerate solutions of the wave equation describing the collision. One, $\psi_0(r_1)F_0(r_2)$, belongs to a continuum while the other, $\psi_1(r_1)X_j(r_2)$, is a discrete function corresponding to a doubly-excited state of H⁻. In terms of the analysis of § 1, $\psi_0(r_2)\mathscr{G}_0(r_1)r_1^{-1}$ corresponds to the continuum wave function ψ_E and $\psi_1(r_2)X_j(r_1)r_1^{-1}$ to the doubly-excited function ϕ. It is important to notice, however, that we are now using a better approximation to ϕ which in § 1, in the notation of this section, is represented by $\psi_1(r_2)\psi_j(r_1)$, where ψ_j is the wave function for the jth excited state of hydrogen. We now sketch out the analysis based on approximate solution of the equations (100) that leads to the same Breit–Wigner formulae as in § 1.

To include now the effect of the coupling $U_{10} = U_{01}$ we must consider

9.3 RESONANCE PHENOMENA—THRESHOLD BEHAVIOUR

the solution of the inhomogeneous equation (100 b). This may be expanded in terms of the eigenfunctions of the corresponding homogeneous equation (102).

Writing
$$G_1 = \sum_j a_j X_j + \int a(E) X_E \, dE, \tag{103}$$

where X_E is a continuum wave function of (102) corresponding to an energy $E = \hbar^2 k_E^2/2m$, we have, on substitution in (100 b),

$$\sum_j (\kappa_j^2 - \kappa^2) a_j X_j - \int (k_E^2 + \kappa^2) a(E) X_E \, dE = U_{10} G_0. \tag{104}$$

This follows because

$$\frac{d^2 X_j}{dr^2} - (\kappa_j^2 + U_{11}) X_j = 0, \tag{105 a}$$

$$\frac{d^2 X_E}{dr^2} + (k_E^2 - U_{11}) X_E = 0. \tag{105 b}$$

Multiplying both sides of (104) by X_j and integrating over all r gives

$$a_j = \int X_j U_{10} G_0 \, dr / (\kappa_j^2 - \kappa^2). \tag{106 a}$$

Similarly we find
$$a_E = - \int X_E U_{10} G_0 \, dr / (k_E^2 + \kappa^2). \tag{106 b}$$

Thus, in the solution we have obtained, singularities occur when κ^2 is equal to one of the κ_j^2 corresponding to the bound eigenfunctions of (102). It is these singularities which lead to the resonance effects we have discussed earlier.

To analyse the matter further we suppose $\kappa^2 \simeq \kappa_j^2$ so a_j is much larger than any of the remaining coefficients in (103). We may then take

$$G_1 \simeq X_j \int X_j U_{10} G_0 \, dr / (\kappa_j^2 - \kappa^2), \tag{107}$$

and substitute this in (100 a) to obtain

$$\frac{d^2 G_0}{dr^2} + (k^2 - U_{00}) G_0 = A U_{01} X_j, \tag{108}$$

where
$$A = \int X_j U_{10} G_0 \, dr / (\kappa_j^2 - \kappa^2). \tag{109}$$

The solution of this inhomogeneous equation, satisfying the boundary condition (99 a), may be expressed in terms of that for the homogeneous equation

$$\frac{d^2 \mathcal{G}_0}{dr^2} + (k^2 - U_{00}) \mathcal{G}_0 = 0 \tag{110}$$

by means of a Green's function $\mathcal{K}(r, r')$. Thus we have

$$G_0(r) = e^{i\eta} \mathcal{G}_0(r) + A \int \mathcal{K}(r, r') U_{01}(r') X_j(r') \, dr', \tag{111}$$

where $\mathscr{G}_0$ has the asymptotic form $\sin(kr+\eta)$. But, from (109),

$$\int X_j U_{10} \mathscr{G}_0 \, dr = (\kappa_j^2 - \kappa^2)A. \tag{112}$$

So, multiplying both sides of (111) by $X_j U_{10}$ and integrating over all r, gives

$$(\kappa_j^2 - \kappa^2)A$$
$$= e^{i\eta} \int X_j U_{10} \mathscr{G}_0 \, dr + A \iint \mathscr{K}(r,r') U_{01}(r') U_{01}(r) X_j(r) X_j(r') \, dr \, dr'. \tag{113}$$

The Green's function $\mathscr{K}(r,r')$ is given by†

$$\mathscr{K}(r,r') = \begin{cases} -k^{-1}\{\mathscr{G}_0(r)\mathscr{G}_0^c(r') + i\mathscr{G}_0(r)\mathscr{G}_0(r')\} & (r' > r) \\ -k^{-1}\{\mathscr{G}_0^c(r)\mathscr{G}_0(r') + i\mathscr{G}_0(r)\mathscr{G}_0(r')\} & (r' < r). \end{cases} \tag{114}$$

$\mathscr{G}_0^c(r)$ is the solution of (110) which has the asymptotic form

$$\cos(kr+\eta).$$

We may therefore write

$$A = e^{i\eta}\lambda_j / (\kappa_j^2 - \kappa^2 - \Delta\kappa_j^2 + i\lambda_j^2/k), \tag{115}$$

where

$$\lambda_j = \int X_j U_{10} \mathscr{G}_0 \, dr, \tag{116a}$$

$$\Delta\kappa_j^2 = \iint \{\operatorname{re}\mathscr{K}(r,r')\} U_{01}(r') U_{01}(r) X_j(r) X_j(r') \, dr \, dr'. \tag{116b}$$

It follows from (114) that, for large r,

$$\mathscr{K}(r,r') \sim -k^{-1} e^{i(kr+\eta)} \mathscr{G}_0(r'), \tag{117}$$

so, from (111), (115), and (117),

$$G_0 \sim e^{i\eta}\sin(kr+\eta) - k^{-1} e^{ikr} e^{2i\eta} \lambda_j^2 / (\kappa_j^2 - \kappa^2 - \Delta\kappa_j^2 + i\lambda_j^2/k)$$
$$= \sin kr + e^{ikr}\left\{\frac{e^{2i\eta}-1}{2i} - e^{2i\eta}\frac{\tfrac{1}{2}\Gamma}{E-E_r+\tfrac{1}{2}i\Gamma}\right\}, \tag{118}$$

where

$$\Gamma = (4m/k\hbar^2)\left\{\int X_j V_{01} \mathscr{G}_0 \, dr\right\}^2, \quad E = k^2\hbar^2/2m = (k_0^2-\kappa^2)\hbar^2/2m,$$
$$E_r = (k_0^2-\kappa_j^2+\Delta\kappa_j^2)\hbar^2/2m, \quad V_{01} = U_{01}\hbar^2/2m. \tag{119}$$

The partial elastic cross-section then becomes

$$Q_0^{\text{el}} = \frac{\pi}{k^2}\left|\frac{i\Gamma}{E-E_r+\tfrac{1}{2}i\Gamma} + e^{-2i\eta} - 1\right|^2. \tag{120}$$

Comparison with (29) of § 1.1 shows that the two expressions would be equal if

$$(4m/k\hbar^2)\left\{\int X_j V_{01} \mathscr{G}_0 \, dr\right\}^2 = 2\pi V_{E\phi}^2. \tag{121}$$

† Mott, N. F. and Massey, H. S. W., *The theory of atomic collisions*, 3rd edn, p. 76 (Clarendon Press, Oxford, 1965).

9.3 RESONANCE PHENOMENA—THRESHOLD BEHAVIOUR

Now

$$\int X_j V_{01} \mathcal{G}_0 \, dr = (e^2/4\pi) \iint X_j(r_1)\psi_1(r_2)r_{12}^{-1}\psi_0(r_2)\mathcal{G}_0(r_1)r_1^{-2} \, d\mathbf{r}_1 d\mathbf{r}_2$$
$$= e^2(\pi k\hbar^2/2m)^{\frac{1}{2}} \iint \phi(r_1,r_2)r_{12}^{-1}\psi_E(r_1,r_2) \, d\mathbf{r}_1 d\mathbf{r}_2 \quad (122)$$

in the notation of § 1, it being remembered that in that section a simpler, less accurate approximation is used for ϕ. The factor $(\pi k\hbar^2/2m)^{\frac{1}{2}}$ arises because ψ_E is normalized per unit energy range so as to have the asymptotic form for large r_1 (see (25))

$$\psi_E(r_1,r_2) \sim \psi_0(r_2)(m/8\pi^3 k\hbar^2)^{\frac{1}{2}} \sin(kr+\eta).$$

We see then that (121) and (29) are equivalent apart from the better approximation used in the former for the doubly-excited wave function ϕ.

It is possible to solve coupled equations such as (100) quite accurately no matter how large the coupling U_{01} may be, so that we have a means of determining the positions and level widths of the resonances to rather greater accuracy. The only requirement is that the integration of the equations can be carried out at intervals of k^2 small compared with $2m\Gamma/\hbar^2$: otherwise the resonances will be missed or, at best, broadened.

There is no difficulty, at least in principle, in extending the analysis to more complicated sets of coupled equations including allowance for electron exchange as discussed in Chapter 8, § 2.2.3.

Before describing the results of calculations of this kind it is worth noting that, if the level shift $\Delta\kappa_j^2 \hbar^2/2m$ due to the coupling is small, the resonance energies may be determined to good approximation by calculating the energies of the bound states for the interaction V_{11}. In the case of more complicated sets of equations the corresponding procedure would be to determine the bound states that arise as eigenfunctions for the reduced set of coupled equations that involve the closed channels only, the coupling with the open channels being ignored. Examples of such calculations and comparison with results obtained by accurate solution of the equations in which this latter coupling is included will be given in § 3.1.

It is also possible, at great cost in computing time, to combine the differential equation and usual variational approach so that the level shift due to the interaction is allowed for in addition to the contribution from the closed channels not otherwise included.

3.1 *Application to the calculation of resonance energies and level widths associated with the doubly-excited states of helium based on the 2s and 2p states of* He^+

The simplest cases of resonances of the kind we have been discussing are those that arise due to the occurrence of the Rydberg series of doubly-excited states based on the 2-quantum excited states of He$^+$. Resonance effects due to these states are observable in the energy-loss spectra of electrons that have produced ionization of neutral helium atoms and also in photo-ionization cross-sections at sufficiently high frequencies. They also give rise to resonance effects in elastic scattering of electrons by He$^+$ ions that are not as yet susceptible to experimental observation. Despite the fact that the elastic scattering resonances in electron collisions with hydrogen atoms are observable, we choose to discuss them later because they depend on the excitation of doubly-excited states of a negative ion that are in many ways less simple and familiar than the states of neutral atoms.

To a good approximation we can expect that the Rydberg series of He$^+$ in question can be obtained by solution of the coupled equations that arise when the wave function describing the collisions of electrons with the He$^+$ ions is represented by an eigenfunction expansion that includes the $1s$, $2s$, and $2p$ states of the ion but no others—the $1s$–$2s$–$2p$ close-coupling approximation discussed in Chapter 8, § 2.2.3.

As discussed in that section the possible states which can arise from the coupling of $1s$, $2s$, and $2p$ orbitals may be distinguished by the respective total orbital and spin quantum numbers L and S and the parity. Corresponding to each value of L, $S = 0$ or 1, giving the respective singlet and triplet states.

Since we are considering an electron incident on a helium ion in the ground state, $L = l$, the angular momentum quantum number for the motion of the electron relative to the ion. The states which couple with this initial state must then also have $L = l$. To distinguish the possibilities we need only use the rule for combination of angular momenta (see also Chap. 8, § 2.2.3).

When $l = L = 0$, the states which couple may be written $2s$, $l_2 = 0$, $2p$, $l_2 = 1$, where l_2 refers to the angular momentum quantum number of the relative motion of the 'scattered' electron relative to the atom in a $2s$ or $2p$ state respectively. The close-coupling approximation in this case therefore involves two equations associated with closed channels. The resonance states obtained from these channels correspond to the doubly excited series $2s\,ns$ and $2p\,np$, respectively. Some information about the identification of any resonance state in these terms may be obtained from the asymptotic forms of the solutions of the equations in the close channels. Thus if we denote the radial functions for the two closed channels as $\mathscr{G}_{2s}$, $\mathscr{G}_{2p}$ respectively and at a resonance energy E_r^j
$$\mathscr{G}_{2s} \sim N_{2s}^j e^{-\kappa_j r}, \qquad \mathscr{G}_{2p} \sim N_{2p}^j e^{-\kappa_j r}, \tag{123}$$

then the relative magnitude of N_{2s}^j, N_{2p}^j is some guide to the relative weight to be assigned to the 2s and 2p series in the actual solution at the resonance.

Further evidence can be obtained by comparison of the energy values with those calculated by perturbation methods based on particular configuration assignments. In any case the zero-order phase shift η_0 will show resonance effects due to interaction with doubly-excited states $2s\,ns$, and $2p\,np$.

When $l\,(=L) = 1$ the states which couple are 2s, $l_2 = 1$; 2p, $l_2 = 0$, and 2p, $l_2 = 2$ corresponding to the series $2s\,np$, $2p\,ns$, $2p\,nd$ respectively. $2p\,np$ does not contribute because such states are of even parity whereas the initial state is odd. In this case there will be four coupled equations. Resonance effects in the first-order phase shift will arise from interaction with states of the three series above.

Tables 9.1–9.4 give the values of the resonance energies for the ^1S, ^3S, $^1P^0$, and ^3P series respectively, calculated by Burke and McVicar† by numerical solution of the appropriate 1s–2s–2p coupled equations, using a high-speed computer. The integrodifferential equations were solved by the non-iterative method of Marriott,‡ which made it possible to obtain solutions at close energy intervals through the resonances. In this way the level widths as well as the resonance energies could be determined. In addition the weighting factors attached to each of the coupled states are given in the form $N_\alpha^j/(\sum N_\alpha^j)^2$, where the subscript distinguishes the different states and the N_α^j are as defined in (123).

Included also in Tables 9.1 and 9.2 are values of the resonance energies uncorrected for level shifts that have been calculated by O'Malley and Geltman† by a variational method. This procedure has the advantage of allowing to some extent for contributions from closed channels in addition to those associated with the 2s and 2p states of He^+. Thus if the interaction with the incident 1s channel is ignored the resonance energies are the eigenvalues of the infinite set of equations representing the closed channels. It has been shown by Hahn, O'Malley, and Spruch§ how a good approximation to these energies may be obtained by a variational method. It is necessary to use a trial wave-function that contains no admixture of the open-channel state. If $\Psi_t^{(0)}(\mathbf{r}_1, \mathbf{r}_2)$ is an initial trial wave-function, then to eliminate terms corresponding to

---

† BURKE, P. G. and McVICAR, D. D., *Proc. phys. Soc.* **86** (1965) 989.
‡ MARRIOTT, R., ibid. **72** (1958) 121.
§ O'MALLEY, T. F. and GELTMAN, S., *Phys. Rev.* **137** (1964) A1344.
‖ HAHN, Y., O'MALLEY, T. F., and SPRUCH, L., ibid. **128** (1962) 932.

the open channel this function is transformed to

$$\Psi_t^{'(1)}(\mathbf{r}_1,\mathbf{r}_2) = (1-P_1-P_2-P_1P_2)\Psi_t^{'(0)}(\mathbf{r}_1,\mathbf{r}_2), \qquad (124)$$

where P_1 and P_2 are projection operators defined as follows. Suppose $\Psi_t^{'(0)}$ is expressed in the form

$$\Psi_t^{'(0)} = \left(\sum + \int\right)\psi_n(\mathbf{r}_2)\chi_n(\mathbf{r}_1), \qquad (125)$$

where the $\psi_n(\mathbf{r}_1)$ are the complete set of eigenfunctions for He^+. Then

$$P_1\Psi_t^{'(0)} = \psi_0(r_2)\chi_0(r_1)$$
$$= \psi_0(r_2)\int \Psi_t^{'(0)}(\mathbf{r}_1,\mathbf{r}_2')\psi_0(r_2')\,\mathrm{d}\mathbf{r}_2', \qquad (126)$$

and similarly for P_2.

Once $\Psi_t^{'(1)}$ is obtained the variational calculation proceeds in the usual way. Thus if $\Psi_t^{'(1)}$ is made up of a linear combination of n functions the extremum condition leads in general to n roots for the energy of the system. According to the Hylleraas–Undheim theorem[†] these roots each represent an upper bound to the respective exact eigenvalue.

O'Malley and Geltman used trial functions of the form

$$\Psi_t^{'(0)} = \Phi_t(r_1,r_2) \pm \Phi_t(r_2,r_1), \qquad (127)$$

where

$$\Phi_t(r_1,r_2) = r_1^L P_L(\cos\theta_1) \times$$
$$\times \left[\psi_{2s}(r_2)\sum_{i=1}^{N} a_i e^{-\alpha_i r_1} + \sum_{\lambda=0}^{2}\left\{\sum_{j\lambda=1}^{N\lambda} b_{j\lambda}P_\lambda(\cos\theta_{12})r_1^\lambda r_2^\lambda e^{-\beta_{j\lambda}r_1-\gamma_{j\lambda}r_2}\right\}\right], \qquad (128)$$

and included as many as twenty-five terms in their calculations.

A further set of values, extending to higher members of the $^1P^0$ and ^3P series have been derived from calculations of the wave function in the $1s$–$2s$–$2p$ close-coupling approximation at energies above the $n=2$ excitation threshold, by extrapolation procedures based on quantum defect analysis (see § 5.1). These agree well with those calculated directly at energies below the threshold when comparison is possible.

Finally, as described in Chapter 8, § 2.5, Burke and Taylor[‡] combined a trial function rather similar in form to (128) with a $1s$–$2s$–$2p$ eigenfunction expansion. The only major difference in the form of Φ_t is the inclusion of terms depending explicitly on the interelectronic separation r_{12} so that correlation effects should be properly allowed for.

Results obtained by these methods are given in Tables 9.1 and 9.2, for the first three ^1S and the first ^3S resonances. There is very close agreement between the resonance energies calculated by the combined method and those, $E_r - \Delta$, uncorrected for level shift, calculated by

[†] HYLLERAAS, E. A. and UNDHEIM, B., *Z Phys.* **65** (1930) 759.
[‡] BURKE, P. G. and TAYLOR, A. J., *Proc. phys. Soc.* **88** (1966) 549.

9.3 RESONANCE PHENOMENA—THRESHOLD BEHAVIOUR

TABLE 9.1

Resonance energies and level widths in helium associated with the ^1S doubly-excited series based on 2-quantum excited states

Resonance energies (eV)			Effective quantum number $n-\mu$	Level width (eV) Γ		Reduced level width (eV) $(n-\mu)^3\Gamma$		N_{2s}/N $(N^2 = N_{2s}^2 + N_{sp}^2)$	N_{2p}/N	Configuration assignment
E_r (calculated) $1s-2s-2p\dagger$ correlation	$(E_r - \Delta)$ (calculated) O'Malley§ and Geltman	E_r (observed)		(calculated) $1s-2s-2p\dagger$	(calculated) $1s-2s-2p\ddagger$ + correlation					
57·865	57·84	57·82‖††	1·343	0·141	0·124	0·341	0·300	0·85	−0·53	$2s^2$
62·81	62·16	62·15‖	2·290	0·019	0·007	0·228	0·084	−1·00	0·06	$2p^2$
63·01	62·96	62·95‖	2·384	0·0325	0·036	0·440	0·487	−0·77	0·63	$2s\,3s$
	64·04		3·339	0·016		0·596		0·94	−0·33	$2p\,3p$
	64·22	64·22‖	3·387	0·003		0·117		0·44	−0·90	$2s\,4s$
	64·68		4·337	0·008		0·653		−0·92	0·39	$2p\,4p$
	64·70		4·393	0·0005		0·042		−0·13	0·99	$2s\,5s$

† BURKE, P. G. and MCVICAR, D. D., *Proc. phys. Soc.* **86** (1965) 989.
‡ BURKE, P. G. and TAYLOR, A. J., *ibid.* **88** (1966) 549.
§ O'MALLEY, T. F. and GELTMAN, S., *Phys. Rev.* **137** (1964) A1344.
‖ RUDD, M. E., *Phys. Rev. Lett.* **13** (1964) 503; **15** (1965) 580.
†† SIMPSON, J. A., MIELCZAREK, S. R., and COOPER, J., *J. opt. Soc. Am.* **54** (1964) 269.

TABLE 9.2

Resonance energies and level widths in helium associated with the ^3S doubly-excited series based on 2-quantum excited states

Resonance energies (eV)			Effective quantum number $n-\mu$	Level width (eV) Γ		Reduced level width (eV) $(n-\mu)^3\Gamma$	N_{2s}/N $(N^2 = N_{2s}^2 + N_{2p}^2)$	N_{2p}/N	Configuration assignment
E_r (calc.) $1s$–$2s$–$2p$†	$1s$–$2s$–$2p$‡ correlation	$(E_r - \Delta)$ (calc.) O'Malley§ and Geltman		$1s$–$2s$–$2p$	$1s$–$2s$–$2p$ +correlation				
62·62	62·615	62·62	2·211	2×10^{-4}	2×10^{-4}	$2\cdot2\times 10^{-3}$	$-0\cdot82$	$0\cdot57$	$2s\ 3s$
63·82		63·76	2·934	$7\cdot5\times 10^{-6}$		$1\cdot9\times 10^{-4}$	$0\cdot79$	$0\cdot61$	$2p\ 3p$
64·08		63·95	3·203	$8\cdot5\times 10^{-5}$		$2\cdot8\times 10^{-3}$	$0\cdot80$	$-0\cdot60$	$2s\ 4s$
64·54		64·56	3·957	$4\cdot2\times 10^{-6}$		$2\cdot6\times 10^{-4}$	$-0\cdot77$	$-0\cdot64$	$2p\ 4p$
64·63		64·68	4·198	$4\cdot2\times 10^{-5}$		$3\cdot1\times 10^{-3}$	$-0\cdot78$	$-0\cdot62$	$2s\ 5s$

† BURKE, P. G. and McVICAR, D. D., *Proc. phys. Soc.* **86** (1965) 989. § O'MALLEY, T. F. and GELTMAN, S., *Phys. Rev.* **137** (1964) A1344.
‡ BURKE, P. G. and TAYLOR, A. J., ibid. **88** (1966) 549.

O'Malley and Geltman.† Apart from the $2p^2\,^1S$ level the $1s$–$2s$–$2p$ close-coupling approximation without inclusion of the additional correlation terms also gives results in close agreement. The discrepancy for the $2p^2\,^1S$ resonance is probably due to the importance of coupling with the $3d^2\,^1S$ configuration, which is not allowed for with a $1s$–$2s$–$2p$ eigenfunction expansion. Comparison may also be made between the level widths calculated with and without inclusion of the trial function which allows for correlation. Again there are only small differences, except for the $2p^2\,^1S$ level.

For other ^1S and ^3S levels and for the $^1P^0$ and ^3P series, comparison of resonance energies calculated using the $1s$–$2s$–$2p$ expansion without inclusion of correlation and by the variational method without inclusion of interaction with the open channel reveals very good agreement.

For the $^1P^0$ and ^3P series values of E_r and Γ derived by quantum defect extrapolation, as described in § 10.7 are also included. These extend to much higher terms in the series and agree quite well with those calculated using the $1s$–$2s$–$2p$ expansion for the lower values of n for which comparison is possible.

It can therefore be assumed that, apart from a few exceptional cases, the $1s$–$2s$–$2p$ expansion method above will give good results.

The assignment of configurations to the different doubly-excited levels raises many points of interest. Turning to Table 9.1 we note that, for the ^1S levels, there is a close approach to degeneracy between the $2s\,ns$ and $2p\,(n-1)p$ levels with the result that there is considerable mixing between them. The assignments are therefore rather poor representations of the real situation. This is borne out by the weighting factor N_{2s}, N_{2p}.

The ^3S series begins with $2s\,3s$, $2s^2$ being excluded by the Pauli principle. It will be seen from Table 9.2 that the level widths are very small compared with those for ^1S. This is because the wave functions for the triplet states are very small when the electrons are close together because of the antisymmetry in the electron coordinates. To produce auto-ionization one of the electrons must acquire a high kinetic energy from its interaction with the other and this is only possible when the two approach closely. It follows that the rate of auto-ionization will be low for the triplet series. Although all the widths are small they are seen to fall into two series, one set having widths of the order ten times larger than the other. The latter have all been assigned configurations of the type $2p\,np$, the former $2s\,ns$. This is consistent with the level

† loc. cit., p. 623.

widths because the main contribution to auto-ionization must come from close to the nucleus where the overlap with the 1s state is large and where high kinetic energy is acquired. This strongly favours $2s\,ns$ as against $2s\,np$.

The position as regards the P levels is especially interesting. There is energy degeneracy in the zero order approximation between the $2s\,np$ and $2p\,ns$ levels and they must be expected to mix strongly under the influence of the electronic interaction. Cooper, Fano, and Prats† were the first to point out that it is more correct to consider mixed configurations $2s\,np \pm 2p\,ns$ in which the two states have nearly equal weight. This applies to both the $^1P^0$ and ^3P series, but whereas the states of the positive series have the larger level widths for the $^1P^0$ series the reverse is the case for ^3P. The reason for this is as follows.

Apart from any symmetry conditions the levels of the negative series would be expected to be higher than the corresponding levels of the positive series for the following reason. The wave functions $\psi_{2s}(r_1)\psi_{np}(\mathbf{r}_2)$ and $\psi_{2p}(\mathbf{r}_1)\psi_{ns}(r_2)$ are nearly equal when both electrons lie close to the nucleus, so the wave functions of the negative states nearly vanish under these circumstances. This reduces the mean interelectronic repulsion compared with the positive states for which the same considerations do not apply. It would also follow by the same argument that the auto-ionization probability and level widths should be relatively small for the negative states.

We must now superpose the symmetry requirements demanded by the Pauli principle. Because of this the first member of the negative series for the $^1P^0$ states is excluded. Thus

$$\psi_{2s}(r_1)\psi_{np}(\mathbf{r}_2) - \psi_{2p}(\mathbf{r}_1)\psi_{ns}(r_2) + \psi_{2s}(r_2)\psi_{np}(\mathbf{r}_1) - \psi_{2p}(\mathbf{r}_2)\psi_{ns}(r_1)$$

vanishes when $n \to 2$. Again, because wave functions ψ_{np} vary very slowly with n at small distances from the nucleus it follows that, for the $^1P^0$ series, the symmetry requirements enhance still further the tendency for the positive states to be higher than the corresponding negative ones and to have larger level widths. Reference to Table 9.3 shows that these features are clearly present.

For the ^3P states, however, the symmetry requirements exclude the first member of the positive series. Thus

$$\psi_{2s}(r_1)\psi_{np}(\mathbf{r}_2) + \psi_{2p}(\mathbf{r}_1)\psi_{ns}(r_2) - \psi_{2s}(r_2)\psi_{np}(\mathbf{r}_1) - \psi_{2p}(\mathbf{r}_2)\psi_{ns}(r_1)$$

vanishes when $n \to 2$. In this case the symmetry requirement tends to act in the opposite way, reducing the tendency of the negative states to

† COOPER, J. W., FANO, U., and PRATS, F., *Phys. Rev. Lett.* **10** (1965) 518.

be lower and have smaller level widths. Reference to Table 9.4 shows that in fact the symmetry requirement is partly dominant—the states of the ^3P negative series have the smaller level widths but lie higher than those of the positive series, though the separation is smaller for the ^3P than for the $^1P^0$ series.

In addition to the mixed positive and negative series there also exists a $2pnd$ series. Because of the smallness of the d wave functions close to the nucleus the terms of this series lie relatively high while the auto-ionization probabilities, and hence level widths, are very small. Both these features are clear from the assignments in Tables 9.3 and 9.4.

3.2. Comparison with experiment

Experimental information about the doubly-excited states of helium comes from observations of electron energy-loss and photo-ionization spectra. The first association of a doubly-excited state with observations of the former type was made by Massey[†] who suggested in 1933 that the line observed by Priestley and Whiddington[‡] (see Chap. 5, § 5.3.3) at an energy loss near 64 eV was due to excitation of one of the low-lying doubly-excited levels. This was followed by calculations,[§] using Born's approximation, of cross-sections for excitation of different doubly-excited helium states by electron impact. It was concluded from the calculations that the state in question was the $2s\,2p\ ^1P$ state. Although at about this time Fano discussed[||] the interaction between discrete and continuum states in relation to the interpretation of Beutler's[††] observations (see Chap. 14, § 7.3) of auto-ionizing lines in the absorption spectra of rare gases, no application of this technique to the collision problem was made at the time. The matter rested until Silverman and Lassettre (see Chap. 5, § 5.3.3) confirmed the existence of a resonance effect in the energy-loss spectrum and improved the precision of observation. The next step in the theoretical interpretation was made by Fano[‡‡] who analysed the position and shape of the level observed by Lassettre and Silverman[§§] in terms of the theory described in § 1. Assuming, on the basis of interpolation across the resonance, that the continuous background is falling uniformly at a rate of 10 per cent per eV over this

---

[†] MASSEY, H. S. W., private communication to Professor R. H. Fowler (1933).
[‡] PRIESTLEY, H. and WHIDDINGTON, R. R., *Proc. Leeds phil. Soc.* **2** (1934) 491.
[§] MASSEY, H. S. W. and MOHR, C. B. O., *Proc. Camb. phil. Soc. math. phys. Sci.* **31** (1935) 604.
[||] FANO, U., *Nuovo Cim.* **12** (1935) 156; see also earlier work of RICE, O. K., *J. chem. Phys.* **1** (1933) 375.   [††] BEUTLER, H., *Z. Phys.* **93** (1935) 177.
[‡‡] FANO, U., *Phys. Rev.* **124** (1961) 1866.
[§§] SILVERMAN, S. M. and LASSETTRE, E. N., *J. chem. Phys.* **40** (1964) 1265.

region Fano obtained a reasonable fit using his profile formula (44) with the following parameters:

$$E = 60\cdot 1 \text{ eV}, \qquad \Gamma = 0\cdot 04 \text{ eV}, \qquad q^2 = 3\cdot 06.$$

Further evidence about this resonance level and about other levels of the $^1P^0$ series has come not only from the very precise electron impact spectra obtained by Simpson and Mielczarek[†] as described in Chapter 5, § 4.2.1 but also from experiments[‡] on photo-ionization using high wavelength resolution. These latter are described in Chapter 14, § 7.1 but the values of the resonance energies obtained by both methods are given in Table 9.3. It will be seen that agreement is very good, both between the experimental results and between these and the calculated values. A more detailed discussion of the resonance profiles in the photo-ionization spectra is given in Chapter 14, § 7.1. It is of interest, however, to note here that the resonance width of the $2s\,2p\,^1P^0$ level derived from the measurements is 0·038 eV, which agrees closely with that deduced by Fano[§] from the electron spectra of Silverman and Lassettre.

The doubly-excited levels of the other series cannot be excited by optical absorption. As described in Chapter 5, § 4.2.1 Simpson, Mielczarek, and Cooper[‖] observed energy losses corresponding to excitation of states with energy a little below that of the $2s\,2p\,^1P$ level. Two of these may be identified with the $2s^2\,^1S$, $2s\,2p\,^3P$ levels, the correspondence with the calculated energies being good. The third, very weak loss, occurring very close to the strong excitation, the $2s\,2p\,^1P$ state, is less easy to identify.

Further evidence concerning the ^1S and ^3P series has been obtained by Rudd[††] who excited the various levels by proton impact as described in Volume IV. Although the detailed discussion of this work will be deferred to that chapter the resonance energies derived from it are also given in Tables 9.1 and 9.4. These include those found by electron impact studies as well as several others that may be identified through their close agreement with the calculated energies of other states of the series.

It will be noted also that all the resonance levels that have been observed are those that have relatively large calculated level widths. Thus no levels of the ^3S series have yet been observed.

The agreement between theory and experiment is already very good.

[†] SIMPSON, J. A. and MIELCZAREK, S. R., ibid. **39** (1963) 1606; SIMPSON, J. A., MIELCZAREK, S. R., and COOPER, J., *J. opt. Soc. Am.* **54** (1964) 269.
[‡] MADDEN, R. P. and CODLING, K., *Phys. Rev. Lett.* **10** (1963) 516; *J. opt. Soc. Am.* **54** (1964) 2683; *Astrophys. J.* **141** (1965) 364.
[§] FANO, U., *Phys. Rev.* **124** (1961) 1866. [‖] loc. cit., p. 321.
[††] RUDD, M. E., *Phys. Rev. Lett.* **13** (1964) 503 and **15** (1965) 580.

Results obtained by photo-ionization experiments which are discussed in Chapter 14 not only extend the region of agreement to line profiles for excitation of the Rydberg series based on 2s or 2p orbitals but also reveal the excitation of series based on states with total quantum number $n > 2$.

4. Resonance energies and level widths of doubly-excited states of H⁻

Although H⁻ is isoelectronic with He, certain differences arise in the structure of the doubly-excited energy level system due to the fact that the mean effective interaction that a hydrogen atom in any excited state exerts on an additional electron falls off faster than r^{-2} at large distances r. It is well known that an interaction that falls off faster than r^{-2} will, in general, support only a finite number of bound states so that, at first sight, it would seem that there could at most be only a finite number of doubly-excited states of H⁻ based on a singly-excited orbital of H of total quantum number n. In fact, as first pointed out by Gailitis and Damburg,† the situation is affected by the energy degeneracy of the H atom states of different angular momenta associated with a given n. The coupling between these states modifies the effective interaction acting on an additional electron so that it falls off as slowly as r^{-2} for large r. If the degeneracy were exact this would lead to an infinite series of doubly-excited states associated with a given n but for certain values only of the total orbital angular-momentum quantum number L. Thus we have such series for

$$n = 2, \quad L = 0, 1, 2;$$
$$n = 3, \quad L = 0, 1, 2, 3, 4;$$
$$n = 4, \quad L = 0, 1, 2, 3, 4, 5, 6.$$

The energy level diagram is as shown in Fig. 9.5.

Even when these series occur they are very different from Rydberg series. To a close approximation the energies of the successive levels below the ionization limit to which they converge form a geometrical progression. Thus if $E_{nL}^{(1)}$ is the energy of the lowest level of a series associated with particular values of n and L the energies of the higher levels are given by $E_{nL}^{(1)}(\alpha, \alpha^2, \alpha^3,....)$, where $\alpha < 1$. Some values for series associated with small values of n are given in Table 9.5.

In fact the degeneracy between the s, p, d,... levels of hydrogen for a given n is not exact because of the Lamb shift. It may be shown that

† GAILITIS, M. and DAMBURG, R., *Proc. phys. Soc.* **82** (1963) 192.

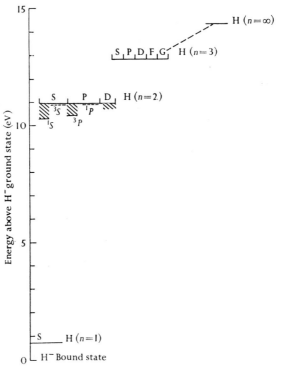

Fig. 9.5. Energy level diagram for H⁻ ions showing the different series converging to states of excited H atoms.

TABLE 9.5

Ratios ($1/\alpha$) of energy separations of successive levels from the series limit for various doubly excited levels in H⁻

				L				
n	0	1	2	3	4	5	6
2	17·429	29·334	4422·18				
3	4·823	5·164	6·134	9·323	62·416		
		16·752	80·552				
4	2·982	3·047	3·197	3·485	4·070	5·608	16·698
	16·210	4·360	4·940	6·494	14·492		
		27·299	18·777	8·516 × 10⁸			
			3226·57				

if $\Delta\epsilon$ is the actual energy difference between nearly degenerate states, the above considerations remain applicable provided the energy $E_n^{(s)}$ of a particular state below the corresponding ionization threshold satisfies

$$E_n^{(s)} \gg \tfrac{1}{2}(L+1)\Delta\epsilon. \qquad (129)$$

9.4 RESONANCE PHENOMENA—THRESHOLD BEHAVIOUR

This means that, for sufficient large values of s, the series of levels associated with n and L must terminate.

As an example for $n = 2$, $\Delta\epsilon$ is about 4×10^{-6} eV. The deepest level of the $L = 0$ series lies 0·6 eV below the ionization limit. Reference to Table 9.5 shows that $\alpha = 0\cdot057$ for this series. It then follows from (129) that the series will terminate after four terms.

To indicate how these results are obtained we consider in more detail the case of $n = 2$, $L = 0$. The $2s$–$2p$ close-coupling equations for this case take the form, for large r,

$$\left(\frac{d^2}{dr^2} - \kappa^2\right) G_{2s} - \frac{6}{r^2} G_{2p} = 0, \tag{130 a}$$

$$\left(\frac{d^2}{dr^2} - \kappa^2 - \frac{2}{r^2}\right) G_{2p} - \frac{6}{r^2} G_{2s} = 0, \tag{130 b}$$

where G_{2s}, G_{2p} satisfy the condition

$$G_{2s,2p} \sim e^{-\kappa r} \beta_{2s,2p} \quad (r \to \infty). \tag{131}$$

It will be seen that the coupling term falls off only as r^{-2}.

The equations may be decoupled as follows. The linear combination

$$G_\gamma = G_{2s} + \gamma G_{2p}$$

satisfies the equation

$$\left(\frac{d^2}{dr^2} - \kappa^2\right)(G_{2s} + \gamma G_{2p}) - \frac{2}{r^2}\{(3+\gamma)G_{2p} + 3\gamma G_{2s}\} = 0. \tag{132}$$

If we choose γ so that

$$3\gamma/(3+\gamma) = 1/\gamma \tag{133}$$

then

$$\left(\frac{d^2}{dr^2} - \kappa^2 - \frac{6\gamma}{r^2}\right) G_\gamma = 0. \tag{134}$$

From (133)

$$\gamma = \{1 \pm \sqrt{(37)}\}/6, \tag{135}$$

so (134) may be written

$$\left(\frac{d^2}{dr^2} - \kappa^2 - \frac{\lambda(\lambda+1)}{r^2}\right) G_\lambda = 0, \tag{136}$$

where

$$\lambda = -\tfrac{1}{2} \pm \sqrt{\{\pm\sqrt{(37)} + \tfrac{5}{4}\}}.$$

This apparently gives four alternative values for λ but the ambiguity due to expression of 6γ in the form $\lambda(\lambda+1)$ does not lead to anything new. We may thus take

$$\lambda = -\tfrac{1}{2} + \{\pm\sqrt{(37)} + \tfrac{5}{4}\}^{\frac{1}{2}}, \tag{137}$$

so

$$\lambda + \tfrac{1}{2} = \{\sqrt{(37)} + \tfrac{5}{4}\}^{\frac{1}{2}}, \tag{138 a}$$

or

$$i\{\sqrt{(37)} - \tfrac{5}{4}\}^{\frac{1}{2}}. \tag{138 b}$$

Consider now the behaviour of the solutions of (136) for small κ. To do this we distinguish three regions:

$$\text{I, } 0 < r \leqslant r_1; \quad \text{II, } r_1 \leqslant r \leqslant r_2; \quad \text{III, } r_2 \leqslant r \to \infty. \tag{139}$$

In regions II and III the equations (130) are valid but in region II we suppose that κ^2 is negligible compared with $\lambda(\lambda+1)r^{-2}$. On the other hand, in region I, while κ^2 is negligible other coupling terms, ignored in (130 a) and (130 b) and hence in (136), must be taken into account. This means that, for $r = r_1$,

$$G_\lambda/G_\lambda' = c, \tag{140}$$

where c is a constant independent of κ but depending on the internal interactions and $G'_\lambda = dG_\lambda/dr$.

In region II,
$$G_\lambda = ar^\lambda + br^{-\lambda-1}, \tag{141}$$
where, in view of (140),
$$a/b = -r_1^{-2\lambda-1}\{r_1+(\lambda+1)c\}/(r_1-\lambda c), \tag{142}$$
and is independent of κ.

Finally, in region III,
$$G_\lambda = (i\kappa r)^{\frac{1}{2}}\{AJ_{\lambda+\frac{1}{2}}(i\kappa r) + BJ_{-\lambda-\frac{1}{2}}(i\kappa r)\}. \tag{143}$$

The Bessel functions† have the asymptotic form
$$J_{\lambda+\frac{1}{2}}(i\kappa r) \sim (2/\pi i\kappa r)^{\frac{1}{2}}\cos(i\kappa r - \tfrac{1}{2}\lambda\pi - \tfrac{1}{2}\pi), \tag{144a}$$
$$J_{-\lambda-\frac{1}{2}}(i\kappa r) \sim (2/\pi i\kappa r)^{\frac{1}{2}}\cos(i\kappa r + \tfrac{1}{2}\lambda\pi), \tag{144b}$$
so that if (143) is to have the asymptotic form (131)
$$B = -iAe^{i\lambda\pi}. \tag{145}$$

When κ is small so that $\kappa r_2 \ll 1$, we may use the series expansions of the Bessel functions†
$$J_{\lambda+\frac{1}{2}}(i\kappa r) \simeq (2i\kappa r/\pi)^{\frac{1}{2}}(2i\kappa r)^\lambda \Gamma(\lambda+1)/\Gamma(2\lambda+2), \tag{146a}$$
$$J_{-\lambda-\frac{1}{2}}(i\kappa r) \simeq \cos\pi\lambda(2i\kappa r/\pi)^{\frac{1}{2}}(2i\kappa r)^{-\lambda-1}2\Gamma(2\lambda+1)/\Gamma(\lambda+1). \tag{146b}$$

Substitution in (143) and comparison with (141) gives
$$\frac{a}{b} = -(2\kappa)^{2\lambda+1}\{\Gamma(\lambda+1)\}^2 \sec\pi\lambda/2\Gamma(2\lambda+2)\Gamma(2\lambda+1). \tag{147}$$

This shows that
$$\kappa^{2\lambda+1} = \text{constant}. \tag{148}$$

When $\lambda = -\tfrac{1}{2}+i\mu$ this condition becomes
$$\kappa^{2i\mu} = \text{constant}. \tag{149}$$

This is satisfied by
$$2\mu\ln\kappa = \pm 2|s|\pi + 2\mu\ln\kappa_0, \quad |s| = 0, 1, 2,...$$
or
$$\kappa^2 = \kappa_0^2 \exp(\pm 2|s|\pi/\mu), \tag{150}$$
showing that if $\kappa_0^2 \hbar^2/2m$ is an allowed energy value then so also is
$$\exp(\pm 2|s|\pi/\mu)\kappa_0^2 \hbar^2/2m.$$

This will be true provided κ_0^2 is such that κ^2 is not too small for (129) to hold nor too large for κr_2 to be no longer small.

In the case we have considered the progression of energy levels, measured below the ionization threshold, will be of the form
$$E_0(1, \alpha, \alpha^2,...),$$
where
$$\alpha = e^{-2\pi/\mu}.$$
Since
$$\mu = \{\sqrt{(37)} - \tfrac{5}{4}\}^{\frac{1}{2}} = (4\cdot 833)^{\frac{1}{2}} = 2\cdot 199,$$
we have
$$1/\alpha = 17\cdot 429$$
as in Table 9.5 for $n = 2, L = 0$.

In Table 9.6 the energies of resonance levels below the $n = 2$ threshold for H⁻ are given. The first set of values was obtained by O'Malley and

† WHITTAKER, E. T. and WATSON, G. N., *Modern analysis*, 4th edn, Chap. xvii (Cambridge University Press, 1927).

9.4   RESONANCE PHENOMENA—THRESHOLD BEHAVIOUR   633

TABLE 9.6

Energies and widths of resonance levels of H^- below the $n = 2$ threshold

State	$E_r - \Delta$ (calc.) (eV)†	E_r (calc.) (eV)		E_r (obs.) (eV)‡	Γ (calc.) (eV)		Γ (obs.) (eV)‖
		1s-2s-2p‡	1s-2s-2p§ +correlation		1s-2s-2p‡	1s-2s-2p§ +correlation	
^1S	9·559	9·61	9·56	9·56	0·109	0·047$_5$	0·043±0·006
	10·178 (10·167)	10·178	10·178		0·0024	0·0022	
^3S	10·149	10·151	10·150		0·189×10^{-4}	0·206×10^{-4}	
	10·202 (10·200$_6$)						
^1P	10·178						
	10·204 (10·203)						
^3P	9·727	9·78		9·71±0·03	0·009		> 0·009
	10·198 (10·187$_5$)						

† O'Malley, T. F. and Geltman, S., loc. cit., p. 623.
‡ Burke, P. G. and Smith, K., loc. cit., p. 634; McEachran, R. P. and Fraser, P. A., loc. cit., p. 634.
§ Burke, P. G. and Taylor, A. J., loc. cit., p. 622.
‖ McGowan, J. W., *Phys. Rev. Lett.* **17** (1966) 1207.

Geltman† using the same variational technique as for the isoelectronic case of He⁺ discussed in § 3.1. For each series they calculated the lowest two energy values. According to the preceding discussion the respective ratios of these values should be as given in Table 9.5 for $n = 2$, $L = 0, 1$. That this holds quite closely may be seen from the bracketed values for the upper energy levels, derived by dividing the variational value for the lowest level by the appropriate factor given by Table 9.5. These compare quite well with the variational values.

Comparison is also made with the resonance energies, including level shift, but not the effects of states with $n > 2$, calculated by the $1s$–$2s$–$2p$ close-coupling approximation.‡ The agreement is not unsatisfactory although the results differ by 0·05 eV from those of O'Malley and Geltman.† Level widths for the ^1S, ^3S, and ^3P states have also been obtained by the close-coupling method but the widths for the ^3S and ^3P are so small as to be beyond determination within the accuracy of the calculation.

Burke and Taylor§ have extended these calculations by combining the $1s$–$2s$–$2p$ close-coupling trial function with an additional trial function including electron correlation effects as described in Chapter 8, § 2.5 and in § 3.1, p. 622 of this chapter. They obtain resonance energies agreeing very closely with those of O'Malley and Geltman† but the main effect of the correlation terms is to reduce the level width for the lowest ^1S state by a factor of more than 2. This very much improves the agreement with observation as we shall see.

A detailed analysis of the experimental data (Chap. 1, § 6.2.2) has been made by McGowan.‖ He considered first the observations made by McGowan, Clarke, and Curley†† who observed the variation with energy of the scattered intensity in a small solid angle around 90°. Reference to Fig. 1.20 shows clear evidence of a resonance near 9·5 eV with some indications only of a second resonance near 9·7 eV. Reference to Table 9.6 shows that we would expect only a ^1S and ^3P resonance state close to these energies. As the contribution from ^3P will vanish at 90° it is reasonable to suppose that the strong resonance arises from ^1S. McGowan made this assumption and then proceeded to derive the resonance energy and level width. To do this it was necessary to allow both for the finite

---

† O'MALLEY, T. F. and GELTMAN, S., *Phys. Rev.* **137** (1964) A1344.

‡ BURKE, P. G. and SMITH, K., *Atomic collision processes*, ed. McDOWELL, M. R. C., p. 89 (North Holland, Amsterdam, 1964); McEACHRAN, R. P. and FRASER, P. A., *Proc. phys. Soc.* **82** (1963) 1038.

§ BURKE, P. G. and TAYLOR, A. J., loc. cit., p. 622.

‖ McGOWAN, J. W., *Phys. Rev. Lett.* **17** (1966) 1207.   †† loc. cit., p. 34.

solid angle of acceptance of scattered electrons and for the width of the electron energy distribution in the actual experiments. Neither of these quantities could be accurately determined but limits could be placed upon them which were sufficient to permit evaluation of E_r^+ and Γ^+ for the ^1S level.

The differential cross-section for scattering into the solid angle $d\omega$ about the angle θ is given by the appropriate generalization of (83) to allow for the two spin states

$$I(\theta)\,d\omega = \tfrac{3}{4}I^- + \tfrac{1}{4}I^+,$$

where

$$I^- = \frac{\pi}{k^2}\left|\sum_{l\neq 1}(2l+1)(e^{2i\eta_l^-}-1)P_l(\cos\theta) + 3\left\{\frac{\Gamma^- e^{2i\eta_1^-}}{E-E_r^- + \tfrac{1}{2}i\Gamma^-} + e^{2i\eta_1^-} - 1\right\}\cos\theta\right|^2,$$

$$I^+ = \frac{\pi}{k^2}\left|\sum_{l\neq 0}(2l+1)(e^{2i\eta_l^+}-1)P_l(\cos\theta) + \frac{\Gamma^+ e^{2i\eta_0^+}}{E-E_r^+ + \tfrac{1}{2}i\Gamma^+} + e^{2i\eta_0^+} - 1\right|^2.$$

Here k is the electron wave number and E its energy. $\eta_l^\pm$ are the phase shifts for the background scattering in the respective singlet and triplet states and $E_r^\pm$, $\Gamma^\pm$ are the resonance energies and level widths for the ^1S and ^3P states, the plus sign referring to the former. To calculate the expected shape of the intensity energy variation, the phase shifts $\eta_0^\pm$ were taken as given by Schwartz (see Chap. 8, Table 8.2) and $\eta_1^\pm$, $\eta_2^\pm$ from $1s$–$2s$–$2p$ close-coupling calculations (see Figs. 8.1 and 8.2). The solid angular range of scattering electron acceptance was taken as given in terms of polar angles by $\Delta\theta = 15°$, $\Delta\phi = 15°$, and the electron energy distribution as gaussian with a half-width of 0·06 eV, a value that is consistent with the width of the main resonance in helium observed with the same equipment. It was found that a good fit to the data could be obtained only if $E_r^+ = 9\cdot56$ eV. Fig. 9.6 illustrates the sensitivity to the assumed value of Γ^+. The results are not sensitive to E_r^+ and Γ^- but these were taken as 9·73 and 0·010 eV respectively. Comparison with the actual observations shows that the best fit is obtained with Γ^+ near 0·045 eV. From further comparison of this type the best fit was obtained with

$$\Gamma^+ = 0\cdot043 \pm 0\cdot006 \text{ eV}.$$

It was then confirmed that change of the half-width of the electron energy spread from 0·06 to 0·08 eV had little effect on the derived value of Γ^+.

Reference to Table 9.6 shows that this value is less than half of that calculated by Burke and Smith from the $1s$–$2s$–$2p$ close-coupling approximation but agrees very much better with that obtained from the more elaborate calculations of Burke and Taylor, which included electron correlation terms in addition.

As far as the ^3P resonance is concerned its magnitude depends very much on the actual solid angle for scattered electron acceptance. From the data of McGowan, Clarke, and Curley, E_r^- is estimated to be $9 \cdot 71 \pm 0 \cdot 03$ eV. If Γ^- is taken as theoretically estimated, $0 \cdot 009$ eV, an unreasonably large acceptance angle $\Delta\theta$ of $25°$ is required to give good

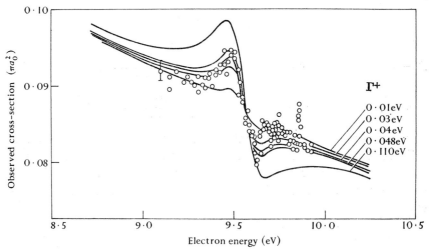

FIG. 9.6. Illustrating the sensitivity of the shape of the resonance in elastic scattering of electrons through $90°$ in atomic hydrogen, to the assumed value of the level width Γ^+, the values of the other parameters being as follows. $E_r^+ = 9 \cdot 56$ eV, $E_r^- = 9 \cdot 73$ eV, $\Gamma^- = 0 \cdot 010$ eV, $\Delta\theta \times 15°$, $\Delta\phi = 15°$, half width of electron energy distribution $= 0 \cdot 06$ eV. ○ observed results.

agreement with the observations. It is therefore considered that the true value of Γ^- is somewhat greater.

Once the various parameters have been determined the variation of the scattered intensity with electron energy at other angles of scattering may be calculated. Fig. 9.7 illustrates results obtained in this way assuming an acceptance solid angle about θ defined by $\Delta\theta = 15° = \Delta\phi$. Included in this figure are the results of the experiments of McGowan, Clarke, and Curley who observed the scattering at $\theta = 90°$, of Kleinpoppen and Raible who observed at $\theta = 94°$, and of Schulz who observed the transmitted beam so that emphasis was placed on small scattering angles. It will be seen that the combined resonance profile varies quite rapidly with angle so that, even in Kleinpoppen and Raible's experiments, the contribution of the ^3P resonance is considerably more marked than in those of McGowan et al., while in Schulz's experiments

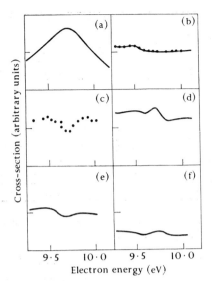

Fig. 9.7. Comparison of observed and calculated resonance profiles for electrons scattered elastically by atomic hydrogen. The energy spread of the incident electrons has been included in the calculations. (a) observed transmission (Schulz). (b) ● observed by McGowan, Clarke, and Curley at 90°; ——— calculated. (c) ● observed by Kleinpoppen and Raible at 94°. (d), (e), and (f) calculated for 70°, 100°, and 160° respectively.

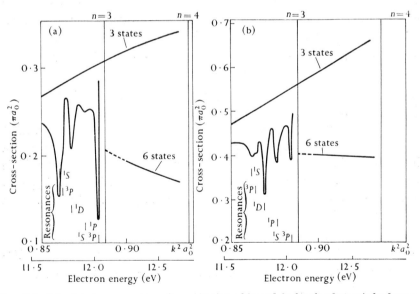

Fig. 9.8. Calculated cross-sections for excitation of $2s$ and $2p$ levels of atomic hydrogen by electron impact showing resonance effects just below the $n = 3$ threshold. (a) $2s$, (b) $2p$. The resonance effects are obtained with the 6 coupled-state approximation. No such effects are obtained with the $1s$–$2s$–$2p$ 3-state approximation.

it is completely dominant. The apparent discrepancies between the three experiments are thereby largely resolved.

Resonance effects due to doubly-excited states of H⁻ associated with 3-quantum excited states of H have been obtained by Burke, Ormonde, Taylor, and Whitaker† in the course of their calculations of the cross-sections for excitation of the $2s$ and $2p$ levels of atomic hydrogen (see Chap. 8, § 4.3) using the six-state ($1s$–$2s$–$2p$–$3s$–$3p$–$3d$) truncated eigenfunction expansion and solving the corresponding coupled equations. Fig. 9.8 shows the existence of the resonance effects in the 3-quantum excitation cross-section and their association with different levels of H.

5. Resonance energies and level widths of doubly and triply excited states of He⁻ (for discussion of experimental techniques and results see Chap. 1, § 6.2.3 and Chap. 5, § 4.2.1)

There seems to be little doubt that the resonance effect that appears as a peak in transmission of electrons by helium at 19·3 eV‡ and in a reduction of scattered intensity at an angle of 72° at the same energy§ is due to the $1s\,2s^{2}\,^{2}S$ state of He⁻.

In the experiments, for example, of Simpson‡ the energy spread of the electron beam was greater than the resonance width, so (89) applies. An increase of transmission and hence a decrease of the total elastic cross-section was observed. It follows that, if L is the angular momentum quantum number of the relative motion of the electron and atom, the phase shift η_L in the absence of resonance effects must satisfy

$$\cos 2\eta_L < 0.$$

All phase shifts at the resonance energy are certainly $< \frac{1}{4}\pi$ for $L > 0$ (see Table 8.6) but η_0 is close to $\frac{1}{2}\pi$ so that $\cos 2\eta_0$ is indeed negative.

In experiments with higher energy resolution‖ (see Chap. 1, Fig. 1.21) the shape of the resonance is more clearly defined and shows up as a high peak of transmission followed by a strong reduction at a little higher energy. This is just what would be expected if, in (84) $q_L < 0$ and $|q_L| < 1$, conditions which, since $q_L = \cot \eta_L$, can only be satisfied for helium at 19·3 eV for $L = 0$.

Further evidence that the resonance occurs for $L = 0$ is provided by the observed angular distributions†† of electrons scattered at energies

---

† BURKE, P. G., ORMONDE, S., TAYLOR, A. J., and WHITAKER, W., *Proc. Phys. Soc* **92** (1967) 319, 336.
‡ SIMPSON, J. A. and FANO, U., *Phys. Rev. Lett.* **11** (1963) 158.
§ SCHULZ, G. J., ibid. **10** (1963) 104.
‖ KUYATT, C. E., SIMPSON, J. A., and MIELCZAREK, S. R., *Phys. Rev.* **138** (1965) A385.
†† KLEINPOPPEN, H. and RAIBLE, V., *Phys. Lett.* **18** (1965) 24.

close to the resonance energy. As discussed in Chapter 5, § 5.3.3 the angular distribution of the resonance scattering, separated out as explained in that section, is isotropic (see Fig. 5.54), which it should be if $L = 0$.

There is no very clear evidence of the existence of other resonance states below the (2^3S) excitation threshold though Kuyatt et al.† found some indication of a weak resonance near 19·45 eV which they tentatively ascribed to $1s\,2s\,2p\,^2P^0$. s–p coupling between singly-excited states of the same total quantum number n is not as strong as in atomic hydrogen but is still close. Burke and Cooper‡ have investigated the bound states that are associated with coupled 2^3S, 2^1S, 2^3P, and 2^1P closed channels, from direct solution of the appropriate coupled equations. They find a ^2S bound state at 19·3 eV close to the prominent resonance in the elastic scattering, but the interaction is not strong enough to produce a ^2P bound state. The identification of the weak observed resonance near 19·45 eV if it proves real is therefore uncertain (see, however, Chap. 8, § 6.2).

Above and between the 2-quantum excitation thresholds a much greater wealth of detailed structure has been observed, particularly in separate inelastic differential cross-sections at small angles (see Chap. 5, § 5.3.3) but also in the total cross-section (see Chap. 1, § 6.2.3).

Between the thresholds the structure is probably due to interactions between the 2-quantum singly-excited states. The peak in the 2^3S cross-section close to its threshold is almost certainly due to the proximity of the ^2S resonance state of He⁻ (see Chap. 8, § 6.2).

At energies beyond the threshold the resonances arise from states of He⁻ that lie close below the thresholds for excitation of 3-, 4-, and 5-quantum singly-excited states. This is clearly seen in Fig. 5.29, which shows as functions of the incident electron energy the observed§ intensities of forward scattering of electrons after exciting each of the four 2-quantum states. There seems to be little doubt that the doubly-excited states of He⁻ which are concerned belong to configurations such as $1s\,3s^2$, etc.

An interesting further feature is the observation of effects that seem to be due to triply-excited states. Thus Kuyatt et al.,† in the course of their experiments on the transmission of electrons through helium, found resonance effects at $57·1\pm0·1$ eV and $58·2\pm0·1$ eV. Evidence for

---

† KUYATT, C. E., SIMPSON, J. A., and MIELCZAREK, S. R., *Phys. Rev.* **138** (1965) A385.
‡ BURKE, P. G. and COOPER, J. W., in course of publication.
§ CHAMBERLAIN, G. E. and HEIDEMAN, H. G. M., *Phys. Rev. Lett.* **15** (1965) 337.

the existence of these resonances in the forward scattering of electrons after exciting the 2-quantum states has also been found by Chamberlain and Heideman.† The first falls about 0·8 eV below the $2s^2\ ^1S$ doubly-excited state of He (see Table 9.1) and the second about 0·3 eV below the $2s\ 2p\ ^3P$ state (see Table 9.4). Fano and Cooper‡ identify these respective He⁻ states as $2s^2\ 2p\ ^2P^0$ and $2s\ 2p^2\ ^2D$.

6. Doubly-excited states of Ne⁻ (for discussion of experimental techniques and results see Chap. 1, § 6.2.4 and Chap. 5, § 4.2.1)

The most prominent resonance features (see Chap. 1, Fig. 1.25) occurring at electron energies below the excitation threshold in neon appear as two sharp decreases in transmission separated by $0·095\pm0·002$ eV. They are located at $16·04\pm0·02$ and $16·13_5\pm0·02$ eV, while the first, $^3P_2$, excited state of Ne lies at 16·619 eV.

The separation between the two levels is nearly the same as that (0·097 eV) between the $^2P_{\frac{1}{2}}$ and $^2P_{\frac{3}{2}}$ levels of Ne⁺. This is probably significant because the lowest excited configuration of Ne $1s^2\ 2s^2\ 2p^5\ 3s$ is based on the ^2P configuration of Ne⁺. The lowest doubly-excited configuration should be obtained by adding a second occupied $3s$ orbital. This is equivalent to adding a closed $3s$ shell to the ^2P levels of Ne⁺ so as to give $^2P_{\frac{1}{2}}$ and $^2P_{\frac{3}{2}}$ states of Ne⁻ with much the same fine structure separation as for Ne⁺. Further support for this interpretation of the pair of resonance levels comes from the following argument.

When allowance is made for fine-structure splitting we must regard a particular auto-ionizing state as determined not only by the orbital angular momentum quantum number l of the incident electron but also its total angular momentum quantum number $j = l\pm\frac{1}{2}$. If a resonance effect occurs for a particular pair of values L, J of l, j (89) becomes

$$\int_{\text{res}} Q\,\mathrm{d}E = \int_{\text{res}} (Q_a+Q_b)\,\mathrm{d}E + \frac{\pi^2}{k^2}(2J+1)\Gamma\cos 2\eta_L.$$

The experimental results with moderate resolution§ show decreases in transmitted intensity at the resonances and hence increases in mean cross-section. This requires $\cos 2\eta_L > 0$. This excludes $L = 0$ as η_0 is between $\frac{3}{2}\pi$ and 2π (see Chap. 6, § 4.3). The ratio of the intensity decreases at the two resonances is close to 2, the one at the lower energy being the deeper. This is consistent with $L = 1$ and $J = \frac{3}{2}, \frac{1}{2}$ for the first and second resonances respectively. Also η_1 for neon at the resonance

---

† CHAMBERLAIN, G. E. and HEIDEMAN, H. G. M., *Phys. Rev. Lett.* **15** (1965) 337.
‡ FANO, U. and COOPER, J. W., *Phys. Rev.* **138** (1965) A400.
§ SIMPSON, J. A. and FANO, U., *Phys. Rev. Lett.* **11** (1963) 158.

9.6 RESONANCE PHENOMENA—THRESHOLD BEHAVIOUR

energy is between $\tfrac{3}{4}\pi$ and π (see Chap. 6, § 4.3) so $\cos 2\eta_1 > 0$ as required.

This choice of resonance level is also consistent with the resonance profile observed in experiments with higher energy resolution† and is confirmed by the angular distribution measurements discussed in Chapter 5, § 5.3.3. When analysed, as for helium, these measurements show that the resonance scattering varies nearly as $\cos^2\theta$ as it must if it is a p-resonance.

The fact that the resonance effects below the threshold are much more clearly seen in helium than in neon can be ascribed to the fact that in neon the p-wave resonance is small compared with the mainly s-wave background, whereas in helium both the resonance and background scattering are s-type.

As described in Chapter 1, § 6.2.4 other possible resonance effects in transmission have been observed at electron energies above the excitation threshold. Identification of the responsible levels of Ne$^-$ is difficult and is still uncertain.

7. Doubly-excited states of Ar$^-$, Kr$^-$, and Xe$^-$

Resonance effects observed in the transmission of electrons through argon, krypton, and xenon have been described in Chapter 1, § 6.2.5. In all cases a pair of resonances is observed about 0·5 eV below the excitation threshold at a separation close to that between the $^2P_{\frac{3}{2}}$ and $^2P_{\frac{1}{2}}$ states of the corresponding ground state positive ions. It seems clear that they are to be interpreted in the same way as for neon.

Evidence of resonances beyond the excitation threshold has been observed for xenon.

8. Doubly-excited states of Hg$^-$

Fano and Cooper‡ have discussed the interpretation of the resonance observed by Kuyatt *et al.*† in the transmission of electrons with energy in the range 4 to 5 eV through mercury vapour.

The nearest levels of excited Hg belong to the $6s\,6p$ configuration and are $^3P_{0,1,2}$ between 4·67 and 5·46 eV and ^1P at 6·7 eV. Levels of other configurations such as $6s^2$, $6s\,6d$, $6p^2$, etc., lie several eV higher, so that it is reasonable to assume that the resonances that are observed at 4·0$_7$, 4·2$_9$, and 4·8$_9$ eV (see Chap. 1, § 6.2.6) belong to the $6s\,6p^2$ configuration of Hg$^-$. This gives rise to four terms ^4P, ^2D, ^2S, and ^2P.

---

† KUYATT, C. E., SIMPSON, J. A., and MIELCZAREK, S. R., *Phys. Rev.* **138** (1965) A385.   ‡ FANO, U. and COOPER, J. W., ibid. **138** (1965) A400.

If LS coupling were strictly adhered to the ^4P terms could be excluded because it would not be possible to attain a quartet state of Hg⁻ by capture of an electron by an Hg atom in its singlet ground state. However, in a heavy atom such as mercury LS coupling is a poor approximation. Indeed Fano and Cooper† suggest that the resonance states are actually, in ascending order of energy, the three fine structure levels $^4P_{\frac{1}{2}}$, $^4P_{\frac{3}{2}}$, and $^4P_{\frac{5}{2}}$. Thus their separations are comparable with the separations of the ^4P levels of the isoelectronic neutral atom Tl. Furthermore, levels identified as ^2D, ^2S, and ^2P terms of the $6s\,6p^2$ configuration from the spectra of isoelectronic atoms and ions lie a few eV higher than the ^4P levels.

The $^4P_{\frac{1}{2}}$ level can only be reached by capture of an s electron, the $^4P_{\frac{3}{2},\frac{5}{2}}$ only from capture of a d electron. In terms of the formulae (84), (89) the shape of the resonance at $4 \cdot 0_7$ eV, involving at first a sharp increase of transmission followed by a smaller dip, can be understood, as well as its relative smallness, if the phase shift for s-wave elastic scattering near resonance lies close to 150° and the width is much less than the resolution of the measuring equipment. The $^4P_{\frac{3}{2},\frac{5}{2}}$ resonances are broad and well resolved, requiring that the d-wave phase shift near resonance should differ from an integral multiple of π by about 10°.

9. One-body or shape resonances

So far we have been concerned with resonance effects that arise through capture of an electron into an intermediate complex with lifetime of order 10^{-13} s at least. These complexes can be considered as doubly- or multiply-excited states of the system of target+electron. It is of interest to re-examine now the relation of this theory to the one-body scattering theory. Thus the resonance states we have been discussing are responsible for fine structure effects in the variation of cross-sections with electron energy. Can we interpret the broad features, such as the maxima of width 1–2 eV in the elastic cross-section of argon, krypton, and xenon, in terms of intermediate states of correspondingly shorter lifetime—10^{-15} s or so—and, if so, what is the nature of such states?

It is best to begin by returning to the problem of the scattering of electrons with zero angular momentum by a structureless centre of force. At a maximum of the zero-order partial cross-section the corresponding phase shift η_0 is an odd integral multiple of $\frac{1}{2}\pi$. The expression

$$M = k \cot \eta_0,$$

† Fano, U. and Cooper, J. W., loc. cit., p. 641.

9.9 RESONANCE PHENOMENA—THRESHOLD BEHAVIOUR

where k is the electron wave number, will therefore vanish at this maximum. Hence we may write, for an electron energy E in the neighbourhood of that E_0 at the maximum,

$$M = (E-E_0)\left(\frac{\partial M}{\partial E}\right)_m, \qquad (151)$$

where $\partial M/\partial E$ is calculated at the maximum. The partial cross-section q_0 is given by

$$q_0 = 4\pi/|ik-M|^2 \qquad (152)$$

so, if we write

$$\left(\frac{\partial M}{\partial E}\right)_m = \frac{2k}{\Gamma}, \qquad (153)$$

we have, at electron energies near E_0,

$$q_0 = \frac{\pi}{k^2}\frac{\Gamma^2}{(E-E_0)^2 + \tfrac{1}{4}\Gamma^2}. \qquad (154)$$

This is of exactly the same form as (65) and suggests that we can regard a broad maximum in the cross-section as arising from capture of an electron into a state of lifetime $\hbar/\Gamma$, where Γ is given, close to the maximum, by (153). We can regard this as useful only when $\hbar/\Gamma$ is appreciably longer than the transit time for an electron through the effective scattering field.

Since

$$\left(\frac{\partial M}{\partial E}\right)_0 = \left[\left(k\,\mathrm{cosec}^2\eta_0\,\frac{\partial \eta_0}{\partial k} + \cot \eta_0\right)\frac{\partial k}{\partial E}\right]_{\eta_0 = \tfrac{1}{2}\pi}$$

$$= k\left(\frac{\partial \eta_0}{\partial k}\right)_m \frac{\partial k}{\partial E}$$

$$= (m/\hbar^2)\left(\frac{\partial \eta_0}{\partial k}\right)_m,$$

(153) shows that the lifetime of the state is given by

$$\tau = \frac{\hbar}{2k}\left(\frac{\partial M}{\partial E}\right)_m = \frac{m}{2k\hbar}\left(\frac{\partial \eta_0}{\partial k}\right)_m = \frac{1}{2v}\left(\frac{\partial \eta_0}{\partial k}\right)_m, \qquad (155)$$

where v is the electron velocity. The relation between $\partial \eta_0/\partial k$ and the time spent by the electron in the neighbourhood of the scatterer was derived by Wigner† using a quite different approach that confirms our interpretation.

Wigner considered the scattering of a wave packet of prescribed angular momentum and velocity v. This packet may, for simplicity, be made up of two beams of nearly equal energies $E \pm \delta E$ and wave

† WIGNER, E. P., *Phys. Rev.* 98 (1955) 145.

numbers $k\pm\delta k$ respectively. The incident packet then has the asymptotic form

$$\psi_{\text{inc}} = r^{-1}[\![\exp[-i\{(k+\delta k)r + \hbar^{-1}(E+\delta E)t\}] + \\ + \exp[-i\{(k-\delta k)r + \hbar^{-1}(E-\delta E)t\}]]\!].$$

The two constituent waves are in phase when

$$r\,\delta k + \hbar^{-1} t\,\delta E = 0,$$

so this locates the centre of the packet at time t.

Scattering of the packet will introduce phase shifts $\eta \pm \delta\eta$ in the waves with respective wave numbers $k\pm\delta k$, so the wave function for the outgoing wave packet will have the asymptotic form

$$\psi_{\text{out}} = r^{-1}[\![\exp[-i\{-(k+\delta k)r + \hbar^{-1}(E+\delta E)t - 2(\eta+\delta\eta)\}] + \\ + \exp[-i\{-(k-\delta k)r + \hbar^{-1}(E-\delta E)t - 2(\eta-\delta\eta)\}]]\!].$$

Its centre will be located where

$$r\,\delta k - \hbar^{-1} t\,\delta E + 2\,\delta\eta = 0,$$

giving
$$r = \hbar^{-1}t\frac{dE}{dk} - 2\frac{\partial\eta}{\partial k} = \hbar^{-1}\frac{dE}{dk}\left(t - 2\hbar\frac{\partial\eta}{\partial E}\right).$$

This shows that, through scattering, the packet is retarded as if it were delayed for a time

$$2\hbar\frac{\partial\eta}{\partial E} = \frac{2}{v}\frac{\partial\eta}{\partial k}.$$

As a typical example of the time delays involved we refer to Fig. 6.6, which shows typical phase shifts calculated, as a function of k, for the optical model potential for argon. The maximum value of $\partial\eta_2/\partial k$, occurring near $ka_0 = 0.8$, is approximately $3a_0$. Since $v = 1.7\times 10^8$ cm/s the delay time is only 2×10^{-16} s. This is so short-lived as to hardly merit description in terms of a transitory complex. It is clear from (155) that $\partial\eta/\partial k$ must be considerably larger than a_0 if the delay time is to be much longer than the unretarded electron transit time across the atom.

It is not difficult to find central scattering potentials that would give rise to resonance capture states, for electrons, which would have much longer lifetimes. These are of the form of a deep attraction at close distances changing to a repulsion at larger distances. Because of the difficulty in penetrating the potential barrier, electrons within the barrier will be reflected back and forth many times before escaping. Quasi-stationary states will exist at those electron energies for which constructive interference occurs between various reflected waves just as if the barrier were infinitely high. With proper choice of parameters specifying the range and strength of the interactions the lifetime of these states can be made as long as desired. Examples of similar phenomena occurring in the collisions between atoms are discussed in Chapter 16, §§ 5.4.4 and 12.1.3.

For the scattering of electrons with non-vanishing angular momentum the centrifugal force provides a long-range repulsive barrier but it does not seem that the field parameters are such as to lead to quasi-stationary states with widths much less than 1 eV. Although an atom is not just a structureless source of central potential we can regard the short-lived states we have been considering as associated with transitory capture to the ground states of the atom. Following Taylor, Nazaroff, and Golebiewski† we shall refer to them as Type II resonance states to distinguish them from the much longer-lived Type I states in which the electron is bound to an excited state of an atom. Type II states will also arise as quasi-stationary states in the fields of excited atoms. Resonance effects due to Type II atomic states are usually referred to as one-body or shape resonances. They can be regarded as responsible for the broad maxima in the behaviour of cross-sections but for most atomic problems the breadth is so large that the resonance interpretation is rather artificial.

Although there is little reliable evidence of the excitation of Type II atomic states with width less than 1 eV, states of even this short life can be important in producing certain types of molecular collisions. There is also some evidence of the existence of Type II states with width as small as 0·1 eV for certain molecules. We shall defer consideration of these molecular phenomena until Chapter 10, § 3.5.2 and Chapter 12, §§ 3.6.1 and 6.

10. The S, T, R, and M matrices—behaviour of cross-sections at thresholds

10.1. *The two-channel S, T, R, and M matrices for states of zero angular momentum*

We now discuss some aspects of generalized scattering theory that may readily be seen to follow from appropriate formulae for the case of scattering by a centre of force or, perhaps more accurately, single-channel scattering. Apart from introducing the formulae in a simple way and indicating how they may be used in practice we shall be particularly concerned with the information provided about the variation with incident

† TAYLOR, H. S., NAZAROFF, G. V., and GOLEBIEWSKI, A., *J. chem. Phys.* **45** (1966) 2872.

electron energy of cross-sections in the neighbourhood of thresholds. For example, we shall consider the variation of the elastic cross-section at energies immediately above and immediately below that which is just sufficient to produce excitation of the first excited state, as well as that of the inelastic cross-section just above this energy.

We limit our discussion at first to the single-channel case with zero angular momentum. At large distances r outside the range of the interaction that produces the scattering the wave function $r^{-1}G_0$ for the electron motion satisfies

$$\frac{d^2 G_0}{dr^2} + k^2 G_0 = 0,$$

where k is the wave number of the electron motion. We may therefore always write
$$G_0 \sim A(k)\sin\{kr+\eta(k)\}, \qquad (156)$$

where η is a *real* phase shift which is a function of k, determined by the condition $G_0(0) = 0$, and A is a normalizing constant which may also be a function of k.

If we take $A(k) = -2ie^{i\eta}$ then
$$G_0 \sim e^{-ikr} - S(k)e^{ikr},$$
where
$$S(k) = e^{2i\eta}. \qquad (157)$$

We note that $|S(k)|^2 = 1$, a result that expresses the conservation of particles—the so-called *unitarity* property. The scattered amplitude is then given in terms of S by

$$f(k) = \frac{1}{2ik}\{S(k)-1\}, \qquad (158)$$

or, writing
$$S(k) = 1 + iT(k), \qquad (159)$$
$$f(k) = T(k)/2k, \qquad (160)$$

and the partial elastic cross-section is given by
$$q_0 = (\pi/k^2)|1-S(k)|^2 = (\pi/k^2)|T(k)|^2. \qquad (161)$$

We note here that
$$|T(k)| \leqslant 2. \qquad (162)$$

Alternatively, we may take
$$A(k) = \sec\eta,$$
so that
$$G_0 \sim \sin kr + R(k)\cos kr,$$
where
$$R(k) = \tan\eta.$$

We note that
$$S = (1+iR)/(1-iR), \qquad (163\,\text{a})$$
$$T = 2R/(1-iR). \qquad (163\,\text{b})$$

9.10 RESONANCE PHENOMENA—THRESHOLD BEHAVIOUR

These formulae are already of interest in connection with the preservation of unitarity in any approximate solution for the scattered amplitude. If $S(k)$ is calculated directly by, say, Born's approximation, it will not in general satisfy the unitarity condition. This means that the derived value of $T(k)$ will not in general satisfy (162). However, if R is first calculated directly by the approximate method to give R_b, say, then approximate values of S_b and T_b given by

$$S_\mathrm{b} = \frac{1+iR_\mathrm{b}}{1-iR_\mathrm{b}}, \qquad T_\mathrm{b} = \frac{2R_\mathrm{b}}{1-iR_\mathrm{b}} \tag{164}$$

will always be consistent with unitarity. It does not, of course, follow that they will be good approximations if R_b is not small, but at least the error will be bounded.

An alternative procedure for ensuring that an approximate method provides results consistent with the conservation of electrons is to calculate directly an approximation η_a for the phase shift η and then take
$$S_\mathrm{a} = e^{2i\eta_\mathrm{a}}, \qquad T_\mathrm{a} = e^{2i\eta_\mathrm{a}}-1. \tag{165}$$

Extrapolation to negative values of k^2, relating the energies of bound states to the scattering phases, is most conveniently carried out by working with $k\cot\eta$ because this remains continuous in passing from $k^2 < 0$ to $k^2 > 0$. This has already been discussed in Chapter 6, § 3.5 in connection with effective range formulae. We write

$$k\cot\eta = M(k) \tag{166}$$

and note that if we take $A(k) = k\operatorname{cosec}\eta$ in (156)

$$G_0 \sim M(k)\sin kr + k\cos kr.$$

Also
$$M(k) = k\{R(k)\}^{-1} \tag{167 a}$$
and
$$T(k) = 2k/\{M(k)-ik\}. \tag{167 b}$$

We now consider how to generalize S, T, R, and M to apply to collisions in which more than one channel is involved. The simplest such case is that of two open channels in both of which the angular momentum of the colliding electron is zero. The eigenfunction expansion for the full collision wave function, ignoring exchange effects, consists of two terms

$$\Psi = \psi_0(r_2)r_1^{-1}G_0(r_1) + \psi_1(r_2)r_1^{-1}G_1(r_1). \tag{168}$$

To describe collisions in which the electron is incident with wave number k_0 on the system in its ground state with wave function ψ_0 the functions G_0, G_1, which vanish at $r_1 = 0$ and which we shall distinguish as $G_{0,0}$, $G_{1,0}$ respectively, will have the asymptotic form

$$G_{0,0}(r) \sim k_0^{-\frac{1}{2}}\{e^{-ik_0 r}-S_{00}\,e^{ik_0 r}\}, \tag{169 a}$$

$$G_{0,1}(r) \sim -k_1^{-\frac{1}{2}}S_{01}\,e^{ik_1 r}, \tag{169 b}$$

where k_1 is the wave number of the outgoing electron motion after excitation of the state with wave function ψ_1. Thus

$$k_1^2 = k_0^2 - \kappa_{01}^2, \tag{170}$$

where $\kappa_{01}^2 \hbar^2/2m$ is the excitation energy. The partial cross-sections for elastic and inelastic collisions are given respectively by

$$q_{00} = \pi |1 - S_{00}|^2/k_0^2, \qquad q_{01} = \pi |S_{01}|^2/k_0^2. \tag{171}$$

The choice of normalizing factors $k_0^{-\frac{1}{2}}$, $k_1^{-\frac{1}{2}}$ in (169) has been made so that the formulae for the cross-sections are direct generalizations of (161).

The wave function Ψ is also capable of describing collisions of an electron incident with wave number k_1 on the target in its excited state, leading either to elastic scattering or a superelastic collision that leaves the target in its ground state. For this type of collision the functions G_0, G_1 must have the respective asymptotic forms

$$G_{1,0} \sim k_0^{-\frac{1}{2}} S_{10} \, e^{ik_0 r}, \tag{172 a}$$

$$G_{1,1} \sim k_1^{-\frac{1}{2}} \{e^{-ik_1 r} - S_{11} \, e^{ik_1 r}\}. \tag{172 b}$$

The partial cross-sections for elastic and superelastic collisions with the excited target are now given by

$$q_{11} = \pi |1 - S_{11}|^2/k_1^2, \qquad q_{10} = \pi |S_{10}|^2/k_1^2. \tag{173}$$

The four quantities S_{00}, S_{01}, S_{10}, S_{11} forming a matrix

$$\begin{pmatrix} S_{00} & S_{01} \\ S_{10} & S_{11} \end{pmatrix} \tag{174}$$

known as the S-matrix $\mathbf{S}$ or S_{ij}, together represent the generalization of the single-channel $S(k)$ for this case. It was pointed out that the condition $|S(k)|^2 = 1$ expresses the conservation of particles. For the more general case this is replaced by the natural generalization, in terms of matrix multiplication,

$$\mathbf{SS^*} = \mathbf{1}, \tag{175}$$

where $\mathbf{1}$ denotes the unit matrix

$$\begin{pmatrix} 1 & 0 \\ 0 & 1 \end{pmatrix}.$$

In terms of suffixes this becomes

$$\sum_j S_{ij} S_{jk}^* = \delta_{ik}, \tag{176}$$

so that
$$|S_{00}|^2 + S_{01} S_{10}^* = 1, \tag{177 a}$$
$$|S_{11}|^2 + S_{10} S_{01}^* = 1, \tag{177 b}$$
$$S_{00} S_{01}^* + S_{10} S_{11}^* = 0, \tag{177 c}$$

9.10 RESONANCE PHENOMENA—THRESHOLD BEHAVIOUR

three relations which may be proved without difficulty.† A matrix satisfying (176) is known as a *unitary matrix*.

The S matrix is also *symmetrical* so that $S_{01} = S_{10}$. It follows that

$$q_{10} = (k_0^2/k_1^2)q_{01}, \qquad (178)$$

the usual detailed-balance relation between cross-sections for inelastic and the corresponding superelastic collisions.

It follows, from the fact that

$$|S_{00}|^2 = 1 - |S_{01}|^2, \qquad (179)$$

that S_{00} cannot be written in the form $e^{2i\eta}$, where η is a real phase shift. For some purposes it is useful to write

$$S_{00} = e^{2i\delta}, \qquad (180)$$

where δ is complex with positive imaginary part so that $|S_{00}|^2 \leqslant 1$. The same applies to S_{11}.

Because the S matrix is unitary it may be written as a matrix product in the form

$$\mathbf{S} = \mathbf{U}^\dagger e^{2i\eta} \mathbf{U}, \qquad (181)$$

where $\mathbf{U}$ is a real orthogonal matrix, satisfying the condition $\mathbf{U}^\dagger \mathbf{U} = 1$, $\mathbf{U}^\dagger$ being the transpose of $\mathbf{U}$, i.e. the matrix derived from $\mathbf{U}$ by transposition of rows and columns. $e^{2i\eta}$ is the diagonal matrix

$$e^{2i\eta} = \begin{pmatrix} e^{2i\zeta^{(0)}} & 0 \\ 0 & e^{2i\zeta^{(1)}} \end{pmatrix}, \qquad (182)$$

where $\zeta^{(0)}$, $\zeta^{(1)}$ are real quantities known as the *eigenphase shifts*. These are such that the two independent sets of solutions of the coupled equation for G_0 and G_1 (see § 3) have the asymptotic forms

$$\left. \begin{array}{l} G_{0,0}^e \sim \sin(k_0 r + \zeta^{(0)}) \\ G_{0,1}^e \sim \chi \sin(k_1 r + \zeta^{(0)}) \end{array} \right\} \text{(I)} \quad \left. \begin{array}{l} G_{1,0}^e \sim \sin(k_0 r + \zeta^{(1)}) \\ G_{1,1}^e \sim -\chi^{-1} \sin(k_1 r + \zeta^{(1)}) \end{array} \right\} \text{(II)} \quad (183)$$

instead of (169) and (172). The matrix $\mathbf{U}$ is then given by

$$\mathbf{U} = \begin{pmatrix} \cos\epsilon & \sin\epsilon \\ -\sin\epsilon & \cos\epsilon \end{pmatrix}, \qquad (184)$$

where the mixing parameter $\chi = (k_1/k_0)^{\frac{1}{2}} \tan\epsilon$.

There is no difficulty now in defining the $\mathbf{T}$ and $\mathbf{R}$ matrices which are the appropriate generalization of $T(k)$ and $R(k)$ for the single-channel case.

Thus

$$\mathbf{S} = 1 + i\mathbf{T}, \qquad (185)$$

and the partial cross-sections q_{rs} ($r, s = 0, 1$) are given by

$$q_{rs} = (\pi/k_r^2)|T_{rs}|^2. \qquad (186)$$

---

† MOTT, N. F. and MASSEY, H. S. W., *The theory of atomic collisions*, 3rd edn, pp. 370–1 (Clarendon Press, Oxford, 1965).

Similarly we find for the **R** matrix, which is such that the two sets of solutions have the asymptotic forms

$$\left.\begin{array}{l} G^r_{0,0} \sim k_0^{-\frac{1}{2}}(\sin k_0 r + R_{00} \cos k_0 r) \\ G^r_{0,1} \sim k_1^{-\frac{1}{2}} R_{01} \cos k_1 r \end{array}\right\} \text{(I)},$$

$$\left.\begin{array}{l} G^r_{1,0} \sim k_0^{-\frac{1}{2}} R_{10} \cos k_0 r \\ G^r_{1,1} \sim k_1^{-\frac{1}{2}}(\sin k_1 r + R_{11} \cos k_1 r) \end{array}\right\} \text{(II)}, \qquad (187)$$

$$\mathbf{S}(1-i\mathbf{R}) = 1+i\mathbf{R}. \qquad (188)$$

Unlike **S**, **R** is real. It can be written in the form

$$\mathbf{R} = \mathbf{U}^\dagger \tan\boldsymbol{\eta}\, \mathbf{U} \qquad (189)$$

where **U** is as in (184) and $\tan\boldsymbol{\eta}$ is the diagonal matrix

$$\begin{pmatrix} \tan \zeta^{(0)} & 0 \\ 0 & \tan \zeta^{(1)} \end{pmatrix}. \qquad (190)$$

The relationship to the single-channel case is therefore very close. In particular we note that, if $\mathbf{R}_a$ is an approximate value of **R** derived in some way, then an approximate S-matrix $\mathbf{S}_a$ given by

$$\mathbf{S}_a(1-i\mathbf{R}_a) = 1+i\mathbf{R}_a \qquad (191)$$

will always be unitary.† When the limits of validity of a particular approximate method are being approached the calculation of $\mathbf{S}_a$ and hence $\mathbf{T}_a$ via $\mathbf{R}_a$ in this way yield results that are at least consistent with the conservation of electrons. This will not be the case in general if $\mathbf{S}_a$ or $\mathbf{T}_a$ are calculated directly by the approximate method concerned.

If $\mathbf{T}_b^I$ is an approximation to the T-matrix calculated by Born's approximation (see Chap. 7) then the corresponding approximation to **R** is given by

$$\mathbf{T}_b^I = 2\mathbf{R}_b \qquad (192)$$

as $\mathbf{R}_b$ and $\mathbf{T}_b$ are both assumed small. A second approximation $\mathbf{T}_b^{II}$ to **T** is then obtained as

$$\mathbf{T}_b^{II} = 2\mathbf{R}_b/(1-i\mathbf{R}_b) = \mathbf{T}_b^I/(1-\tfrac{1}{2}i\mathbf{T}_b^I). \qquad (193)$$

A second alternative is to work with approximations to the eigenphase shifts in terms of which

$$i\mathbf{T}_b^{III} = e^{2i\mathbf{R}_b} - 1. \qquad (194)$$

The second term possessing a matrix exponent can be interpreted in terms of a power series or in terms of a suitable diagonal matrix **B** such that

$$\mathbf{R}_b = \mathbf{U}_b^\dagger \mathbf{B} \mathbf{U}_b, \qquad (195)$$

where $\mathbf{U}_b$ is unitary. We then have

$$e^{2i\mathbf{R}_b} = \mathbf{U}_b^\dagger e^{2i\mathbf{B}} \mathbf{U}_b, \qquad (196)$$

† Provided $\mathbf{R}_a$ is hermitian as it will normally be for simple approximations.

where
$$e^{2i\mathbf{B}} = \begin{pmatrix} e^{2iB_0} & 0 \\ 0 & e^{2iB_1} \end{pmatrix}. \tag{197}$$

The possibility of using methods of this type for ensuring the satisfaction of conservation conditions in approximate calculations for multi-channel problems was first pointed out by Percival.† A little later, Seaton‡ examined the usefulness of the method in providing better approximations by dealing with some exactly soluble special cases of coupled equations. Applications have been made to a number of cases but,§ as pointed out earlier, while the methods always ensure unitarity it is difficult to tell how accurate they are within this limitation when the approximate eigenphases are not small—at any rate the results will never grossly exaggerate the cross-sections.

There remains now to consider the generalization‖ of $M(k)$ from which we can hope to investigate cross-sections in the neighbourhood of a threshold. To do this we introduce pairs of functions that have the asymptotic form

$$\left.\begin{array}{l} G^m_{0,0} \sim k_0^{-\frac{1}{2}}(M_{00} \sin k_0 r + k_0 \cos k_0 r) \\ G^m_{0,1} \sim k_1^{-\frac{1}{2}} M_{01} \sin k_1 r \end{array}\right\} \text{(I)},$$
$$\left.\begin{array}{l} G^m_{1,0} \sim k_0^{-\frac{1}{2}} M_{10} \sin k_0 r \\ G^m_{1,1} \sim k_1^{-\frac{1}{2}}(M_{11} \sin k_1 r + k_1 \cos k_1 r) \end{array}\right\} \text{(II)}. \tag{198}$$

We then find, in order that the pairs of functions (198) may be obtained from linear combination of the pairs (187), that

$$k_0^{-1} M_{00} R_{10} + k_1^{-1} M_{10} R_{11} = 0,$$
$$k_0^{-1} M_{01} R_{10} + k_1^{-1} M_{11} R_{11} = 1,$$
$$k_0^{-1} M_{00} R_{00} + k_1^{-1} M_{10} R_{01} = 1,$$
$$k_0^{-1} M_{01} R_{00} + k_1^{-1} M_{11} R_{01} = 0. \tag{199}$$

These can be expressed in matrix form as

$$\mathbf{R} \begin{pmatrix} k_0^{-1} M_{00} & k_1^{-1} M_{10} \\ k_0^{-1} M_{01} & k_1^{-1} M_{11} \end{pmatrix} = \mathbf{1}, \tag{200}$$

so
$$\mathbf{R}^{-1} = \begin{pmatrix} k_0^{-1} M_{00} & k_1^{-1} M_{10} \\ k_0^{-1} M_{01} & k_1^{-1} M_{11} \end{pmatrix}. \tag{201}$$

If we denote by $\mathbf{k}$ the diagonal matrix

$$\mathbf{k} = \begin{pmatrix} k_0 & 0 \\ 0 & k_1 \end{pmatrix}$$

---

† PERCIVAL, I. C., *Proc. phys. Soc.* **76** (1960) 206.
‡ SEATON, M. J., ibid. **77** (1961) 174.
§ BURKE, V. M. and SEATON, M. J., ibid. **77** (1961) 199; SOMERVILLE, W. B., ibid. **80** (1962) 806. ‖ Ross, M. H. and SHAW, G. L., *Ann. Phys.* **13** (1961) 147.

then
$$\mathbf{k}^{\frac{1}{2}}\mathbf{R}^{-1}\mathbf{k}^{\frac{1}{2}} = \mathbf{M}, \tag{202}$$

where
$$\mathbf{M} = \begin{pmatrix} M_{00} & M_{10} \\ M_{01} & M_{11} \end{pmatrix}. \tag{203}$$

and is symmetrical. In the one-channel case since $\{R(k)\}^{-1} = \cot\eta$, (202) reduces to (166).

The relation between the **T** and **M** matrices, which is very important for our purpose, is now given by

$$\mathbf{T} = 2\mathbf{k}^{\frac{1}{2}}(\mathbf{M}-i\mathbf{k})^{-1}\mathbf{k}^{\frac{1}{2}}, \tag{204}$$

which reduces, as it must, to (167 b) in the one-channel case.

The importance of the **M**-matrix is that, as might be expected of the generalization of $k \cot \eta$, it is continuous across a threshold and, provided all the interactions are of short range, can be expanded about any chosen value K_0^2 of k_0^2 in the form

$$\mathbf{M}(k_0^2) = \mathbf{M}_0 + \mathbf{M}_1(k_0^2 - K_0^2) + \dots. \tag{205}$$

$\mathbf{M}_0, \mathbf{M}_1, \mathbf{M}_2$ are matrices that depend only on K_0^2 and the expansion is valid whether k_0^2 is greater or less than K_0^2.

Using (204) we can study how T_{00} varies in going through the threshold where $k_1^2 = 0$, $k_0^2 = \kappa_{01}^2$. From (204) we have

$$T_{00} = 2k_0(M_{11} - ik_1)D^{-1}, \tag{206}$$

where D is the determinant of the matrix $\mathbf{M} - i\mathbf{k}$, i.e.

$$D = \begin{vmatrix} M_{00} - ik_0 & M_{10} \\ M_{01} & M_{11} - ik_1 \end{vmatrix} \tag{207}$$
$$= (M_{00} - ik_0)(M_{11} - ik_1) - M_{10}^2.$$

Just above the threshold

$$|T_{00}|^2 = |T_{00}(k_1 = 0)|^2 \{1 - 2Bk_1\}, \tag{208}$$

where
$$B = M_{10}^2 k_0 \{(M_{00}^2 + k_0^2)M_{11}^2 + M_{10}^4 - 2M_{10}^2 M_{00} M_{11}\}^{-1}$$
$$= M_{10}^2 k_0^{-1}(s^2 + M_{11}^2)^{-1},$$

with
$$s = k_0^{-1}\{M_{00} M_{11} - M_{10}^2\}. \tag{209}$$

The **M**-matrix elements are calculated at the threshold $k_1 = 0$. On the other hand, just below the threshold, where $k_1 \to i\kappa_1$,

$$|T_{00}|^2 = |T_{00}(k_1 = 0)|^2 \{1 - 2B\kappa_1/c\}, \tag{210}$$

where
$$c = M_{11}/s. \tag{211}$$

Since $B > 0$ it follows that the partial elastic cross-section falls initially as the energy increases above the excitation threshold. It will rise or fall initially as the energy decreases below the threshold according as $c <$ or > 0. In either case it will possess an infinite derivative with respect to k_0 at the threshold for $k_1 = (k_0^2 - \kappa_{01}^2)^{\frac{1}{2}}$. Fig. 9.9 illustrates the two possible shapes of the variation through the threshold. This is seen to be a cusp or a step according as $c >$ or < 0.

Since
$$T_{01} = 2(k_0 k_1)^{\frac{1}{2}} M_{10} D^{-1} \tag{212}$$
we see that initially the inelastic cross-section increases from the threshold as k_1.

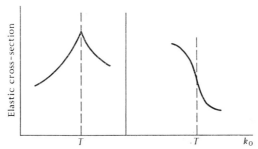

Fig. 9.9. Possible forms of variation with wave number k_0 of the elastic cross-section in passing through an inelastic excitation threshold (T).

Reference to (206) shows that, below the threshold, T_{00} may be written in the form
$$T_{00} = 2/(\alpha - i), \tag{213}$$
where
$$\alpha = k_0^{-1}\{M_{00} - M_{01}^2/(M_{11} + \kappa_1)\}.$$
This shows the existence of a single resonance at $\alpha = 0$ with half-maxima at $\alpha = \pm 1$. Using the effective range expansion of the M-matrix elements Ross and Shaw† show that the full width at half-maximum in k_0-space is given by
$$2M_{01}^2 k_0/(M_{00}^2 - k_0^2),$$
provided κ_1 is small and $M_{01}^2 \ll M_{00}^2$.

10.2. *Generalization‡ to coupling between states of different angular momentum*

There is no difficulty in generalizing the analysis of § 10.1 to apply to coupling between two states of different angular momenta. Thus, if l_0, l_1 are the angular-momentum quantum numbers of the electron before and after impact, we need only, as far as S, T, and R are concerned, replace $\sin k_0 r$, $\cos k_0 r$ by $k_0 r$ times the respective spherical Bessel functions $j_l(k_0 r)$, $j_{-l}(k_0 r)$ (see Chap. 6, § 3.3) and similarly for $\sin k_1 r$, $\cos k_1 r$. For the M-matrix, however, we note that in the one-

† loc. cit., p. 651.
‡ MOTT, N. F. and MASSEY, H. S. W., *The theory of atomic collisions*, 3rd edn, p. 384 (Clarendon Press, Oxford, 1965).

channel case with angular-momentum quantum number l the effective-range expansion applies to $k^{2l+1}\cot\eta_l$. The appropriate generalization for the two-channel case is

$$\mathbf{M} = \mathbf{k}^{l+\frac{1}{2}}\mathbf{R}^{-1}\mathbf{k}^{l+\frac{1}{2}}, \tag{214}$$

where $\mathbf{k}^{l+\frac{1}{2}}$ is to be interpreted as the diagonal matrix

$$\begin{pmatrix} k_0^{l_0+\frac{1}{2}} & 0 \\ 0 & k_1^{l_1+\frac{1}{2}} \end{pmatrix}. \tag{215}$$

The relation (204) between $\mathbf{T}$ and $\mathbf{M}$ now takes the form, in matrix notation

$$\mathbf{T} = 2\mathbf{k}^{l+\frac{1}{2}}(\mathbf{M}-i\mathbf{k}^{2l+1})^{-1}\mathbf{k}^{l+\frac{1}{2}}. \tag{216}$$

The behaviour near a threshold in this case is considerably modified.

We have
$$T_{00} = 2k_0^{2l_0+1}(M_{11}-ik_1^{2l_1+1})D^{-1}, \tag{217}$$

where
$$D = (M_{00}-ik_0^{2l_0+1})(M_{11}-ik_1^{2l_1+1})-M_{10}^2. \tag{218}$$

The corresponding cross-section $\pi|T_{00}|^2/k_0^2$ will no longer have an infinite derivative at the threshold since $l_1 \geqslant 1$, so that the cusp or step illustrated in Fig. 9.9 does not arise.

As far as the threshold behaviour of the excitation cross-section is concerned we have

$$T_{01} = 2k_0^{l_0+\frac{1}{2}}k_1^{l_1+\frac{1}{2}}M_{01}D^{-1}. \tag{219}$$

Hence q_{01} will increase from zero at the threshold as $k_1^{2l_1+1}$.

It must be re-emphasized that these results only follow on the assumption that the interactions, including that responsible for the coupling between the two states, are of short range or fall off exponentially with distance at large distances. Some of the modifications introduced when these conditions are not satisfied are discussed in §§ 10.4 and 10.6.

10.3. *Generalization to any number of channels*†

Generalization of the analysis of §§ 10.1, 10.2 to apply to a collision problem involving an unlimited number of channels is immediate, provided we assume that no channels involve two or more free electrons at infinity. Ionization is therefore excluded. We shall make some further brief reference to this in § 10.8.

We associate a given channel j with a set γ_j of quantum numbers of constants of the motion such as the total angular momentum J, its z-component J_z, the parity and, if spin-orbit and spin-spin interaction is neglected, the total spin s and its z-component s_z. Corresponding to this set there will be an eigenfunction $\mathscr{Y}_j(\gamma)$. We then take as a generalization of (169)

$$G_{ij}^{\gamma}(r) \sim k_j^{-\frac{1}{2}}\{\delta_{ij}(k_j r)^{-1}j_{l_j}(k_j r) - S_{ij}(k_j r)^{-1}j_{-l_j}(k_j r)\}\mathscr{Y}_j(\gamma). \tag{220}$$

† See Mott, N. F. and Massey, H. S. W., ibid., Chap. xiv (Clarendon Press, Oxford, 1965).

The partial cross-section for a transition from channel i to channel j is then given by
$$q_{ij} = \frac{4\pi}{k_i^2} \frac{2J+1}{2s_i+1} |T_{ij}(\gamma)|^2, \qquad (221)$$
where
$$T_{ij} = \delta_{ij} - S_{ij} \qquad (222)$$
and s_i is the spin of the initial channel, supposed unpolarized.

If all the channels are open **S** is a unitary matrix and the **R** matrix, defined by an obvious generalization of (187), is related to it as in (188). On the other hand, if some of the channels are closed, we may select the submatrices ^{0}**S**, ^{0}**R**, which refer to the open channels only. Whereas **S** is then no longer unitary, ^{0}**S** is unitary and ^{0}**R** is related to it as in (188).

The **M** matrix is defined as in (198) and is related to the **T** matrix by the matrix relation (216). The possibility of expansion of the **M** matrix so defined in powers of k^2 depends on all the interactions being of short range or falling off exponentially.

10.4. Modifications introduced by the presence of a Coulomb field

Although the asymptotic form of the wave function for an electron moving in a Coulomb field differs from that for a shorter range field through the logarithmic phase factor (see Chap. 6, § 3.9) there is no difficulty in defining the **S**, **T**, and **R** matrices so as to allow for this. Thus if the interaction behaves as $-Ze^2/r$ for large r we need only replace the spherical Bessel functions $j_{\pm l}(kr)$ by the appropriate functions $G_l^c(k,r)$, $H_l^c(k,r)$ (see Chap. 6, § 3.9 (93)), where
$$G_l^c(k,r) \sim \sin(kr - \tfrac{1}{2}l\pi - \alpha \ln 2kr + \eta_l^c),$$
$$H_l^c(k,r) \sim \cos(kr - \tfrac{1}{2}l\pi - \alpha \ln 2kr + \eta_l^c). \qquad (223)$$

Here $\alpha = -Ze^2/\hbar v$ and
$$\exp(2i\eta_l^c) = \Gamma(l+1+i\alpha)/\Gamma(l+1-i\alpha). \qquad (224)$$

The relation between the **M** and **R** matrices needs to be redefined in order that **M** should be an analytic function of k^2 for $k^2 \geq 0$. Gailitis[†] showed that we should take[‡]
$$\mathbf{M} = k^{l+\tfrac{1}{2}}(2l+1)!!\,\mathbf{C}_l \mathbf{R}^{-1}\mathbf{C}_l(2l+1)!!\,k^{l+\tfrac{1}{2}} +$$
$$+\{(2l+1)!!\}^2 k^{2l+1}\mathbf{p}_l\{-\ln|\alpha|+\mathbf{g}(\alpha)\}, \qquad (225)$$
where
$$C_l = \frac{2^l C_0}{(2l+1)!} \prod_{s=1}^{l}(s^2+\alpha^2)^{\tfrac{1}{2}}, \qquad C_0 = \{2\pi\alpha/(e^{2\pi\alpha}-1)\}^{\tfrac{1}{2}},$$
$$p_l = 2\alpha C_l^2/C_0^2, \qquad g(\alpha) = -\gamma + \sum_1^{\infty} \frac{\alpha^2}{s(s^2+\alpha^2)}. \qquad (226)$$

[†] GAILITIS, M., *Zh. éksp. teor. Fiz.* **44** (1963) 1974; *Soviet Phys. JETP* **17** (1963) 1328.
[‡] The matrices involving functions of l and α are to be interpreted as diagonal matrices in the same way as $k^{l+\tfrac{1}{2}}$ in (214). Also $(2l+1)!! = 1.3.5\ldots(2l+1)$.

For a single channel with $l = 0$ (225) reduces to

$$M = kC_0^2 R^{-1} + 2\alpha k\{-\ln|\alpha| + g(\alpha)\}. \tag{227}$$

Since in this case $R^{-1} = \cot\sigma_0$, where σ_0 is the additional phase shift due to modification of the Coulomb field at small distances (see Chap. 6, § 3.10), we have

$$M = C_0^2 k \cot\sigma_0 + \beta f(k), \tag{228}$$

where

$$\beta = 2mZe^2/\hbar^2$$

and

$$f(k) = -\gamma - \ln|\alpha| + \sum_{s=0} \frac{\alpha^2}{s(s^2+\alpha^2)}. \tag{229}$$

Reference to Chapter 6, § 3.10 shows that (228) is just the form used as a basis for an effective range expansion when the interaction behaves asymptotically like $-Ze^2/r$.

Using the relation between the **T** and **R** matrices we have

$$\mathbf{T} = 2k^{l+\frac{1}{2}}(2l+1)!! \, \mathbf{C}_l[\mathbf{M} - \{(2l+1)!!\}^2 p_l k^{2l+1}\tau]^{-1}\mathbf{C}_l(2l+1)!! \, k^{l+\frac{1}{2}}, \tag{230}$$

where

$$\tau = -\ln\alpha + g(\alpha) + i\pi/(e^{2\pi\alpha}-1).$$

When $Z \to 0$, $\alpha \to 0$,

$$(2l+1)!! \, C_l \to 1, \qquad \{(2l+1)!!\}^2 p_l \tau \to i,$$

and (216) is regained.

We note first that the cross-section for excitation involving the opening of a new channel with $l = l_1$, rises from the threshold as

$$k_1^{2l_1+1}C_{l_1}^2.$$

For an attractive Coulomb interaction $\alpha < 0$ and

$$C_{l_1}^2 \simeq \pi\{(2l_1+1)!\}^{-2}\beta_1^{2l_1+1}k_1^{-2l_1-1}, \tag{231}$$

so the cross-section is finite at the threshold for all l_1 (contrast with (219)).

On the other hand, for a repulsive Coulomb interaction $\alpha > 0$ and

$$C_{l_1}^2 k_1^{2l_1+1} \simeq \pi\{(2l_1+1)!\}^2 \beta_1^{2l_1+1} e^{-\pi\beta/k_1}, \tag{232}$$

so the cross-section and all its derivatives vanish at the threshold for all l_1.

The behaviour of the cross-sections in old channels just below the threshold for a new channel may also be analysed as in § 10.2.

As might be expected, because an attractive Coulomb field can support an infinite number of bound states, these cross-sections exhibit an infinite set of resonances converging on the threshold. The cross-section averaged over these resonances tends to a limit at the threshold greater than the corresponding limit of the cross-section from above

9.10 RESONANCE PHENOMENA—THRESHOLD BEHAVIOUR

the threshold. Nevertheless the total cross-section remains constant in passing through the threshold.

10.5. *Generalization of the quantum-defect method to coupled channels*

In Chapter 6, § 3.10 we described a method for extrapolating from positive to negative values of k^2 for single-channel problems involving interactions that have the asymptotic form of an attractive Coulomb field $-Ze^2/r$. This involved a relationship between the quantum defects for a series of bound states and the phase shifts for the corresponding continuum states which is a consequence of the effective range formula based on (228).

This may be extended in principle to coupled channels.† The most useful starting-point from which to generalize is the formula ((108) of Chap. 6):

$$\tan\{\pi\nu_l(k^2)\}+A_l(k^2)Y_l(k^2) = 0, \qquad (233)$$

where the energy of a particular term in the series of bound states is given by

$$E_{nl} = -2\pi^2 me^4/h^2\nu_{nl}^2 \qquad (234)$$

with $\nu_{nl} = n-\mu_{nl}$, μ_{nl} being the quantum defect.

$A_l(k^2)$ is given by

$$A_l(k^2) = \prod_{s=0}^{l}\{1+4s^2k^2/\beta^2\}, \qquad (235)$$

where $\beta = 2mZe^2/\hbar^2$.

We may now generalize (233) to the matrix equation

$$\tan\{\pi\mathbf{\nu_l}(k^2)\}+\mathbf{A}_l^{\frac{1}{2}}\mathbf{Y}\mathbf{A}_l^{\frac{1}{2}} = 0, \qquad (236)$$

which gives the eigenvalues of the matrix $\mathbf{k}^2$. If these eigenvalues may be obtained from spectroscopic data the matrix may be determined for these values of k^2 (< 0). Extrapolation to other values of k^2 is then carried out using the fact that the elements of $\mathbf{Y}$ can all be expressed in the form

$$Y_{ij} = \frac{\sum a_m^{ij} k_0^{2m}}{\sum b_s^{ij} k_0^s}, \qquad (237)$$

only the first few coefficients being important.

Having obtained $\mathbf{Y}$ in this way it still remains to obtain the relation between $\mathbf{\nu}_l(k^2)$ and the phase shifts that determine the scattering cross-sections. In the special case of two coupled states the elastic scattering phase shift σ_l for $k_0^2 > 0$, $k_1^2 < 0$, i.e. below the excitation threshold, is given by

$$\tan\sigma_l = -\tan\pi\nu_{l,0}-A_{l,0}Y_{01}^2/\{Y_{11}+A_{l,1}^{-1}\tan\pi\nu_{l,1}\}, \qquad (238)$$

---

† BELY, O., MOORES, P., and SEATON, M. J., *Atomic collision processes*, ed. McDOWELL, M. R. C., p. 304 (North Holland, Amsterdam, 1964); SEATON, M. J., *Proc. phys. Soc.* **88** (1966) 801.

where $\nu_{l,0}$, $\nu_{l,1}$ are the quantum defects extrapolated to $k_0^2 > 0$. It is noteworthy that $\tan \sigma_l \to -\tan \pi \nu_{l0} = \tan \pi \mu_{l0}$ as $k_0^2 \to 0$.

As an illustration of this method in operation Seaton[†] has discussed the two $^3D_1$ series in neutral Ca, $4s\,n_1d$, which converges to the $4s\,^2S_{\frac{1}{2}}$ level of Ca$^+$ and $3d\,n_2s$ which converges to the $3d\,^2D$ level of Ca$^+$. The former series is strongly perturbed by interaction with the latter because the $3d\,5s$ level falls between $4s\,8d$ and $4s\,9d$.

TABLE 9.7

Spectroscopic data for ^3D levels of calcium

			n	T_n (cm^{-1})	$\nu_{n,0}$	$\nu_{n,1}$
Ca I	$4s\,3d$	$^3D_1$	3	20335·3	1·9463	1·6046
	$4s\,4d$	$^3D_1$	4	37748·2	3·0813	2·0864
	$4s\,5d$	$^3D_1$	5	42743·1	4·0890	2·3300
	$4s\,6d$	$^3D_1$	6	45049·1	5·0771	2·4755
	$4s\,7d$	$^3D_1$	7	46302·2	6·0441	2·5669
	$4s\,8d$	$^3D_1$	8	47036·3	6·9531	2·6254
	$3d\,5s$	$^3D_1$	9	47456·1	7·7017	2·6608
	$4s\,9d$	$^3D_1$	10	47753·3	8·4066	2·6866
	$4s\,10d$	$^3D_1$	11	48032·0	9·2804	2·7116
	$4s\,11d$	^3D	12	48259·2	10·238	2·733
	$4s\,12d$	^3D	13	48434·8	11·222	2·749
	$4s\,13d$	^3D	14	48570·7	12·216	2·762
	$4s\,14d$	^3D	15	48676·6	13·203	2·772
	$4s\,15d$	^3D	16	48762·4	14·206	2·781
	$4s\,16d$	^3D	17	48830·7	15·193	2·787
Ca II	$4s$	$^2S_{\frac{1}{2}}$		49306·1		2·8353
	$3d$	$^2D_{\frac{3}{2}}$		62956·3		

Table 9.7 summarizes the spectroscopic data[‡] and the quantities $\nu_{n,0}$, $\nu_{n,1}$ for the first seventeen terms of the series as well as the series limits. In preparing this table the quantum number n is assigned simply on the energy order of the term. $\nu_{n,0}$, $\nu_{n,1}$ are then obtained from the formula

$$E_n^{0,1} = -\frac{2\pi^2 m e^4}{h^2 \nu_{n,0,1}^2},$$

where E_n^0, E_n^1 are the energies below the respective series limits.[§]

Fig. 9.10 shows the quantum defect $\mu_{n,0}$ given by $n - \nu_{n,0}$. The effect of the interaction of the two series is shown by the rapid rise in $\mu_{n,0}$ between $n = 8$ and 11.

From the numbers given in Table 9.7 the **Y**-matrix elements are

[†] SEATON, M. J., *Proc. phys. Soc.* **88** (1966) 815.
[‡] MOORE, C. E., *Atomic energy levels*, Natn. Bur. Stand. Circular No. 467 (Printing Office, Washington, 1949).
[§] GARTON, W. R. S. and CODLING, K., *Proc. phys. Soc.* **86** (1965) 1067.

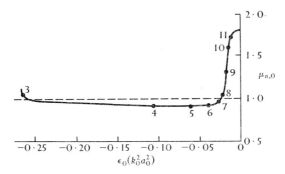

Fig. 9.10. Quantum defects $\mu_{n,0}$ for the $^3D_1$ series in calcium.

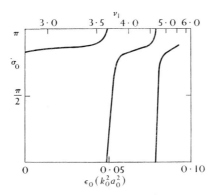

Fig. 9.11. Phase shifts σ_0 for scattering of electrons by Ca^+ $4s$ when the total system is in a ^3D state. The integral multiple of π associated with the phase shift is taken to be such that the phase lies between 0 and π. The resonance effects are apparent.

determined. By fitting the levels $n = 3, 4, 5, 8, 9, 10$ we find

$$Y_{00} = -0.27839 + 0.08977\alpha_0^{-2},$$
$$Y_{11} = 1.51411 - 6.38428\alpha_0^{-2},$$
$$Y_{01} = \pm(0.5466 - 1.76480\alpha_0^{-2}),$$

where $\alpha_0 = \beta/2k_0$.

Fig. 9.11 illustrates the phase shift σ_0 at energies below the excitation threshold in the ^3D state. The resonances stand out clearly.

Moores[†] has carried out a similar analysis for the $^1P^0$ series $4s\,np$ and $3d\,np$ of neutral Ca in order to obtain wave functions for the $^1P^0$ continuum state of Ca which include the corresponding resonance effects. These wave functions were then used to calculate photo-ionization

† Moores, D. L., Proc. phys. Soc. 88 (1966) 843.

cross-sections for Ca including auto-ionization effects arising from the transition
$$3d\,np \to 4s\,kp.$$

Fig. 9.12 shows the variation of the quantum defect μ and the phase shift σ up to the $3d$ threshold. The resonance effects in the phase shift are clearly seen as well as the very similar behaviour of the quantum defect between $n = 6$ and 7—the $3d\,4p$ level falls between $4s\,6p$ and

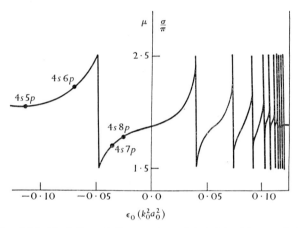

FIG. 9.12. Variation of the quantum defect (μ) of terms of the principal ^1P series of calcium with energy below Ca$^+$ $4s$ ionization threshold and of the corresponding phase shift (σ) for scattering of electrons by Ca$^+$ $4s$ ions with energy between this threshold and that for Ca$^+$ $3d$.

$4s\,7p$. The application of these results to the calculation of the photo-ionization cross-sections for Ca is discussed in Chapter 14, § 7.4.5.

10.6. *Threshold effects of orbital degeneracy*

We have already pointed out in § 4 that the existence of degeneracy between states of the hydrogen atom or of hydrogen-like ions that possess the same total quantum number n leads to the existence of resonance states that otherwise would not appear. It is to be expected that, in addition, threshold behaviour may be modified in other directions also.

In fact this is only so for collisions with neutral hydrogen atoms because the long-range Coulomb field dominates the situation for the positive ions. Gailitis and Damburg† have carried out the analysis for hydrogen and we shall discuss its application to the behaviour of the

† GAILITIS, M. and DAMBURG, R., ibid. **82** (1963) 192.

9.10 RESONANCE PHENOMENA—THRESHOLD BEHAVIOUR

elastic and inelastic cross-sections near the threshold for excitation of the 2-quantum states.

The analysis follows on much the same lines as that outlined in § 4 for the states with the total orbital angular-momentum quantum number $L = 0$. The only difference is that we include the free wave function G_{1s} associated with the atom in its ground state in addition to the G_{2s}, G_{2p} functions associated with the 2s and 2p states respectively.

At large distances r, G_{1s} is not coupled to G_{2s} and G_{2p} so that the latter functions still satisfy (130 a), (130 b) while in addition

$$\left(\frac{d^2}{dr^2} + k^2\right)G_{1s} = 0. \tag{239}$$

The three equations may be written in matrix form as

$$\left(\frac{d^2}{dr^2}\mathbf{1} + \mathbf{k}^2\right)\mathbf{G} = \alpha \mathbf{G} r^{-2}, \tag{240}$$

where $\mathbf{G}$, $\boldsymbol{\alpha}$, and $\mathbf{k}^2$ are the respective matrices

$$(G_{1s} \ G_{2s} \ G_{2p}), \quad \begin{pmatrix} 0 & 0 & 0 \\ 0 & 0 & -6 \\ 0 & -6 & 2 \end{pmatrix}, \quad \text{and} \quad \begin{pmatrix} k^2 & 0 & 0 \\ 0 & -\kappa^2 & 0 \\ 0 & 0 & -\kappa^2 \end{pmatrix}.$$

We have shown in § 4 how the equations for G_{2s}, G_{2p} may be separated by taking

$$G_\gamma = G_{2s} + \gamma G_{2p}, \tag{241}$$

where
$$\gamma = \{1 \pm \sqrt{(37)}\}/6,$$

so that
$$\left\{\frac{d^2}{dr^2} - \kappa^2 - \frac{\lambda(\lambda+1)}{r^2}\right\}G_\gamma = 0, \tag{242}$$

with
$$\lambda = -\tfrac{1}{2} + \sqrt{\{\pm\sqrt{(37)} + \tfrac{5}{4}\}}.$$

There is no difficulty in extending this to include the uncoupled equation for G_{1s}. Again, using matrix notation, we have

$$\left\{\frac{d^2}{dr^2}\mathbf{1} + \mathbf{k}^2 - \frac{\boldsymbol{\lambda}(\boldsymbol{\lambda}+1)}{r^2}\right\}\mathbf{A}\mathbf{G} = 0, \tag{243}$$

where

$$\boldsymbol{\lambda}(\boldsymbol{\lambda}+1) = \begin{pmatrix} 0 & 0 & 0 \\ 0 & 1+\sqrt{(37)} & 0 \\ 0 & 0 & 1-\sqrt{(37)} \end{pmatrix}, \tag{244}$$

$$\mathbf{A} = \begin{pmatrix} 1 & 0 & 0 \\ 0 & -6/a & -6/b \\ 0 & \dfrac{1+\sqrt{(37)}}{a} & \dfrac{1-\sqrt{(37)}}{b} \end{pmatrix} \tag{245}$$

with $a = \sqrt{[\sqrt{(148)}\{\sqrt{(37)}+1\}]}$, $b = \sqrt{[\sqrt{(148)}\{\sqrt{(37)}-1\}]}$. It is to be noted that
$$\mathbf{A}^{-1}\alpha\mathbf{A} = \lambda(\lambda+1) \qquad (246)$$
so that $\mathbf{A}$ is a matrix that diagonalizes α.

Damburg and Gailitis† then show that the $\mathbf{M}$ matrix that may be expanded in powers of k^2 through the excitation threshold is related to the $\mathbf{T}$ matrix by

$$\mathbf{T} = i(1-e^{\frac{1}{2}i\pi l}\mathbf{A}e^{-i\pi\lambda}\mathbf{A}^{-1}e^{\frac{1}{2}i\pi l})+e^{\frac{1}{2}i\pi l}\mathbf{A}e^{-\frac{1}{2}i\pi\lambda}k^{\lambda+\frac{1}{2}} \times$$
$$\times\{2/(\mathbf{M}-i\sec\pi\lambda\, e^{-i\pi\lambda}k^{2\lambda+1})\}k^{\lambda+\frac{1}{2}}e^{-\frac{1}{2}i\pi\lambda}\mathbf{A}^{-1}e^{-\frac{1}{2}i\pi l}. \qquad (247)$$

When $\mathbf{A} = 1$ and $\lambda = 1$ this reduces to the usual form (216).

The first two terms of $\mathbf{T}$ do not depend on the energy and cancel except for transitions between the 2-quantum states, that is to say they contribute only to T_{12} and T_{21}. This means that the variation with energy of the cross-sections, for elastic collisions with ground state atoms and for inelastic collisions involving excitation of the 2-quantum states from the ground state, are determined wholly by the third term in (247).

In particular we find that, since $\mathbf{M}$ is approximately constant in a small energy range about the threshold, the elements of the $\mathbf{T}$ matrix near the threshold that determine the cross-sections for excitation of the $2s$ and $2p$ states are given respectively by

$$T_{01} = -6\sqrt{2}[M_{01}(k_1 = 0)\sqrt{[\sqrt{(37)}\{\sqrt{(37)}-1\}]}]^{-1} \times$$
$$\times \exp[\sqrt{\{\sqrt{(37)}-\tfrac{5}{4}\}}\{i(\ln k_1 - \tfrac{1}{2}\pi) + \tfrac{1}{2}\pi\}],$$
$$T_{02} = -\sqrt{2}[M_{02}(k_1 = 0)\sqrt{[\sqrt{(37)}/\{\sqrt{(37)}-1\}]}]^{-1} \times$$
$$\times \exp[\sqrt{\{\sqrt{(37)}-\tfrac{5}{4}\}}(i\ln k_1 + \pi)]. \qquad (248)$$

It follows that $|T_{01}|^2$ and $|T_{02}|^2$ both tend to finite constants as $k_1 \to 0$. This is to be contrasted with the behaviour expected if no long-range forces were present. Then according to (212) and (219) $|T_{01}|^2$, $|T_{02}|^2$ would behave like k_1 and k_1^3 respectively for small k_1.

Evidence in support of these predictions is afforded by the experiments of Chamberlain, Smith, and Heddle‡ on the excitation of the $2p$ state by electrons with energy close to the threshold. These experiments are described in Chapter 4, § 1.4.1. Reference to Fig. 4.16 shows that the evidence strongly favours the existence of a finite cross-section at the threshold.

---

† loc. cit., p. 660.
‡ CHAMBERLAIN, G. E., SMITH, S. J., and HEDDLE, D. W. O., *Phys. Rev. Lett.* **12** (1964) 647.

10.7. *Quantum defect extrapolation for collisions with* He^+

Although the Coulomb field dominates the threshold behaviour for collisions of electrons with He^+ ions, the existence of orbital degeneracy must be taken account of in performing extrapolation across an inelastic threshold. Bely[†] has carried out such an extrapolation across the $n = 2$ threshold using the appropriate extension of the quantum defect method, which may be derived readily by use of the matrix **A**. The initial data used were those of Burke and McVicar who calculated the **R** matrix in the $1s$–$2s$–$2p$ close-coupling approximation for electron energies above the $n = 2$ threshold. Extrapolation was then carried out to energies below this threshold to obtain the elastic scattering phase shifts. From these, resonance energies and level widths could be derived.

Results obtained for the $^1P^0$ and ^3P levels are given in Tables 9.3 and 9.4 respectively. It will be seen that there is very good agreement with the results of the direct $1s$–$2s$–$2p$ close-coupling calculations in the same energy region, except for the narrowest triplet levels. In addition the extrapolation method locates higher levels of the series as well as giving their widths, even when very small.

If this extrapolation is carried out without allowance for orbital degeneracy substantially different results are obtained, particularly for level widths.

10.8. *Threshold law for ionization*

We have made it clear that the analysis of § 10.1–7 only applies to collisions in which as a result of the impact of an electron with the atom energy transfer to the atom is not sufficient to produce ionization—there must be no final states in which more than one unbound electron is present. The analysis therefore does not apply to ionization. It is possible to extend the analysis to include this possibility but many complications are thereby introduced. Rudge and Seaton,[‡] extending and rendering more rigorous earlier calculations by Peterkop,[§] have discussed the asymptotic form of the solution of the wave equation for a system consisting of a proton and two unbound electrons when all three particles are at large distances apart. From this analysis they were able to deduce that the ionization cross-section Q_i for a neutral atom should rise linearly from the threshold, i.e.

$$Q_i \simeq A(E - E_i), \tag{249}$$

[†] BELY, O., *Proc. phys. Soc.* **88** (1966) 833.
[‡] RUDGE, M. R. H. and SEATON, M. J., *Proc. R. Soc.* **A283** (1965) 262.
[§] PETERKOP, R. K., *Latv. PSR Zināt. Akad. Vest.* **9** (1960) 79; *Zh. éksp. teor. Fiz.* **43** (1962) 616.

where E is the incident and E_i the ionization energy. An earlier, classical calculation by Wannier† led to

$$Q_i \simeq B(E-E_i)^{1\cdot 127}.$$

Experimental results to date are not yet definite enough (see Chap. 3, § 2.5.3.2) to check the validity of (249). As explained in Chapter 3, one difficulty is that it is uncertain for what range of energy (249) is valid and no information is available about the magnitude of the constant A. A further possible experimental complication is that electrons with energy just below the ionization threshold may excite the atoms with which they collide to states very close to this threshold. Such atoms may then be ionized by thermal photons, so extending the apparent ionization threshold to lower energies.

For further ionization of an ion, already n-fold ionized, the threshold behaviour should be as given by

$$Q_i = C(E-E_i)^{n+1}, \qquad (250)$$

but as explained in Chapter 3, § 2.4.4 the complex structure near the threshold for these collisions makes it very difficult to prove.

† WANNIER, G. H., *Phys. Rev.* **90** (1953) 817.

AUTHOR INDEX

Abdelnabi, I., 301, 302, 303, 306.
Aberth, W., 16.
Abrines, R., 587, 589, 590, 593.
Adler, F. P., 70.
Akesson, N., 7.
Allen, H. W., 65.
Allis, W. P., 49, 65, 78, 408, 410.
Allison, S. K., 159.
Altschuler, S., 445.
Amaldi, E., 417, 418.
Anderson, E. M., 562.
Anderson, J. M., 84.
Anderson, N., 115.
Andrick, D., 324, 327, 346, 348.
Appleyard, E. T. S., 178, 182, 183, 214, 222, 223.
Arnot, F. L., 107, 310, 324, 325.
Arthurs, A. M., 166, 475.
Asundi, R. K., 101, 124, 126, 130.
Aubrey, B. B., 13, 14, 27, 28, 576.

Bacon, F. M., 145.
Bailey, V. A., 24, 45, 66, 82, 83, 87, 88, 90, 91, 308.
Baines, G. O., 310.
Baker, F. A., 149, 150, 151, 153.
Balling, L. C., 365, 366, 367, 368, 369, 370, 568.
Bandel, H. W., 19, 31, 32, 36, 38, 83, 84, 544, 545.
Bates, D. R., 202, 454, 495.
Bauer, E., 420.
Bederson, B., 12, 13, 14, 27, 28, 267, 332, 358, 360, 576.
Bely, O., 496, 657, 663.
Bennett, W. R., 230.
Berkowitz, L., 110.
Berman, A. S., 322, 327, 482.
Bethe, H., 388, 399, 436, 499, 540.
Betts, D. D., 307, 309.
Beutler, H., 136, 139, 627.
Bewilogua, L., 499.
Blaha, M., 496.
Blais, N. C., 130.
Blatt, J. M., 388, 607.
Bleakney, W., 99, 103, 104, 105, 123, 124, 126, 130.
Boeckner, C., 93, 95.
Boerboom, A. J. H., 130.
Boersch, H., 311, 312, 313, 315.
Boksenberg, A., 102, 105, 109, 124, 126, 129.
Bolton, H. C., 328.

Bortner, R. T. E., 61.
Bothe, W., 163.
Bowe, J. C., 61, 62, 81, 82, 84, 85, 87, 88, 90, 91, 92, 308.
Boyd, R. L. F., 107, 471.
Boyd, T. J. M., 496.
Bracewell, R., 521.
Brackmann, R. T., 10, 12, 26, 107, 108, 111, 123, 124, 126, 129, 130, 193, 196, 197, 232, 251, 332, 471, 472, 518, 519, 521.
Bradbury, N. E., 58, 62, 64, 81.
Brandt, A. F., 417, 418.
Branscomb, L. M., 107, 153, 154, 155, 156, 158.
Bransden, B. H., 12.
Brattain, W. H., 284.
Breit, G., 142, 532, 601, 607.
Bricout, P., 181.
Brink, G. O., 16, 107, 111, 126, 130, 139, 140.
Brode, R. B., 8, 9, 24, 29, 95, 96, 362.
Bronco, C. J., 190.
Brown, S. C., 67, 70, 71, 72, 73, 74, 78, 80, 83, 84, 85, 86, 89, 295, 296, 297, 301, 302, 303, 306, 554, 556.
Browne, H. N., 420.
Bruche, E., 24, 25, 27, 557.
Bullard, E. C., 322, 323, 324, 325, 333, 546.
Bunyan, P. J., 424.
Burgess, A., 472, 473, 475, 477, 588, 589.
Burhop, E. H. S., 159, 166, 167, 475.
Burke, P. G., 497, 498, 514, 528, 535, 538, 550, 552, 565, 574, 576, 621, 622, 623, 624, 633, 634, 635, 636, 638, 639, 663.
Burke, V. M., 651.
Burns, J. F., 115, 116, 130, 135, 136, 137, 138.
Bush, V., 378.
Byrne, J., 369.

Caldwell, S. H., 378.
Callaway, J., 541.
Caplinger, E., 107, 130.
Caren, R. P., 308.
Čermák, V., 171, 237, 238, 239, 254, 255, 256.
Chamberlain, G. E., 195, 196, 198, 255, 314, 315, 317, 318, 320, 321, 348, 529, 639, 640, 662.
Chen, C. L., 70, 83, 84, 85, 86, 90, 92, 93, 94, 95, 564.
Chibisov, M. J., 500.

AUTHOR INDEX

Childs, E. C., 331, 336.
Christoph, W., 495.
Chupka, W. A., 110.
Clark, J. C., 160, 161, 166.
Clarke, E. M., 34, 112, 118, 130, 131, 634, 636, 637.
Cloutier, G. G., 130.
Codling, K., 316, 628, 658.
Compton, A. H., 159.
Compton, K. T., 99.
Condon, E. V., 274.
Cooper, J., 321, 623, 628.
Cooper, J. W., 495, 550, 552, 609, 612, 626, 639, 640, 641, 642.
Craggs, J. D., 124, 126.
Cravath, A. M., 58.
Crompton, R. W., 56, 57, 58, 67, 81, 82, 83, 544.
Crown, J. C., 567.
Curley, E. K., 34, 634, 636, 637.
Czyzak, S. J., 496, 578, 579.

Dalgarno, A., 12, 202, 443, 483, 484, 523.
Daly, N. R., 143, 144.
Damburg, R., 629, 660, 662.
Damgaard, A., 202.
Dance, D. F., 153, 154, 155, 157, 158, 270, 271, 272, 273, 275, 477, 536.
Davies, D. E., 307.
Davies, D. K., 292, 301, 303.
Davydov, V. B., 49.
De Heer, F. J., 103, 124, 130.
Dehmelt, H. G., 363, 366, 370, 568.
Deichsel, H., 352, 353, 355, 425.
De Vos, 191.
Dibeler, V. H., 107, 130.
Dirac, P. A. M., 599.
Ditchburn, R. W., 107.
Dolder, K. T., 145, 147, 148, 270, 272, 273, 474.
Dorman, F. H., 116, 118, 130, 133, 261.
Dorrestein, R., 234, 235, 254, 258, 259, 304, 305.
Dougal, A. A., 86.
Dowell, J. T., 235, 236, 254, 255, 258, 259, 260, 261.
Druyvesteyn, M. J., 49, 307, 308.
Dunlop, S. H., 301, 303.
Dunn, G. H., 102, 123, 124, 125.
Duveneck, F. B., 160, 588.
Dymond, E. G., 337.

Eckart, C., 478.
Eggleton, P. P., 115.
Ehrhardt, H., 324, 327, 346, 347, 348, 552.
Einstein, A., 438.
Elford, M. T., 56, 57, 58, 67, 81, 82, 93, 544.

Elsasser, W., 459.
El Wakeil, A. N., 567.
Engelhardt, A. G., 299, 307, 308, 309.
Englander-Golden, P., 12, 102, 103, 124, 126, 129.
Errett, D., 308.
Erskine, G. A., 527.

Fano, U., 31, 321, 594, 609, 612, 626, 627, 628, 638, 640, 641, 642.
Farago, P. S., 369, 370, 371.
Fehsenfeld, F. C., 76, 77, 83, 84, 89.
Feldman, P., 240, 241, 261, 262.
Feltsan, P. V., 223.
Fermi, E., 415.
Feshbach, H., 615.
Fink, R. W., 167.
Fineman, M. A., 471.
Fiquet-Fayard, F., 107, 130, 140, 141.
Fischer, O., 182.
Fisher, L., 496.
Fisher, L. H., 293.
Fite, W. L., 10, 26, 107, 108, 109, 110, 111, 123, 124, 126, 129, 130, 193, 195, 196, 197, 232, 251, 332, 471, 472, 518, 519, 520, 521, 529.
Flammersfeld, A., 160, 162.
Flavin, R. K., 77, 95, 564.
Fleming, R. J., 31, 255, 279, 282, 288.
Flower, D. R., 586.
Fock, V., 590.
Foner, S. N., 112, 130, 138.
Fowler, R. G., 190.
Fox, M. A., 491, 492, 493.
Fox, R. E., 106, 114, 115, 118, 119, 125, 130, 132, 133, 134, 135, 138, 140, 235, 254, 255, 304, 305, 317, 319.
Francis, S. A., 311.
Franck, J., 309, 310.
Frank, N. H., 438.
Franken, P., 366, 370, 568.
Fraser, P. A., 633, 634.
Frisch, S., 183, 216, 219.
Frost, D. C., 114, 130.
Frost, L. S., 59, 63, 64, 82, 84, 87, 88, 89, 90, 91, 92, 207, 303, 305, 306, 544, 553.
Füchtbauer, Chr., 417, 418.
Fundaminsky, A., 454, 495.
Fundingsland, O. T., 67, 70, 73, 83, 85, 89, 555, 556.
Funk, H., 107.

Gabathuler, E., 202, 203.
Gaede, W., 102.
Gagge, A. P., 328, 329, 342.
Gailitis, M., 399, 629, 655, 660, 662.
Garton, W. R. S., 658.
Gascoigne, J., 56.
Gaspar, R., 562.

Geiger, J., 311, 312, 313, 329, 330, 331, 434, 486, 495.
Geltman, S., 131, 463, 465, 500, 621, 622, 623, 624, 625, 633, 634.
Gerjuoy, E., 505, 521, 542.
Gilardini, A. L., 85, 86.
Gilbody, H. B., 332, 519, 520.
Gold, A., 495.
Golden, D. E., 19, 31, 32, 36, 37, 39, 83, 84, 293, 544, 545.
Goldstein, L., 83, 84, 85, 86, 459.
Golebiewski, A., 645.
Gossler, F., 417, 418.
Goule, L., 72, 73, 74, 83, 84.
Green, G. W., 107, 160, 161, 165, 166, 167.
Green, L. C., 478.
Greenstein, J. L., 532.
Grove, D. J., 114, 118.
Guggenheimer, K., 139.

Hadeishi, T., 247, 257, 258.
Hafner, H., 195, 226, 586.
Haft, G., 495.
Hagstrum, H. D., 141, 142, 143, 144, 152.
Hahn, Y., 621.
Hanle, W., 176, 179, 181, 182, 220.
Hansen, H., 160.
Hansen, W. W., 160, 162, 165, 166, 588.
Hanson, R. J., 365, 366, 370, 568.
Harries, W., 279.
Harris, L. P., 95, 96.
Harrison, H., 103, 123, 124, 126, 129.
Harrison, M. F. A., 145, 152, 153, 158, 270, 271, 272, 273, 275, 474, 477, 536.
Harrower, G. A., 358.
Harworth, K., 160.
Hasted, J. B., 149, 150, 151, 153.
Haurwitz, E. S., 479, 550.
Heddle, D. W. O., 173, 174, 176, 177, 178, 179, 180, 188, 189, 193, 195, 198, 201, 202, 204, 206, 207, 209, 210, 211, 213, 214, 215, 216, 217, 484, 492, 493, 529, 583, 662.
Heideman, H. G. M., 188, 207, 209, 210, 212, 213, 222, 224, 255, 318, 320, 321, 639, 640.
Heil, H., 103, 124, 130, 138, 139.
Heisenberg, W., 499.
Hellwig, H., 311.
Hendrickson, C. G., 223.
Hendrie, J. M., 243.
Henry, R. J. W., 574, 576.
Herreng, P., 308.
Hertz, G., 279, 309, 310.
Heylen, L. E. D., 301, 302, 303, 307, 308.
Hickam, W. M., 114, 118, 130, 138.
Higginson, G. S., 31, 255, 279, 282, 288.
Hiller, L. A., 589.

Hils, D., 232, 234, 250, 251, 252.
Hinteregger, H. E., 245.
Hirshfield, J. L., 78, 80, 83, 84.
Hobart, J., 366, 370, 568.
Holt, H. K., 231, 236, 242, 245, 246, 253, 254, 255, 256, 257, 268.
Holtsmark, J., 402, 403, 404, 405.
Holzwarth, G., 421, 423.
Hooper, J. W., 145.
Hornbeck, J. A., 60, 62, 63.
Hughes, A. H., 38.
Hughes, A. L., 325, 431, 548, 549.
Hughes, R. H., 178, 179, 214, 215, 223.
Huizinga, W. J., 167.
Hulthèn, L., 395, 396.
Hummer, D. G., 232, 251, 477.
Huxley, L. G. H., 55, 66.
Hylleraas, E. A., 482, 543, 622.
Hyman, H., 230.

Inn, E. C. Y., 110.
Ishii, H., 102, 103.

Jackson, J. D., 388.
Jeffreys, B., 449, 451.
Jeffreys, H., 392, 449, 451.
Jobe, J. D., 191, 199, 204, 205, 206, 480, 481, 494.
John, T. L., 12.
Jones, E. A., 482.
Jones, H., 310.
Jones, T. J., 99, 123, 124, 126, 130.
Jongerius, H. M., 183, 184, 185, 186, 187, 188, 216, 219, 220.
Jopson, R. C., 167.
Jory, R. L., 67, 81, 82, 83, 544.
Jost, K., 352, 355, 425.
Jursa, A. S., 110.

Kaneko, Y., 107, 130, 139.
Kanomata, I., 107, 130.
Karule, E. M., 534, 562, 563, 564, 565, 566, 568, 569.
Kay, R. B., 178, 179, 214, 215.
Keesing, R. G. W., 115, 179, 193, 209, 210, 211, 213, 214, 216, 217, 583.
Kelly, D. C., 74.
Kerr, L. W., 282.
Kerwin, L., 113, 130, 143, 144.
Kessler, J., 323, 336, 337, 353, 355, 425.
Khashaba, S., 528.
Kieffer, L. J., 102, 123, 124, 125.
Kindlmann, P. J., 230.
Kingston, A. E., 496.
Kinney, J. D., 111, 124, 126, 130, 139, 497.
Kipper, A. Y., 532.
Kirkpatrick, P., 160, 161, 165, 166, 167.
Kistemaker, J., 103, 124, 130.

Kistiakowsky, B. B., 110.
Kjeldaas, T., 114, 118, 130, 138.
Kleinpoppen, H., 33, 35, 195, 196, 198, 199, 226, 232, 251, 586, 613, 636, 637, 638.
Knox, R. S., 495.
Kohn, W., 394, 395, 396, 502, 516, 524.
Kollath, R., 24, 25, 27, 36, 38, 83, 84, 89, 93, 328, 329, 333, 545, 546, 557, 558, 559.
Korchevoi, Y. P., 95.
Koschmieder, H., 232, 251.
Krall, N. A., 505, 521, 542.
Krasnow, M. E., 322, 327, 338, 482, 491.
Krotkov, R., 231, 237, 242, 245, 246, 253, 254, 255, 256, 257, 268.
Krueger, T. K., 496, 578, 579.
Krüger, H., 195, 196, 198, 199, 226, 586.
Kruithof, A. A., 307.
Kuprianov, S. E., 145, 148, 151, 236, 253, 256, 258, 259.
Kurepa, M. V., 124, 126, 130.
Kuyatt, C. E., 35, 36, 40, 41, 42, 311, 320, 321, 486, 638, 639, 641.

La Bahn, R. W., 541.
Ladenburg, R., 248, 249.
Lahmani, M., 107, 130, 140, 141.
Lamar, E. S., 49.
Lamb, W. E., 178, 214, 242.
Lamkin, J. C., 512, 513.
Langer, R. E., 392.
Lassettre, E. N., 311, 312, 314, 322, 323, 324, 327, 328, 338, 339, 341, 345, 346, 481, 482, 483, 484, 485, 486, 487, 488, 489, 491, 493, 627, 628.
Latypov, Z. Z., 145, 148, 151.
Latyscheff, G. D., 263, 289, 290.
Lauer, J. E., 496.
Lawrence, E. O., 112, 134.
Lawson, J., 541, 543, 565, 566.
Le Blanc, F. J., 110.
Ledsham, F., 496.
Lee, A. H., 310.
Leech, J. W., 495.
Lees, J. H., 178, 202, 203, 484, 485.
Leiby, C. C., 83, 84.
Leipunsky, A. I., 263, 289, 290.
Lewis, B. A., 487, 498.
Lewis, J. T., 443.
Lewis, M. N., 478.
Lewis, T. J., 301, 302, 303, 307, 308.
Lichten, W., 234, 242, 243, 250, 252, 253, 357, 529.
Lin, C. C., 190, 199, 201, 202, 205, 206, 484, 485, 486, 492, 493.
Lindeman, H., 195, 196, 198, 220.
Lindner, H., 323, 336, 337, 425.
Lineberger, W. C., 145, 146, 148, 152.

Lippert, W., 331.
Liska, J. W., 123, 124, 126, 130.
Llewellyn Jones, F., 292, 301, 303.
Loeb, L. B., 58.
Longmire, M. S., 493.
Löwdin, P. D., 541.
Lowke, J. J., 62, 63, 64.
Lucas, C. B., 176, 177, 178, 179, 180, 188, 189, 193, 201, 214, 215.

McCallum, S. P., 301, 303, 307.
McCoyd, G. C., 496.
McCrea, D., 496.
McCue, J. J. G., 160, 167.
McDaniel, E. W., 145.
MacDonald, A. D., 295, 296, 303, 306, 307, 309.
McDowell, C. A., 114, 130.
McDowell, M. R. C., 211, 214, 216, 499, 500, 541.
McEachran, R. P., 633, 634.
McFarland, R. H., 111, 124, 126, 130, 139, 178, 179, 191, 192, 201, 204, 207, 208, 209, 210, 211, 213, 214, 215, 216, 217, 497, 583.
McGowan, J. W., 34, 119, 120, 121, 130, 131, 132, 143, 144, 471, 633, 634, 636, 637.
McKirgan, T. V. M., 496.
McMillen, J. H., 38, 325, 331, 413, 431, 548, 549, 565.
McVicar, D. D., 535, 538, 621, 623, 624, 663.
Madden, R. P., 316, 628.
Maier-Leibnitz, H., 31, 278, 280, 281, 283, 288, 301, 302, 303, 305, 307, 309, 550.
Maiman, T. H., 178, 214.
Malamud, H., 332.
Mann, J. B., 130.
Margenau, H., 67, 68, 70, 74, 75, 415.
Marino, L. L., 17, 18, 25, 26, 27, 28, 107, 124, 129, 267, 268, 471, 518, 519.
Mark, H., 167.
Marmet, P., 104, 113, 130.
Marriott, R., 479, 550, 621.
Massey, H. S. W., 301, 302, 303, 306, 322, 323, 324, 325, 331, 333, 336, 349, 350, 383, 384, 385, 390, 392, 393, 396, 399, 418, 422, 423, 428, 430, 431, 436, 445, 448, 453, 454, 457, 458, 466, 495, 502, 504, 507, 512, 515, 516, 524, 525, 527, 528, 541, 543, 546, 549, 550, 560, 565, 566, 618, 627, 649, 653, 654.
Mayers, D. F., 423.
Mazur, J., 589.
Meinke, C., 103.
Meister, H. J., 348, 421, 423.
Mercer, G. N., 230.

AUTHOR INDEX

Meyer, V. D., 493.
Meyerand, R. G., 77, 95, 564.
Mielczarek, S. R., 35, 36, 41, 42, 311, 314, 315, 317, 318, 320, 321, 623, 628, 638, 639, 641.
Milatz, J. M. W., 246, 247, 253, 257.
Milford, S. N., 496.
Miller, F. L., 190, 199, 201, 202, 205, 206, 484, 485, 492, 493.
Milne, J. G. C., 307.
Mohler, F. L., 93, 95.
Mohr, C. B. O., 326, 337, 338, 339, 340, 342, 413, 422, 423, 458, 466, 558, 560, 627.
Moiseiwitsch, B. L., 166, 475, 515, 516, 549, 550.
Möllenstedt, G., 329, 331.
Molnar, J. P., 263.
Moores, D. L., 659.
Moores, P., 657.
Morgan, C. G., 292, 301, 303.
Morgulis, N. D., 95.
Morrison, J. D., 104, 116, 118, 130, 133, 136, 137, 138, 261.
Morse, P. M., 49, 408, 409, 410, 479, 498, 549.
Mott, N. F., 3, 349, 350, 378, 383, 384, 385, 390, 392, 393, 396, 399, 418, 422, 423, 428, 430, 431, 436, 445, 448, 453, 454, 475, 502, 504, 507, 524, 525, 589, 618, 649, 653, 654.
Motz, J. W., 160, 162, 165, 166.
Moustafa, H. R., 124.
Mulder, M. M., 478.

Nakano, H., 32, 38, 39.
Nakayama, K., 102, 103.
Nall, B. H., 112, 130, 138.
Nazaroff, G. V., 645.
Newton, A. S., 171, 237, 258, 259, 260.
Neynaber, R. H., 10, 17, 18, 25, 26, 27, 28, 107, 124, 129, 130, 267, 268, 269, 270, 332, 471, 518, 519, 521.
Nicholson, A. J. C., 118, 136, 261.
Nicoll, F. H., 326, 337, 338, 339, 340, 342, 418, 558.
Nielsen, R. A., 58, 62, 64, 81, 84, 85, 87, 88, 308.
Nolan, J. F., 93, 296, 297, 298, 309, 569.
Normand, C. E., 18, 24, 25, 36, 38, 542, 545.
Nottingham, W. B., 112, 134, 135.
Novick, R., 240, 241, 261, 262.

Obed'kov, V. D., 566.
Ochkur, V. I., 455, 470, 528, 533.
Oldeman, J., 220.
Olmsted, J., 171, 237, 258, 259, 260, 261.

O'Malley, T. F., 49, 389, 390, 395, 544, 547, 553, 554, 621, 622, 623, 624, 625, 633, 634.
Omidvar, K., 496, 532.
Oppenheimer, J. R., 453, 583.
Ormonde, S., 528, 533, 550, 552, 638.
Ornstein, L. D., 195, 196, 198, 229, 246, 247, 257.
Osterbrock, D. E., 581.

Pack, J. L., 59, 60, 63, 64, 81, 82, 84, 85, 87, 88, 90, 91, 93, 264, 265, 267, 298, 308.
Palmer, R. R., 338.
Parker, J. H., 56, 57, 58, 82, 88, 308.
Parratt, L. G., 168.
Pawsey, J. L., 521.
Peach, G., 487, 488, 489, 490, 496.
Penney, W. G., 583.
Percival, I. C., 583, 584, 586, 587, 589, 590, 593, 651.
Perel, J., 12, 29, 267, 358, 360.
Perlman, H. S., 166.
Peterkop, R. K., 463, 464, 465, 562, 565, 568, 569, 663.
Peterson, J. R., 111, 124, 130.
Phelps, A. V., 59, 60, 63, 64, 67, 70, 73, 81, 82, 83, 84, 85, 86, 87, 88, 89, 90, 91, 92, 93, 173, 207, 263, 264, 265, 267, 297, 298, 299, 301, 303, 304, 305, 306, 307, 308, 309, 544, 550, 552, 553, 554, 556, 569.
Philbrick, J. W., 255, 311, 316, 317, 318, 319.
Pietenpol, J. L., 240.
Pipkin, F. M., 365, 366, 370.
Placious, R. C., 160, 162, 165, 166.
Plumlee, R. J., 149.
Pockman, L. T., 160, 166, 167.
Pomilla, F. R., 496.
Pottie, R. F., 103.
Prasad, S. S., 470.
Prats, F., 626.
Presnyakov, L., 530, 531.
Priestley, H., 310, 313, 627.

Racah, G., 584.
Raether, M., 93, 94, 95, 564.
Raible, V., 33, 35, 613, 636, 637, 638.
Ramsauer, C., 7, 8, 9, 24, 25, 27, 36, 38, 45, 47, 83, 84, 87, 89, 92, 93, 328, 329, 333, 545, 546, 557, 558, 559.
Rapp, D., 102, 103, 123, 126, 129.
Reder, F. H., 296, 297, 301, 302, 303, 306.
Redhead, P. A., 150, 151, 152, 153.
Reese, R. M., 17, 130.
Reich, G., 103.
Reich, H. J., 311.
Reichert, E., 327, 331, 336, 337, 352, 353, 355, 425.

Reimers, H. J., 417, 418.
Retherford, R. C., 242.
Rice, O. K., 594, 627.
Riemann, H., 308.
Rinehart, E. A., 190.
Roberts, J. E., 310.
Roehling, D., 95, 96.
Roschdestwensky, D., 248.
Rose, D. J., 71, 296.
Rosenberg, L., 389, 395.
Ross, M. H., 651, 653.
Ross, P. A., 161.
Rothe, E. W., 17, 18, 25, 26, 27, 28, 103, 107, 109, 124, 126, 129, 130, 267, 268.
Rubin, K., 267, 358, 359, 360.
Rubinow, S. I., 396.
Rudd, M. E., 623, 628.
Rudge, M. R. H., 463, 464, 465, 469, 473, 475, 498, 523, 663.
Rundel, R. D., 153, 158.
Rusch, M., 24, 25.
Russek, A., 567.
Rutherford, E., 297, 397, 398.

St. John, R. M., 190, 191, 199, 201, 202, 204, 205, 206, 480, 481, 484, 485, 492, 493, 494.
Salmona, S., 563, 564, 568.
Sands, R., 366, 370, 568.
Saraph, H. E., 577, 578, 579.
Schaffernicht, W., 176, 179, 181, 182, 220.
Schey, H. M., 528.
Schiff, H. I., 130.
Schonfelder, J. L., 421, 424.
Schram, B. L., 103, 106, 124, 126, 130.
Schrödinger, E., 460.
Schultz, S., 234, 242, 243, 250, 252, 253, 357, 529.
Schulz, G. J., 21, 22, 23, 24, 31, 32, 33, 34, 35, 40, 42, 235, 254, 255, 284, 287, 304, 305, 311, 316, 317, 318, 319, 613, 636, 637, 638.
Schulz, P., 417, 418.
Schutten, J., 124.
Schwartz, C., 515, 516, 517, 635.
Schwartz, S. B., 473, 498.
Scott, B., 103, 124, 130, 138, 139.
Seaton, M. J., 12, 400, 450, 452, 463, 464, 465, 469, 479, 496, 532, 533, 535, 536, 563, 564, 568, 569, 570, 571, 572, 573, 574, 575, 576, 577, 580, 581, 583, 584, 586, 651, 657, 658, 663.
Segrè, E., 417, 418.
Seiler, R., 283.
Shang-Yi, C., 417, 418.
Shaw, C. H., 168.
Shaw, G. L., 651, 653.
Shemming, J., 577, 575, 579.

Shen, S. T., 262.
Shevera, V. S., 227, 229.
Shimon, L. L., 175, 224, 226, 227, 229.
Shortley, G. H., 274.
Shpenik, O. B., 186, 216, 219.
Shull, H., 541.
Siebertz, K., 216, 219.
Siegmann, H. C., 369, 370, 371.
Silverman, S. M., 322, 327, 338, 339, 341, 345, 346, 482, 483, 485, 487, 488, 489, 491, 627, 628.
Simpson, J. A., 20, 31, 32, 35, 36, 40, 41, 42, 115, 311, 314, 315, 316, 317, 318, 320, 321, 487, 623, 638, 639, 640, 641.
Skerbele, A. M., 311, 312, 314, 324, 484.
Skinner, H. W. B., 178, 182, 183, 214, 222, 223.
Slater, J. C., 408, 410, 438, 560.
Sloan, I. H., 487, 488, 490, 535.
Smernov, B. M., 500.
Smick, A. E., 160, 161, 166.
Smit, C., 188, 204, 208, 209, 210, 211, 212, 213.
Smit, J. A., 188, 207, 209, 210, 212, 213, 216, 301, 302, 306.
Smith, A. C. H., 107, 110, 130, 270, 271, 272, 273, 275, 477, 536.
Smith, K., 533, 535, 538, 574, 576, 633, 634, 635.
Smith, L. G., 99, 104, 130.
Smith, P. T., 102, 103, 104, 106, 123, 124, 126, 129, 130, 138, 139, 283, 301, 303, 487, 490, 497, 542.
Smith, S. J., 195, 198, 529, 662.
Smokina, E. S., 562.
Smyth, H. D., 107.
Sneddon, I. N., 378.
Sobelman, I., 530, 531.
Soltysik, E. A., 175, 191, 201, 204, 214, 215.
Somerville, W. B., 651.
Sommerfeld, A., 440.
Sosnikov, A. K., 224.
Spitzer, L., 532.
Spruch, L., 389, 395, 621.
Stanton, H. E., 130.
Stanton, R. L., 190.
Stebbings, R. F., 193, 196, 197, 232, 250, 251, 332, 472, 519, 520, 529.
Steidl, H., 352, 353, 355.
Sternheimer, R. M., 556.
Stevenson, A., 61.
Stewart, A. L., 202, 483, 484.
Stewart, D. T., 202, 203.
Stickel, W., 311, 313.
Stone, P. M., 566.
Stone, W. G., 61.

Street, K., 171, 237, 258, 259, 260.
Stuber, F. A., 130.
Sunshine, G., 13, 14, 27, 28, 576, 577.
Sweeney, J. P., 190.
Swift, C. D., 167.

Tait, J. H., 497, 498.
Tanaka, Y., 110.
Tate, J. T., 99, 106, 126, 130, 138, 139, 338.
Taylor, A. J., 514, 517, 518, 528, 533, 534, 550, 552, 622, 623, 624, 633, 634, 636, 638.
Taylor, H. S., 645.
Taylor, J. E., 310.
Teller, F., 142, 532.
Temkin, A., 509, 510, 512, 513, 515, 516, 524, 534, 565, 566, 567.
Thieme, O., 178, 182, 202, 203, 206, 207, 484, 485.
Thomas, L. H., 588.
Thomas, M. N., 163.
Thompson, D. G., 555, 556.
Thomson, J. J., 588.
Thonemann, P. C., 145, 270, 273, 474.
Tisone, G., 153, 154, 155, 156, 158, 500.
Townsend, J. S., 24, 45, 53, 54, 56, 66, 82, 83, 87, 88, 90, 91, 291, 301, 303, 307, 308.
Tozer, B. A., 124, 126.
Trefftz, E., 498.
Trujillo, S. M., 17, 18, 25, 26, 27, 28, 107, 124, 129, 130, 267, 268, 471, 518, 519, 576.
Tsi-Zé, N., 417, 418.
Tully, J. A., 477, 496.
Tunitskii, N. N., 145, 148, 151.

Ulmer, R., 195, 196, 198, 199.
Undheim, B., 622.

Vainshtein, L., 530, 531.
Valentine, N. A., 587, 589, 590.
van Atta, L. C., 311.
van der Wiel, M. J., 103, 124, 130.
van Regemorter, H., 496.
Van Vleck, J. H., 442, 553.
van Voorhis, C. C., 99.
Varnerin, L. J., 71.
Veldre, V. J., 534.
Vetterlein, P., 310.
Vinkalns, I., 534, 566.
Vinti, J. P., 443.

Volkova, L. M., 223.
von Hippel, A., 107.
Voshall, R. E., 90, 91.
Vriens, L., 589.

Wahl, J. J., 496.
Wainfan, N., 245.
Walker, W. C., 245.
Wall, F. T., 589.
Wallace, J., 541, 543.
Wannier, G. H., 131, 664.
Warren, R. W., 56, 57, 58, 82, 88, 308.
Watanabe, K., 245.
Watson, G. N., 384, 452, 632.
Watson, W. W., 415.
Weaver, L. D., 178, 179, 214, 215.
Webb, G. M., 431, 548, 549.
Webster, D. L., 160, 161, 165, 166, 167, 588.
Weigmann, H., 160, 162.
Weisskopf, V. P., 607.
Weissler, G. L., 245.
West, H. D., 190.
Wexler, B., 230.
Wheeler, J. A., 482.
Whiddington, R., 310, 313, 627.
Whitaker, W., 528, 638.
Whittaker, E. T., 384, 452, 632.
Wien, W., 311.
Wigner, E. P., 261, 601, 607, 643.
Wild, J. P., 532.
Wilkinson, D., 541, 543.
Williams, E. J., 164, 588.
Williamson, J. H., 499, 500, 541.
Willmann, K., 347, 348, 552.
Woll, J. W., 478.
Womer, R. L., 311, 338.
Woodroofe, E. G., 310.
Woudenberg, J. P. M., 246, 253.
Wu, T. Y., 262.

Yakhontova, V. E., 202, 207, 208, 209, 210, 211, 212, 213.
Yarnold, G. D., 328.
Young, L. A., 479, 549.

Zaazou, A. A., 55, 66.
Zapesochnyi, I. P., 175, 183, 186, 190, 202, 206, 211, 216, 219, 223, 224, 225, 226, 227, 229.
Ziesel, J. P., 107, 130, 140.

SUBJECT INDEX

Absorption:
cross-section for inverse bremsstrahlung, 440.

Ag:
cross-section for K-shell ionization of:
measurement of, 166.
theory of, 475

Alkali metal atoms, theory of large cross-section for elastic scattering of electrons by, 407.

Amplitude of elastic scattering:
imaginary part of, and total cross-section, 505.
real part of, and dispersion relations, 506.

Angular distribution:
of elastically scattered electrons in Ar, 334–5, 343; Cd, 336; H, 333; He, 334–5, 346–8; Hg, 334–5, 336, 337, 344; Kr, 334–5; Ne, 334–5; Xe, 334–5, Zn, 336.
of inelastically scattered electrons:
after excitation of He, 341–2, 345–6.
after ionizing collisions in Ar, 343; H_2, 345; He, 341, 345; Hg, 349.
in Ar, 343; H_2, 345; He, 338, 340–2, 385–6, 348–9; Hg, 349.
theory of:
for He excitation, 552.
for inelastic scattering, by complex atoms, 495; by H atoms, 499.
for ionizing collisions, 466–7.
of optical radiation, dependence on polarization, 171, 194.
theory of for scattering:
by atomic fields, 431.
by centre of force, 381.
of high-energy electrons, 429.
of low-energy electrons in rare gases, 410; Hg, 412; H_2S, 413; PH_3, 413.

Anomalous dispersion method for measuring metastable atom concentrations, 348–9.

Appearance potentials of Ar^{2+}, Kr^{2+}, Xe^{2+}, 151.

Ar:
angular distribution of electrons scattered elastically in, 334, 335, 343.
theory of, 411.
angular distribution of electrons scattered inelastically in, theory of, using distorted wave approximation, 560.

characteristic energy of electrons in, 87, 308.
drift velocity of electrons in, 87.
at large F/p values, 299.
elastic scattering of electrons in, theory of:
effective range expansion, 553.
for fast electrons using Born's approximation, 431.
for slow electrons, 402.
energy-loss spectrum of electrons scattered in, 318.
excitation of metastable states of, 259–60.
inelastic scattering of electrons in, 343.
ionization cross-sections of electrons in, 123, 124, 126–8.
ionization, multiple, of:
threshold potentials for, 134.
through Auger effect, 140.
magnetic deflection factor for electrons in, 90.
metastable ion production in, 141, 143.
momentum-transfer cross-section for electrons in, 89, 554.
phase shifts for electron scattering in, 403.
polarizability of, 556.
polarization of and the Ramsauer–Townsend effect in, 402.
scattering length for electrons in:
estimated from pressure shift in alkali metal spectra, 418.
theory of, 553.
simultaneous excitation and ionization cross-sections for, 230.
total collision cross-sections for electrons in:
broad features of, 25.
fine structure of, 42.
velocity distribution of electrons in, 89.

Ar^+, Ar^{++}, ionization cross-sections of, 151.

Ar^-, doubly excited auto-ionizing states of, 641.

Atom form factor, 430.

Atomic beams:
gas-discharge source of, 108, 110.
measurement of total collision cross-sections using, 9, 10, 12.
oven source of, 108.
produced by charge exchange, 11.

Atomic field:
Hartree and Hartree–Fock, 376.
static, 376.
statistical, 378.

Au:
angular distribution of electrons after elastic scattering in, 337.
cross-section for ionization of K-shell of, 166; of L-shell of, 167.
polarization of electrons in scattering by, 423.
relativistic theory of, 423.

Auger effect:
in collision of metastable atoms with surfaces, 142.
multiple ionization due to, 140.

Auto-ionizing states:
of excited two-electron systems, 597.
of He, 316; Li, 261–2; of metastable atoms, 239–42.

Balanced filter monochromator for X-radiation, 161.

Be, ionization of and Born's approximation, 496.

Be II, excitation of and Coulomb–Born approximation, 496.

Bethe approximation:
excitation and ionization of atoms by electrons, 437.
impact parameter formulation, 452.

Born's approximation and:
angular distribution after ionizing collisions of electrons in H, 466.
detachment of electrons from H^-, 498.
elastic scattering of electrons by complex atoms, 429; H, 426; He^+, 428.
elastic scattering of high-energy electrons by atoms, 429.
energy distribution after ionizing collisions, of electrons, 467.
excitation by electron impact, 434; of H, 458; $He(^1D)$, 493; $He(^1P)$, 430–6; $He(^1S)$, 491–2; $Na(3^1P)$, 495–6; Xe, 494.
form factor, 430.
impact parameter formulation, 448.
inner-shell ionization of AgK, 475; NaK, 475.
ionization cross-sections, by electrons, of atoms, 434; of Be, 496; of H, 461; of He, 486–8.
oscillator strengths, 444.
scattering by Coulomb field, 398.
spherical and truncated spherical average formulae, 465.

Born-exchange approximation, 465.

Born–Oppenheimer approximation, 454.
excitation of triplet states of He using, 494.
ionization of H using, 464.
truncated form of, 464.

Bremsstrahlung, cross-section for, 440.

C:
elastic scattering of electrons in, 575.
energy of ground state of, 574.
excitation of, by electrons, 575.

C^+, excitation of, by electrons, 577.

Ca:
Auger effect and multiple ionization of, 140.
ionization cross-section of, 129.

Ca II, excitation of, and Coulomb–Born approximation, 496.

Cascade processes and optical excitation function measurement, 173; for He, 204–5; Hg, 221.

Cd:
angular distribution of electrons elastically scattered in, 336.
energy variation of total electron scattering cross-section in (broad features), 29.
optical excitation function for, 227; resonance line of, 229.

Channel, in scattering theory, 504.

Characteristic energy of electrons in gases, 54.
measurement of in Ar, 88, 308; He, 82.

Classical scattering theory:
for Coulomb field, 398.
for electron-atom collisions, 587; for ionizing collisions, 588.
for phase-shift determination, 391.

Close-coupling (truncated eigenfunction) approximation, 508.
application to:
calculation of resonant parameters, 615; for doubly excited states of He, 620; H^-, 634.
collision of electrons with atoms and ions with incomplete p shells, 571.
elastic scattering of electrons by Cs, K, Li, Na, 562; H, 514–9; He^+, 535–6.
excitation of Cs, 569; H, 508, 515, 518, 528; He^+, 536–8; He, 550–2; K, 569; Li, 569, Na, 569.

Collision strengths, definition of, 571.
for electron collisions with ions, 577.
with $2p^2$, $2p^3$, $2p^4$, $3p^2$ configuration, 578.
for transitions between fine structure levels of ions, 550–1.

Conductivity, electric, of plasma in an alternating field, 67.
  application to measurement of momentum-transfer cross-section, 69.
    for He, 70; Ne, 70.
  resonant frequency of a cavity and, 71.
Coulomb–Born approximation for elastic scattering of electrons by He^+, 428.
  for excitation of Be II, 496; Ca II, 496; H-like ions, 477; He^+, 473; Mg II, 496; N V, 496; Ne VIII, 496, O VI, 496.
  for ionization of He^+, 473; and isoelectronic ions, 474.
Cross-section:
  concept of, 2.
  detachment, 153–8.
  differential, 5.
    measurement of, for electrons, 321–46.
    theory of, calculation of, 381.
  diffusion, 47.
  energy loss, and resonance effects, 316, 321.
  excitation, 97.
    measurement of:
      by diffusion method, 280–4.
      by optical excitation, 169–72.
      for metastable state, 231–45.
  ionization, 98.
    energy variation of, 125.
    fine structure in energy dependence of, 112–18.
    inner shell, 158–68.
    measurement of, 99–112.
    momentum-transfer (diffusion), 47.
      measurement of, 67, 74, 78.
    multiple, 125.
    of positive ions, 144–52.
    simultaneously with excitation, 228–30.
    single, 125.
    total, apparent, 99.
      measurement, by collision chamber method, 99–104; by crossed-beam method, 106–12.
    results of measurements of, 123–4.
  scattering, of electrons into polarized states, 349–58.
  spin-exchange, 357–69.
  total, for scattering by centre of force, 379.
    calculation of 379–84.
Crossed-beam method of measuring:
  angular distribution of electrons scattered in H, 332.
  cross-section:
    detachment, of H^-, 153–7.
    excitation of H metastable states, 231–45.

  of $He(2^3S)$ states, 267–70.
  of He^+, 270–6.
  excitation, optical, 193–5.
  ionization (total apparent), 106–12.
  of positive ions, 145–9.
  total collision, 9–12.
Cs:
  Auger effect and multiple ionization of, 140.
  drift velocity of electrons in, 93.
  elastic scattering of electrons by, 562.
  excitation of resonance transitions, close-coupling approximation for, 569.
  ionization cross-section for, measurement of, 124.
  fine structure near threshold, 138.
  polarizability of, 562.
  relativistic effects in scattering of electrons by, 420.
  simultaneous excitation and ionization cross-section for, 229, 309.
  spin-exchange cross-section for electrons in, 370.
  theory of, 568.
  total electron cross-sections of, broad features of, 30.
Cyclotron resonance method for measurement of momentum-transfer cross-sections, 74; in Ar, 89; Cs, 95; He, 84; Ne, 86.

Detachment from negative ions by electron impact, 153.
  cross-section for:
    Born's approximation for, 499.
    effect of ionic charge on, 500.
    measurement of, 153–7.
Diffusion:
  of electron through gases, 45, 278–80.
  length, characteristic, 294.
Dipole distortion in electron atom scattering, 510.
  for alkali atoms, 562; Ar, 555; H, 524; He^+, 534; Ne, 555.
Dispersion relations for real part of forward elastic scattered amplitude, 506.
  for electrons in H, 519; He, 542.
Distorted-wave method, 525; application to calculation of:
  angular distributions of inelastic electron scattering in Ar, Ne, 560.
  cross-sections for collision of electrons with atoms and ions with incomplete p shells, 573.
    for excitation of $H(2s, 2p)$, 528; H, highly excited states, 33; He, two-quantum states, 549.
    for ionization of H, 533.

Drift velocity of electrons in gases, 45.
for large F/p values, 296–9.
magnetic, in He, 83.
measurement of, in Ar, 87, 299; Cs, 93; He, 81; Kr, 91; Ne, 85; Xe, 91.
errors in, 62.
method of measurement:
by electrical-shutter method, 58.
by Hornbeck's method, 58.
from glow discharge study, 64.
proportional-counter method, 62.

Effective range expansion, 388.
for electron scattering by Ar, 553; He, 544, 547; Kr, 553; Ne, 553; Xe, 553.
in a modified Coulomb field, 399.
including effect of long-range forces, 389.
relation to bound states, 391.
Electrical breakdown:
in microwave fields, 293–6.
in steady fields, 291–3.
Electrical shutter method for measurement of electron drift velocity, 58.
Electron distribution in microwave cavity, 295.
Electron gun, 272, 322–31.
Electron-optical focusing effect, 35.
Electron-trap method for studying excitation cross-sections, 384–8.
Electron velvet, 113–14.
Energy analyser for electron scattering measurements:
electron-trap method, 284–8.
electrostatic, 38, 120–2.
spherical lens type, 20, 324.
trochoidal type, 104–6.
using retarding potential difference (RPD) method, 21; 114–16.
high resolution, for study of scattering of fast electrons through small angles, 329–31.
Energy, characteristic, of electrons in He, 82; Ar, 88, 308.
Energy distribution of electrons:
after diffusion through gases, 280–4.
after ionizing collisions, 467.
forward scattered in Ar, 318; He, 315; Ne, 317.
scattered in He, 338, 340; Hg, 339.
width of, 121–2.
Exchange-adiabatic method, 514.
application to electron scattering in Ar, 556; H, 515, 518; He, 541; He$^+$, 535–7; Li, 565; Na, 567; Ne, 556.
Exchange of electrons on impact, 453.
Born–Oppenheimer theory of, 454.
theory of:
for excitation of He triplet states, 479.

in scattering by H, 455; He, 455.
intermultiplet transitions, 453.
Excitation coefficient, 291.
Excitation cross-section, 97.
of He$^+$ ions, 270–6, 473.
measurement of:
by diffusion method, 280–4.
by electron-trap method, 384–8.
for metastable atomic states, 231–45.
optical:
absolute measurement of, in alkali metals, 223–7; He, 188–93, 202–7; Hg, 179–88, 218, 220; inert gases, 223.
for H, 194–9.
measurement of, using:
bulk gas or vapour, 179–93.
crossed-beam method, 193–5.

Fe VIII, Fe XIV, Fe XV, Fe XVI:
calculation by Coulomb–Born approximation of excitation cross-section for, 496.
calculation by Coulomb–Born approximation of ionization cross-section for, 498.
Fine structure in energy-dependence of cross-sections for:
electron-atom collisions, 6.
methods of observation of, 19.
results of measurements in Ar, 42; H, 31; He, 31, 35; Hg, 43; Kr, 42; Ne, 41; Xe, 42.
inelastic scattering of electrons in Ar, 318; He, 319; Ne, 317.
ionization, 112.
by Auger effect, 140.
near threshold, in Cs, 138; Hg, 134–5; K, 139; Xe, 135–8.
using Morrison's differential analysis method, 116–18.
using RPD method, 114–16.
Forward scattering amplitudes, for electrons in He, 543.
Free paths, of electrons in gases, 46.

H:
elastic scattering of electrons in:
angular distribution for, 333.
dispersion relations for, 519.
scattering phases for, 515, 518.
Temkin's analysis of, 509.
excitation cross-section of:
calculation of:
using Born's approximation, 458, 496.
using various theories, 523, 528, 532, 533.

H (*cont.*):
   excitation cross-section of:
      measurement of:
         for metastable states, 231–4, 242–5, 251–2.
         for optical states, 194–7, 198–9.
      polarizability of, 513.
      sources of, 108.
      total cross-section of:
         fine structure in energy variation of, 23, 31.
         for electrons, 10, 18, 25.

H⁻:
   detachment cross-section of, by electron impact, 153–8.
   resonance energies and level widths of doubly-excited states of, 629–38.

H₂, angular distribution of electrons after ionizing collisions in, 345.

He:
   angular distribution of electrons scattered:
      in elastic collisions in, 334–5; 340, 411.
      in ionizing collisions in, 340, 345.
      with excitation of, 338, 340.
         2^3S state (theoretical), 552.
         2^1S–2^3S transition, calculated by close-coupling approximation, 552.
         resonant states, 345–52.
   auto-ionizing states:
      effect of, in photo-ionization, 628.
      resonance energies and level widths for, 621–7.
   characteristic energy of electrons in, 82.
   de-excitation of He(2^1S) to He(2^3S), cross-section for, 263–7.
   diamagnetic susceptibility of, 478.
   drift velocity of electrons in, 81, 83.
   elastic scattering of electrons in, 407.
      at high energy, calculated by Born's approximation, 431.
      at low energy, 539.
         by exchange-adiabatic method, 541.
         by polarized-orbital method, 540.
      dispersion relations for, 542.
   energy-loss spectrum of electrons:
      in exciting auto-ionizing states, 316.
      in scattering in forward direction, 319.
      in scattering through a fixed angle, 314–15.
   excitation cross-sections in:
      absolute, 188–93; 202–7.
      calculation of, using Born and Born–Oppenheimer approximations, 447.
      for ^1P states (comparison of theory and experiment), 480–6.
      for ^1S–^1D transitions, calculated using Born's approximation, 493.
      for ^1S–^1S transitions, calculated using Born's approximation, 491–2.
   ionization coefficients in:
      measured by microwave breakdown, 297.
      measured by steady field electrical breakdown, 294.
      Townsend's first coefficient and, 303.
   ionization cross-section of, 123–4, 126, 128.
      calculated by Born's approximation, 487; and modified Born's approximation, 488–91.
      relation to photo-ionization cross-section, 487.
   metastable-state excitation cross-section in, 234–9, 245, 253–7, 306–7.
      calculated by close-coupling method, 551.
      calculated by distorted-wave method, 550.
   optical excitation cross-sections for, 188–93, 199–202.
      dependence on principal quantum number, 206.
      fine structure, in energy dependence of, 207–13.
   oscillator strengths, generalized, for, 481.
   polarizability of, 541.
   polarization of impact radiation in, 188–90, 191–3, 214–17.
   radiative transition probabilities in, 203.
   relativistic effects in scattering of electrons in, 420.
   scattering length for electrons in, 544.
      estimated from pressure shift in alkali metal spectra, 418.
   total cross-section for scattering of electrons in:
      broad features in observed energy variation of, 26.
      fine structure in observed energy variation of, 31.
      from He(2^3S) atoms, 267–70.
   wave functions:
      for S states of, 478–9.
      Hylleraas, simple form, 479.
         six-term function, 482, 543.
      Shull–Löwdin, 541.
      Wheeler, for 2^1P state, 482.

He⁺:
   elastic cross-sections for electron scattering by, below excitation threshold, 535.

excitation cross-section for electrons by, 270–6.
    for $2s$ state, comparison of theory and experiment, 477.
    theory of, Coulomb–Born approximation, 473.
    theory of $2s$ and $2p$ excitation, 538.
  ionization cross-section of, 151–2.
    theory of, by Coulomb–Born and related approximations, 473.
He^-, resonance energies and level widths of, 638.
Hg:
  angular distribution of electrons after scattering in:
    elastic collisions, 334–7.
    excitation, 339, 344.
    ionizing collisions, 344.
    theory of, for slow electrons in, 411.
  energy levels of, 218.
  ionization cross-section of, for electrons, 123–4, 126, 129.
    fine structure of, near threshold, 134–5.
  ionization potentials for multiple ionization in, 134.
  optical excitation cross-sections for, 179–87, 216–24.
    excitation of resonant line of, 220–2.
    fine structure in, energy variation of, 186, 218–20.
  polarization of impact radiation in, 182–3, 222–4.
  theory of, in slow electron scattering, 421.
  relativistic effects in scattering of electrons by, 420–2, 424.
  scattering length for electrons in, from pressure shift of alkali metal spectra, 418.
  spin polarization of elastically scattered electrons in, 355–8.
  superelastic collisions of electrons with metastable atoms of, 289–90.
  total electron cross-section for, 29.
Hg^+, ionization cross-section of, 151.
Hg^{++}, ionization cross-section of, 151.

Impact parameter, formulation of collision theory, 447.
  application to transitions between closely coupled states, 452; to $2s$–$2p$ transitions of H, 532; to resonance transitions in Cs, K, Li, Na, 569.
Impact radiation:
  cross-section for production of, 196–71.
    measurement of, 179–93; in He, 188–95, 197–214; in Hg, 183–7, 216–22.
  polarization of, 171–2, 178–9.
    measurement of, in alkali metals, 223–7; in He, 214–17; in Hg, 222–4; in inert gases, 223.
    theory of, 581.
    near threshold for ^6Li, ^7Li, ^{23}Na resonant lines, 586.
Imprisonment of resonance radiation, effect of in measurement of optical excitation functions, 173–5.
Inelastic collisions, effect on electron velocity distribution function, 299–300.
Inner shell ionization by electron impact, 158–68.
  cross-section for:
    measurement of, using thin targets, 160–5; double, 168; K-shell of Ag, Au, Ni, Pb, Sn, W, 166; L-shell of Au, 167.
    theory of, in K-shells of Ag and Ni, 475.
Ion current, measurement of, 100.
Ion-trap method, for studying positive ion ionization, 149–51.
Ionization coefficient, 290–3; in He, 294, 297, 303.
Ionization cross-section, 98.
  energy variation of, 125.
  fine structure in energy dependence of, 112–18.
  inner shell, 158–168.
  measurement of, 99–112.
  multiple, 125.
  of metastable atoms by collisions of the second kind, 237–9.
  of positive ions, 144–52.
  single, 125.
  simultaneously with excitation, 228–30.
  theory of:
    for positive ions, 474.
    using Born's approximation for Be, 496; H, 469; He, 473; Li, 496; Mg, 496; Na, 496, Ne, 496.
    using classical theory, 588.
    using Coulomb–Born approximation for Fe XV, Fe XVII, O IV, 498.
  total, apparent, 99.
  measurement of:
    by collision-chamber method, 99–104.
    by crossed-beam method, 106–12.
    fine structure effects near threshold in Cs, 138; Hg, 134–5; K, 139; Xe, 135–8.
    Morrison's differential analysis method, 116–18.
    results of, for Ar, H, He, Na, Ne, Xe, 126.

Ionization cross-section (*cont.*):
  measurement of:
    RPD method, 114–16.
    threshold behaviour of, 118–22, 131–4.
    with high-energy resolution, 112–14.

Ionization potentials for multiple ionization in Ar, Hg, Kr, Xe, 134.

Ionized atoms, cross-section for electron collisions with, 263–76.

Jeffreys's approximation, 392.

K:
  elastic electron scattering cross-section of, calculated by close-coupling method, 562.
  excitation cross-section for electrons in, 225:
    calculated by close-coupling method, 568.
  ionization cross-section for electrons in, 151–2.
    fine structure in, near threshold, 124, 139.
  multiple ionization through Auger effect in, 140.
  polarizability of, 562.
  polarized electron beam, produced by spin exchange from, 369–72.
  spin exchange cross-section for electrons in:
    measured, 362, 370.
    theory of, 568.
  total cross-section for electron scattering in:
    broad features of observed cross-section, 30.
    measurement of, 12.

Kr:
  angular distribution of electrons after scattering in:
    measured, 334–5.
    theoretical, 411.
  drift velocity of electrons in, 91.
  elastic scattering (theoretical):
    and effective range formula, 553.
    of fast electrons in, 431.
    of slow electrons in, 406.
    using exchange–adiabatic approximation, 556.
  excitation of metastable states of, 259–60.
  ionization cross-section of, by electrons, 123–4, 128–9, 134.
  metastable ion production in, by electrons, 141.
  momentum transfer cross-section for electrons in, 92, 554.
  polarization of, and the Ramsauer–Townsend effect in, 406.
  relativistic effects in calculation of electron scattering by, 420.
  scattering length for electrons in, 553.
    from pressure shift in alkali-metal spectra, 418.
  total scattering cross-section of electrons in:
    broad features of energy variation of, 25.
    fine structure in energy variation of, 42.

Kr^+, Kr^{++}, ionization cross-section of, 151.
Kr^-, doubly-excited auto-ionizing states of, 641.

Langer's approximation, 392.

Level width, 601.
  of doubly-excited states of He, 621–7.
  partial, 607.

Li:
  elastic scattering of electrons by, theory of:
    using close-coupling method, 562.
    using exchange-adiabatic method, 566.
    using polarized orbital method, 567.
  excitation of auto-ionizing metastable states of, 239–42.
    resonance terms of, 496, 568.
  ionization cross-section of, 123–4, 128.
    theory of, 496.
  polarizability of, 562.
  polarization of impact radiation from, 227–8.
  polarization of resonance lines, theory of, 586.
  spin-exchange cross-section for electrons in, 568.
  total cross-section for electrons in, 11.

Li^+, ionization cross-section of, 151–2.

Line width:
  of cyclotron resonance line for electrons in gases, 75.
  application to measurement of momentum-transfer cross-sections, 74.

M matrix, 646–64.
  effect of orbital degeneracy on, 662.
  generalization to any number of channels, 655.
  modification due to Coulomb field, 655.
  single-channel case, 646.
  threshold behaviour and, 652, 654.
  two-channel case:
    generalization to any angular momentum, 654.
    with zero angular momentum, 651.

Magnetic field, effect of on ionization cross-section measurements, 101–2, 119.
Magnetohydrodynamical conversion of heat to electricity (MHD), 560.
McLeod gauge, precautions needed in absolute pressure measurements by, 103.
Metastable atoms:
auto-ionizing states of, cross-section for, 239–42; of Li, 261–2.
concentration of:
measured by anomalous dispersion method, 248–9.
measured by optical absorption method, 246–8, 268–5.
electron ejection from surfaces by, 234–6, 242–5.
production coefficient for, 304.
quenching of, in electric field, 231, 242–5.
Metastable ions:
detection of, 272–3.
production of, 141–4, 149.
Metastable states of atoms:
deactivation of 2^1S to 2^3S states of He, 263–7.
excitation cross-section of, for Ar, 259–60; H(2s), 250–2; He, 253–7, 550; Kr, 259–60; Ne, 257–8; Xe, 259–61.
by crossed-beam method, 231–4, 236–7.
direct measurement of, 231–45.
phase-sensitive detection of, 235.
Method, experimental:
atomic beam, for studying spin-exchange collisions, 358–63.
collision chamber:
for measurement of ionization cross-sections, 99–104.
for measurement of optical excitation cross-sections, 180–2, 184–5, 189–93.
crossed beam:
for angular distribution of electron scattering, 332.
for detachment of electrons from negative ions, 153–8.
for ionization cross-section measurements, 106–12, 119–22.
for ionization of positive ions, 149–51.
for optical excitation cross-section measurements, 193–5.
for spin polarization of elastically scattered electrons, 353–5.
for total cross-section of electrons, 9–18.
diffusion, for studying excitation cross-sections by electrons, 284.
electron trap for studying excitation by electron impact, 284–8.

for angular distribution studies of scattering of electrons, 322–8.
for angular distribution studies of scattering of very slow electrons, 328–32.
for auto-ionizing metastable state studies, 241–2.
for characteristic energy measurements of electrons in gases, 58–64.
for detachment of electrons from negative ions, 153–8.
for double elastic scattering of electrons, 353–5.
for drift velocity measurements of electrons, 58–62.
at large F/p values, 296–7.
by electrical-shutter method, 58–60.
by Hornbeck's method, 60–1.
using proportional counters, 61–2.
for electron energy analysis, 309–16, 326.
by retarding potential difference method, 114–16.
by spherical energy analyser, 20.
for inner shell ionization measurements, 160–5.
for ionization coefficient measurement, 291–3.
for ionization cross-section measurements:
by collision chamber, 99–104.
by crossed beam, 104–6.
for ionization rate measurement, in microwave discharge, 293–6.
for metastable atom concentration measurement, 246–9, 263–5.
for metastable atom flux measurements, 234–8, 242–5.
for metastable atom production rate studies, 303–4.
for momentum-transfer cross-section measurement:
by microwave afterglow, 67–73.
from effect of magnetic field on resonance of a gas-filled cavity, 78–9.
using cyclotron resonance, 74–8.
for optical excitation cross-section measurements:
using collision chamber, 180–2, 184–5, 188–93.
using crossed beams, 193–5.
for polarization of impact radiation studies, 182–3, 188–90, 192–3.
for polarized electron beam production by spin exchange, 369–72.
of phase-sensitive detection for study of excitation of He^+ ions, 270–3.
optical pumping, 363–9.

N:
    elastic scattering of electrons in, calculated by close-coupling approximation, 575.
    energy of ground terms of, 574.
    ionization cross-section of, 128.
    source of, 110.
    total cross-section for electrons:
        broad features of, 28.
        measurement of, 18.
N^+, ionization cross-section of, 151–2.
N V, theory of excitation of, by Coulomb–Born approximation, 496, 492.

Na:
    elastic scattering of electrons in:
        calculated by close-coupling method, 562.
        calculated by exchange-adiabatic method, 567.
        calculated by polarized-orbital method, 567.
    excitation (optical) cross-sections for, by electrons, 225–6.
    excitation of 3^1P state of, and Born's approximation, 495.
    excitation of resonance transitions in, 568; and Born's approximation, 496.
    ionization cross-section of, by electrons, 126, 128; and Born's approximation, 496.
    ionization and simultaneous excitation by electrons in, 229.
    multiple ionization of, due to Auger effect, 140.
    polarizability of, 562.
    polarization of impact radiation from, 226, 228.
    resonance lines from (theory), 586.
    scattering length for electrons in, 564.
    spin exchange cross-sections for electrons in, 370; theory of, 568.
    total electron scattering cross-section in:
        measurement of, 12.
        observed broad features of energy dependence of, 30.
Na^+, ionization cross-section of, 151–2.

Ne:
    angular distribution of electrons scattered elastically in, 334–5; theory of, 411.
    angular distribution of electrons scattered inelastically in, theory of, 560.
    drift velocity of electrons in, 84.
    elastic scattering of electrons in (theory), 407.
        and effective range formula, 553.
        and exchange-adiabatic approximation, 556.
    excitation of metastable states of, 257–8.
    ionization cross-section of, 123–4, 126, 128; theory of, by Born's approximation, 496.
    multiple ionization of, due to Auger effect, 140.
    polarizability of, 556.
    resonance effects in inelastic scattering in, 317.
    scattering length, measurement of from pressure shift in alkali metal spectra, 418; theory of, 553.
    total cross-section for electron scattering in:
        broad features of energy dependence, of, 26.
        fine structure of energy dependence of, 40.
Ne^+, ionization cross-section of, 151–2.
Ne^-, doubly excited auto-ionizing states of, 640.
Ne VII, theory of excitation of, and Coulomb–Born approximation, 496.
Negative ions, detachment of electrons from, by electron impact, 153–8.

Ni:
    K-shell ionization cross-section for, 166.
    theory of, 475.
    relativistic effects in, 475.

O:
    elastic scattering of electrons, theory of, by close-coupling approximation, 575.
    energy of ground terms of, 574.
    ionization cross-section of, 128.
    source of, 108.
    total cross-section for scattering of electrons in:
        measurement of, 13, 17.
        observed cross-section; broad features of, 12.
O IV, ionization of, by Coulomb–Born approximation, 498.
O VI, excitation of, by Coulomb–Born approximation, 496.
Ochkur's approximation and electron exchange, 455.
    and ionization of Be, 496; H, 470; He, 490; Li, Mg, Na, 496.
Optical absorption method for measuring metastable atom concentrations, 246–8.
Optical excitation cross-section:
    absolute, for alkali metals, 223–8; He, 202–7; Hg, 218, 220; inert gases, 223.

measurement of, 172–8.
fine structure in, for Hg, 186.
for H, 198–9.
principles of, 169–72.
correction for cascade processes in, 173.
effect of imprisonment of resonance radiation, 173–5.
importance of pressure dependence effects, 173–8.
using bulk gas or vapour, 179–93.
using crossed-beam method, 193–5.
Optical excitation functions:
of H, 196–7; He, 188–95, 197–214; Hg, 183–7, 216, 218–22.
dependence of, on principal quantum number, for He, 206.
dependence of, on transition type, 199–202; 208–13.
fine structure of, for alkali metals, 223–7, 229; for He, 206; for Hg, 218–20; for inert gases, 223.
Optical pumping method for studying spin-exchange collisions, 363–9.
Optical radiation, intensity measurement of:
use of G.M. counter, 194–5.
polarization of, 171.
measurement of, 182–3, 188–90, 192–3.
Optical transition probabilities, 438.
matrix elements for, 438.
alternative forms for, 443.
for transitions in H, 460.
Oscillator strengths:
generalized and Born's approximation, 444; for transitions to ^1P states of He, 484.
optical, 439.
for Ar, Ne, Kr, 495.
differential, 439.
sum rules for, 441, 445.

Path length of electrons in ionization cross-section measurements, effect of collimating field on, 101.
Pb cross-section for K-shell ionization in, 166.
Phase-sensitive detection, use in measurement of:
angular distribution of elastically scattered electrons, 332.
detachment cross-section of H^-, 153–7.
excitation cross-section for metastable states, 231–5.
excitation cross-section for optical states, 194–5; of He^+, 270–6.
ionization (total, apparent) cross-sections, 106–12.

ionization cross-sections of positive ions, 145–9.
total electron scattering cross-sections, 9, 10, 12, 17.
observed fine structure in, 33.
PH_3, angular distribution of electron scattering in, 413.
Phase-shifts for scattering:
approximate, when small, 392, 430.
and analysis of elastic scattering below threshold for excitation, 504.
and angular distribution of slow electrons scattered by atoms, 411.
and bound states, relation to number of, 390.
and quantum defect, 400.
at zero energy, 387.
by central force, 380.
by modified Coulomb field, 399.
classical approximation for, 391.
dependence of, on energy and angular momentum, 385.
effective range expansion for, 388.
for electron scattering by Ar, 403; Kr, 406; Xe, 406.
resonance behaviour of, 600.
variational method for determining, 393.
wave-number dependence of, 643.
zero-order, for electrons below threshold, theory of, by various methods, 541–2; for H, 514; He^+, 535.
Photo-ionization, cross-section for, 439.
Polarizability of alkali metal atoms, 561–2; Ar, 556; H, 513; He, 541; Ne, 556.
Polarization of atoms in collisions:
and the effective range expansion, 389.
and the Ramsauer–Townsend effect in Ar, 402; Kr, 406.
Polarization of elastically scattered electrons, 349–55.
theory of, 419; by Hg, 421; Au, 422.
Polarization of impact radiation, 171–2.
angular distribution of the radiation, 171, 194.
measurement of, 178–93.
of alkali metals, 226; H, 195–7; He, 188–93, 214–17; Hg, 182–3, 222–4.
theory of, 581.
for resonance lines of ^6Li, ^7Li, ^{23}Na, 586.
including electron spin, 583.
including nuclear spin, 586.
Polarized electron beam, produced by spin exchange, 369–72.
Polarized orbital approximation for Cs, 567; H, 514–15; He, 540; He^+, 535–7; Li, 566; Na, 567.

Positive ions, ionization of, 144–52.
  by crossed-beam method, 144–9.
  by ion-trap method, 149–52.
  results of measurements of in Ar^+, Ar^{++}, He^+, Hg^+, Hg^{++}, Kr^+, Kr^{++}, Li^+, N^+, Ne^+, Kr^+, Xe^{++}, 151–2.

Post and prior interactions, 455.

Pressure:
  dependence on, of Hg (λ 5461) intensity, 186.
  dependent effects in optical excitation cross-section measurements, 173–8.
  dependent effects in measurement of polarization of impact radiation, 178–80.
  precautions needed in measurement of, by McLeod gauge, 102–3.

Pressure shift of high series terms of alkali metal spectra by foreign gas atoms, 414.
  and scattering length for foreign gas atoms, 416.

Probability of ionization, 99.

Proportional counters:
  for electron drift velocity measurements, 61.
  for X-ray intensity measurements, 167.

Quantum defect, 400.
  and collision strength for excitation of transitions in p^x configurations, 580.
  allowance for orbital degeneracy, case of He^+, 663.
  generalization to coupled channels, 657.
    application to Ca^+, 658.
  and phase shifts in a modified Coulomb field, 400.
  for elastic scattering of electrons by He^+, 353.

R matrix:
  for any number of channels, 655.
  for single-channel cases, 646.
  for two-channel cases with zero angular momentum, 650.
  for two-channel cases with any angular momentum, 654.
  modification due to Coulomb field, 655.

Radiative transition probabilities in He, 203.

Radiative capture, cross-section for, 440.

Ramsauer–Townsend effect, 24, 44, 45.
  and electron velocity distribution in Ar, 49, 59; Kr, Xe, 49.
  theory of, 401.

Rb:
  cross-section for electron spin-exchange in, 470.
  excitation of, 225–6.
  ionization of, 124, 129.
  simultaneous ionization and excitation, 229.

Relativistic effects in scattering of electrons by atoms, 419.

Resonance effects:
  angular distribution of electrons scattered after exciting resonances in He, elastic, 346–50; inelastic, 348–9, 351–2.
  in elastic scattering of electrons in H, 33.
  in excitation of unbound states from ground states (theory), 602.
  in inelastic scattering, 316–21.
  in inelastically forward scattered electrons in Ar, 318; He, 319; Ne, 317.
  in observed total cross-section of electrons in Ar, 42; H, 31; He, 35; Kr, Xe, 42.
  perturbation theory of:
    in terms of one discrete and one continuum, 597; two or more continua, 607–10.
  one-level formula for, 601.

Resonance shift and line-width in optical pumping experiments, to study electron spin exchange cross-sections, 363–9.

Resonances:
  shape, or one body, 642.
  Type I and Type II, 645.

Resonant frequency of a cavity containing plasma, 72.
  application to measurement of momentum transfer cross-section of electrons in He, 84.
  in a magnetic field, 78.

Resonant parameters:
  accurate calculation of, 615; for H^-, 629; for He^-, 638–9; for Ne^-, 640.
  variation of, along a Rydberg series, 611.

Retarding potential difference (RPD) method, 21, 114–16.
  applied to measurement of excitation cross-section of metastable states, 236.

Rutherford scattering formula, 397;
  and Born's approximation, 398, 430.

S-matrix:
  generalization to any number of channels, 654.
  single-channel case, 646.
  two-channel case and zero angular momentum, 648.

two-channel case and any angular momentum, 654.
Scattering of electrons by atoms, theory of:
angular distribution for slow electrons in H_2S, PH_3, 413; Hg, 511; rare gases, 411.
angular distribution for elastic scattering at high energy, 429; for Ar, Cs, He, Kr, 431.
by centre of force, 379.
by Coulomb field, 397.
by modified Coulomb field, 399.
elastic, by H atoms:
using generalized variation method, 502.
using truncated eigenfunction (close-coupling) expansion, 508.
using $1s$–$2s$–$2p$ close-coupling expansion for, 508.
Temkin's analysis of s-scattering of, 509.
polarized orbital approximation for, 514.
exchange-adiabatic approximation for, 514.
for Ar atoms, 402.
Scattering length of Ar, He, Hg, Kr, N_2, Ne, Xe, from pressure shift produced in alkali metal spectra, 418.
and effective range expansion, 388.
upper bound for, 395.
variational methods for determining, 394.
Secondary electrons, emission of from surfaces due to metastable atoms, 142.
and spread of energy of electron beams, 113.
in ionization cross-section measurements, elimination of effects of, 101.
Shape profile parameter (profile index), 601.
Shape resonance, 642.
Sn, cross-section for K-shell ionization of, 166.
Spin-exchange collisions, 357–72.
and radio-emission of 21-cm line, theory of, 522.
production of polarized electron beams by, 369–72.
Spin-polarization of electrons following elastic scattering, 349–57.
Superelastic collisions, 263; with $He(2^3S)$ atoms, 267–70; with metastable Hg atoms, 289–90.
Surface ionization probability, 111.
Swarm experiments, for measurement of inelastic scattering cross-sections, 277–309.

T matrix:
modification of by Coulomb field, 656.
by orbital degeneracy, 662.
single-channel case, 646.
two-channel case for zero angular momentum, 650.
behaviour at threshold, 653.
case of any angular momentum, 654.
Threshold behaviour of cross-sections for:
excitation of Xe metastable states, 261.
excitation, theory of, 653–4.
in presence of Coulomb field, 656.
effect of orbital degeneracy on, 662.
ionization, 118–22, 131–4; of H, 132; theory of, 663.
multiple ionization of Ar, 134; Hg, 134; Kr, 133–4; Xe, 134.
Tl, total electron cross-section in, observed broad features of, 30.
Total cross-section for electron scattering, measurement of, 7–24.
by atomic-beam method 9–18.
by observation of beam attenuation, 12–18.
by observation of scattered electrons, 10–12.
by Ramsauer's method, 7–9.
fine structure in energy variation of cross-sections, 19–24.
Total ionization cross-section for electrons, apparent, 99.
measurement of, by collision-chamber method, 99–104.
by crossed-beam method, 106–12.
positive ion current analysis of, 104–6.
Townsend coefficients, 291; for He, 309.
Transmission of electrons in Ar, 42; H, 31; He, 35; Hg, 45; Kr, 42; Ne, 40; Xe, 42.
Truncated eigenfunction expansion (close-coupling) approximation, 506.

Variational methods for determining phases η_l, 393.
Kohn, 394; Hulthèn, 395; Rubinow, 396.
and eigenfunction expansion, 503.
for determining bounds on scattering length, 395.
generalized, for calculation of $\eta_0^\pm$ for H: below threshold, 502.
above threshold, 524.
results of calculation of $\eta_0^\pm$ for H by, 514; $\eta_1^\pm$, $\eta_2^\pm$ for H by, 516.

Variational methods (*cont.*):
  results of calculation:
    of $2s$, $2p$ excitation of H by, 528.
    of resonance energies and level widths of He by, 621; H^-, 634.
Velocity analysis of scattered electrons, 310–12.
Velocity distribution of electrons in gases, 49; in Ar, 49; He, 49, 302; Ne, 49.
  derivation of formula for, in atomic gases, 49.
    allowance for energy of motion of gas atoms, 53.
    in presence of magnetic field:
      alternating, 75.
      uniform, 65.
    under conditions when inelastic collisions are important, 299–300.

W, cross-section for K-shell ionization of, 166.
Width of electron-energy distribution in ionization cross-section measurements, 121–2.

X-ray intensity measurements:
  by ionization chamber, 161.
  by proportional counter, 167.
  by scintillation counter, 162–3.
X-ray production cross-section for Ag K, 165–6; Ag L_{III}, 167; Au K, 162, 166; Au L_I, 167; Au L_{II}, 167; Au L_{III}, 167; Cu K, 165; Hg K, 165; Ni $K\alpha$, 161, 165–6; Pb K, 166; Sn K, 162–4, 166; W K, 165.
X-ray satellite lines, excitation of, 167–8.

Xe:
  angular distribution of electrons scattered elastically in, 334–5.
    theory of, for slow electrons, 411.
  drift velocity of electrons in, 91.
  elastic scattering of electrons in, theory of:
    for slow electrons, 406.
      effective range formula, 553.
    for fast electrons, by Born's approximation, 411.
  excitation of metastable states of, 259–61.
  ionization cross-section of, for electrons, 124, 129, 135–8.
  metastable ion production by electrons in, 141, 144.
  momentum-transfer cross-sections for electrons in, 92; theory of, 554.
  multiple ionization, due to Auger effect in, 134.
    threshold potentials for, 134.
  scattering length of, from pressure shift in alkali metal spectra, 418.
  scattering length, theory of, 553.
  total electron cross-section, observed:
    broad features of, 25.
    fine structure of, 42.
Xe$^+$, ionization cross-section of, 151.
Xe^{++}, ionization cross-section of, 151.
Xe$^-$, doubly excited auto-ionizing states of, 641.

Zn: angular distribution of electrons scattered elastically in, **336**.
  total electron cross-section, observed broad features of, 29.